Clinical Virology Manual

Clinical Virology Manual

Steven Specter, Ph.D.
Associate Professor

Gerald J. Lancz, Ph.D.
Associate Professor

Department of Medical Microbiology and Immunology
University of South Florida College of Medicine
Tampa, Florida

Elsevier
New York • Amsterdam • London

Published by:
Elsevier Science Publishing Company, Inc.
52 Vanderbilt Avenue, New York, New York 10017

Sole distributors outside the United States and Canada:
Elsevier Science Publishers B.V.
P.O. Box 1000 AE Amsterdam, the Netherlands

Library of Congress Cataloging in Publication Data

Clinical virology manual.

Includes bibliographies and index.
1. Diagnostic virology—Handbooks, manuals, etc.
I. Specter, Steven. II. Lancz, Gerald J.
QR387.C48 1986 616′.0194 86-16560
ISBN 0-444-01085-8

Current printing (last digit)
10 9 8 7 6 5 4 3 2 1

Manufactured in the United States of America

Contents

Preface

Clinical virology is an area that is undergoing rapid expansion. As a service for patient care, the utility of the clinical virology laboratory has increased significantly in the past decade. Due to the availability of commercial test kits, sophisticated yet simple diagnostic reagents, and the standardization of laboratory assays, accurate, reliable and, in many instances, rapid protocols are currently available for the diagnosis of a variety of viral agents producing human infections. Thus, the demands (on both the physician and the clinical laboratory virologist) for the diagnosis of viral infections will continue to increase. With this in mind, this volume is written as both an aid to the clinician and as a guide for the clinical laboratory.

This manual has three sections. The first describes laboratory procedures to detect viruses. The initial chapters deal with quality control in the laboratory and specimen handling, areas that are critical for an effective diagnostic laboratory. This is followed by individual chapters that provide information or a detailed protocol on how to set up and test samples for viral diagnosis using this technique. Both classical and the newer, more experimental techniques are described in detail.

The second section focuses on the viral agents. Viruses are grouped into chapters based on a target organ-system categorization. In this way, viruses producing infection in a particular organ or tissue are discussed and compared in a single chapter. This approach more accurately reflects the problems and choices faced by the attending physician and clinical technician for the diagnosis of a viral infection. Each chapter includes information relating basic, pathogenic, immunologic, and protective measures concerning each virus group, as well as information on its isolation, propagation, and diagnosis. This section also includes a chapter on *Chlamydia*. There are two reasons for including this family: The clinical laboratory often isolates and diagnoses *Chlamydia*, and the techniques used in its isolation and diagnosis are used in other instances.

The third section is designed to be used for reference. Here we supply information about Federal Reference Laboratories at the Centers for Disease Control and their role in the diagnosis of viral infection. The diagnostic and regulatory activities of state health laboratories and services available at individual hospital laboratories are provided in survey form. This listing is somewhat incomplete in that it contains information provided in response to an initial questionnaire and follow-up.

The aim and scope of this volume is service: to the physician, as a source of basic and clinical information regarding viruses and viral diseases, and to the laboratories, as a reference source to aid in the diagnosis of virus infection by providing detailed information on individual techniques and the impetus to expand services offered.

Steven Specter
Gerald J. Lancz

Contributors

George R. Anderson, D.V.M.
Deputy Chief for Laboratories
Bureau of Laboratory and Epidemiological Services
Michigan Department of Public Health
Lansing, Michigan

Harold C. Ballew, M.D.
Virologist
Virology Training Section
Division of Laboratory Training and Consultation
Laboratory Program Office
Centers for Disease Control
Atlanta, Georgia

Thomas A. Brawner, Ph.D.
Senior Scientist
Department of Infectious Diseases and Immunology
Abbott Laboratories
North Chicago, Illinois

George H. Burgoyne, Ph.D.
Acting Chief
Division of Vaccine Production
Bureau of Laboratory and Epidemiological Services
Michigan Department of Public Health
Lansing, Michigan

Francis W. Doane, M.A.
Professor, Department of Medical Microbiology
Faculty of Medicine
University of Toronto
Toronto, Ontario, Canada

John J. Docherty, Ph.D.
Associate Professor
Virology Laboratory
Department of Molecular and Cellular Biology
The Pennsylvania State University
University Park, Pennsylvania

R. Gordon Douglas, Jr., M.D.
E. Hugh Luckey Distinguished Professor of Medicine
Professor and Chairman
Department of Medicine
Cornell University Medical College
New York;
Physician-in-chief
The New York Hospital
New York, New York

M. Nixon Ellis, Ph.D.
Senior Research Scientist
Department of Virology
The Wellcome Research Laboratories
Research Triangle Park, North Carolina

Mario R. Escobar, Ph.D.
Professor of Pathology
Department of Virology
Medical College of Virginia
Virginia Commonwealth University
Richmond, Virginia

David A. Fuccillo, Ph.D.
Director of Virology
Microbiological Associates Inc.
Bethesda, Maryland

Robert C. Gallo, M.D.
Chief, Laboratory of Tumor Cell Biology
National Cancer Institute
Bethesda, Maryland

P. S. Gardner, M.D.
Retired, Division of Microbiological Reagents and Quality Control
Central Public Health Laboratory
Colindale, London, England

Helen Gay, M.S.
Senior Research Technician
Investigative Cytology Laboratory
Memorial Sloan Kettering Cancer Center
New York, New York

Kenneth L. Herrmann, M.D.
Assistant Director for Laboratory Science
Division of Viral Diseases
Center for Infectious Diseases
Centers for Disease Control
Atlanta, Georgia

G. D. Hsiung, Ph.D.
Professor, Department of Laboratory Medicine
Yale University School of Medicine
New Haven;
Director, Virology Reference Laboratory
Veterans Administration Medical Center
West Haven, Connecticut

Bruce H. Keswick, Ph.D.
Assistant Professor
Program in Infectious Diseases and Clinical Microbiology
University of Texas Medical School at Houston
Houston, Texas

Bryan L. Kiehl, Ph.D.
Director of Product Development
Cytotech Inc.
San Diego, California

Gerald J. Lancz, Ph.D.
Associate Professor
Department of Medical Microbiology and Immunology
University of South Florida College of Medicine
Tampa, Florida

Marie L. Landry, M.D.
Assistant Professor
Departments of Laboratory Medicine and Medicine
Yale University School of Medicine
New Haven;
Associate Director
Virology Reference Laboratory
Veterans Administration Medical Center
West Haven, Connecticut

Ira C. Lee, Ph.D.
Senior Scientist
Difco Research and Development Center
Ann Arbor, Michigan

David A. Lennette, Ph.D.
Codirector
Virolab, Inc.
Berkeley, California

Evelyne T. Lennette, Ph.D.
Codirector
Virolab, Inc.
Berkeley, California

A. L. Lewis, D.V.M.
Biological Administrator
Epidemiology Research Center
State of Florida
Department of Health and Rehabilitative Services
Office of Laboratory Services
Tampa, Florida

Leroy C. McLaren, Ph.D.
Professor, Department of Microbiology
Director, Clinical Virology Laboratory
University of New Mexico School of Medicine
Albuquerque, New Mexico

Isa K. Mushahwar, Ph.D.
Head, Department of Infectious Diseases and Immunology
Abbott Laboratories
North Chicago, Illinois

James H. Nakano, Ph.D.
Chief, Poxvirus Laboratory and World Health Organization Collaborating Center for Smallpox and Other Poxvirus Infections
Virus Exanthems and Herpesvirus Branch
Division of Viral Diseases
Centers for Disease Control
Atlanta, Georgia

Paul E. Palumbo, M.D.
Assistant Professor of Pediatrics
Department of Pediatrics
Cornell University Medical College
New York;
Assistant Attending in Pediatrics
The New York Hospital
New York, New York

Larry K. Pickering, M.D.
Professor of Pediatrics
Program in Infectious Diseases and Clinical Microbiology
University of Texas Medical School at Houston
Houston, Texas

Fred Rapp, Ph.D.
Professor and Chairman
Department of Microbiology
The Pennsylvania State University College of Medicine
Hershey, Pennsylvania

Charles A. Reed, A.B.
Research Associate
Supervisor of Clinical Virology
Department of Pediatrics
Washington University School of Medicine
St. Louis, Missouri

Douglas D. Richman, M.D.
Associate Professor of Pathology and Medicine
University of California, San Diego, School of Medicine
San Diego;
Staff Pathologist
Clinical Virology Section
Veterans Administration Hospital
La Jolla, California

Julius Schachter, Ph.D.
Professor of Epidemiology
Department of Laboratory Medicine
University of California, San Francisco, School of Medicine
San Francisco;
Director, World Health Organization Collaborating Center for Reference and Research on Chlamydia
San Francisco, California

Jörg Schüpbach, M.D.
Institute for Immunology and Virology
University of Zurich
Zurich, Switzerland

John L. Sever, M.D., Ph.D.
Chief, Infectious Diseases Branch
National Institute of Neurological and Communicative Diseases and Stroke
National Institutes of Health
Bethesda, Maryland

Keerti V. Shah, M.D., Ph.D.
Professor and Associate Chairman
Department of Immunology and Infectious Diseases
The Johns Hopkins University School of Hygiene and Public Health
Baltimore, Maryland

Isabel C. Shekarchi, Ph.D.
Research Scientist
Microbiological Associates, Inc.
Bethesda, Maryland

Robert E. Shope, M.D.
Professor of Epidemiology
Yale Arbovirus Research Unit
Department of Epidemiology and Public Health
Yale University School of Medicine
New Haven, Connecticut

Holly A. Smith
Microbiologist, Medical Virology Section
Laboratory of Clinical Investigation
National Institute of Allergy and Infectious Diseases
National Institutes of Health
Bethesda, Maryland

Roger D. Smith, M.D.
Mary M. Emory Professor of Pathology
Director, Department of Pathology and Laboratory Medicine
University of Cincinnati Medical Center
Cincinnati, Ohio

Thomas F. Smith, Ph.D.
Professor of Microbiology and of Laboratory Medicine
Mayo Medical School
Director of Virus Laboratory
Section of Clinical Microbiology
Department of Laboratory Medicine
Mayo Clinic and Mayo Foundation
Rochester, Minnesota

Steven Specter, Ph.D.
Associate Professor
Department of Medical Microbiology and Immunology
University of South Florida College of Medicine
Tampa, Florida

Stephen E. Straus, M.D.
Head, Medical Virology Section
Laboratory of Clinical Investigation
National Institute of Allergy and Infectious Diseases
National Institutes of Health
Bethesda, Maryland

Karen Sutherland, C.T. (A.S.C.P.)
Diagnostic Cytology Section
Department of Pathology and Laboratory Medicine
University of Cincinnati Medical Center
Cincinnati, Ohio

Fulvia diMarzo Veronese, Ph.D.
Scientist 2
Bionetics Research Inc.
Rockville, Maryland

F. M. Wellings, Sc.D.
Director, Epidemiology Research Center
State of Florida
Department of Health and Rehabilitation Services
Office of Laboratory Services
Tampa, Florida

Heinz Zeichhardt
Private Lecturer
Institute for Clinical and Experimental Virology
The Free University of Berlin
Hindenburgdamm, Berlin, Germany

Clinical Virology Manual

Section 1

Laboratory Procedures for Detecting Viruses

1

Quality Control in Clinical Virology

Ira C. Lee

1.1 INTRODUCTION

Quality control in the clinical laboratory consists of a set of procedures designed to help ensure delivery to the medical staff of laboratory results that are consistent and accurate. These results must be supplied in a timely fashion while the data are still clinically relevant. Quality control should not hamper the timely delivery of these results, and must be cost-effective in these times of operating budget restraints. Additionally, laboratory personnel should regard quality control as a reflection of their personal commitment to patient care, rather than as an enforced burden or necessary evil.

Clinical virology is a relatively new department in the clinical laboratory. In fact, many large medical centers do not have such a service or offer limited service, such as referred viral serology. Consequently, quality control procedures for clinical virology may not be as formalized as the procedures for more established departments, such as microbiology, hematology, or chemistry. However, many of the quality control procedures used by these other areas are applicable, either directly or with modification, to clinical virology. Examples are media quality control procedures from microbiology and adaptation of instrument quality control procedures from the clinical chemistry laboratory. Some quality control procedures, such as screening fetal calf serum for virus inhibitory activity are, of course, unique to clinical virology. This chapter will attempt to provide guidelines that are adapted from quality control procedures used by various departments, and includes those unique to clinical virology.

Quality control should be regarded as a holistic activity. In many hospitals, quality control is an intradepartmental discipline with interdepartmental or medical staff interaction occurring only in times of difficulties, proficiency testing, or inspections. Ideally, quality control should encompass all activities from the time of specimen procurement through the final report and should include all departments involved. Good communication between the clinical virologist, pathologist, and other department heads, and medical staff aids greatly in achieving this goal.

1.2 FACILITY DESIGN AND MAINTENANCE

The physical plant for a clinical virology laboratory should be designed minimally with two basic concepts in mind:

1. Biohazard control, to minimize risk to laboratory personnel and the general public
2. Protection of all cultures used in viral isolation from contamination by environmental agents

Although other concepts, such as facility design to maximize work efficiency while maintaining pleasant work conditions certainly bear consideration, this section will deal with the two basic concepts cited.

Service capabilities of clinical virology laboratory facilities vary greatly. At one extreme are limited viral serology services, which essentially are adjuncts to a serology department. Intermediate to this are virology laboratories that are physically incorporated into microbiology departments and may attempt limited isolations along with serology. At the other extreme are full service clinical virology laboratories. These facilities represent a major expenditure to a hospital and often must justify their development and continued existence by serving as reference centers, because virology testing in-house is often low volume, compared with microbiology or chemistry.

The following describes some of the salient features of a facility designed specifically for clinical virology, and its maintenance:

1. The facility should be physically separated from the microbiology laboratory and not share common air returns or equipment, such as biological safety hoods or incubators.
2. The environment should be controlled so that ambient temperature is in the 22°–26°C range and relative humidity is 30–50%.
3. The facility should be under negative pressure with respect to the rest of the laboratory area.
4. Internally, the clinical virology laboratory can be divided into positive and negative air pressure areas. The positive areas are used for tissue culture and media preparation, whereas, the negative areas are designated for viral isolation or serology, because they deal with viable pathogens.
5. All surfaces, such as work benches, floors, and walls should be composed of materials that can be decontaminated easily and do not trap dust and foamites.
6. Good standard microbiology procedures appropriate to the biosafety level of the viruses handled must be observed. These procedures are described in detail elsewhere (Richardson and Barkley, 1984; Warren, 1981), and include daily decontamination of all work surfaces, proper laboratory attire, use of safety pipetting devices, and a minimization of aerosol generation.
7. Biological safety hoods should be available for both tissue culture and viral isolation. Ideally, separate hoods or sets of hoods should be provided, with tissue culture hoods being housed in positive pressure rooms, and viral isolation hoods in negative pressure areas. The hood rooms should have minimal furnishings or equipment, as these serve to trap foamites or may require decontamination in event of pathogen spillage or aerosolization. Hood rooms should not have common air ducting and the exhaust from hoods in which pathogens are handled should be externally vented, as an additional safety precaution in event of a failure of the HEPA filter in the hood. If only one safety hood is available, tissue culture work should be scheduled before viral inoculations and the hood decontaminated between these operations.
8. The facility must be properly maintained. Biohazardous wastes must be properly disposed, floors must be disinfected periodically, and the air pressure balance must be routinely checked. Common laboratory safety checks, such as for electrical safety, fire control, and chemical or physical hazards must be ob-

served. All instruments should be in good working order, as will be described later.

1.3 WRITTEN PROCEDURES

Two sets of written procedures are of importance to all departments in the clinical laboratory. The first set is intended for use by the medical staff, providing all details necessary to order a test in a given department. The second set of procedures, set up as procedure manuals, are intended for intradepartmental use providing details necessary for performance of each test.

The procedures used by the medical staff to order tests should be drawn up by the clinical virologist and pathologist, and reviewed by a medical committee. The parameters that must be included are:

1. Purpose and limitations of the test
2. Hours the test is performed
3. Test turnaround times
4. Types and quantity of specimen required
5. Specimen transport and holding instructions

These instructions should be circulated to all wards or nursing stations and, if desired, to all staff physicians. They serve as guidelines to ensure that tests are ordered correctly and that appropriate specimens are collected and handled properly. Because clinical virology is a relatively new discipline, nurses and physicians often do not have complete updated training in this area. This inadequacy extends to laboratory personnel as well, and reinforces the need to disseminate information with regard to ordering specific tests.

The instructions for test ordering need not be ironclad, but should be informative guidelines with some flexibility. A particular clinical situation may warrant performance of a test beyond the routine workday, or the submission of an atypical specimen. The degree of flexibility allowed must be decided by consultation between the clinical virologist, pathologist, and the ordering physician, as well as factors such as availability of manpower.

In many instances, a specimen that is submitted may be tested by several departments. In order to avoid confusion and specimen mishandling, the department heads and supervisors should decide the priorities for aliquotting specimens and, thereby, insure handling that is acceptable to all departments involved. These consensus decisions should be codified and made available to all departments. If consensus cannot be reached for a particular specimen type, the pathologist or medical staff should be consulted to arbitrate. Similarly, if an inadequate specimen is submitted, the ordering physician may be asked to prioritize the tests to be performed.

Test ordering information provided to the medical staff may be incorporated into the laboratory procedure manual as the first portion of each test procedure. It is not advisable to provide the entire test procedure to the medical staff, as portions subsequent to the test ordering information are irrelevant and may serve only to confuse floor personnel.

The procedure manual used in the laboratory should be complete enough in detail so that an inexperienced technologist or student could perform the procedure without additional information. One copy of the manual should be readily available to bench personnel, while another copy should be stored separately in event of laboratory accidents.

Information provided in the procedure manual must be properly referenced with current citations, and updated as improved techniques become available. The entire manual must be reviewed periodically by the supervisor, clinical virologist, and pathologist, and these reviews documented by signature. These periodic reviews should not coincide with imminent state or College of American Pathologists (CAP) inspections. Minor changes in procedure may be inserted into the text, but these changes should be approved and initiated by the supervisor or clinical virologist. All changes, major or minor, should be reviewed with:

1. The clinical virology laboratory staff if the change only affects intralaboratory operations
2. The entire laboratory staff or, if impractical, all department supervisors, if changes affect shared specimens
3. Medical staff if the procedural changes concern specimen types, processing, or times of test availability

1.4 QUALITY CONTROL FOR SPECIMEN COLLECTION, TRANSPORT, AND EVALUATION

Timely collection of appropriate specimens and their correct handling are of great importance if false-negative viral cultures are to be avoided. Unfortunately, many medical personnel are unaware of the specimen types suitable for isolation of specific viruses, and that viruses in clinical specimens are generally more labile than bacteria or fungi in clinical specimens. Additionally, many viral specimens are submitted after the prodromal period, the time when virus is often present for many viral infections. These facts reinforce the need for in-house seminars and broad dissemination of information regarding appropriate viral specimens and how they should be handled.

The types of viral specimens that are appropriate for specific viruses have been exhaustively catalogued in this text (see Chapter 2) and elsewhere (Ballew et al., 1977; Lennette and Schmidt, 1979).

It is not recommended that disinfectants or lubricants be used prior to obtaining a viral specimen, as these solutions contain physical agents that can inactivate viruses. Excessively bloody or purulent lesions may be wiped with sterile gauze as needed before obtaining viral or chlamydial specimens.

The relative scarcity of clinical virology laboratories and the logistics of tissue culture inoculation dictate that specimens for viral isolation be held for periods ranging from several hours to days. Viral pathogens that are enveloped such as respiratory syncitial virus (RSV) and cytomegalovirus (CMV) are extremely labile to both room temperature storage and freeze-thaw cycles (Ballew et al., 1982), whereas, nonenveloped viruses such as enteroviruses tolerate these conditions very well. As a general rule, viral specimens held for short periods (e.g., less than overnight) are refrigerated; those held for longer periods are frozen at −70°C. Freezing at −70°C is often difficult, as many physicians' offices and small laboratories do not possess ultralow freezers, and the −20°C freezers available may have automatic defrost cycles, which essentially subject viruses to multiple freeze–thaw cycles. Recent studies have shown that CMV may be held for up to 7 days at refrigerator temperatures (Stagno et al., 1980). A reasonable recommendation, therefore, would be that viral specimens be refrigerated and not frozen unless inordinately long storage times are expected.

1.4.1 Transport Media

The composition and type of viral transport media also can dramatically affect viral isolation rates. In general, the media should have the following characteristics:

1. Balanced salt solution (isotonic) at physiologic pH
2. Contains a substance that will stabilize virus, such as gelatin, fetal calf serum, or bovine serum albumin
3. Swabs may be composed of dacron or rayon, although toxicity with these materials may vary greatly with the manufacturer. Calcium alginate swabs have been reported to be toxic to herpes simplex virus (HSV) (Crane et al., 1980).
4. Antibiotics may be included depending

on whether the transport device and medium are to be frozen or held at room temperature. The particular antibiotic to be employed is important: gentamicin, for example, is stable at room temperature, whereas, fungizone degrades rapidly at this temperature.

Specimens submitted for chlamydial isolation should be transported separately in appropriate media with a stabilizing agent. Penicillin should not be included in the medium, as *Chlamydia trachomatis* is susceptible in vitro to this antibiotic.

1.4.2 Submission of Smears

The submission of smears for fluorescent or immunoperoxidase staining must also be quality controlled. Recent availability of high-caliber monoclonal reagents have made direct immunostaining an attractive alternate to viral and chlamydial cultures, although the rapidity of staining techniques is still offset by the subjectivity of test interpretations. Criteria for acceptance of a smear include:

1. The smear should be from an anatomic site appropriate for the pathogen. Smears should be made by rolling the swab across a cleaned glass slide.
2. The amount of blood or purulent discharge on a smear preparation needs to be evaluated for acceptability due to the problems associated with nonspecific staining of background debris.
3. The smear diameter should be limited to a reasonable size, such as 5–15 mm, as larger smears are wasteful of reagents, and require excessive time to examine.
4. The smear should contain a representative number of cells on a total smear or per field basis. Reasonable numbers would be 50–100 total cells, or 5 cells per low power field.

1.4.3 Specimen for Serology

Sera submitted for viral serology usually present less of a problem in quality control than do cultures or smears. However, the following should be considered:

1. Excessively hemolyzed, lipemic, or bacterially contaminated sera should be rejected.
2. Sera should be heat-inactivated, depending on the specific test(s) to be performed.
3. Paired sera should be submitted for assay of antiviral IgG levels. A single serum sample may be of little diagnostic value, as it often only reflects immune status rather than identifying an acute illness. The physician and clinical virologist or serologist should discuss the times when blood samples should be drawn during the disease course.
4. IgM assays should be properly controlled for interference by rheumatoid factor.
5. IgG assays for TORCH (toxoplasma, rubella, CMV, and HSV) organisms should compare antibody levels in both the infant's and the mother's serum.

In the event that a specimen is rejected for any of the criteria just cited, the ward or attending physician must be informed. This is preferably done with an oral report followed by a written one, stating the reason for rejection of the specimen. Again, extenuating circumstances may warrant acceptance of a substandard specimen, but this must be done after consultation between the clinical virologist, pathologist, and attending physician. Acceptance of this specimen, although substandard, should be documented. The attending physician must be informed of the possible ramifications of an improperly submitted specimen. Routine submission of substandard specimens by individuals should be monitored.

1.5 TISSUE CULTURE AND MEDIA

Tissue culture remains as the mainstay of nonserologic viral diagnosis, even with the advent of promising new techniques, such as antigen capture immunoassays and DNA probes. Thus, adequate quality control for commercially purchased or for in-house preparation of tissue culture cells is of great importance.

With a given cell line, there may be significant variations in sensitivity to virus infections dependent on the particular cell subline or clone, and the passage number. It is relevant, therefore, to accurately record pertinent information regarding cell lines purchased for immediate use or those passaged routinely for in-house use by the clinical virology laboratory. This information should include:

1. Source (e.g., commercial company or cell repository)
2. Individual cell line (e.g., WI-38, ATCC CCL75)
3. Animal and organ of origin (e.g., human lung)
4. Ploidy and morphology (e.g., diploid fibroblast)
5. Passage number at receipt (e.g., passage 19)

Diploid or continuous cell lines routinely passaged for use in the clinical virology laboratory should have the following additional information recorded in a daily cell log:

1. Current passage number (e.g., passage 24)
2. Configuration (e.g., two 25-cm^2 flasks)
3. Confluency (e.g., 90%)
4. Comments on cell condition (e.g., OK, medium acidic)
5. Action taken (e.g., one flask split at 1:7 ratio, second flask held)

Loss of a cell line routinely passaged for use can lead to a severe disruption of work flow in a clinical virology laboratory; therefore, provisions must be made for back-up cells in the event of contamination or laboratory accident. These back-up systems may include:

1. Use of paired stock flasks. When the flasks reach confluency, only one is passaged, while the second flask is held as a back-up until the new flask displays good growth
2. Carrying of a parallel set of stock flasks by a different technologist using a separate set of tissue culture reagents and glassware
3. Freezing and storage of low passage cells at −70°C or in liquid nitrogen; critical cell lines should be stored frozen in two separate freezers or facilities

Similarly, a commercial supplier may fail to ship cells or occasionally may provide cells of poor quality. If possible, contingency arrangements should be made with second supplier. Alternatively, the clinical virology laboratory might consider receiving half their allotment of a particular cell line from each of two different suppliers on a routine basis.

Cells purchased from a commercial supplier should be certified to be free from mycoplasma, fungal, and bacterial contamination. They should be examined on receipt for signs of contamination; for example, primary monkey kidney cells must be examined for evidence of foamy agent or simian viruses, such as SV-5 and SV-40. Overtly contaminated cells should be destroyed immediately. If there is suspicion of minimal contamination that may be due to another cause (e.g., medium acidity), the purchased cells should be held and refed separately from in-house lines until a definite decision is reached, in order to lessen the risk of crosscontamination.

Tissue culture lines carried in-house should be submitted to the following inspections:

1. Daily examination of stock flasks for growth rate and evidence of contamination: Use of a cell log aids in charting the expected growth rate of a healthy cell line.
2. Use of antibiotic-free medium is of ques-

tionable value in a clinical virology laboratory. Use of such medium on a periodic basis, however, will aid in detecting minimal contamination.

3. Freedom from mycoplasma contamination should be monitored periodically (e.g., monthly) by Hoechst stain, (Chen, 1977) 3T6 culture (Douglas and Dell'Orco, 1979) or plating on mycoplasma medium.
4. The sensitivity to viral infections may be checked by conducting periodic $TCID_{50}$ experiments with a stock of reference virus.
5. The quality control tests described in 2, 3 and 4 should be performed whenever there is a suspicion of contamination, such as unduly acid or alkaline medium, turbidity, or alteration in growth kinetics.

Other quality control procedures that may aid in minimizing risk of contamination include:

1. Laboratory personnel with any evidence of infectious disease should not be allowed to handle tissue culture.
2. Separate laboratory apparel, cell culture reagents, and glassware should be used for tissue culture as opposed to specimen inoculated cells. Each technologist should have a separate set of reagents and glassware.
3. If tissue culture tasks and specimen inoculation are to be performed daily by a single technologist, then the cell culture tasks should be performed first, whenever feasible.
4. Cell lines should be handled separately and the biological safety cabinet should be decontaminated in between. On a daily basis the "cleanest" cell lines should be manipulated prior to other cell lines that may have minimal contamination or carry passenger viruses, such as the simian agents.

The following procedures should be employed to quality control tissue culture media used for growth and maintenance of cells:

1. Following the filter sterilization of medium, aliquots should be sampled and either incubated to allow microbial growth or should be plated on media that supports bacterial or fungal growth. These samples should be examined daily for 5 days and should be free of contamination before that lot of medium is released for use. Presence of antibiotics in the medium may mask minimal contamination.
2. Sample aliquots of all other medium components, such as fetal calf serum and L-glutamine should also be checked as above before release for use.
3. New lots of medium that have passed this sterility check should be monitored for their ability to support cell growth.
4. Fetal calf serum is a key component that should be checked for its ability to support cell growth, as well as for absence of virus inhibitory activity.
5. Purchase of large lots of fetal calf serum and other medium components will help minimize the work load of medium quality control. Care should be taken that large lot purchase does not result in reagent expiration before their use.
6. Manufacturer's specifications for medium component storage and the expiration dates should be observed and all aliquots of a given medium component must bear the expiration date. Outdated medium and reagents should be discarded.

1.6 REFERENCE VIRUS STOCKS

Reference virus stock cultures may be obtained from the ATCC (American type culture collection) or other reference institutions such as the Centers for Disease Control and National Institutes of Health.

These reference stocks may serve several purposes in a clinical virology laboratory quality control program:

1. The reference stocks may be used to demonstrate typical cytopathic effects on susceptible host cell lines. Staff technologists should use these reference virus stocks to periodically review the characteristic cytopathology produced. These stocks may also be used to train medical technology and medical students. Fixed and stained infected tissue cultures (Gurtler et al., 1982) serve as a ready supply of material to demonstrate typical cytopathic effects while reducing biohazard risk to students.
2. Tissue culture cells infected with reference virus stocks can also be used to quality control both nonspecific (e.g., Giemsa) and specific stains (e.g., fluorescent, immunoperoxidase) used in clinical virology. These infected cell cultures serve to verify expected staining characteristics on both an intralot and interlot basis.
3. Titered reference virus stocks stored at −70°C or in liquid nitrogen, and used to periodically check the sensitivity of tissue culture lines to each viral agent. Conversely, these reference viruses can be used as an antigen source for interrun controls for antigen capture immunoassays.

1.7 REAGENTS, STAINS, AND KITS

The initial step in the quality control of reagents, stains, and kits (generally referred to from here on as reagents) is to establish a practice of ordering from reputable manufacturers or dealers with reliable transportation systems. Dealers should have a reasonable turnover in stocks so that the laboratory can utilize the kit prior to its expiration date. The reagents should have been warehoused and transported under the environmental conditions specified by the manufacturer. Establishing a standing order for reagents can serve to assure timely arrival of fresh reagents if storage in the laboratory poses a problem.

Upon receipt, the reagents should be checked for obvious breakage or contamination, as well as deviation from specified temperature storage conditions. The quantity, source, lot number, and date of receipt should be entered in a log, and the latter three pieces of information recorded as needed on all aliquots or components if the initial lot is subdivided. Again, one should adhere to the manufacturer's storage specifications. These recommendations by the manufacturers, however, should not be regarded as absolutes guaranteeing the "freshness" of reagents. General guides for storage of different types of reagents have been described (Bartlett, 1980) and should be consulted. Reagents bought in bulk and intended for long-term storage up to expiration should be monitored at receipt and periodically with standardized reference materials. An inventory of all reagents received in-house should be maintained.

Inorganic chemicals present somewhat of a problem, because expiration dates often are not specified by the manufacturer. However, quality control can be maintained by observing specified storage conditions for chemicals, such as in amber glass bottles or under desiccant. New lots of reagents prepared from chemicals should be compared to the old, and comparable performances noted and documented before the new lot is put into service.

When new lots of any reagent are opened, the date should be noted on the container. New lots should perform comparably with old lots. The degree of variability that is acceptable will be discussed in a later section.

It is advisable to obtain reference materials for use to test various reagents for interrun and interlot variation. Examples of these reference materials would be patient sera of known titers, or fixed and frozen cells infected with reference viruses. Unused sera or antigen from state or national (e.g., CAP) proficiency surveys are excellent as reference materials to monitor and control test variation, as these materials are generally well standardized and preserved.

Two notes of caution should be mentioned on the use of kits. Different components of a kit may have different storage conditions and expiration dates. These components, therefore, should be separated and stored under the recommended conditions when received. The lot number for the kit should be noted on each component if not already present. Individual components of a kit may be calibrated against each other. Therefore, components from different lots of a type of kit should not be interchanged.

A common question is how often controls supplied with kits or other reagents should be run. Here are some considerations:

1. The frequency of running a control reflects the interrun variation of the test. Tests with greater intrinsic variability need to be more tightly controlled.
2. One should also consider how often the test is run. If, for instance, several runs of a test are conducted daily with the same lot of reagents, one set of controls may be adequate. On the other hand, infrequently run tests probably should always have controls run with them, as proficiency and technical judgment can fade with disuse. In the same vein, one should consider sending tests that are run very infrequently to a reference laboratory.
3. The cost-effectiveness of running controls must also be balanced against their benefits. Excessive use of controls can reach a point of diminishing returns.

1.8 INSTRUMENTS

Clinical laboratory instruments should be subjected to routine preventive maintenance and also surveyed on a regular basis for satisfactory performance. Consistent patient results cannot be expected with unreliable equipment. Furthermore, improperly maintained defective instruments can pose a threat to the safety of laboratory personnel.

The minimum preventive maintenance performed should follow the manufacturer's recommendations. Note that these recommendations assume normal usage and applications for the instrument. Excessive use or special applications may warrant additional preventive maintenance. This work can be performed by appropriate workshops in the hospital physical plant, or by an in-house instrument specialist. Work beyond the scope of in-house capabilities can be performed on manufacturer's service contracts. Logs should be kept of the routine preventive maintenance, and service manuals should be readily available to perform routine trouble-shooting.

A portion of routine preventive maintenance, such as cleaning of microscope objectives, can be performed by laboratory personnel. Logs of this routine intralaboratory maintenance work can be combined with routine performance checks, and charts documenting this work attached to the instrument. Information on such charts should include:

1. Instrument name and serial number
2. Performance checks or maintenance to be done
3. Performance ranges
4. Date work is done with technologist's initials

Failure in instrument performance can be logged in a central log. Remedial action taken to correct this failure should be noted and initialed by the clinical virologist or supervisor.

The following are some recommendations for routine laboratory maintenance and performance checks on instruments commonly found in the clinical virology laboratory:

Incubators: Daily temperature, CO_2, and humidity checks; weekly decontamination of interior

Biosafety hoods: Daily cleaning of ultraviolet lamp and check of air pressure; decontamination of work surface before and after each use; annual checks for air velocity and filter integrity; paraldehyde decontamination, as applicable

Microscopes: Daily cleaning of objectives and stage; log of lamp usage with mercury vapor bulbs; annual overhaul of microscope company

Refrigerators and freezers: Daily temperature check; annual check of compressor and refrigerant levels

Water baths and heat blocks: Daily temperature check; weekly decontamination

Centrifuges: Weekly decontamination; quarterly speed calibrations with tachometer; annual inspection of motor and drive system

Autoclaves: Daily temperature check and spore strip testing

Rotators: Daily rpm check

Spectrophotometers: Absorbance and linearity check with each run

pH Meters: Single reference buffer check before each use; multiple point check monthly

Pipetting devices: Gravimetric volume check monthly; annual overhaul

Microdiluters: Before each use

1.9 PROFICIENCY TESTING

Proficiency testing may be broadly divided into those initiated intralaboratory, and those applied by external agencies or laboratories. A mixture of these two types of test aid in insuring that accurate results are generated by the clinical virology laboratory.

Typically, the most frequently encountered intralaboratory quality control tests are the regular use of controls that are run with the specific tests. These serve to establish ranges or validate results on an intra- or interrun basis. Every effort must be made to insure that these controls are treated in the same manner as patient samples, so as not to introduce bias. However, it is often human nature that controls are treated with special deference, especially when controls have not given expected values in a recent series of runs. Therefore, it is very useful for the clinical virologist or supervisor to introduce blind controls on a regular basis. These should be indistinguishable from patient specimens. This should not be construed by the technologist as an attempt to police their work, but rather as additional proof of their consistent competency. The bench technologists may reciprocally ask the clinical virologist or supervisor to view (with them) some blind specimens set up by the technologists.

The type of externally applied proficiency testing with which most laboratories are familiar are the formal tests issued by state laboratories or the CAP. (Laboratory accreditation is contingent on satisfactory performance in these formal proficiency tests). Until recently, the proficiency testing by these institutions has been limited to viral serology, such as rubella and HSV titers. The CAP has recently introduced samples comprised in part of smears and specimens for viral isolation, Another type of externally applied proficiency testing may be done on a formal or informal basis with other hospital and clinical virology laboratories. Directors of the laboratories may agree to exchange unknown specimens on a routine schedule. These specimens are then inserted into routine test runs as blind unknowns.

All categories of proficiency testing, be

they routine controls, blind in-house samples, or externally supplied samples, must be adequately documented. The clinical virologist or supervisor should review these results on a routine basis and approve acceptable performance by initialing the results. In the event of a failure of the proficiency test, it is the duty of these department heads to determine the cause of the quality control failures and to rectify them. This is most simply accomplished by having a log with the following information:

1. Type of test failing quality control
2. Date of failure
3. Name of technologist performing the test
4. Reagent lots associated with the failure
5. Probable reason for failure
6. Rectifying action taken
7. Supervisory signature

All such quality control failures should be reviewed with the pathologist so that he or she is aware of them. Discussion between the clinical virologist and the pathologist should be targeted at identifying the trends in quality control failure. The trends may be associated with certain reagent lots, instruments, or technologists. Identification of these trends, thus, can serve to prevent future occurrences of quality control failures.

1.10 STATISTICS

Clinical virology laboratories generate both qualitative and quantitative patient results. Quantitative results tend to come from serologic tests, whereas, viral isolations and stains yield qualitative results.

Generally, traditional viral serology has not been statistics oriented. However, new instrumentation and technology, such as enzyme immunoassays have militated the introduction of criteria to decide if an analytic run is within control limits. The former standard for rejection of a quantitative analysis was to reject a run if any control value fell outside two or three standard deviations from the mean (Howanitz and Howanitz, 1983). Many analytic laboratories have now adopted the "CAP modified" approach (Haven et al., 1980), in which a run is rejected when:

1. Patient results appear incorrect, regardless of the control results
2. Any single control value deviates more than 3 standard deviations from the mean
3. One control value is greater than 2 S.D. from the mean on two consecutive runs
4. Two controls are used and both control values are greater than 2 S.D. from the mean

Statistical analysis of qualitative results cannot be so precisely defined. Viral isolation rates vary according to geographic region, age and socioeconomic class of the patient population, and introduction of new viral antigenic variants. A practical approach to determining if a clinical virology laboratory's isolation rate is acceptable would be to compare these rates with numbers generated by the CDC, state and regional laboratories, or clinical virology laboratories in area hospitals having similar patient populations. If a laboratory's isolation rate differs significantly from these institutions, then measures to determine a cause for these lower isolation rates, such as improper specimen transport or mycoplasma contamination of cell lines, should be undertaken.

Inconsistency of isolation rates can also be calculated on an intralaboratory basis. Monthly calculations of isolation rates allow the clinical virologist to determine if performance is typical. In comparison of these monthly isolation rates, seasonal variations such as onset of the influenza season should be considered.

REFERENCES

Ballew, H., Forrester, F.T., and Lyerla, H.C. 1982. Laboratory Methods for Diagnosing Respiratory Virus Infections. Atlanta, GA: Centers for Disease Control, pp. 38, 49.

Ballew, H., Forrester, F.T., Lyerla, H.C., Valleca, W.M., and Bird, B.R. 1977. Laboratory Diagnosis of Viral Diseases. Atlanta, GA: Centers for Disease Control, p. 10.

Bartlett, R.C. 1980. Quality control in clinical microbiology. In E.H. Lennette, A. Balows, W.J. Hausler, Jr., and J.P. Truant (eds.), Manual of Clinical Microbiology. Washington, D.C.: American Society for Microbiology, pp. 20–21.

Chen T.R. 1977. In situ detection of mycoplasma contamination in cell cultures by fluorescent Hoechst 33258 stain. Exp. Cell Res. 104:255–262.

Crane, L.R., Gutterman, P.A., Chapel, T., and Lerner, M. 1980. Incubation of swab materials with herpes simplex virus. J. Infect. Dis. 141:531.

Douglas, W.H.J., and Dell'Orco, R.T. 1979. Physical aspects of a tissue culture laboratory. In W.B. Jakoby and I.H. Pastan (eds.) Methods in Enzymology, Vol. 58, Cell Culture. N.Y.: Academic Press, p. 24.

Gurtler, J., Ballew, H., and Smith, T. 1982. Cell culture medium for preserving cytopathic effects in cell cultures. Lab. Med. 13:244–245.

Haven, G.T., Lawson, N.S., and Ross, J.W. 1980. Quality control outline. Pathologist 34:619–624.

Howanitz, P.J., and Howanitz, J.H. 1983. Quality control for the clinical laboratory. Clin. Lab. Med. 3:541–551.

Lennette, E.H., and Schmidt, N.J. (eds.). 1979. Diagnostic Procedures for Viral, Rickettsial and Chlamydial Infections. Washington, D.C.: American Public Health Association.

Richardson, H.J., and Barkley, W.E. (eds.). 1984. Biosafety in Microbiological and Biomedical Laboratories. Washington, D.C.: Centers for Disease Control and National Institutes of Health. U.S. Government Printing Office, pp. 10–23.

Stagno, S., Pass, R.F., Reynolds, D.W., Moore, M.A., Nahmias, A.J., and Alford, C.A. 1980. Comparative study of diagnostic procedures for congenital cytomegalovirus infection. Pediatrics 65:251–257.

Warren, E. 1981. Laboratory safety. In J.A. Washington (ed.), Laboratory Procedures in Clinical Microbiology. N.Y.: Springer-Verlag, pp. 729–744.

Specimen Requirements
Selection, Collection Transport, and Processing

Thomas F. Smith

2.1 INTRODUCTION

The diagnosis of most viral infections has been based on clinical grounds rather than by specific laboratory tests. The extensive time lapse between submission of specimens and the availability of results, and the lack of effective forms of treatment or prevention of viral disease have been major reasons for this situation. This trend has changed dramatically, especially in the last few years. For example, herpes simplex virus (HSV), once thought to produce only benign lip lesions, is now recognized to be a major cause of sexually transmitted disease with possible transmission to newborns (Tummon et al., 1981). Further, HSV has emerged as an important viral infection today in terms of prevalence, morbidity, and mortality in that it is capable of producing a wide spectrum of clinical manifestations, including keratitis, gingivostomatitis, encephalitis, and oftentimes disseminated disease in immunosuppressed individuals (Felman and Sonnabend, 1979; Smith, 1983).

Recently, the diagnosis of HSV infections has become accessible to microbiology laboratories without preexisting virologic capabilities, with a commercially available system called Cultureset™, which uses a peroxidase–antiperoxidase antigen detection protocol (Rubin and Rogers, 1984). Moreover, a rapid 5-hour enzyme-linked immunosorbent assay (ELISA) for the direct detection of HSV in clinical specimens is now available (Morgan and Smith, 1984). Once the laboratory diagnosis of HSV infection has been established, specific antiviral drug therapy may be instituted in many clinical situations (Koch-Weser et al., 1983). For at least one very important viral infection, therefore, prompt laboratory diagnosis and treatment can be implemented at the primary level of medical care. This trend is sure to follow for other viral infections, such as cytomegalovirus (CMV), now that monoclonal antibodies have been developed and can be purchased for use in the laboratory (Gleaves et al., 1984).

The success rate of the diagnostic laboratory is dependent on frequent communication between the laboratory and the physician. This is especially true in an era of evolving technology. No technique, however, regardless of how rapid, is useful unless the quality of specimen is adequate regarding the source, method of transport, and means of processing. Finally, the ever increasing cost of operating a clinical laboratory obligates the laboratory director to suggest to clinicians the selection of the most probable specimen(s) that would yield a definitive laboratory diagnosis of a viral infection.

2.2 SPECIMEN SELECTION

Comprehensive tables listing the selection of viral specimens based on clinical infections and the suspected viral agents have been published (Lennette and Lennette, 1981; Chernesky et al., 1982; Smith, 1984). A specific guide for most clinical situations is presented (Table 2.1).

Several viruses may cause respiratory tract infections. In almost all cases, a throat or nasal swab/washing should suffice. With other viral infections, such as those involving the central nervous system (CNS) or those associated with congenital disease, several types of specimens should be submitted.

2.3 COLLECTION

2.3.1 Timing

Specimens should be collected early in the acute phase of infection. However, the duration of viral shedding depends on the type of virus and systemic involvement, as well as other factors. For example, the duration of respiratory syncytial virus (RSV) shedding is usually 3 to 7 days, but with a range of 1 to 36 days (Hall et al., 1976a;b). Further, viral excretion was of longer duration in those with lower respiratory tract disease than in those with clinical manifestations limited to the upper respiratory tract (Hall et al., 1976a).

Interpretation of the significance of enterovirus isolates from stool specimens is confounded by the prolonged shedding (6 to 8 weeks) of these viruses, especially in children (Wilfert et al., 1983). Virus is commonly present for a shorter time in the oropharynx, usually resulting in successful isolation from specimens collected during the first 5 to 7 days of illness.

The immunocompetence of a patient with a viral illness has a significant effect on the excretion pattern of the infectious agent. For example, high concentrations of HSV ($>10^4$ plaque forming units) were detected from lesions for more than 3 weeks (Daniels et al., 1975). The mean duration of HSV shedding from immunocompetent men and women with genital infection was 11.4 days (Corey et al., 1983). Different echoviruses were recovered from the cerebral spinal fluid (CSF) of children with agammaglobulinemia for periods varying from 2 months to 3 years (Wilfert et al., 1977). Varicella-zoster virus (VZV) could be recovered from the buffy coat of blood specimens from patients with malignant disease for at least 8 days after the onset of cutaneous lesions, but could not be isolated from patients with typical varicella and no underlying malignancy (Myers, 1979). Generally, VZV can be recovered from lesions for up to 7 days after the initial vesicles appear, but from blood only during the late incubation period or days 1 to 4 of the acute illness (Feldman and Epp, 1979). Isolation of an enterovirus from the CSF of immunocompromised patients usually is successful only within 2 to 3 days after onset of the CNS manifestation when few leukocytes are present in the CSF (Lerner et al., 1978).

Some viruses, such as CMV, rubella, adeno-, and enteroviruses can be excreted from various sites in the asymptomatic individual for months and even years. However, collection of the specimen at the appropriate time is meaningful in regard to association with the current disease process. For example, a CMV-positive urine specimen collected during the first 3 weeks of life from an infant with congenital disease is strong laboratory evidence supporting the viral etiology of the congenital anomalies. Conversely, isolation of CMV from the same source after 3 weeks would not discriminate between congenitally or postna-

Table 2.1. Specimen Information for Diagnostic Virology Services

Clinical Presentation and Common Virus Infection	Serology Available	Specimens for Culture	Collection Procedure	Transport Device	Comments
Respiratory tract					
Adenovirus	Yes	Throat swab, sputum	*Throat swab:* Swab inflamed area of throat *Sputum:* Instruct patient to produce sputum, in response to deep cough	Culturette, Virocult, or other swab (not calcium alginate) extracted in transport medium	Specimens usually collected from patients with lower respiratory tract disease
Enterovirus	No				
HSV	Yes				
Influenza virus	Yes				
Mumps virus	Yes				
Parainfluenza virus	Yes				
Respiratory syncytial virus	Yes				
Rhinovirus	No				
Rash					
Maculopapular					
Adenovirus	Yes	Throat, rectal swab	*Rectal swab:* Insert swab 3–5 cm into rectum to obtain fecal material		Differentiated diagnosis of viral vs. bacterial infections important for therapeutic considerations
Enterovirus	No				
Rubella virus	Yes				
Measles (rubeola)	Yes				
Vesicular					
Coxsackievirus A or echovirus	No	Swab of lesion or vesicular fluid	Rupture vesicle and scrape cells from base of lesion		Tzanck smear (cells from lesion stained with a Wright's or Giemsa preparation) and direct fluorescent antibody stains are rapid, but usually only 40%–60% sensitive for diagnosis of HSV, VZV. Rapid tests available for detection and reliable serotyping of HSV
HSV	Yes				
VZV	Yes				
Vaccinia and other poxviruses	No		Aspirate vesicular fluid from several fresh lesions	Syringe with 26-gauge needle	

Table 2.1. (*continued*)

Clinical Presentation and Common Virus Infection	Serology Available	Specimens for Culture	Collection Procedure	Transport Device	Comments
Central nervous system					
Arbovirus	Yes	CSF, throat, and rectal swabs, urine (mumps), brain tissue for biopsy	*CSF:* Specimen collection by physician or specifically trained personnel	Sterile, screw-capped tube	
Enterovirus	No				
HSV	Yes				
Mumps	Yes		*Urine:* Collect clean voided urine	Sterile, screw-capped tube	Incidence of mumps virus infections is rapidly decreasing
Infectious mononucleosis, immunodeficiency, and hepatitis					
EBV	Yes	Does not replicate in usual cell cultures	*Blood:* 5 ml of heparinized blood	Heparinized blood collection tube	First line test is the Monospot test for heterophile antibodies indicative of EBV infection. Rapid tests available for detection CMV in cell cultures after 16 hr
CMV	Yes	Urine, blood			
Hepatitis viruses, A, B, and non-A, non-B	Antigen and/or antibody (no test for non-A, non-B)	Serum for antigen or antibody			

Gastrointestinal					
Adenovirus	Yes	Stool specimen or rectal swab	*Blood:* 5 ml of heparinized blood	*Stool:* sterile screw-capped container	Assays for Norwalk agent infection are performed in only a few research laboratories. Rotavirus antigen can be rapidly detected by ELISA methodology.
Norwalk-like agents	No				
Rotavirus	Antigen detection				Etiology of adenovirus in gastrointestinal tract disease is being evaluated. Rapid tests available for detection and reliability of serotyping of HSV
Genital					
HSV	Yes	Swab of lesion or vesicular fluid	*Blood:* 5 ml of heparinized blood		
Congenital and perinatal					
CMV	Yes	Urine, blood, swab of lesion	*Blood:* 5 ml of heparinized blood		Tests for IgM class antibody, rather than IgG antibodies, necessary for serologic diagnosis
Enterovirus	No				
HSV	Yes				

Abbreviations: CSF, cerebrospinal fluid; HSV, herpes simplex virus; VZV, varicella-zoster virus; EBV, Epstein–Barr virus.
Table adapted from Chernesky et al., 1982.

tally acquired CMV infection (Spector, 1983).

Adeno- and enteroviruses can be excreted in stool specimens for several days to weeks, even though the ability of these viruses to cause gastroenteritis is debated by some. Nevertheless, both viruses are recognized pathogens causing acute upper respiratory tract disease (Kepfer et al., 1974).

2.3.2 Source

2.3.2.a Throat, Nasopharyngeal Swab, Nasal Washing

Throat or nasal washing may be more productive for viral isolation than throat or nasal swabs, but few comparisons have been reported. In one study, for example, nasal wash specimens were collected from children, but throat or combined nose and throat swabs were obtained only from adults (Hall and Douglas, 1976). The convenience of using a swab by medical personnel and the willingness of the patient to allow collection of this specimen (compared with washings) are important factors in this choice. Swabs are considered superior to nasal wash because several problems are associated with the latter. Nasal wash specimens submitted for fluorescent antibody detection of viral antigens often contain debris, such as mucus, squamous cells, leukocytes, and erythrocytes. Also, the number of cells obtained by nasal wash was smaller than that obtained by nasal and oropharyngeal swabs combined (Blumenfeld et al., 1984).

2.3.2.b Sputum

Recovery of a virus from this source does not necessarily reflect lower respiratory tract infection, but may be due to "contamination" of the specimen with the agent present in the throat. Alternatively, the recovery or cytologic detection of a virus from a transtracheal aspirate or transbronchial biopsy indicates lower respiratory tract involvement (Blumenfeld et al., 1984). Although Kimball et al. (1983) obtained an isolation rate of 20% from patients diagnosed as having radiologically confirmed pneumonia in one study, they suggested that not all of the viruses recovered from this source may be of clinical importance in the etiology of lower respiratory tract disease in their study population. Thus, of all of the viral isolates: influenza virus (H_3N_2), six; respiratory syncytial virus, two; HSV, nine; and rhinovirus, three, the investigators felt that only the influenza and RSV (total, 8%) caused lower respiratory tract disease in their patients. On this basis, they speculated that sputum specimens may be of particular value in the laboratory diagnosis of lower respiratory tract disease due to viruses. Unfortunately, because of the severity of illness in their study population, these investigators were unable to obtain throat washings for comparison with the sputum specimens.

Interestingly, HSV and CMV are capable of growth in human alveolar macrophages, thus, bronchopulmonary lavage specimens yielding these cells may be useful as a relatively simple technique to make the laboratory diagnosis of lower respiratory tract infection due to these viruses without resorting to an open-lung biopsy procedure (Drew et al., 1979; Stover et al., 1984).

2.3.2.c Rectal Swabs and Stool Specimens

The use of feces as a specimen for virus isolation has been reduced with the realization that viruses that are noncultivatable in cell cultures (rotavirus, Norwalk-like agent, and perhaps some adenoviruses) are responsible for most cases of viral gastroenteritis. Enteroviruses, however, may be recovered from the stool but not the CSF in cases of CNS disease. The etiologic association of such isolates from the gastrointestinal tract becomes somewhat more tenuous than from CSF, however, owing to the common excretion of these agents from the stool subsequent to respiratory tract involvement (Bowen et al., 1983). In addition, enteroviruses can be recovered from the gastrointestinal tract of unaffected children as frequently as from those children with respiratory tract disease. There is a higher rate of virus isolation from stool specimens but these are less convenient than rectal swabs for the diagnosis of viral gastroenteritis (Mintz and Drew, 1980).

Viruses surrounded by a lipid envelope generally are not found in an active form in stool specimens; however, CMV is now recognized as a cause of intestinal disease and has been recovered from that source (Foucar et al., 1981).

2.3.2.d Urine

Many viruses are excreted in the urine during the incubation period of infection. Mumps, adenovirus, and CMV are commonly recovered from urine by the laboratory after symptoms develop. Importantly, mumps virus can be isolated from urine when specimens from other sites are negative, as in cases of CNS disease.

Urine specimens submitted for the laboratory diagnosis of CMV infection can be inoculated directly into cell cultures because the virus is not concentrated in urine sediment after low-speed centrifugation (Lee and Balfour, 1977). Urine is the best single specimen for recovery of CMV, but the virus may only be isolated using a throat swab (Henson et al., 1972; Glenn, 1981). Because excretion of CMV is found to be somewhat sporadic, multiple specimens may be required. In third-trimester pregnant women, cervical excretions appreciably exceeded urinary shedding of virus, the rates being 11.6% (48 of 404) and 6.3% (29 of 463), respectively (Reynolds et al., 1973). As expected, HSV was recovered more frequently from specimens of the cervical canal compared with urine in pregnant women with acute genital infections (Kawana et al., 1982).

2.3.2.e Dermal Lesions

HSV (70%), VZV (29%), coxsackievirus type A (1%), and perhaps some echoviruses are the principal agents that can be recovered on a routine basis from dermal lesions (Desada-Tous et al., 1977; Smith, 1983). The sensitivity of any laboratory test for HSV varies with the stage of the lesion. For example, HSV was recovered from 94% of vesicular lesions, 87% of pustular lesions, 70% of ulcers, and 27% of crusted lesions (Moseby et al., 1981). Similarly, smears prepared with cells obtained from vesicles for Papanicolaou, crystal violet, or immunofluorescence staining were superior to cells obtained from ulcers for the diagnosis of HSV infections.

2.3.2.f Cerebrospinal Fluid

Nonpolio enteroviruses and mumps virus are the most common isolates from the CSF. HSV, an important cause of CNS disease, rarely has been isolated from the CSF. Those instances have usually been associated with meningitis due to HSV type 2 (Rubin, 1983). Although isolation rates of viruses from CSF specimens generally are low (<4%), this source has been particularly productive for the recovery of enterovirus (67–80%) (Rubin, 1983). CMV, VZV, and adenoviruses are rare isolates from CSF in compromised hosts. Togaviruses, although present in CSF, usually are not collected for testing because they do not replicate well in the cell cultures commonly used for routine viral diagnosis.

2.3.2.g Eye

HSV (66%) and adenovirus (34%) are the viruses commonly associated with eye infections (Smith, 1984); however, enteroviruses have been associated with hemorrhagic conjunctivitis (Pal et al., 1983).

2.3.2.h Blood

Isolation of viruses from the blood provides evidence of acute phase infection and symptomatic disease (Neiman et al., 1977). Specific separation procedures for the collection of leukocyte fractions of the blood will yield higher isolation rates of viruses, such as CMV and VZV, compared with buffy coat preparations, especially from specimens of neutropenic patients (Howell et al., 1979). On the other hand, enteroviruses were isolated from 14 of 31 frozen serum specimens obtained from hospitalized patients with enterovirus infection that had been documented previously by recovery of virus from stool, throat, or CSF (Moseby et al., 1981). Similarly, whole blood was used to document congenital infection due to echovirus type 11 (Jones et al., 1980).

2.3.2.i Tissue

Generally, lung and other tissue from the respiratory tract and brain tissue are the only specimens that yield viruses in cell cultures. Occasionally, liver and, rarely, spleen tissue have yielded CMV or HSV. Of 95 viral isolates from tissue over an 8-year period (isolation rate, 3.6%), 82 (86%) of these were CMV, HSV, parainfluenza, influenza, rhino- and adenovirus from respiratory tissues (Smith, 1983). Certainly, selection of certain tissue specimens will improve the recovery rate of viruses from this source. For example, of 105 open-lung biopsy specimens, obtained mostly from immunosuppressed adults, CMV was recovered from 20 (19%), influenza virus type A from one. Generally, however, the recovery rate is less than 10% and the majority of isolates are CMV from lung specimens. The low rate of isolation may stem from the release of viral inhibitors in tissue after homogenization. Alternatively, enzymatic digestion of tissue fragments have provided higher rates of viral isolation compared with homogenized specimens (Shope et al., 1972).

2.4 TRANSPORT

2.4.1 Swabs

Swab tips consisting of rayon, cotton, dacron, or polyester are not toxic to HSV; however, contact of this virus with calcium alginate results in rapid inactivation (Crane et al., 1980).

2.4.2 Stability of Viruses

In general, viruses that are enveloped, such as the herpesviruses and the myxo- and paramyxoviruses (especially RSV), are relatively labile compared with those without envelopes. Nevertheless, these viruses survive transit for at least 24 to 48 hours if maintained at 4°C (Levin et al., 1984). HSV can survive for as long as 2 hours on the surface of skin, 3 hours on cloth, and 4 hours on plastic (Turner et al., 1982).

2.4.3 Transport System

Comprehensive studies of viral transport media are difficult to find in the scientific literature. The reason is likely the substantial expense of comparing two or more types of media for transport of swabs from the physician's office to the laboratory. As a result, viral transport devices are often selected on the basis of convenience. For example, at the Mayo Clinic, Culturettes™, (Marion Scientific, Kansas City, MO) have been used for over 15 years. This swab consists of a plastic tube containing a sterile rayon-tipped applicator and an ampule of modified Stuart's transport medium, which was originally formulated about 35 years ago specifically to prolong the viability of *Neisseria gonorrhoeae* in transit. These swabs have allowed recovery of commonly isolated viruses in clinical laboratories at rates ranging up to 50% (Smith, 1979).

An advantage for our institution is that one swab, the Culturette™, may be used for specimen collection and transport of organisms appropriate for diagnostic use in bacteriology, mycology, parasitology, and virology.

Two other transport media [Hanks' balanced salt solution (HBSS) and Leibovitz–Emory medium (LEM)] were compared with Stuart's and incorporated into the ampules of Culturettes by Marion Scientific. The swabs were coded and provided to pediatricians with instructions to insert all three swabs simultaneously into the oropharynx or on vesicles on the dermal surface, and then transport the specimen to the laboratory by routine means (transport time, 30 minutes to 21 hours). Of 80 isolates from 200 children (40% isolation rate), 72 (90%) were recovered in HBSS, 64 (80%) in Stuart's, and 63 (79%) in LEM. Although the greatest number of isolates were recovered in HBSS, the differences in isolation rates among the three media were not statistically significant (Huntoon et al., 1981). It would be necessary to use several hundred patients in such a study in order to demonstrate any substantial difference in the performance of the media.

The incorporation of protein, such as serum, albumin, or gelatin into transport media has been advocated as a means of stabilizing viruses during transit to the laboratory. Recently, we participated in a four-site evaluation of a laboratory test kit for rapid diagnosis of genital HSV infections. All four laboratories processed specimens sent to their specific laboratories with request for HSV culture. Two laboratories used a transport medium containing 0.5% gelatin and another incorporated 2% fetal bovine serum as a stabilization agent. Our laboratory used the Culturette™ during the study, and almost all specimens were submitted from another reference laboratory that required a minimum of 24 hours transit time between collection and receipt at the Mayo Clinic laboratory. Transport time for specimens submitted to the other three laboratories was routinely shorter than 24 hours. The addition of protein to stabilize HSV during transit provided no apparent positive effect on the overall isolation rate of virus (Table 2.2). Admittedly, this is a comparison of apples with oranges, because the specimens are not common, however, the methodology used at all investigational sites was similar.

Comparative studies of media for viral transport have been outlined in detail elsewhere. Most of these studies used laboratory-passaged strains of virus in assessment of different formulations of transport media (Smith 1979; 1983; Chernesky et al., 1982). Two of these studies indicated that HSV survives equally well in Stuart's or HBSS compared with protein-containing medium (Rodin et al., 1971; Yeager et al., 1979). Thus, at least for HSV, it would seem that the type of transport medium is not of paramount importance for virus isolation. The presence of protein, no doubt, helps to stabilize viruses during a freeze–thaw cycle.

Virocult (Medical Wire and Equipment Co., Cleveland, OH) is another commercially available self-contained viral collection and transportation system that consists of a sterile pack containing a green color-coded collection swab and plastic transtube. The transtube contains a small sponge saturated with 1.0 ml of a buffered phosphate (pH 7.2), solution containing D-glucose, lactalbumin hydrolysate, chloramphenicol, and cycloheximide. At ambient temperature it has a shelf-life of 2 years. Surprisingly, the recovery rate of HSV from specimens submitted with the Virocult system was 15.0% and 15.7% from samples transported in tryptose phosphate broth (TPB). The investigators concluded that Virocult was at least equal in efficiency to TPB and offered the advantage of commercial availability, extended shelf-life at ambient temperature, and ease of use (Perez et al., 1984). In another evaluation, this transport system provided a HSV iso-

Table 2.2. Isolation of Herpes Simplex Virus

Type of Laboratory	Transport Medium	Number of Isolates/ Number of Specimens	Recovery (percent)
Private virology reference	Hanks' Balanced Salt Solution + 0.5% gelatin	33/124	26.6
Medical center and reference laboratory (Mayo Clinic)	Culturette	60/179	33.5
Community hospital laboratory	Eagles Minimal Essential Medium + 2% fetal bovine serum	32/110	29.1
Medical school hospital laboratory	Veal infusion broth + 0.5% gelatin	63/214	29.4

lation rate of 22.4% from 2000 clinical specimens, a result consistent with most other transport media (Johnson et al., 1984). Interestingly, however, HSV could be successfully isolated after a holding period of 12 days in the Virocult transport tube. Overall results from their study indicated that shipping times of up to 2 to 3 days from outlying areas usually result in satisfactory survival of the virus.

A new type of commercial transport system for viruses, the Transporter Tube (Bartels Immunodiagnostics, Bellevue, WA), consists of a plastic centrifuge tube containing a monolayer of human diploid fibroblast monolayers around the lower conical portion of the centrifuge tube with 2 ml of medium. This system was tested in parallel with sucrose-phosphate-glutamate (SPG) for the recovery of viruses from specimens transported an average distance of 73 miles to the laboratory. The Transporter Tubes were kept at ambient temperature and the SPG extracts were refrigerated at 1°–10°C on cold packs. Ninety-two (91%) of 101 viral isolates were recovered in cell culture from the Transporter Tube compared with 82 (81%) from SPG. Twenty-five (24.7%) of the viral isolates were detected by cytopathic effects (CPE) that developed 1 to 4 days earlier with specimens transported in the Transporter Tube than specimens transported in SPG. Of the 101 total virus isolates in the study, however, nine were recovered exclusively in SPG and only 19 in Transporter Tubes. Overall, the authors felt that the Transporter Tubes system was superior to SPG because of the enhanced development of CPE and the facilitated transport of clinical material at ambient temperature.

2.4.4 Storage of Specimen

Viruses, such as adeno- and enteroviruses, that do not have structurally labile lipid envelopes survive freeze–thaw procedures with relatively little loss of viral titer. Conversely, a single freeze–thaw cycle decreased the infectivity titer of HSV by 100-fold. Even storage at room temperature for 1 to 30 days significantly reduced the infectivity of this virus. In contrast, none of 65 samples stored for this period of time at 4°C in HBSS or TPB media showed more than a tenfold loss in infectivity. For short-term (<5 days) transit or storage of most viral suspensions, therefore, the specimen should be held at 4°C, rather than frozen (Chernesky et al., 1982).

2.5 PROCESSING SPECIMENS

2.5.1 Inoculation

Generally, 0.2–0.3 ml are inoculated onto human diploid fibroblast (HDF) cells, primary monkey kidney, and a continuous cell line, such as HeLa or HEp-2 cells. Primary rabbit kidney cells may be used in place of HDF cells for the recovery of HSV (Moore, 1984) and rhabdosarcoma (RD) and buffalo green monkey (BGM) kidney cells have been found to yield more enteroviruses compared with the usual cell systems.

Removal of liquid medium from cell monolayers before inoculation of a specimen, in order to allow for adsorption of viral particles to the cells (1 hour) probably enhances the rate of recovery of these viruses; however, comparative data from clinical specimens have not been published. Based on studies that demonstrated that low-speed centrifugation (2000–3000 × *g*) of specimen inoculum onto cell culture monolayers increased the efficiency of infection of *Chlamydia* spp. (Reeve et al., 1975), murine CMV (Osborn and Walker, 1969), and the AD169 strain of human CMV (Hudson et al., 1976), we applied this technique to urine specimens for the diagnosis of human strains of CMV (Gleaves et al., 1984). In this procedure, 0.2 ml of urine is inoculated onto monolayers of MRC-5 cells seeded in 1-dram shell vials containing a circular coverslip. The inoculated shell vials are then centrifuged at 700 × *g* for 1 hour, medium is added back and the vials, incubated at 36°C for 16 hours. At this time, the coverslip containing the cell monolayer is tested for the presence of early antigen of CMV by immunofluorescence. This rapid test reduced the detection time for CMV from an average of 8 days in conventional tube cell cultures to 1 day (16 hours). The sensitivity of the rapid test was 100% when compared with conventional cell culture isolation. If the specimens were not centrifuged, however, the sensitivity was only 37.5%. This technique has been applied for the rapid diagnosis of HSV infections in the laboratory, and seems applicable to other viruses, as well (Gleaves et al., 1985). The only limitation of the methodology is the availability of specific monoclonal or polyclonal antibodies for use as a probe for early antigens synthesized by a virus.

Inoculation of specimens directly into cell cultures at the bedside has yielded 500-fold higher titers of agents, such as RSV (Hall and Douglas, 1975). However, a recent study of 135 samples inoculated at bedside or held for 3 hours at 4°C before transport to the laboratory showed no difference in the rate of RSV recovery. Of 51 positive specimens, 44 (86%) were positive by both inoculation procedures, three (6%) were recovered only from specimens inoculated at bedside, and four (8%) were positive for RSV only when the specimen was inoculated in the laboratory (Bromberg et al., 1984).

2.5.2 Examination

Conventional test tube cell cultures are examined at 40 × to 100 × magnifications for the presence of typical viral CPE. In the near future, the need for this type of visual examination could be reduced, as new rapid and sensitive techniques become available for each virus. For example, an ELISA test could detect HSV prior to the development of CPE because the assay is performed on lysates of infected cells (Morgan and Smith, 1984). Yet other rapid techniques for the detection of viruses are not dependent on the use of cell cultures, and appear to be extremely sensitive (Richman et al., 1984). However, with most rapid techniques, especially those performed directly on the clinical specimen, the sensitivity is often less than 100% compared with conventional cell cultures (Chow and Merigan, 1983). The most successful and sensitive techniques have combined initial viral amplification in cell cultures with subsequent application of rapid assay a few hours later (Richman et al., 1984).

Even with the practical advent of rapid techniques in clinical virology, therefore, we must still preserve the maximal infectivity of viruses in specimens by giving careful attention to the principles of proper specimen selection, collection and transport, and inoculation in cell cultures.

2.6 SEROLOGIC DETERMINATION

Although techniques and high quality reagents have reduced the time required for the diagnosis of many viral infections to just a few hours, the need for some serologic tests is still apparent. Agents such as Epstein–Barr virus (EBV), rubella, measles, arboviruses (Togaviruses), and hepatitis viruses are not capable of replicating in the usual cell cultures with the development of recognizable CPE. Serology remains the test of choice for the laboratory diagnosis of these virus infections. Further, the assessment of immune status of individuals to rubella, CMV (renal transplant recipients), and VZV (children with neoplastic disease), has remained an important diagnostic function of viral laboratories.

The detection of viral-specific IgM class antibodies in the acute phase serum of patients and, in contrast to IgG antibodies, indicates a recent primary infection to a particular agent. Inclusion of the proper controls and separation methods is necessary to provide specific, reliable results (Smith, 1983). Thus, sensitive techniques for IgM determination using a single specimen provide a necessary complement to rapid techniques for demonstrating viral antigens directly in specimens or after amplification in cell culture.

REFERENCES

Blumenfeld, W., Wager, E., and Hadley, W.K. 1984. Use of transbronchial biopsy for diagnosis of opportunistic pulmonary infections in acquired immunodeficiency syndrome (AIDS). Am. J. Clin. Pathol. 81:1–5.

Bowen, G.S., Fisher, M.C., Deforest, A., Thompson, C.M. Jr., Kleger, B., and Friedman, H. 1983. Epidemic of meningitis and febrile illness in neonates caused by ECHO type 11 virus in Philadelphia. Pediatr. Infect. Dis. 2:359–363.

Bromberg, K., Daidone, B., Clarke, L., and Sierra, M.F. 1984. Comparison of immediate and delayed inoculation of HEp-2 cells for isolation of respiratory syncytial virus. J. Clin. Microbiol. 20:123–124.

Chernesky, M.A., Ray, C.G., and Smith, T.F. 1982. Laboratory diagnosis of viral infections. In W.L. Drew (coordinating ed.), Cumitech 15. Washington, D.C.: American Society for Microbiologists.

Chow, S., and Merigan, T.C. 1983. Rapid detection and quantitation of human cytomegalovirus in urine through DNA hybridization. N. Engl. J. Med. 308:921–925.

Corey, L., Adams, H.G., Brown, Z.A., and Holmes, K.K. 1983. Genital herpes simplex virus infections: Clinical manifestations, course, and complications. Ann. Intern. Med. 98:958–972.

Corey, L., and Holmes, K.K. 1983. Genital herpes simplex virus infections: Current concepts in diagnosis, therapy, and prevention. Ann. Intern. Med. 98:973–983.

Crane, L.R., Gutterman, P.A., Chapel, T., and Lerner, A.M. 1980. Incubation of swab materials with herpes simplex virus. J. Infect. Dis. 141:531.

Daniels, C.A., LeGoff, S.G., and Notkins, A.L. 1975. Shedding of infectious virus/antibody complexes from vesicular lesions of patients with recurrent herpes labialis. Lancet ii:524–528.

Deseda-Tous, J., Byalt, P.H., Cheny, J.D. 1977. Vesicular lesions in adults due to echovirus 11 infections. Arch. Dermatol. 113:1705–1706.

Drew, W.L., Mintz, L., Hao, R., and Finley,

T.N. 1979. Growth of herpes simplex and cytomegalovirus in cultured human alveolar macrophages. Ann. Rev. Resp. Dis. 287:291.

Feldman, S., and Epp, E. 1979. Detection of viremia during incubation of varicella. J. Pediatr. 94:746–748.

Felman, Y.M., and Sonnabend, J.A. 1979. Herpes simplex virus infections. N.Y. State J. Med. 79:179–185.

Foucar, E., Mukai, K., Foucar, K., Sutherland, D.E.R., and Van Buren, C.T. 1981. Colon ulceration in lethal cytomegalovirus infection. Am. J. Clin. Pathol. 76:788–801.

Gleaves, C.A., Smith, T.F., Shuster, E.A., and Pearson, G.R. 1984. Rapid detection of cytomegalovirus in MRC-5 cells inoculated with urine specimens by using low-speed centrifugation and monoclonal antibody to an early antigen. J. Clin. Microbiol. 19:917–919.

Gleaves, C.A., Wilson, D.J., Wold, A.D., and Smith, T.F. 1985. Detection and serotyping of herpes simplex virus in MRC-5 cells using centrifugation and monoclonal antibodies 16 h postinoculation. J. Clin. Microbiol. 21:29–32.

Glenn, J. 1981. Cytomegalovirus infections following renal transplantation. Rev. Infect. Dis. 3:1151–1178.

Hall, C.B., and Douglas, R.G. Jr. 1975. Clinically useful method for the isolation of respiratory syncytial virus. J. Infect. Dis. 131:1–5.

Hall, C.B., and Douglas, R.G. Jr. 1976. Respiratory syncytial virus and influenzae. Arch. Am. J. Dis. Child. 130:615–620.

Hall, C.B., Douglas, R.G. Jr., and Geiman, J.M. 1976a. Respiratory syncytial virus infections in infants: Quantitation and duration of shedding. J. Pediatr. 89:11–15.

Hall, C.B., Geiman, J.M., Biggar, R., Kotok, D.I., Hogan, P.M., and Douglas, R.G., Jr. 1976b. Respiratory syncytial virus infections within families. N. Engl. J. Med. 294:414–419.

Henson, D., Siegel, S.E., Fuccillo, D.A., Matthew, E., and Levine, A.S. 1972, Cytomegalovirus infections during acute childhood leukemia. J. Infect. Dis. 126:469–481.

Howell, C.L., Miller, M.J., and Martin, W.J. 1979. Comparison of rates of virus isolation from leukocyte populations separated from blood by conventional and Ficoll-Paque/Macrodex methods. J. Clin. Microbiol. 10:533–537.

Hudson, J.B., Misra, V., and Mosmann, T.F. 1976. Cytomegalovirus infectivity: Analysis of the phenomenon of antifungal enhancement of infectivity. Virology 72:235–243.

Huntoon, C.J., House, R.F. Jr., and Smith, T.F. 1981. Recovery of viruses from three transport media incorporated into Culturettes. Arch. Pathol. Lab. Med. 105:436–437.

Johnson, F.B., Leavitt, R.W., and Richards, D.F. 1984. Evaluation of the virocult transport tube for isolation of herpes simplex virus from clinical specimens. J. Clin. Microbiol. 20:120–122.

Jones, M.J., Kolb, M., Votava, H.J., Johnson, R.L., and Smith, T.F. 1980. Intrauterine echovirus type 11 infection. Mayo Clin. Proc. 55:509–512.

Kawana, T., Kawogoe, K., Takizawa, K., Chen, J.T., Kawaguchi, T., and Sakamoto, S. 1982. Clinical and virologic studies on female genital herpes. Obstet. Gynecol. 60:456–461.

Kepfer, P.D., Hable, K.A., and Smith, T.F. 1974. Viral isolation rates during summer from children with acute upper respiratory tract disease and healthy children. Am. J. Clin. Pathol. 61:1–5.

Kim, H.W., Wyatt, R.G., Fernie, B.F., Brandt, C.D., Arrobio, J. O., Jeffries, B.C., and Parrott, R.H. 1983. Respiratory syncytial virus detection by immunofluorescence in nasal secretions with monoclonal antibodies against selected surface and internal proteins. J. Clin. Microbiol. 18:1399–1404.

Kimball, A.M., Foy, H.M., Cooney, M.K., Allan, I.D., Mattock, M., and Plorde, J.J. 1983. Isolation of respiratory syncytial and influenza viruses from the sputum of patients hospitalized with pneumonia. J. Infect. Dis. 147:181–184.

Koch-Weser, J., Hirsch, M.S., and Schooley, R.T., 1983. Treatment of *Herpesvirus* infections. N. Engl. J. Med. 309:963–970.

Lee, M.S., and Balfour, H.H. Jr. 1977. Optimal method for recovery of cytomegalovirus from urine of renal transplant patients. Transplantation 24:228–230.

Lennette, D.A., and Lennette, E.T. 1981. A User's Guide to the Diagnostic Virology Laboratory. Baltimore: University Park Press.

Lerner, M., Silverman, S.H., Rausen, A.R., Haughton, P., and Winter, J.W. 1978. Viral meningitis. N.Y. State J. Med. 78:746–750.

Levin, M.J., Leventhal, S., and Master, H.A. 1984. Factors influencing quantitative isolation of varicella-zoster virus. J. Clin. Microbiol. 19:880–883.

Mintz, L., and Drew, W.L. 1980. Relation of culture site to the recovery of nonpolio enteroviruses. Am. J. Clin. Pathol. 74:324–326.

Moore, D.F. 1984. Comparison of human fibroblast cells and primary rabbit kidney cells for isolation of herpes simplex virus. J. Clin. Microbiol. 19:548–549.

Morgan, M.A., and Smith, T.F. 1984. Evaluation of an enzyme-linked immunosorbent assay for the detection of herpes simplex virus antigen. J. Clin. Microbiol. 19:730–732.

Moseby, R.C., Corey, L., Benjamin, D., Winter, C., and Remington, M.L. 1981. Comparison of viral isolation, direct immunofluorescence, and indirect immunoperoxidase techniques for detection of genital herpes simplex virus infection. J. Clin. Microbiol. 13:913–918.

Myers, M.G. 1979. Viremia caused by varicella-zoster virus: Association with malignant progressive varicella. J. Infect. Dis. 140:229–233.

Neiman, P.E., Reeves, W., Ray, G., Flourney, N., Lerner, K.G., Sale, G.E., and Thomas, E.D. 1977. A prospective analysis of interstitial pneumonia and opportunistic viral infection among recipients of allogeneic bone marrow grafts. J. Infect. Dis. 136:754–767.

Osborn, J.E., and Walker, D.L. 1969. Enhancement of infectivity of murine cytomegalovirus *in vitro* by centrifugal inoculation. J. Virol. 2:853–858.

Pal, S.R., Szucs, and Melnick, J.L. 1983. Rapid immunofluorescence diagnosis of acute hemorrhagic conjunctivitis caused by enterovirus 70. Intervirology. 20:19–22.

Perez, T.R., Mosman, P.L., and Juchau, S.V. 1984. Experience with Virocult as a viral collection and transportation system. Diagn. Microbiol. Infect. Dis. 2:7–9.

Prather, S.L., Jenista, J.A., and Menegus, M.A. The isolation of nonpolio enteroviruses from serum. Diagn. Microbiol. Infect. Dis. 2:353–357.

Reeve, P., Owen, J., and Oriel, J.D. 1975. Laboratory procedures for the isolation of *Chlamydia trachomatis* from the human genital tract. J. Clin. Pathol. 28:910–914.

Reynolds, D.W., Stagno, S., Hosty, T.S., Tiller, M., and Alford, C.A. Jr. 1973. Maternal cytomegalovirus excretion and perinatal infection. N. Engl. J. Med. 289:1–5.

Richman, D.D., Cleveland, P.H., Redfield, D.C., Oxman, M.N., and Wahl, G.M. 1984. Rapid viral diagnosis. J. Infect. Dis. 149:298–310.

Rodin, P., Hare, M.J., Barwell, C.F., and Withers, M.J. 1971. Transport of herpes simplex virus in Stuart's medium. Br. J. Vener. Dis. 47:198–199.

Rubin, S.J. 1983. Detection of viruses in spinal fluid. Am. J. Med. 75:124–128.

Rubin, S.J., and Rogers, S. 1984. Comparison of Cultureset and primary rabbit kidney cell culture for the detection of herpes simplex virus. J. Clin. Microbiol. 19:920–922.

Shope, T.C., Klein-Robbenhaar, J., and Miller, G. 1972. Fatal encephalitis due to *Herpesvirus hominis*: Use of intact brain cells for isolation of virus. J. Infect Dis. 125:542–544.

Smith, T.F. 1979. Specimen requirements, transport, and recovery of viruses in cell cultures. In D. Lennette, S. Specter, and K. Thompson (eds.), Diagnosis of Viral Infections: The Role of the Clinical Laboratory. Baltimore: University Park Press.

Smith, T.F. 1983. Clinical uses of the diagnostic virology laboratory. Med. Clin. North Am. 67:935–951.

Smith, T.F. 1984. Diagnostic virology in the community hospital. Postgrad. Med. 75:215–223.

Smith, T.F., Holley, K.E., Keyes, T.F., and

Macasaet, F.F. 1975. Cytomegalovirus studies of autopsy tissue. I. Virus isolation. Am. J. Clin. Pathol. 63:854–858.

Spector, S.A., 1983. Transmission of cytomegalovirus among infants in hospital documented by restriction-endonclease-digestion analysis. Lancet i:378–380.

Stover, D.E., Zaman, M.B., Hajdu, S.I., Lange, M., Gold, J., and Armstrong, D. 1984. Bronchoalveolar lavage in the diagnosis of diffuse pulmonary infiltrates in the immunosuppressed host. Ann. Intern. Med. 101:1–7.

Tummon, I.S., Dudley, D.K.L., and Walters, J.H. 1981. Genital herpes simplex. Can. Med. Assoc. J. 125:23–29.

Turner, R., Shehab, Z., Osborne, K., and Hendley, J.O. 1982. Shedding and survival of herpes simplex virus from "fever blisters." Pediatrics 70:547–549.

Wilfert, C.M., Buckley, R.H., Mohanakumar, T., Griffith, J.F., Katz, S.L., Whisnant, J.K., Eggleston, P.A., Moore, M., Treadwell, E., Oxman, M.N., and Rosen, F.S. 1977. Persistent and fatal central-nervous system echovirus infections in patients with agammaglobulinemia. N. Engl. J. Med. 296:1485–1489.

Wilfert, C.M., Nusinoff Lehrman, S., and Katz, S.L. 1983. Enteroviruses and meningitis. Pediatr. Infect. Dis. 2:333–341.

Yeager, A.S., Morris, J.E., and Prober, C.G. 1979. Storage and transport of cultures for herpes simplex virus, type 2. Am. J. Clin. Pathol. 72:977–979.

Primary Isolation of Viruses

Marie L. Landry and G. D. Hsiung

3.1 INTRODUCTION

Viruses are obligate intracellular parasites and, therefore, require living cells in which to replicate. This is very different from the cultivation of bacteria, for which nutrient broth or agar plates suffice. The "living cells" essential for virus isolation and assay can be in the form of cultured cells, embryonated eggs, or laboratory animals, most frequently newborn mice. The variety of methods and host systems employed for the isolation of different viruses from clinical specimens reflects the fact that the optimum growth conditions for each virus may differ tremendously. If an insensitive host system is inoculated with a specimen containing a particular virus, or if suboptimal growth conditions exist, the virus probably will not be isolated and a false-negative result will be obtained. Due to limitations in resources and personnel, not all available isolation systems can be maintained in every laboratory. Depending on the patient population the laboratory serves, a decision can be made to select the best host systems and methods needed to optimize the isolation of those viruses causing the most morbidity in that group of patients.

In this chapter, the host systems available for virus isolation, as illustrated in Figure 3.1, will be reviewed. Although embryonated eggs and laboratory animals are very useful for the isolation of certain viruses, cell cultures are the sole isolation

Figure 3.1. Host systems for virus isolation, (A) tissue culture method (CPE); (B) embryonated eggs; (C) newborn mice; (D) tissue culture method (plaques). Reproduced with permission from Hsiung, G.D. 1982. Diagnostic Virology. New Haven: Yale University Press, p. 18.

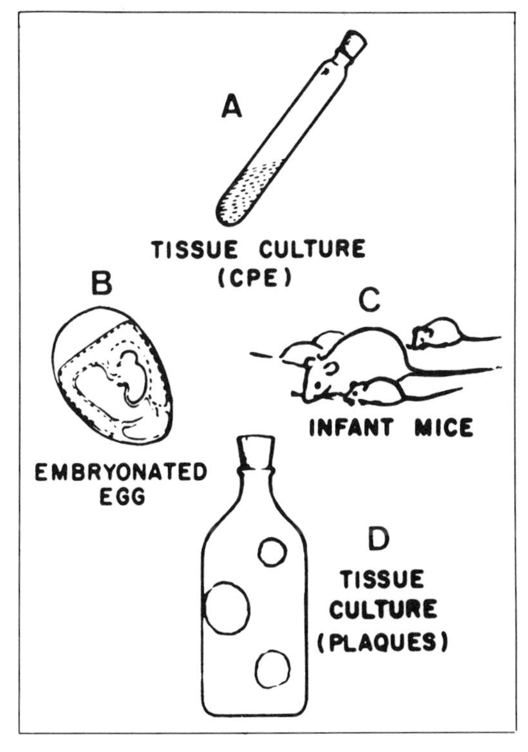

Table 3.1. Cell Cultures Commonly Used in a Clinical Virology Laboratory

Cell Culture	Origin	Number of Subpassages
Primary	Kidney tissues from monkeys, rabbits, etc. Embryos from chickens, guinea pigs, etc.	1 or 2
Diploid (limited passage)	Human embryonic lung/or human newborn foreskin	20–50
Heteroploid	Human epidermoid carcinoma of larynx (HEp-2), of cervix (HeLa), of lung (A549)	Infinite

system utilized in many, if not most, clinical virology laboratories and remain the mainstay of viral diagnosis.

3.2 VIRUS ISOLATION IN CELL CULTURES

3.2.1 Background

The discovery by Enders, Weller, and Robbins in the late 1940s that poliovirus replicates in cultivated mammalian cells derived from nonnervous tissues revolutionized and simplified procedures for the isolation of viruses (Enders et al., 1949). Until that time intact animals or embryonated eggs were the common systems used. After that landmark discovery, cell cultures were prepared for virus studies from a wide variety of animal and human tissues and as a result, in the years following, most of the common viruses we are familiar with today were discovered.

3.2.2 Types of Cell Culture

Cell cultures are generally separated into three types (Table 3.1): Primary cells, which are prepared directly from animal or human tissues and can be subcultured only one or two passages; diploid cell cultures, which usually are derived from human tissues, either fetal or newborn, and can be subcultured 20 to 50 times before senescence; and continuous cell lines, which can be established from human or animal tissues, from tumors, or following the spontaneous transformation of normal tissues. These have a heteroploid karyotype and can be subcultured an infinite number of times.

3.2.3 Variation in Susceptibility to Different Viruses

Cell cultures vary greatly in their susceptibility to different viruses (Table 3.2). If a virus is inoculated into an insusceptible cell culture, the virus will not be able to replicate and a negative result will be obtained. When small amounts of virus are present in a clinical sample, a positive result may be obtained only when the most sensitive systems are used. Therefore, it is critical that those caring for the patients inform the laboratory of the clinical syndrome and/or virus(es) suspected, so that the most sensitive cell cultures can be used and appropriate detection methods employed.

Table 3.2. Variation in Sensitivity of Cell Cultures to Infection by Viruses Commonly Isolated in a Clinical Virology Laboratory

	Cell Culture[a]			
Virus	PMK	HDF	HEp-2/A549	RK
RNA VIRUS				
Enterovirus	+ + +	+ +	+/−	−
Rhinovirus	+	+ + +	+	−
Myxovirus	+ + +	+	−	−
Respiratory syncytial	−	+	+ + +	−
DNA VIRUS				
Adenovirus	+	+ +	+ + +	−
HSV	+	+ +	+ +	+ + +
VZV	+	+ + +	−	−
CMV	−	+ + +	−	−

Abbreviations: PMK, primary monkey kidney; HDF, human diploid fibroblast; HEp-2/A549, human heteroploid cell lines; RK, primary rabbit kidney.
[a] Degree of sensitivity: + + +, highly sensitive; + +, moderately sensitive; +, low sensitivity; −, nonsensitive.

3.2.4 Virus Isolation Methods

3.2.4.a Obtaining and Processing Specimens

Although this area has been reviewed in the previous chapter, it should be reiterated that without appropriate specimens that are properly collected and promptly transported to the laboratory, the subsequent time and effort spent in isolation attempts will be wasted. Accomplishing this is an important task of the clinical virology laboratory and requires continuing communication with and education of the clinicians.

3.2.4.b Supplies and Equipment Needed

The materials needed for the isolation of viruses in cell culture (Table 3.3) are those necessary for the safe handling and processing of cell cultures, maintenance and

Table 3.3. Supplies and Equipment Needed for Isolation of Viruses in Cell Culture

	Supplies and Equipment Needed
Processing of cell cultures	Laminar flow hood, pipettes, automatic pipetting device, pipette jar and discard can, disposable gloves, disinfectant, and sterile glass- and plasticware
Maintenance of cell cultures	Culture media, serum, antibiotics, 4°C refrigerator, test tube racks, and/or rotating drum, 35°C incubators, CO_2 incubator, waterbath, and upright and inverted microscopes
Preservation and storage of viruses	Freezer vials, ultralow temperature freezer (−70°C), and DMSO as stabilizer.

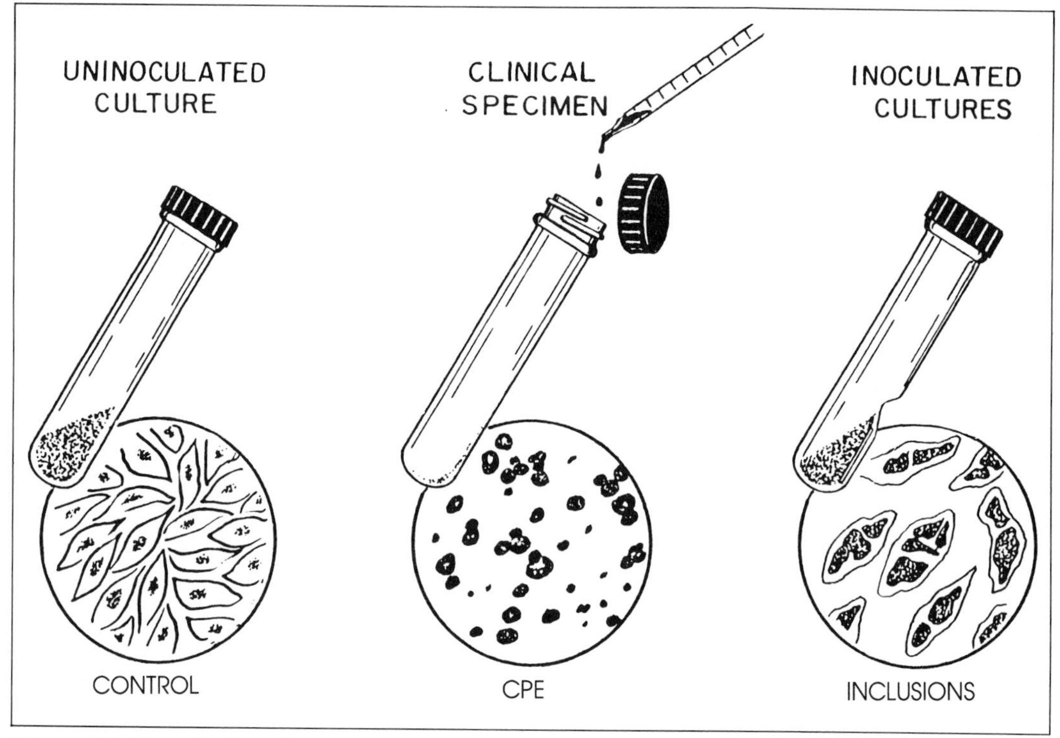

Figure 3.2. Inoculation of cell cultures: Uninoculated cell culture control; cell culture inoculated with a clinical specimen showing virus-induced cytopathic effect (CPE); inoculated Leighton tube containing coverslip which, after fixation and staining, shows viral induced inclusions.

observation of cell cultures, and preservation and storage of clinical specimens and virus isolates. Although the clinical virologist is primarily interested in virus isolation, maintaining different cell cultures in healthy condition is absolutely necessary in order to ensure good results. A wide variety of cell cultures are available commercially and can be purchased and delivered weekly or biweekly, depending on the needs of the laboratory. It may be elected to prepare some cell cultures in the laboratory from available animal or human tissues (e.g., rabbit kidney, guinea pig embryo, human newborn foreskin), or passage certain cell lines [e.g., HEp-2, human diploid fibroblast (HDF) cell strains] for reasons of availability, economy, or quality. The preparation and maintenance of cell cultures can be found in several reference books cited in the reference list (Hsiung and Green, 1978; Lennette and Schmidt, 1979; Hsiung, 1982).

3.2.4.c Inoculation and Incubation

Generally, cell cultures of several different types are inoculated, such as HDF, a human heteroploid cell line, (e.g., HEp-2, A549), and a primary monkey kidney cell culture.

The cell type(s) most susceptible to the suspected viruses in the clinical specimen should be included. Ideally, only healthy, freshly prepared, young cell cultures should be used, because aged cells are less sensitive to virus infection. All cell cultures should be examined under the microscope before inoculation to ensure that the cells are in good condition. Although techniques may vary somewhat for different viruses, in general, the following procedures apply:

1. Pour off culture media and inoculate specimens, 0.1–0.3 ml, into each culture tube (Figure 3.2). In selected instances, Leighton tube cultures containing cover-

slips can be inoculated in a similar manner, the coverslips later removed for fixing and staining, and examined for virus-induced inclusions (see chapter 4). An uninoculated culture should be kept in parallel for comparison.

2. Allow specimen to adsorb in the incubator at 35°C for 30 to 60 minutes. Then, 1.0–1.5 ml of maintenance medium should be added and the inoculated cultures returned to the incubator. Inoculated cultures can be placed in a rotating drum if available, which is optimal for the isolation of respiratory viruses, especially rhinoviruses, and results in earlier appearance of cytopathic effect (CPE) for many viruses. If stationary racks are used to conserve space, it is critical that culture tubes be positioned so that the cell monolayer is bathed in nutrient medium, otherwise the cells will degenerate, especially at the edge of the monolayer.
3. Check inoculated culture tubes at least every other day for virus-induced CPE and compare with uninoculated control tubes from the same lot of cell cultures.
4. Certain specimens, such as urine and stool, frequently will be toxic to the cell cultures and this toxicity can be confused with virus-induced cytopathology. With such specimens it is a good practice to check inoculated tubes within 24 hours of inoculation and refeed with fresh medium if necessary. If toxic effects are extensive, it may be necessary to subpassage the inoculated cells in order to dilute toxic factors and provide viable cells for virus growth.
5. Inoculated cultures and the uninoculated controls should be kept for observation for virus-induced effects for at least 2 weeks, during which time cell cultures may need to be refed to maintain the cells in good condition. Some cultures, such as HEp-2 cells, may require refeeding or subculturing every few days. Great care must be taken when refeeding cultures that cross contamination from one specimen to another does not occur. Separate pipettes should be used for separate specimens.
6. When virus-induced effects occur (see below), passage infected culture fluid (especially in doubtful cases) into a fresh culture of the same cell type to ensure recovery of virus for further identification of the isolate. For certain cell-associated viruses, such as cytomegalovirus (CMV) or varicella-zoster virus (VZV), it is necessary to trypsinize and passage intact infected cells (Taylor-Robinson, 1959). Adenovirus can be subcultured after freezing and thawing infected cells, which disrupts the cells and releases intracellular virus (Rowe and Hartley, 1962).
7. For certain fastidious viruses, or when the amount of infectious virus in the specimen is low, blind passage (i.e., subculture of the inoculated culture in the absence of virus-induced effects) into a set of fresh culture tubes may be necessary before virus growth can be detected.

3.2.4.d Detection of Virus Induced Effects

Cytopathic effects. Many viruses can be identified by the characteristic cellular changes they induce in susceptible cell cultures. These can be visualized under the light microscope. Examples of CPE characteristic for each particular virus are shown in Figure 3.3 and described in greater detail in the sections on individual viruses. Degree of CPE is usually graded from + to + + + + as it progresses to involve less than 25% of the cell monolayer (+), to 50% (+ +), 75% (+ + +), and finally 100% (+ + + +). There are two important points that should be emphasized regarding CPE induced by virus:

1. The *rate* at which the CPE progresses may help to distinguish similar viruses; for example, HSV progresses rapidly to

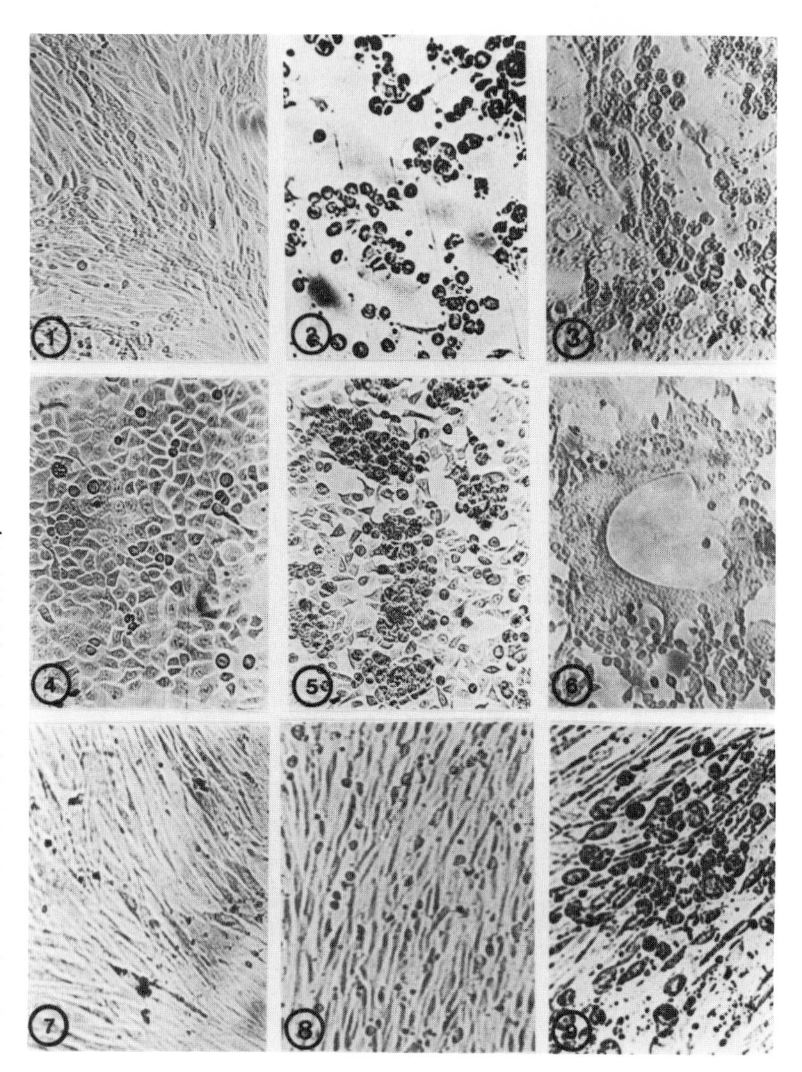

Figure 3.3. Examples of characteristic cytopathic effects (CPE) of different viruses: (1) Uninfected rhesus monkey kidney (RhMK) cells; (2) poliovirus CPE in RhMK cells; (3) influenza B virus CPE in RhMK cells; (4) uninfected HEp-2 cells; (5) adenovirus CPE in HEp-2 cells; (6) respiratory syncytial virus CPE in HEp-2 cells; (7) uninfected human diploid fibroblasts (HDF); (8) rhinovirus CPE in HDF cells; and (9) cytomegalovirus CPE in HDF cells.

involve the entire monolayer of several cell systems (Figure 3.4), whereas, two other herpesviruses, CMV and VZV, grow primarily in HDF cells and progress slowly over a number of days or weeks (Weller et al., 1958, Weller and Hanshaw, 1962).

2. The *type* of cell culture(s) in which the virus replicates is important; that is, although the CPE may be similar within a virus group, the susceptibility of different cell types to different viruses may differ greatly. For example, both polio and echovirus induce similar CPE in primary rhesus monkey kidney (RhMK), however, echovirus does not induce CPE in HEp-2 cells, thus, allowing presumptive identification (Figure 3.5).

It should be cautioned that virus-induced CPE must be distinguished from "nonspecific" CPE caused by toxicity of specimens, contamination with bacteria or fungi, or old cells. A subculture into fresh cells should amplify virus effects and dilute toxic effects. With experience, the appearance of the cellular changes, taken together with the susceptible cell systems, the specimen source, and clinical disease, usually allow a presumptive diagnosis to be made as soon as the virus-induced cellular changes occur.

Hemadsorption. Ordinarily, influenza and parainfluenza viruses do not induce any distinctive cellular changes; however, they do possess hemagglutinins, which have an affinity for red blood cells (RBC). When a freshly obtained guinea pig RBC suspension is added to the infected cultures, the RBC adsorb onto the infected cells, resulting in a hemadsorption phenomenon as shown in Figure 3.6(2). When a culture shows positive hemadsorption, the culture fluid is subcultured into a fresh culture to confirm the virus isolation and to permit further identification. Caution should be taken when aged guinea pig RBC are used, however, because nonspecific hemadsorption often occurs in an uninoculated culture [Figure 3.6(3)] and should be distinguished from that resulting from a specific viral infection (Dowdle and Robinson, 1966). Furthermore, the hemadsorption test is usually performed at 4° or 22°C because the RBC will elute when incubated at 37°C. It should be noted that not all viruses that agglutinate RBC can adsorb them onto infected cell monolayers, because hemadsorption is a property of those viruses that bud from the host cell membrane during maturation. The technique is further described in Chapter 13.

Interference. Certain viruses that do not readily induce CPE in infected cultures can be detected by their ability to interfere with

Figure 3.4. Cell susceptibility and rate of progression of CPE of two herpesviruses: herpes simplex virus type 1 (HSV-1) and human cytomegalovirus (CMV). (A) Uninfected WI-38 cells; (B) extensive HSV-1 CPE in WI-38 cells, 2 days postinoculation; (C) CMV in WI-38 cells, 1 week postinoculation; (D) uninfected rabbit kidney cells (RK); (E) extensive HSV-1 CPE in RK cells, 1 day postinoculation; (F) absence of CMV CPE in RK cells, 2 weeks postinoculation. Modified from Hsiung, G.D. 1982. Diagnostic Virology, Figure 79. New Haven: Yale University Press, p. 206.

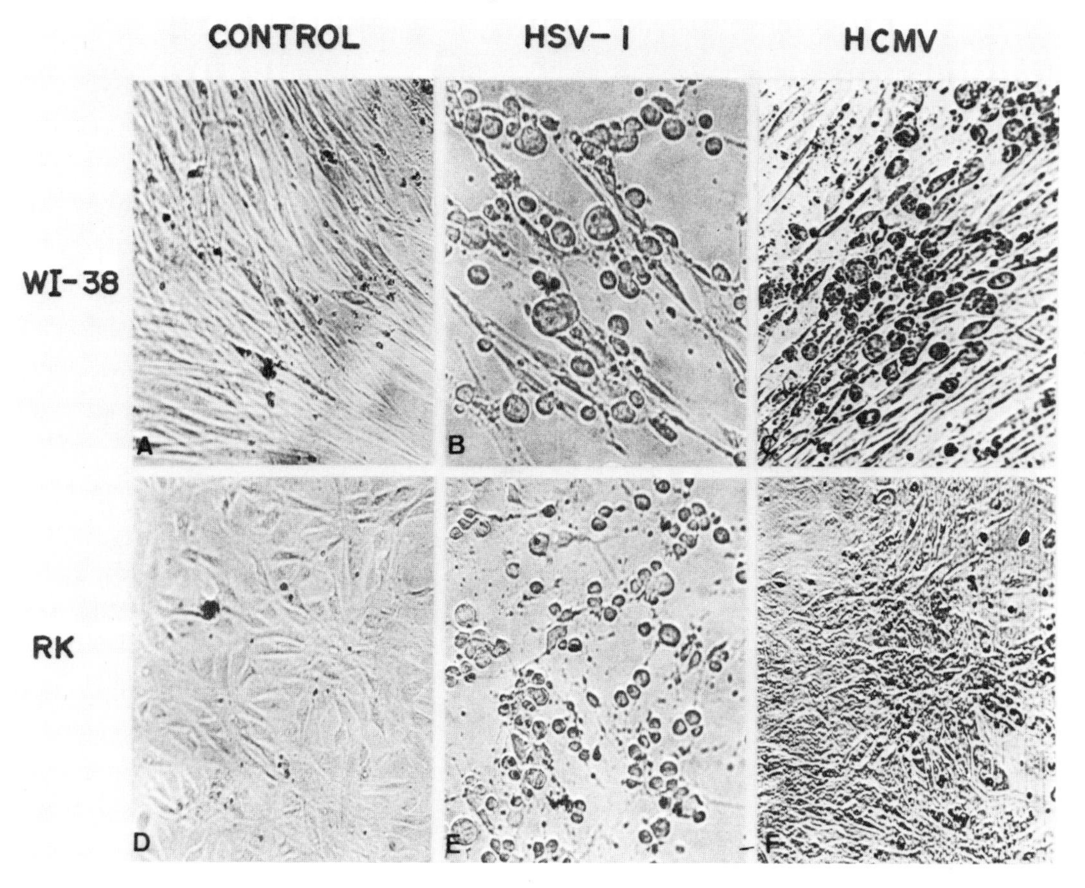

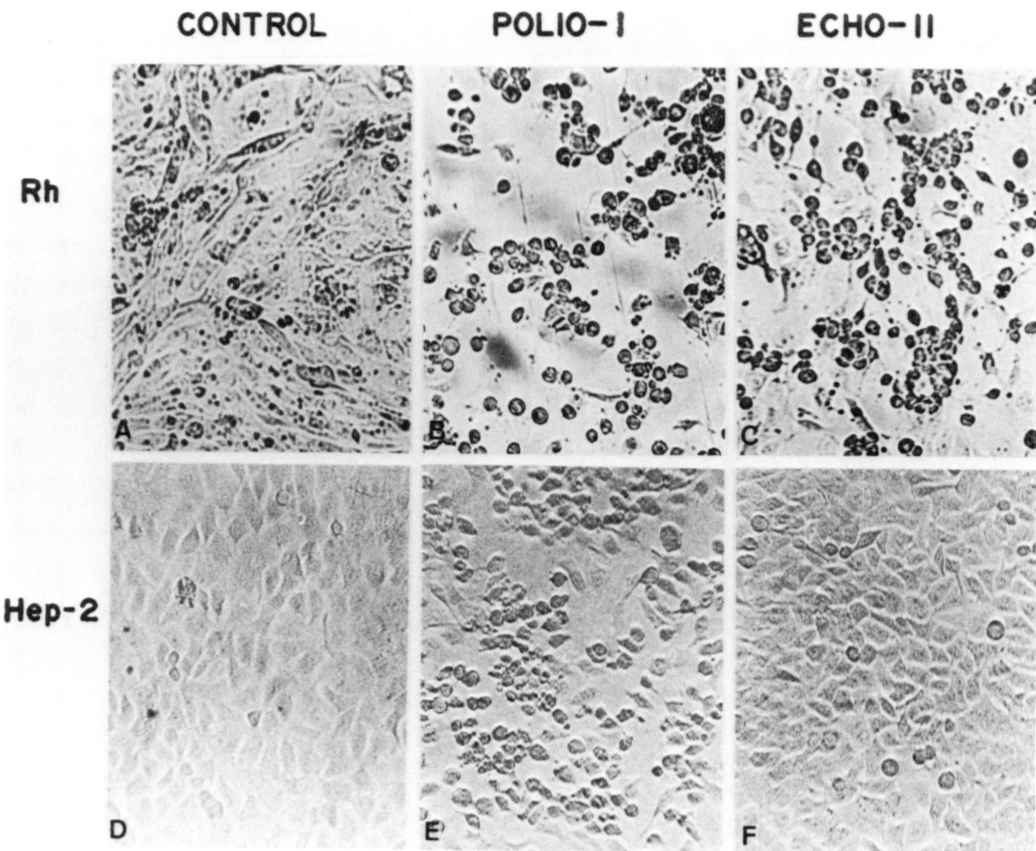

Figure 3.5. Differential susceptibility of cell cultures to enteroviruses. (A) Uninfected rhesus monkey kidney (RhMK); (B) poliovirus infected RhMK cells showing advanced CPE; (C) echovirus infected RhMK cells showing advanced CPE; (D) uninfected HEp-2 cells; (E) poliovirus infected HEp-2 cells showing advanced CPE; (F) echovirus infected HEp-2 cells showing absence of CPE. Reproduced with permission from Hsiung, G.D. 1982. Diagnostic Virology, Figure 20. New Haven: Yale University Press, p. 94.

Figure 3.6. Hemadsorption of guinea pig red blood cells by parainfluenza virus in monkey kidney cells (MK). (1) Uninfected MK cells; (2) specific hemadsorption in parainfluenza infected MK cells; (3) nonspecific hemadsorption seen with aged red blood cells in uninfected cell cultures. Modified from Hsiung, G.D. 1982. Diagnostic Virology, Figures 43 and 44. New Haven: Yale University Press, p. 145.

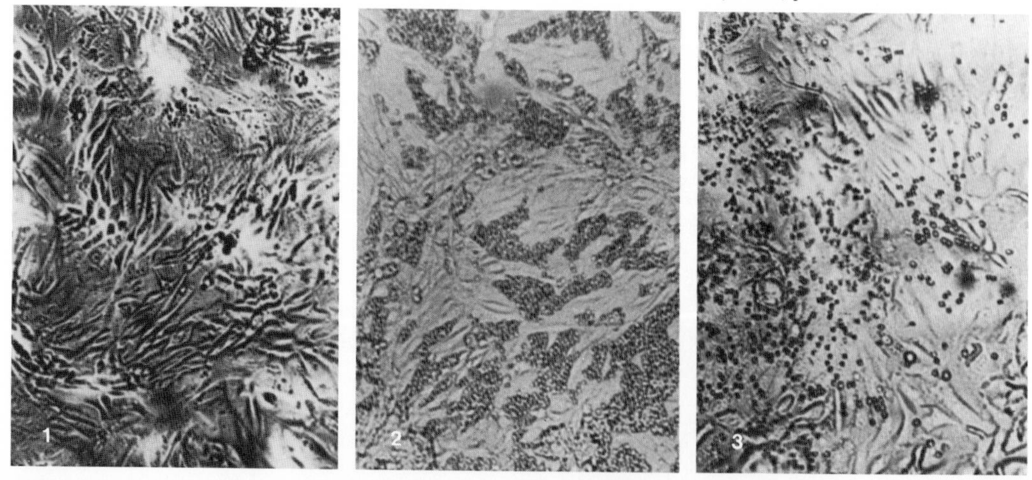

the growth of a second virus inoculated into the same culture. This test has been used most frequently in detection of rubella virus and is referred to as the "interference phenomenon." African green monkey kidney cell cultures infected with rubella virus commonly do not show CPE. After a 10-day incubation, a standard dose of a challenge virus, such as echovirus 11, is inoculated into the same tube, as well as into a control tube without rubella virus, and incubated for 2 or more additional days. The infection of these cells by rubella virus will inhibit the replication of a superinfection by echovirus 11. Thus, echovirus CPE will not be observed in rubella virus-infected cultures, due to the presence of an interfering agent.

3.2.4.e Additional Methods for Virus Isolation

Explant culture or cocultivation. Clinical specimens, such as fresh tissue cells, can be grown out (explanted) or cocultivated with cells susceptible to the suspected agent in the specimen and observed for the development of CPE. The cocultivation method is especially useful for recovery of viruses from buffy coat cells, using the following method:

1. Mince tissues finely in 0.25% trypsin
2. Centrifuge at low speed to pellet the cells
3. Remove the trypsin
4. Resuspend in Hanks' balanced salt solution (HBSS) with antibiotics to a 10 or 20% suspension
5. For cocultivation, inoculate 0.1–0.2 ml onto cell monolayers
6. For explant culture, resuspend in growth medium with 20% calf serum and disperse into petri dishes or flasks

Organ culture. For certain fastidious viruses, such as rhinovirus, coronavirus, or rotavirus, isolation can be accomplished using whole or part of an organ in vitro, such as human embryonic trachea or intestine, allowing preservation of architecture and/or function (Tyrrell and Bynoe, 1965; McIntosh et al., 1967; Wyatt et al., 1974). Virus growth may be detected by subculture of nutrient fluids onto monolayers, or by observation of cessation of ciliary activity and eventual degeneration of epithelial cells. This technique is tedious and is not routinely used in a clinical laboratory. Detailed procedures can be found in the textbooks in the reference list.

3.2.5 Virus Assay and Identification

3.2.5.a Virus Infectivity Assay by the End Point of CPE

At times it is necessary to quantitate the amount of infectious virus present in a specimen or a cell culture. The specimen can be assayed by determining the highest dilution of the fluid that produces CPE (or hemadsorption or interference) in 50% of the cell cultures inoculated; this endpoint is the 50% tissue culture infectious dose ($TCID_{50}$), as follows:

1. Add 0.9 ml of HBSS to seven sterile test tubes
2. Add 0.1 ml of virus suspension to the first dilution tube, mix thoroughly and transfer 0.1 ml of the mixture to the next tube
3. With a separate 1-ml pipette, mix the suspension and transfer 0.1 ml to the next tube; continue this process with all seven tubes
4. Inoculate 0.1 ml of each dilution of the virus suspension into a tube of a sensitive cell culture, four tubes per dilution (use a separate pipette for each dilution)
5. Determine the assay endpoint by the method of Reed and Muench (1938) (See Table 3.4)

Calculation: 50% end point

$$= \frac{\%\text{ with CPE} > 50\% - 50\%}{\%\text{ with CPE} > 50\% - < 50\%}$$

$$= \frac{83 - 50}{83 - 40}$$

$$= 0.7$$

Table 3.4. An Example of Determination of $TCID_{50}$

	Number of Cultures		Total Cultures			
Virus Dilution	With CPE	No CPE	With CPE	No CPE	CPE Ratio	% Cultures with CPE
10^{-3}	4	0	9	0	9/9	100
10^{-4}	3	1	5	1	5/6	83
10^{-5}	2	2	2	3	2/5	40
10^{-6}	0	4	0	7	0/7	0

Therefore a virus dilution of $10^{4.7}$ per 0.1 ml represents one $TCID_{50}$, i.e., at that dilution 50% of the inoculated cultures will become infected. A dilution of $10^{2.7}$ per 0.1 ml of virus suspension will contain 100 $TCID_{50}$ in a volume of 0.1 ml.

3.2.5.b Plaque Formation

Many viruses that produce CPE, and also certain viruses that do not produce detectable CPE under fluid medium in cell cultures, may be detected by their ability to form plaques in cell monolayers under a solid medium, such as agar, agarose, or starch (Porterfield, 1960; Hsiung, 1961). Virus plaques are colorless areas of infected cells, which do not take up the vital stain neutral red when it is incorporated into the overlay medium (Figure 3.7). Alternatively, monolayers can be stained with crystal violet after the medium is removed to visualize plaques (Figure 3.8). Virus particles from a focus of infection are localized by the solid overlay medium and the virus spreads from infected cells to adjacent cells resulting in discrete foci of infection. Dif-

Figure 3.7. Examples of plaque morphology of different enteroviruses in RhMK cultures. *Left to right:* Poliovirus type 3, echovirus type 8, and coxsackievirus B4.

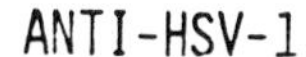

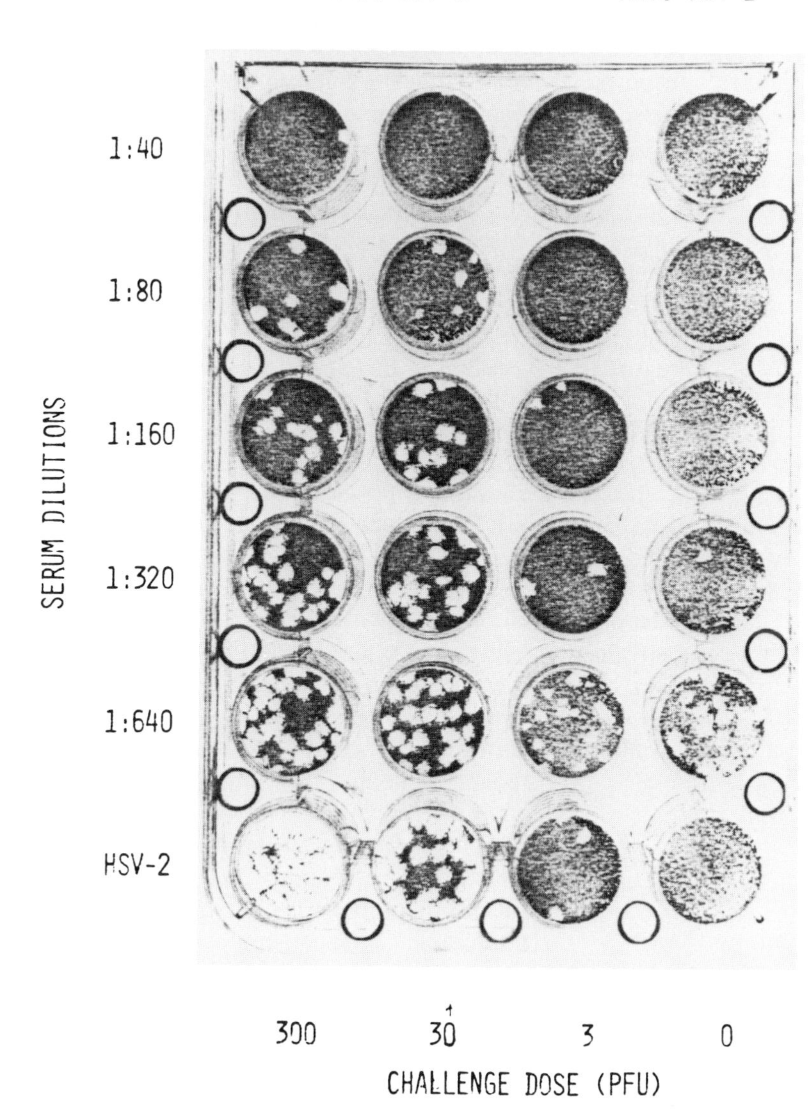

Figure 3.8. Plaque reduction neutralization of herpes simplex virus type 2 (HSV-2) using antiserum against HSV-2. Inhibition of 90 percent of the plaques induced by HSV-2 (challenge dose, 30 PFU) is seen at the 1:40 dilution of anti–HSV-1 serum and up to the 1:320 dilution of anti–HSV-2 serum.

ferent enteroviruses induce plaques of varying size and shape in much the same manner that different bacteria produce characteristic colonies (Hsiung and Melnick, 1957). Characteristic plaques can be helpful in detecting mixed infections of two viruses within the same virus group, such as poliovirus and echovirus, which induce the same type of CPE but show two distinct sizes and shapes of plaques (Figure 3.7). Because one infectious unit is capable of initiating one plaque, this technique can be used both for accurate quantitative assay of virus infectivity and for purification of virus strains. However, plaque assays are not commonly used in a clinical laboratory; therefore, detailed procedures are not included in this chapter and the reader is referred to the reference list at the end of this chapter.

3.2.5.c Identification of Virus Isolates

As mentioned above, presumptive identification usually can be made on the basis of characteristic virus induced effects (e.g., type of CPE or hemadsorption) and selective cell susceptibility. For final identification, either immunofluorescence with specific antisera or neutralization of virus-induced effects in cell culture can be performed. These tests will be discussed in greater detail in subsequent chapters. The plaque reduction neutralization test is illustrated in Figure 3.8. Occasionally a new isolate cannot be identified by the standard serologic tests, and it may be necessary to study the more basic properties of the new agent (Hsiung, 1982). These include the nucleic acid type, size and shape of the isolate, as well as the presence or absence of a lipid envelope. Nucleic acid type can be determined by exposure to 5′-bromodeoxyuridine (BrdU), an inhibitor of DNA viruses, followed by assay of virus infectivity. Virus size can be estimated by passing infected culture fluids through a series of filters or by electron microscopy (EM), and the presence of a lipid envelope can be determined by exposure of the virus to ether, then reinoculation into culture to determine if infectivity has been lost. The morphologic properties of the infecting virus can also be determined by EM.

3.2.6 Advantages and Limitations

The advantages of cell culture for virus diagnosis include relative ease compared with animal inoculation, broad spectrum and sensitivity when compared with other available diagnostic methods, and the recovery of unknown or unexpected infectious virus(es) that may be present in the specimen. It is limited by the difficulty in maintaining cell cultures, by the sometimes variable quality of cultures, and by the decreased sensitivity of cell lines after repeated subculture. Contamination with adventitious agents, such as endogenous viral agents and mycoplasma, occur which can inhibit the growth of viruses in clinical specimens (Hsiung, 1968; Smith, 1970; Stanbrige, 1971; Chu et al., 1973). Endogenous viruses that are latent in the tissue culture can be reactivated during cultivation and cause CPE or hemadsorption and, thus, be confused with virus isolated from the patient's specimen. In addition, some common viruses as yet do not produce identifiable effects in readily available cell cultures, for example, hepatitis viruses, rotavirus, some group A coxsackieviruses and togaviruses, so that other methods of detection are necessary. To get the best results from primary isolation in cell culture, it is most important to maintain healthy cell cultures and have a spectrum of cell types available. In general, a primary monkey kidney cell culture, an HDF cell strain, and a human heteroploid cell line (e.g., as HEp-2 or A549) constitute a satisfactory combination. If the isolation of a particular virus is a high priority, the most sensitive system available should be selected.

Another problem in isolating certain viruses, especially of the myxo- and paramyxovirus groups, is the presence of inhibitory substances and/or antibodies in calf serum used in the cell culture media (Krizanova and Rathova, 1969). Ideally, maintenance media for inoculated cultures should be serum-free; however, serum is required for long-term maintenance of cells. Using fetal or agamma calf serum reduces this problem, but adds to expense. To date, no completely satisfactory, chemically defined medium is available.

3.2.7 Viruses Commonly Isolated in a Clinical Laboratory

3.2.7.a Herpes Simplex Viruses Types 1 and 2

Both herpes-simplex virus type 1 and type 2 (HSV-1, HSV-2) infect a wide variety of cell cultures and animals. Because cell culture isolation is very sensitive and rapid, inoculation of eggs or animals is no longer used. Primary cell cultures, such as rabbit kidney (RK), human embryonic kidney (HEK), and guinea pig embryo (GPE) cells, as well as HDF, are all very sensitive to HSV infection. Continuous cell lines of baby hamster kidney (BHK), African green monkey kidney (Vero), and HEp-2 cells have been used for isolation of HSV. HDF are the most widely used, although some investigators have found non-primate cells, RK or GPE, to be more sensitive (Douglas and Couch, 1970; Landry et al., 1982). Others have prepared HDF in their own laboratories from human placenta, newborn foreskin, or embryonic tissues and found them to be highly sensitive. Comparison of sensitivities of different cultures is shown in Table 3.5. Vesicular fluids, throat swabs, and genital lesions are the most common sources for virus isolation. HSV produces a rapid degeneration of cells, often appearing within 24 hours of inoculation of the cell culture (Figure 3.4). Over 90 percent of positives will be identified within 3 to 5 days (Herrmann, 1972). Occasionally, CPE develops later and, rarely, will be detected after blind passage.

The CPE begins as clusters of enlarged, rounded, refractile cells and spreads to involve the entire monolayer, usually within 48 hours. Another type of CPE also can be seen: the formation of multinucleated giant cells. Subcultures are performed by passaging 0.2 ml of supernatant fluid to a fresh culture tube.

Identification of virus as HSV can be done by immunofluorescence or neutralization. Commercial kits are available that use immunoperoxidase stains to identify HSV infected cells (Fayram et al, 1983; Sewell et al, 1984). Isolates can be differentiated as type 1 or 2 by several means, including immunofluorescence with monoclonal antibodies (Balkovic and Hsiung, 1985), ability to replicate in chick embryo cell monolayers (Nordlund et al., 1977), sensitivity to bromovinyldeoxyuridine (BVDU) (Mayo, 1982), or by restriction enzyme analysis (Lonsdale, 1978).

3.2.7.b Varicella-Zoster Virus

HDF are the most sensitive cells for the isolation of VZV, although the virus also has been isolated using human epithelial cells, primary MK cells, and occasionally GPE cells. Vesicle fluid and lesion swabs are the usual sources for VZV isolation. The virus is quite labile; therefore, prompt inoculation into cell culture is desirable. In contrast with HSV, VZV does not cause disease in newborn mice or embryonated eggs.

Cytopathology starts as foci of rounded enlarged cells, as seen with HSV, however, the onset and progression are much slower

Table 3.5. Comparison of Sensitivity and Rapidity of CPE Induced by Herpes Simplex Virus in Different Cell Culture Systems

HSV dose ($TCID_{50}$)	Cell Culture/Days Postinoculation CPE Observed											
	RK			HDF			HEp-2/A549			Vero		
	1[a]	2	4	1	2	4	1	2	4	1	2	4
10000	+ + +	+ + +	+ + +[b]	0	+	+ +	0	+	+ +	0	+ +	+ +
1000	+ +	+ +	+ + +	0	+	+ +	0	0	+ +	0	+	+ +
100	+	+	+ + +	0	0	+	0	0	0	0	0	0
10	0	+	+ +	0	0	0	0	0	0	0	0	0
1	0	0	+	0	0	0	0	0	0	0	0	0

Abbreviations: RK, primary rabbit kidney cells; HDF, human diploid fibroblast, WI-38/MRC-5; HEp-2, human heteroploid cell line derived from carcinoma of the larynx; Vero, African green monkey kidney cell line; $TCID_{50}$, tissue culture infectious dose, 50 percent.

[a] Days postinoculation.

[b] Degree of CPE: +, 25 percent of cells show CPE; + +, 50 percent of cells show CPE; + + +, 75 percent of cells show CPE; + + + +, 100 percent of cells show CPE.

and the foci of CPE tend to progress linearly along the axis of the cells similar to CMV. CPE first appears 4 to 7 days after inoculation but may take 2 or 3 weeks. The virus is cell-associated and subpassages are performed by trypsinization and passage of infected intact cells to fresh monolayers of cells. Stocks of VZV should be maintained as suspensions of viable cells frozen at or below −70°C. Final identification is by immunofluorescence, neutralization, or complement fixation.

3.2.7.c Cytomegalovirus

HDF are the single most successful culture system for the isolation of CMV. The source of the fibroblasts can be either human embryonic tissues or newborn foreskin. The latter, however, lose their sensitivity after the tenth to 15th passage. Virus can be isolated from a variety of body secretions including urine, saliva, tears, milk, semen, stools, vaginal or cervical secretions, and blood elements. Clinical specimens can produce CPE within a few days to many weeks, depending on the amount of virus in the specimen. Characteristic CPE consists of foci of enlarged, refractile cells that slowly enlarge over weeks and often do not involve the entire monolayer [Figure 3.3(9)]. Thus, it is important that the monolayers be maintained in good condition for 4 to 6 weeks. On the other hand, when a large quantity of CMV is inoculated, one may see generalized rounding at 24 hours. Incubation of cultures at 36°C instead of 33°C results in more rapid onset of CPE and higher isolation rates (Gregory and Menegus, 1983). For subculture, early passage of intact infected cells is essential. Alternatively, degenerating monolayers can be dispersed and then dispensed onto fresh uninfected cells. To maintain stocks of CMV, viable cells should be frozen in modified Eagle's medium with 20% calf serum and 10% dimethyl sulfoxide (DMSO). Identification of isolates can be accomplished with immunofluorescence or complement fixation tests.

3.2.7.d Adenovirus

In general, human adenoviruses produce CPE in continuous human cell lines, such as HEp-2 and A549, in HEK and HDF cell cultures. Each of these cell systems has its disadvantages: the continuous cell lines may be difficult to maintain; HEK often are not readily available and are expensive; the HDF are less sensitive and the changes are not characteristic. Non-human cells, such as rhesus monkey kidney (RhMK), are of variable sensitivity and virus growth is slower. Throat swabs, nasal swabs, eye swabs, and stool are good sources of virus, the choice depending on the clinical syndrome. Characteristic CPE consists of grape-like clusters of rounded cells [Figure 3.3(5)], which appear in 2 to 7 days with types 1, 2, 3, 5, 6, and 7. Other adenovirus types may require 4 weeks or blind passage. Adenovirus remains cell-associated, similar to VZV and CMV; however, adenovirus is nonenveloped and stable to freezing and thawing. Therefore, two to three cycles of freezing at −70°C and thawing disrupts the cells and releases intranuclear infectious virus. A number of adenovirus types associated with diarrhea have been detected by electron microscopy, but do not grow readily in cell culture (see Chapter 22).

Identification of isolates can be done by immunofluorescence using antihexon serum for the adenovirus group or complement fixation test. Isolates can be separated into four subgroups by agglutination with rat and rhesus erythrocytes; hemagglutination inhibition (HAI) and neutralization tests will identify virus types.

3.2.7.e Enteroviruses

The first 67 types of enteroviruses have been divided into subgroups based on their growth characteristics in cell cultures and

Table 3.6 Host Susceptibility for Enteroviruses

Virus Type	Serotypes	Cell Cultures RhMK	HDF	HEp-2	Newborn Mice
Poliovirus	1–3	+ +[a]	+ +	+ +	–
Coxsackie B	1–6	+ +	+ +	+ +	+ +
Coxsackie A	1–24	+/–	+/–	–	+ +
Echovirus	1–34	+ +	+	–	–

Abbreviations: RhMK, primary rhesus monkey kidney; HDF, human diploid fibroblast; HEp-2, human heteroploid cell line.
[a] Degree of sensitivity: + +, sensitive; +, less sensitive; –, not sensitive.

pathogenicity in newborn mice. (Table 3.6). However, many exceptions occur and newly recognized virus types are now classified simply as enteroviruses. In general, enteroviruses grow best in epithelial cells of primate origin. Polio- and coxsackie B viruses grow well in primary MK and HEp-2 cells, echovirus grows well in primary MK but not HEp-2 cells, and the universal host for coxsackie group A is the newborn mouse; however, some strains grow in HDF, HEK, MK, RD (a rhabdomyosarcoma cell line), or GPE cells. Enteroviruses can be recovered from feces, throat swabs, CSF, blood, vesicle fluid, conjunctival swabs, and urine. Characteristically, infected monolayer cells round up, become refractile, shrink, degenerate, then detach from the surface of the culture vessel [Figure 3.3(2)]. The supernatant fluid can be subpassaged. Final identification of enteroviruses can be made in micro- or macroneutralization tests in cell cultures using antiserum pools.

3.2.7.f Rhinovirus

Rhinoviruses are classified as picornaviruses along with the enteroviruses, but can be separated from the latter by their sensitivity to low pH. Unlike enteroviruses, therefore, they cannot survive passage through the stomach and are not found in the gut. They can be isolated in cells of human origin (usually HDF), although some laboratories have used HEK, and occasionally primary MK. Varying sensitivity of different lots of cells can be a problem. Many rhinovirus types were originally isolated in organ cultures of human embryonic trachea. Sources of virus include throat swabs and nasal swabs. Cultivation at 33°C in a roller drum apparatus is optimal (Gwaltney and Jordan, 1964). CPE may occur from the first to the third week of incubation. The CPE is similar to the enteroviruses, starts as foci of rounded cells, and spreads gradually [Figure 3.3(8)]. CPE may not progress and may even disappear; if it is not progressing, subpassage of supernatant fluids from infected cells should be performed. Identification of isolates is by characteristic CPE and inactivation at pH 3.

3.2.7.g Influenza

Primary MK is the most widely used cell culture for isolation of influenza, although the host range may be increased by addition of trypsin to the media (Frank et al., 1979). Influenza A is reliably isolated in eggs and usually in MK as well, though strains differ (Smith and Reichrath, 1974). Influenza B is more readily isolated in MK than in eggs. Throat swabs, nasal swabs, and nasal washings are good sources for virus and should be collected early in illness. Serum components may inhibit influenza virus from replicating. Therefore, serum should be removed from cell cultures by rinsing with HBSS before inoculation and cultures should be maintained in serum-free media after inoculation. Incubation at 33°C in a roller drum is optimal for isolation. The presence of virus is generally detected by hemadsorption of guinea pig RBC onto infected monolayers [Figure 3.6(2)]. CPE may be seen with influenza B [Figure 3.3(3)], and often with influenza A, but it occurs later than the detection of virus by hemadsorption. Subcultures can be performed by passaging the supernatant fluids. Isolates can be identified as influenza by immunofluorescence or complement fixation, and can be identified as influenza A or B by immunofluorescence using monoclonal antibodies. Strain differences are determined by hemagglutination inhibition.

3.2.7.h Parainfluenza

Primary MK is the most sensitive system for isolation of these viruses. Parainfluenza grows poorly or not at all in hens' eggs. HEK, HDF, and HEp-2 are less sensitive. Throat and nasal swabs are good sources for virus. Cell cultures should be washed with HBSS before inoculation and refed with media without serum. Incubation at 33°–36°C in a roller drum is optimal. The presence of virus is detected by hemadsorption [Figure 3.6(2)], which occurs before CPE. Parainfluenza type 2 may produce syncytia, especially in HEp-2 cells. On subculture parainfluenza type 3 also may induce syncytia formation. In those instances when high levels of virus are present, hemadsorption may be detected in the infected cultures within a few days; with specimens containing less virus, 10 days or more of incubation may be necessary. Identification is by hemadsorption inhibition (HAI).

3.2.7.i Respiratory Syncytial Virus

Respiratory syncytial virus (RSV) grows best in continuous cell lines, such as HEp-2 or A549 cells, in which it produces its characteristic syncytia [Figure 3.3(6)]. However, different cell lines vary in their sensitivity to the virus. HDF cells support its growth but are less sensitive and the cytopathology is not characteristic. RSV is found in respiratory secretions from the nose and oropharynx. Identification is by immunofluorescence, neutralization, or complement fixation tests.

3.3 VIRUS ISOLATION IN EMBRYONATED EGGS

3.3.1 Background

The chick embryo is a highly sensitive host for the primary isolation of several virus types. Some influenza A virus strains are more readily isolated in embryonated eggs than in cell culture, although the reverse is true for influenza B virus strains. During "influenza season," it can be helpful to have embryonated eggs available, especially for the isolation of influenza A viruses. The chick embryo is also a sensitive host for the isolation of mumps, virus, however, MK cells are a sensitive host for the rapid isolation of this agent. For primary isolation of influenza and mumps, inoculation of the amniotic sac is necessary. Subsequently, virus isolates can be adapted to grow in the allantoic sac.

The chorioallantoic membrane (CAM) is highly susceptible to pock formation by HSV and poxviruses. With the availability of numerous sensitive cell cultures for the isolation of HSV and vaccinia viruses, and the eradication of smallpox, CAM inoculation is now rarely employed in the clinical laboratory. However, HSV and poxviruses can be differentiated by the morphologic characteristics of the pocks that they produce in this system.

3.3.2 Maintenance and Source of Eggs

Fertile hens' eggs for virus isolation should be obtained from flocks that are free of infection (Newcastle disease virus and mycoplasma can be particularly troublesome). Eggs should be incubated at 37°C in an atmosphere of 40–70% humidity to ensure

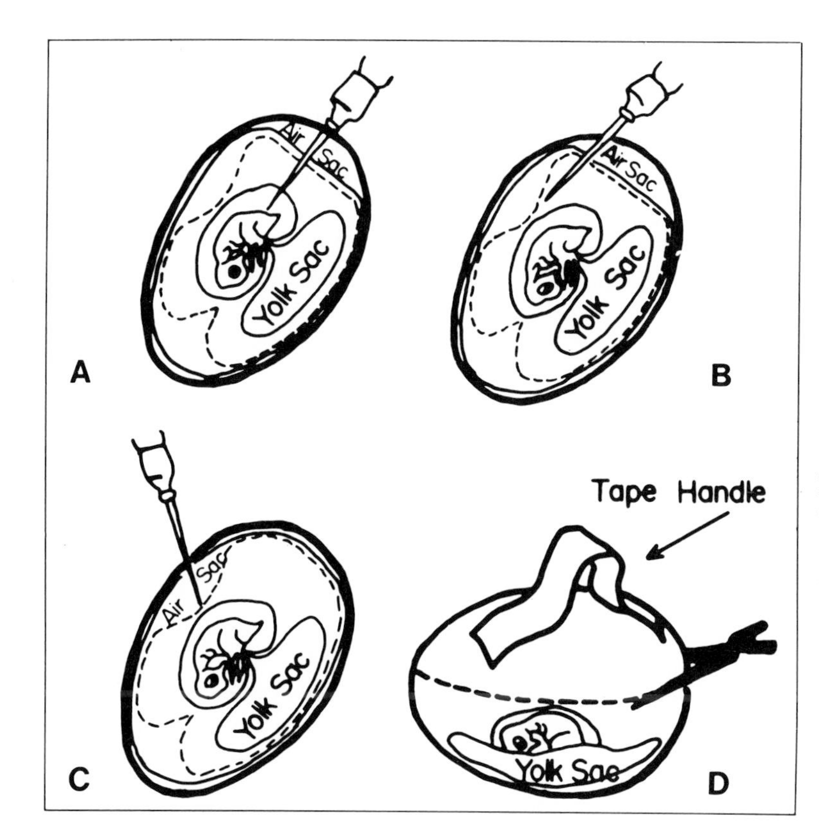

Figure 3.9. Embryonated egg inoculation and harvesting. (A) Amniotic cavity inoculation; (B) allantoic cavity inoculation. (C) chorioallantoic membrane inoculation (note artificial air sac); (D) chorioallantoic membrane harvesting. Reproduced with permission from Hsiung, G.D. 1982. Diagnostic Virology, Figure 5. New Haven: Yale University Press, p. 22.

proper development of the air sac. To prevent adhesions of the embryonic membranes and to keep the embryo centralized, the eggs should be turned two to four times each day. After 4 to 5 days of incubation the eggs are candled to determine which ones are fertile and contain developing embryos.

3.3.3 Inoculation and Incubation[1]

3.3.3.a Amniotic or Allantoic Cavity Inoculation

1. Use 7- to 13-day-old embryonated eggs (optimum for mumps, 7 to 8 days; for influenza, 10 to 13 days)
2. Candle the eggs to locate the embryo and detect movement; make a puncture through the shell over the air sac
3. Inoculate 0.1–0.2 ml of the specimen into the amniotic sac, which is entered with a quick stabbing motion, and/or allantoic cavity as the needle is withdrawn using a 1¾ in.-long, 23-gauge needle; use three or four eggs per specimen. When the needle is in the amniotic cavity, gentle pressure will move the embryo [Figure 3.9(A)]. (These procedures should be performed while the egg is illuminated on a candler)
4. Seal the hole in the shell with Scotch tape
5. Incubate eggs at 35°–37°C with air sac uppermost
6. Candle inoculated eggs daily. Discard those that die within 24 hours after inoculation

3.3.3.b Chorioallantoic Membrane Inoculation

1. Candle eggs containing 9- to 12-day-old developing chicken embryos; mark an

[1] Parts excerpted from Diagnostic Virology by G.D. Hsiung, Yale University Press, 1982.

area free from large blood vessels on the side where the embryo is located and the area over the air sac
2. Drill two slits in the eggshell, one on the side and the other over the air sac
3. Puncture the shell membrane under the slits with a sterile needle; care should be taken not to damage the chorioallantoic membrane (CAM)
4. With a rubber bulb, gently apply suction at the hole over the air sac. If this procedure is carried out while the egg is being candled, one can see the CAM drop and a new, artificial air sac form
5. Place 0.05 ml inoculum on the dropped membrane by inserting the needle through the side slit to a depth of about 5 mm (Fig. 3.9(C). Withdraw the needle very slowly. Seal the opening with Scotch tape or wax and incubate at 35°C–37°C with the artificial air sac uppermost
6. Candle the inoculated eggs daily; discard those that die within 24 hours after inoculation

3.3.4 Harvesting, Assay, and Identification of Isolates

3.3.4.a Amniotic and Allantoic Fluids

1. Harvest amniotic fluids and allantoic fluids separately, 2 to 4 days after inoculation for influenza and 5 to 7 days for mumps. The procedure for harvesting egg fluids is as follows:
    - Chill eggs at 4°C overnight, or for 30 minutes at −20°C then 2 to 4 hours in the refrigerator, in order to clot the blood
    - Open the egg shell over the air sac
    - Cut out and remove the overlying shell membrane and CAM with scissors
    - Aspirate amniotic and allantoic fluids separately with sterile capillary pipettes
2. Carry out a spot hemagglutination test by mixing 0.025 ml allantoic or amniotic fluids, both undilute and a 1:5 dilution in phosphate buffered saline (PBS) with 0.025 ml of a 0.5% suspension of guinea pig or chick RBC. Allow this mixture to stand at room temperature for 35 to 45 minutes before reading. If a hemagglutinating virus is present, the RBC will not settle to the bottom but will remain in suspension. Nonspecific hemagglutination can be caused by bacterial contamination
3. Blind passage of egg fluids in eggs may be necessary before virus is detected
4. Isolates are identified by the inhibition of hemagglutination using specific antisera (HI test)

3.3.4.b Chorioallantoic Membrane

1. Place the inoculated egg on a holder so that the slit through which the inoculum was delivered faces upward
2. Make a tape handle over the area of inoculation [Figure 3.9(D)]
3. Cut off the top half of the eggshell, including the infected area, and gently remove the CAM, which is attached to the shell
4. Place the infected CAM in a Petri dish with a few milliliters of phosphate buffered saline; spread the membrane flat against the bottom of the dish; place the dish on a dark surface to facilitate counting of pocks
5. Variola, vaccinia, monkeypox, cowpox viruses, and HSV types 1 and 2 can be differentiated by the morphologic characteristics of the pocks they produce on the CAM. In doubtful cases, the pocks can be ground and examined by EM, or used as antigens for complement fixation or agar gel precipitation tests.

3.4 VIRUS ISOLATION IN MICE

3.4.1 Background

Although suckling mice are the definitive host for certain viruses, isolation of virus in mice is rarely performed in a clinical laboratory today. Group A and B coxsackieviruses were originally isolated in suckling mice and were differentiated by the pathologic lesions and illnesses that they produce (Melnick, 1983). Coxsackie B viruses, however, are readily isolated in cell culture. Although no universal cell system for the isolation of group A viruses has been found, several cell systems such as RD cells (Schmidt et al., 1975), GPE cells (Landry et al., 1981), and HDF support the growth of a number of the coxsackie A serotypes.

The suckling mouse is also the universal host for togaviruses, many of which are etiologic agents of encephalitis. Several cell lines and cell cultures derived from insects are now available for many of these viruses (Shope and Sather, 1979).

3.4.2 Inoculation and Observation for Illness

Pregnant mice are obtained and the entire litter of mice is inoculated with the clinical specimen within 24 to 48 hours of birth, as described below (Figure 3.10). Older mice are less susceptible to infection. Mice should be obtained from pretested virus-free mouse colonies (Sendai virus, coronavirus, and many others can be troublesome). Materials needed include animal facilities for housing noninfected and infected animals, needles, syringes, scissors, and forceps.

1. Inoculate newborn mice within 24 to 48 hours of birth, with 0.01–0.02 ml/mouse intracerebrally, and/or with 0.03–0.05 ml/mouse intraperitoneally, using a 27-gauge ⅜-in. long needle and a ½-cc syringe (Figure 3.10)

2. Check inoculated mice twice daily for signs of illness, paralysis, or death (Figure 3.10)

Figure 3.10. Newborn mouse inoculation. (A) Intracerebral inoculation; (B) intraperitoneal inoculation; (C) flaccid hindlimb paralysis as seen with coxsackie A virus infection; (D) spastic paralysis secondary to coxsackie B infection. Modified from Hsiung, G.D. 1982. Diagnostic Virology, Figure 4. New Haven: Yale University Press, p. 22.

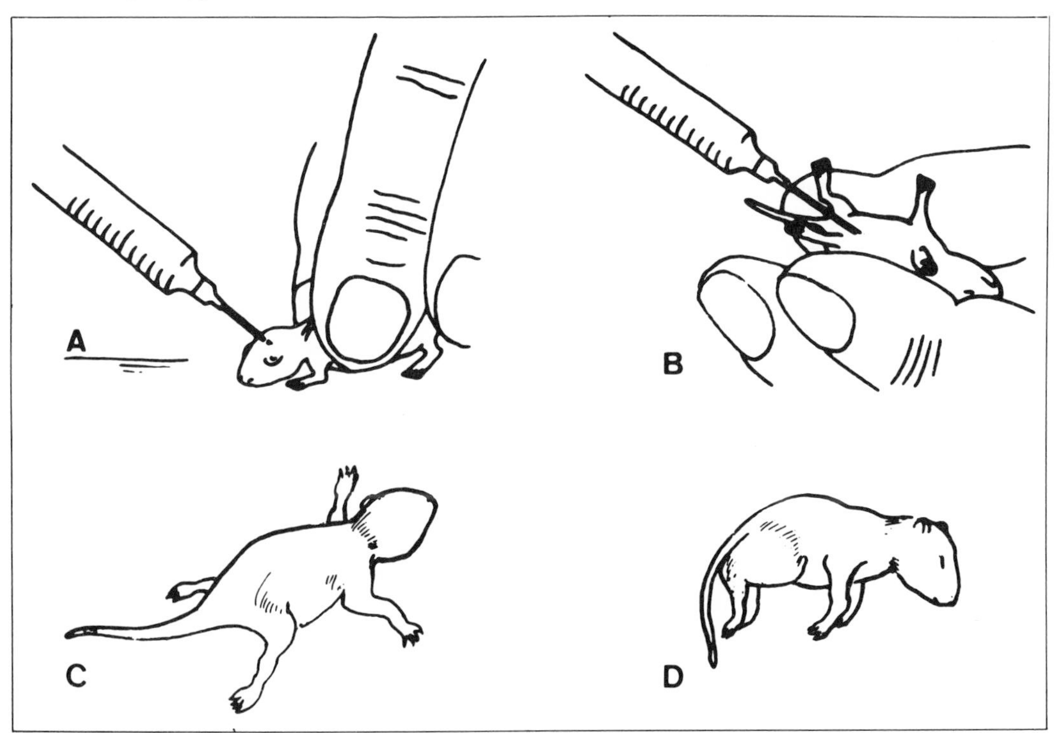

3.4.3 Harvesting and Processing Infected Tissues[2]

1. Harvest mouse brain or skeletal muscle when animals are paralyzed or when other symptoms appear
2. Make 10% tissue suspension and inoculate into additional mice and/or appropriate cell cultures, if the latter are available, for further study

3.4.4 Identification of Isolates

Mouse protection neutralization of virus-induced death or inhibition of CPE in cell culture can be performed for final identification. It is important to recognize that the mice may also become ill due to their own endogenous viruses and that cannibalism may occur.

When working with togaviruses, it should be appreciated that the isolation and identification of certain of these viruses can be associated with risks to the laboratory personnel (Hanson et al., 1967). Unless appropriate precautions can be taken, no attempts should be made to isolate these viruses outside of the appropriate reference facilities.

[2] Parts excerpted from Diagnostic Virology by G.D. Hsiung, Yale University Press, 1982.

REFERENCES

Balkovic, E.S. and Hsiung, G.D. 1985. Comparison of immunofluorescence using commercial monoclonal antibodies with biochemical and biological techniques for typing clinical herpes simplex virus isolates. J. Clin. Microbiol. 22:870–872.

Chu, F.C., Johnson, J.B., Orr, H.C., Probst, P.G., and Petricciani, J.C. 1973. Bacterial virus contamination of fetal bovine sera. In Vitro 9:31–34.

Douglas, R.G., Jr., and Couch, R.B. 1970. A prospective study of chronic herpes simplex virus infection and recurrent herpes labialis in humans. J. Immunol. 104:289–295.

Dowdle, W.R., and Robinson, R.Q. 1966. Nonspecific hemadsorption by rhesus monkey kidney cells. Proc. Soc. Exp. Biol. Med. 121:193–198.

Enders, J.F., Weller, T.H., and Robbins, F.C. 1949. Cultivation of the Lansing strain of poliomyelitis virus in cultures of various human embryonic tissues. Science 109:85–87.

Fayram, S.L., Aarnaes, S., and de la Maza, L.M. 1983. Comparison of Cultureset to a conventional tissue culture-fluorescent antibody technique for isolation and identification of herpes simplex virus. J. Clin. Microbiol. 18:215–216.

Frank, A.L., Couch, R.B., Griffis, C.A., and Baxter, B.D. 1979. Comparison of different tissue cultures for isolation and quantitation of influenza and parainfluenza viruses. J. Clin. Microbiol. 10:32–36.

Gregory, W.W. and Menegus, M.A. 1983. Effect of incubation temperature on isolation of cytomegalovirus from fresh clinical specimens. J. Clin. Microbiol. 18:1003–1005.

Gwaltney, J.M., Jr., and Jordan, W.S., Jr. 1964. Rhinoviruses and respiratory disease. Bacteriol. Rev. 28:409–422.

Hanson, R.P., Sulkin, S.E., Buescher, E.L., Hammon, W.McD., McKinney, R.W., and Work, T.H. 1967. Arbovirus infections of laboratory workers. Science 158:1283–1286.

Herrmann, E.C., Jr. 1972. Rates of isolation of viruses from a wide spectrum of clinical specimens. Am. J. Clin. Pathol. 57:188–194.

Hsiung, G.D. 1961. Production of an inhibitor substance by DA myxovirus. Auto-inhibition of plaque formation. Soc. Exp. Biol. Med. 108:357–360.

Hsiung, G.D. 1968. Latent virus infections in primate tissues with special reference to simian viruses. Bact. Rev. 32:185–205.

Hsiung, G.D. 1982. Diagnostic Virology, 3rd ed. New Haven: Yale University Press.

Hsiung, G.D., and Green, R.H., eds. 1978. Virology and Rickettsiology, Vol. 1, Parts 1 and 2. Handbook Series in Clinical Laboratory Science, Section H. West Palm Beach, FL: CRC Press.

Hsiung, G.D., and Melnick, J.L. 1957. Morphologic characteristics of plaques produced on monkey kidney monolayer cultures by enteric viruses (poliomyelitis, coxsackie and ECHO groups). J. Immunol. 78:128–136.

Krizanova, O., and Rathova, V. 1969. Serum inhibitors of myxoviruses. Curr. Top. Microbiol. Immunol. 47:125–151.

Landry, M.L., Madore, H.P., Fong, C.K.Y., and Hsiung, G.D. 1981. Use of guinea pig embryo cell cultures for isolation and propagation of group A coxsackieviruses. J, Clin. Microbiol. 13:588–593.

Landry, M.L., Mayo, D.R., and Hsiung, G.D. 1982. Comparison of guinea pig embryo cells, rabbit kidney cells and human embryonic fibroblast cell strains for the isolation of herpes simplex virus. J. Clin. Microbiol. 15:842–847.

Lennette, E.H., and Schmidt, N.J., eds. 1979. Diagnostic Procedures for Viral, Rickettsial and Chlamydial Infections, 5th ed. Washington, D.C.: American Public Health Association.

Lonsdale, D.M. 1978. A rapid technique for distinguishing herpes simplex virus type 1 from type 2 by restriction enzyme technology. Lancet i:849–851.

McIntosh, K., Dees, J.H., Becker, W.B., Kapikian, A.Z., and Chanock, R.M. 1967. Recovery in tracheal organ culture of novel viruses from patients with respiratory disease. Proc. Natl. Acad. Sci. USA. 57:933–940.

Mayo, D.R. 1982. Differentiation of herpes simplex virus types 1 and 2 by sensitivity to (E)-5-(2-Bromovinyl)-2′-deoxyuridine. J. Clin. Microbiol. 15:733–736.

Melnick, J.L. 1983. Portraits of viruses: The picornaviruses. Intervirology 20:61–100.

Nordlund, J.J., Anderson, C., Hsiung, G.D., and Tenser, R.B. 1977. The use of temperature sensitivity and selective cell culture system for differentiation of herpes simplex virus types 1 and 2 in a clinical laboratory. Proc. Soc. Exp. Biol. Med. 155:118–123.

Porterfield, J.S. 1960. A simple plaque-inhibition test for the study of arthropod-borne viruses. Bull. W.H.O. 22:373.

Reed, L.J., and Muench, H.A. 1938. A simple method of estimating fifty percent end points. Am. J. Hyg. 27:493–497.

Rowe, W.P., and Hartley, J.W. 1962. A general review of adenoviruses. Ann. N.Y. Acad. Sci. 101:466–474.

Schmidt, N.J., Ho, H.H., and Lennette, E.H. 1975. Propagation and isolation of group A coxsackieviruses in RD cells. J. Clin. Microbiol. 2:183–185.

Sewell, D.L., Horn, S.A., Silbeck, P.W. 1984. Comparison of Cultureset and Bartels Immunodiagnostics with conventional tissue culture for isolation and identification of herpes simplex virus. J. Clin. Microbiol. 19:705–706.

Shope, R.E., and Sather, G.E. 1979. Arboviruses. In E.H. Lennette and N.J. Schmidt (eds.), Diagnostic Procedures for Viral, Rickettsial and Chlamydial Infections. Washington, D.C.: American Public Health Association.

Smith, K.O. 1970. Adventitious viruses in cell cultures. Prog. Med. Virol. 12:302–336.

Smith, T.F., and Reichrath, L. 1974. Comparative recovery of 1972–1973 influenza virus isolates in embryonated eggs and primary rhesus monkey kidney cell cultures after one freeze–thaw cycle. Am. J. Clin. Pathol. 61:579–584.

Stanbridge, E. 1971. Mycoplasmas and cell cultures. Bact. Rev. 35:206–227.

Taylor-Robinson, D. 1959. Chicken pox and herpes zoster. III. Tissue culture studies. Br. J. Exp. Pathol. 40:521–532.

Tyrrell, D.A.J., and Bynoe, M.L. 1965. Cultivation of a novel type of common cold virus in organ cultures. Br. Med. J. 1:1467–1470.

Weller, T.H., and Hanshaw, J.B. 1962. Virologic and clinical observation on cytomegalic inclusion disease. N. Engl. J. Med. 266:1233–1244.

Weller, T.H., Witton, M.M., and Bell, E.J. 1958. The etiologic agents of varicella and herpes zoster: Isolation, propagation, and cultural characteristics *in vitro*. J. Exp. Med. 108:843–852.

Wyatt, R.D., Kapikian, A.Z., Thornhill, T.S., Sereno, M.M., Kim, H.W., and Chanock, R.M. 1974. In vitro cultivation in human fetal intestinal organ culture of a reovirus-like agent associated with nonbacterial gastroenteritis in infants and children. J. Infect. Dis. 130:523–528.

The Cytopathology of Virus Infections

Roger D. Smith and Karen Sutherland

4.1 INTRODUCTION

Virus-infected cells that exfoliate or are scraped from the skin or mucous membranes may contain readily identifiable morphologic changes that permit rapid diagnosis. In many instances, the cytologic alterations may be so distinctive as to be pathognomonic for infection with a specific agent. In others, the changes may point to a virus group or merely raise a suspicion of infection to be confirmed by other means. The purpose of this chapter is to describe the methods used to obtain and prepare cells for cytologic examination, and to illustrate characteristic changes encountered in common virus infections. Most of the methods described are used routinely by diagnostic cytology laboratories and are best applied by the cytotechnologist or pathologist who analyzes cytologic material on a daily basis.

4.2 PREPARATION AND STAINING

The proper collection, fixation, preparation, and staining of specimens for cytology is essential. Rapid fixation in 95% alcohol or with cytology spray fixative before the smear dries is imperative for accurate interpretation. The purpose of fixation is to maintain the existing form and structure of the cellular elements and to achieve consistent staining characteristics and identifiable structures. Improper specimen preparation will decrease the diagnostic accuracy and may lead to false-positive results. Air drying causes nuclear swelling and distortion, cytoplasmic vacuolization, and atypical staining. These changes can mimic nuclear and/or cytoplasmic alterations seen with some virus infections or can distort the characteristic details to such an extent that viral cytopathology cannot be identified. For example, the ground-glass nuclear appearance seen in early herpes simplex virus (HSV) infection can be confused with the smudgy nuclear detail seen in air-dried specimens. Aqueous fixatives (e.g. formalin) result in poor staining and irregular condensation of chromatin, which can be mistaken for a nuclear inclusion.

The preparatory methods utilized for the microscopic examination of cytologic specimens can be divided into four categories: direct smears, preparation by cytocentrifugation, membrane filter preparation, and cell block preparation (Bales and Durfee, 1979).

4.2.1 Direct Smears

Direct scraping of vesicular or bullous lesions of the skin and mucous membranes is the simplest method of cell collection for the identification of viral changes. The scrapings from the base and edges of the suspected lesion should be smeared evenly onto a clean glass slide and immediately fixed with 95% alcohol or cytology spray fixative. Once fixed, the smears are almost indefinitely stable at room temperature and can be stored, mailed, or transported to a cytology laboratory for further processing. The slides are then stained with a modified Papanicolaou staining technique. Proper sampling is important because the crusted or eczematous areas often fail to show the diagnostic cellular features. If a specific lesion, such as an ulcer or unroofed vesicle, is present the base and edges of the lesion should be thoroughly scraped with a spatula, tongue blade, or endoscopic brush to insure proper sampling.

4.2.1.a Modified Papanicolaou Stain

The modified Papanicolaou stain technique is used to detect cytologic changes due to viral infection. The procedure for staining is as follows:

1. Ten dips in 95% ETOH
2. Ten dips in 70% ETOH
3. Ten dips in 50% ETOH
4. Ten dips in distilled H_2O
5. Two minutes in hematoxylin [Gill hematoxylin, consisting of 2190 ml distilled H_2O, 750 ml ethylene glycol, 6 g hematoxylin (C.I. #75290), 0.6 g sodium iodate, 528 g aluminum sulfate, and 60 ml glacial acetic acid]
6. One minute under running tap water
7. One minute in Scott's tap water substitute (consisting of 10 g anhydrous magnesium sulfate, 2 g sodium bicarbonate, and 1 L distilled water)
8. Ten dips in tap water
9. Ten dips in 50% ETOH
10. Ten dips in 95% ETOH
11. One and one-half minutes in OG-6[1]
12. Ten dips in 95% ETOH
13. Ten dips in 95% ETOH
14. Three minutes in EA-65[1]
15. Ten dips in 95% ETOH
16. Ten dips in 95% ETOH
17. Ten dips in 95% ETOH
18. Ten dips in 100% ETOH
19. Ten dips in 100% ETOH
20. Ten dips in 100% ETOH/Hemo-De[2] (equal amounts)
21. Ten dips in Hemo-De, in three consecutive dishes
22. Coverslip slide with Permount media[3]

4.2.2 Cytocentrifugation and Filtration

New cytocentrifugation and filtration techniques have been developed which concentrate small numbers of cells suspended in fluids and are the preferred method for preparation of samples from urine, cerebrospinal, and body cavity fluids. To determine which of the two techniques should be used, one needs to consider the expected number of cells in the fluid. Fluids containing many cells should be centrifuged. The specimen is placed in a centrifuge tube and spun at 1500 rpm for 10 minutes. The supernatant fluid is decanted leaving a volume of 2 ml in the tube and the sediment is resuspended. The specimen is then prepared using the standard cytocentrifugation technique (Barrett, 1976). If little or no sediment is present, then the filtration technique (Gill, 1976) is the method of choice. The use of membrane filters should be limited to cases where the number of cells is low and where additional cell sampling presents a problem. Cytospin preparations offer several advantages. The cells are evenly dispersed on the

[1] From Harleco Co., Gibbstown, New Jersey.
[2] Clearing Agent from Fisher Scientific Co., Philadelphia, Pennsylvania.
[3] From Fisher Scientific Co., Philadelphia, Pennsylvania.

slide (monolayer), little or no background artifact is present, and the cell preparation can be utilized easily for other diagnostic procedures, such as special stains, immunofluorescence, immunoperoxidase, and electron microscopy. Cell block preparations require a histopathology laboratory and are used when there is an abundance of cellular material that can be embedded in paraffin and sectioned for histologic examination.

4.3 VIRUS CYTOPATHOLOGY

The eye and respiratory, genital, and urinary tracts are locations that readily yield cytologic material for rapid viral diagnosis. Characteristic cytologic changes depend on the cytopathic effect of a virus in infected cells, which need not include all cells of the involved organ. In practice, however, it is most useful to consider cytologic alterations in the context of organ system affected and clinical presentation. Therefore, the following discussion and illustrations are organized according to organ system.

4.3.1 Viral Infections of the Respiratory Tract

Smears of cells obtained by nasal and throat swabs, tracheal aspirates, sputum, bronchial washings and brushings, and pulmonary lavage may exhibit cytologic alterations that are diagnostic of virus infection (Table 4.1). In adults, the most frequently encountered virus that is readily detectable by cytology is cytomegalovirus (CMV) (Warner, 1964). Particularly in patients who are receiving immunosuppressive therapy for transplantation or cancer chemother-

Table 4.1. Cytopathology of Respiratory Viral Infections

Virus	Clinical Presentation	Cytologic Findings
Adeno	URI Pneumonia	Small multiple eosinophilic IN inclusions (early); large, single, dense basophilic IN inclusion (late)
CMV	Pneumonia	Cytomegaly; large, single, amphophilic IN inclusions; small PAS-positive IC inclusion
HSV	Tracheobronchitis	Large ground-glass nucleus (early); eosinophilic IN inclusions (late); multinuclearity with nuclear molding
Parainfluenza	Bronchitis pneumonia	Cytomegaly, single nucleus, small eosinophilic IC inclusions
RSV	Tracheobronchitis pneumonia	Large multinucleated cells; IC basophilic inclusions with prominent halos
Measles	Prodromal	Mulberry-like clusters of lymphocytic nuclei in nasal secretions
	Pneumonia	Multinucleated giant cells with IN and IC eosinophilic inclusions
Nonspecific (many viruses)	Bronchitis pneumonia	Ciliocytophthoria

Abbreviations: IN, intranuclear; IC, intracytoplasmic.

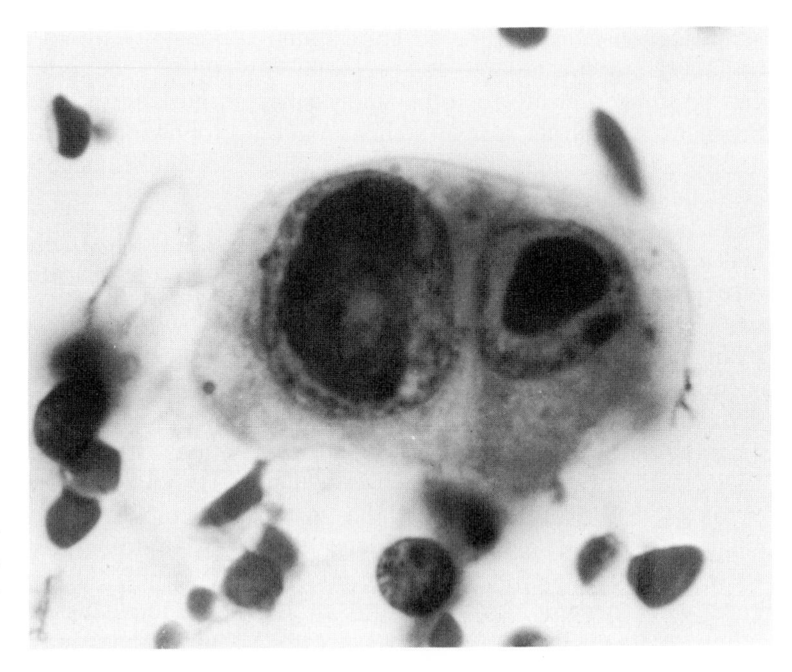

Figure 4.1. CMV in bronchial brushing. Large intranuclear inclusions in binucleated cell and small cytoplasmic inclusions. Pap stain, ×800

apy, the rapid cytologic identification of characteristic inclusions may be a great asset in patient management, particularly because CMV isolation in tissue culture takes many days and often weeks. Because CMV infection involves the lungs, a deep specimen containing pulmonary macrophages is needed. Patients with CMV pneumonia rarely produce abundant sputum, thus requiring bronchial washings and brushings or pulmonary lavage to obtain an adequate specimen. The characteristic cytologic changes are seen in pulmonary macrophages or in cells lining the alveoli. They most often exhibit a single nucleus but occasionally two or more nuclei that are four

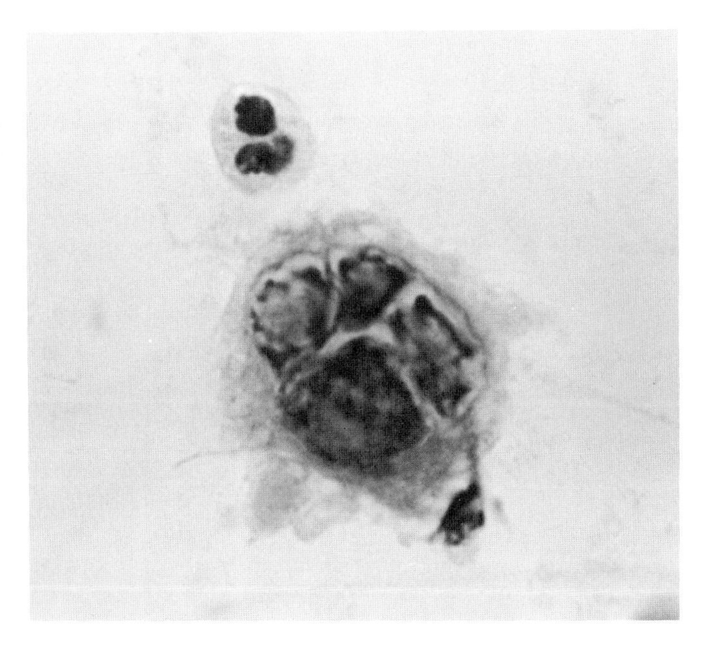

Figure 4.2. HSV in sputum. Epithelial cells showing the ground-glass nuclear appearance of the early stage. Pap stain, ×800

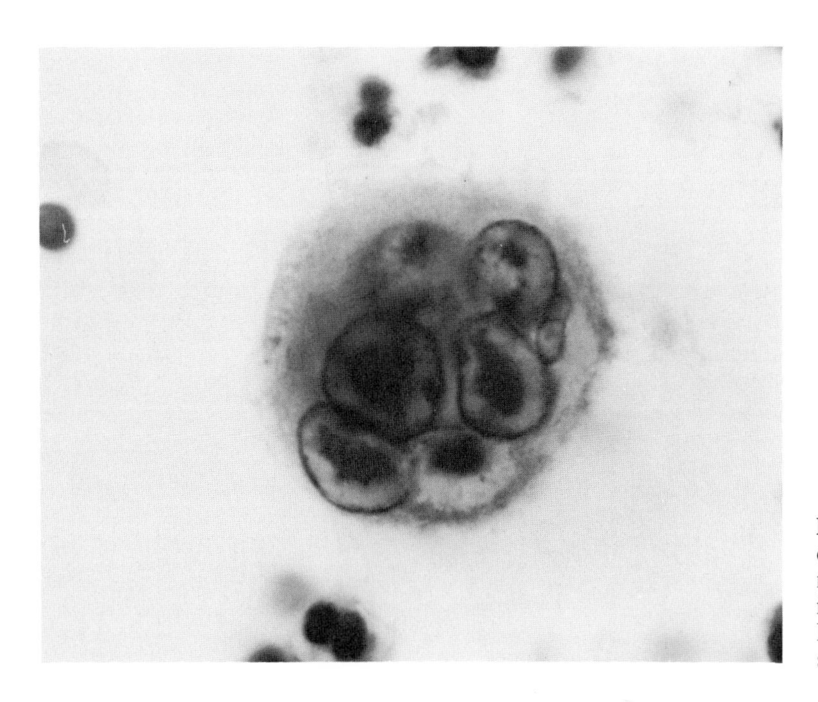

Figure 4.3. HSV in bronchial brushings. Multinucleated cell with molding of nuclei and typical HSV intranuclear inclusions. Pap stain, ×800

to six times their normal size are present (Figure 4.1). Early in the infection these nuclei contain amphoteric or basophilic inclusions that are granular, and the inclusions become condensed and surrounded by a halo in the later stages of infection. Smaller, more eosinophilic oval cytoplasmic inclusions that are periodic acid–Schiff reaction (PAS)-positive are often, but not invariably, present. At later stages of the infection, cytoplasmic inclusions may predominate with an empty or collapsed nucleus (Figure 4.8).

Characteristic inclusion-bearing cells may be observed in sputum and bronchial washings in HSV tracheobronchitis, which is encountered often in immunosuppressed and burn patients (Vernon, 1982). Inclusion-bearing cells tend to be multinucleated and contain either eosinophilic intranuclear inclusions that are centrally located and surrounded by a halo or, at an earlier stage of infection, these cells contain ground-glass inclusions that stain poorly (Figure 4.2). The chromatin often appears as a basophilic ring condensed at the periphery of the nuclear membrane. Cytoplasmic inclusions are not present. When there are multiple nuclei, they are frequently molded or indented by each other (Figure 4.3). Because HSV tends to produce cellular necrosis, the background of these smears usually contains an abundance of cellular debris.

Adenovirus infections of the upper respiratory tract may be identified by smears of secretions from the nasopharynx. Adenovirus pneumonia may be diagnosed by finding typical intranuclear inclusions in bronchial or epithelial cells obtained by bronchoscopy. At an early stage (Figure 4.16) the nucleus contains multiple small, rounded eosinophilic inclusions, each surrounded by a halo. At a later stage, a single larger, dense, basophilic intranuclear inclusion is seen (Figure 4.17).

Respiratory syncytial virus (RSV) and parainfluenza viruses frequently cause bronchitis and pneumonia in infants and young children. They can be rapidly diagnosed by identifying characteristic cytologic changes in respiratory epithelial cells obtained by nasopharyngeal swabs and tracheal aspirates (Naib et al., 1968). Parainfluenza can be differentiated from RSV by finding large cells containing a single nucleus and multiple small eosinophilic inclusions. In RSV infection, the epithelial cells

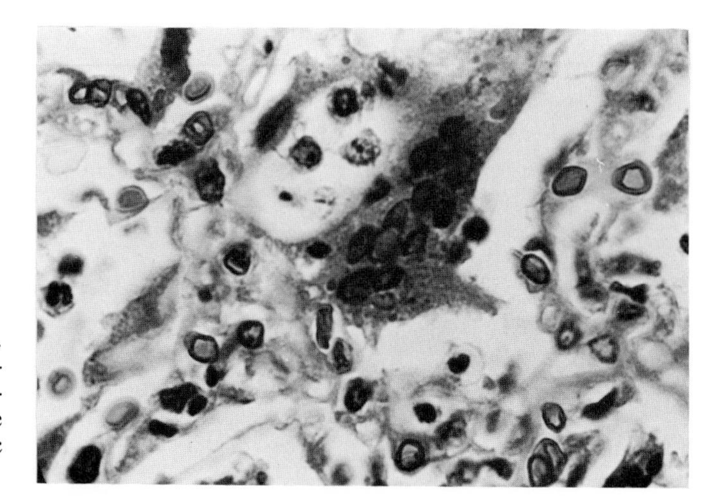

Figure 4.4. Measles in lung tissue. Paraffin embedded tissue of measles pneumonia showing a multinucleated giant cell with multiple intranuclear and intracytoplasmic inclusions. H&E stain, ×550

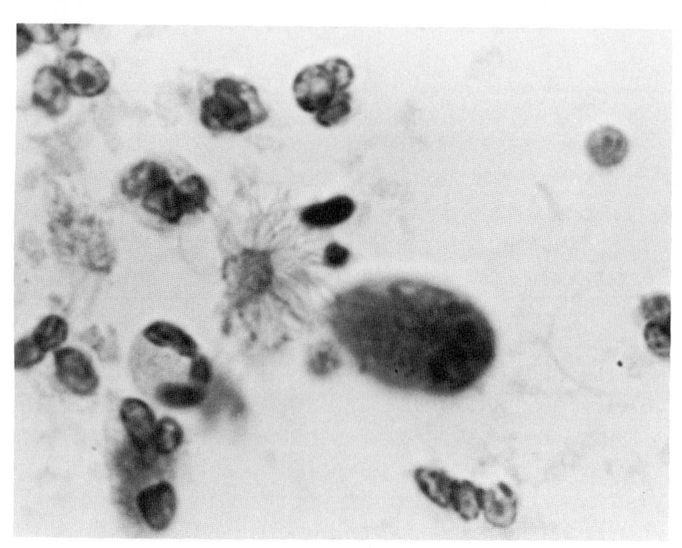

Figure 4.5. Ciliocytophthoria in sputum. Single karyorrhexic nucleus with degenerated cytoplasm containing small inclusion bodies and a tuft of detached cilia. Pap stain, ×550

are large and multinucleate and the cytoplasm contains multiple basophilic inclusions with prominent halos.

Measles can be detected during the prodrome by finding mulberry-like clusters of lymphocytes having up to 50 nuclei in smears of nasal secretions (Tomkins and Macanlay, 1955). Measles giant cells (Figure 4.4) are multinucleate respiratory epithelial cells with intranuclear and cytoplasmic inclusions. These cells may appear in the sputum of patients with measles pneumonia.

A nonspecific change referred to as ciliocytophthoria is found in various inflammatory diseases of the respiratory tract and, in particular, virus infections (Pierce and Hirsch, 1958). The ciliated bronchial epithelial cells undergo a degenerative process in which a pinching off occurs between the cytoplasm and nucleus, resulting in detached tufts of cilia, and a degenerating nucleus and cytoplasm (Figure 4.5). The degenerated cytoplasm may contain small, round eosinophilic inclusion bodies (Takahashi, 1981). Ciliocytophthoria occurs most frequently with influenza, parainfluenza, and adenovirus infection, but may also occur in bronchiectasis and other nonviral inflammatory conditions. It is seen more frequently in sputum specimens than in those obtained by bronchoscopy.

4.3.2 Virus Infections of the Urinary Tract

Although many viruses that cause systemic infections have been isolated from the urine, those most readily diagnosed by urine cytology are CMV, HSV, and a member of the papovavirus group, designated the BK virus (BKV) (Table 4.2). In each of these infections, epithelial cells of the urinary tract from the renal tubules to the bladder and urethra, detach and enter the urine. These cells often contain characteristic inclusions. Because of the relatively small number of cells in a large fluid volume, filtration or cytocentrifugation is necessary to obtain a suitable preparation (Schumann et al., 1977). Each of these infections occurs most often, but not exclusively, in immunosuppressed patients (commonly renal transplant recipients), and may coexist as mixed infections. CMV infected urothelial cells were first described in the urine of newborn infants with cytomegalic inclusion disease (Fetterman, 1952). Cytologic examination of the urinary sediment is indicated with any infant suspected of having neonatal CMV infection. Positive cytology is seen in approximately 50% of neonates that will subsequently have CMV-positive cultures (Hanshaw et al., 1968). CMV is the most frequently encountered viral infection in renal transplant recipients. In one study it was found in 31% of 2354 cytologically examined routine urine samples obtained from 91 patients (Traystman et al., 1980). Cellular changes include cytomegaly of the urothelial cells with typical large intranuclear inclusions surrounded by a clear halo and smaller eosinophilic cytoplasmic inclusions. This cytopathology most often involves single cells (Figure 4.1). Although the classical large, dense, intranuclear inclusion is the easiest to identify, occasional cells may be binucleated and some cells may have large dense eosinophilic cytoplasmic inclusions with little or no evidence of an intranuclear inclusion (Figure 4.6).

BKV was first isolated from the urine of a 39-year-old man who developed ureteral stenosis 4 months after renal transplantation (Gardner et al., 1971). The urine sediment contained abnormal transitional cells with dense intranuclear inclusions composed of crystalline arrays of papova virions as revealed by electron microscopy. BKV was later isolated from many asymp-

Table 4.2. Virus Infections of the Urinary Tract

Virus	Clinical Presentation	Cytologic Findings
Adeno	Hemorrhagic cystitis	Dense basophilic IN inclusions in transitional cells
BK (human papova)	Urethral stenosis in renal transplant and asymptomatic immunosuppressed patients	Large full mucoid IN inclusions (early); dense full basophilic inclusions bulging from cytoplasm (late)
CMV	Asymptomatic immunosuppressed patients	Large basophilic IN inclusions surrounded by halo; multiple eosinophilic IC inclusions
HSV	Generalized infection or local cystitis; may be contaminant from herpes genitalis	Ground-glass nuclei (early), eosinophilic IN inclusions (late); mutinuclearity, may be part of tubular cast
Measles	Measles with exanthema	Multinucleated giant cells with eosinophilic IN and IC inclusions

Abbreviations: IN, intranuclear; IC, intracytoplasmic.

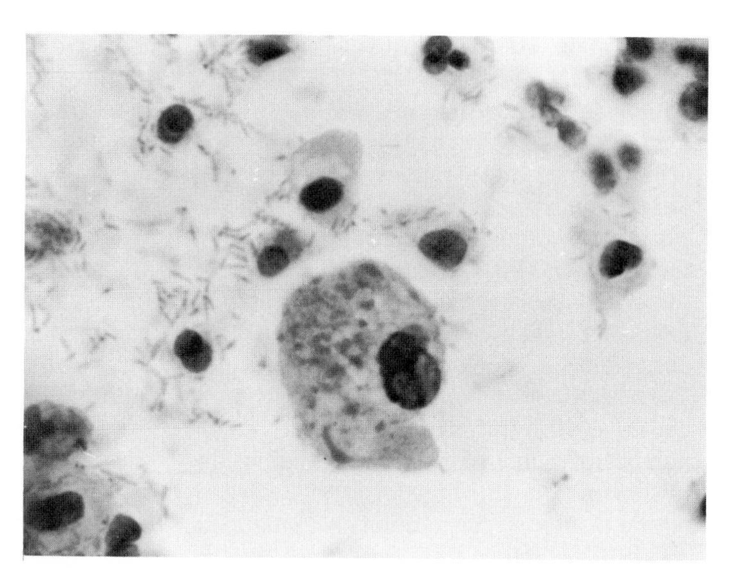

Figure 4.6. CMV in urine. Multiple intracytoplasmic inclusions with no significant intranuclear inclusions in late stage of infection. The nucleus is degenerated and appears empty. Pap stain, ×750

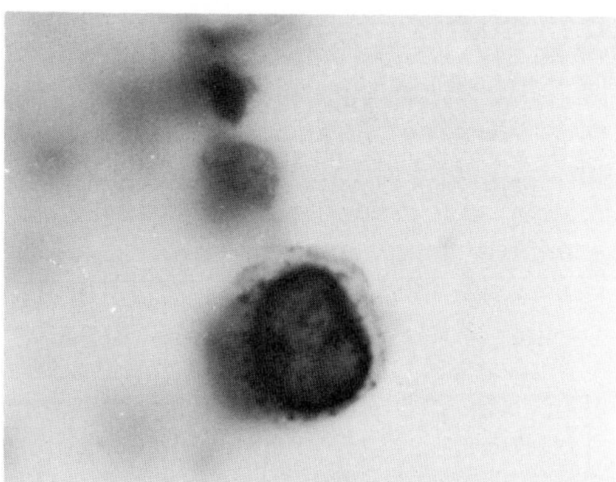

Figure 4.7. BKV in urine. Urothelial cell with the early changes of BKV infection. The enlarged nucleus has a condensed nuclear membrane with a mucoid inclusion. Pap stain, ×750

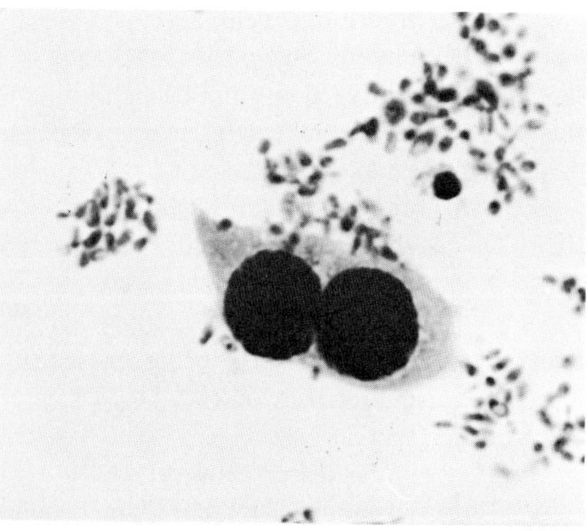

Figure 4.8. BKV in urine. Binucleated transitional cell with a dense homogeneous intranuclear inclusions. Pap stain, ×550

tomatic transplant recipients (Coleman, 1975) and two other patients with ureteral stenosis (Coleman et al., 1978). At an early stage, the most recognizable cytologic change due to BKV infection is the enlarged nucleus in an epithelial cell that contains a mucoid inclusion filling the nucleus (Figure 4.7). A more homogeneous, densely basophilic inclusion is detached at a later stage (Figure 4.8). This is sometimes referred to as a "decoy cell." Frequently, the nucleus appears to be bulging from the cytoplasm or thrusting from it, giving it a comet-like effect (Figure 4.9). Although most involved cells have single nuclei, occasional binucleated forms are seen with both types of inclusions present. At a later stage, the inclusion shrinks from the nuclear membrane leaving an incomplete thin halo. The intranuclear inclusions of BKV can be distinguished from those of CMV by the complete and consistent halo around the CMV inclusion and the lack of cytoplasmic inclusions in BKV (compare Figure 4.1 with 4.8).

Cytologic changes of HSV in the urinary sediment are similar to those described in the respiratory tract. They include multinuclear syncytial cells with enlarged ground-glass nuclei, seen at an early stage of infection, and typical eosinophilic intranuclear inclusions surrounded by a halo at a later stage. Elongated clumps of infected epithelial cells, probably of tubular origin, may contain inclusions in varying stages of development (Figure 4.10). Urinary sediment cells characteristic of HSV may occur in a generalized HSV infection involving the kidney, or a localized cystitis; they also may result from herpes genitalis, particularly when there is vaginal involvement during the infection (Masukawa et al., 1972). Other cytologic changes that may be observed in urinary sediment cells include intranuclear inclusions of adenovirus associated with acute hemorrhagic cystitis in children (Numazaki et al., 1973) and inclusion-bearing cells in the urine of patients with measles (Bolande, 1961).

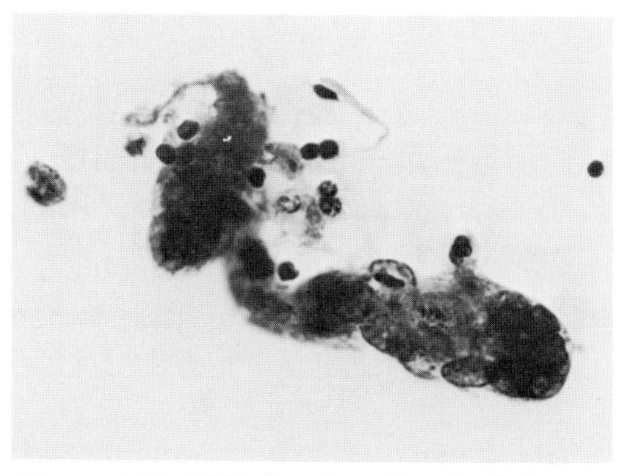

Figure 4.10. HSV in urine. Renal tubular cast showing characteristic HSV inclusions and nuclear molding. Pap stain, ×800

4.3.3 Virus Infections of the Genital Tract

Cytologic recognition of typical viral changes in cells of routine "pap" smears is the most readily available and cost-effective method of detecting genital herpes infections (Table 4.3). This is important in abating the spread of genital HSV, as well as epidemiologic studies that indicate a putative etiologic involvement of HSV-2 in human neoplasia. Cytologic recognition of HSV is of critical importance in directing the management of pregnancy near term. The overall incidence of HSV in routine vaginal smears has been reported at approximately 0.3% (Naib, 1980), although this figure varies greatly depending on the patient population. The sensitivity of cytology for detecting HSV infection depends somewhat on the location of the herpetic lesions and the adequacy of the sample. In one study (Vontver et al., 1979), 41% of 69 cases with external lesions that were virus isolation-positive had positive smears, but the rate was 23% with women that had only cervical lesions. Similar results were reported in a study of 76 patients with genital HSV comparing virus isolation, immunofluorescence, immunoperoxidase, and cytology as means of making a diagnosis (Moseley et al., 1981). The overall positive cytology rate was 37.6%, but was 47.9% for cases with vaginal or cutaneous lesions. Significantly, there were a number of cases in which the cytology was positive but virus isolation in

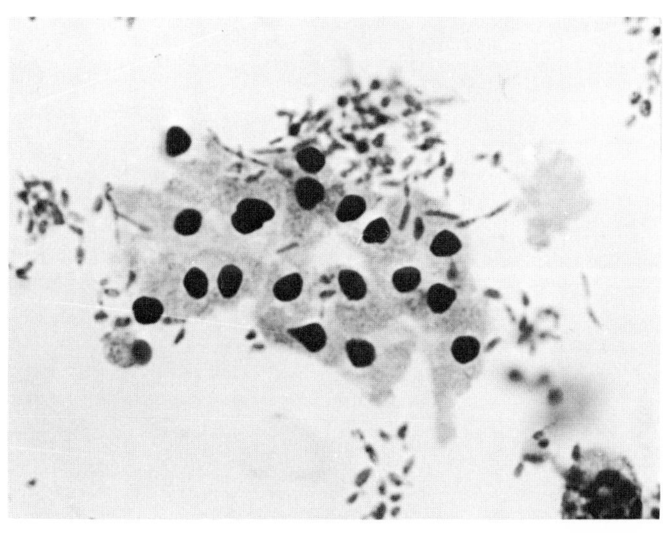

Figure 4.9. BKV in urine. Urothelial cells each containing a dense intranuclear inclusion, some of which have a comet-like configuration. Pap stain, ×550

Table 4.3. Virus Infections of the Genital Tract

Virus	Clinical Presentation	Cytologic Findings
HSV (types I and II)	Herpes genitalis	Ground-glass nuclei (early); eosinophilic IN inclusions (late) with peripheral chromatin condensation; multinuclearity
Papilloma	Condyloma acuminatum; cervical dysplasia	Enlarged hyperchromatic nucleus; rare basophilic IN inclusions; perivascular cytoplasmic clearing and vacuolar degeneration (koilocytotic change)
Molluscum contagiosum	Vaginal, penile, or perineal papule with central umbilication	Large dense staining IC inclusions displacing nucleus; squamous cell often bean-shaped

Abbreviations: IN, intranuclear; IC, intracytoplasmic.

tissue culture from the same sample was negative. This discrepancy is repeatedly encountered in reported studies comparing virus isolation and cytopathology with various viruses at different sites, and indicates that the greatest diagnostic yield with virus infections amenable to cytologic diagnosis is from the combination of cytology and virus isolation.

The identifiable cytologic changes in genital HSV infection are identical to those described with infections of the respiratory and urinary tracts. At an early stage, the enlarged nuclei have a bland ground-glass appearance with the chromatin displaced to the periphery, resulting in an apparent thickening of the nuclear membrane (Figure 4.11). At a later stage, the nucleus contains an eosinophilic inclusion surrounded by a clear halo. Multinuclear cells are common with up to ten nuclei, which often exhibit molding. Inclusions in multinucleated cells may all be at the same stage (Figure 4.12) or may exhibit different stages of develop-

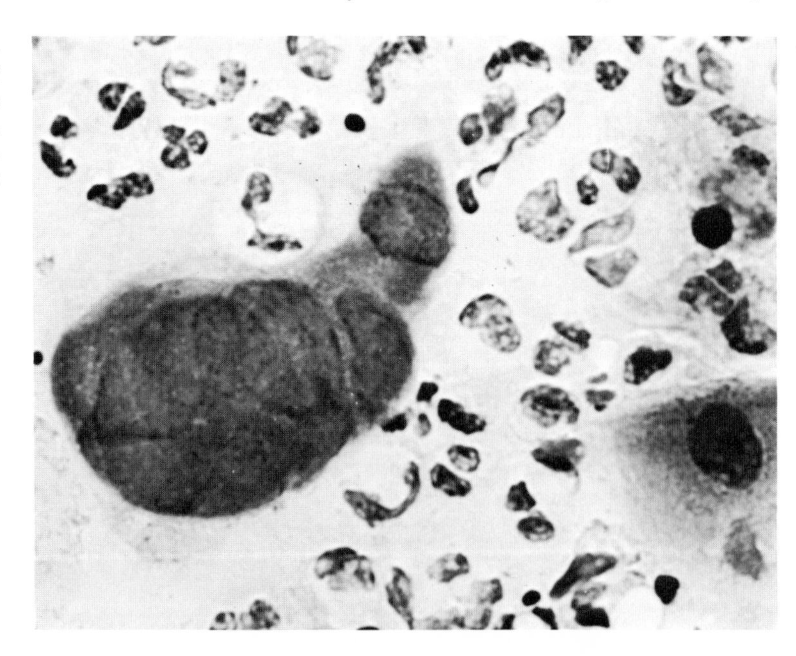

Figure 4.11. HSV in genital smear. Early stage of infection showing the molded nuclei with ground-glass nuclear appearance. Pap stain ×800

ment (Figure 4.13). HSV-1 and -2 produce identical morphologic changes and cannot be differentiated on the basis of cytology. Although it has been reported that primary HSV infection can be differentiated from recurrent or secondary infection by a predominance of bland ground-glass nuclei in primary infection (Ng et al., 1970), this has not been confirmed in subsequent studies (Naib, 1980).

Infection of the cervical or vaginal mucosa and the skin of the perineum with a human papilloma virus (HPV) may result in proliferation of epithelial cells forming a vegetative papillary growth known as a condyloma acuminatum or venereal wart. Atypical cellular changes often accompany the proliferative process and may result in cellular alterations similar to those of malignant cells (Meisels et al., 1981). HPV has

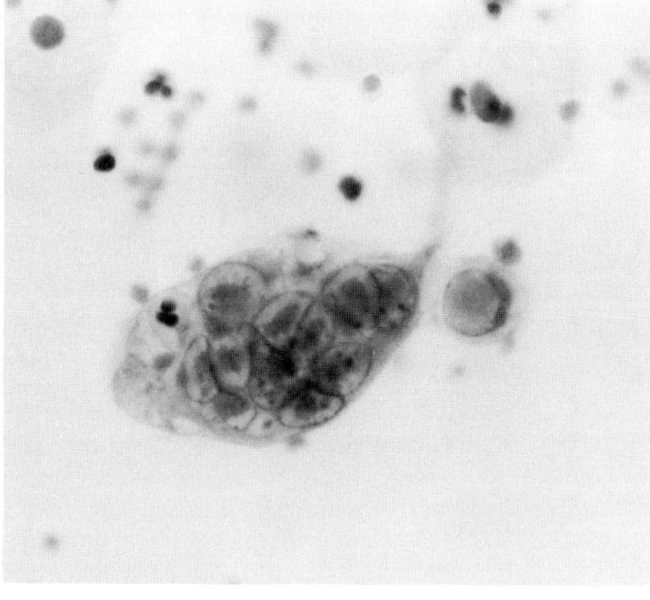

Figure 4.12. HSV in genital smear. Multinucleated giant cells with intranuclear inclusions surrounded by halos. Pap stain, ×720

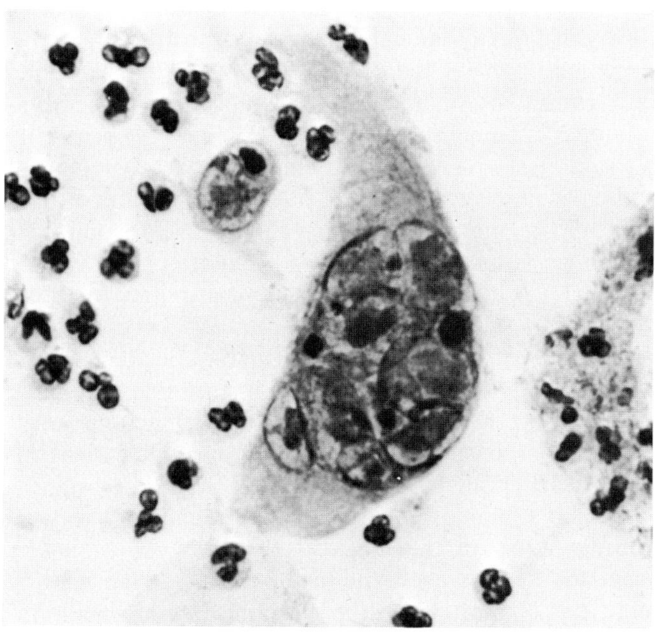

Figure 4.13. HSV in genital smear. Note inclusions at different stages. Pap stain, ×720

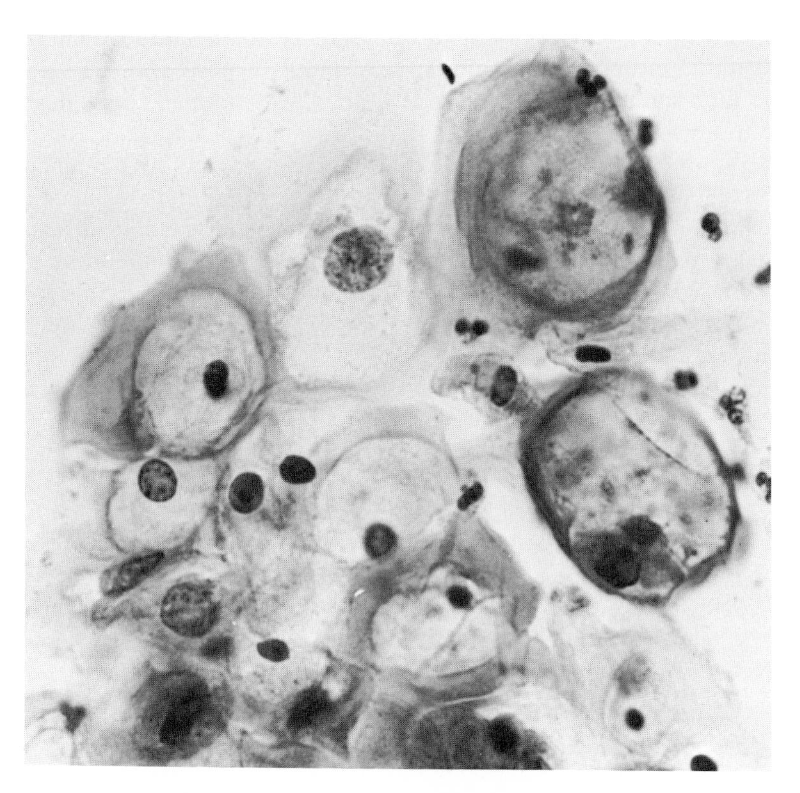

Figure 4.14. Papilloma virus in genital smear. Swollen squamous cells exhibiting koilocytotic change characteristic of a condylomatous lesions. Pap stain, ×720

been associated with premalignant cervical dysplasia and carcinoma, and most recently identified by immunohistology and/or molecular probes in association with squamous cell carcinoma of the vulva (Zachow et al., 1982; Pilotti et al., 1984). At the present time, cytology is the only practical way of identifying an HPV infection of the genital tract when a characteristic gross condylomatous lesion is not observed. The changes attributable to virus infection involve squamous cells, which appear swollen and have a perinuclear halo with poor cytoplasmic keratinization resulting in irregular staining. This produces a picture referred to as koilocytotic change (Figure 4.14). The nucleus is frequently enlarged and occasionally contains a poorly defined basophilic inclusion which, by electron microscopy, is composed of virus particles and fibrillar material (Caras-Cordeo et al., 1981). Although the incidence of papilloma virus infection and the accuracy of diagnosis based on finding koilocytotic change by cytology have not yet been fully documented, the identification of papilloma virus infection of the genital tract is likely to be important in preventing sexual transmission of HPV. Women who have genital HPV infection should have follow-up with periodic cytology to detect early atypical changes that may indicate malignant transformation.

Other viruses that have been identified by characteristic cytologic findings in vaginal smear include CMV and the poxvirus, which causes molluscum contagiosum (Brown et al., 1981). Molluscum contagiosum is a benign cutaneous infection most often observed in children and young adults, which is easily transmitted by direct contact. Circumstantial evidence suggests the infection is transmitted between young adults during sexual intercourse. Although the lesions have a characteristic appearnce consisting of a small, firm papule with a centralized umbilication on the skin or vaginal mucosa, virus isolation techniques to confirm the diagnosis are not yet available. However, the histopathology necessitating biopsy and the morphologic changes in individual cells as observed in a pap stained smear are diagnostic. The cytologic changes consist of large, dense staining cytoplasmic inclusions occupying the entire squamous cell and resulting in peripheral displacement of a flattened nucleus (Figure 4.15). The cells frequently assume a bean shape with the nucleus displaced to the concave aspect.

4.3.4 Virus Infections of the Eye

In many common ocular lesions, and particularly in those involving the cornea, a rapid diagnosis is essential in order to initiate therapy and avoid progressive corneal damage. Although biopsy and virus isolation are the usual definitive procedures for the diagnosis of virus infections involving the cornea and conjunctiva, exfoliative cytology offers a simple, inexpensive, and rapid means of diagnosing adenovirus keratoconjunctivitis and keratitis due to herpes viruses (Naib et al., 1967; Schumann et al., 1980) (Table 4.4). In addition, characteristic cytologic changes consisting of multinucleated giant cells in measles keratoconjunctivitis and the large dense cytoplasmic inclusions of molluscum contagiosum in conjunctival and eyelid lesions may also yield a definitive diagnosis, although these infections as isolated ophthalmic disease are rarely encountered. Finally, cytology is most useful for the diagnosis of chlamydial conjunctivitis, which may be clinically difficult to differentiate from the disease caused by adenovirus but can be diagnosed by the finding of characteristic cytoplasmic inclusions caused by the chlamydial infection (Gupta et al., 1979).

Specimens containing conjunctival or corneal cells should be collected by a physician, preferably an ophthalmologist, by swab or superficial conjunctival or corneal scraping. Corneal scraping requires examination with a slit lamp microscope for localization of the lesion. The collected material is immediately spread on an alcohol-moistened slide and, after partial evaporation, the slide is placed in 95% ethyl alcohol for proper fixation. Because there are usually very few cells and little fluid substrate, air-drying of the smears is a frequent problem, but can be avoided by immediate fixation of the specimen on the slide.

In adenovirus infections, the conjunc-

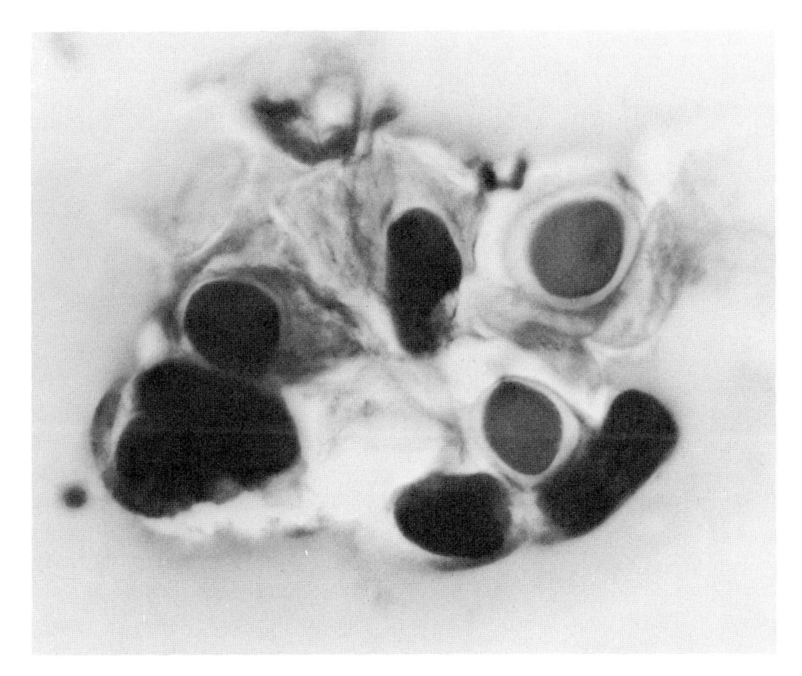

Figure 4.15. Molluscum contagiosum in genital smear. Molluscum bodies from a vaginal smear showing large intracytoplasmic inclusions displacing the nucleus. Pap stain, $\times 720$

Table 4.4. Virus Infections of the Eye

Virus	Clinical Presentation	Cytologic Findings
Adeno	Acute (epidemic) keratoconjunctivitis and conjunctivitis with pharyngiitis	Multiple eosinophilic IN inclusions (early); dense central basophilic inclusions surrounded by halo (late)
HSV	Corneal vesicle or ulcer; may be isolated ophthalmic lesion or with other HSV vesicles	Multinucleated cells with eosinophilic IN inclusions surrounded by halo; nuclei has ground-glass appearance (early stage)
Molluscum contagiosum	Reddish papular 5-mm lesions of eyelid or conjunctiva	Large dense basophilic IC inclusions displacing nucleus
Varicella zoster	Vesicular eruptions in dermatome involving eye (shingles) or accompanying chicken pox	Multinucleated cells with IN eosinophilic inclusions
Chlamydia	Granular conjunctiva with corneal ulcerations (trachoma) or conjunctivitis only (inclusion conjunctivitis)	Enlarged corneal (trachoma only) and conjunctival cells with numerous IC basophilic inclusions surrounded by individual halos

Abbreviations: IN, intranuclear; IC, intracytoplasmic.

tival or corneal cells are mixed with lymphocytes and plasma cells, and contain distinctive intranuclear inclusions. In the early stages of infection (Figure 4.16), the intranuclear inclusions are multiple, small, and sometimes granular and eosinophilic. At a later stage, the small inclusions coalesce as a single dense basophilic body, usually centrally located and surrounded by a clear halo (Figure 4.17).

Keratoconjunctivitis caused by HSV can occur as a primary lesion or as part of a systemic infection. Superficial scrapings from the margin of the ulcerated area will usually contain multinucleated cells with characteristic large eosinophilic intranu-

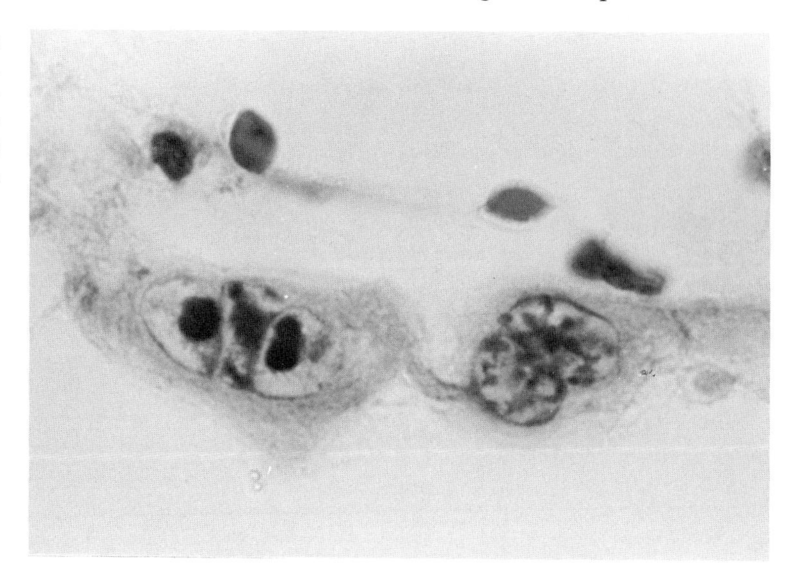

Figure 4.16. Adenovirus in conjunctival scraping. Multiple small intranuclear inclusions characteristic of the early stage of infection. Pap stain, ×800

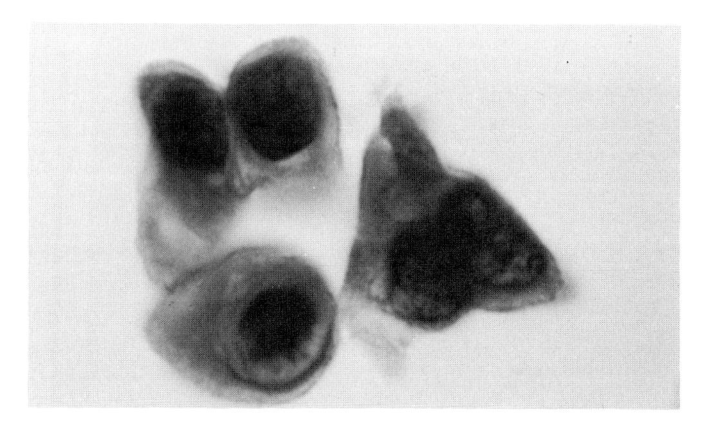

Figure 4.17. Adenovirus in conjunctival scraping. Conjunctival cells showing a centrally located intranuclear inclusion surrounded by a clear halo seen in late stages of the infection. Pap stain, ×800

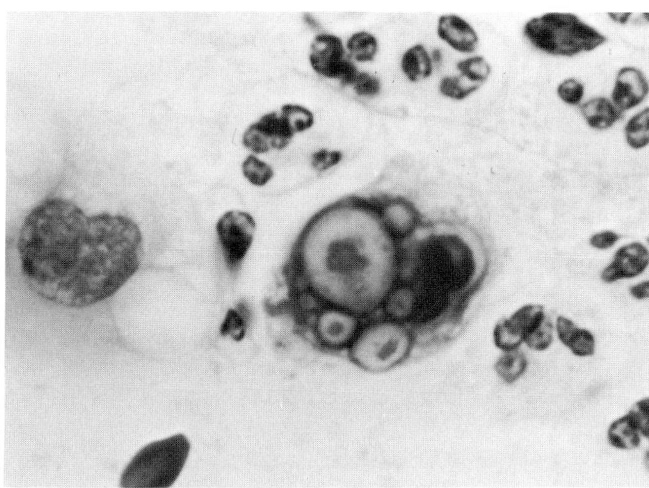

Figure 4.18. Chlamydia in conjunctival scraping. Epithelial cells with intracytoplasmic inclusions surrounded by distinct halos. Pap stain, ×720

clear inclusions surrounded by prominent halos. Early in the infection, as with HSV at other sites, scrapings will reveal enlarged nuclei having a ground-glass appearance.

Herpes zoster keratitis is well recognized clinically and usually does not require additional diagnostic confirmation, such as cytology. However, scrapings of the lesions will yield cells similar to the ones found in HSV infection, although it is reported that syncytia and intranuclear inclusions are less prominent than with HSV (Naib, 1967). The acute conjunctivitis that occurs, usually during the prodromal of measles, is also associated with characteristic cytologic findings seen in conjunctival smears or scrapings. The characteristic cells may contain up to 100 round nuclei surrounded by an abundant cytoplasm in which there are numerous eosinophilic inclusions. Occasionally, similar eosinophilic inclusions can be found within the nuclei. These findings can precede the appearance of the typical exanthem by 2 to 3 days.

The morphologic changes in conjunctival and corneal cells due to virus infection must be differentiated from the cytologic changes due to chlamydial infections causing trachoma and inclusion conjunctivitis. In both these diseases, the epithelial cells generally are enlarged and have abundant cytoplasm containing clusters of basophilic intracytoplasmic inclusions, each surrounded by a large individual halo (Figure 4.18). In trachoma, the corneal cells are involved, whereas, only conjunctival cells show the changes in the more benign chlamydial conjunctivitis. The presence of cells containing cytoplasmic inclusions in cytologic examination of specimens from the eye suggests chlamydial, rather than viral, infection.

REFERENCES

Bales, C., and Durfee, G. 1979. Cytologic techniques. In L. Koss (ed.), Diagnostic Cytology, Vol. 2, 3rd ed. Philadelphia: J.B. Lippincott Co., pp. 1187–1266.

Barrett, D. 1976. Cytocentrifugation technique. In C. Keebler, J. Reagan, and G.L. Wied (eds.), Compendium on Cytopreparatory Techniques, 4th ed. Chicago: Tutorials of Cytology, pp. 80–83.

Bolande, R.P. 1961. Significance and nature of inclusion-bearing cells in the urine of patients with measles. N. Engl. J. Med. 265:919–923.

Brown, S.T., Nalley, J.F., and Kraus, S.N. 1981. Molluscum contagiosum. Sex. Transm. Dis. 8:227–234.

Caras-Cordero, M., Morin, C., Roy, M., Fortier, M., and Meisels, A. 1981. Origin of the koilocytes in condylomata of the human cervix. Ultrastructural study. Acta Cytol. 25:383–392.

Coleman, D.V. 1975. The cytodiagnosis of human polyoma infection. Acta Cytol. 9:93–96.

Coleman, D.V., MacKenzie, E.F.D., Gardner, S.D., Poulding, J.M., et al. 1978. Human polyomavirus (BK) infection and ureteric stenosis in renal allograft recipients. J. Clin. Pathol. 31:338–347.

Fetterman, G.H. 1952. New laboratory aid in clinical diagnosis of inclusion disease of infancy. Am. J. Clin. Pathol. 22:424–427.

Gardner, S.D., Field, A.M., Coleman, D.V., and Hulme, B. 1971. New human papovavirus (BK) isolated from urine after renal transplantation. Lancet i:1253–1257.

Gill, G. 1976. Methods of cell collection on membrane filters. In C. Keebler, J. Reagan, and G.L. Wied (eds.), Compendium on Cytopreparatory Techniques, 4th ed. Chicago: Tutorials of Cytology, pp. 34–44.

Gupta, P.A., Lee, E.F., Erozan, Y.S., Frost, J.K., Geddes, S.T., and Donovan, P.A. 1979. Cytologic investigations in chlamydia infection. Acta Cytol. 23:315–320.

Hanshaw, J.B., Steinfeld, H.J., and White, C.J. 1968. Fluorescent–antibody test for cytomegalovirus macroglobulin. N. Engl. J. Med. 279:566–570.

Masukawa, T., Jarancis, J.C., Rytel, M., and Mattingly, R.F. 1972. Herpes genitalis virus isolation from human bladder urine. Acta Cytol. 16:416–428.

Meisels, A., Ray, M., Fortier, M., Morin, C., Cassas-Cordero, M., Shah, K.V., and Turgeon, H. 1981. Human papillomavirus infection of the cervix. Acta Cytol. 25:7–16.

Moseley, R.C., Corey, L., Benjamin, D., Winter, C., and Remington, M.L. 1981. Comparison of viral isolation, direct immunofluorescence, and direct immunoperoxidase techniques for detection of genital herpes simplex virus infection. J. Clin. Microbiol. 13:913–918.

Naib, Z.M., Clepper, A.S., and Elliott, S.R. 1967. Exfoliative cytology as an aid in diagnosis of ophthalmic lesions. Acta Cytol. 11:295–303.

Naib, Z., Stewart, J., Dowdle, W., Casey, H., Marine, W., and Nahmias, A. 1968. Cytologic features of viral respiratory tract infections. Acta Cytol. 12:162–171.

Naib, Z.M. 1980. Exfoliative cytology in the rapid diagnosis of herpes simplex infection. In A.J. Nahmias, W.R. Dowdle, and R.F. Schinaza (eds.), The Human Herpes Virus: An Interdisciplinary Perspective. Amsterdam: Elsevier/North Holland Biomedical Press, pp. 381–386.

Ng, A.B.P., Reagan, J.W., and Yen, S.S. 1970. Herpes genitalis: Clinical and cytopathological experience with 256 patients. Obstet. Gynecol. 36:645–651.

Numazaki, Y., Kumasaka, T., Yano, N., Yamanaka, M., Miyazawe, T., Takai, S., and Ishida, N. 1973. Further study on acute hemorrhagic cystitic due to adenovirus type 11. N. Engl. J. Med. 289:344–347.

Pierce, C.H., and Hirsch, J.G. 1958. Ciliocytophthoria relationship to viral respiratory infections in humans. Proc. Soc. Exp. Biol. Med. 98:489–492.

Pilotti, S., Rilke, F., Shah, K.V., Torre, G.D., and DePalo, G. 1984. Immunohistochemical and ultrastructural evidence of papilloma virus infection associated with in situ and microinvasive squamous cell carcinoma of the vulva. Am. J. Surg. Pathol. 8:751–761.

Schumann, G.B., Berring, S., and Hill, R. 1977. Use of the cytocentrifuge for the detection of cytomegalovirus inclusions in the urine of renal allograft patients. Acta Cytol. 21:168–172.

Schumann, G.B., O'Dowd, G.J., and Spinnler,

P.A. 1980. Eye cytology. Lab. Med. 11:533–540.

Takahashi, M. 1981. Color Atlas of Cancer Cytology. Tokyo–New York: Igaku-Shoin, p. 291.

Tomkins, V., and Macanlay, J.C. 1955. A characteristic cell in nasal secretions during prodromal measles. J. Am. Med. Assoc. 157:711–712.

Traystman, M.D., Gupta, P.K., Shah, K.V., Reissig, M., et al. 1980. Identification of viruses in the urine of renal transplant recipients by cytomorphology. Acta Cytol. 24:501–510.

Vernon, S.E. 1982. Cytologic features on nonfatal herpes virus tracheobronchitis. Acta Cytol. 26:237–242.

Vontver, L.A., Reeves, W.C., Rathay, M., Corey, L., Remington, M.A., Tolentino, E., Schweid, A., and Holmes, K.K. 1979. Clinical course and diagnosis of genital herpes virus infection and evaluation of topical surfactant therapy. Am. J. Obstet. Gynecol. 133:548–554.

Warner, N.E., McGrew, E.A., and Nanos, S. 1964. Cytologic study of sputum in cytomegalic inclusion disease. Acta Cytol. 8:311–315.

Zachow, K.R., Ostrow, R.S., Bender, M., Watts, S., Okagaki, T., Pass, F., and Faras, A.J. 1982. Detection of human papillomavirus DNA in anogenital neoplasms. Nature 300:771–773.

5

Electron Microscopy and Immunoelectron Microscopy

Frances W. Doane

5.1 INTRODUCTION

Virus diagnosis by electron microscopy (EM) relies on the detection and identification of viruses on the basis of their characteristic morphology. As long ago as 1948 it was realized that the electron microscope could be used to distinguish one virus from another and, soon after the first commercial electron microscopes became available, they were being used to distinguish the large brick-shaped smallpox virus from the smaller, round varicella-zoster virus (Nagler and Rake, 1948; van Rooyen and Scott, 1948). With the introduction of the simple but effective negative staining technique for revealing viral ultrastructure (Brenner and Horne, 1959) it became possible to observe greater structural detail than had been possible with the earlier shadowing technique. An increasing number of virologists ventured into this field when it became apparent that the electron microscope could be an important tool in rapid diagnosis of virus infections (Peters et al., 1962; Nagington, 1964; Cruickshank et al., 1966; Chambers and Evans, 1966; Doane et al., 1967, 1969; Joncas et al., 1969). In more recent years immunoelectron microscopy (IEM), involving the visualization by EM of virus–antibody complexes, has increased the sensitivity of virus detection, providing a means of serotyping viruses directly on the EM specimen grid (Almeida et al., 1963; Almeida and Waterson, 1969; Doane, 1974).

This chapter presents a number of EM methods that we and our colleagues have found to be useful in diagnostic virology. For more extensive dissertations the reader should consult one or more of the following sources (Doane and Anderson, 1977, 1986; Chernesky, 1979; Hsiung et al., 1979; Kjeldsberg, 1980; Field, 1982; Palmer and Martin, 1982; McLean and Wong, 1984).

5.1.1 Advantages

A major advantage of virus diagnosis by EM is the ability to visualize the virus. The virologist who employs other methods in the diagnostic armamentarium is denied this gratification. Instead, he must rely on indicator systems—indirect clues involving cytopathology, lysing blood cells, color changes, or radioactivity counts—that signify the presence of a virus. By identifying viruses directly, on the basis of their morphology, it is possible to perform an examination without a preconceived concept of the etiologic agent, in contrast with those assays that require a specific viral probe (e.g., a particular viral antibody or nucleic acid sequence). Speed is another major advantage provided by EM. With the commonly used negative staining technique, a clinical specimen can be processed within minutes after collection. In many cases, virus morphology alone is sufficiently characteristic to permit family identification. Finally, by depending on the morphology of a virus for identification, its level of infectivity is of no consequence in making that identification. Thus, EM methods are applicable for the detection of inactivated viruses or of fastidious viruses, such as hepatitis B, Norwalk, or papovaviruses.

5.1.2 Limitations

The increasing use of EM in virus laboratories attests to the value of this instrument in diagnostic virology. But it would be unwise to enter this field without a realistic understanding of its limitations. Probably the most serious limitation of the electron microscope as a tool in diagnostic virology is its inability to examine multiple specimens coincidentally. Thus, it is not possible to examine every clinical specimen by EM. Consequently, only selected specimens can be examined by this technique.

A second limitation of the electron microscope is one shared by other detection methods, viz. there must be a minimum number of virus particles present in the specimen in order to be detected on the EM specimen grid. Some clinical specimens, such as vesicle fluid from herpetic or poxvirus lesions or feces from rotavirus-infected patients, usually have a very high virus content and are ideal EM specimens. Others, such as throat washings and urine, may contain too few virus particles to be detected by EM. For this reason many laboratories initially inoculate all specimens into cell culture systems. This serves to screen out the negatives, and amplify the virus content in the positives. The EM can then be utilized at this stage, to identify the cell culture isolates.

5.1.3 Special Facilities

A high-resolution transmission electron microscope (TEM) and appropriately trained personnel are essential elements for diagnostic virology by EM.[1] One should consider the facility of operation and available service when purchasing a TEM. Alternatively, the virus laboratory can depend entirely on the services of an external EM facility, thereby avoiding maintenance responsibilities.

The principal EM methodology used in diagnostic virology is the negative staining technique, which requires little special equipment. The EM specimen grids can be coated with plastic alone, but it is preferable to stabilize the plastic with carbon, a procedure that requires a simple vacuum evaporator. Plastic/carbon grids can be prepared in large quantities, and stored for several weeks.

If thin sectioning techniques are to be employed, a fairly extensive collection of equipment and reagents must be available, including oven, ultramicrotome, glass or diamond knives, fixatives, embedding media, and stains. In addition, someone must be available to cut thin sections (1/10–1/20 μm), a skill that is not easily acquired. For this reason, most virus laboratories will prefer to rely on a neighboring pathology facility for tissue fixation, embedding, and sectioning. It then remains for the virologist to examine prepared sections by EM.

Rarely will one perform EM without the need—or desire—to produce micrographs. Consequently, the operator must have a supply of photographic materials, and develop a technical capability in photographic processing.

[1] A list of EM courses offered in North America is available through the Electron Microscopy Society of America, c/o Dr. Judy Murphy, Center for Electron Microscopy, Southern Illinois University, Carbondale, Illinois 62901, U.S.A.

5.2 CLINICAL SPECIMENS SUITABLE FOR VIRUS DETECTION BY ELECTRON MICROSCOPY

2.1 Vesicle Fluid and Crusts

Vesicle fluid and scrapings from poxvirus and herpetic skin lesions are ideal EM specimens, as they often contain large quantities of virus that can be detected readily (Fig. 5.1) (Peters et al., 1962; Nagington, 1964; Cruickshank et al., 1966; Macrae et al., 1969; Blank et al., 1970). Vesicular fluid can be collected in a capillary tube or a fine-bore needle attached to a small syringe. It is expelled onto a glass slide and mixed with a drop of filtered, distilled water.

A scalpel blade scraping of the base of a vesicle can be smeared on a glass slide and moistened with a small quantity of water. Crusts removed from dried vesicles can be minced using a scalpel blade in a drop of water on a slide.

Specimens to be negatively stained are added to the EM grid either by touching a coated grid to the fluid on the slide or by processing the suspension by the agar diffusion method (see below).

5.2.2 Respiratory Tract Secretions

On occasion, these specimens contain a sufficient quantity of virus (e.g., paramyxoviruses) to be detected by direct EM (Doane et al., 1967; Joncas et al., 1969), although they usually require some form of amplification, such as ultracentrifugation, IEM, or passage in cell culture. A small amount of specimen is placed on a glass slide. The specimen, if viscous, can be diluted slightly with distilled water prior to negative staining.

5.2.3 Cerebrospinal Fluid

Evans and Melnick (1949) reported the detection by EM of herpesvirus in cerebrospinal fluid (CSF) obtained from a patient with herpes zoster. One of the first clinical specimens we examined was a CSF from a patient with mumps encephalitis; a paramyxovirus was detected by negative staining (Doane et al., 1967). In general,

Figure 5.1. Negatively stained herpes varicella zoster virus seen in smear from vesicle scraping. In this particular field all of the virus particles exhibit a stain-penetrated core surrounded by a hexagonal-shaped capsid and a tightly adhering envelope. Bar = 100 nm

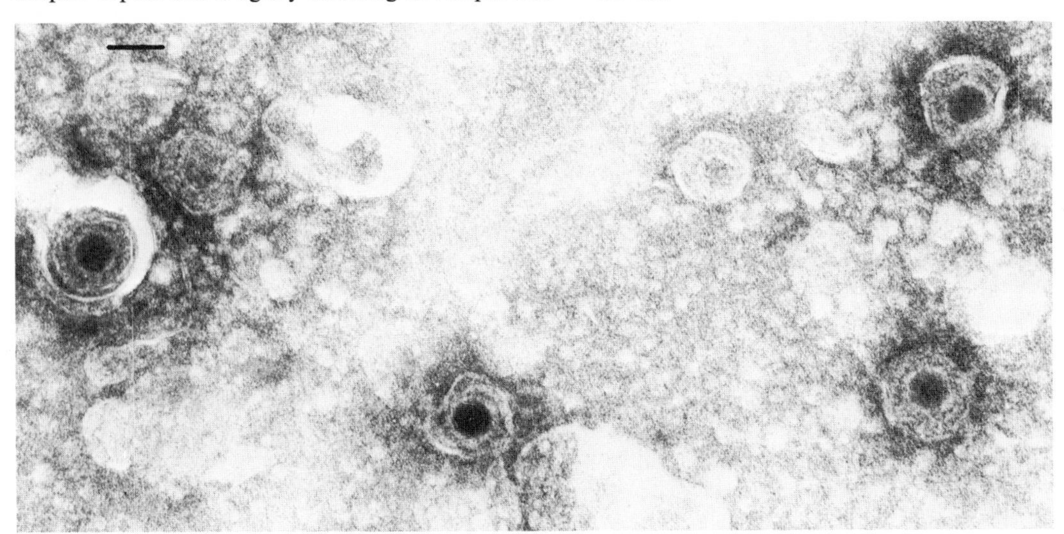

concentration of virus content should be attempted, for example, Airfuge ultracentrifugation (see below).

5.2.4 Feces

The negative staining technique can be used to detect many different viruses that are found in feces (Fig. 5.2) (Tyrrell and Kapikian, 1982), including rotaviruses (Flewett et al., 1973; Bishop et al., 1973; Middleton et al., 1974), coronaviruses (Caul et al., 1975; Mathan et al., 1975), adenoviruses (Anderson and Doane, 1972; Flewett et al., 1974), astroviruses (Madeley and Cosgrove, 1975), caliciviruses (Madeley and Cosgrove, 1976; Flewett and Davies, 1976), hepatitis A virus (Feinstone et al., 1973), and a variety of "small round viruses," in the 18- to 34-nm range, identified

Figure 5.2. Isometric viruses found in negatively stained human feces. (A) Adenoviruses; (B) rotaviruses. Each family can be identified by morphologic characteristics, such as virion size and the appearance of the capsid(s) and capsomers. Bar = 100 nm Micrographs kindly provided by M. Szymanski. Reproduced from Doane and Anderson 1977, with permission.

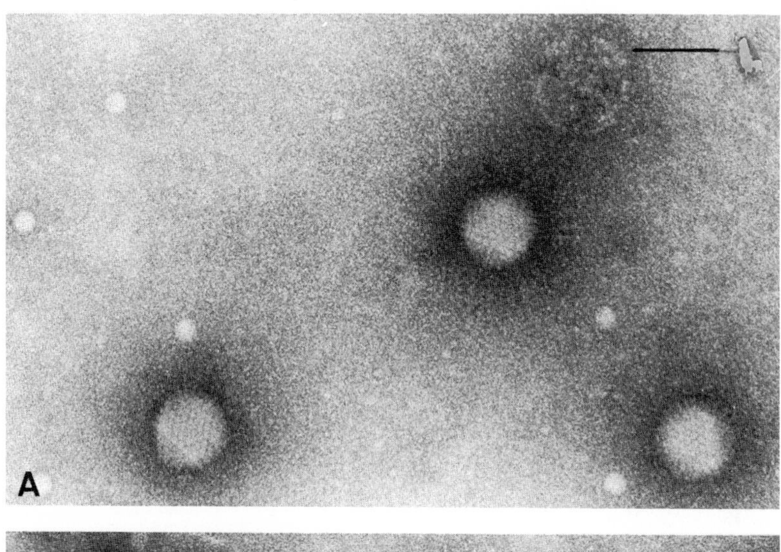

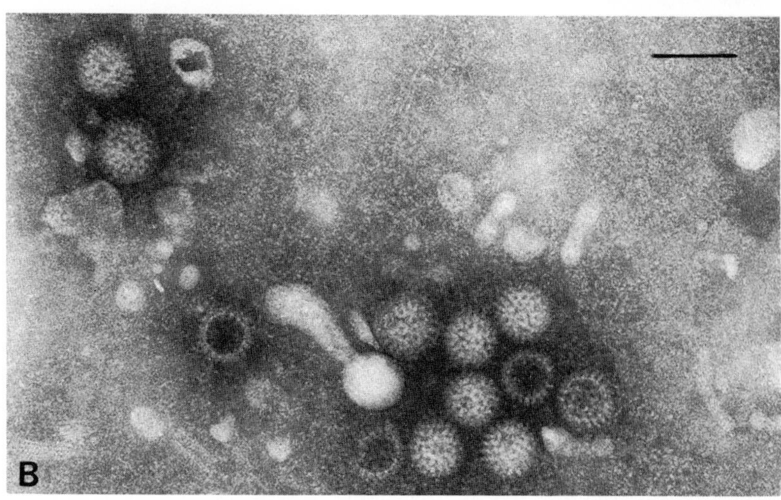

as picornaviruses and/or parvoviruses (for reviews, see Bishop, 1982; Field, 1982). Some investigators routinely prepare 10–20% suspensions of feces, which they clarify free of bacteria and debris using light centrifugation (approximately 1000 × *g*), then ultracentrifuge at 50–100,000 × *g* to pellet any virus present (Flewett et al., 1974; Davies, 1982). A simple and effective method is to mix a small amount of feces with 1% ammonium acetate, and then transfer the mixture to a grid for negative staining (Middleton et al., 1977).

5.2.5 Urine

Herpesvirus particles have been demonstrated by negative staining in the urine of patients excreting cytomegalovirus (Montplaiser et al., 1972; Lee et al., 1978). Papovaviruses of the polyomavirus genus have been detected in the urine of immunosuppressed renal transplant patients (Gardner et al., 1971; Field, 1982) and in the urine of pregnant women (Coleman et al., 1977).

5.2.6 Blood

Electron microscopy has provided a sensitive and rapid method for demonstrating hepatitis B virus in serum, both by direct EM and by IEM (Figure 5.3) (Bayer et al., 1968; Almeida et al., 1969; Hirschman et al., 1969; Dane et al., 1970).

5.2.7 Tissues

Any number of different tissues have been examined by EM for the presence of viruses. Brain biopsy tissue from patients with suspected herpes simplex virus infection can be ground with a tissue grinder or mortar and pestle in filtered distilled water to produce a suspension that is then negatively stained. By this method, herpesvirus particles have been identified within a few minutes after receipt of the specimen (Chia and Spence, personal communication; Szy-

Figure 5.3. Hepatitis B virus particles from chronic carrier. Patient's serum was incubated for 1 hour with rabbit anti-HBs prior to negative staining. A single isometric virus particle is seen amidst the spherical and tubular forms of HBsAg. Bar = 100 nm Micrograph kindly provided by M. Fauvel. Reproduced from Doane and Anderson 1977, with permission.

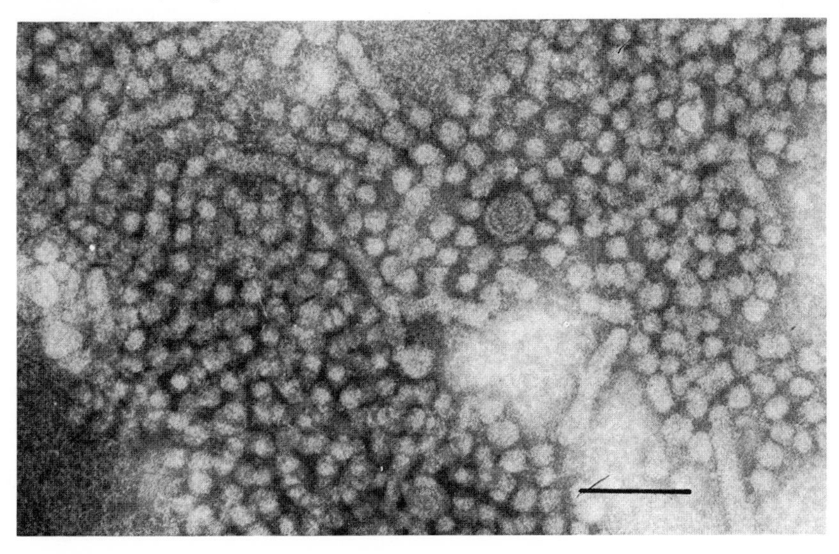

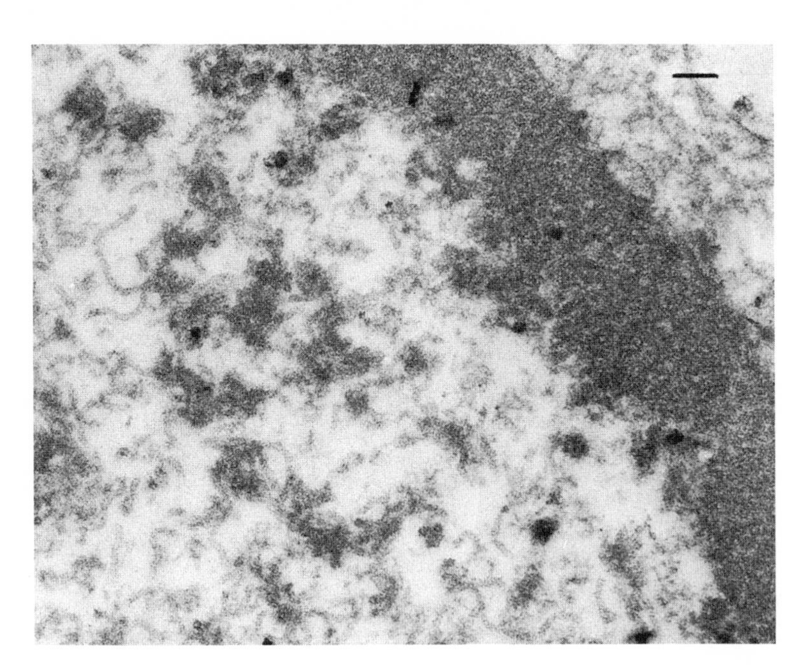

Figure 5.4. Thin section (in region of nucleus) of brain biopsy from patient with SSPE. Note margination of chromatin across upper right corner, and worm-like viral nucleocapsids scattered throughout the nucleoplasm. Bar = 100 nm Reproduced from Doane and Anderson 1977, with permission.

manski and Middleton, personal communication). Although this method is undoubtedly the most rapid, it may not be as sensitive as thin sectioning (Field, 1982). Standard fixation and embedding procedures, or a rapid embedding method (described later) have been used to demonstrate virus in the brain of patients with herpes simplex encephalitis (Harland et al., 1967; Roy and Wolman, 1969; subacute sclerosing panencephalitis (SSPE) (Figure 5.4) (Bouteille et al., 1965; Herndon and Rubinstein, 1968), and progressive multifocal leukoencephalopathy (PML) (Zu Rhein and Chou, 1965).

5.3 NEGATIVE STAINING METHODS

Several commonly used negative staining methods are presented below. They vary in complexity and in sensitivity. Except where noted, they use essentially the same materials. Prepared specimen grids should be exposed to ultraviolet (UV) radiation (approximately 700–1000 $\mu W/cm^2$ at a distance of 4–6 in) for at least 10 minutes prior to EM examination.

5.3.1 Materials

300-mesh copper specimen grids, coated with a Formvar or Parlodion film (preferably stabilized with a thin layer of carbon)

Negative stain: 2% phosphotungstic acid (PTA) adjusted to pH 6.5 with 1N KOH; or 1% uranyl acetate, pH 4.0; or 3% ammonium molybdate, pH 7.2; stain must be prepared in distilled water sterilized by filtration to remove bacteria

Fine-bore pasteur pipettes

Filter paper

EM forceps

UV lamp

5.3.2 Direct Application Method

As the name implies, this is the simplest of the negative staining methods. It cannot be used for specimens that contain a high concentration of salt.

1. Place a small drop of specimen on a coated copper grid held with EM forceps
2. Add a drop of negative stain
3. Touch the fluid with the torn edge of filter paper, leaving only a moist layer on the grid
4. Air dry (1–2 minutes)

5.3.3 Water Drop Method

This is a simple method for removing salt from specimens that contain a high salt concentration which, if allowed to dry on the EM grid, will produce crystals that tend to hide virus particles. It yields a nicely stained preparation, but it requires a starting concentration of approximately 10^9 virus particles per milliliter (Doane et al., 1969).

1. On a waxed surface (e.g., Parafilm®) place a drop of filtered distilled water
2. Place a small drop of specimen on top of the water drop
3. Briefly touch the coated surface of a specimen grid to the top of the drop
4. Add negative stain and air-dry as described above in Direct Application Method

5.3.4 Agar Diffusion Method

This is a modification of the method of Kelen et al. (1971). It is useful for specimens containing a high salt concentration, especially when the virus concentration is low. The limit of detectability is approximately 10^7 virus particles per milliliter (Anderson and Doane, 1972).

Special materials required (Cooke Engineering Co., Alexandria, VA)

Flexible 96-well microtiter plates with wells filled approximately $\frac{3}{4}$ with 1% agar or agarose; covered with transparent sealing tape and stored at 4°C until needed.

1. For each specimen to be examined, cut a pair of agar-filled cups from the plate; remove the sealing tape and dry the agar surface at room temperature for approximately 5 minutes
2. Place a coated specimen grid on the surface of each cup
3. Add a drop of specimen to each grid
4. Air-dry at room temperature (30–60 minutes)
5. Add a drop of negative stain; remove the grid and air-dry as described for Direct Application Method.

5.3.5 Pseudoreplica Method

Although more involved than the Agar Diffusion Method, this method utilizes the same principles (Smith, 1967; Lee et al., 1978; Boerner et al., 1981).

Special materials required

10 mm × 10 mm × 5 mm cubes of 2% agar (or agarose)

Glass slides

0.5% Formvar in ethylene dichloride

1. Place a cube of agar on a glass slide; place a drop of specimen in the center of the cube
2. Air-dry the specimen (10–15 minutes)
3. When the surface appears dry, flood with 1–2 drops of Formvar solution; drain off any excess with absorbent paper
4. When the Formvar has dried completely, trim the block slightly on all four sides and move it to the very end of the slide
5. Slowly dip the end of the slide into a container of negative stain, at a slight angle, until the Formvar film floats off

6. Place a bare 300-mesh copper grid in the center of the floating film
7. Turn up the corner of a piece of filter paper that is two to three times the size of the floating film; holding this corner, gently place the paper on top of the film; as soon as the paper becomes wet quickly flip the paper (plus film and grid) 180° out onto paper towelling.

5.3.6 Airfuge Ultracentrifugation

Hammond et al. (1981) routinely employ this method for all clinical specimens submitted to the virus laboratory.

Special materials required

Airfuge ultracentrifuge with EM-90 rotor

Parafilm

Filtered distilled water

2.5% glutaraldehyde or 10% formalin

1. Place a coated specimen grid into the end of each of the six rotor sectors
2. Add 90 μl of specimen into each sector
3. Centrifuge at top speed (approximately 90,000 rpm) for 30 minutes
4. To stain each grid, invert it briefly on a drop of negative stain resting on Parafilm
5. Air dry
6. After use, decontaminate the rotor and rotor cover by immersing in glutaraldehyde or formalin for 15–30 minutes; rinse in tap water, brushing the sectors with a cotton tipped swab; soak briefly in 90% alcohol; air dry

5.4 IMMUNOELECTRON MICROSCOPY METHODS

Immunoelectron microscopy, like other immunoassays, involves the use of specific antibody as a probe or detector of viral antigen. Unlike other immunoassays, however, IEM permits the direct visualization of the resultant immune complex. In reading IEM assays one looks for antibody-trapped negatively stained virus particles or viral antigen.

Immunoelectron microscopy has a variety of applications in virology (Doane, 1974). It was used effectively to identify rubella virus, and provided the first morphologic characterization of this elusive virus (Best et al., 1967). Similarly, hepatitis B virus ("Australia antigen") was first identified by IEM (Bayer et al., 1968). It can also be used to increase the sensitivity of EM detection. We have found that the presence of antibody in our EM detection system can increase the sensitivity for enteroviruses by 100-fold (Anderson and Doane, 1973). The sensitivity can be further increased by using a label of colloidal gold coupled either to antibody or to protein A. Although still under development, the immunogold method has been used with fluid specimens to detect low concentrations of viral antigens associated with rotavirus, poliovirus and hepatitis B virus (Stannard et al., 1982; Beesley and Betts, 1985; Hopley and Doane, 1985).

IEM has also been used to serotype viruses, including enteroviruses (Anderson and Doane, 1973; Petrovicova and Juck, 1977), adenoviruses (Luton, 1973; Vassall and Ray, 1974; Svensson and von Bonsdorff, 1982), papovaviruses (Gardner et al., 1971; Penny et al., 1972; Penny and Narayan, 1973), myxoviruses (Kelen and McLeod, 1974; Edwards et al., 1975), and rotaviruses (Gerna et al., 1984).

Finally, IEM can be used to assay specific antibody. This approach, which has been extensively studied by Kapikian et al. (1975, 1976, 1979) uses a predetermined concentration of reference virus mixed with dilutions of acute and convalescent sera. The amount of antibody present is determined by measuring the width of the antibody halo surrounding the aggregated virus particles. Improved visibility and greater sensitivity can be achieved by using gold-labeled protein A to detect viral antibody (Hopley and Doane, 1985).

5.4.1 Direct Immunoelectron Microscopy Method

Materials

Plastic/carbon-coated copper specimen grids, 300-mesh

Microtiter plate

Antibody (monoclonal, polyclonal)

Flexible microtiter plates (see agar diffusion method)

Negative stain

1. Prepare antiserum (or virus) dilutions in a microtiter plate; prepare 1:1 mixtures of virus and antiserum in cups; incubate at 37°C for 1 hour
2. Process the mixtures by the agar diffusion method
3. Examine grid for the presence of virus-antibody aggregates

The original method described by Almeida and Waterson (1969) recommended an overnight refrigeration of the virus-antiserum mixture after step 1, followed by centrifugation at 10,000–15,000 rpm for 30 minutes. The pellet is then resuspended in distilled water and negatively stained. In direct IEM studies on enteroviruses the short method listed above was found to be equally sensitive to the overnight procedure (F. Lee, unpublished results).

5.4.2 Serum-in-Agar Method

This is a modification of the agar diffusion method, incorporating antiserum in the agar itself (Anderson and Doane, 1973). We have found it to be the most sensitive of the IEM methods described here (Pegg-Feige and Doane, 1984).

Materials

Plastic/carbon-coated copper specimen grids, 300 mesh

Microtiter plates (flexible plastic)

Antibody (monoclonal, polyclonal)

Negative stain

1. Prepare a 1% molten solution of agar or agarose by heating to boiling point
2. Allow the molten agar to cool to approximately 45°C and add antiserum to a concentration that results in aggregation of virus particles
3. Pipette the mixture into microtiter cups (approximately $\frac{3}{4}$ full); allow to solidify at room temperature. (Plates can now be covered with plastic sealing tape and stored at 4°C)
4. For use, cut a pair of cups from the plate, remove the tape, and air-dry the agar surface (approximately 5 minutes at room temperature)
5. Place a coated specimen grid on the surface of each cup
6. Add a drop of specimen to each grid
7. Air-dry at room temperature (30–60 minutes)
8. Add a drop of negative stain; remove the grid and air-dry as described for direct application method
9. Examine grid for the presence of virus-antibody aggregates (Figure 5.5)

5.4.3 Solid Phase Immunoelectron Microscopy

Originally described by Derrick (1973) for use with plant viruses, this method also has been used for detection of a number of human viruses. The presence of antibody on the grid has been shown to increase the sensitivity of detection of enteroviruses approximately 60-fold over direct EM (Pegg-Feige and Doane, 1984). For the detection of rotaviruses in stool, Svensson et al. (1983) found solid phase immunoelectron microscopy (SPIEM) to be approximately 30 times more sensitive than direct EM and ten times more sensitive than enzyme-linked immunosorbent assays (ELISA).

Materials

Formvar/carbon grids, 300-mesh, pretreated with UV light (1700 μW/cm^2 for 30 minutes prior to use)

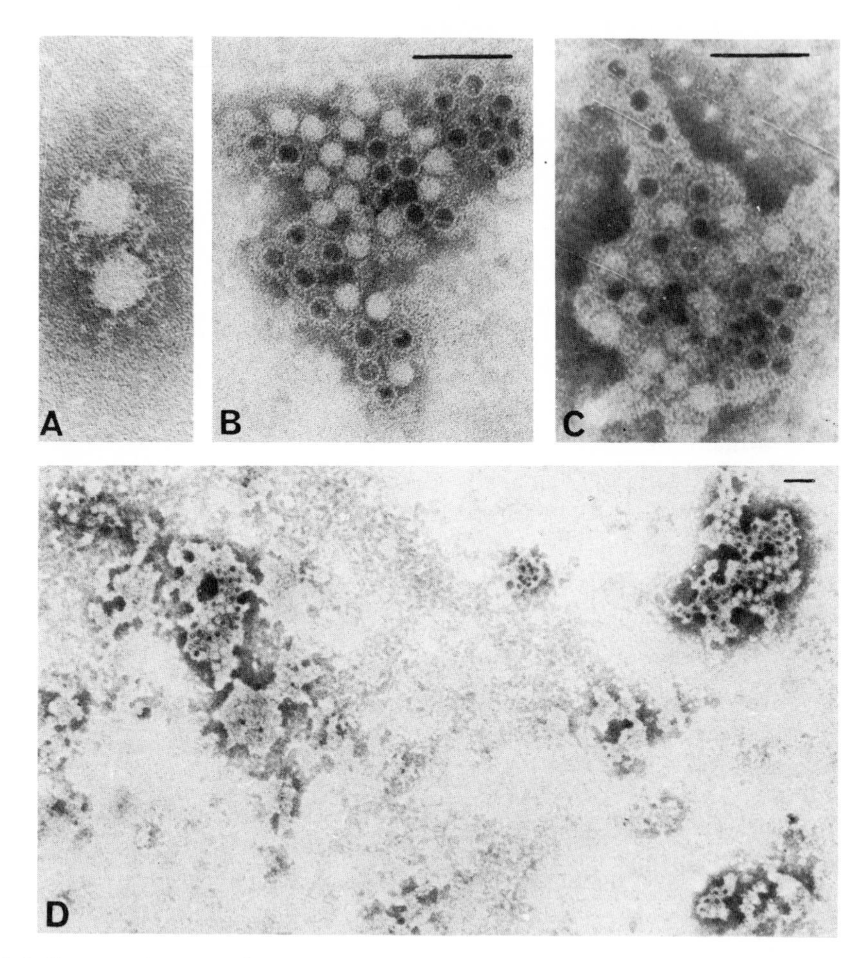

Figure 5.5. Immunoelectron microscopy of enterovirus, using the serum-in-agar method. (A) At high magnification antibody molecules can be seen bridging adjacent virus particles. (Courtesy of F. Lee.) (B) Immune complex formed with dilute antiserum shows little surrounding antibody. (C) Complex formed with lower dilution of antiserum has a dense surrounding halo of antibody. (D) Immune complexes may be large enough to be detected at low magnifications. Bars = 100 nm Courtesy of N. Anderson. Reproduced from Doane 1974, with permission.

Antiserum
Protein A, 1 mg/ml
Parafilm
0.05M Tris buffer (pH 7.2)
Negative stain

The procedure given below is one developed for rotavirus (Pegg-Feige and Doane 1983). It is recommended that optimum test conditions be determined for individual viruses, prior to routine use of SPIEM.

1. Invert a freshly pretreated grid on a drop of protein A on Parafilm; leave at room temperature for 10 minutes
2. Drain grid briefly by touching the edge to a paper towel; transfer grid across three separate drops of Tris buffer (total, 1–2 minutes).
3. Invert on a drop of a 1/100 dilution of antiserum to rotavirus for 10 minutes
4. Drain briefly; transfer across three separate drops of Tris buffer (1–2 minutes)
5. Invert on a drop of specimen at room temperature for 30 minutes

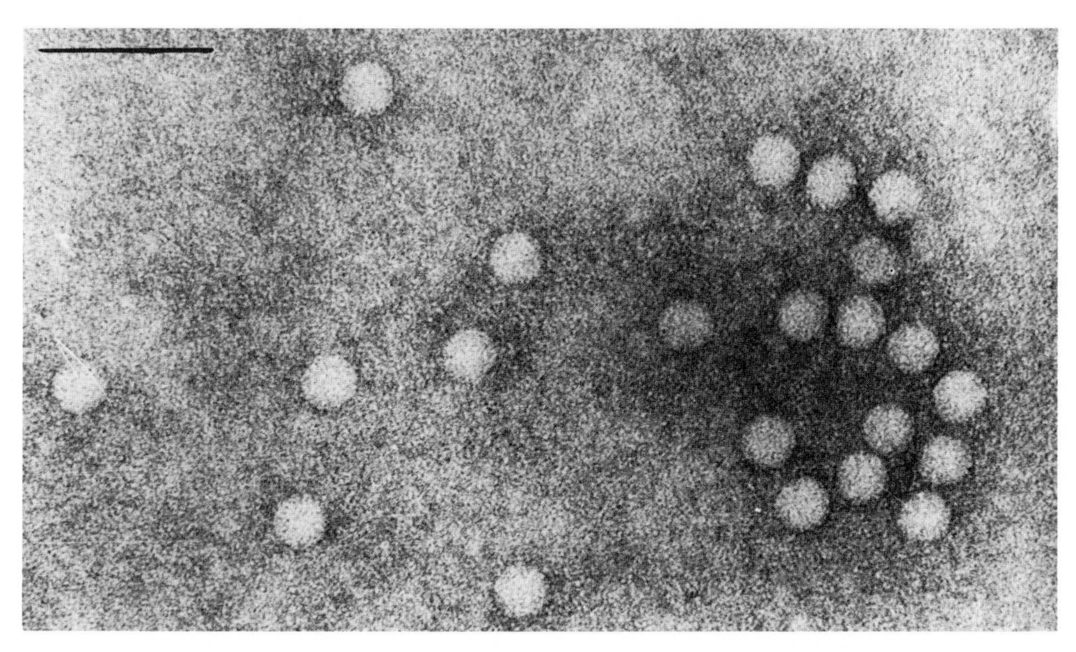

Figure 5.6. Immunoelectron microscopy of enterovirus, using SPIEM. A Formvar/carbon-coated grid was exposed to protein A, then to poliovirus antiserum, prior to exposure to the specimen. Bar = 100 nm Kindly provided by K. Pegg-Feige.

6. Rinse on three drops of Tris buffer (1–2 minutes)
7. Float briefly on a drop of negative stain; air dry
8. Examine grid for trapped virus (Figure 5-6)

Grids coated with protein A and/or antiserum are most effective when used shortly after preparation. They can be stored at 4°C, but with a resultant 50% reduction in trapping efficiency after 4–5 weeks.

5.4.4 Routine Immunoelectron Microscopy Screening for Clinical Specimens or Cell Culture Isolates

By incorporating immune serum globulin in the serum-in-agar method, a broad, nonselective virus detection system is produced (Juneau, 1979; Berthiaume et al., 1981). Enteroviruses detected by this method can be further identified by using the serum-in-agar method with pooled reference antisera (Doane and Anderson, unpublished results).

Materials

Serum-in-agar cups containing pooled human immune serum globulin, or pooled reference antisera, or individual antisera.

1. Process the specimen by the serum-in-agar method, using cups containing pooled human immune serum globulin at a final dilution of 1/50; identify the virus family on the basis of virus particle morphology
2. To serotype the identified virus, process by the serum-in-agar method, but use pooled or individual reference antisera; identify the virus type by carefully comparing test grids with control grids (virus alone, and virus plus normal serum)

5.5 THIN SECTIONING METHODS

Any standard EM embedding procedure can be used to process tissue samples or cell pellets (Weakley, 1981). However, these procedures usually take 1 to 3 days to complete and, for a rapid diagnosis, a shortened embedding method such as the one given below (Doane et al., 1974) is advantageous. Preservation of ultrastructure and specimen contrast are slightly reduced by this method, when compared with standard methods.

Materials

2.5% glutaraldehyde in 0.13 *M* phosphate buffer, pH 7.3

1% osmium tetroxide in 0.13 *M* phosphate buffer, pH 7.3

Acetone

Epoxy embedding medium

Bare 200-mesh copper grids

Miscellaneous equipment and reagents for preparing thin sections (e.g., oven, microtome, knives, stains)

1. Fix 1-mm cubes of tissue in glutaraldehyde for 15 minutes at 4°C
2. Rinse in three changes of phosphate buffer (1 minute each)
3. Fix in osmium tetroxide for 15 minutes at room temperature
4. Dehydrate through acetone as follows: two changes of 70% (total, 5 minutes); three changes of 100% (total, 5 minutes)
5. Place in 1:1 mixture of 100% acetone and embedding medium for 10 minutes at room temperature
6. Transfer through two changes of 100% embedding medium (5 minutes each)
7. Place in embedding capsule in fresh embedding medium and heat at 95°C for 60 minutes
8. Cool block; cut and stain sections as usual

5.6 PROCESSING CELL CULTURE ISOLATES FOR ELECTRON MICROSCOPY

Because it is not possible to examine a large number of clinical specimens by EM on a daily basis, many virus laboratories routinely select only those specimens that have induced a cytopathic effect in inoculated cell cultures. The viral isolates can then be identified by EM, either by negative staining (Figures 5.7, 5.8) or thin sectioning techniques (Figure 5.9).

5.6.1 Negative Staining

Materials

Filtered distilled water

Pasteur pipettes

Negative stain

Agar diffusion cups with specimen grids

1. Withdraw medium from culture to be examined; put medium aside temporarily
2. Add 2–3 drops of water to culture; resuspend cells
3. After 2–3 minutes, negatively stain cell lysate by the agar diffusion method or by the serum-in-agar IEM method.
4. If no virus is found in the lysate, process the temporarily stored medium by Airfuge ultracentrifugation and negative staining or by the serum-in-agar method

5.6.2 Thin Sectioning

Materials

Hematocrit centrifuge

1.3 mm × 75 mm capillary tubes

Parafilm

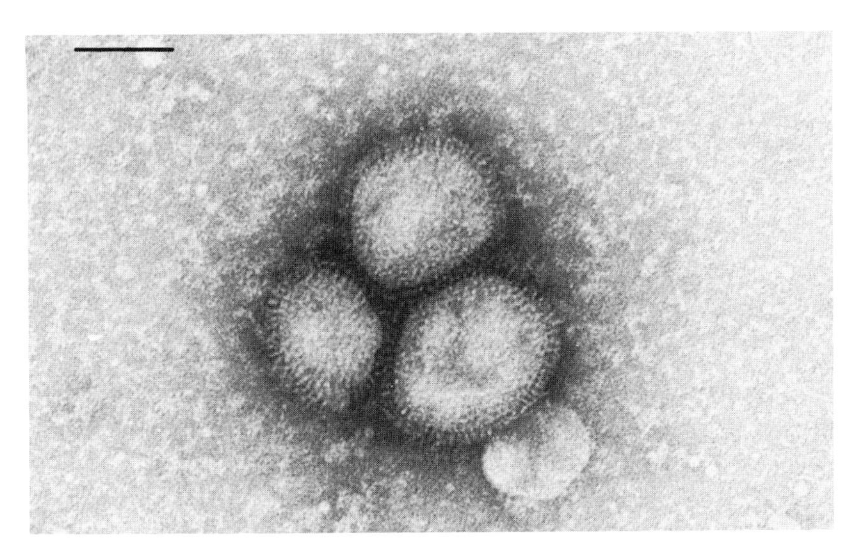

Figure 5.7. Three influenza virus particles from negatively stained cell culture lysate. Characteristic morphologic features include coarse spikes and an intact envelope (not easily penetrated by stain). Bar = 100 nm

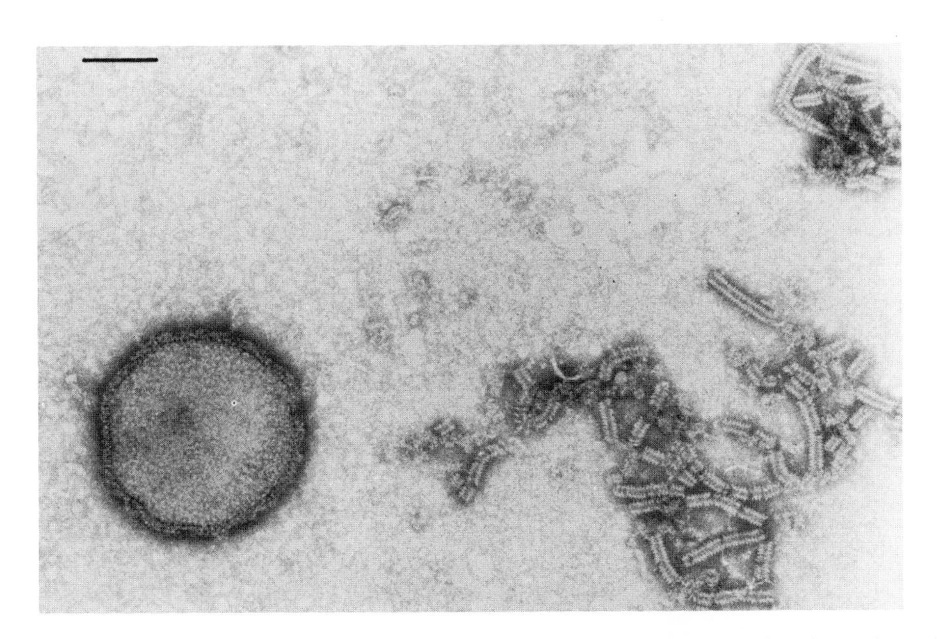

Figure 5.8. Simian paramyxovirus found in "normal" monkey kidney cell culture (negatively stained). An intact virus particle with a fine fringed envelope is seen at left. At right are clusters of viral nucleocapsids exhibiting the herring bone configuration that is characteristic of paramyxoviruses. Bar = 100 nm

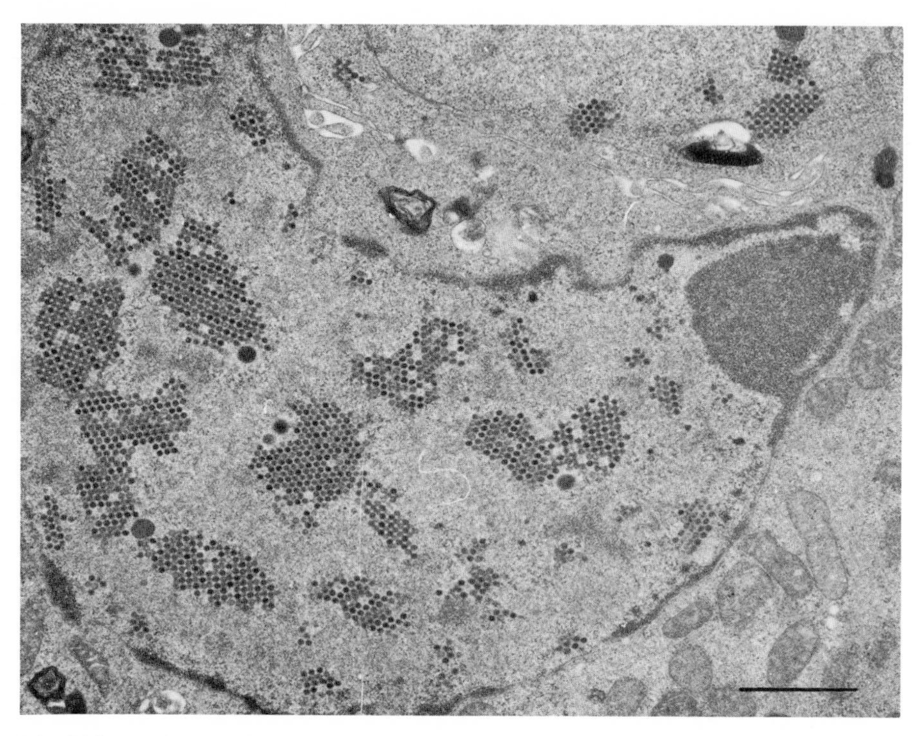

Figure 5.9. Thin section of adenovirus-infected cell culture. Crystalline arrays of virus particles are seen in the nucleus. Bar = 1 μm

Plasticine
Paper clip
2.5% glutaraldehyde in 0.13 *M* phosphate buffer, pH 7.3

1. Gently resuspend cells from culture into overlying medium, using a rubber policeman or a Pasteur pipette.
2. Transfer suspension to a small conical-tipped centrifuge tube; centrifuge at 1500 rpm for 3 minutes
3. Withdraw the medium and replace with 2–3 drops of glutaraldehyde; transfer to a sheet of Parafilm
4. Draw the suspension into a capillary tube; seal one end with a small plug of plasticine
5. Centrifuge in a hematocrit centrifuge for 3 minutes at 12,500 rpm; the cells now form a compact pellet immediately above the plasticine
6. Score the glass tube and break at a distance 6–7 mm above the cell pellet
7. Invert the tube so that the open end is directed toward a drop of glutaraldehyde on Parafilm; use a straightened paper clip to push against the plasticine plug, forcing the cell pellet into the fixative
8. Transfer the cell pellet to a vial containing fresh fixative; process by a standard embedding method or by the rapid embedding method

ACKNOWLEDGMENT

The assistance of Nan Anderson and Joan Stubberfield is gratefully acknowledged.

REFERENCES

Almeida, J., Cinader, B., and Howatson, A. 1963. The structure of antigen-antibody complexes. J. Exp. Med. 118:327–340.

Almeida, J.D., Zuckerman, A.J., Taylor, P.E., and Waterson, A.P. 1969. Immune electron microscopy of the Australia-SH (serum hepatitis) antigen. Microbios 2:695–698.

Almeida, J.D., and Waterson, A.P. 1969. The morphology of virus-antibody interaction. Adv. Virus Res. 15:307–338.

Anderson, N., and Doane, F.W. 1972. Agar diffusion method for negative staining of microbial suspensions in salt solutions. Appl. Microbiol. 24:495–496.

Anderson, N., and Doane, F.W. 1973. Specific identification of enteroviruses by immunoelectron microscopy using a serum-in-agar diffusion method. Canad. J. Microbiol. 19:585–589.

Bayer, M.E., Blumberg, B.S., and Werner, B. 1968. Particles associated with Australia antigen in the sera of patients with leukemia, Down's syndrome and hepatitis. Nature (London) 218:1057–1059.

Beesley, J.E., and Betts, M.P. 1985. Virus diagnosis: a novel use for the protein A-gold probe. Med. Lab. Sci. 42:161–165.

Berthiaume, L., Alain, R., McLaughlin, B., Payment, P., and Trepanier, P. 1981. Rapid detection of human viruses in faeces by a simple and routine immune electron microscopy technique. J. Gen. Virol. 55:223–227.

Best, J.M., Banatvala, J.E., Almeida, J.D., and Waterson, A.P. 1967. Morphological characteristics of rubella virus. Lancet ii:237–239.

Bishop, R.F. 1982. Other small virus-like particles in humans. In D.A. Tyrrell and A.Z. Kapikian (eds.), Virus Infections of the Gastrointestinal Tract. New York: Marcel Dekker, pp. 195–201.

Bishop, R.F., Davidson, G.P., Holmes, I.H., and Ruck, B.J. 1973. Virus particles in epithelial cells of duodenal mucosa from children with acute non-bacterial gastroenteritis. Lancet ii:1281–1283.

Blank, H., Davis, C., and Collins, C. 1970. Electron microscopy for the diagnosis of cutaneous viral infections. Br. J. Dermatol. 83:69–80.

Boerner, C.F., Lee, F.K., Wickliffe, C.L., Nahmias, A.J., Cavanagh, H.D., and Strauss, S.E. 1981. Electron microscopy for the diagnosis of ocular viral infections. Ophthalmology 88:1377–1380.

Bouteille, M., Fontaine, C., Verdrenne, C., and Delarue, J. 1965. Sur un cas d'encephalite subaigue a inclusions. Etude anatomo-clinique et ultrastructurale. Rev. Neurol. 118:454–458.

Brenner, S., and Horne, R.W. 1959. A negative staining method for high resolution electron microscopy of viruses. Biochim. Biophys. Acta 34:103–110.

Caul, E.O., Paver, W.K., and Clarke, S.K.R. 1975. Coronavirus particles in faeces from patients with gastroenteritis. Lancet i:1192.

Chambers, V.C. Ito, Y., and Evans, C.A. 1966. Technique for visualizing papovaviruses in tumors and in tissue cultures. J. Bacteriol. 91:2090–2092.

Chernesky, M.A. 1979. The role of electron microscopy in diagnostic virology. In D. Lennette, S. Specter, and K. Thompson (eds.), Diagnosis of Viral Infections: The Role of the Clinical Laboratory. Baltimore: University Park Press, pp. 125–142.

Chia, K., and Spence, L.P. 1983. Personal communication.

Coleman, D.V., Russel, W.J.I., Hodgson, J., Tun, P., and Mowbray, J.F. 1977. Human papovavirus in Papanicolaou smears of urinary sediment detected by transmission electron microscopy. J. Clin. Pathol. 30:1015–1020.

Cruickshank, J.G., Bedson, H.S., and Watson, D.H. 1966. Electron microscopy in rapid diagnosis of smallpox. Lancet ii:527–530.

Dane, D.S., Cameron, C.H., and Briggs, M. 1970. Virus-like particles in serum of patients with Australia-antigen–associated hepatitis. Lancet i:695–698.

Davies, H.A. 1982. Electron microscopy and immune electron microscopy for detection of gastroenteritis viruses. In D.A. Tyrrell and A.Z. Kapikian (eds.), Virus Infections of the Gastrointestinal Tract. New York: Marcel Dekker, pp. 37–49.

Derrick, K.S. 1973. Quantitative assay for plant viruses using serologically specific electron microscopy. Virology 56:652–653.

Doane, F.W. 1974. Identification of viruses by immunoelectron microscopy. In E. Kurstak and R. Morisset (eds.), Viral Immunodiagnosis. New York: Academic Press, pp. 237–255.

Doane, F.W., and Anderson, N. 1977. Electron and immunoelectron microscopic procedures for diagnosis of viral infections. In E. Kurstak and C. Kurstak (eds.), Comparative Diagnosis of Viral Diseases, Vol. II, part B. New York: Academic Press, pp. 505–539.

Doane, F.W., and Anderson, N. 1986. Electron Microscopy in Diagnostic Virology. A Practical Guide and Atlas. New York: Cambridge University Press (in press).

Doane, F.W., Anderson, N., Chao, J., and Noonan, A. 1974. Two-hour embedding procedure for intracellular detection of viruses by electron microscopy. Appl. Microbiol. 27:407–410.

Doane, F.W., Anderson, N., Chatiyanonda, K., Banatyne, R.M., McLean, D.M., and Rhodes, A.J. 1967. Rapid laboratory diagnosis of paramyxovirus infections by electron microscopy. Lancet ii:751–753.

Doane, F.W., Anderson, N., Zbitnew, A., and Rhodes, A.J. 1969. Application of electron microscopy to the diagnosis of virus infections. Canad. Med. Assoc. J. 100:1043–1049.

Edwards, E.A., Valters, W.A., Boehm, L.G., and Rosenbaum, M.J. 1975. Visualization by immune electron microscopy of viruses associated with acute respiratory disease. J. Immunol. Meth. 8:159–167.

Evans, A.S., and Melnick, J.L. 1949. Electron microscope studies of the vesicle and spinal fluids from a case of herpes zoster. Proc. Soc. Exp. Biol. Med. 71:283–286.

Feinstone, S.M., Kapikian, A.Z., and Purcell, R.H. 1973. Hepatitis A: Detection by immune electron microscopy of a virus-like antigen associated with acute illness. Science 182:1026–1028.

Field, A.M. 1982. Diagnostic virology using electron microscopic techniques. Adv. Virus Res. 27:1–69.

Flewett, T.H., and Davies, H. 1976. Caliciviruses of man. Lancet i:311.

Flewett, T.H., Bryden, A.S., and Davies, H. 1973. Virus particles in gastroenteritis. Lancet ii:1497.

Flewett, T.H., Bryden, A.S., and Davies, H. 1974. Diagnostic electron microscopy of faeces. I. The viral flora of the faeces as seen by electron microscopy. J. Clin. Pathol. 27:603–608.

Gardner, S.D., Field, A.M., Coleman, D.V., and Hulme, B. 1971. A new human papovavirus (BK) isolated from urine after renal transplantation. Lancet i:1253–1257.

Gerna, G., Passarani, N., Battaglia, M., and Percivalle, E. 1984. Rapid serotyping of human rotavirus strains by solid-phase immune electron microscopy. J. Clin. Microbiol. 19:273–278.

Hammond, G.W., Hazelton, P.R., Chuang, I., and Klisko, B. 1981. Improved detection of viruses by electron microscopy after direct ultracentrifuge preparation of specimens. J Clin Microbiol. 14:210–221.

Harland, W.A., Adams, H.J., and McSeveney, D. 1967. Herpes simplex particles in acute necrotising encephalitis. Lancet ii:581–582.

Herndon, R.M., and Rubinstein, L.J. 1968. Light and electron microscopy observations on the development of viral particles in the inclusions of Dawson's encephalitis (subacute sclerosing panencephalitis). Neurology 18 (part 2):8–20.

Hirschman, R.J., Schulman, N.R., Barker, L.F., and Smith, K.O. 1969. Virus-like particles in sera of patients with infectious and serum hepatitis. J. Am. Med. Assoc. 208:1667–1670.

Hopley, J.F.A., and Doane, F.W. 1985. Development of a sensitive protein A-gold immunoelectron microscopy method for detecting viral antigens in fluid specimens. J. Virol. Meth. 12:135–147.

Hsiung, G.D., Fong, C.K.Y., and August, M.J. 1979. The use of electron microscopy for diagnosis of virus infections: An overview. Prog. Med. Virol. 25:133–159.

Joncas, J.H., Berthiaume, L., Williams, R., and Beaudry, P. 1969. Diagnosis of viral respiratory infections by electron microscopy. Lancet i:956–959.

Juneau, M.L. 1979. Role of the electron microscope in the clinical diagnosis of viral infections from patients' stools. Canad. J. Med. Technol. 41:53–57.

Kapikian, A.Z., Dienstag, J.L., and Purcell, R.H. 1976. Immune electron microscopy as a method for the detection, identification, and characterization of agents not cultivable in an *in vitro* system. In N.R. Rose and H. Friedman (eds.), Manual of Clinical Immunology, Washington: American Society of Microbiologists, pp. 467–480.

Kapikian, A.Z., Feinstone, S.M., Purcell, R.H., Wyatt, R.G., Thornhill, T.S., Kalica, A.R., and Chanock, R.M. 1975. Detection and identification by immune electron microscopy of fastidious agents associated with respiratory illness, acute nonbacterial gastroenteritis, and hepatitis A. Persp. Virol. 9:9–47.

Kapikian, A.Z., Yolken, R.H., Greenberg, H.B., Wyatt, R.G., Kalica, A.R. Chanock, R.M., and Kim, H.W. 1979. Gastroenteritis viruses. In E.H. Lennette and N.J. Schmidt (eds.), Diagnostic Procedures for Viral, Rickettsial and Chlamydial Infections. Washington: American Public Health Association, pp. 927–995.

Kelen, A.E., Hathaway, A.E., and McLeod, D.A. 1971. Rapid detection of Australia/SH antigen and antibody by a simple and sensitive technique of immunoelectron microscopy. Canad. J. Microbiol. 17:993–1000.

Kelen, A.E., and McLeod, D.A. 1974. Differentiation of myxoviruses by electronmicroscopy and immunoelectronmicroscopy. In E. Kurstak and R. Morisset (eds.), Viral Immunodiagnosis. New York: Academic Press, pp. 257–275.

Kjeldsberg, E. 1980. Application of electron microscopy in viral diagnosis. Path. Res. Pract. 167:3–21.

Lee, F.K., Nahmias, A.J., and Stagno, S. 1978. Rapid diagnosis of cytomegalovirus infection in infants by electron microscopy. N. Engl. J. Med. 299:1266–1270.

Luton, P. 1973. Rapid adenovirus typing by immunoelectron microscopy. J. Clin. Pathol. 26:914–917.

Macrae, A.D., Field, A.M., McDonald, J.R., Meurisse, E.V., and Porter, A.A. 1969. Laboratory differential diagnosis of vesicular skin rashes. Lancet ii:313–316.

Madeley, C.R., and Cosgrove, B.P. 1975. 28 nm particles in faeces in infantile gastroenteritis. Lancet ii:451–452.

Madeley, C.R., and Cosgrove, B.P. 1976. Calicivirus in man. Lancet i:199.

Mathan, M., Mathan, V.I., Swaminathan, S.P., Yesudoss, S., and Baker, S.J. 1975. Pleomorphic virus-like particles in human faeces. Lancet i:1068–1069.

McLean, D.M., and Wong, K.K. 1984. Same-Day Diagnosis of Human Virus Infections. Boca Raton, FL: CRC Press.

Middleton, P.J. Szymanski, M.T., Abbott, G.D., Bortolussi, R., and Hamilton, J.R. 1974. Orbivirus acute gastroenteritis of infancy. Lancet i:1241–1244.

Middleton, P.J., Szymanski, M.T., and Petric, M. 1977. Viruses associated with acute gastroenteritis in young children. Am. J. Dis. Child. 131:733–737.

Montplaisir, S., Belloncik, S., Leduc, N.P., Onji, P.A., Martineau, B., and Kurstak, E. 1972. Electron microscopy in the rapid diagnosis of cytomegalovirus: Ultrastructural observations and comparison of methods of diagnosis. J. Infect. Dis. 125:533–538.

Nagington, J. 1964. Electron microscopy in differential diagnosis of variola, vaccinia and varicella. Br. Med. J. 2:1499–1500.

Nagler, F.P.O., and Rake, G. 1948. The use of the electron microscope in diagnosis of variola, vaccinia and varicella. J. Bacteriol. 55:45–51.

Palmer, E.L., and Martin, M.L. 1982. An Atlas of Mammalian Viruses. Boca Raton, FL: CRC Press.

Pegg-Feige, K., and Doane, F.W. 1983. Effects of specimen support film in solid phase immunoelectron microscopy. J. Virol. Meth. 7:315–319.

Pegg-Feige, K., and Doane, F.W. 1984. Solid phase immunoelectron microscopy for rapid diagnosis of enteroviruses. 42nd Annual Proceedings of the Electron Microscopy Society of America, G.W. Bailey, ed.

Penny, J.B., and Narayan, O. 1973. Studies of the antigenic relationships of the new human papovaviruses by electron microscopy agglutination. Infect. Immun. 8:299–300.

Penny, J.B., Weiner, L.P., Herndon, R.M., Narayan, O., and Johnson, R.T. 1972. Virions from progressive multifocal leukoencephalopathy: Rapid serological identification by electron microscopy. Science 178:60–62.

Peters, D., Nielsen, G., and Bayer, M.E. 1962. Variola. Dtsch. Med. Wschr. 87:2240–2246.

Petrovicova, A., and Juck, A.S. 1977. Serotyping of coxsackieviruses by immune electron microscopy. Acta Virol 21:165–167.

Roy, S., and Wolman, L. 1969. Electron microscopic observations on the virus particles in herpes simplex encephalitis. J. Clin. Pathol. 22:51–59.

Smith, K.O. 1967. Identification of viruses by electron microscopy. In H. Busch (ed.), Methods in Cancer Research, Vol. 1. New York: Academic Press, pp. 545–572.

Stannard, L.M., Lennon, M., Hodgkiss, M., and Smuts, H. 1985. An electron microscopic demonstration of immune complexes of hepatitis B e-antigen using colloidal gold as a marker. J. Med. Virol. 9:165–175.

Svensson, L., and von Bonsdorff, C.D. 1982. Solid-phase immune electron microscopy (SPIEM) by use of protein A and its application for characterization of selected adenovirus serotypes. J. Med. Virol. 10:243–253.

Svensson, L., Grandien, M., and Pettersson, C.A. 1983. Comparison of solid-phase immune electron microscopy by use of protein A with direct electron microscopy and enzyme-linked immunosorbent assay for detection of rotavirus in stool. J. Clin. Microbiol. 18:1244–1249.

Szymanski, M., and Middleton, P.J., 1983. Personal communication.

Tyrell, D.A., and Kapikian, A.Z., eds. 1982. Virus Infections of the Gastrointestinal Tract. New York: Marcel Dekker.

van Rooyen, C.E., and Scott, G.D. 1948. Smallpox diagnosis with special reference to electron microscopy. Canad. J. Pub. Health 39:467–477.

Vassall, J.H., and Ray, C.G. 1974. Serotyping of adenoviruses using immune electron microscopy. Appl. Microbiol. 28:623–627.

Weakley, B.S. 1981. A Beginner's Handbook in Biological Electron Microscopy. Edinburgh: Churchill Livingstone.

Zu Rhein, G.M., and Chou, S.M. 1964. Particles resembling papovaviruses in human cerebral demyelinating disease. Science 148:1447–1479.

The Interference Assay

Charles A. Reed

6.1 HISTORY

For many decades, rubella had been looked upon as a rather unimportant and mild disease of childhood. It was commonly referred to as German measles or the 3-day measles, and seemed to cause few complications. Rubella had first been described in the German medical literature by DeBergen in 1752 and Orlow in 1758 (Emminghaus, 1870). But it was not until 1941 that the teratogenic effects of this virus, when acquired during pregnancy, became well documented. It was then that an Australian ophthalmologist, Gregg (1941), reported a high incidence of cataracts and other anomalies among offspring of mothers who had contracted rubella early in pregnancy during a rubella epidemic in Australia. A few years later, Swan et al. (1943, 1946) reported more cases of congenital defects with later observations related especially to rubella. Since then, many reports have appeared in the literature confirming these findings and reporting many other congenital defects and complications of pregnancy following maternal rubella. Through the years the pattern of events with their resulting effects have become known as "the rubella syndrome." Many retrospective studies and other reports in the literature are in agreement with the earlier findings (Skinner, 1961; also see Chapter 30).

With the advent of newer virologic techniques, more precisely defined chemical media, and a multiplicity of cell lines developed in the late 1940s and early 1950s, the world of virology exploded with the isolation and identification of many viruses (Melnick et al., 1979). However, it was not until 1962 that two groups of investigators succeeded in isolating the rubella virus in two different systems. Weller and Neva were able to isolate the virus in primary human amnion cell cultures, which demonstrated a visible cytopathic effect (CPE) with time (Weller and Neva, 1962). Parkman et al. (1962) were able to demonstrate the presence of the virus in primary African green monkey kidney (AGMK) cell cultures, which produced no rubella virus-associated CPE. However, it was shown that rubella virus infection interfered with the CPE production of a challenge virus (e.g., ECHO-11). This was the first demonstration of the interference assay that could be used to identify rubella virus subsequent to its isolation in cell culture. At about the same time, Veronelli et al. (1962), reported similar findings using a continuous Rhesus monkey cell line, LLC-MK_2, which had been shown to be useful for viral research (Hull et al., 1962; Veronelli et al., 1962).

Parkman et al. continued their research with the rubella virus interference phenomenon and it was shown that infection in cell culture interfered with CPE production by many enteroviruses, myxoviruses, and arboviruses in AGMK cell cultures (Parkman et al., 1964a). Thirteen different cell lines were tested and it was found that viral replication occurred in 11 of these cell lines, but the interference of the CPE by a challenge virus varied with the different lines. In addition, their studies showed that rubella virus is heat labile and that specimens should be chilled or frozen after collection. In continuing studies, Parkman et al. demonstrated that rubella viral replication and neutralization tests showed similar results when the AGMK and LLC-MK_2 cell lines were compared (Parkman et al., 1964b). All of the techniques and methods were put to the test when the United States was swept by a major epidemic of rubella in 1964–1965 and most virology laboratories started to culture for the rubella virus. With the licensure of a rubella vaccine in 1969, however, the demands on the clinical virus lab for culturing rubella virus have been significantly reduced.

6.2 SPECIMEN COLLECTION, TRANSPORT, AND STORAGE

As with all viral isolation, the process starts with the proper collection and handling of the specimen (see Chapter 2). Once inoculated onto tissue culture, the remaining specimen should be frozen and stored at −70°C.

For rubella identification by the interference assay the following specimens are appropriate (Sanders, 1978; Herrman, 1979).

1. Respiratory: Nasal and pharyngeal swabs from children and adults. Throat washings may be suitable from adults. These should be collected no later than 5 days after the onset of the rash.
2. Blood: Heparinized blood should be collected as near the onset of the rash as possible and as early as 7 days after possible exposure.
3. Cerebrospinal fluid: CSF is most useful in cases involving suspected congenital rubella.
4. Urine: Urine should be collected as aseptically as possible. Urine has been shown to contain virus for months after birth of a congenitally infected infant (Cooper and Krugman, 1966).
5. Tissues: All tissues and body fluids may be collected and tested when appropriate. These would include ocular tissues and fluids, autopsy or biopsy tissues, placenta, amniotic fluid, and any fetal tissues that are accessable.

6.3 SPECIMEN PREPARATION AND INOCULATION

Specimens collected on swabs should be expressed into 2 ml of maintenance medium if not collected directly into a transport broth. All specimens—with the exception of CSF—should be treated with the appropriate antimicrobial agents per each laboratory's preference. Tissue material should be ground and prepared as a 10% suspension in the same maintenance medium that is to be added to the final cell culture. Our laboratory utilizes Earle's basic salt solution (EBSS), Eagle's minimum essential medium (EMEM) and 2% fetal calf serum as maintenance medium. Adjust the pH of all specimens to 7.0 to 7.2.

Select the cell line to be inoculated. Primary AGMK is the most widely used, but it has drawbacks that will be discussed later. LLC-MK_2 is used in our laboratory if problems arise with the AGMK. Remove the medium from the selected tube cultures, using four tubes for each specimen. Inoculate 0.25 ml of the specimen directly onto each cell culture followed by a 1-hour adsorption period at 36°C (stationary rack). Add 1.5 ml of fresh maintenance medium to each tube and incubate in a roller drum. Observe each tube daily and record any changes noted.

6.4 ECHO-11 CHALLENGE AND NEUTRALIZAITON

Rubella virus may not produce visible CPE, but may be detectable using the property of viral interference. For example, if rubella virus is growing in AGMK cells, these cells become resistant to ECHO-11 virus, to which they are ordinarily highly susceptible. As a result, no CPE is produced following infection by ECHO-11. The following procedure is our adaptation of the procedures of Parkman et al. (1964b):

1. Prepare a stock virus pool of ECHO-11 in LLC-MK_2; aliquot into tubes (1 ml per tube) and store at −70°C
2. Titer this challenge virus (ECHO-11) in LLC-MK_2 cells to an endpoint using the Reed–Muench (1938) method
3. Inoculate specimens submitted that are suspected to contain rubella virus (including all specimens from newborns with congenital malformations) into four tubes of AGMK cells of the same lot
4. At 10 days postinoculation remove the medium from two inoculated tubes and two control tubes of AGMK cells.
    - Add 100 $TCID_{50}$ in 0.1 ml of ECHO-11 challenge virus into one inoculated tube and one control tube each. Add 10 $TCID_{50}$ in 0.1 ml of ECHO-11 challenge virus into each of the remaining inoculated and control tubes. These dilutions of virus stock are made in EMEM supplemented with 2% fetal calf serum
    - Incubate in a stationary rack for 1 hour at 36°C; manually rotate the tubes every 10 to 15 minutes
    - Add 1.5 ml EMEM with 2% FCS at the end of this incubation period
    - Incubate in a stationary rack at 36°C
5. Examine tubes daily; interference is present if extensive CPE develops in the control cell cultures but not in the cell cultures inoculated with the clinical specimen

Note: This method allows for detection of low titer viral replication. If the 10 $TCID_{50}$ challenge tube is negative for ECHO-11, the unchallenged tubes should be harvested and passed to 4 new tubes. Repeat steps 3 and 4 above.

6. Interference-neutralization test:
    - Harvest material (by one freeze–thaw cycle) from the two inoculated tubes that were not subjected to challenge virus
    - Place 0.15 ml of the harvested material in a sterile screw cap tube (16 × 75 mm) and add 0.15 ml rubella antiserum at a dilution to represent 50 neutralizing antibody units per 0.1 ml
    - As a control, incubate the harvest material without antiserum
    - Incubate samples in a stationary rack for 1 hour at 36°C
    - Inoculate 0.2 ml of the virus antiserum mixture and 0.1 ml of the untreated virus into one tube each of AGMK cells; incubate at 36°C in a stationary rack. Duplicate tubes may be used.
    - After 5 days inoculate each tube with 100 $TCID_{50}$ ECHO-11; if duplicate tubes were prepared, one set may be challenged with 10 $TCID_{50}$ of ECHO-11; the virus is confirmed as rubella virus if viral interference is prevented by antiserum treatment, and is again demonstrated in the control tube

Note: Several different log dilutions of virus can be employed at step six if the virus titer is too high for the neutralizing capacity of the antibody.

6.5 PROS AND CONS

The "rubella-challenge-interference" system is probably the most rapid system of isolating the virus from a patient specimen in a single passage. It can be achieved within 14 days on primary AGMK. Other cell lines that are capable of replicating rubella virus usually require more than one blind passage before a visible CPE is noted. During the period of time required for virus replication the physical condition of the cell sheet is critical and must be visually compared very carefully with uninfected control tubes. Most laboratories today do not pass the specimen from tubes of a set that has shown a positive ECHO-11 CPE. The specimen is reported negative after 14 days.

However, because laboratories may experience contamination with indigenous resident simian viruses in purchased AGMK cell lines from time to time, these cell lines should be examined daily for simian virus associated CPE and tested for the presence of these simian viruses. If the cell cultures are positive for contaminating agents they should not be used because misleading results may occur. If specimens have already been inoculated into tubes of that lot, reinoculate a new lot of tested cells from the specimen sample that has been held in frozen storage. This problem is not encountered if the LLC-MK_2 line is used, however, LLC-MK_2 cells are not as sensitive for the primary isolation of rubella virus. The LLC-MK_2 line can be used for the neutralization test subsequent to the primary isolation of the virus.

Because of the expense of the cell cultures and laboratory personnel time that are required to isolate, challenge and then confirm rubella virus isolation, this protocol has been questioned especially if a serologic test will give the physician the necessary information. However, virus isolation and identification by interference assay is still the standard for proof of current rubella virus infection.

REFERENCES

Cooper, L.Z., and Krugman, S. 1966. Diagnosis and management: Congenital rubella. Pediatrics 37:335–338.

Emminghaus, H. 1870. Uber rubeolen. Jahrb. Kinderheilkd. 4:47–59.

Gregg, N.M. 1941. Congenital cataract following German measles in the mother. Trans. Ophthal. Soc. Aust. 3:35–46.

Herrmann, K. 1979. Rubella virus. In E.H. Lennette and J. Schmidt (eds.), Diagnostic Procedures for Viral, Rickettsial and Chlamydial Infections, 5th ed. Washington, D.C.: American Public Health Association, pp. 725–766.

Hull, R.H., Cherry, W.R., and Tritch, O.J. 1962. Growth characteristics of monkey kidney cell strains LLC-MK_1, LLC-MK_2 and LLC-MK_2 (NCTC-3196) and their utility in virus research. J. Exp. Med. 115:903–917.

Melnick, J.L., Wenner, H.A., and Phillips, C.A. 1979. Enteroviruses. In E.H. Lennette and N.J. Schmidt (eds.), Diagnostic Procedures for Viral, Rickettsial and Chlamydial Infections, 5th ed. Washington, D.C.: American Public Health Association, pp. 471–534.

Parkman, P.D., Buescher, E.L., and Artenstein, M.S. 1962. Recovery of rubella virus from army recruits. Proc. Soc. Exp. Biol. Med. 111:225–230.

Parkman, P.D., Buescher, E.L., Artenstein, M.S., McCown, J.M., Mundon, F.K., and Druzd, A.D. 1964a. Studies of rubella I. Properties of the virus. J. Immunol. 93:595–607.

Parkman, P.D., Mundon, F.K., McCown, J.M., and Buescher, E.L. 1964b. Studies of rubella II. Neutralization of the virus. J. Immunol. 93:608–617.

Reed, L.J., and Muench, H. 1938. A simple method of estimating fifty percent end points. Am. J. Hyg. 27:493–497.

Sanders, C.V., Jr. 1978. Diagnostic virology. In H. Rothschild, F. Allison, Jr., and C. Howe (eds.), Human Diseases Caused by Viruses.

New York: Oxford University Press, pp. 259–281.

Skinner, C.W. 1961. The rubella problem. Am. J. Dis. Child. 101:78–86.

Swan, C., Tostevin, A.L., Moore, B., Mayo, H., and Black, G.H.B. 1943. Congenital defects in infants following infectious disease during pregnancy. Med. J. Australia 2:201–210.

Swan, C., Tostevin, A.L., and Black, G.H.B. 1946. Final observations on congenital defects in infants following infectious diseases during pregnancy with special reference to rubella. Med. J. Australia 2:889–908.

Veronelli, J.A., Maassab, H.F. and Hennessy A.V. 1962. Isolation in tissue culture of an interfering agent from patients with rubella. Proc. Soc. Exp. Biol. Med. 111:472–476.

Weller, T.H., and Neva, F.A. 1962. Propagation in tissue culture of cytopathic agents from patients with rubella-like illness. Proc. Soc. Exp. Biol. Med. 111:215–225.

7

Immunofluorescence

P. S. Gardner

7.1 HISTORY

Developments of all kinds tend to come in fits and starts and the moment of take-off depends not on the actual merits of a discovery but on local factors and public acceptance. It took the automobile a relatively long time to fully develop: It depended on the development of roads, public acceptance (the walker in front with a red flag had to disappear) and readily available fuel supplies. It also depended on producing the goods in sufficient quantity and economically and, here, men like Ford came into their own. Perhaps in the field of rapid diagnosis Coons should be not only looked on as the originator of the fluorescent antibody technique (FAT), which he first described as early as 1941, but as the father of the techniques on which all modern rapid diagnostic methods are based, though obviously this could not have come into his original concept.

The period between 1940 and 1970 is seeded with the names of people, all of whom made a contribution to rapid diagnosis by immunofluorescence (IF); however, the method did not receive popular acclaim. The reasons for this need now to be considered. It was in 1941 that Coons et al. employed immune serum globulin labeled with a fluorescent dye to locate the corresponding antigens. From this start Beigelsen et al. (1959) investigated herpetic skin lesions and Kaufmann (1960) examined corneal scrapings for herpes, both groups detecting antigen at the lesion site. Another example of IF use for detecting antigen in lesions was in the case of influenza A where the pioneer was Liu (1955, 1956), working first with ferret and then with human nasal washings.

The first legitimization of the FAT came from the World Health Organization (WHO) in the case of rabies. The somewhat poor results of the then conventional Seller's stain for Negri bodies prompted Goldwasser et al. (1958) to use a direct FAT and they showed it to be effective. By 1966 (and reaffirmed in 1973), WHO expert committees recommended that in the hands of suitably trained technicians the FAT was the best test currently available for the rapid diagnosis of rabies. This is still true today.

In the late 1960s and 1970s, investigators in Newcastle upon Tyne decided to explore the possibility of using IF for the rapid diagnosis of respiratory virus infections together with superficial herpetic lesions, and within a short time showed that all the frequently occurring respiratory viruses could be diagnosed without difficulty. These viruses included respiratory syncytial virus (RSV) (Gardner and McQuillin, 1968), influenza A (McQuillin et al., 1970), the parainfluenza viruses (Gardner et al., 1971), adenoviruses (Gardner et al., 1972), influenza B (Kerr et al., 1975), and measles (McQuillin et al., 1976). Superficial herpetic lesions were also investigated successfully (Gardner et al., 1968). It became possible, therefore, to run a routine diagnostic service in Newcastle based on the use of IF. However, this depended on the production of specific antisera, a task that required multiple absorptions to produce a satisfactory product (Gardner and McQuillin, 1974, 1980). It was this tedious task that deterred manufacturers from attempting to enter the field, and also because, by this time, the FAT had fallen into disrepute by some investigators using their unassessed homemade antisera with disastrous results. It was the painstaking research by the Newcastle workers that showed that IF could become an efficient, sensitive technique, provided suitable reagents were available.

British manufacturers now looked more closely at the potential of the FAT and one in particular started to produce reagents of good quality. About this time the European Group for Rapid Virus Diagnosis was formed, took up the cause of rapid virus diagnosis, and established a close liaison with the WHO. Reference laboratories were established to ensure that IF reagents to be used for rapid virus diagnosis reached approved standards. Two memoranda were issued by the WHO (1977, 1978), as well as the findings of a Scientific Group (1981), which established the standards required for the reagents and clearly stated that IF was the method of choice for the diagnosis of acute respiratory virus infection. It was also suggested that when an electron microscope was not available, IF was still the method of choice for the diagnosis of suspected herpetic lesions. The further improvements, both in the production and the purification of reagents, will be discussed elsewhere in this chapter.

The evolution of the FAT in the United Kingdom and Europe and the slow development of commercially prepared quality assessed reagents, especially in the U.S.A., led to the divergent development of rapid methods with U.S. investigators favoring the enzyme-linked immunosorbent assay (ELISA). However, the formation of the Pan American Group for Rapid Viral Diagnosis and the close liaison that has developed with its European counterpart has resulted in a convergence of ideas and methods.

7.2 MATERIALS

7.2.1 Specimens

No technique, however simple or sophisticated, will achieve results if the specimen taken is inadequate. The original success of the FAT for detection of viruses in respiratory specimens was attributable to the taking of specimens by suction. A fine catheter is passed through each nostril into the nasopharynx. The catheter is attached to a mucous trap that is connected to the source of suction. The catheter is passed with suction on and the specimen is collected in the mucous trap. This procedure is fully described by Gardner and McQuillin (1980). Though this method was originally devised for taking specimens from young infants who were the major clientele it is also suitable for older children and adults (I will personally vouch for the latter statement). Other specimens that are normally suggested for the respiratory tract are throat swabs, nasal washings, gargles, or sputa, all of which are about 20% less effective than suction into a mucous trap. The investigation of superficial herpetic lesions presented no special problems. The clinical material is scraped from the appropriate place (eye or skin). For the best results vesicular lesions of the skin should be selected and cells at the base of the lesion scraped; older crusted lesions show increased nonspecific activity and the results of FAT staining, therefore, are more difficult to interpret. The technician at the bedside teases this material to separate cells in a drop of buffered saline on a slide using needles. The preparation is then allowed to air-dry.

7.2.2 Reagents

The introduction to this chapter stressed the need for high quality reagents and an adequate system for controlling such products. To achieve this, certain guidelines have been laid down and were published by the WHO (Almeida et al., 1979). These are quoted below, slightly modified to take in recent developments.

7.2.3 Quality Control of Specific Antisera or Antiglobulins

Infected and control cells should be tested to determine the optimum dilution of the antiserum that will detect the antigen. With a previously standardized and appropriate antispecies conjugate, this antiserum must be diluted at least four times greater than that concentration revealing any nonspecific fluorescence with uninfected cells. Recommended control cell cultures should include a continuous human line, a continuous monkey kidney line, a diploid cell strain, and a primary monkey kidney line.

A limited titration for final evaluation of the antiserum should be made using positive clinical material from an appropriate site for the virus being tested.

Antisera should be tested at their optimum dilutions on negative pharyngeal secretions and, where possible, on other negative human clinical specimens.

Antisera should be tested on cell cultures infected with a representative collection of viruses [e.g., RSV and parainfluenza 1, 2, 3, 4a and 4b, mumps, influenza A and B, measles, herpes simplex virus, cytomegalovirus (CMV), and at least two enteroviruses]. There should not be any nonspecific reaction using cell cultures infected with other viruses and these antisera at their optimum dilutions, and the antisera should not show nonspecific activity at their optimum dilutions when tested against heterologous virus antigens.

7.2.4 Quality Control of Antispecies Conjugate

Dilutions of antiglobulin conjugate along with an appropriate antiserum should be tested on positive tissue cultures and positive human specimens to find the optimum dilution for use. A preliminary optimum dilution may be obtained on tissue culture, but must be modified for final use on material taken directly from the patient.

The dilutions are then tested at their optimum for any nonspecific reaction on cell lines likely to be used in the laboratory, including those mentioned above for antisera.

Antiglobulin conjugates should then be tested for nonspecific reactions at the optimum dilution on negative human material, including nasopharyngeal secretions, lung material, and biopsy material.

By appointing a small number of reference laboratories for IF reagents that used the above criteria, the WHO ensured that high quality reagents for rapid virus diagnosis were available throughout the U.K., members of the European Group for Rapid Laboratory-Virus Diagnosis, and for all WHO-supported viral laboratories.

Antispecies conjugates have not presented any real problem, provided they have been assessed as described above, and most commercially available conjugates have proved suitable. There have been a number of new developments, however, regarding the production of antisera. The trend had been to move away from the uneconomical small animals wherever possible, and investigators at Wellcome Research Laboratories discovered that high titered antisera against RSV and parainfluenza virus type 3 could be produced in calves. Both viruses cause natural infections of this species. The high titers produced in this way enabled the manufacturers to prepare an antiserum that, when diluted, was devoid of nonspecific activity. However, relatively few antisera could be produced in this way, because few human pathogens cause natural infections in animals used to raise these immunoglobulins. A further advance came when the same group of workers at Wellcome Research Laboratories produced antiglobulins in the yolk sac of eggs after laying hens had previously been inoculated. This proved to be an almost unlimited supply of high titered antiglobulin without the need for elaborate purification (Polson et al., 1980). If the titer dropped the appropriate hen could be boosted. This method is being used successfully for routine production of reagents used for diagnosis (Gardner and Kaye, 1982). The principle source of quality assessed specific antiglobulins and antisera suitable for use in the FAT are those issued by Wellcome Diagnostics (Dartford, U.K.). Table 7.1 summarizes the present reagent situation. There is an urgent need for parainfluenza type 2 and mumps antiglobulins.

The role of monoclonal antibodies also requires mention. These have many potential advantages, such as unlimited supply, reproducibility, high titers, and no nonspecific activity. However, their precision itself may be a disadvantage in not necessarily being able to detect every wild type virus and a "cocktail" of monoclonal antibodies would then be necessary. There would also need to be a different dimension in their quality assessment than that presently used with the conventional polyclonal antibodies. Large numbers of wild type strains from different years should be tested to determine if any of these strains would be undetected by a particular monoclonal antibody (Gardner, 1982). Though their value in many fields of diagnostic virology can be recognized, currently their use in rapid diagnosis by IF shows no distinct advantages over polyclonal antibodies, and their quality assessment could present some logistic difficulties.

7.2.5 The Physics of Immunofluorescence and the Microscope

The basic principle of IF is that certain dyes (fluorochromes) become excited when they are stimulated by light of short wavelengths, in the ultraviolet (UV) and violet–blue end of the spectrum. To revert to their resting state the fluorochrome molecules emit light of a longer wavelength, which then is visible. This is known as fluorescence. The fluorochrome most frequently used in IF is fluorescein isothiocyanate, although other dyes (e.g., rhodamine isothiocyanate) also can be used. For simplicity fluorescein only will be considered. Fluorescein has a peak wavelength for absorbing light, thereby, achieving maximum stimulation at 490 nm. The peak emission wavelength is 517 nm. The successful utilization of the IF technique rests with the ability to separate these two wavelengths.

Table 7.1. Currently Available Commercially Produced, Quality Assessed Antisera and Globulins Suitable for Immunofluorescence

Bovine Antisera	Egg Globulins	Rabbit Antisera
Respiratory syncytial virus	Influenza A	Herpes virus hominis
Influenza A	Influenza B	
Measles	Parainfluenza type 1	
Rotavirus	Parainfluenza type 3	
Parainfluenza type 3	Adenovirus	
	Rotavirus	

The art of producing good IF is compromise. A light source is needed (usually an HBO 200 bulb), which stimulates fluorescein near its optimum at 495 nm. Then an exciting filter system capable of blocking out all light above 500 nm is used so that maximum stimulation is achieved. Then a barrier filter will be used, which allows only light over 517 nm through and which will be seen as an apple-green color. Ordinary glass filters cannot achieve sharp cut-off points and, therefore, cannot achieve maximum stimulation. Recently, more effective fluorescence has been achieved by using interference filters that can transmit excitation light up to 490 nm and should be used with a suitable barrier filter that transmits light over 500 nm. Interference filters are layers of glass in which thin layers of metallic salt compounds are deposited in vacuum. If the vacuum coating process is repeated using different materials a multilayer filter is obtained. Light that does not pass through interference filters is reflected. Variation of the thickness and refractive index of the vacuum-deposited layers makes it possible to select specific wavelengths of transmission and reflection that are close to the wavelength of maximum absorption of the fluorochrome to be used. A further property of fluorochromes is their ability to be coupled with proteins. If they are coupled with antibodies they can be used to detect antigens by irradiating such stained specimens under the fluorescence microscope. There are two kinds of microscope that can be used for fluorescence microscopy: transmitted light microscope and incident light microscope. The first method is now rarely used and should be considered obsolete. I would strongly recommend the incident light microscope for all routine purposes. The principles of its use are illustrated in Figure 7.1. The simplicity of the theory is

Figure 7.1. Incident light fluorescence.

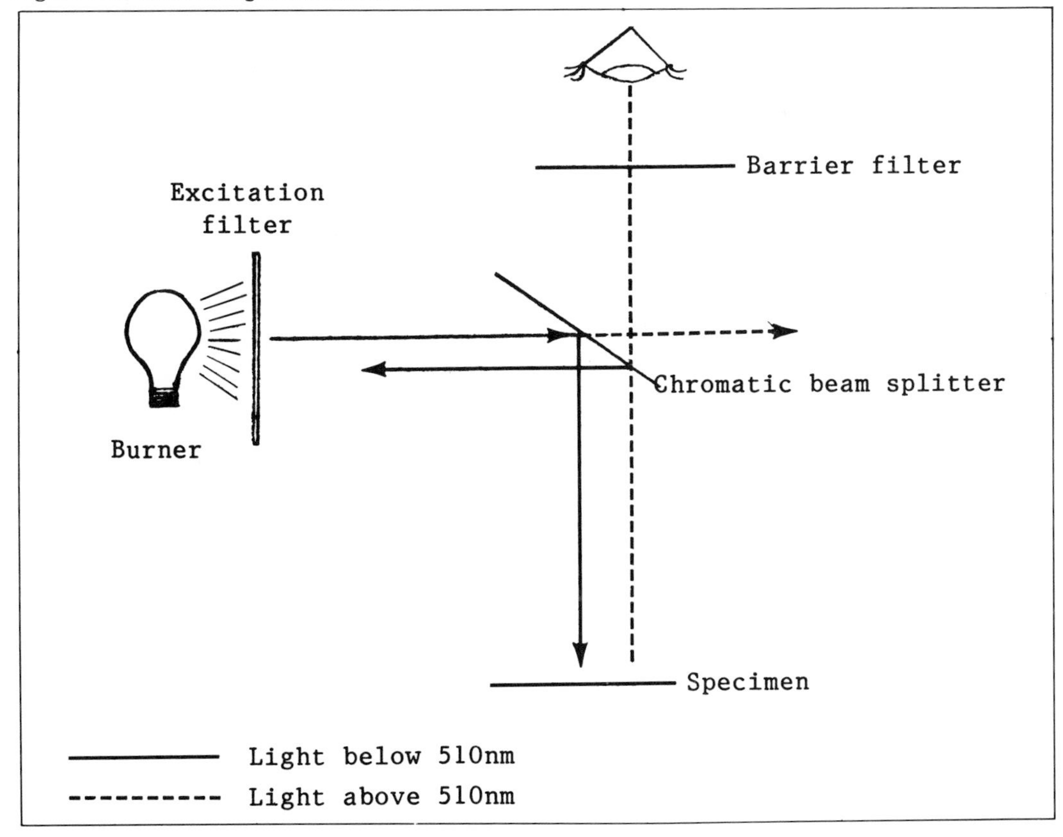

borne out in practice. A more detailed review of the theory of IF and fluorescent microscopy has been recently presented (Gardner and McQuillin, 1980).

7.3 METHODS

There have been many debates on the advantages of both the direct and indirect methods for IF. Generally, Europeans and the WHO favor the indirect method, as it required only one or two antispecies conjugates to be used against many different primary antisera. This was probably valid when antisera were difficult to make and when many were home-made. Now it should probably be a decision by manufacturers. They may still feel that when batteries of antisera are used, such as for the detection of respiratory viruses, it is easier to prepare reagents for the indirect method, whereas, in those situations when only one antigen is usually being sought (e.g., rabies or herpes simplex) a directly labeled antisera would be advantageous. There is still a belief that the indirect test is more sensitive than the direct by a factor of ten, because of the extra molecules of fluorescein involved (Pressman et al., 1958). Figure 7.2 illustrates the principles of both tests.

7.3.1 Laboratory Technique

Nasopharyngeal secretions are transported on ice as quickly as possible to the laboratory. Cells are separated by centrifugation at 350 × *g* at 4°C. The supernatant is removed and the cell pellet, which will be used for rapid diagnosis, is resuspended in 3–4 ml of phosphate buffered saline (PBS) and gently pipetted with a wide-bore pasteur pipette. Thick fragments of mucous that will not break up can be discarded. The cell suspension is transferred to a test tube and a further 4 ml of PBS added. The contents are mixed and recentrifuged at 350 × *g* for 10 minutes at 4°C. The deposit is resuspended in PBS to dilute any remaining mucous and to make sufficient numbers of slide preparations to diagnose that particular specimen and to build up a library of positive and negative slides for teaching and quality assessment purposes.

The slides may be precoated with Teflon.[1] This simplifies the task of making suitable slides on which to place drops of suspension. One drop of cell suspension is spread evenly in each square and is allowed to air-dry. The preparation is then fixed in acetone for 10 minutes at 4°C. Slides are now ready for staining. If required for a slide library they should be stored in airtight boxes below −30°C.

7.3.2 Practical Details of the Indirect Method of Staining

The direct method will not be described separately, as it consists of steps 1–4 and 9 of the indirect test described here, except that the appropriate antiviral antiserum is fluorescein labeled.

1. Slides balanced on parallel sticks are placed in suitable boxes containing wet cotton wool to maintain a moist atmosphere and to prevent cells from deteriorating
2. A drop of appropriate antiserum is spread onto a fixed preparation
3. The slide is incubated at 37°C for 30 minutes to allow the antigen–antibody reaction to take place
4. The antiserum is washed off with PBS and then immersed for three washes of 10 minutes in PBS contained in staining troughs; this 30-minute wash is important to remove all traces of unbound antibody
5. The slides are allowed to air-dry
6. A drop of the appropriate antispecies conjugate is now added and spread onto the preparation (a further reaction will take place if the antibody has previously

[1] Slides precoated with Teflon can be obtained from either Flow Laboratories Ltd. or C.A. Hendley (Essex) Ltd., Oakwood Hill Industrial Estate, Loughton, Essex IG10 3TZ.

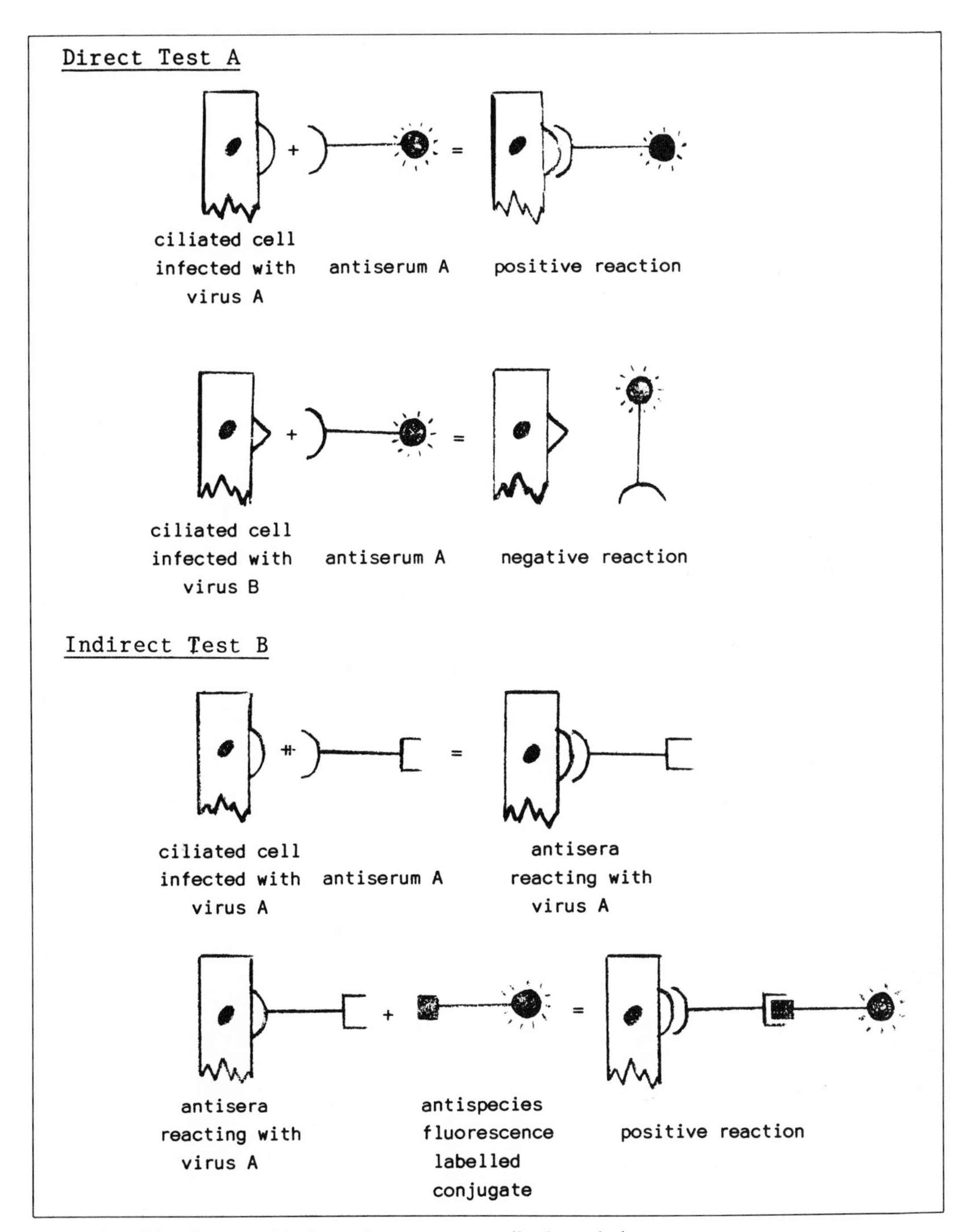

Figure 7.2. The direct and indirect fluorescence antibody technique.

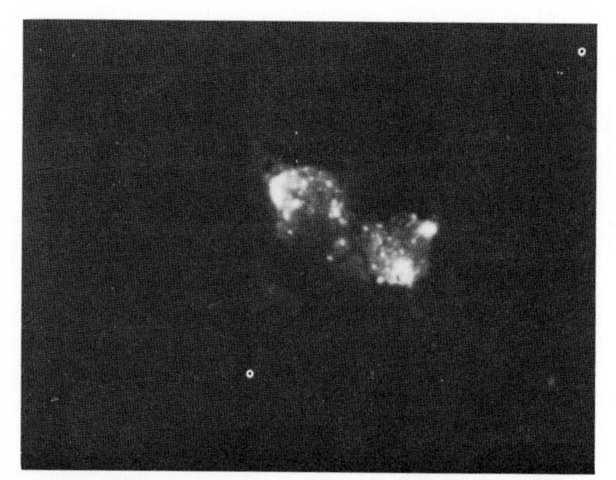

Figure 7.3. RSV in cells of a nasopharyngeal secretion. × 1100

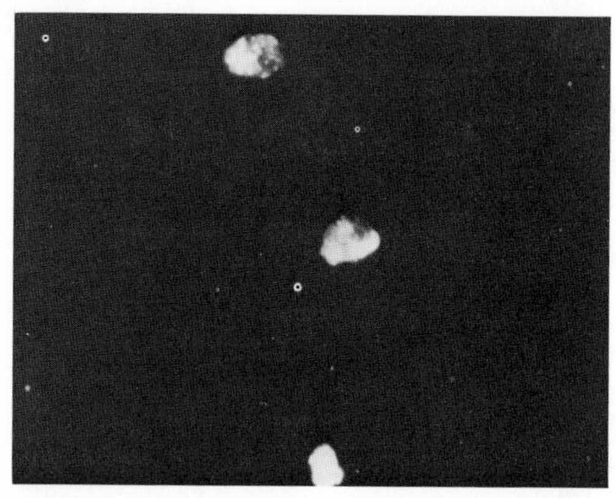

Figure 7.4. Influenza A in cells of a nasopharyngeal secretion. × 1100

Figure 7.5. Adenovirus in cells of a nasopharyngeal secretion. × 1100

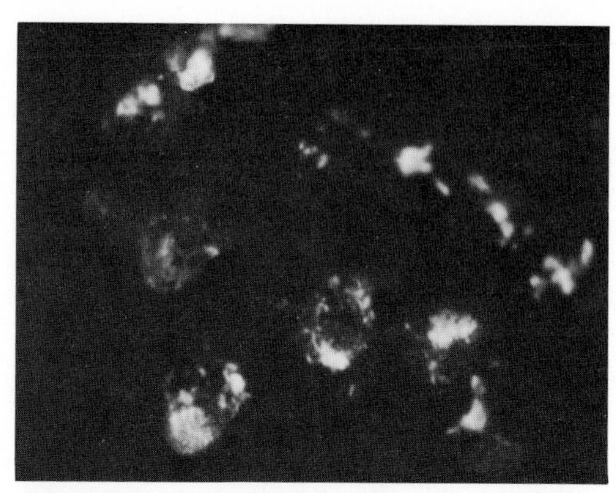

Figure 7.6. Parainfluenza 1 in cells of a nasopharyngeal secretion. × 1100

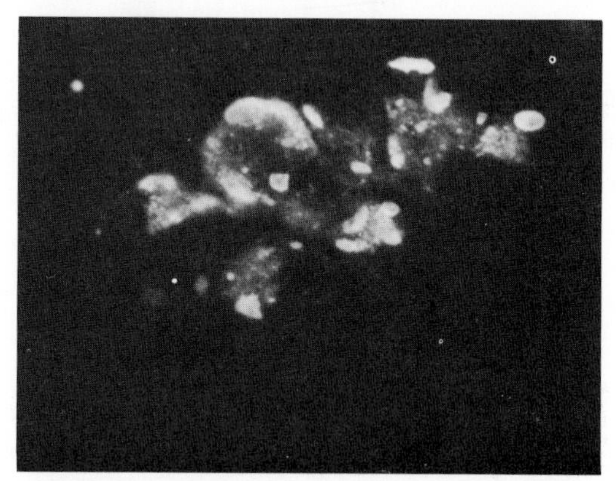

Figure 7.7. Impression smear of lung showing RS virus infected cells. × 1100

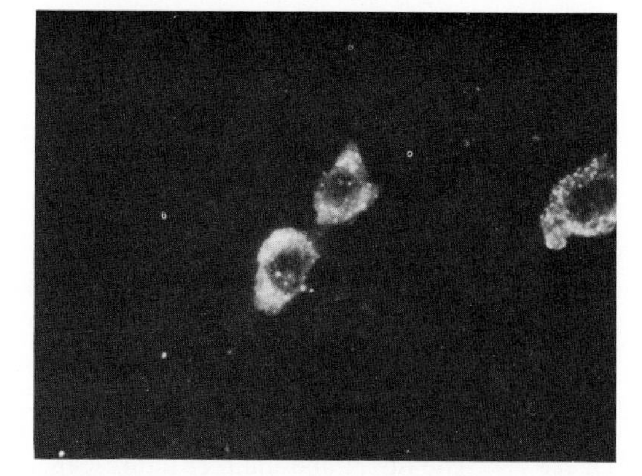

Figure 7.8. Impression smear of lung showing cells infected with Parainfluenza virus type 3. × 1100

been bound); this stage again takes place at 37°C for 30 minutes in a moist atmosphere

7. The conjugate is removed by washing in exactly the same way as in step 4; to get a clean preparation it is vital to remove all traces of unbound conjugate
8. The slide is immersed in distilled water for 2 minutes to remove PBS; if this is not done the slide will be unreadable because of crystal deposit
9. Finally, the slide is air-dried or dried on a warm plate at no more than 30°C; there is no need to mount a coverslip onto the slide; nonfluorescent immersion oil can be put directly on the dry preparation and read in blue light at a magnification of 500

7.4 VIRUSES FOR WHICH FLUORESCENT ANTIBODY TECHNIQUE IS AVAILABLE

Viruses available from Wellcome Diagnostic Ltd. are listed in Table 7.1. It will be helpful to illustrate the ease of identification of virus antigen in secretions and other respiratory tract material. Figure 7.3 shows the IF identification of RSV in the nasopharyngeal secretion of an infant with bronchiolitis and Figure 7.4 shows a similar preparation containing influenza A infected cells of a toddler presenting with a febrile convulsion.

Figure 7.5 shows the characteristic intranuclear fluorescence of adenovirus in the nasopharyngeal cells of a child with a severe pharyngitis and Figure 7.6 shows parainfluenza virus type 1 infection in a case of croup. The importance of these illustrations is that all the secretions were diagnosed as containing a specific virus within 3 hours of the patients' admission to a hospital, and if there had been need to a further scale down of the time schedule this could have been done.

In the U.K., respiratory infections are still a frequent cause of infant mortality and a large proportion of this is due to viruses. Autopsy specimens can rapidly be examined by IF and a cause of death found when conventional techniques will be time consuming and often unsatisfactory. Infants die principally of bronchiolitis and pneumonia (Aherne et al., 1970) and, though many viruses may play a part, the principle viral agent is RSV. Figure 7.7 shows a stained area of an impression smear of lung taken from a child with RS bronchiolitis and Figure 7.8 shows another impression smear of a lung from an infant with a parainfluenza virus type 3 pneumonia. The sudden infant death syndrome (SIDS) has been a tragic and perplexing problem for many years and it appears that about 33% of these deaths are associated with a respiratory virus, usually RSV. In these cases the virologic and histologic picture is identical with that of bronchiolitis (Downham et al., 1978). Superficial herpes simplex infection of the skin and eye can also be investigated. Figure 7.9 illustrates the ease of making such a diagnosis from a skin scraping.

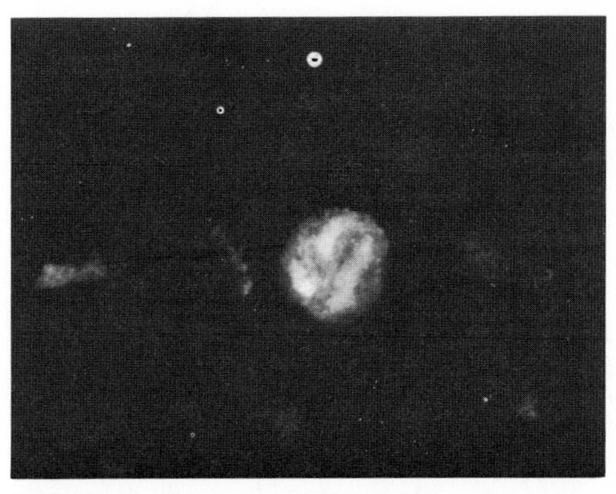

Figure 7.9. Herpes virus hominis in a skin scraping. × 1100

7.5 APPLICATION OF FLUORESCENT ANTIBODY TECHNIQUE AND ITS FUTURE

The previous section gave a pictorial display of the FAT. Having seen the beauty of the technique, its advantages and its pitfalls must be examined. But before doing this, one should refer back to the statement in the introduction that most modern techniques were derived from the FAT. The patient's specimen fixed to the slide was, in fact, the first solid phase to be used and actually contained the diagnostic material; IF was solely used as a novel indicator system. The two main developments from the FAT were ELISA and RIA: the only difference being a change of indicators. Radioimmunoassay (RIA) has become the most sensitive technique for many diagnostic tests, although it does not lend itself to nationwide use as a routine test. This is because the reagents are expensive, labile, and potentially dangerous. Also, it is difficult to assess their quality because of their short half-life. Therefore, RIA is a technique used by larger specialist and reference laboratories. ELISA and IF have developed into the present-day routine test systems for diagnostic virology.

Modern diagnostic virology requires rapid answers, and specific IgM detection for rubella, CMV, and hepatitis A are examples where the ELISA system has been accepted as a conventional technique. IF could be used, but endpoints are too difficult to define. The detection of viruses in feces, especially rotavirus, has been a province in which the ELISA system is effective, as well as for the detection of hepatitis B surface antigen. Although IF may not play a major role in the diagnosis of these virus infections, it is the technique of choice for the diagnosis of respiratory tract infections.

Before comparing the sensitivity of IF with ELISA for the detection of respiratory tract infections, it should be noted that IF has been shown to be more sensitive than virus isolation (Gardner et al., 1970). This is due to a coating of infected cells by antibody produced by the patient, which allows virus to be detected but not cultured. This was demonstrated by using a double staining technique (Gardner and McQuillin, 1978). The fluorescence seen was dull and only occurred in the convalescent stages of the illness (i.e., after 4 to 5 days and, on occasions, for as long as 14 days). On theoretical grounds one would anticipate that both the ELISA and RIA system would be less sensitive than IF, as their reaction would be blocked by coating antibody. The technique of IF would allow the dully fluorescent cells still to be observed.

A number of investigations comparing the sensitivity of ELISA and the FAT have been reported. Most studies indicated that FAT resulted in a small increase in sensitivity compared with ELISA, especially for the detection of RSV (Hornsleth et al., 1982; McIntosh, 1984). A comparative study by the author's laboratory and the National Bacteriological Laboratory, Stockholm (Dr. M. Grandien), yet to be fully reported, showed a similar result (Table 7.2). A larger trial by Grandien in the National Bacteriological Laboratory, Stockholm, involving 1267 tests again confirmed that for RSV ELISA was 10–15% less sensitive than FAT, though for influenza A there was essentially no difference between the two techniques. The intranuclear fluorescence of influenza A makes recognition of a coated cell in the convalescent stage of illness more problematic, so the two techniques approach equal sensitivity. Another factor that makes the FAT a more sensitive technique is that a firm diagnosis is dependent upon identifying a few cells only that contain fluorescence of the right color and with the correct antigen distribution.

One of the main criticisms of the FAT has been its labor-intensiveness and the need for highly trained staff for reading the stained specimens. One cannot deny that the method is labor-intensive but, compared with the isolation of virus, it is highly efficient. In fact, the many detailed viral epidemiologic reports by investigators in Newcastle (Martin et al., 1978) would not have been feasible without the FAT. Even during the largest epidemics of respiratory virus infection, it was rare to receive more than 20 specimens a day, which would represent admissions to all Tyneside hospitals

Table 7.2. Immunofluoresence Results

Investigating Laboratory	Influenza A	Influenza B	RSV	Adenovirus	Parainfluenza 3	Negatives	Total
Division of Microbiological Reagents and Quality Control, Colindale, London	15[a]	1	18	1	2	41	78
ELISA result							
National Bacteriological Laboratory, Stockholm	14	1	14	1	2	46	78

[a] Number of positive specimens.

covering a population of two million. The reading of such a small number of specimens would only represent approximately 1 hour of work for two trained people. WHO virologists were so impressed by the efficiency of the method that they have organized courses worldwide on rapid diagnosis, in which IF was taught especially for the diagnosis of respiratory virus infections. This has been a practical exercise, not an academic one. All the trainees on returning to their home laboratories were supplied with quality assessed reagents and were able to investigate the respiratory problems of their areas. To ensure consistency, reference laboratories in Europe checked samples of difficult slides for these newly established laboratory services. There appeared to be no problems with the method and much satisfaction was experienced by participants who brought rapid virus diagnosis to areas of the world, which would have seemed an impossibility only a few years ago.

Two similar experiences are worth recording. First, in the U.K. (Report to the MRC, 1978) ten laboratories with no experience using IF techniques took part in a trial of the incidence and importance of RSV infections. A second study for the incidence of RSV and other respiratory viruses was conducted by members of the European Group for Rapid Virus Diagnosis in eight centers, some with little or no previous experience using IF techniques (Orstavik et al., 1980, 1984). Both trials were highly successful.

Once reliable reagents and suitable microscopes are available the mystique of IF rapidly disappears after minimal training. Within such training programs, however, one should point out clearly the pitfalls of the technique. Table 7.3 illustrates the major problem areas that may occur with the reagents, the techniques, and the interpretation of results. Most of these are self-evident and need little further explanation, as this has been discussed previously under the quality assessment of reagents. Additional problems may occur and are listed under the general heading of "specimens." Table 7.4 catalogs the principal problems. Most of the points have been covered elsewhere and others are self-evident. Tables 7.3 and 7.4, therefore, are offered as a summary of the major problems of the technique.

Evaluating the pitfalls in this way tends to overemphasize the problems that in practice are relatively few and far between. Perhaps the focus should be more on the advantages, which include rapidity, specificity, sensitivity, reproducibility, and time and cost saving in comparison with conventional isolation techniques. Besides these advantages the method has certain

Table 7.3. Pitfalls in IF Diagnosis

Reagents
Antiglobulins, including conjugates
specificity
optimal titers
deposits
insufficient absorption
Diluents
pH
crystal deposits
Immersion oil
autofluorescence
Counterstain
autofluorescence
loss of specific fluorescence by wrong concentration
Technique
errors in handling
quality of preparation
over fixation
under washing
wrong antiglobulin
wrong conjugate
incorrect illumination
wrong filter system
Difficulties in interpretation
Nonspecific fluorescence due to:
contaminants
autofluorescence
mucous
reactions between specimens and antiglobulins
Criteria for diagnosis
intracellular fluorescence
color
experience

Table 7.4. Pitfalls due to the Specimen in IF Diagnosis

Respiratory
Type of specimen
nasopharyngeal specimen best
Stage of illness
Type of illness
Contaminants
Skin and eye scrapings
Type of lesion
Site sampled
Contaminants
Autopsy material
Problems in delay
Contaminants
Sampling errors
Biopsy material
Sampling errors
Quality available

other boons that are of great clinical importance. A major problem of pediatric wards is crossinfection, the extent of which has been reported previously for a number of respiratory viruses (Sims et al., 1975; Weightman et al., 1974; Ditchburn et al., 1971). Approximately 50% of all crossinfections involved the lower respiratory tract, were severe, and occasionally were fatal. One need only visualize the example of a toddler with a febrile convulsion admitted to hospital having lost both his convulsion and his pyrexia but still excreting the influenza A virus that caused the convulsion. He moves around the ward infecting patients who are immunologically compromised. Now that we have the tools to diagnose immediately all respiratory admissions within 2 to 3 hours, such spread can be limited by early diagnosis and subsequent patient isolation.

In a similar vein, a number of patients in pediatric wards are immunosuppressed with leukemia. Many studies now confirm that acute lymphocytic leukemia can be controlled and even cured, but the patient may die of intercurrent infection which is often viral (Craft et al., 1977, 1978). These children must be protected and IF could be a great help in diagnosis, thus, leading to measures to limit the spread of virus infection. These children also lack an immune response and IF can readily diagnose their virus infection. Measles is especially lethal in the leukemic child and often the illness presents in atypical fashion. Figure 7.10 shows a typical giant cell using IF. Similarly, RSV infections are not often associated with severe illness in those outside infancy. Respiratory syncytial virus may be excreted for many months by the leukemic child with infected cells present in large numbers in the secretions during this period but with no evidence of antibody coating of such cells, confirming the absence of any local immune response.

We have seen how the diagnosis of virus infections in the immunosuppressed child can be facilitated by IF, especially when the clinical features are atypical. The same is true for the normal child who may react in a variety of ways to the common virus infections. It has been pointed out that febrile convulsions may be a sign of influenza A infections (Brocklebank et al., 1972). During influenza B outbreaks, abdominal pain, often a prominent symptom, simulates acute appendicitis. In this instance influenza B can be speedily diagnosed by IF (Kerr et al., 1975).

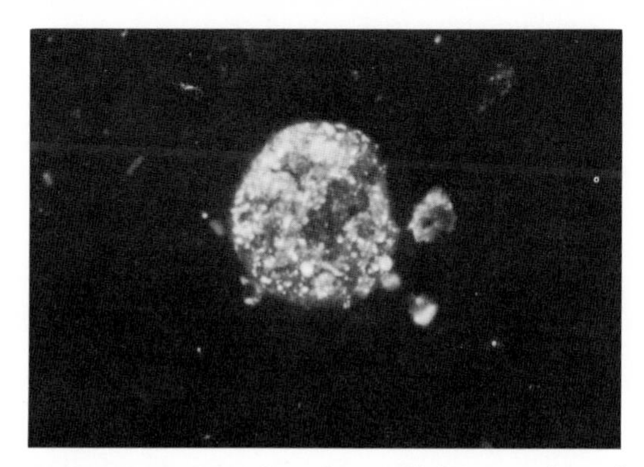

Figure 7.10. Measles giant cell in secretion of patient with acute lymphatic leukemia. × 1100.

Although virology laboratories have mushroomed over the last few years there are many areas even in the U.K. that do not have a virus laboratory. The FAT lends itself well for diagnosis at a distance. A hospital without a diagnostic virology service, perhaps situated many miles from a virus diagnostic center, can take secretions or specimens, prepare and fix slides, and send them by post or other transport to a virus laboratory. The reference laboratory stains the slides, reads them, and supplies an answer by telephone, which is rapid diagnosis in an area that previously offered no virus diagnostic service. An example of a routine diagnostic service run in this way has been reported (Downham et al., 1974).

These days the "in" word is "automation," and "labor intensiveness" is "passe." This summarizes the dilemma of fluorescence microscopy. This chapter has attempted to present the advantages and disadvantages of the FAT. Some investigators (Halonen et al., 1983) have tried to introduce automation into IF by time resolved fluorometry. This technique is based on having a fluorescent probe with a long life time, excited by a short pulse, and specific fluorescence is measured after a selected delay time. During the delay time background fluorescence, which is said to have a shorter delay, is eliminated and specific fluorescence only is detected. The rare earth metal europium has a long decay time of 100–1000 μsec and also has a large difference between excitation and emission wavelength, resulting in a further reduction of background noise. The specimens are read in a single photon counting fluorometer equipped with a xenon flash lamp. The excitation wavelength is 340 nm and the length of the excitation pulse is 1 μsec. After a delay of 400 μsec, single photon emission is counted for 500 μsec at 613 nm. After a further delay of 100 μsec a new cycle begins. The cycle is repeated 1000 times in overall counting time of 1 second. The work of Halonen et al. (1983) is the only report of this method to be used and at this stage it is too early to judge its full potential. Particularly, one is interested in the power of this machine to distinguish unequivocally between specific and nonspecific fluorescence. One also wonders if a need exists for such an elaborate, sophisticated, and expensive apparatus for approximately 1 hour reading a day. There is also a feeling of security of being able to see with your own eyes the characteristics of fluorescence in the right cell, of correct distribution, and apple-green in color. If a machine can do this infallibly so be it. But until that time comes let us continue to use apparatus and reagents that all can understand and leave a little creative joy in the work of the virus diagnostician.

REFERENCES

Aherne, W., Bird, T., Court, S.D.M., Gardner, P.S., and McQuillin, J. 1970. Pathological changes in virus infections of the lower respiratory tract in children. J. Clin. Path. 23:7–18.

Almeida, J.D., Atanasiu, P., Bradley, D.W., Gardner, P.S., Maynard, J., Schuurs, A.W., Voller, A., and Yolken, R.H. 1979. Manual for rapid viral diagnosis. WHO Offset Publication No. 47.

Biegeleisen, J.Z. Jr., Scott, L.V., and Lewis, V. Jr. 1959. Rapid diagnosis of herpes simplex virus infection with fluorescent antibody. Science 129:640–641.

Brocklebank, J.T., Court, S.D.M., McQuillin, J., and Gardner, P.S. 1972. Influenza A infection in children. Lancet ii:497.

Coons, A.H., Creech, H.J., and Jones, R.N. 1941. Immunological properties of an antibody containing a fluorescent group. Proc. Soc. Exp. Biol. Med. 47:200–202.

Craft, A.W., Reid, M.M., Bruce, E., Kernahan, J., and Gardner, P.S. 1977. The role of infection in the death of children with acute lymphoblastic leukaemia. Arch. Dis. Child. 52:752–757.

Craft, A.W., Gardner, P.S., Jackson, E., Kernahan, J., Noble, T.C., Reid, M.M., and Walker, W. 1978. Virus infections in acute childhood lymphoblastic leukaemia. Arch. Dis. Child. 53:836.

Ditchburn, R.K., McQuillin, J., Gardner, P.S., and Court, S.D.M. 1971. Respiratory syncytial virus in hospital cross-infection. Br. Med. J. 3:671–673.

Downham, M.A.P.S., Elderkin, F.M., Platt, J.W., McQuillin, J., and Gardner, P.S. 1974. Rapid virus diagnosis in paediatric units by a postal service: Respiratory syncytial viral infection in Cumberland. Arch. Dis. Child. 49:467–471.

Downham, M.A.P.S., Scott, D., Gardner, P.S., McQuillin, J., and Stanton, A.N. 1978. Respiratory viruses and cot death. Br. Med. J. 2:12–13.

Gardner, P.S., and McQuillin, J. 1968. Application of the immunofluorescent antibody technique in the rapid diagnosis of respiratory syncytial virus infection. Br. Med. J. 3:340–343.

Gardner, P.S., McQuillin, J., Black, M.M., and Richardson, J. 1968. The rapid diagnosis of herpes virus hominis infections in superficial lesions by immunofluorescent antibody techniques. Br. Med. J. 4:89–92.

Gardner, P.S., McQuillin, J., and McGuckin, R. 1970. The late detection of respiratory syncytial virus in cells of respiratory tract by immunofluorescence. J. Hyg. Camb. 68:575–580.

Gardner, P.S., McQuillin, J., McGuckin, R., and Ditchburn, R.K. 1971. Observations on the clinical and immunofluorescent diagnosis of parainfluenza virus infections. Br. Med. J. 2:7–12.

Gardner, P.S., McGuckin, R., and McQuillin, J. 1972. Adenovirus demonstrated by immunofluorescence. Br. Med. J. 3:175.

Gardner, P.S., and McQuillin, J. 1974. Rapid Virus Diagnosis, Application of Immunofluorescence, 1st ed. London: Butterworths.

Gardner, P.S., and McQuillin, J. 1978. The coating of RS virus infected cells in the respiratory tract by immunoglobulins. J. Med. Virol. 2:77–87.

Gardner, P.S., and McQuillin, J. 1980. Rapid Virus Diagnosis, Application of Immunofluorescence, 2nd ed. London: Butterworths.

Gardner, P.S. 1982. Monoclonal antibodies in relation to viruses and their possible clinical application. Proc. Royal Soc. Edinburgh 81B:277–291.

Goldwasser, R.A., Kissling, R.E., Carski, T.R., and Hosty, T.S. 1958. Fluorescent antibody staining of street and fixed rabies virus antigens. Proc. Soc. Exp. Biol. Med. 98:219–223.

Halonen, P., Meurman, O., Lovgren, T., Hemmile, I., and Soini, E. 1983. Detection of viral antigens by time-resolved fluoroimmunoassay. In P.A. Bachmann (ed.), New Developments in Diagnostic Virology. Berlin, Heidelberg, New York: Springer-Verlag.

Hornsleth, A., Friis, B., Anderson, P., and Brenoe, E. 1982. Detection of respiratory syncytial virus in nasopharyngeal secretions by ELISA: Comparison with fluorescent antibody techniques. J. Med. Virol. 10:273–281.

Kaufman, H.E. 1960. The diagnosis of corneal

herpes simplex infection by fluorescent antibody staining. Arch. Opthal. (Chicago) 64:382–384.

Kerr, A.A., Downham, M.A.P.S., McQuillin, J., and Gardner, P.S. 1975. Gastric "flu." Lancet i:291–295.

Liu, C. 1955. Studies on influenza infection in ferrets by means of fluorescein-labeled antibody. I. The pathogenesis and diagnosis of the disease. J. Exp. Med. 101:665–676.

Liu, C. 1956. Rapid diagnosis of human influenza infection from nasal smears by means of fluorescein-labelled antibody. Proc. Soc. Exp. Biol. Med. 92:883–887.

Martin, A.J., Gardner, P.S., and McQuillin, J. 1978. Epidemiology of respiratory viral infection among hospitalized children over a six year period in North East England. Lancet ii:1035–1038.

McIntosh, K. 1984. Editorial: FA for rapid diagnosis of respiratory viruses. Pan American Group for Rapid Virus Diagnosis. Vol. 9, January 1984. Publication of the Pan American Group.

McQuillin, J., Gardner, P.S., and McGuckin, R. 1970. Rapid diagnosis of influenza by immunofluorescent techniques. Lancet ii:690–694.

McQuillin, J., Bell, T.M., Gardner, P.S., and Downham, M.A.P.S. 1976. Application of immunofluorescence to a study of measles. Arch. Dis. Child. 51:411–419.

Orstavik, I., Grandien, M., Halonen, P., Arstila, P., Mordhurst, C.H., Hornsleth, A., Popow-Kraupp, T., McQuillin, J., and Gardner, P.S. 1980. Rapid immunofluorescence diagnosis of respiratory syncytial virus infections among children in European countries. Lancet ii:32.

Orstavik, I., Grandien, M., Halonen, P., Mordhurst, C.H., Hornsleth, A., Popow-Kraupp, T., McQuillin, J., Gardner, P.S., Almeida, J., Bricout, F., and Marques, A. 1984. Epidemiological finds by rapid immunofluorescence diagnosis of respiratory virus infections among children in different European countries. WHO Bull. 62.

Polson, A.M., von Wechmar, B., and van Regenmortel, M.H.V. 1980. Isolation of viral IgY antibodies from yolks of immunized hens. Immunol. Commun. 9:475–493.

Pressman, D., Yagi, Y., and Hiramoto, R. 1958. A comparison of fluorescein and I^{1131} as labels for determining the "in vivo" localization of anti-tissue antibodies. Int. Arch. Allergy 12:125–136.

Report of the MRC Subcommittee on RSV Vaccines. 1978. Respiratory syncytial virus infection in industrial, urban and rural areas as a cause of admission to hospital. Br. Med. J. 2:796–798.

Sims, D.G., Downham, M.A.P.S., Webb, J.K.G., Gardner, P.S., and Weightman, D. 1975. Hospital cross-infection on children's wards with respiratory syncytial virus and the role of adult carriage. Acta Paediat. Scand. 64:541–545.

Weightman, D., Downham, M.A.P.S., and Gardner, P.S. 1974. Introduction of a cross-infection rate in children's wards, its application to respiratory virus infections. J. Hyg. Camb. 73:53–60.

World Health Organization. 1966. Expert Committee on Rabies (fifth report). Technical Report Series No. 321.

World Health Organization. 1973. Expert Committee on Rabies (sixth report). Technical Report Series No. 523.

WHO Bulletin. 1977. Laboratory techniques for rapid diagnosis of viral infections: A memorandum. WHO Bull. 55:33–37.

WHO Bulletin. 1978. Progress in the rapid diagnosis of viral infections: A memorandum. WHO Bull. 56:241–244.

WHO Scientific Group. 1981. Rapid laboratory techniques for the diagnosis of viral infections. Report of a Scientific Group. Technical Report Series No. 661.

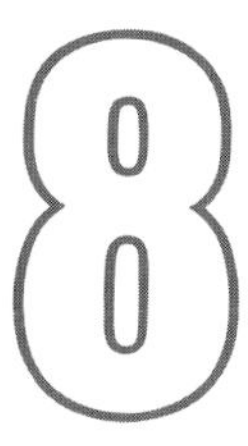

Radioimmunoassay

Isa K. Mushahwar and Thomas A. Brawner

8.1 INTRODUCTION

Ever since the introduction of radioimmunoassay (RIA) techniques (Yalow and Berson, 1960) for the determination of endogenous human plasma insulin, RIA has become the essential key analytical method in many sciences. Besides endocrinology, RIA has been a valuable tool in such fields as enzymology (Kolb and Grodsky, 1970), hematology (Rutland, 1984), toxicology (Shimada et al., 1983), pharmacology (Robinson and Smith, 1983), parasitology (Avraham et al., 1982), neurochemistry (Dowse et al., 1983), microbiology (Zollinger et al., 1976), plant pathology (Ghabrial and Shepherd, 1980), diagnostic medicine (Matsui et al., 1982), mycology (Poor and Cutler, 1979), and virology (Witkor et al., 1972). Thus, RIA has contributed to the revolution of biology during the past 25 years in many other disciplines besides endocrinology. This trend will continue to have a positive impact on public health in the future through diagnostic and epidemiologic studies in the areas of infectious diseases and cancer.

8.2 COMPETITIVE BINDING RADIOIMMUNOASSAY

The original RIA described by Yalow and Berson (1960) and subsequently by other investigators (Faiman and Ryan, 1967; Odell et al., 1967; Goodfriend and Ball, 1969) was a competitive binding RIA, where the competition between an unlabeled antigen and a radiolabeled counterpart for a limited amount of antibody is monitored. This is illustrated as follows:

$$AG^* + AB + AG^0 \rightarrow AG^*AB + AG^0AB$$

where AG^* represents radiolabeled antigen; AG^0 is the unlabeled antigen, AB is the antibody; AG^*AB is a complex of radiolabeled antigen with antibody and AG^0AB is a complex of unlabeled antigen with antibody. The higher the concentration of AG^0, the lower the concentration of radioactive AG^*AB and the higher the concentration of free AG^*. The reaction is designed to take place in a solution giving a mixture of bound and free radiolabeled antigen. Separation of antibody bound antigen from the free antigen is achieved by electrophoretic or a variety of immunoprecipitation techniques (Morgan and Lazarow, 1962; Hales and Randle, 1963).

8.3 SOLID PHASE RADIOIMMUNOASSAYS

A simpler and widely used variation of the competitive binding RIA is a solid phase

RIA, which is also based on competitive inhibition (Catt and Tregear, 1967) utilizing radiolabeled antigen and solid phase antibody for the separation of bound and free antigen as shown:

$$SP.AB + AG^* + AG^0 \rightarrow \begin{matrix} SP.AB.AG^* \\ SP.AB.AG^0 \end{matrix} + AG^* + AG^0$$

where the count rate of radioactive antigen (AG*), bound through antibody linkage to the surface of the solid phase (SP.AB) to form the radiolabeled antigen–antibody complex (SP.AB.AG*) is in direct proportion to the quantity of competitive unlabeled antigen (AG^0) in the reaction mixture. Removal of the reaction mixture and washing the solid phase (polystyrene tube) serves as an effective and highly reproducible method for separating bound and free antigen. Soon after the introduction of solid phase assays, it became apparent that RIA procedures based on principles other than competitive inhibition could be developed. These were the direct solid phase sandwich RIA (Wide et al., 1971). These methods were found applicable to all biological substances that have a minimum of two binding sites and, thus, can be utilized to assay for both antigen and antibodies. Soon, a direct solid phase sandwich RIA utilizing, for the first time, radiolabeled specific immunoglobulin as a probe for macromolecular viral antigens was introduced (Ling and Overby, 1972). The potential advantages of using radiolabeled antibody over radiolabeled antigen (e.g., viruses) is the greater availability, ease of labeling and immobilization of immunoglobulins.

The last decade has witnessed a rapid increase in the variety of solid phase RIA developed for the detection and quantification of a multitude of antigens and their corresponding antibodies (Overby and Mushahwar, 1979; Mushahwar and Overby, 1983). For the most part, these assays have been used extensively in highly specialized research environments. Most importantly, the vast majority of these tests never made

Table 8.1. Hepatitis

Hepatitis Marker	Reference
Hepatitis B virus (HBV)	
HBsAg	Hollinger et al., 1971 Ling and Overby, 1972 Purcell et al., 1973
Anti-HBc IgG	Neurath et al., 1978 Overby and Ling, 1976 Purcell et al., 1973 Vyas and Roberts, 1977
Anti-HBs	Ginsberg et al., 1973
Anti-HBe	Blum et al., 1979 Frosner et al., 1978 Mushahwar et al., 1978 Neurath et al., 1979
HBeAg	Blum et al., 1979 Frosner et al., 1978 Mushahwar et al., 1978 Neurath et al., 1979
Anti-HBc IgM	Chau et al., 1983
HBcAg	Overby and Ling, 1976 Purcell et al., 1973 Vyas and Roberts, 1977
Hepatitis D virus (HDV)	
HDAg	Rizzetto et al., 1980 Mushahwar and Decker, 1983
Anti-HD IgG	Mushahwar and Decker, 1983 Rizzetto et al., 1980
Anti-HD IgM	Smedile et al., 1982
Hepatitis A virus (HAV)	
HAV	Decker et al., 1979 Hollinger et al., 1975 Hollinger and Maynard, 1976 Purcell et al., 1976
Anti-HAV IgG	Decker et al., 1979 Hollinger and Maynard, 1976 Purcell et al., 1976 Safford et al., 1980

Table 8.1. (*continued*)

Hepatitis Marker	Reference
Hepatitis A virus (HAV)	
Anti-HAV IgM	Bradley et al., 1977
	Decker et al., 1981
	Devine et al., 1979
	Flehmig et al., 1979
	Lemon et al., 1980
	Roggendorf et al., 1980
Anti-HAV IgA	Overby et al., 1981

the transition from a research procedure to a commercially available and standardized procedure (Hill and Matsen, 1983). Because the hepatitis viruses have been widely characterized with highly specific and commercially available RIA, the authors will use these models to describe the various direct and indirect (competitive) methods that employ solid phase sandwich RIA to detect the antigens and antibodies of viral hepatitis (Table 8.1).

8.4 HEPATITIS B SEROLOGIC MARKERS

Three distinct viral antigens and their respective antibodies are used for the diagnosis of hepatitis B virus (HBV) infections: hepatitis B surface antigen (HBsAg) and its antibody (anti-HBs), hepatitis B core antigen (HBcAg) and its antibody (anti-HBc); and hepatitis B e antigen (HBeAg) and its antibody (anti-HBe). These antigens and antibodies occur in serum sequentially during the course of disease and recovery (Mushahwar et al., 1981). Because HBsAg is a defective particle without nucleic acid and is produced in large quantities, it is an easily detectable marker of an ongoing HBV infection. Seroconversion to anti-HBs is an indication of recovery and eventual immunity. HBcAg is found in the internal core of the DNA-containing virus and has not been found to occur free in serum. Anti-HBc rises with the onset of viremia and persists through recovery and immunity. This antibody is sometimes the only detectable marker for exposure to the virus. HBeAg rises in serum at the same time as HBsAg, and its presence is associated with acute disease, chronic disease, and the presence of infectious virus. Seroconversion to anti-HBe signifies better clinical prognosis and a lower level of viremia.

8.5 DIRECT SOLID PHASE SANDWICH ASSAYS

8.5.1 Radiometric Assays for Hepatitis B Virus Antigens and Antibodies

The direct solid phase sandwich RIA procedure for antigen detection is divided into three main operational stages, as follows: adsorption of unlabeled antibody onto the solid phase, binding of the antigen by the adsorbed antibody, and detection of bound antigen by reaction with radiolabeled and highly specific antibody. The first successful direct solid phase RIA for the detection of viral antigens was the AUSRIA prototype (Ling and Overby, 1972) for the detection of HBsAg. In this system highly specific human anti-HBs was adsorbed to polystyrene tubes. The serum to be tested for HBsAg was incubated with the anti-HBs coated tube giving a complex of anti-HBs and HBsAg in the solid phase. The radioactive probe in the next reaction was affinity purified guinea pig anti-HBs labeled with radioactive iodine (^{125}I).

A direct solid phase RIA for HBeAg similar to AUSRIA was developed (Mushahwar et al., 1978) employing radiolabeled specific antibodies as illustrated in Figure 8.1. Briefly, 6-mm diameter polystyrene beads are coated with anti-HBe by incubating a dilution of antiserum with the beads at pH 9.0 for 24 hours at room temperature. These antibody coated beads are then used as the solid phase antibody. Anti-

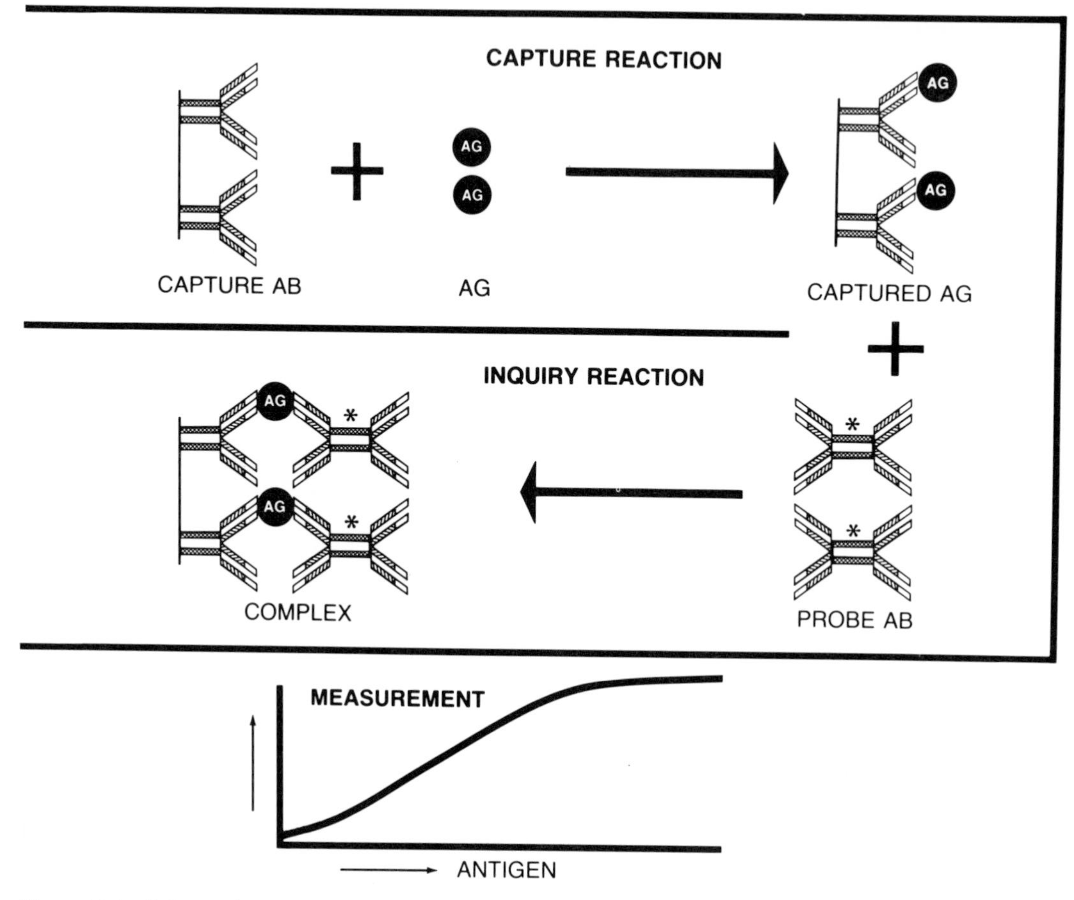

Figure 8.1. Schematic representation of a direct solid phase assay designed to detect HBeAg. The procedure involves capture of the antigen by antibody attached to the solid phase (*top*) and subsequent binding of labeled antibody to exposed sites on the captured antigen (*middle*). The amount of antigen measured is in direct proportion to the count rate of the washed solid phase (*bottom*).

HBe IgG preparations are radiolabeled with ^{125}I by the chloramine-T method (Greenwood et al., 1963) often resulting in a specific radioactivity of 18–25 μCi/μg IgG. The ^{125}I-labeled anti-HBe solution is diluted to approximately 3 μCi/ml in a diluent containing 50% fetal calf serum. For HBeAg detection, serum samples of 0.2 ml are incubated overnight at room temperature with the antibody coated bead. These are then washed with distilled water, and the beads are further incubated in 0.2 ml of ^{125}I-labeled anti-HBe solution at 45°C for 4 hours. The beads are then washed and counted for ^{125}I-labeled anti-HBe uptake. The resulting count rate of the multiple layered beads are directly proportional to HBeAg concentration in the sample. Human convalescent antisera containing anti-HBe are a source of reagents for the diagnostic RIA described, without recourse to animal immunizations with purified preparations of HBeAg. The manufacturers package insert for HBeAg determination describes the procedures in a stepwise manner.

The presence or absence of HBeAg is determined by comparing the counts per minute (cpm) of the unknown specimen to a predetermined cutoff value (Mushahwar

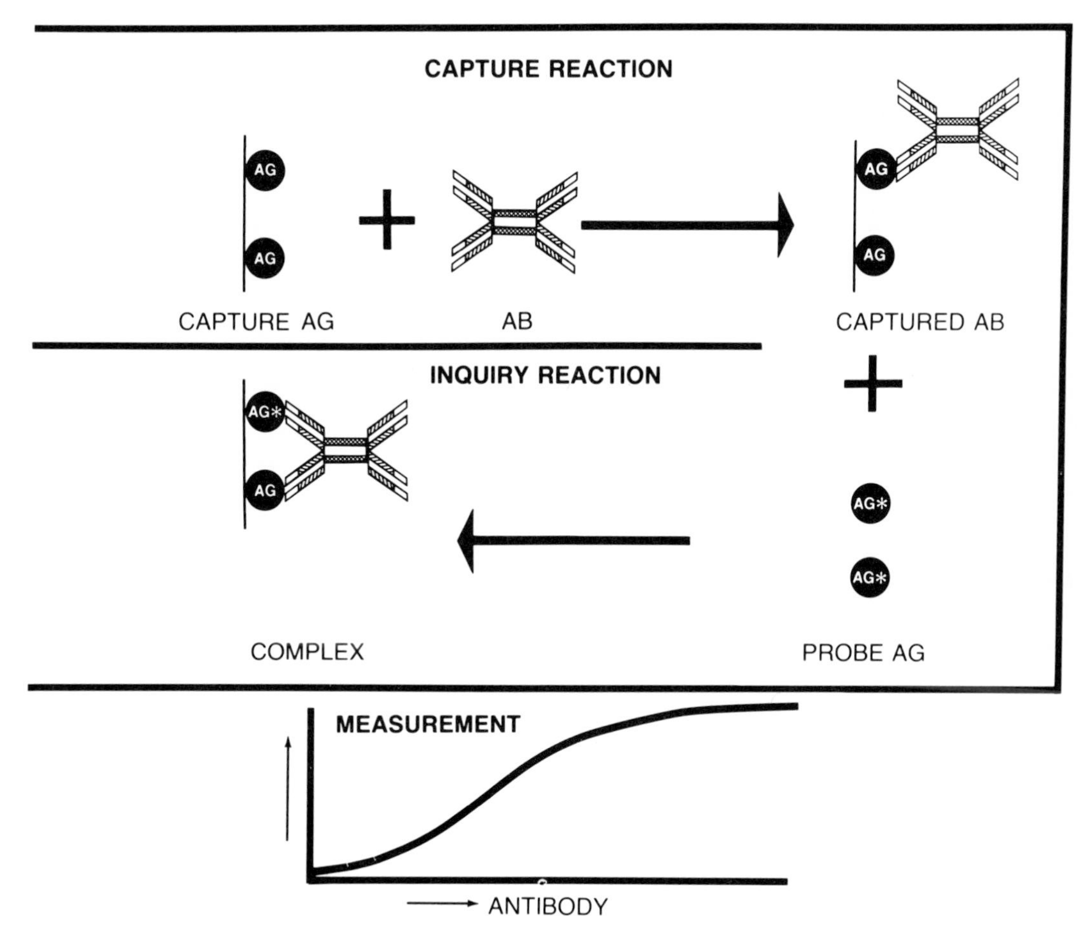

Figure 8.2. Schematic representation of direct, solid phase assay to detect anti-HBs. The two-step procedure involves capture of specific antibody by antigen bound to the solid phase (*top*) and binding of labeled antigen to specific, available sites on the captured antibody (*middle*). The amount of antibody measured is in direct proportion to the count rate of the washed solid phase (*bottom*).

et al., 1978). Specimens whose count rates are equal to or greater than this cutoff value are considered to be reactive for HBeAg.

For a test run to be valid the mean value for the positive control should be at least four times the negative control mean. It is recommended by the manufacturer that three negative and three positive HBeAg controls should be assayed with each run. A similar direct solid phase sandwich RIA for anti-HBs detection has been developed (Ginsberg et al. 1973). The principle is illustrated in Figure 8.2.

Highly purified HBsAg is bound to a polystyrene bead by adsorption to produce the solid phase antigen. The serum specimen to be assayed for anti-HBs is incubated with the solid phase antigen. If anti-HBs is present, it will complex with the antigen. ^{125}I-HBsAg is added in the next step, which will bind to the antibody already trapped on the bead, forming an antigen–antibody–^{125}I-HBsAg complex. The amount of radioactivity on the bead is in proportion to the concentration of anti-HBs in the serum.

The step-by-step procedure is described in the manufacturer's package insert. An example of this is the AUSAB test

(Abbott Laboratories) for anti-HBs. It is recommended that seven negative and three positive controls should be assayed with each run of unknowns.

Again, as in the case of HBeAg detection, reactive and nonreactive specimens are determined by relating the net counts per minute of unknowns to a cutoff value calculated by multiplying the negative control mean by the factor 2.1. Specimens with values greater than the cutoff are reactive for anti-HBs.

Table 8.2. Commercial Hepatitis Radioimmunoassay Kits

Test	Serologic Marker	Product Name	Solid Phase Configuration	Manufacturer
Antigen	HBsAg detection	AUSRIA II	6-mm bead	Abbott Laboratories
		Heparia	6-mm bead	North American Biologicals, Inc.
		Rialyze	Gel column	Ames Co., Division Miles Laboratories
		AUK-3	6-mm bead	Sorin Biomedica
		Riasure	Microbeads	Electro-Nucleonics Laboratories, Inc.
		Clinical Assays	Plastic tube	Connought Laboratories (Travenol Laboratories)
		NML HBsAg RIA	6-mm bead	Nuclear-Medical Laboratories
	HBeAg	ABBOTT-HBe	6-mm bead	Abbott Laboratories
		EBK	6-mm bead	Sorin Biomedica
Antibody	Anti-HBs	AUSAB	6-mm bead	Abbott Laboratories
		AB-AUK	6-mm bead	Sorin Biomedica
		Hepab	6-mm bead	North American Biologicals, Inc.
		Clinical Assays	Plastic tube	Connought Laboratories (Travenol Laboratories)
	Anti-HBc (Total)	CORAB	6-mm bead	Abbott Laboratories
		AB-COREK	6-mm bead	Sorin Biomedica
	Anti-HBc (IgM)	CORAB-M	6-mm bead	Abbott Laboratories
		CORE-IGMK	6-mm bead	Sorin Biomedica
	Anti-HAV (IgM)	HAVAB-M	6-mm bead	Abbott Laboratories
	Anti-HBe	ABBOTT-HBe	6-mm bead	Abbott Laboratories
		EBK	6-mm bead	Sorin Biomedica
	Anti-HAV (Total)	HAVAB	6-mm bead	Abbott Laboratories
	Anti-Delta (Total)	ABBOTT-ANTI-DELTA	6-mm bead	Abbott Laboratories
	Anti-Rubella IgG	Gamma Coat	Plastic tube	Travenol–Genentech Diagnostics

8.6 INDIRECT (COMPETITIVE) RADIOMETRIC ASSAYS FOR HEPATITIS B VIRUS ANTIBODIES

In these assays, radiolabeled antibodies are utilized as probes for the detection of serum antibodies in either a one-step or a two-step procedure. The solid phase is an appropriate antigen coated polystyrene tube or bead. The radiolabeled probe will have fewer antigen binding sites for the reaction when serum antibodies are present. Hence, the final count rate of the solid phase will be inversely proportional to the amount of antibody in the serum specimen being tested. These techniques have been applied to the detection of anti-HBc (Overby and Ling, 1976) and anti-HBe (Mushahwar et al., 1978).

8.6.1 One-Step Procedure

The anti-HBc assay is illustrated in Figure 8.3. The solid phase reagent is a polystyrene bead coated with purified HBcAg.

Figure 8.3. Schematic representation of an indirect, competitive solid phase assay for anti-HBc utilizing labeled antibodies. The two-step procedure involves capture of antibody by an antigen (HBcAg) coated solid phase (*top*) and binding of labeled antibody to available sites on the bound antigen (*middle*). The count rate of the washed solid phase will be inversely proportional to the amount of antibody in the assay specimen (*bottom*).

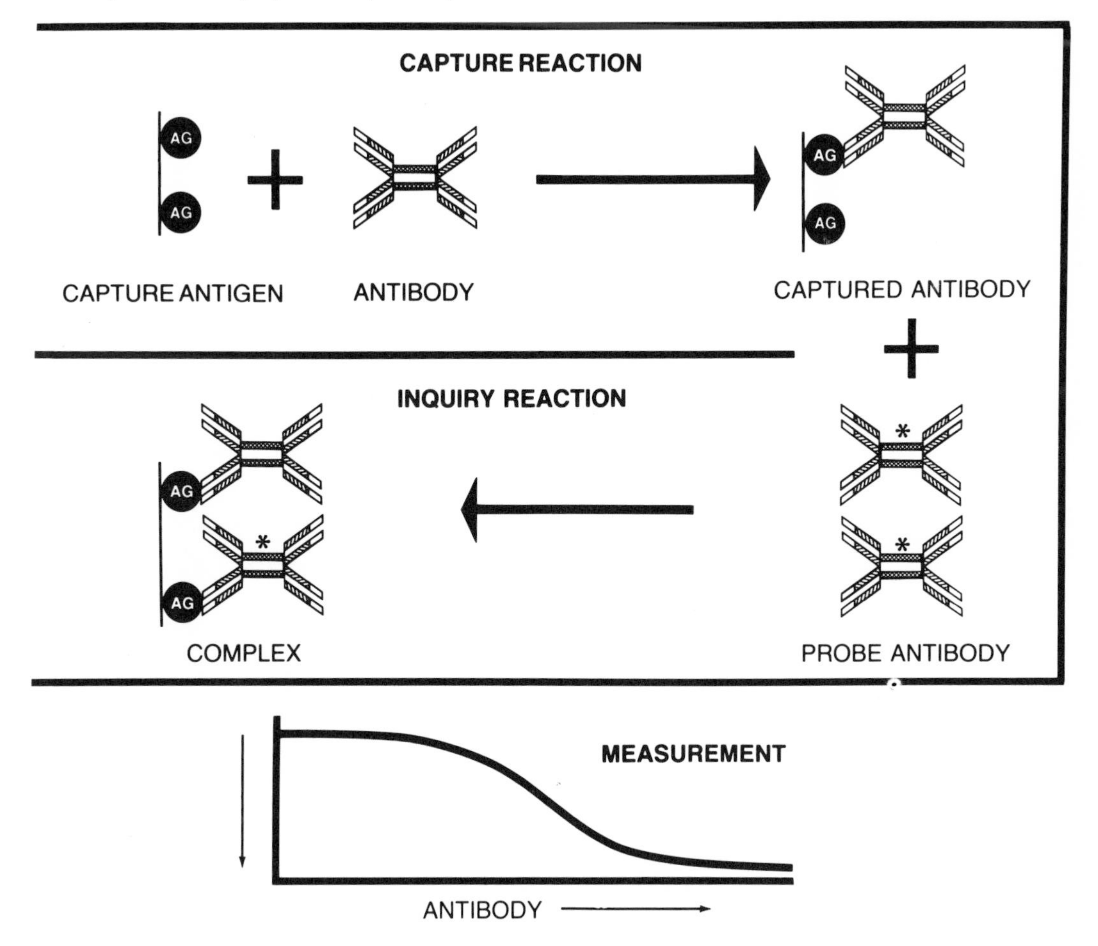

The bead is first reacted with the serum specimen to be assayed for anti-HBc. The resulting HBcAg–anti-HBc complex is challenged with purified human ^{125}I–anti-HBc IgG in a second step. The level of radioactivity bound to the bead is inversely proportional to the amount of anti-HBc in the test specimen. A competitive binding RIA is available commercially (Table 8.2). The step-by-step procedure is described by the manufacturer's package insert.

Five positive and five negative controls are tested with each run of unknown specimens. The presence or absence of anti-HBc is determined by comparing the net counts per minute of the specimen to a cutoff value, calculated as the sum of the negative control and positive control means divided by 2. Specimens whose count rates are equal to or lower than the cutoff value are considered reactive for anti-HBc.

8.6.2 Two-Step Procedure

The anti-HBe assay has been described in detail (Mushahwar et al., 1978). This was the first successful indirect solid phase RIA for the detection of serum anti-HBe utilizing

Figure 8.4. Schematic representation of a competitive solid phase sandwich assay for anti-HBe antibodies. The solid phase, coated with anti-HBe, is incubated with a standardized amount of HBeAg and a sample of the patient's serum. The anti-HBe in the patient's serum combines with the antigen and reduces the amount of antigen binding to the solid phase bound anti-HBe (*top*). The sandwich is formed by the addition of labeled anti-HBe (*middle*). The amount of labeled anti-HBe bound is inversely proportional to the amount of specific antibody present in the patient's serum (*bottom*).

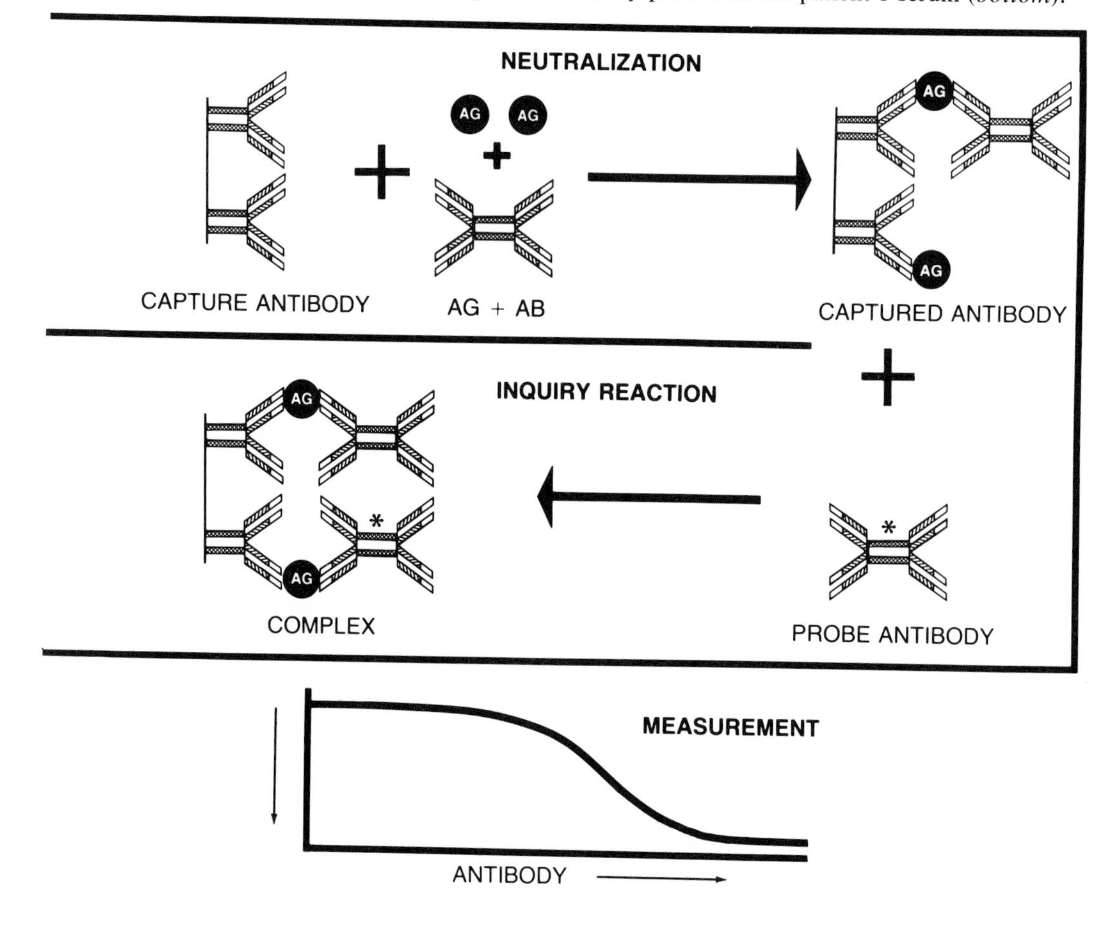

a two-step neutralization procedure. The principle of the assay is illustrated in Figure 8.4. In this assay, 0.1 ml of the patient's serum is mixed with an equal volume of standardized HBeAg-positive serum (neutralizing reagent). The mixture is incubated overnight at room temperature with the solid phase, a polystyrene bead coated with high-titered human anti-HBe serum. After washing, the bead is incubated with ^{125}I-labeled anti-HBe IgG for 4 hours at 45°C, as described under the direct solid phase HBeAg assay. The quantity of neutralizing reagent in the initial incubation has been selected to give over 10,000 counts per minute on the bead in the presence of a serum that is negative for anti-HBe. A 50% or more reduction in the count rate indicates the presence of anti-HBe in the test sample. This assay is available commercially (Table 8.2). The step-by-step procedure is described by the manufacturer's package insert. Three negative and three positive anti-HBe controls are used each time the test is run for anti-HBe analysis.

The presence or absence of anti-HBe is determined by comparing the count rate of the specimen to a cutoff value calculated from the sum of the negative control and positive control divided by 2. Specimens whose count rates are equal to or less than the cutoff value are considered reactive for anti-HBe.

8.7 ASSAY FOR IgM CLASS ANTIBODIES

Virus-specific IgM antibody has been confirmed as a prominent early immune response in many viral infections. Because it is relatively short lived, it is a good marker for acute disease (Chau et al., 1983). A reliable and reproducible RIA for the detection of hepatitis A virus IgM antibody (anti-HAV IgM) has been described (Decker et al., 1979). The anti-HAV IgM assay is based on the following reactions and is illustrated in Figure 8.5.

1. SP-Abμ + IgM → Sp-Abμ-IgM
2. SP-Abμ-IgM + HAV → SP-Abμ-IgM-HAV
3. SP-Abμ-IgM.HAV + *Ab-HAV → SP-Abμ-IgM HAV.*Ab-HAV

Where SP-Abμ is a polystyrene surface coated with μ chain-specific goat antihuman antibody. In step one, if anti-HAV IgM is present in a patient's serum, it will be bound by the μ chain-specific solid phase antibody. In step two, HAV will be attached to this complex to form the SP–Abμ–IgM–HAV. This complex is then detected in step three by incubation with the probe antibody, *Ab-HAV, an ^{125}I-labeled human anti-HAV IgG. The resulting count rate of the multiple layer product SP–Abμ–IgM–HAV.*Ab–HAV is in proportion to anti-HAV IgM concentration in the patient's serum. This test was shown to be highly specific. No crossreactions were observed in the presence of high-titered anti-HAV IgG, when hepatitis A convalescent serum samples were tested. This assay is commercially available. The step-by-step procedure is described by the manufacturer's package insert.

Two negative and three positive controls are used each time the test is run for anti-HAV IgM analysis. The presence or absence of anti-HAV IgM is determined by comparing the count rate of the specimen to a predetermined (Decker et al., 1979) cutoff value.

8.8 OTHER VIRAL MARKERS

Several recent reviews (Overby and Mushahwar, 1979; Mushahwar and Overby, 1983) have surveyed the application of RIA for rapid diagnosis of viral, bacterial, and fungal diseases. The following review describes the use of RIA to identify viral antigens or antiviral antibodies appearing in the most recent literature. Tables 8.3 and 8.4 summarize the information presented in this review.

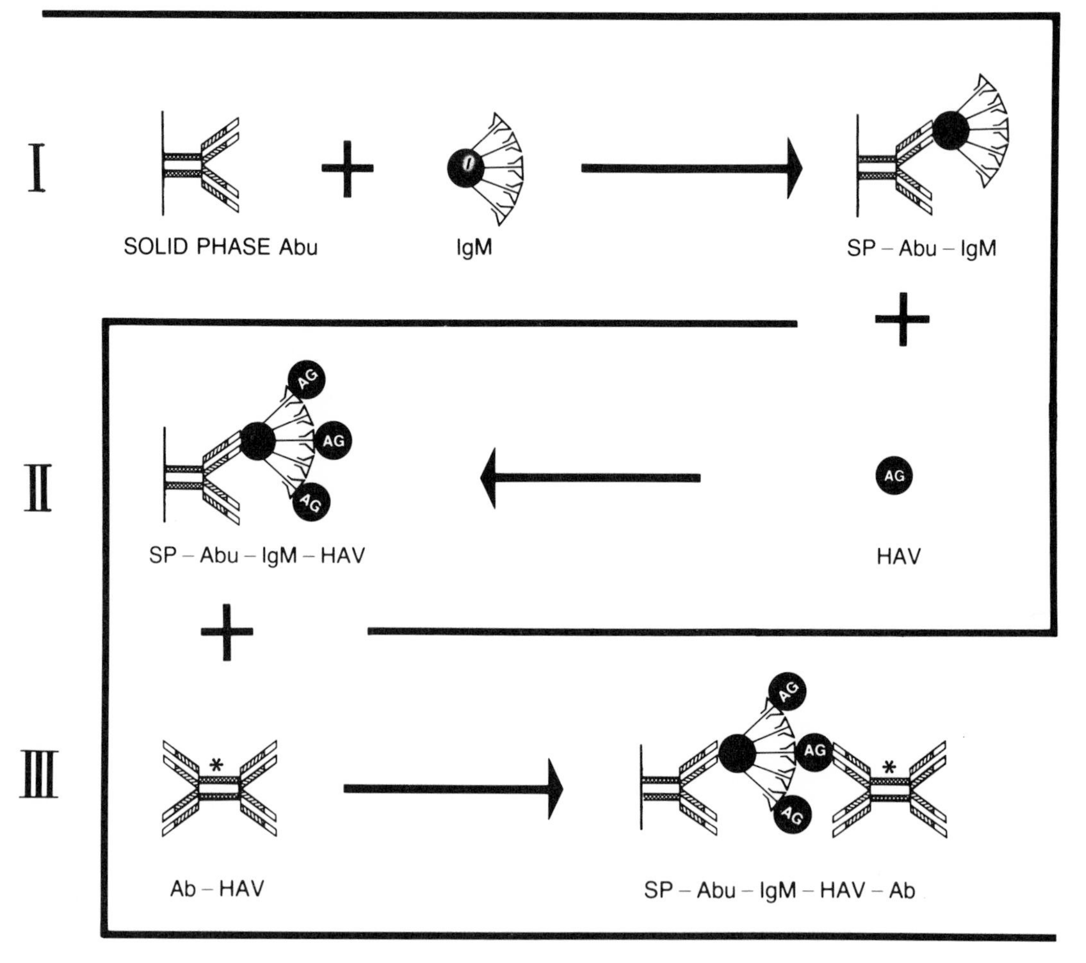

Figure 8.5. Schematic representation of a direct assay for IgM class anti-HAV antibody. The solid phase is coated with anti-human antibody specific for the μ chain. The three-step procedure involves capture of IgM antibodies by the solid phase μ chain specific antibody (*top*), subsequent binding of HAV to the captured antibody (*middle*) and binding of radiolabeled antibody to the solid-phase bound complex (*bottom*). The amount of antibody in the serum sample is directly proportional to the count rate of the washed solid phase.

8.8.1 Enteroviruses

8.8.1.a Enterovirus Antibody Detection

A solid phase IgM capture RIA designed to detect antibody specifically directed against coxsackievirus B4 and B5 has been described (Morgan-Capner and McSorley, 1983). Positive results were obtained with sera from cases of heterologous enterovirus infection. These results may be due to anamnestic IgM response or shared, common antigenic determinants. The authors note that specificity may be improved by the development of monoclonal detector antibodies.

IgM class, as well as IgG class antibodies to coxsackievirus A7, A9, A16, B2, B4, B5 or echovirus 4, 17 or 25, could be detected using RIA procedures (Torfason et al., 1984). Immunoglobulin G titers were demonstrated to be higher in the convalescent phase sera. Modification of the indirect IgM test resulted in an antibody capture assay that demonstrated type specificity. The authors note that the antigen must be

Table 8.3. Antigen Detection

Organism	Radiolabeled Reagent	Sample	Separation Technique	Reference
Rotavirus	Ab	Stool	Polystyrene bead	Sarkkinen et al., 1979
	Ab	Stool	Polystyrene bead	Haikala et al., 1983
	Ab	Stool	Microtiter plate	Cukor et al., 1978
	Ab	Stool	Microtiter plate	Cukor et al., 1984
	Ab	Stool	Polystyrene tube	Middleton et al., 1977
	Ab	Stool	Polystyrene bead	Sarkkinen et al., 1980
Norwalk Agent	Ab	Stool	Microtiter plate	Greenberg et al., 1978
Adenovirus	Ab	Stool	Polystyrene bead	Halonen et al., 1980
	Ab	NP[a]	Polystyrene bead	Sarkkinen et al., 1981
	Ab	NP	Polystyrene bead	Vesikari et al., 1982
Herpes simplex	Protein A	Virus[b]	Microtiter plate	Marsden et al., 1984
	Protein A	Cell[c]	Filter paper disc	Cleveland et al., 1979
	Ab	Virus[b]	Polystyrene tube	Enlander et al., 1976
	Ab	Cell[c]	Glass vials	Forghani et al., 1974
Respiratory viruses	Ab	NP	Polystyrene bead	Sarkkinen et al., 1981
(RSV)	Ab	NP	Polystyrene bead	Meurman et al., 1984
	Ab	NP	Polystyrene bead	Vesikari et al., 1982
Paraflu type 2	Ab	NP	Polystyrene bead	Sarkkinen et al., 1981
Paraflu type 3	Ab	NP	Polystyrene bead	Vesikari et al., 1982
Influenza	AMP	NP	Polystyrene bead	Coonrod et al., 1984
	Ab	NP	Polystyrene bead	Sarkkinen et al., 1981
	Ab	NP	Polystyrene bead	Vesikari et al., 1982

[a] Nasopharyngeal sample.
[b] Virus from cell culture.
[c] Antigens in cell culture.

carefully standardized in order to maintain the type specificity.

8.8.2 Togavirus

8.8.2.a Anti-Alphaviruses Antibody

A previous review (Mushahwar and Overby, 1983) has detailed the sensitive and specific procedures used to identify antibody to Venezuelan, Western, and Eastern equine encephalitis viruses (Jarling et al., 1978).

8.8.2.b Anti-Flaviviruses Antibody

Trent et al., (1976) have described a solid phase assay for the detection of IgG and IgM antibodies to specific St. Louis encephalitis (SLE) virus structural protein. Other investigators (Wolff et al., 1981) have extended the solid phase system to use the crude antigens now commonly used to diagnose SLE infections. Their results indicated that the procedure was as sensitive as conventional serologic tests but not as specific. The procedure was capable of differentiating SLE from similar clinical infections with alphaviruses. However, infections caused by related flaviviruses could not be accurately differentiated.

8.8.2.c Anti-Rubella Virus Antibody

The ability to diagnose acute infections with serologic methods becomes more important when the clinical signs and symptoms are subtle. The principle of diagnosing rubella

Table 8.4. Antibody Detection

Organism	Ab Species Identified	Radiolabled Reagent	Separation Technique	Reference
Enteroviruses	IgM	Ab	Polystyrene bead	Morgan-Capner and McSorley, 1983
	IgG, IgM	Ab, Ag		Torfason et al., 1984
Togaviruses				
Rubella	IgM	Ab	Microtiter plate	Kangro et al., 1981
	IgM	Ab	Polystyrene bead	Mortimer et al., 1981
	IgG, IgM	Ab	Polystyrene bead	Meurman, 1978
	IgG, IgM	Ab	Polystyrene bead	Kalimo et al., 1976
	IgG, IgM	Ab	Polystyrene bead	Meurman & Granfors, 1977
	IgM	Ab	Polystyrene bead	Tedder et al., 1982
	IgM	Ab	Red cells	Sexton et al., 1982
St. Louis encephalitis virus	IgG, IgM	Ab	Polyvinyl microtiter plate	Wolff et al., 1981
	IgG	Ab	Bead	Trent et al., 1976
Rotavirus	IgG, A, M	Ab	Polystyrene bead	Sarkkinen et al., 1979
	IgG, A, M	Ab	Filter paper	Watanabe & Holmes, 1977
	IgA	Ab	Microtiter plate	Cukor et al., 1979
Herpesviruses				
Cytomegalovirus	IgG	Ab	Polyvinyl microtiter plate	Kimmel et al., 1980
	IgM	Ab	Microtiter plate, polystyrene well	Kangro et al., 1984
Herpes simplex	IgG	Ab	Microtiter plate	Dreesman et al., 1979
	IgG	Ab	Infected cells	Smith et al., 1974
	IgG, A, M	Ab	Plastic coated bead	Patterson et al., 1978
	IgG, IgM	Ab	Polystyrene bead	Kalimo et al., 1977
	IgG, IgM	Ab	Polystyrene bead	Kalimo et al., 1977
	IgG	Ab	Polyvinyl chloride plate	Matson et al., 1983
	IgG	Ab	Polyvinyl chloride plate	Adler-Storthz et al., 1983
Varicella-Zoster	IgG	Ab	Polyvinyl microtiter plate	Friedman et al., 1979
	IgG, IgM	Ab	Polyvinyl microtiter plate	Arvin & Koropchak, 1980
	IgG	Ab	Microtiter plate	Benzie-Campbell et al., 1981
	IgM	Ab	Polystyrene tube and beads	Tedder et al., 1981
Adenovirus	IgA	Ab	Polystyrene bead	Halonen et al., 1979

virus infections by the detection of virus-specific IgM antibody in a serum sample is now well accepted. Solid phase RIA has been shown to detect IgM, as well as IgG, antibodies to rubella virus. The adsorption of purified rubella virus to polystyrene beads resulted in a highly sensitive test capable of detecting IgM antibodies (Meurman et al., 1977; Meurman, 1978; Mortimer et al., 1981) or IgM or IgG antibodies (Kalimo et al., 1976). The development of a solid phase, IgM capture procedure (Mortimer et al., 1981; Tedder et al., 1982) resulted in a high degree of sensitivity and a decreased probability of interference by rheumatoid factor. The use of monoclonal antibody directed against the hemagglutinin protein of rubella virus resulted in a strong and specific reaction when used in the MACRIA assay (Tedder et al., 1982). Evaluation of clinical samples with this system may result in greater correlation with more traditional hemagglutination assays used to evaluate the immune status. Other solid phase systems, such as microtiter plates (Kangro et al., 1981) and the use of red cells (Sexton et al., 1982) have been successfully employed to detect IgM antibodies to rubella virus.

8.8.3 Rotavirus

8.8.3.a Antigen Detection

The use of solid phase detection systems for the identification of rotavirus antigens in stool specimens has become increasingly common. The assays have taken three general forms, as follows: capture of antigen on microtiter plates (Kalica et al., 1977; Cukor et al., 1978, 1984; Greenberg et al., 1978) on polystyrene beads (Sarkkinen et al., 1979a; Haikala et al., 1983), or using polystyrene tubes (Middleton et al., 1977).

Comparison of RIA procedures (Sarkkinen et al., 1979) to latex agglutination (Haikala et al., 1983) indicate that RIA is more sensitive and less subject to interference by particulate material. The key to a sensitive and specific test for antigen identification is the selection of immune reagents of the highest quality. Cukor et al. (1978) noted that screening of antiserum by conventional procedures, such as complement fixation (CF), is not suitable or appropriate for selecting antibody capable of identifying rotavirus in stool samples with great sensitivity. The use of monoclonal antibody directed against the common group-specific antigen has resulted in a sensitive and specific test (Cukor et al. 1984).

8.8.3.b Antibody Detection

Sarkkinen et al. (1979b) described a solid phase RIA for the detection of human rotavirus-specific IgG, IgM, and IgA antibodies. This system, using viral antigens to capture specific antibodies, was found to be more sensitive than the reference CF test. Cukor et al. (1979) have described a procedure for the detection of IgA in human milk. The RIA was capable of detecting antibody for longer periods of time than enzyme immunoassay (ELISA), immunofluorescence assay (IF), or neutralization assays.

8.8.4 Norwalk Virus

8.8.4.a Antigen Detection

A second viral agent has been linked to acute episodes of diarrhea or vomiting (Cukor and Blacklow, 1984). The Norwalk virus is a small, 27-nm nonenveloped particle. Although the particle was visualized in 1972, major advances were not made until an RIA capable of detecting the agent was reported (Greenberg et al., 1978). The use of a solid phase system employing microtiter plates allows a large number of patient specimens to be screened.

8.8.5 Adenovirus

8.8.5.a Antigen Detection

Adenovirus infections may result in virus shedding from two different sites. Virus shed in feces has been identified in a manner similar to that described for rotavirus. Halonen et al. (1980) described a simple yet sensitive method for identifying adenovirus antigen in stool suspensions. Confirmatory tests have shown the procedure to be sensitive, while maintaining high specificity.

Identification of adenovirus antigens in nasopharyngeal secretions by RIA has been demonstrated (Sarkkinen et al., 1981). These investigators demonstrated complete correlation with reference to the IF assay.

8.8.6 Cytomegalovirus

8.8.6.a Antibody Detection

The presence or absence of IgG antibodies to cytomegalovirus (CMV) was determined (Kimmel et al., 1980) using an indirect RIA detection system. The procedure was capable of accurately detecting anti-CMV antibody in the presence of antibodies to other members of the herpesvirus family. A modification of the original RIA incorporating a primary 1:100 dilution resulted in a rapid screening procedure with excellent correlation with serum titration experiments (Kimmel et al., 1980).

Kangro et al. (1984) compared RIA procedures with ELISA and found comparable sensitivity with sera from adults. The RIA procedures were more sensitive, however, when cord serum was used. Results indicate that IFA procedures were less sensitive than RIA or ELISA. The identification of IgM antibodies with an indirect solid phase RIA has been described (Griffiths and Kangro, 1984). The procedure was shown to be highly specific for detecting congenital infection and sensitive for identifying primary CMV infection in pregnant women.

8.8.7 Herpes Simplex Virus

8.8.7.a Antibody Detection

Early application of RIA procedures for the detection of specific antibodies to herpes simplex virus (HSV) used virus infected, fixed monolayers to measure naturally occurring immunoglobulins (Smith et al., 1974). Extension of this procedure involved adsorption of viral antigens to the surface of a solid phase with results comparable to using monolayers (Smith et al., 1974). Refinement of serum dilution procedures and incubation times resulted in greater sensitivity (Patterson and Smith, 1973).

Improvements in the procedures, such as attachment of viral antigens to polystyrene beads (Kalimo et al., 1977c), use of specific viral antigens (Kalimo et al., 1977a) or removal of crossreacting antigens by sample adsorption (Patterson et al., 1978) resulted in lower background, greater sensitivity, and specificity.

Crossreactivity between members of the HSV group was observed to be a problem during the early phases of assay development (Kalimo et al., 1977b). Adsorption of patient serum with potential crossreacting organisms or using a solid phase coated with specific antigens resolve major problems of crossreactivity with other herpesviruses (Patterson et al., 1978; Smith and Kennell, 1981). The identification and use of specific HSV-1 and -2 glycoproteins attached to the solid phase (Matson et al. 1983) allowed identification of an immune response to either HSV-1 or HSV-2 with high levels of sensitivity and specificity (Forghani et al., 1975; Adler-Storthz et al., 1983).

8.8.7.b Antigen Detection

Generally, the identification of HSV in clinical samples has required viral isolation in cell culture and subsequent identification by reaction with fluorescent or enzyme tagged antibody. RIA technology has been applied to the detection of HSV antigens in cell culture (Forghani et al., 1974; Cleveland et al., 1979) virus containing fluids (Enlander et al., 1976) or clinical specimens (Forghani et al., 1974). Identification of viral antigens was shown to be highly sensitive and specific when clinical specimens were examined (Forghani et al., 1974).

Commonly, labeled antibodies have been used to detect specific antigens.

Staphylococcal protein A, labeled with ^{125}I, has been used to detect immune complexes (Cleveland et al., 1979). This generic approach lends itself to modification for the identification of other viral antigens.

8.8.8 Varicella-Zoster Virus

8.8.8.a Anti-Varicella-Zoster Virus Antibody

Several investigators have described methods for the detection of IgM (Tedder et al., 1981), IgG (Friedman et al., 1979; Benzie-Campbell et al., 1981; Richman et al., 1981), or IgG and IgM antibodies to varicella-zoster virus (VZV) (Arvin and Koropchak, 1980). All of the assays were shown to be more specific and sensitive than the standard CF procedures.

8.8.9 Respiratory Viruses

8.8.9.a Influenza Antigen Detection

Coonrod et al. (1984) observed that viral antigen detection was less sensitive during the first 3 days after onset of infection, and antigen could be detected longer than infectious virus. A comparison of two antigen detection systems (Sarkkinen et al., 1981) indicates that the sensitivity and specificity of RIA and IF are equivalent.

Solid phase assays for the detection of influenza A viral antigens from clinical samples have been described (Sarkkinen et al., 1981; Coonrod et al., 1984).

8.8.9.b Other Respiratory Viruses

Identification of viral antigens in the appropriate clinical specimen is the most direct method of identifying a causative agent. Sarkkinen et al. (1981) reported the use of a solid phase antigen capture RIA procedure to identify the presence of respiratory syncytial virus (RSV), parainfluenza type 2, or adenovirus in nasopharyngeal secretions. Agreement between the RIA and IF assays was very good for all viruses. Correlation between IF and RIA was independent of the sample fraction, either mucus or cells, assayed for all viruses except RSV. Lower sensitivity was observed when only the mucus fraction (devoid of cells) was tested in the case of RSV.

The ability of RIA procedures to identify viral antigens in nasopharyngeal samples using RIA and a comparison with diagnosis by other serologic methods has been described (Meurman et al., 1984). These results indicate that the ability to identify viral antigen in the sample was dependent on the time after the onset of symptoms and the age of the patient. Antigen detection was more sensitive in specimens taken from children under 6 months of age when compared with older children.

8.9 SUMMARY

Since its introduction, 25 years ago, as a highly sensitive, specific, and reproducible method for the determination of insulin levels in plasma (Yalow and Berson, 1960), RIA has been a valuable tool for the measurement of many molecules. Over the past 15 years, RIA has proven to be one of the key serologic procedures in diagnostic medicine and has been utilized in many clinical and epidemiologic investigations. The wide use of RIA in the field of infectious diseases has improved considerably our diagnostic accuracy and efficiency and enabled us to avoid diagnostic errors based on the use of less sensitive immunologic procedures.

REFERENCES

Adler-Storthz, K., Matson, D.O., Adam, E., and Dreesman, G.R. 1983. A micro solid-phase radioimmunoassay for detection of herpesvirus type-specific antibody: Specificity and sensitivity. J. Virol. Meth. 6:85–97.

Arvin, A.M., and Koropchak, C.M. 1980. Immunoglobulins M and G to *Varicella-zoster* virus measured by solid-phase radioimmunoassay: Antibody responses to varicella and *Herpes zoster* infection. J. Clin. Microbiol. 12:367–374.

Avraham, H., Golenser, J., Gazitt, Y., Spira, D. T., and Sulitzeanu, D. 1982. A highly sensitive solid-phase radioimmunoassay for the assay of *Plasmodium falciparum* antigens and antibodies. J. Immunol. Meth. 53:61–68.

Benzie-Campbell, A., Kangro, H.O., and Heath, R.B. 1981. The development and evaluation of a solid-phase radioimmunoassay (RIA) procedure for the determination of susceptibility to varicella. J. Virol. Meth. 2:149–158.

Blum, H.E., Dölken, G., and Gerok, W. 1979. Solid-phase radioimmunoassay for hepatitis B e-antigen. Klin. Wochenschr. 57:1129.

Bradley, D.W., Maynard, J.E., Hindman, S.H., Hornbeck, C.L., Fields, H.A., McCaustland, K.A., and Cook, E.H., Jr. 1977. Serodiagnosis of viral hepatitis A: Detection of acute-phase immunoglobulin M antihepatitis A virus by radioimmunoassay. J. Clin. Microbiol. 5:521–530.

Catt, K.J., and Tregear, G.W. 1967. Solid-phase radioimmunoassay in antibody coated tubes. Science 158:1570–1572.

Chau, K.H., Hargie, M.P., Decker, R.H., Mushahwar, I.K., and Overby, L.R. 1983. Serodiagnosis of recent hepatitis B infection by IgM Class Anti-HBc. Hepatology 3:142–149.

Cleveland, P.H., Richman, D.D., Oxman, M.N., Wickham, M.G., Binder, P.S., and Worthen, D.M. 1979. Immobilization of viral antigens on filter paper for a ^{125}I-staphylococcal protein A immunoassay: A rapid and sensitive technique for detection of herpes simplex virus antigens and antiviral antibodies. J. Immunol. Meth. 29:369–386.

Coonrod, J.D., Betts, R.F., Linnemann, C.C., Jr., and Hsu, L.C. 1984. Etiological diagnosis of influenza A virus by enzymatic radioimmunoassay. J. Clin. Microbiol. 19:361–365.

Cukor, G., Berry, M.K., and Blacklow, N.R. 1978. Simplified radioimmunoassay for detection of human rotavirus in stools. J. Infect. Dis. 138:906–910.

Cukor, G., and Blacklow, N.R. 1984. Human viral gastroenteritis. Microbiol. Rev. 48:157–179.

Cukor, G., Perron, D.M., Hudson, R., and Blacklow, N.R. 1984. Detection of rotavirus in human stools by using monoclonal antibody. J. Clin. Microbiol. 19:888–892.

Cukor, G., Blacklow, N.R., Capozza, F.E., Panjvani, Z.F.K., and Bednarek, F. 1979. Persistence of antibodies to rotavirus in human milk. J. Clin. Microbiol. 9:93–96.

Decker, R.H., Kosakowski, S.M., Vanderbilt, A.S., Ling, C.-M., and Overby, L.R. 1981. Diagnosis of acute hepatitis A by HAVAB-M, a direct radioimmunoassay for IgM anti-HAV. Am. J. Clin. Pathol. 76:140–147.

Decker, R.H., Overby, L.R., Ling, C.-M., Frösner, G., Deinhardt, F., and Boggs, J. 1979. Serology of transmission of hepatitis A in humans. J. Infect. Dis. 139:74–82.

Devine, R.E., Sit, F., and Larke, R. 1979. Laboratory diagnosis of acute hepatitis A virus (HAV) infection by detection of HAV-specific IgM antibody using radioimmunoassay. Can. J. Pub. Health 70:58.

Dowse, C.A., Carnegie, P.R., Linthium, D.S., and Bernard, C.C.A. 1983. Solid-phase radioimmunoassay for human myelin basic protein using a monoclonal antibody. J. Neuroimmunol. 5:135–144.

Dreesman, G.R., Matson, D.O., Courtney, R.J., Adam, E., and Melnick, J.L. 1979. Detection of herpes virus type-specific antibody by a micro solid-phase radioimmunometric assay. Intervirology 12:115–119.

Enlander, D., Remedios, L.V.D., Weber, P.M., and Drew, L. 1976. Radioimmunoassay for herpes simplex virus. J. Immunol. Meth. 10:357–362.

Faiman, C., and Ryan, R.J. 1967. Radioimmunoassay for human follicle-stimulating hormone. J. Clin. Endocrinol. 27:444–447.

Flehmig, B., Ranke, M., Berthold, H., and Gerth, H.-J. 1979. A solid-phase radioimmunoassay for detection of IgM antibodies to hepatitis A virus. J. Infect. Dis. 140:169–175.

Forghani, B., Schmidt, N.J., and Lennette, E.H. 1974. Solid-phase radioimmunoassay for identification of *Herpesvirus hominis* types 1 and 2 from clinical materials. Appl. Microbiol. 28:661–667.

Forghani, B., Schmidt, N.J., and Lennette, E.H. 1975. Solid-phase radioimmunoassay for typing herpes simplex viral antibodies in human sera. J. Clin. Microbiol. 2:410–418.

Friedman, M.G., Leventon-Kriss, S., and Sarov, I. 1979. Sensitive solid-phase radioimmunoassay for detection of human immunoglobulin G antibodies to varicella-zoster virus. J. Clin. Microbiol. 9:1–10.

Frösner, G.G., Brodersen, M., Papaevangelou, G., Sugg. Y., Haas, H., Mushahwar, I.K., Ling, C.-M., Overby, L.R., and Deinhardt, F. 1978. Detection of HBeAg and anti-HBe in acute hepatitis B by a sensitive radioimmunoassay. J. Med. Virol. 3:67–76.

Ghabrial, S.A., and Shepherd, R.J. 1980. A sensitive radioimmunosorbent assay for the detection of plant viruses. J. Gen. Virol. 48:311–317.

Ginsberg, A.L., Conrad, M.E., Bancroft, W.H., Ling, C.M., and Overby, L.R. 1973. Antibody to Australia antigen: Detection with a simple radioimmune assay, incidence in military populations, and role in the prevention of hepatitis B with gammaglobulin. J. Lab. Clin. Med. 82:317–325.

Goodfriend, T.L., and Ball, D., 1969. Radioimmunoassay of bradykinin: Chemical modification to enable use of radioactive iodine. J. Lab. Clin. Med. 73:501–511.

Greenberg, H.B., Wyatt, R.G., Valdesco, J. Kalica, A.R., London, W.T., Chanock, R.M., and Kapikian, A.Z. 1978. Solid-phase microtiter radioimmunoassay for detection of the Norwalk strain of acute nonbacterial epidemic gastroenteritis virus and its antibodies. J. Med. Virol. 2:97–108.

Greenwood, F.C., Hunter, W.M., and Glover, J.S. 1963. The preparation of ^{131}I-labeled human growth hormone of high specific radioactivity. Biochem. J. 89:114–123.

Griffiths, P.O. and Kangro, H.O. 1984. A user's guide to the indirect solid-phase radioimmunoassay for the detection of cytomegalovirus-specific IgM antibodies. J. Virol. Meth. 8:271–282.

Haikala, O., Kokkonen, J.O., Leinonen, M.K., Nurmi, T., Mantyjarvi, R., and Sarkkinen, H.K. 1983. Rapid detection of rotavirus in stool by latex agglutination: Comparison with radioimmunoassay and electron microscopy and clinical evaluation of the test. J. Med. Virol. 11:91–97.

Hales, C.N. and Randle, P.J. 1963. Immunoassay of insulin with insulin-antibody precipitate. Biochem. J. 88:137–146.

Halonen, P., Bennich, H., Torfason, E., Karlsson, T., Ziola, B., Matikainen, M.-T., Hjertsson, E., and Wesslen, T. 1979a. Solid-phase radioimmunoassay of serum immunoglobulin A antibodies to respiratory syncytial virus and adenovirus. J. Clin. Microbiol. 10:192–197.

Halonen, P., Meurman, O., Matikainen, M.T., Torfason, E., and Bennick, H. 1979b. IgG antibody response in acute rubella determined by solid-phase radioimmunoassay. J. Hyg. (Camb.) 83:69–75.

Halonen, P., Sarkkinen, H., Arstila, P., Hjertsson, E., and Torfason, E. 1980. Four-layer radioimmunoassay for detection of adenovirus in stool. J. Clin. Microbiol. 11:614–617.

Hill, H.R., and Matsen, J.M. 1983. Enzyme-linked immunosorbent assay and radioimmunoassay in the serologic diagnosis of infectious diseases. J. Infect. Dis. 147:258–263.

Hollinger, F.B., Vorndam, V., and Dreesman, G.R. 1971. Assay of Australia antigen and antibody employing double-antibody and solid-phase radioimmunoassay techniques and comparison with the passive hemagglutination methods. J. Immunol. 107:1099–1111.

Hollinger, F.B., Bradley, D.W., Maynard, J.E., Dreesman, G.R., and Melnick, J.L. 1975. Detection of hepatitis A viral antigen by radioimmunoassay. J. Immunol. 115:1464–1466.

Hollinger, F.B. and Maynard, J.E. 1976. Recent diagnostic techniques for detecting hepatitis A virus and antibody. Rush Presbyt. St. Luke's Med. Bull. 15:93–103.

Jahrling, P.B., Hesse, R.A., and Metzger, J.F. 1978. Radioimmunoassay for quantitation of antibodies to alphaviruses with staphylococcal protein A. J. Clin. Microbiol. 8:54–60.

Kalica, A.R., Purcell, R.H., Sereno, M.M.,

Wyatt, R.G., Kim, H.W., Chanock, R.M., and Kapikian, A.Z. 1977. A microtiter solid-phase radioimmunoassay for detection of the human reovirus-like agent in stools. J. Immunol. 118:1275–1279.

Kalimo, K.O.K., Meurman, O.H., Halonen, P.E., Ziola, B.R., Viljanen, M.K., Granfors, K., and Toivanen, P. 1976. Solid-phase radioimmunoassay of rubella virus immunoglobulin G and immunoglobulin M antibodies. J. Clin. Microbiol. 4:117–123.

Kalimo, K.O.K., Martilla, R.J., Granfors, K., and Viljanen, M.K. 1977a. Solid-phase radioimmunoassay of human immunoglobulin M and immunoglobulin G antibodies against herpes simplex virus type 1 capsid, envelope and excreted antigens. Infect. Immun. 15:883–889.

Kalimo, K.O.K., Martilla, R.J., Ziola, B.R., Matikainen, M.T., and Panelius, M. 1977b. Radioimmunoassay of herpes simplex and measles virus antibodies in serum and CSF of patients without infectious or demyelinating diseases of the central nervous system. J. Med. Virol. 10:431–438.

Kalimo, K.O.K., Ziola, B.R., Viljanen, M.K., Granfors, K., and Toivanen, P. 1977c. Solid-phase radioimmunoassay of herpes simplex virus IgG and IgM antibodies. J. Immunol. Meth. 14:183–195.

Kangro, H.O., Booth, J.C., Bakir, T.M.F., Tryhorn, Y., and Sutherland, S. 1984. Detection of IgM Antibodies against cytomegalovirus: Comparison of two radioimmunoassays, enzyme-linked immunosorbent assay and immunofluorescent antibody test. J. Med. Virol. 14:73–80.

Kangro, H.O., Jackson, C., and Heath, R.B. 1981. Comparison of radioimmunoassay and the gel filtration technique for routine diagnosis of rubella during pregnancy. J. Hyg. (Camb.) 87:249–255.

Kimmel, N., Friedman, M.G., and Sarov, I. 1980. Detection of human cytomegalovirus-specific IgG antibodies by a sensitive solid-phase radioimmunoassay and by the rapid screening test. J. Med. Virol. 5:195–203.

Kolb, H.J., and Grodsky, G.M. 1970. Biological and immunological activity of fructose 1,6-di-phosphatase. Application of a quantitative displacement radioimmunoassay. Biochemistry 9:4900–4906.

Lemon, S.M., Brown, C.D., Brooks, D.S., Simms, T.E., and Bancroft, W.H. 1980. Specific immunoglobulin M response to hepatitis A virus determined by solid-phase radioimmunoassay. Infect. Immun. 28:927–936.

Ling, C.M. and Overby, L.R. 1972. Prevalence of hepatitis B virus antigens as revealed by direct radioimmune assay with ^{125}I-antibody. J. Immunol. 109:834–841.

Marsden, H.S., Buckmaster, A., Palfreyman, J.W., Hope, R.G., and Minson, A.C. 1984. Characterization of the 92,000-dalton glycoprotein induced by herpes simplex type 2. J. Virol. 50:547–554.

Matson, D.O., Adler-Storthz, K., Adam, E., and Dreesman, G.R. 1983. A micro solid-phase radioimmunoassay for detection of herpesvirus type-specific antibody: Parameters involved in standardization. J. Virol. Meth. 6:71–83.

Matsui, A., Psacharopoulos, H.T., and Mowat, M.P. 1982. Radioimmunoassay of serum glycocholic acid, standard laboratory tests of liver function and liver biopsy findings: Comparative study of children with liver disease. J. Clin. Pathol. 35:1011–1017.

Meurman, O.H. 1978. Antibody in patients with rubella infection determined by passive hemagglutination, hemagglutination inhibiters, complement fixation, and solid-phase radioimmunoassay tests. Infect. Immunol. 19:369–372.

Meurman, O., and Granfors, K. 1977. Completion of hemagglutination inhibition test by solid-phase radioimmunoassay test in routine diagnostic rubella serology. Med. Biol. 55:241–244.

Meurman, O.H., Viljanen, M.K., and Granfors, K. 1977. Solid-phase radioimmunoassay of rubella virus immunoglobulin M antibodies: Comparison with sucrose density gradient centrifugation test. J. Clin. Microbiol. 5:257–262.

Meurman, O., Sarkkinen, H., Ruuskanen, O., Hanniners, P., and Halonen, P. 1984. Diagnosis of respiratory syncytial virus infection in children: Comparison of viral antigen detection and serology. J. Med. Virol. 14:61–65.

Middleton, P.J., Holdaway, M.D., Petric, M., Szymanski, M.T., and Tam, J.S. 1977.

Solid-phase radioimmunoassay for the detection of rotavirus. Infect. Immun. 16:439–444.

Morgan, C.R., and Lazarow, A. 1962. Immunoassay of insulin using a two-antibody system. Proc. Soc. Exp. Biol. Med. 110:29–32.

Morgan-Capner, P., and McSorley, C. 1983. Antibody capture radioimmunoassay (MACRIA) for coxsackievirus B4 and B5-specific IgM. J. Hyg. (Camb.) 90:333–349.

Mortimer, P.P., Tedder, R.S., Hambling, M.H., Shafi, M.S., Burkhardt, F., and Schilt, U. 1981. Antibody capture radioimmunoassay for anti-rubella IgM. J. Hyg. (Camb.) 86:139–153.

Mushahwar, I.K. and Decker, R.H. 1983. Prevalence of anti-delta in various HBsAg positive populations. In G. Verme, F. Bonino, and M. Rizzetto (eds.), Viral hepatitis and delta infection. New York: Alan R. Liss, p. 269.

Mushahwar, I.K., Dienstag, J.L., Polesky, H.F., McGrath, L.C., Decker, R.H., and Overby, L.R. 1981. Interpretation of various serological profiles of hepatitis B virus infection. Am. J. Clin. Pathol. 76:773–777.

Mushahwar, I.K., and Overby, L.R. 1983. Radioimmune assays for diagnosis of infectious diseases. In F.S. Ashkar (ed.), Radiobioassays. Boca Raton, FL: CRC Press. pp. 167–194.

Mushahwar, I.K., Overby, L.R., Frösner, G., Deinhardt, F., and Ling, C.-M. 1978. Prevalence of hepatitis B e-antigen and its antibody as detected by radioimmunoassays. J. Med. Virol. 2:77–87.

Neurath, A.R., Szmuness, W., Stevens, C.E., Strick, N., and Harley, E.J. 1978. Radioimmunoassay and some properties of human antibodies to hepatitis B core antigen. J. Gen. Virol. 38:549–559.

Neurath, A.R., Strick, N., Szmuness, W., Stevens, C.E., and Harley, E.J. 1979. Radioimmunoassay of hepatitis B e-antigen (HBeAg); Identification for HBeAg not associated with immunoglobulins, J. Gen. Virol. 42:493.

Odell, W.D., Ross, G.T., and Rayford, P.L. 1967. Radioimmunoassay for luteinizing hormone in human plasma or serum: Physiological studies. J. Clin. Invest. 46:248–255.

Overby, L.R., and Ling, C.-M. 1976. Radioimmune assay for anti-core as evidence for exposure to hepatitis B virus. Rush Presbyt. St. Luke's Med. Bull. 15:83–92.

Overby, L.R., Ling, C.-M., Decker, R.H., Mushahwar, I.K., and Chau, K. 1981. Serodiagnostic profiles of viral hepatitis. Viral Hepatitis 1981 International Symposium. W. Szmuness, H.J. Alter, and J.E. Maynard (eds.). Philadelphia: Franklin Institute Press. pp. 169–182.

Overby, L.R., and Mushahwar, I.K. 1979. Radioimmune assays. In M.W. Rytel (ed.), Rapid Diagnosis in Infectious Disease. Boca Raton, FL: CRC Press. 39–69.

Patterson, W.R., Rawls, W.E., and Smith, K.O. 1978. Differentiation of serum antibodies to herpesvirus types 1 and 2 by radioimmunoassay. Proc. Soc. Exp. Biol. Med. 157:273–277.

Patterson, W.R., and Smith, K.O. 1973. Improvement of radioimmunoassay for measurement of viral antibody in human sera. J. Clin. Microbiol. 2:130–133.

Poor, A.H., and Cutler, J.E. 1979. Partially purified antibodies used in a solid-phase radioimmunoassay for detecting candidal antigenemia. J. Clin. Microbiol. 9:362–368.

Purcell, R.H., Wong, D.C., Moritsugo, Y., Dienstag, J.L., Routenberg, J.A., and Boggs, J.D. 1976. A microtiter solid-phase radioimmunoassay for hepatitis A antigen and antibody. J. Immunol. 116:349–356.

Purcell, R.H., Wong, D.C., Alter, H.J., and Holland, P.V. 1973. Microtiter solid phase radioimmunoassay for hepatitis B antigen. Appl. Microbiol. 26:478–484.

Richman, D.D., Cleveland, P.H., Oxman, M.N., and Zaia, J.A. 1981. A rapid radioimmunoassay using ^{125}I-labeled staphylococcal protein A for antibody to varicella-zoster virus. J. Infect. Dis. 143:693–699.

Rizzetto, M., Gocke, D.J., Verme, G., Shih, J.W.-K., Purcell, R.H., and Gerin, J.L. 1979. Incidence and significance of antibodies to delta antigen in hepatitis B virus infection. Lancet ii:986–990.

Rizzetto, M., Shih, J.W.-K., and Gerin, J.L. 1980. The hepatitis B virus-associated delta antigen: Isolation from liver, development of solid-phase radioimmunoassays for delta antigen and anti-delta and partial characterization of delta antigen. J. Immunol. 125:318–324.

Robinson, K., and Smith, R.N. 1983. Methadone radioimmunoassay: Two simple methods. J. Pharm. Pharmacol. 35:566–569.

Roggendorf, M., Frösner, G.G., Deinhardt, F., and Scheidt, R. 1980. Comparison of solid-phase test systems for demonstrating antibodies against hepatitis A virus (anti-HAV) of the IgM-class. J. Med. Virol. 5:47–62.

Rutland, P.C. 1984. The development of a radioimmunoassay for the measurement of fetal haemoglobin and its use in determining the distribution of HbF in the British population. Med. Lab. Sci. 41:84–85.

Safford, S.E.S., Needleman, S.B., and Decker, R.H. 1980. Radioimmunoassay for detection of antibody to hepatitis A virus. Am. J. Clin. Pathol. 74:25–31.

Sarkkinen, H.K., Halonen, P.E., and Arstila, P.P. 1979a. Comparison of four-layer radioimmunoassay and electron microscopy for detection of human rotavirus. J. Med. Virol. 4:255–260.

Sarkkinen, H.K., Halonen, P.E., and Salmi, A.A. 1981. Detection of influenza A virus by radioimmunoassay and enzyme immunoassay from nasopharyngeal specimens. J. Med. Virol. 7:213–220.

Sarkkinen, H.K., Meurman, O.H., and Halonen, P.E. 1979b. Solid-phase radioimmunoassay of IgA, IgG and IgM antibodies to human rotavirus. J. Med. Virol. 3:281–289.

Sarkkinen, H.K., Tuokko, H., and Halonen, P.E. 1980. Comparison of enzyme-immunoassay and radioimmunoassay for detection of human rotaviruses and adenoviruses from stool specimens. J. Virol. Meth. 1:331–341.

Sexton, S.A., Hodgsen, J., and Morgan-Capner, P. 1982. The detection of rubella-specific IgM by an immunosorbent assay with solid-phase attachment of red cells (SPARC). J. Hyg. (Camb.) 88:453–461.

Shimada, N., Ushioda, K., Nagatsuka, S., Ueda, T., and Yokoshima, T. 1983. Comparison between radioreceptor assay and RIA for the determination of dihydroergotoxine in rabbit plasma samples. J. Immunol. Meth. 65:191–198.

Smedile, A., Lavarini, C., Crivelli, O., Raimondo, G., Fassone, M., and Rizzetto, M. 1982. Radioimmunoassay detection of IgM antibodies to the HBV-associated delta antigen: Clinical significance in delta infection. J. Med. Virol. 9:131–138.

Smith, K.O., Gehle, W.D., and McCracken, A.W. 1974. Radioimmunoassay techniques for detecting naturally occuring viral antibody in human sera. J. Immunol. Meth. 5:337–344.

Smith, K.O., and Kennell, W. 1981. Differentiation of members of the human herpesviridae family by radioimmunoassay. Infect. Immun. 33:491–497.

Tedder, R.S., Mortimer, P.P., and Lord, R.B. 1981. Detection of antibody to varicella-zoster virus by competitive and IgM-antibody capture immunoassay. J. Med. Virol. 8:89–101.

Tedder, R.S., Yao, J.L., and Anderson, M.J. 1982. The production of monoclonal antibodies to rubella hemagglutinin and their use in antibody-capture assays for rubella-specific IgM. J. Hyg. (Camb.) 88:335–350.

Trent, D.W., Harvey, C.L., Quereshi, A., and LeStourgeon, D. 1976. Solid-phase radioimmunoassay for antibodies to flavivirus structural and nonstructural proteins. Infect. Immun. 13:1325–1333.

Torfason, E.G., Frisk, G., and Diderholm, H. 1984. Indirect and reverse radioimmunoassays and their apparent specificities in the detection of antibodies to enteroviruses in human sera. J. Med. Virol. 13:13–31.

Torfason, E.G., Källander, C., and Halonen, P. 1981. Solid-phase radioimmunoassay of serum IgG, IgM and IgA antibodies to cytomegalovirus. J. Med. Virol. 7:85–96.

Vesikari, T., Kuusela, A.-L., Sarkkinen, H.K., and Halonen, P.E. 1982. Clinical evaluation of radioimmunoassay of nasopharyngeal secretions and serology for diagnosis of viral infections in children hospitalized for respiratory infections. Ped. Infect. Dis. 1:391–394.

Vyas, G.N., and Roberts, I.M. 1977. Radioimmunoassay of hepatitis B core antigen and antibody with autologous reagents. Vox Sang. 33:369.

Watanabe, H., and Holmes, I.H. 1977. Filter-paper solid-phase RIA for human rotavirus surface immunoglobulins. J. Clin. Microbiol. 6:319–324.

Wide, L., Kirkham, K.E., and Hunger, W.M., eds. 1971. Solid phase antigen–antibody systems. In Radioimmunoassay Methods. Edinburgh: Churchill-Livingstone, p. 405.

Witkor, R.J., Koprowski, H., and Dixon, F. 1972. Radioimmunoassay procedure for rabies binding antibodies. J. Immunol. 109:464–470.

Wolff, K.L., Muth, D.J., Hudson, B.W., and Trent, D.W. 1981. Evaluation of the solid-phase radioimmunoassay for diagnosis of St. Louis encephalitis infection in humans. J. Clin. Microbiol. 14:135–140.

Yalow, R.S., and Berson, S.A. 1960. Immunoassay of endogenous plasma insulin in man. J. Clin. Invest. 39:1157–1175.

Zollinger, W.D., Dalrymple, J.M., and Artenstein, M.S. 1976. Analysis of parameters affecting the solid-phase radioimmunoassay quantitation of antibody to meningococcal antigens. J. Immunol. 117:1788–1798.

Enzyme Immunoassay

Isabel C. Shekarchi and John L. Sever

9.1 INTRODUCTION

The enzyme-linked immunosorbent assay (ELISA) was first described by Avrameas and Guilbert (1971), Engvall and Perlmann (1971, 1972) and Van Weeman and Schuurs (1971). It is related to two older technologies, the fluorescence immunoassay (FIA) and the radioimmunoassay (RIA) described elsewhere in this manual. The principles of the three tests are similar in that antigen or antibody immobilized on a solid phase is used to separate free antigen or antibody from a specimen and the presence of the complex is determined by fluorescence (FIA), radioactivity (RIA), or a color change in an appropriate substrate (ELISA). The ELISA incorporates most of the desireable features of FIA and RIA and provides the advantage of long shelf-life and no radioactivity. The test is technically easy to perform requiring simple dilution, incubation, and washing protocols. There is no need for secondary reactions such as agglutination or complement fixation and quantitation, based on a color change due to interaction of the enzyme and substrate, can be done visually or with a colorimeter or spectrophotometer. The assay is rapid and easily automated and its reagents are safe and relatively inexpensive with a long shelf-life. ELISA has been shown to be more sensitive than hemagglutination and complement fixation and equal in sensitivity to RIA. In recent years ELISA has been applied to wide areas of research and has been thoroughly covered by reviews (Yolken, 1980; Benjamin, 1979; Schuurs and Van Weeman, 1977; Voller et al., 1982; Avrameas, 1983), books (Wardley and Crowther, 1982; Maggio, 1979; Malvano, 1979; Engvall and Pesce, 1978), and conferences (Sever and Madden, 1977).

ELISA has proved to be a useful tool in detection and quantitation of viral antigen and antibody. The indirect method for assay of antibody is diagramed in Figure 9.1 (1) and the double antibody sandwich assay for antigen is shown in Figure 9.1 (2). Theoretically, these methods, or modifications of them, can be used to identify any viral agent for which a specific immune serum can be prepared. These methods also can detect and titrate antibody to any viral antigen that can be grown in cell cultures or in an animal host. The remainder of this chapter will discuss the technical application of the heterologous ELISA as applied to the detection of viral antigens and antibodies.

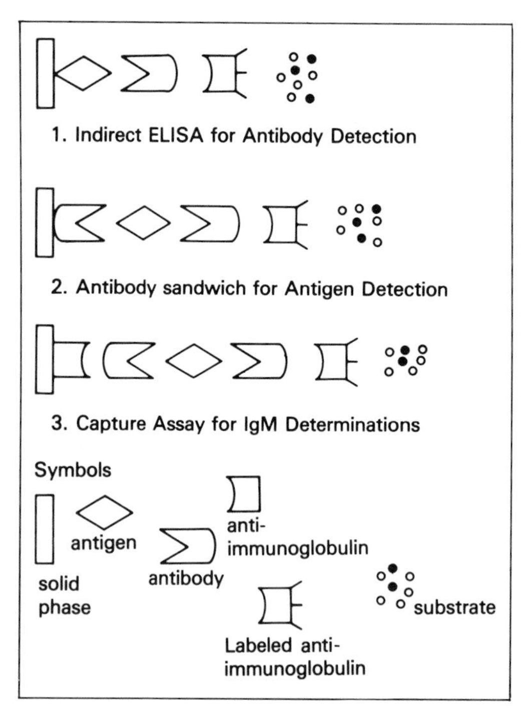

Figure 9.1. Diagram of the bonding sequence of various ELISA assay methods used in virology. In all of these procedures an incubation step follows the addition of each reactant and unbound material is removed by careful washing after each incubation.

9.2 MATERIALS

9.2.1 Solid Phase

In ELISA, the specific immunologic reaction takes place on a solid phase carrier to which one of the reactants, antigen or antibody, is bound. Microtiter plates (Voller et al., 1980), tubes (Ruitenberg et al., 1974; Leinikki and Passila, 1977), discs (Halbert and Anken, 1977), beads (Smith and Gehle, 1977), and sticks (Shekarchi et al., 1982) of glass, plastic, paper, or rubber have been used. Although all of these carriers have been successfully coated for ELISA purposes, microtiter plates of polystyrene or polyvinyl chloride are most widely used. Problems due to variations in bonding between different types of plastic, between lots and batches of plates, and even between wells on the same plate, have been reported (Shekarchi et al., 1984). In the newer plates, designed specifically for ELISA, variation has been greatly reduced but differences in lots do occur; therefore, bonding characteristics for each antigen should be checked with each new plate lot (Bidwell et al., 1977). Because not all proteins bond equally to all plastics, selection of the carrier for each new antigen must be made on an individual basis (Shekarchi et al., 1984). The reactant must retain its immunologic activity after bonding and the bond to the solid phase must be stable, reproducible, and uniform. Generally, γ-irradiated and polyvinyl chloride plates tend to have a greater protein bonding capacity than untreated polystyrene. Nonspecific bonding to these plates tends to reduce their effectiveness when using crude antigen. For crude antigens, such as cell lysates, untreated polystyrene gives less background.

Antigen or antibody diluted in coating buffer is attached to the solid phase usually by electrostatic bonding. Carbonate buffer at pH 9.6 is routinely used for most antigens and immunoglobulins, although phosphate buffered saline (PBS) at pH 7.4 to 7.8 has also been reported as a coating buffer. Purified reactants are more sensitive to differences in pH, temperature, and other bonding conditions.

Covalent bonding procedures in which the carrier surface is chemically altered to allow for a more stable bond between the reactant and the solid phase have been described (Neurath and Strick 1981; Hendry and Herrmann 1980).

9.2.2 Antigen

The type and degree of purity required for ELISA antigens is dependent on the information sought by the test (Kenny and Dunsmoor, 1983). Cell lysates, purified virions, viral subunits, or single proteins are used.

Cell lysates of viral agents are often satisfactory ELISA antigens, especially if a broad spectrum antigen providing the greatest number of antigenic determinants in one preparation is needed. Such preparations have been used in ELISA assays for measles (Forghani and Schmidt, 1979), cytomegalovirus (CMV) (Castellano et al., 1977) and mumps (Leinikki et al., 1979). These relatively crude antigens are economical and convenient. A basic procedure for preparation of cell lysates follows:

1. Virus is grown in stationary cell cultures until cytopathic effect (CPE) is 75–100%
2. Cells are washed and scraped from culture vessel into cold PBS
3. They are pelleted by 700 × *g* for 30 minutes centrifugation and resuspended in about 1/500 the original volume of PBS
4. Cells are disrupted by sonication or freezing and thawing and crude debris is removed by low speed centrifugation
5. The supernatant is then centrifuged at 25,000 rpm for 30 minutes in a Spinco SW 27.1 rotor
6. The pellet is resuspended in a small volume of PBS
7. The dilution to be subsequently used is determined by block titration and the remaining antigen is appropriately aliquoted for storage at −70°C

These antigens should always be used in conjunction with a control antigen consisting of uninfected cell cultures processed the same way as the infected cells.

Cell lysate antigens are not satisfactory for all viral agents or where greater specificity is required. Differential centrifugation on sucrose or other gradients for purification of virus and viral components has been described for rubella (Forghani and Schmidt, 1979; Gravell et al., 1977), Epstein–Barr virus (EBV) (Hopkins et al., 1982), and other agents. Control antigens for purified agents are not always available because uninfected cell cultures processed by the purification techniques often result in preparations with no protein to bond. Blocks of positive and negative sera are used with these antigens as controls.

Antigens commercially prepared for hemagglutination and complement fixation assays have been used for ELISA with mixed success. It has been suggested that residual amounts of tween in some commercial preparations blocked satisfactory bonding of the antigen.

Virus infected cells grown in microtiter plates and fixed with acetone (Nerurkar et al., 1983), or dried to the solid phase have also been used in ELISA assays (Saunders, 1977). Whatever method is used for antigen preparation, it is important that they be free of serum or other protein supplements from the culture medium and that they contain no tween, triton x, or other wetting agents that would compete for binding sites on the solid phase.

Proper concentration of the antigen to be bound to the solid phase must be determined because too little antigen reduces sensitivity and low levels of antibody are not detected. Too much antigen often results in high background with negative samples. For most assays 10 μg protein/ml of carbonate buffer is satisfactory. See Method section for titration of antigen to establish the working concentration.

9.2.3 Antibodies

Antibodies to a specific viral product or agent are bound to the solid phase in ELISA assays for the detection of viral antigens, such as viral neuraminidase in nasopharyngeal washings (Yolken et al., 1980), herpes simplex virus from vaginal swabs (Nerurkar et al., 1984), or rotavirus in stool specimens (Yolken and Stopa, 1979). Antibody bound to the solid phase is also used in ELISA assays for the detection of viral specific IgM (Forghani et al., 1983; Naot and Remington 1980). Antibodies to most classes of animal and human immunoglobulins, as well as to specific infectious agents, are commercially available either as whole serum or immunoglobulins. Monoclonal antibodies are also available on a more limited basis. These products should be of high titer and free of activity that would produce false results. Monoclonal or affinity purified antibodies (see 9.2.4.a) are therefore preferred. Antibody preparations in carbonate buffer pH 9.6 bond easily to irradiated polystyrene and polyvinyl chloride microtiter plates. Optimal antibody concentration is determined by block titration as described in the Method section.

9.2.4 Conjugates and Substrates

To detect the immunologic reaction in ELISA one of the reactants, usually antibody, is labeled with an enzyme. This enzyme label on the second antibody reacts with a substrate to produce a color change at the reaction site. Specificity of the conjugate generally is broad (anti-human IgG, anti-human IgM, etc.) so that it can be used in a variety of assays to detect antibodies directed to different infectious agents. Although specific antiviral antibodies can be labeled, in most ELISA assays for viral antigen, a second unlabeled antibody prepared in a different animal species is used. The conjugate is directed against the IgG of the second animal species. This technique adds sensitivity to the test, although it may also increase background.

9.2.4.a Purification of Antibodies

Antisera that are to be used as conjugates should be of high titer and should be purified enough to give the specificity required by the tests in which they are used. Immunoglobulins are separated from other serum components by ammonium or sodium sulfate precipitation.

Ammonium Sulfate $(NH_4)_2SO_4$ Precipitation of Immunoglobulins

1. Prepare saturate $(NH_4)_2SO_4$ at room temperature (about 54 g in 70 ml H_2O); adjust to pH 7.8 with 2M sodium hydroxide before use
2. Add dropwise 5 ml of saturate $(NH_4)_2SO_4$ to 10 ml of serum; after each drop shake to remove precipitate; finally, precipitate will remain
3. Continue shaking for 2 hours at room temperature
4. Centrifuge at room temperature for 30 minutes at 1400 × *g*
5. Discard supernatant; dissolve precipitate and make up to 10 ml with saline
6. Repeat steps 1 through 4
7. Dissolve precipitate in 4 ml PBS
8. Dialyse against several changes of PBS pH 7.4 at 4°C to remove sulfate; check for sulfate ions by adding a few drops of 10% barium chloride to a small sample of dialysate
9. When all ammonium sulfate has been removed, check for purity of antibody and for protein concentration

Immunoglobulin Purification by Affinity Chromatography

Preparation of immunosorbent

1. Swell 2 g CNBr-activated Sepharose 4B(Pharmacia) in 200 ml of 0.001 *M* HCl in a beaker (about 15 minutes)
2. Put in a small column (0.9 × 15 cm) and wash with carbonate buffer (C buffer, pH 8.2, 0.1 *M*, which consists of $NaHCO_3$ 8.4 g, NaCl 29.2 g, plus distilled H_2O to make 1 L)
3. To the washed gel in a small beaker add 5 mg of protein in C buffer (e.g., purified human γ-globulin 5 mg/ml)
4. Stir gently at room temperature for 2 hours (do not use a stirring bar)
5. Put in a small column (0.9 × 15 cm) and wash as follows:
   - PBS 20–50 ml
   - 8 *M* urea in PBS—SLOWLY! at least 2 hours gradually increasing urea concentration to 8 *M*
   - PBS—to get rid of the urea
   - Glycine HCl pH 2.5—until effluent pH is 2.5
   - PBS + 0.05 *M* 2 amino ethanol
   - PBS (+0.02% azide if not to be used immediately)

Purification on immunosorbent column

1. Wash immunosorbent with PBS in a small (0.9 × 15 cm) column
2. Add 1 ml of anti-globulin to gel; let stand 15 minutes at room temperature
3. Wash with PBS (~50 ml)
4. Elute with glycine HCl using the smallest possible void volume; collect in 0.2 *M* Tris-HCl—1 drop/tube
5. Pool the tubes containing the eluted peak and neutralize immediately with 0.2 *M* Tris
6. Concentrate to a final protein concentration of 5 mg/ml

Affinity purified immunoglobulin is conjugated to one of several enzymes used in ELISA (Johnson et al., 1980; O'Sullivan and Marks, 1981; Nilsson et al., 1981). Conjugation methods for alkaline phosphatase and for horseradish peroxidase (HRP) are presented below. Also included are methods for preparation of substrates and reaction characteristics.

9.2.4.b Preparation of Alkaline Phosphatase Conjugates

One-step Glutaraldehyde Method (Avrameas and Ternynck, 1969)

1. Centrifuge alkaline phosphase suspension (Sigma type VII) containing 5 mg of enzyme. Discard supernatant
2. Add 2 mg protein (antibody) to be labeled in 1 ml PBS and mix
3. Dialyse against several changes of PBS at 4°C overnight
4. Add 25% electron microscopy (EM) grade glutaraldehyde suspension to give a final concentration of 0.2%
5. Hold at room temperature with occasional gentle mixing for 2 hours
6. Dialyse against several changes of PBS at 4°C overnight
7. Dilute conjugate to 4 ml with Tris buffer (pH 8.0) containing 1% bovine serum albumin and add 0.2% sodium azide. Store in the dark at 4°C

Alkaline Phosphatase Substrates

Para nitrophenylphosphate (PNP)

1. Prepare diethanolamine (DEA) buffer, pH 9.8, as follows:
   - Mix 97 ml diethanolamine in 800 ml H_2O
   - Add 200 mg sodium azide (NaN_3) and 100 mg magnesium chloride ($MgCl_2$ $6H_2O$)
   - Adjust to pH 9.8 with 1 *M* HCl
   - Make up volume to 1 L with H_2O; store at 4°C in dark [for use, mix 5 mg PNP (Sigma 104–105 phosphatase substrate) in 5 ml DEA buffer; use within 30 minutes and protect from light; stop enzyme-substrate reaction with 3 *M* NaOH]
2. Positive reaction yields a yellow color
3. Reading wavelength—405 nm

4 Methylumbelliferylphosphate (4MUP)

1. Prepare DEA buffer as above
2. Stock 4MUP (10 mg MUP in 40 ml DEA; store in the dark at 4°C) (For use, mix 1 ml 4MUP stock with 9 ml DEA buffer; protect from light; Read with fluorometer; 365 nm excitation; 450 nm emission)

9.2.4.c Preparation of Horse Radish Peroxidase (HRP) Conjugates

One-step Glutaraldehyde Method (Avrameas, 1969)

1. Prepare 5 mg protein (antibody) in 1 ml of 0.1 *M* phosphate buffer pH 6.8
2. Add 12 mg HRP (Sigma Type IV RZ~3.0) and mix
3. Add 0.05 ml of 1% aqueous glutaraldehyde dropwise while stirring
4. Let stand at room temperature for 2 hours
5. Dialyse at 4°C against three changes of PBS over 12 to 24 hours
6. Centrifuge at 4°C to remove any precipitate present

7. Remove free enzyme (see below) and store

Two-step Glutaraldehyde Method (Avrameas and Ternynck, 1969; 1971)

1. Dissolve 10 mg HRP (Type IV RZ ~3.0) in 0.2 ml 0.1 *M* phosphate buffer pH 6.8 containing 1.25% EM glutaraldehyde
2. Allow to stand overnight at room temperature
3. Remove excess glutaraldehyde by dialysis against 0.15 *M* normal saline or on a sephadex G25 column equilibrated with saline—the brown fraction is the activated peroxidase
4. Concentrate to 1 ml and add 1 ml of 0.15 *M* saline containing 5 mg of the protein (antibody) to be labeled
5. Add 0.1 ml 1 *M* carbonate-bicarbonate buffer pH 9.5
6. Let stand at 4°C overnight
7. Add 0.1 ml of 0.2 *M* lysine and let stand 2 hours at room temperature
8. Dialyse against several changes of PBS at 4°C
9. Remove free enzyme (see below) and store

Schiff Base Method (Nakane and Kawaoi, 1974; Wilson and Nakane, 1978)

1. Dissolve 5 mg HRP (Type IV RZ~3.0) in 1 ml fresh 0.3 *M* sodium bicarbonate pH 8.1
2. Add 0.1 ml fresh 1% 2,4,dinitro-fluorobenzene in absolute alcohol
3. Mix gently at room temperature for 1 hour
4. Add 1 ml of 0.4–0.8 *M* sodium periodate in distilled H_2O
5. Mix gently for 30 minutes; solution should be yellow–green (if not, discard and start over with new supply of $NaIO_4$)
6. Add 1 ml 0.16 *M* ethylene glycol in distilled H_2O
7. Mix gently at room temperature for 1 hour
8. Dialyse against three changes (1 L each) of 0.01 *M* sodium carbonate buffer pH 9.5 at 4°C overnight
9. Add 5 mg protein (antibody) to 3 ml of HRPO-aldehyde solution; antibody may be in 1 ml carbonate buffer or in powder form free of salts
10. Mix gently at room temperature for 2 to 3 hours
11. Add 5 mg sodium borohydride
12. Let stand at 4°C 4 to 12 hours
13. Dialyse against PBS 4°C
14. Remove any precipitate by centrifugation
15. Remove free enzyme and store (see below)

Purification and Storage of HRP Conjugates

1. To remove free enzyme, precipitate the conjugate by adding dropwise an equal amount of cold, neutral, saturated ammonium sulfate with stirring
2. Centrifuge and wash the precipitate twice with cold half-saturate ammonium sulfate
3. Redissolve the precipitate in PBS and dialyse against several changes of PBS to remove the ammonium sulfate
4. Store at 4°C in aliquots containing 50% glycerol and 1% bovine serum albumin; they may also be frozen at −20°C or lower

Peroxidase Substrates

1. Add 50 mg *5-amino salicylic acid* to 50 ml 0.2 *M* phosphate buffer pH 6.8
2. Heat at 60°C in a water bath until clear solution is obtained
3 Add 2 mg charcoal and filter
4. When cool, add 0.5 ml of 1% hydrogen peroxide solution
5. Use within 1 hour and discard unused reagent (Stop reaction with 0.5% NaN_3 or 0.2 *M* NaOH; positive reaction yields a dark purple–brown color; reading wavelength—450 nm)

Ortho phenylenediamine (*OPD*)

1. Prepare phosphate-citrate buffer pH 5.0 (PC buffer, as follows: stock A 0.1 M citric acid 19.2 g/L; stock B 0.2 M sodium phosphate NaHPO 28.4 g/L; mix 24.3 ml stock A with 25.7 ml stock B, add 50 ml H_2O)
2. For use, mix 40 mg OPD in 100 ml of PC buffer and add 40 μl of 30% H_2O_2; use immediately! Reagent is light-sensitive. (Stop reaction with 2.5 M H_2SO_4; positive reaction is orange to reddish brown; reading wavelength—450–492 nm)

2,2′ azino-di-[3 ethyl-benzthiazoline sulfonate (6)](ABTS)

1. Prepare: stock 0.05 *M* citrate pH 4; 9.6 g citric acid in 1 L H_2O; adjust to pH 4 with 1 *M* NaOH
2. Prepare: stock 40 mM ABTS; 548.7 mg ABTS in 25 ml H_2O; store at 4°C
3. Prepare: stock 0.5 *M* H_2O_2; add 0.5 ml 8 *M* solution to 7.5 ml H_2O; store at 4°C (For use, mix 0.05 ml ABTS stock, 0.04 ml H_2O_2 stock, 10 ml citrate stock; stop reaction with 0.1 *M* HF pH 3.3; positive reaction is dark green; reading wavelength—405–414)

9.2.5 Test Specimens

ELISA for viral antigens or antibody can be performed using serum, cerebrospinal fluid (CSF), feces, milk, or washings from nasal, ear, or genital swabs. Pretreatment for nonspecific inhibitors is not usually required and nonreactive specimen is removed from the test by washing.

Rheumatoid factor is a problem in assays for specific IgM. Absorption of IgG from the serum with protein A, or aggregated IgG (Leinikki et al., 1978) greatly reduces false-positive results due to rheumatoid factor. Absorption is also useful in detecting low levels of IgM masked by competition from high levels of specific IgG.

Fecal suspensions occasionally cause capture antibody to be stripped from the solid phase. This problem may be circumvented by covalently bonding the antibody to the solid phase and by adjusting the pH of the sample suspension.

9.2.6 Buffers

9.2.6.a Coating Buffer pH 9.6

Sodium carbonate Na_2CO_3 1.59 g

Sodium bicarbonate $NaHCO_3$ 2.93 g

Sodium azide NaN_3 0.2 g

Make up to 1 L with distilled water; store at 4°C

9.2.6.b PBS Tween pH 7.4–7.6

Sodium chloride NaCl 8g

Potassium phosphate (monobasic) KH_2PO_4 0.2 g

Sodium Phosphate (dibasic) Na_2HPO_4 $12H_2O$ 2.9g

Make up to 1 L with distilled water and add 0.5 ml Tween 20; store at 4°C

9.2.6.c Sample and Conjugate Diluent

PBS 90 ml

10% BSA stock 10 ml

EDTA 0.05%

9.3 ELISA TEST METHODS

All ELISA procedures, whether the actual test or titrating reagents, have certain common methods. Washing to remove unbound reactants is probably the most important. It should be performed so that all unabsorbed reagents are removed completely. Simple emersion of the plate in wash solution is not satisfactory. Each well should be filled and emptied several times, either manually with a pipette or mechanically with a commercial washing device. The wash fluid is usually PBS containing 0.05% tween 20, although distilled water, PBS, or saline have been used with or without added tween, BSA, or serum. High salt concentration or pH below

7.2 may cause disruption of the antigen–antibody bond. Following the wash step, the empty plate should be vigorously tapped on a towel to remove excess wash solution.

Length and temperature of incubation, as well as concentration of reactants, may be varied to suit the schedule of laboratory and economy of reagents; for consistent results parameters, once set, must be followed. To assure that incubation temperature is the same in all wells and evaporation is kept at a minimum, plates may be covered and floated in a water bath or incubated in a closed plastic box containing a wet paper towel.

Optimal coating concentration for each antigen–antibody system and optimal conjugate concentration should be determined to perform sensitive and specific ELISA assays. The methods outlined below are suggested for use with flat bottom 96 well microtiter plates.

9.3.1 Determination of Optimal Conjugate Concentration

Conjugates, whether commercial or laboratory prepared, should be titrated before use.

1. Dissolve human γ-globulin (HGG) in coating buffer to a concentration of 100 ng/ml (stock HGG 1 mg in 20 ml with buffer; mix 50 μl stock in 25 ml coating buffer)
2. Add 200 μl of diluted HGG to each of 48 wells; cover, incubate at 4°C overnight
3. Empty wells and fill with PBS-Tween; let stand 10 minutes, empty and fill two more times; empty and tap to remove all wash solution
4. Dilute enzyme–anti-HGG conjugate 1:200, 1:400, 1:600, 1:800, 1:1000, 1:1200, 1:1600, and 1:2000 in conjugate diluent
5. Test each conjugate dilution in four sensitized wells; to each well add 100 μl of the diluted conjugate
6. Cover and incubate in a moist chamber 37°C for one hour
7. Wash as above (step 3)
8. Prepare substrate
9. Add 100 μl to each test well and to wells with HGG only
10. Incubate in moist chamber 37°C 30 min
11. Stop reaction and read results, either visually or with a spectrophotometer

The dilution giving a reading close to an optical density of 1.0 indicates the range to be tested with the specific ELISA system being used.

9.3.2 Determination of Optimal Antigen Dilution

1. Dilute antigen and control antigen in coating buffer 1:100, 1:200, 1:400, and 1:800
2. Across the plate (12 wells) add 200 μl per well of one antigen dilution; to the next row add the corresponding dilution of control antigen; repeat these two rows for the other antigen dilutions; cover
3. Incubate at 4°C overnight
4. Empty wells and fill with PBS Tween; after 10 minutes empty and fill the wells two more times; empty the final wash and tap plate dry
5. Dilute a positive and negative antigen specific antiserum 1:100, 1:200, 1:400, 1:800, and 1:1600; to one well of each antigen and control antigen dilution add 100 μl of the serum dilutions (eight vertical wells each contain same serum dilution); two rows of eight wells will serve as conjugate and substrate controls
6. Cover and incubate in a moist chamber, 37°C for 90 minutes
7. Wash as above (step 4)
8. Dilute conjugate in serum diluent and add 100 μl to all wells except eight for substrate control

9. Cover and incubate in a moist chamber 37°C for 60 minutes
10. Wash as above (step 4)
11. Prepare substrate for conjugate used and add 100 μl to each well
12. Cover and incubate in a moist chamber 37°C for 30 minutes
13. Stop with 50 μl of appropriate stop reagent
14. Read results visually or spectrophotometrically; the highest dilution of antigen that gives an optical density reading of 1.0 with the selected serum dilution of positive serum and a reading ≤0.1 with the same dilution of negative serum is selected

Block titrations of two or three dilutions of antigen against two or three dilutions of conjugate close to the optimal determined by the above methods with a single dilution of one positive and one negative sample will indicate optimal concentration of both for a specific test. The final dilutions to be used can be slightly varied without a significant effect on the sensitivity or specificity of the test. This may be useful, if it is necessary to conserve any of the reagents.

9.3.3 Determination of Optimal Antibody Dilution

1. Dilute antibody 1:500, 1:000, 1:1500, and 1:2000 in coating buffer
2. For each antibody dilution add 200 μl per well to 12 wells (across the plate

3–4. As in 9.3.2

5. Prepare antigen dilutions such as 400, 200, 100, 50, and 25 plaque forming units of virus
6. Add the antigen dilutions to the sensitized wells so that all antibody dilutions are tested, 100 μl per well

7–8. As in 9.3.2, steps 6 and 7

9. Add labeled specific antibody (conjugate) 100 μl per well. If labeled specific antibody is not available, use a specific antibody from a different animal source and a conjugate suitable to that animal source.

10–14. As in 9.3.2, steps 9–14

15. Select antibody dilution which offers the most sensitive detection of virus antigen

9.3.4 ELISA for Detection of Antiviral Antibody

1. Sensitize the plates; add antigen and control antigen (optimal dilution in coating buffer) to alternate rows 200 μl per well

2–3. As in 9.3.2, steps 3 and 4

4. Add sample diluted in sample diluent. Include samples of known positive and negative control specimens. Include conjugate and substrate control wells—diluent only at this step. 100 μl per well.
5. As in 9.3.2, steps 6–14

9.3.5 ELISA for Detection of Viral Antigens

1. Sensitize the plates; add optimal dilution of antibody in coating buffer 200 μl per well

2–3. As in 9.3.2, steps 3 and 4

4. Add samples and known positive and negative controls diluted in sample diluent; test plate should also include wells for conjugate and substrate control; 100 μl per well

5–6. As in 9.3.2, steps 6 and 7

7. Add optimal dilution of enzyme labeled specific antiviral antibody (Alternate: if labeled specific antiviral antibody is not available, antibody to the virus prepared in a different species from the coating antibody may be used; add optimal dilution to test wells and include a control well without test sample; 100 μl per well

cover and incubate 37° 60′ then proceed as in 9.3.2 steps 7–14)

8–16. As in 9.3.2, steps 9–4. See 9.4 for quantitation.

9.3.6 Detection of Antiviral IgM

Two methods for detection of IgM are currently used. The first is a modification of section 9.3.4 above. Before the serum sample is added to the virus coated well, IgG is removed by mixing the serum with protein A (Leinikki et al., 1978) or by sucrose gradient centrifugation or by column chromatography. We prefer absorption with staffinoc, a commercially available preparation containing staphylococcus protein A and streptococci, which removes all IgG, including IgG_3, and IgA (Kronvall et al., 1979). The assay is performed as described using a class specific anti-IgM conjugate. To assay the quality of the IgG removal, the assay is also performed on absorbed and unabsorbed serum using an anti-IgG conjugate.

The other method is a capture assay similar to 9.3.5 (Naot and Remington, 1980), see Figure 9.1(3). The solid phase is coated with anti-species (e.g., human) IgM. The specimen is added and IgM is bound by the attached antibody. After washing away unbound antibody the viral antigen is added. Labeled antiviral antibody is used to detect the presence of virus and, therefore, antiviral IgM in the specimen. Recent reports (Forghani et al., 1983) indicate that removal of rheumatoid factor by absorption of the serum with anti-IgG or aggregated IgG is also necessary for the capture assay.

9.4 QUANTITATION

The presence and quantity of bound enzyme conjugate is indicated by a color change in the substrate and is read by eye or by a spectrophotometer. There is no standard method for reporting ELISA results. On the basis of visual reading, results may be reported as positive or negative with the cutoff value equal to background color. Endpoint titer, which may be read visually or by photometer, is reported as the last sample dilution with color greater than the negative control or background.

Optical density (OD) readings are less subjective and allow interpolation of titer or relative ELISA units from a single sample dilution. By one technique three or four standards of known value are run along with the samples at a single predetermined dilution. Absorbance of the standards plotted against the known values (e.g., titer) define the curve on which test samples can be evaluated (Parker et al., 1976). By another technique, absorbance yielded by a test sample is compared with absorbance yielded by a positive and negative control sample at the same dilution. The difference between positive and negative readings is divided into units, for example, 100 for high positive and 0 for negative. Readings between the two are reported in relative units (Siegel and Remington, 1983).

Results have also been reported as optical density readings (Voller et al., 1980), effective dose (ED) (Leinikki and Passila, 1977), or area under the curve (Murphy et al., 1980).

The use of photometric readers has made analysis of samples in a single dilution practical and in addition, automation of reading and data manipulations has many advantages. Analysis of a single dilution saves technician time, reagents, and the amount of samples required for testing. It has also been reported that the single dilution method lowers variability and increases reproducibility of test results (Siegel and Remington, 1983).

9.5 INSTRUMENTS

A wide variety of automated or semiautomated equipment for ELISA assays is now available. Included are single and multiple pipetting devices, aspirating and washing machines, and spectrophotometers with or without programmable data reduction (Sever, 1983). The amount of automated equipment that might be useful in a laboratory is dependent on the number and variety of tests to be run each day. The pipetting and washing devices are particularly useful, regardless of the number of tests run. Pipetting equipment should be carefully selected for accuracy and precision. Disposable tips should fit well and periodic checks for accuracy should be made. Aspirating and/or washing devices need monitoring at all times and should be constructed so that this is possible. Each cycle, all wells should be filled and the wash jet should not be so strong that bound reagents are dislodged or wash solution is splashed from one well to the next. Aspiration of reagents from the wells must be thorough but gentle. Strong air flow can cause drying and dislodging of reactants. Hand emptying and tapping is preferred in our laboratory. The spectrophotometer provides a rapid evaluation of the test which is less subjective than visual readings and a simple data reduction program, which can be applied to a number of different assays, is particularly useful. Fully automated systems are developed for large volume use and usually are not applicable in a research or small facility.

Kits for determination of viral antigen and antibody by ELISA and instruments dedicated to their performance are commercially available. The following is a partial list of manufacturers of such products:

Hepatitis A and B, rubella, rotavirus: Abbott Laboratories, N. Chicago, IL 60064

CMV, herpes simplex, varicella zoster, measles, mumps: Calbiochem-Behring Corp., 109333 N. Torrey Pines Rd., La Jolla, CA 92037

Hepatitis A and B, rubella, EBV: Cordis Laboratories, Inc., Box 523580, Miami, FL 33152

Rubella, CMV, herpes, human T-cell leukemia virus: Litton Bionetics, 2020 Bridge View Drive, Charleston, SC 29405

Rubella, herpes simplex, CMV, mumps: Whittaker, M.A. Bioproducts, Bldg. 100, Biggs Ford Rd., Walkersville, MD 21793

9.6 APPLICATIONS OF ELISA IN VIROLOGY

Use of ELISA for detection of viral antigen and antibodies is so widespread that to list specific applications would be a massive task. A literature search for the past 5 years on almost any known virus will reveal at least a few attempts to apply ELISA to its study.

9.7 COMMENTS AND SOURCES OF ERROR

The ELISA assay is deceptively simple but, unless attention is given to every detail, sensitivity and reproducibility will be lost. The interaction of solid phase, antigen, antibody, enzyme conjugate, and substrate under a defined set of conditions, time, temperature, and concentration, has been standardized to react in a measured way and variation in any one element can change the outcome of the assay.

The solid phase should be selected and periodically monitored for bonding capacity because plates which vary only by batch or lot may have altered bonding for a given protein.

Reagents must be properly prepared and stored to avoid contamination or deterioration. Improper pH or osmolality of reagents due to improper dilution or mixing may impair the performance of the system. For use, reagents and wash solution should be warmed to room temperature; for storage they should be refrigerated. Freezing alkaline phosphatase conjugates causes deterioration but peroxidase conjugate may be frozen.

Because test volumes are small, accurate and consistent pipetting is essential. Good equipment properly handled and regularly monitored should be used. Correct dilution and complete mixing of reactants and reagents is important at each step.

Washing should be thorough but not rough enough to remove bound material. Tapping the plate on a dry towel will remove excess wash solution which could dilute the next reagent added.

Incubation temperatures determined for a specific test system must be maintained because higher or lower temperatures will distort the results. For this reason, incubation at room temperature, which may show wide fluctuations, is not reliable.

The assays should be protected from light during the substrate incubation because most substrates are light sensitive. They should also be protected from exposure to sodium hypochlorite vapors which emanate from waste disposal vessels because the vapor inactivates the enzyme.

ELISA technology is still in a state of rapid development. New carriers, bonding techniques, conjugates, substrates, and instruments continue to be introduced. Standardization of test procedures and quantitation methods rarely extend beyond the individual laboratories. Production and monitoring of commercial kits and widespread use of ELISA in clinical diagnostic laboratories will be greatly enhanced by some standardization.

ELISA is a safe, simple and rapid method to detect viral antigen and antibody. It is easily automated and the potential for its becoming a universal tool is attractive. With the development of pools of standard antigens, antisera and conjugates, quantitation, and interlaboratory communication of results will be simplified.

REFERENCES

Avrameas, S. 1969. Coupling of enzymes to proteins with glutaraldehyde. Use of the conjugates for detection of antigens and antibodies. Immunochemistry 6:43–52.

Avrameas, S. 1983. Enzyme immunoassays and related techniques: Development and limitations. Curr. Topics in Microbiol. Immunol. 104:93–99.

Avrameas, S., and Guilbert, B. 1971. A method for quantitative determination of cellular immunoglobulins by enzyme-labelled antibodies. Eur. J. Immunol. 1:394–396.

Avrameas, S., and Ternynck, T. 1969. The cross-linking of proteins with glutaraldehyde and its use for the preparation of immunosorbents. Immunochemistry 6:53–66.

Avrameas, S., and Ternynck, T. 1971. Peroxidase labelled antibody and Fab conjugates with enhanced intracellular penetration. Immunochemistry 8:1175–1179.

Benjamin, D.R. 1979. Immunoenzymatic methods. In E.H. Lennette and N.J. Schmidt (eds.) Diagnostic Procedures for Viral Rickettsial and Chlamydial Infections. Washington, D.C.: APHA Inc, pp. 153–170.

Bidwell, D.E., Bartlett, A., and Voller, A. 1977. Enzyme immunoassays for viral diseases. J. Infect. Dis. 136:S274–S278.

Castellano, G.A., Hazzard, G.T., Madden, D.L., and Sever, J.L. 1977. Comparison of the enzyme-linked immunosorbent assay and the indirect hemagglutination test for detection of antibody to cytomegalovirus. J. Infect. Dis. 136:S337–S340.

Engvall, E., and Perlman, P. 1971. Enzyme-linked immunosorbent assay (ELISA). Quantitative assay of immunoglobulin G. Immunochemistry 8:871–874.

Engvall, E., and Perlman, P. 1972. Enzyme-linked immunosorbent assay. III. Quantitation of specific antibodies by enzyme-labelled anti-immunoglobulin in antigen-coated tubes. J. Immunol. 109:129–136.

Engvall, E., and Pesce, A.J., eds. 1978. Quantitative Enzyme Immunoassay. Blackwell Scientific Publications. Scand. J. Immunol. 8:Suppl. 7.

Forghani, B., and Schmidt, N.J. 1979. Antigen

requirements, sensitivity and specificity of enzyme immunoassays for measles and rubella viral antibodies. J. Clin. Microbiol. 9:657–664.

Forghani, B., Myoraku, C., and Schmidt, N. 1983. Use of monoclonal antibodies to human immunoglobulin M in "capture" assays for measles and rubella immunoglobulin M. J. Clin. Microbiol. 18:652–657.

Gravell, M., Dorsett, P.H., Gutenson, O., and Ley, A.C., 1977. Detection of antibody to rubella virus by enzyme-linked immunosorbent assay. J. Infect. Dis. 136:S300–S303.

Halbert, S.P., and Anken, M. 1977. Detection of hepatitis B surface antigen (HB_S Ag) with use of alkaline phosphatase labeled antibody to HB_SAg. J. Infect. Dis. 136:S318–S323.

Hendry, R.M., and Herrmann, J.E. 1980. Immobilization of antibodies on nylon for use in enzyme-linked immunoassay. J. Immunol. Meth. 35:285–296.

Hopkins, R.F. III, Witmer, T.J., Neubauer, R.H., and Rabin, H. 1982. Detection of antibodies to Epstein–Barr virus antigens by enzyme-linked immunosorbent assay. J. Infect. Dis. 146:734–740.

Johnson, R.B., Libby, R.M., and Nakamura, R.M. 1980. Comparison of glucose oxidase and peroxidase as labels for antibody in enzyme-linked immunosorbent assay. J. Immunoassay 1:27–37.

Kenny, G.E., and Dunsmoor, C.L. 1983. Principles, problems and strategies in the use of antigenic mixtures for enzyme-linked immunosorbent assay. J. Clin. Microbiol. 17:655–665.

Kronvall, G., Simmons, E.B., Myhre, E.B., and Jonsson, S. 1979. Specific absorption of human serum albumin, immunoglobulin A, and immunoglobulin G with selected strains of group A and group G streptococci. Infect. Immun. 25:1–10.

Leinikki, P.O., and Passila, S. 1977. Quantitative, semiautomated enzyme-linked immunosorbent assay for viral antibodies. J. Infect. Dis. 136:S294–S299.

Leinikki, P.O., Shekarchi, I., Dorsett, P., and Sever, J.L. 1978. Determination of virus-specific IgM antibodies by using ELISA: Elimination of false-positive results with protein-A sepharose absorption and subsequent IgM antibody assay. J. Lab. Clin. Med. 92:849–857.

Leinikki, P.O., Shekarchi, I., Tzan, N., Madden, D.L., and Sever, J.L. 1979. Evaluation of enzyme-linked immunosorbent assay (ELISA) for mumps virus antibodies. Proc. Soc. Exp. Biol. Med. 160:363–367.

Maggio, E.T., ed. 1979. Enzyme-Immunoassay. Boca Raton, FL: CRC Press.

Malvano, R., ed. 1979. Immunoenzymatic Assay Techniques (Developments in Clinical Biochemistry, Vol. 1). Hague: Martinus Nijhoff Publishers.

Murphy, B.R., Tierney, E.L., Barbour, B.A., Yolken, R.H., Alling, D.W., Holley, H.P., Mayner, R.E., and Chanock, R.M. 1980. Use of enzyme-linked immunosorbent assay to detect serum antibody responses of volunteers who received attenuated influenza A virus vaccines. Infect. Immun. 29:342–347.

Nakane, P.K., and Kawaoi, A. 1974. Peroxidase-labeled antibody. A new method of conjugation. J. Histochem. Cytochem. 22:1084–1091.

Naot, Y., and Remington, J.S. 1980. An enzyme-linked immunosorbent assay for detection of IgM antibodies to *Toxoplasma gondii*: Use for diagnosis of acute acquired toxoplasmosis. J. Infect. Dis. 142:757–766.

Nerurkar, L.S., Jacob, A.J., Madden, D., and Sever, J.L. 1983. Detection of genital herpes simplex infection by a tissue culture-fluorescent-antibody technique with biotin–avidin. J. Clin. Microbiol. 17:149–154.

Nerurkar, L.S., Namba, M., Brashears, G., Jacob, A.J., Lee, Y.J., and Sever, J.L. 1984. Rapid detection of herpes simplex virus in clinical specimens by use of a capture biotin-streptavidin enzyme-linked immunosorbent assay. J. Clin. Microbiol. 20:109–114.

Neurath, A.R., and Strick, N. 1981. Enzyme-linked fluorescence immunoassays using β-galactosidase and antibodies covalently bound to polystyrene plates. J. Virol. Meth. 3:155–165.

Nilsson, P., Bergquist, N.R., and Grundy, M.S. 1981. A technique for preparing defined conjugates of horseradish peroxidase and immunoglobulin. J. Immunol. Meth. 41:81–93.

O'Sullivan, M.J., and Marks, V. 1981. Methods for the preparation of enzyme-antibody *conjugates* for use in enzyme immunoassay. Enzymology 73:147–166.

Parker, J.C., O'Beirne, A., and Collins, M.J. 1979. Sensitivity of enzyme-linked immunosorbent assay, complement fixation and hemagglutination inhibition serological tests for detection of sendai virus antibody in laboratory mice. J. Clin. Microbiol. 9:444–447.

Ruitenberg, E.J., Steerenberg, P.A., Brosi, B.J.M., and Buys, J. 1974. Serodiagnosis of *Trichinella spiralis* infections in pigs by enzyme-linked immunosorbent assays. Bull. WHO 51:108–109.

Saunders, G.C. 1977. Development and evaluation of an enzyme-labeled antibody (ELA) test for the rapid detection of hog cholera antibodies. Am. J. Vet. Res. 38:21–25.

Schuurs, A.H.W.M., and Van Weeman, B.K. 1977. Enzyme-immunoassay. Clin. Chim. Acta 81:1–40.

Sever, J.L. 1983. Automated systems in viral diagnosis. In M. Cooper, P.H. Hofschneider, H. Koprowski, F. Melchers, R. Rott, H.G. Schweiger, P.K. Vogt, and R. Zinkernagel (eds.), Current Topics in Microbiology. Berlin: Springer-Verlag, pp. 57–75.

Sever, J.L., and Madden, D.L., eds. 1977. Enzyme-linked Immunosorbent Assay (ELISA) for Infectious Agents. J. Infect. Dis. 136:S257–S340.

Shekarchi, I.C., Sever, J.L., Ward, L.A., and Madden, D.L. 1982. Microsticks as solid-phase carriers for enzyme-linked immunosorbent assays. J. Clin. Microbiol. 16:1012–1018.

Shekarchi, I.C., Sever, J.L., Lee, Y.J., Castellano, G., and Madden, D.L. 1984. Evaluation of various plastic microtiter plates with measles, toxoplasma and gamma globulin antigens in enzyme-linked immunosorbent assays. J. Clin. Microbiol. 19:89–96.

Siegel, J.P., and Remington, J.S. 1983. Comparison of methods for quantitating antigen-specific immunoglobulin M antibody with a reverse enzyme-linked immunosorbent assay. J. Clin. Microbiol. 18:63–70.

Smith, K.O., and Gehle, W.D. 1977. Magnetic transfer devices for use in solid-phase radioimmunoassays and enzyme-linked immunosorbent assays. J. Infect. Dis. 136:S329–S336.

Van Weeman, B.K., and Schuurs, A.H.W.M. 1971. Immunoassay using haptoenzyme conjugates. FEBS Lett. 24:77–81.

Voller, A., Bidwell, D.E., and Bartlett, A. 1980. Microplate enzyme immunoassays for the immunodiagnosis of virus infections. In N.R. Rose and H. Friedman (eds.), Manual of Clinical Immunology. Washington, D.C.: American Society of Microbiologists, pp. 359–371.

Voller, A., Bidwell, D.E., and Bartlett, A. 1982. ELISA techniques in virology. In C.R. Howard (ed.), New Developments in Practical Virology. New York: Alan R. Liss, pp. 59–81.

Wardley, R.C. and Crowther, J.R. eds. 1982. The ELISA: Enzyme-linked Immunosorbent Assay in Veterinary Research and Diagnosis. (Current Topics in Veterinary Medicine and Animal Science, Vol 22). Hague: Martinus Nijhoff Publishers.

Wilson, M.B., and Nakane, P.K. 1978. Recent developments in the periodate method of conjugating horseradish peroxidase. In W. Knapp, K. Holubar, and G. Wick (eds.), Immunofluorescence and Related Staining Techniques. Elsevier/North Holland, Biomedical Press, pp. 215–224.

Yolken, R.H. 1980. Enzyme-linked immunosorbent assay (ELISA): A practical tool for rapid diagnosis of viruses and other infectious agents. Yale J. Biol. Med. 53:85–92.

Yolken, R., Torsch, V., Berg, R., Murphy, B., and Lee, Y.C. 1980. Fluorometric assay for measurement of viral neuraminidase-application to the rapid detection of influenza virus in nasal wash specimens. J. Infect. Dis. 142:516–523.

Yolken, R., and Stopa, P. 1979. Enzyme-linked fluorescence assay: Ultrasensitive solid-phase assay for detection of human rotavirus. J. Clin. Microbiol. 10:317–321.

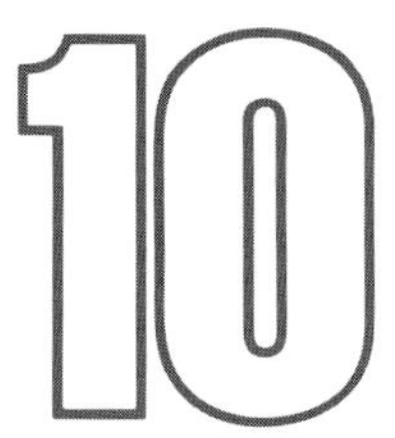

Immunoperoxidase Detection of Viral Antigens in Cells

Helen Gay and John J. Docherty

10.1 INTRODUCTION

Immunocytochemical staining, a sensitive and specific method for the localization of antigens with labeled antibodies, has been used extensively to detect viral antigens. Coons introduced immunocytochemistry in 1942 by using fluorescein labeled anti-pneumococcal III antibody to detect pneumococcal antigens in the livers and the spleens of experimentally infected mice (Coons et al., 1941; Coons et al., 1942). In the following years, immunofluorescence was used to detect many different bacterial and viral antigens in vivo and in vitro (Coons et al., 1950; Kaplan et al., 1950; Kurstak, 1971). Information gathered from such studies advanced our knowledge of the structure and function of viral proteins, transforming mechanisms of viruses, and viral infectious cycles. Indeed, immunofluorescence (IF) continues to be used successfully in a variety of ways, in both the research and clinical laboratory.

As with any procedure there has been a continuing quest to improve on the original procedures of Coons. By the mid 1960s a procedure began to emerge that used the basic methodology of previously described fluorescent methods but the fluor was replaced by an enzyme. When used to detect virus antigens, a precipitate formed marking the location of the antibody–antigen reaction. This immunoenzymatic procedure, which was reported by Ram et al. (1966) used acid phosphatase conjugated antibodies to localize tissue antigens. Because of rapid loss of activity of that antibody–enzyme conjugate (Nakane and Pierce, 1967), however, acid phosphatase was soon replaced by the more stable enzyme horseradish peroxidase (HRP). Introduced by Nakane and Pierce, immunoperoxidase staining involves reacting the antigen with HRP conjugated antibody, then visualizing the complex by exposing the tissue to a solution containing appropriate electron transfer substrates, such as 3,3′-diaminobenzidine tetrahydrochloride (DAB) and hydrogen peroxide (Nakane and Pierce, 1967). The bound enzyme first reduces the hydrogen peroxide to water and then oxidizes DAB. The resulting oxidized DAB polymerizes to form a nondiffusing, insoluble dark brown precipitate that settles at the site of the antigen (Pearse, 1972). This immunoenzymatic staining method overcame many of the weaknesses of immunofluorescence. The peroxidase stained preparations, once mounted, are essentially permanent, can be viewed with an ordinary light microscope, and avoid cellular autofluorescence (Spendlove, 1967). The peroxidase enzyme is available in pure form, is stable after chemical conjugation and because it is small it readily penetrates tissue (Sternberger, 1974). The practical advantages offered by the immunoperoxidase staining method have made it an attractive procedure for the detection of viral antigens. Peroxidase, however, does have one major drawback. It is endogenous to some mammalian tissues (Blain et al., 1975; Straus, 1971; Streefkerk, 1972; Weir et al., 1974), which results in nonspecific staining that can interfere with the interpretations of immunoperoxidase staining test results. This problem can be solved by eliminating endogenous peroxidase by pretreatment of tissues with peroxidase inactivating reagents (Straus, 1971, 1979; Streefkerk, 1972; Weir et al., 1974). Most of the endogenous enzyme activity is eliminated by this treatment, but some investigators have reported that this treatment also damages antigens (Fink et al., 1979; Straus, 1979).

During the years of refinements of both the IF and immunoperoxidase staining techniques, improvements were introduced that greatly increased the sensitivity of these staining methods. Coons had first introduced IF as a direct stain in which the fluorochrome was directly attached to the specific—or primary—antibody (Coons et al., 1941, 1942). Twelve years later, Weller and Coons showed that staining sensitivity could be increased by tenfold if the tissue was first treated with unlabeled specific antibody and then treated with fluorescein labeled anti-immunoglobulin (e.g., anti-IgG) antibody (Weller and Coons, 1954). Through this indirect staining, several anti-IgG molecules could attach to a single bound primary antibody, thus, greatly amplifying the detection of the antigen (Sternberger, 1974). In addition, indirect staining made the technique more general because a single preparation of fluorochrome labeled anti-IgG was used to detect antibodies raised in the same species to different antigens and avoided the necessity of labeling each specific antibody.

The indirect staining method has been applied to immunoperoxidase staining, but two problems have been realized. First, the chemical conjugation process inactivates some of the HRP that is coupled to the antibody; second, some of the antibody remains unconjugated. Both the unconjugated antibody and the antibody conjugated to inactive enzyme compete with antibody molecules labeled with active enzyme for the antigenic sites and, thus, decrease the sensitivity of the assay. In 1970, Sternberger et al. (1970) introduced an immunoperoxidase procedure that required no chemical conjugation of enzyme to antibody. Peroxidase was bound to an antiperoxidase antibody via specific antibody–antigen interactions. Viral antigens were localized by this unlabeled antibody–enzyme technique, schematically presented in Figure 10.1, when the specimen was treated sequentially with the following reagents:

1. Antibody to a specific antigen raised in species A (primary antibody)
2. Anti-species-A serum raised in species B (bridge antibody) applied in sufficient excess that one binding site of the antibody attaches to the primary antibody and the other binding site remains free
3. A purified antibody–enzyme complex made up of anti-HRP raised in species A that had been combined with HRP
4. DAB and peroxide

This method, known as the unlabeled peroxidase anti-peroxidase (PXAPX) technique, is 20 times more sensitive than the indirect immunoperoxidase staining method and 100 to 1000 times more sensitive than IF (Pearson et al., 1979). Sensitivity is greatly increased because each viral antigen is attached to at least one, if not more, active enzyme molecules.

Immunoperoxidase has been used to detect many different viruses including enteroviruses (Herrmann et al., 1974), influenza (Gardner et al., 1978), respiratory syn-

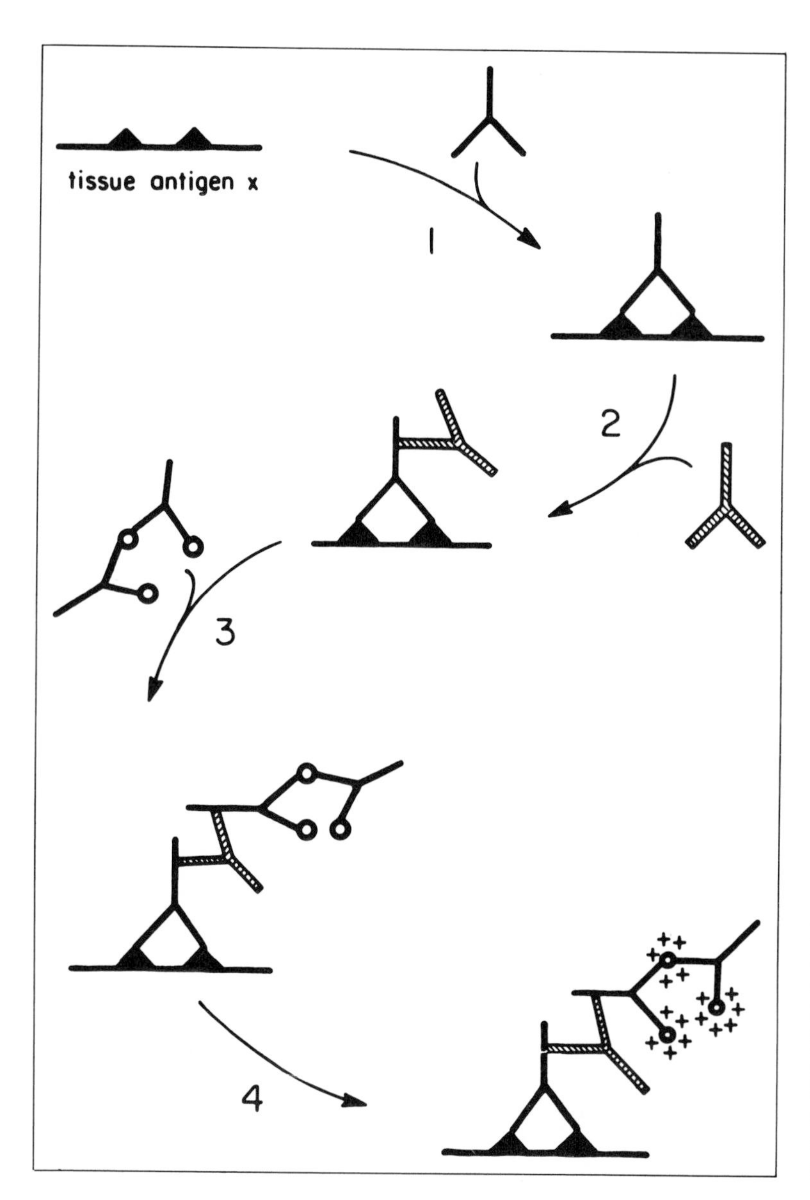

Figure 10.1. Schematic representation of the unlabeled peroxidase anti-peroxidase method of antigen localization. (1) Primary antibody; (2) bridge antibody; (3) peroxidase–anti-peroxidase complex; (4) DAB and peroxide.

cytial virus (RSV; Gardner et al., 1978), hepatitis A (Shimizu et al., 1978), mouse mammary tumor virus (Keydar et al., 1978), papillomavirus (Kurman et al., 1981; Syrjanen and Pyrhonen, 1982), and herpes simplex virus (HSV; Kurchak et al., 1977; Pearson et al., 1979). Our experience has been primarily with the application of the PXAPX method to detect HSV in tissue culture or tissue scrapings from human genitalia. We have also used the procedure to successfully detect a single HSV polypeptide in infected cells using polyclonal primary antibody prepared against that single viral peptide. We will attempt to provide the reader with some of our practical experience using immunoenzymatic PXAPX staining to detect viral antigens. It is not our intent to belabor the pros and cons of one procedure versus another, but to describe the PXAPX method routinely used in our laboratory. This procedure can serve as a base and may be adapted and modified to the specifications of any laboratory or virus, as conditions dictate.

10.2 STAINING CONSIDERATIONS

Successful staining by the PXAPX method requires consideration of several critical factors. The first, and one of the most important, considerations is the choice of tissue fixative. Detection of the antigens cannot occur by any immunocytochemical method if they have been destroyed or greatly altered by the fixative. An appropriate yet effective fixative for the antigen must first be determined and may vary for different viruses. Secondly, optimum concentrations of the three different antisera used must be established. This is true whether the serum was prepared in the research laboratory or obtained from a commercial source. Establishment of appropriate dilutions is carried out by block titering the three antisera and recording those combinations that produce optimum results. Once approximate values are obtained we frequently adjust to final concentrations by titrating one of the three (frequently the primary antibody) while holding the other two constant and subjectively evaluating the quality of the stain for optimum results. Optimum results are defined as maximum staining intensity of the antigen with minimal staining of tissue that does not contain the antigen. This may be repeated for the bridging antibody and the antiperoxidase antibody. The dilutions and incubations used in the procedure detailed below were optimized for the detection of HSV-2 in acetone (Pearson et al., 1979) or 70% ethanol (Steinkamp and Crissman, 1974) fixed tissue culture cells or human specimens. The staining procedure is easily adaptable to other viral systems, and the specific dilutions and incubation times described here can be used as a starting point toward optimizing the procedure for other viruses.

10.3 MATERIALS FOR IMMUNOPEROXIDASE STAIN

The materials listed below are for the PXAPX staining system in which the antiviral serum and antiperoxidase serum have been raised in rabbits and the bridge antibody has been raised in a goat against rabbit immunoglobulins. Other combinations of hosts can be used but it is essential that consistency be maintained (i.e., that the two hosts are not too closely related, and that the antiviral and antiperoxidase antibodies have been raised in the same host). Normal, nonimmune serum of the host in which the bridge antibody was raised (in this case, goat) is used to block nonspecific binding of antibodies and is layered on the samples before addition of each of the three different antibodies. The nonimmune serum contains an insignificant amount of crossreacting antibody that does not interfere with the specific binding of any of the antibodies.

1. Sera
   - Nonimmune goat serum (NGS): prepared in our laboratory but can be commercially obtained
   - Antiviral serum raised in rabbit: prepared in our laboratory but can be commercially obtained
   - Preimmune serum from rabbit: prepared in our laboratory but can be commercially obtained
   - Anti-rabbit serum raised in goat: commercially obtained, antibody protein concentration 15 mg/ml
   - Rabbit peroxidase–antiperoxidase (PXAPX): commercially obtained, antibody protein concentration 3 mg/ml
2. Phosphate buffered saline (PBS) (0.01 *M* sodium phosphate pH 7.2; 0.15 *M* NaCl)
3. Absolute methanol
4. Hydrogen peroxide (fresh)
5. 3,3′-diaminobenzidine tetrahydrochloride (DAB)
6. Absolute ethanol, 90% ethanol and 70% ethanol
7. Xylene
8. Permount

10.4 METHODS

Phosphate buffered saline is used throughout for all dilutions and washes. All procedures are performed at room temperature (~22°C).

1. If samples are immersed in fixative, rinse with PBS (5 minutes) before beginning to stain; this is not necessary if samples have been previously fixed and air-dried (we routinely fix in acetone at 4°C for 5–10 minutes, air dry, and store desiccated at 4°C for up to 21 days before staining)
2. Cover samples with a few drops of a 1:30 dilution of NGS for 5 minutes; if the samples are dry at the beginning of the staining procedure prewet by dipping into PBS before adding the NGS
3. Drain NGS by tipping the sample onto a paper towel at about a 90° degree angle; this removes the liquid that has accumulated at the edges of the coverslip or slide
4. Overlay samples with a few drops of an appropriate dilution of anti-viral serum and incubate for 1 hour in a humid chamber; the optimum dilution for anti–HSV-2 serum raised in rabbits in this laboratory has ranged from 1:200 to 1:2000 (Controls include the following: positive control—virus infected cells plus immune serum; negative controls—virus infected cells plus preimmune serum, virus infected cells plus PBS in place of the primary antibody, uninfected cells plus immune serum)
5. Wash samples twice (5 minutes each)
6. Cover samples with a few drops of diluted NGS for 5 minutes, then drain
7. Overlay each sample with a few drops of a 1:10 dilution of goat anti-rabbit serum; incubate for 20 minutes and drain
8. Wash samples twice (5 minutes each)
9. Immerse samples into absolute methanol for 15 minutes
10. Wash samples three times (5 minutes each)
11. Cover samples with a few drops of the diluted NGS for 5 minutes, then drain
12. Apply a few drops of rabbit PXAPX diluted 1:100 in 1% NGS–PBS (1 ml NGS + 99 ml PBS) for 20 minutes, then drain
13. Wash samples twice (5 minutes each)
14. Stain samples by immersing them in the disclosing reaction for 7 minutes protected from light; the disclosing reaction consists of 0.025% DAB in PBS, freshly prepared and filtered (0.45 micron filter) to which hydrogen peroxide is added to a final concentration of 0.005%; allow the DAB to be stirred rapidly for at least 30 minutes before filtering and adding peroxide
15. Wash stained samples twice (5 minutes each)
16. Mount the samples (dehydrate the stained samples by immersing them sequentially for 1 minute in 70% ethanol, 90% ethanol, absolute ethanol, and then xylene; add a drop of permount and place a glass coverslip on top)

10.5 RESULTS

Immunoperoxidase stained samples contain dark brown deposits at the site of the antigen (if DAB has been used as a substrate) and very little staining elsewhere [Figure 10.2(A),(C)]. Control specimens, in which preimmune serum or PBS was used in place of the primary serum, will often show some diffuse light-brown nonspecific staining throughout the cell or tissue specimen [Figure 10.2(B),(D)]. As demonstrated in Figure 10.2 the immunoperoxidase stain worked equally well in detecting HSV-2 antigens in either HEp-2 or VERO cells. The intensity of the stain can be regulated to varying degrees by the length of time the sample is in the disclosing reaction. As the staining time is increased, however, DAB can be oxidized by atmospheric oxygen, resulting in progressive deposition of product over the entire sample. This reduces the contrast between stained and unstained areas in the sample and may make interpretation of the results difficult. Manipulation of the DAB and H_2O_2 concentrations can result in faster or slower substrate conversion time and is frequently adjusted in a subjective manner to meet the demands or constraints of a particular system. It is possible, however, to overstain the sample so that normal tissue is intensely stained and cannot be distinguished from infected tissue.

Figure 10.2. Immunoperoxidase stained HSV-2 infected VERO and HEp-2 cells. (A) HSV-2 infected VERO cells + anti–HSV-2 serum; (B) HSV-2 infected VERO cells + preimmune serum; (C) HSV-2 infected HEp-2 cells + anti–HSV-2 serum; (D) HSV-2 infected HEp-2 cells + preimmune serum.

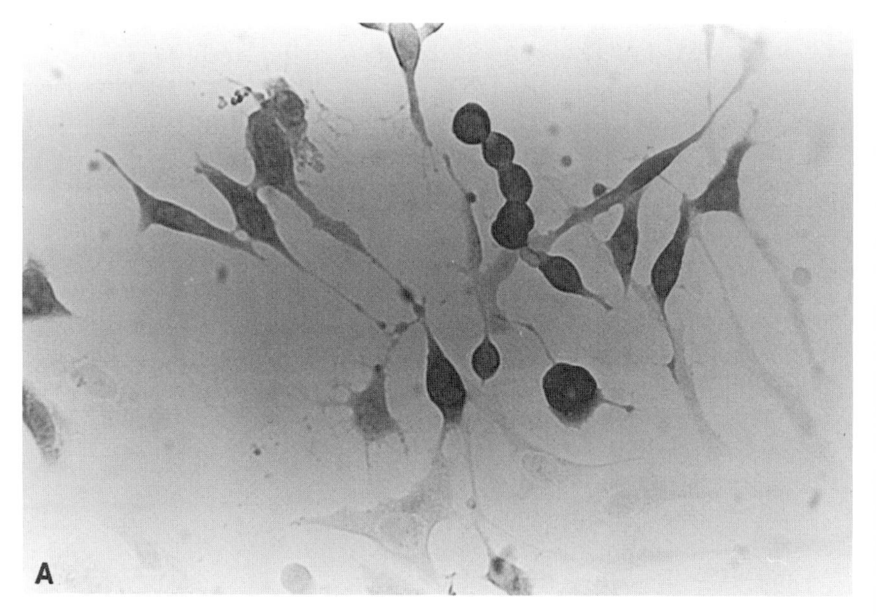

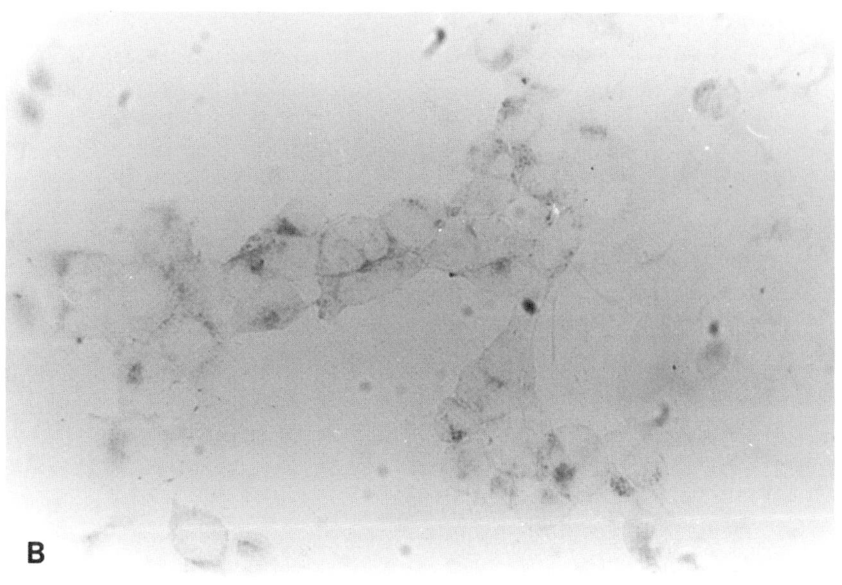

An added attraction of the PXAPX method is that it can easily be interfaced with other staining procedures that may assist in viewing specific virus induced cellular changes. In this regard, we have successfully interfaced the procedure with the Papanicolaou (PAP) stain by merely taking the sample through a standard PAP procedure after it was stained by PXAPX (Pearson et al., 1979). In Figure 10.3(A) and (B) the combination PXAPX–PAP procedure was used to stain HSV-2 infected VERO cells [Figure 10.3(A)] and the genital specimen from a female patient with a confirmed HSV-2 infection [Figure 10.3(B)]. Although the black and white photograph does not readily reveal color differences, the HSV-2 PXAPX-positive cells are stained dark brown [Figure 10.3(A),(B), arrows]. The uninfected cells in the human specimen, PAP stained, are light orange, blue, or pink with dark blue–purple nuclei. In the center

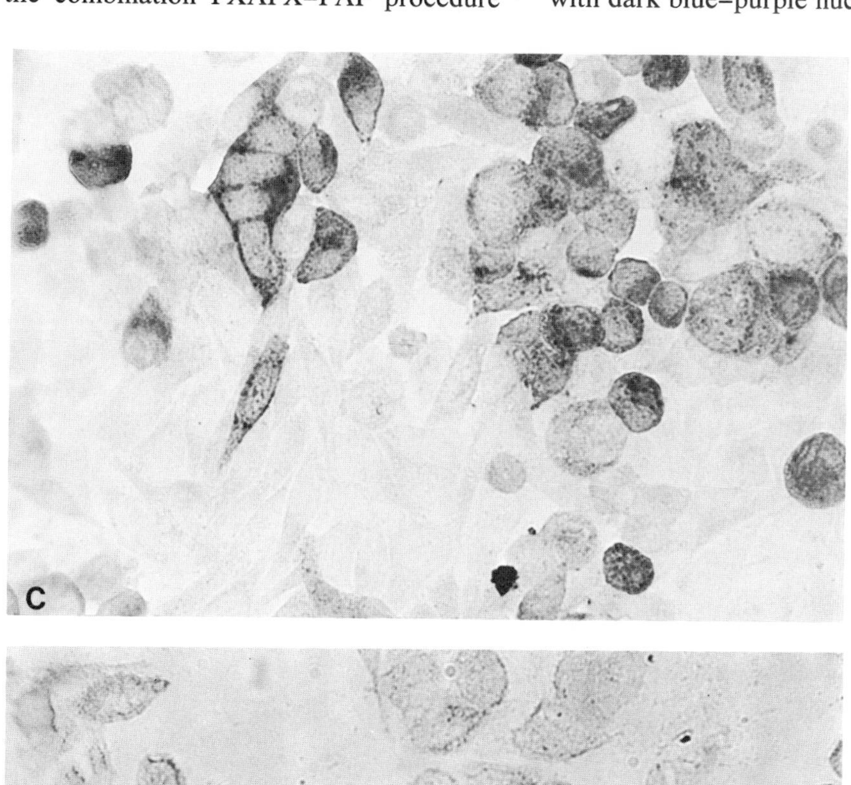

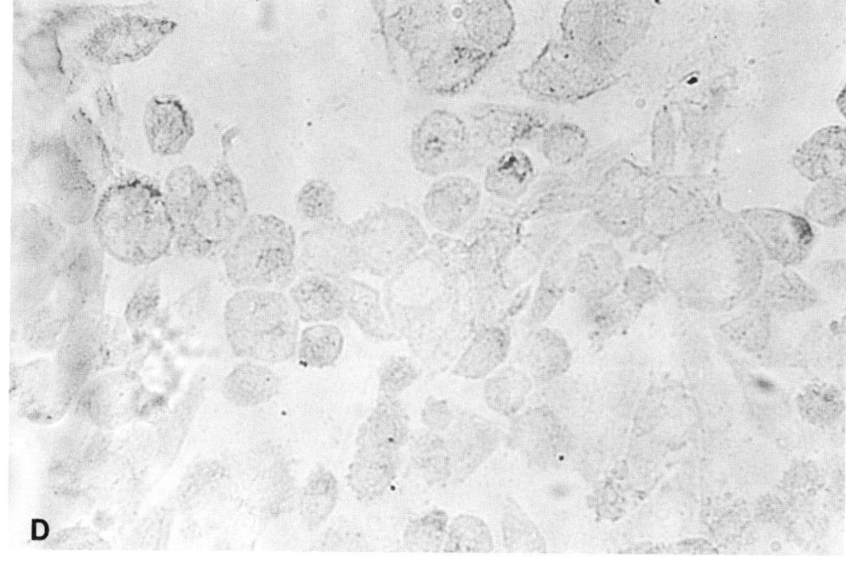

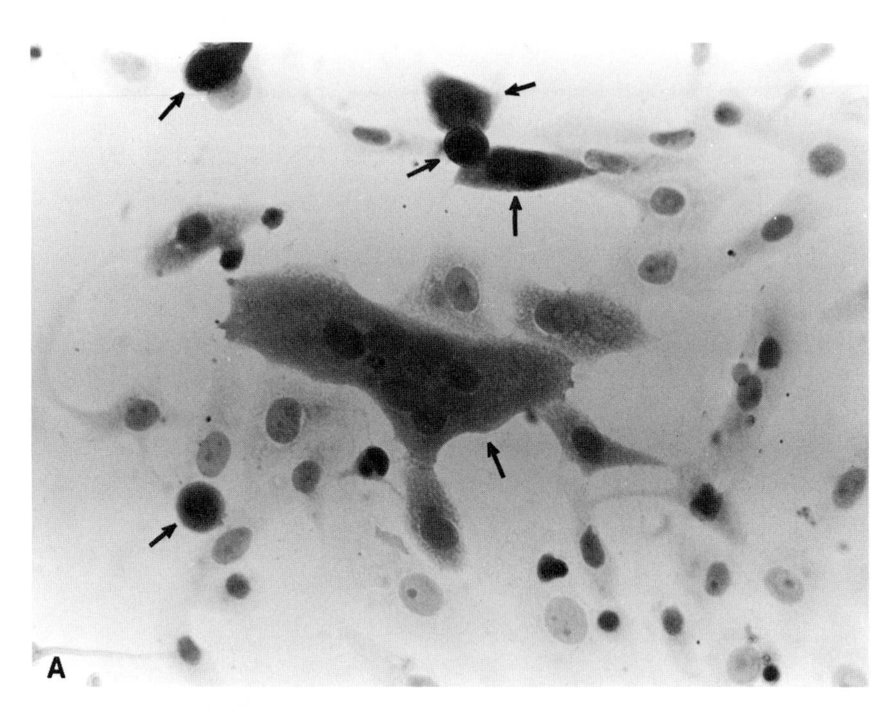

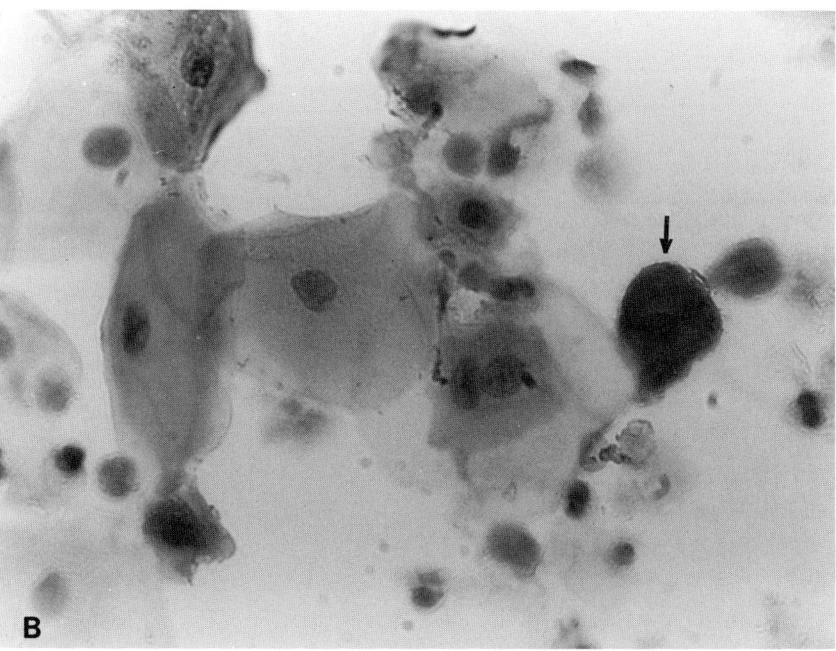

Figure 10.3. Immunoperoxidase–Papanicolaou stained VERO cells and human cells. Primary antibody in (A), (B), and (C) was rabbit anti–HSV-2. (A) HSV-2 infected VERO cells; (B) human cells with confirmed HSV-2 infection; (C) human cells with confirmed HSV-2 infection stained only by the immunoperoxidase method.

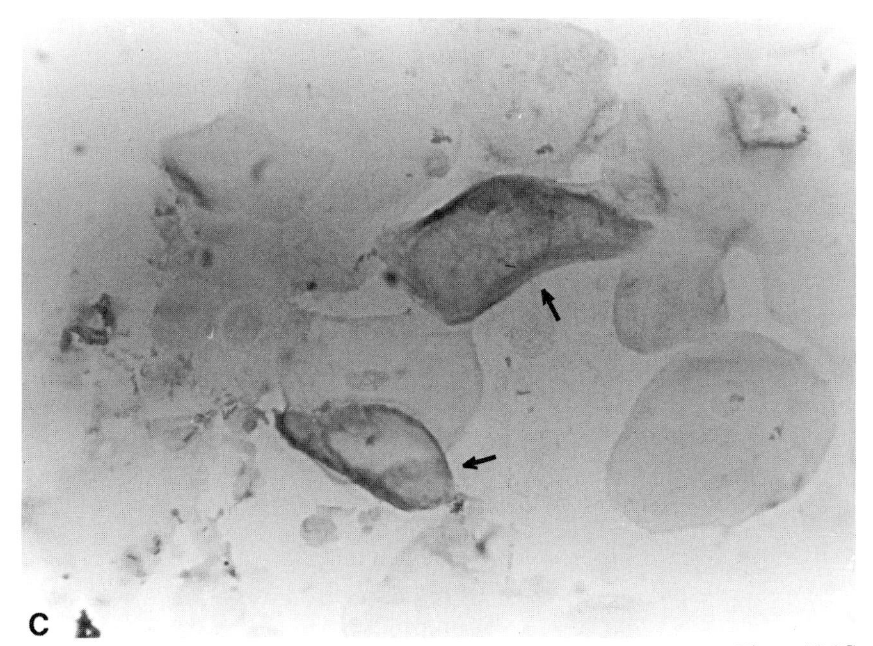

Figure 10.3C

of Figure 10.3(A) is an HSV syncytium that is characterized by multiple dark purple nuclei surrounded by a brown cytoplasm that is antigen-positive. For comparative purposes Figure 10.3(C) is a sample from the same patient as in Figure 10.3(B), but Figure 10.3(C) is only PXAPX stained revealing two PXAPX-positive syncytia (arrows). Uninfected cells from this specimen were light tan and the nuclei were not visible [Figure 10.3(C)]. Although we chose to use a common laboratory staining procedure (i.e., PAP), many other staining procedures could be used in conjunction with the PXAPX method should the staining of specific cellular elements, viral inclusions, or pathologic changes be required.

10.6 DISCUSSION

It has been our experience that immunoperoxidase staining produces consistent, sensitive, and reliable results. Some of the advantages of the procedure are that it is straightforward and uncomplicated, the reagents needed are commonly found in the laboratory or are commercially available, no extraordinary equipment is required to carry out the procedure or view the results, and the stain is permanent. We have found this last trait to be particularly valuable for review of data several months after staining.

As with most procedures, the PXAPX method described has inherent weaknesses. The disadvantages include the rather long period of time required to complete the PXAPX procedure, which precludes its use when rapid results are required. Indirect immunoperoxidase is more rapid than the PXAPX procedure, however, and it can be considered for use when speed is a greater consideration than sensitivity (Benjamin, 1975). Additionally, DAB is a suspected carcinogen and care must be taken when handling and disposing this material. Indeed, considering the carcinogenic properties of DAB, one may wisely consider the use of 4-chloro-1-napthol (Hawkes et al., 1982) or p-phenylenediamine dihydrochloride/pyrocatechol (Hanker et al., 1977) in its place.

Perhaps the biggest drawback to immunoperoxidase staining that we have found is background staining caused by the presence of endogenous peroxidase in mammalian mucosecretions (Blain et al., 1975) and inflammatory cells (Straus, 1971). This was particularly troublesome because most of our studies have been of HSV infections of the female genital tract. This location is rich in mucosecretions and, during an active herpetic infection, inflammatory cells. Because of these two sources of endogenous peroxidase, nonspecific staining can make the detection of a positive reaction extremely difficult or impossible. Consequently, this endogenous enzyme must be inactivated in order to obtain satisfactory peroxidase staining. Fortunately there are several methods to accomplish this and pretreatment with methanol appears satisfactory (Straus, 1971, 1979; Streefkerk, 1972; Weir et al., 1974). However, some reports suggest that such treatment may alter antigenic structure (Fink et al., 1979; Straus, 1979) and we have found this to be true for HSV antigens. Indeed, if an HSV positive sample is treated with methanol prior to the PXAPX procedure the HSV antigens are altered to such an extent that they are not recognized by the primary antibody. This may result in a false-negative, a highly undesirable result in either the clinical or research laboratory. When we were confronted with this apparent impasse we reasoned that while the methanol adversely affected the viral antigens, it may not affect the antibodies after they have reacted with the HSV antigens. Therefore, we ran a series of experiments in which methanol treatment was attempted at each stage of the procedure after the reaction with the primary antibody (Figure 10.1). These studies eventually demonstrated that methanol treatment after goat anti-rabbit IgG, but before rabbit PXAPX, preserved the viral antigen primary antibody reaction, inactivated endogenous enzyme, and did not interfere with the detection of viral antigen (Pearson et al., 1979).

Nonetheless, because the possibility for antigenic alteration exists by such pretreatment, we have recently begun to use an immunoenzymatic staining method that uses a non-mammalian enzyme, glucose oxidase (Gay et al., 1984). Essentially, the procedure is the same as the PXAPX method, but glucose oxidase is used in place of peroxidase. Because glucose oxidase is not present in mammalian cells, this modification has allowed us to eliminate methanol, yet preserve the sensitivity and permanency of the PXAPX method (Campbell and Bhatnager, 1976; Clark et al., 1982; Gay et al., 1984; Suffin et al., 1979).

The PXAPX method provides a permanent preparation and is an extremely sensitive method of detecting viral antigens. It has been used for several different viruses and undoubtedly will gain wider acceptance with time. Currently, it is more commonly used in the research laboratory, but is gaining wider acceptance as evidenced by the appearance of commercially prepared peroxidase kits for use by the clinical laboratory for the diagnosis of viral diseases.

REFERENCES

Benjamin D.R. 1975. Use of immunoperoxidase for rapid viral diagnosis. In D. Schlessinger (ed.), Microbiology—1975. Washington, D.C.: American Society for Microbiology, pp. 89–96.

Blain, J.A., Heald, P.J., Mack, A.E., and Shaw, C.E. 1975. Peroxidase in human cervical mucus during the menstrual cycle. Contraception 11:677–680.

Campbell, G.T., and Bhatnager, A.S. 1976. Simultaneous visualization by light microscopy of two pituitary hormones in a single tissue section using a combination of indirect immunohistochemical methods. J. Histochem. Cytochem. 24:448–452.

Clark, C.A., Downs, E.C., and Primus, F.J. 1982. An unlabeled antibody method using glucose oxidase–antiglucose oxidase complexes (GAG): A sensitive alternative to immunoperoxidase for the detection of tissue antigens. J. Histochem. Cytochem. 30:27–34.

Coons, A.H., Creech, H.J., and Jones, R.N. 1941. Immunological properties of an antibody containing a fluorescent group. Proc. Soc. Exp. Biol. Med. 47:200–202.

Coons, A.H., Creech, H.J., Jones, R.N., and Berliner, E. 1942. The demonstration of pneumococcal antigens in tissues by the use of the fluorescent antibody. J. Immunol. 45:159–170.

Coons, A.H., Snyder, J.C., Cheever, F.S., and Murray, E.S. 1950. Localization of antigen in tissue cells. IV. Antigens of rickettsiae and mumps virus. J. Exp. Med. 91:31–38.

Fink, B., Loepfe, E., and Wyler, R. 1979. Demonstration of viral antigens in cyrostat sections by a new immunoperoxidase procedure eliminating endogenous peroxidase activity. J. Histochem. Cytochem. 27:686–688.

Gardner, P.S., Grandien, M., and McQuillin, J. 1978. Comparison of immunofluorescence and immunoperoxidase methods for viral diagnosis at a distance: A WHO collaborative study. Bull. WHO 56(1):105–110.

Gay, H., Clark, W.R., and Docherty, J.J. 1984. Detection of herpes simplex virus infection using glucose oxidase–antiglucose oxidase immunoenzymatic stain. J. Histochem. Cytochem. 32:447–451.

Hanker, J.S., Yates, P.E., Metz, C.B., and Rustioni, A. 1977. A new specific, sensitive and non-carcinogenic reagent for the demonstration of horseradish peroxidase. Histochem. J. 9:789–792.

Hawkes, R., Niday, E., and Gordon, J. 1982. A dot-immunoblotting assay for monoclonal and other antibodies. Analyt. Biochem. 119:142–147.

Herrmann, J.E., Morse, S.A., and Collins, F. 1974. Comparison of techniques and immunoreagents used for indirect immunofluorescence and immunoperoxidase identification of enteroviruses. Infect. Immun. 10:220–226.

Kaplan, M.H., Coons, A.H., and Deane, H.W. 1950. Localization of antigen in tissue cells. III. Cellular distribution of pneumococcal polysaccharides types II and III in the mouse. J. Exp. Med. 91:15–30.

Keydar, I., Mesa-Tejada, R., Ramanarayanan, M., Ohno, T., Fenoglio, C., Hu, R., and Spiegelman, S. 1978. Detection of viral proteins in mouse mammary tumors by immunoperoxidase staining of paraffin sections. Proc. Natl. Acad. Sci. USA 75:1524–1528.

Kurchak, M., Dubbs, D.R., and Kit, S. 1977. Detection of herpes simplex virus-related antigens in the nuclei and cytoplasm of biochemically transformed cells with peroxidase/anti-peroxidase immunological staining and indirect immunofluorescence. Intl. J. Cancer. 20:371–380.

Kurman, R.J., Shah, K.H., Lancaster, W.D., and Jensen, A.B. 1981. Immunoperoxidase localization of papillomavirus antigens in cervical dysplasia and vulvar condylomas. Gynecology 140:931–935.

Kurstak, E. 1971. The immunoperoxidase technique: Localization of viral antigens in cells. In K. Maramorosch and H. Koprowski (eds.), Methods in Virology, Vol. 5. New York: Academic Press, pp. 423–444.

Nakane, P.K., and Pierce, G.B. 1967. Enzyme-labeled antibodies: preparation and application for the localization of antigens. J. Histochem. Cytochem. 14:929–931.

Pearse, A.G.E. 1972. Histochemistry: Theoretical and Applied, Vol. 2. Edinburgh: Churchill Livingston.

Pearson, N.S., Fleagle, G., and Docherty, J.J. 1979. Detection of herpes simplex virus infection of female genitalia by the peroxidase–antiperoxidase method alone or in conjunction with the Papanicolaou stain. J. Clin. Microbiol. 10:737–746.

Ram, J.S., Nakane, P.K., Rawlinson, D.G., and Pierce, G.B. 1966. Enzyme-labeled antibodies for ultrastructural studies. Fed. Proc. 25:732.

Shimizu, Y.K., Mathiesen, L.R., Lorenz, D., Drucker, J., Feinstone, S.M., Wagner, J.A., and Purcell, R.H. 1978. Localization of hepatitis A antigen in liver tissue by peroxidase-conjugated antibody method: Light and electron microscopic studies. J. Immunol. 121:1671–1679.

Spendlove, R.S. 1967. Microscopic techniques. In K. Maramorosch and H. Koprowski (eds.), Methods of Virology, Vol. 3. New York: Academic Press, pp. 482–520.

Steinkamp, J.A., and Crissman, H.A. 1974. Automated analysis of deoxyribonucleic acid, protein and nuclear to cytoplasmic relationships in tumor cells and gynecologic specimens. J. Histochem. Cytochem. 22:616–621.

Sternberger, L.A. 1974. Immunocytochemistry. Englewood Cliffs, N.J.: Prentice-Hall.

Sternberger, L.A., Hardy, P.H., Cuculis, J.J., and Meyer, H.G. 1970. The unlabeled antibody enzyme method of immunohistochemistry. Preparation and properties of soluble antigen–antibody complex (horseradish peroxidase–antihorseradish peroxidase) and its use in identification of spirochetes. J. Histochem. Cytochem. 18:315–333.

Straus, W. 1971. Inhibition of peroxidase by methanol and by methanol-nitroferricyanide for use in immunoperoxidase procedures. J. Histochem. Cytochem. 19:682–688.

Straus, W. 1979. Peroxidase procedures. Technical problems encountered during their application. J. Histochem. Cytochem. 27:1349–1351.

Streefkerk, J.G. 1972. Inhibition of erythrocyte pseudoperoxides activity by treatment with hydrogen peroxidase following methanol. J. Histochem. Cytochem. 20:829–831.

Suffin, S.C., Much, K.B., Young, J.C., Lewin, K., and Porter, D.D. 1979. Improvement of the glucose oxidase immunoenzymatic technic. Am. J. Clin. Pathol. 71:492–496.

Syrjanen, K.J., and Pyrhonen, S. 1982. Demonstration of human papilloma virus antigen in the condylomatous lesions of the uterine cervix by immunoperoxidase technique. Gynecol. Obstet. Invest. 14:90–96.

Weir, E. E., Pretlow, T.G., Pitts, A., and Williams, E.E. 1974. Destruction of endogenous peroxidase activity in order to locate cellular antigens by peroxidase-labeled antibodies. J. Histochem. Cytochem. 22:51–54.

Weller, T.H., and Coons, A.H. 1954. Fluorescent antibody studies with agents of varicella and herpes zoster propagated in vitro. Proc. Soc. Exp. Biol. Med. 86:789–794.

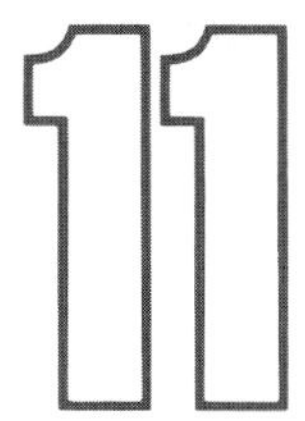

Complement Fixation Test

F. M. Wellings and A. L. Lewis

11.1 INTRODUCTION

Techniques gain or lose their status in all fields of endeavor as methodologies are improved or as new ones are introduced. In the early years of clinical virology the neutralization test served as the standard against which the specificity of other methods was evaluated. However, refinement of the complement fixation (CF) test and purification of reagents subsequently made this test the standard. Over the past decade there has been a definite trend to replace the CF test with more direct, sensitive, and rapid techniques, such as radioimmunoassay (RIA) and the enzyme-linked immunosorbent assay (ELISA). Even so, the CF test will continue to be of importance in the virologist's armamentarium for many years to come because of its value, not only for antibody studies, but for virus identification and typing, as well.

11.1.1 Background

The CF test was used extensively in syphilis serology after being introduced by Wasserman et al. in 1909. For approximately 30 years, application of the CF test in the field of virology was utilized in veterinary pursuits, such as the study of vaccinia in calves, foot-and-mouth disease, psittacosis, the equine arthropod borne encephalitides (Ciuca, 1929; Rice, 1948a, b; Bankowski et al., 1953). It required the efforts of many investigators over a span of roughly three decades to adapt the CF test for routine use in the diagnosis of virus disease in humans.

The major problem encountered by the early investigators was a lack of specificity due to the use of crude antigens prepared from infected tissue. The low virus titers and/or the volume of extraneous material present in these preparations resulted in extensive nonspecific and anticomplementary reactions (Howitt, 1937; Sosa-Martinez and Lennette, 1955). The latter could be removed by adequate adsorption techniques; however, the nonspecificity remained. Kidd and Friedewald (1942) showed that nonspecific reactions, attributed to the presence of a natural antibody in normal serum, could be removed by heating the serum for 30 minutes at 60°–65°C. High-speed centrifugation (30,000 × *g*) of the antigen was also effective. Thromboplastin activity of many tissue suspensions was shown by Maltaner (1946) to be directly correlated with nonspecific reactions that were enhanced by inactivated normal serum. He identified a cephalin-like substance in tissue extracts as the factor responsible for the nonspecific reactions. Lichter (1953) showed that nonspecific complement fixation occurred frequently with sera that were positive in the Wasserman test. Thus, nonspecificity stemmed from a variety of causes requiring a variety of treatments for removal.

11.1.2 Problems Associated with Obtaining Complement Fixation Antigen Preparations

Some viruses replicate to high titer in mouse brain, which made this material a good source for CF test antigen. However, interfering substances had to be removed if specificity was to be achieved. High-speed centrifugation of CF antigen preparations was useful (Kidd and Friedewald, 1942; Havens et al., 1943; Lennette et al., 1956a) and was improved with subsequent Seitz filtration (Casals and Palacios, 1941). This treatment did reduce the virus titer. Repeated freeze–thaw cycles followed by centrifugation produced a more specific antigen preparation without the concomitant loss in titer.

Howitt (1937) was the first to use lipid extraction for the preparation of CF antigens for the equine arthropod borne encephalitides, lymphocytic choriomeningitis, and St. Louis encephalitis (SLE) viruses. The procedure underwent many refinements over the years (DeBoer and Cox, 1947; España and Hammon, 1948; Casals, 1949), but the sucrose–acetone method devised by Clark and Casals (1958) is used routinely in arbovirology because it preserves both the hemagglutinin and CF antigen for many of the arboviruses (Hammon and Sather, 1969). Intracerebral inoculation of newborn mice (less than 3 days old) results in an antigen preparation with a low lipid content provided the brains are harvested within 1 week, because lipid deposition in the mouse brain is minimal during the first week of life. Other mammalian tissues usually fail to produce high titered antigens. One exception is the spleen of guinea pigs infected with lymphocytic choriomeningitis virus (Smadel et al., 1939). Additional sources of CF antigen have been developed using embryonated eggs and tissue culture.

The propagation of viruses in embryonated chicken eggs stemmed from the early work of Woodruff and Goodpasture (1931), Goodpasture et al. (1931), and Stevenson and Butler (1933). Beveridge and Burnet (1946) provided methodologies that led the way for the use of chick embryos in the production of high titered CF antigens for certain viruses (French, 1952; Whitney et al., 1953; Sosa-Martinez and Lennette, 1955; Lennette et al., 1956a, b). This host became the standard for CF antigen production for several of the orthomyxoviruses and paramyxoviruses because of the high titered type specific antigens obtained.

The introduction of cell culture techniques in virology provided an important new host system for the propagation of antigens that could be used in CF reactions. Initially, cell cultures were used principally for virus cultivation (Enders et al., 1949; Robbins et al., 1950), quantitation (Dulbecco, 1952), and isolation of viruses from human specimens. (Robbins et al., 1951; Rowe et al., 1953; Enders and Peebles, 1954; Hilleman and Werner, 1954; Huebner et al., 1954; Henle et al., 1955). The early utilization of cell culture fluids as a source of CF antigens posed the same types of problems encountered with homogenized tissues (i.e., low titers) (Ruckle and Rogers, 1957; Schmidt, 1957; Girardi et al., 1958; Weller and Witton, 1958; Taylor-Robinson and Downie, 1959), nonspecific and anticomplementary reactions discussed above (Svedmyr et al., 1952; Black and Melnick, 1954). Early attempts at concentrating the antigens led to either increased anticomplementary activity (Svedmyr et al., 1953) or reduced specificity when antigens were heated to remove the anticomplementary activity (LeBouvier et al., 1954; Black and Melnick, 1955; Schmidt and Lennette, 1956). Various approaches to produce a better antigen that might enhance the CF reaction were tested. These included reducing the amount of growth medium used in stationary cell cultures (Schmidt et al., 1957), growing cells in roller bottles (Churcher et al., 1959), or using cells grown in suspension culture (Westwood et al., 1960; Suggs et al., 1961; Halonen et al., 1967). In addition, the composition of the maintenance medium, the multiplicity of infection, and the site of the CF antigen in infected cell cultures (i.e., intra- or extracellular), were shown to affect the quantity and quality of CF antigens produced (Schmidt, 1969).

11.1.3 Basic Principles of the Complement Fixation Test

Initially, CF tests were conducted in test tubes, which required the preparation of large quantities of reagents. Various protocols were used, reflecting the personal bias of the individual virologist. In the late 1950s and early 1960s the Communicable Disease Center, currently known as the Centers for Disease Control—U.S. Department of Health and Human Services (Atlanta, Georgia) developed a standardized CF test. With the introduction of commercially prepared reagents and the availability of a standard technique, more laboratories began performing the CF test. The development of microtechniques (Takatsy, 1950; Sever, 1962) led to the adaptation of the Laboratory Branch Complement Fixation Test (LBCF) to this methodology (Casey, 1965), which requires very small amounts of reagents. This has been generally accepted as the standard. Over the years this LBCF test has been revised twice, once in 1974 and again in 1981. With the availability of commercially prepared antigens and the standard LBCF test, results obtained by various laboratories are now comparable.

Although the CF test is considered to be a relatively simple test, it is a very exacting procedure because of the five variables involved. In essence, the test consists of two antigen–antibody reactions, one of which is the indicator system. The first reaction, between a known virus and the test serum or an unknown virus and a specific antiserum, takes place in the presence of a predetermined amount of complement. As the reaction proceeds, complement is removed from solution by becoming an integral part of the antigen–antibody complex (i.e., complement is "fixed"). The second antigen–antibody reaction consists of reacting sheep red blood cells (RBC) with hemolysin (rabbit serum containing antibodies to the RBC). This results in sensitizing the RBC to free (unfixed) complement. Therefore, when this antigen–antibody complex is added to the initial reactants the sensitized RBC will lyse only if free complement is present. If a known antiserum is allowed to react with an unknown virus preparation in the presence of complement and the subsequently added sensitized RBC fail to lyse, the virus is identified because complement was fixed when the antigen reacted with the virus-specific antibody in the serum. In the case of an unknown serum, the antibody titer can be determined based on the highest dilution of serum in which complement was fixed.

11.2 GENERAL CONSIDERATIONS

11.2.1 Sensitivity and Specificity

The value of any diagnostic test is directly related to the sensitivity and the specificity of the test. Unfortunately, in the CF test, these two qualities appear to be incompatible in that the more sensitive the test the less specific it is. However, currently available antigens reduced the problem of specificity. But both sensitivity and specificity are affected by the relative concentration of the various reagents used in the test.

The importance of accurate measurements cannot be overemphasized, particularly in the micro-CF technique because of the minute amount of each reagent used. For instance, if one uses dropper pipettes that are calibrated to deliver 0.025 ml or 0.05 ml, care must be exercised to maintain a perpendicular position over the center of the well so that the reagents are delivered accurately. The loss of even a small portion of the drop on the lip of the well will lead to inaccuracies in the test. It is readily apparent that with five reactants in the test, there may be a compounding of errors; that is, the system's error is greater than the sum of the individual errors. In laboratories where large numbers of specimens are tested, use of an automatic microdiluting mechanism not only decreases the amount of time required to perform the test but it also increases accuracy provided the microdiluter is well maintained and calibrated. If the CF test is performed properly, it has been shown to be both sensitive and specific.

11.2.2 Standardization of Reagents

Accuracy is not only vital in the performance of the CF test, per se, but it is equally important in the standardization of the various components. Inaccuracies in the titration of hemolysin, antigen or complement, or in the standardization of the sheep RBC may be compounded by errors during the test performance. Washing and standardization of the RBC must be done carefully to avoid RBC lysis or enhanced sensitivity of the cell membrane to complement. Whether one uses the packed cell or the spectrophotometric method for standardization of the RBC concentration, lysis of even a few cells may produce inaccuracies in the test.

Lennette (1969) indicates that it is prudent to titrate complement in the presence of test antigen at the concentration to be used in the test, because different antigens have different degrees of activity in their ability to fix complement. If this is not taken into consideration the amount of complement used in the test may be insufficient. This is most critical in CF techniques utilizing exactly two units of complement. The LBCF test, which utilizes five times the amount of complement required to lyse 50% of the sensitized RBC, would not require complement be titrated in the presence of the antigen. However, the test does require the inclusion of three dilutions of complement, as controls in the antigen titration, to detect anticomplementary problems. Finally, complement controls are included in the CF test, per se.

11.2.3 Test Sera

The reliability of any test results is dependent on the quality of the specimens submitted. Thus, if sera are not collected and handled properly and obtained at specified time intervals the results are useless. Sera for CF tests must be collected in sterile tubes containing no preservatives or anticoagulants. Care must be taken to avoid hemolysis. Sufficient blood must be drawn to provide at least 1–2 ml of serum after the RBC are removed. This is readily accomplished if 5–10 ml of blood is permitted to clot in the tube, which is maintained at a 45° angle. Refrigeration for a few hours before the serum is decanted is often beneficial. If RBC are present in the decanted serum, they should be removed by centrifugation before the serum is frozen in order to prevent cell lysis. The serum must be handled aseptically to avoid bacterial contamination, which may produce false-positive reactions. Although filtration of the serum may ameliorate most problems caused by contamination, in some instances it is ineffective. Grossly hemolysed sera must be treated to remove the hemoglobin if an accurate test is to be achieved. Hemoglobin may be removed by molecular sieving through a G-75 or G-100 Sephadex column, a process that does not alter the antibody content of the sera.

These expensive and time consuming treatments can be avoided by proper specimen collection and handling. An acute phase specimen obtained during the first week of illness is necessary if the required fourfold rise in antibody titer is to be demonstrated between the acute and convalescent sera, the latter being obtained approximately 2 weeks after the former. This is particularly critical in those diseases in which CF titers may peak at relatively low levels. A single convalescent specimen is usually sufficient, but in immunologically compromised hosts, antibody production may be delayed. In such cases, a third serum obtained 4 to 6 weeks after the acute phase serum should be tested in conjunction with the acute and convalescent sera. It should be noted that in rickettsial infections, the CF antibodies may fail to appear or may show a very low titer if the patient received early and intensive antibiotic treatment. There is little that can be done under these circumstances except to resort to more sophisticated techniques to remove IgG and repeat the test to determine whether or not the low titer(s) was due to IgM, indicative of a recent infection. One method to remove IgG is to absorb the serum with *Staphylococcus aureus*, protein A. This material is commercially available bound to a matrix packed in a column, which simplifies the procedure. In any situation in which the timing of specimen collection may be responsible for a questionable result, testing of a third specimen is advisable. A fourfold drop in titer in the third specimen may occur in some instances, which is also acceptable as a diagnostic criterion, provided, of course, all serum samples were tested at the same time.

Sera vary extensively in their anticomplementary activity. This activity may be present at the time of collection or develop during storage. Often this activity may be destroyed by heat inactivation (56°C for 30 minutes) of the sera. If this is ineffective, we have used the following treatment in our laboratory. To 0.15 ml of each serum pair, add 0.05 ml of undiluted complement. This is incubated at 37°C for 30 minutes, 1.0 ml of veronal buffer diluent (VBD) is then added to yield a 1:8 dilution and the sera are again inactivated by heating at 56°C for 30 minutes (Hawkes, 1979). If the serum continues to be anticomplementary, an additional serum specimen should be requested and tested. This may not result in a valid test because some individuals have persistent anticomplementary factors in their serum but the additional serum should be tested in case the anticomplementary component may have been transiently present. Usually, anticomplementary reactions occur in the lower serum dilutions (i.e., 1:8 to 1:16); thus, if the initial titer is in that range and the antibody titer shown by a convalescent serum is in the range of 1:64 and above, the diagnostic fourfold rise in titer would be demonstrated. Because of these types of problems paired sera must be included in the same CF test to be valid.

11.2.4 Complement Fixation Antigens

There are two general types of CF antigens. One is the intact virion referred to as the V antigen and the other is a mixture of virion components and virus directed proteins that are released on disintegration of the particle or an infected cell. This is the S (soluble) antigen that was first identified by Craigie and Wishart (1936 a, b) while working with vaccinia virus. The antigen was obtained either by filtration or high-speed centrifugation and had both heat labile and heat stable antigens that were type specific. S antigen was initially described as a small molecular-weight ribonucleoprotein (Hoyle, 1952; Ada and Perry, 1954; Schafer and Zillig, 1954; Paucker et al., 1956); but eventually it was demonstrated that the S antigen was a mixture of molecules (Westwood et al., 1965; Cohen and Wilcox, 1966). S antigens have been demonstrated for a number of viruses and have been found to vary in their stability and specificity (Chambers et al., 1950; Black and Melnick, 1955; Ende et al., 1957).

The S antigen is of primary importance in the mumps CF test because anti-S–antibodies rise early and disappear within a few months, whereas, the anti-V–antibodies rise later and persist for years (Henle et al., 1948). Thus, a recent infection is indicated if an acute serum is positive for the S-antigen and negative for the V-antigen. This finding should be confirmed by testing the acute and convalescent sera in the same test to demonstrate the required fourfold rise in antibody titer (Hoyle, 1952).

When the influenza viruses are treated with ether two antigenic moieties are released, an S-antigen and the hemagglutinin (Hoyle, 1952). Standard ether treated virus preparations release uniform amounts of these antigens, but in preparations containing increased numbers of defective virus particles the amount of S antigen is decreased relative to the hemagglutinin (Lief and Henle, 1956 a, b). The S-antigen, which must be isolated from the hemagglutinin for use in the CF test, is type-specific for each influenza virus type, A, B, and C. It is derived from the noninfectious nucleocapsid and is an antigen that is not associated with the development of immunity. The V-antigen is the strain-specific antigen that stimulates the formation of protective antibody.

Although many laboratories may prefer to prepare their own antigens, particularly research laboratories that have animal breeding facilities, extensive cell culture capabilities, and egg incubators, the availability of commercially prepared CF antigens makes it unnecessary. Table 11.1 lists the sources of many of the commercially available CF antigens.[1]

11.2.5 Interpretation of Complement Fixation Test Results

Complement fixation and other laboratory test results should be "viewed" in the larger context that includes patient medical history, to avoid problematic interpretations that may arise. For example, this is of particular importance in arbovirus serology where antigenic crossreaction is the norm. If a patient never had been infected or immunized with a group A or group B arbovirus, a fourfold rise in CF titer against a particular antigen would be considered diagnostic. However, an individual who had been previously infected with a group B dengue virus, for instance, could show an anamnestic response to the dengue infection, following infection with another group B virus, such as SLE virus. This so-called "original antigenic sin" may result in a rapid rise of the group B IgG antibody. If this occurs before the first serum is drawn, it would preclude demonstration of the diagnostic fourfold rise in antibody titer. An immunization with yellow fever vaccine could have the same effect. As previously discussed, the sera may be treated and then tested for the presence of IgM.

Interpretation of test results, particularly in the arbovirus battery may be greatly facilitated by a travel history. There are literally hundreds of arthropod borne viruses (arboviruses), many of which are indigenous to specific geographic areas. Even in the United States, the LaCross virus, a type of California encephalitis virus, appears to be limited to certain geographic areas, centered mainly around the Great Lakes area and in certain areas of North Carolina and Georgia. This distribution may be more apparent than real because of the limited amount of viral diagnostic studies performed in some states. In Florida, however, where extensive arbovirus serologies are performed, only one verifiable illness due to an infection with LaCross virus has been identified. This occurred in a Florida county near the Georgia border. This may indicate that the virus is moving into new areas, such as has happened with the introduction of three additional dengue virus types into the Caribbean Islands over the past decade. Keeping abreast of such changes is a prerequisite in a virus diagnostic laboratory so

[1] Information concerning unlisted antigens may be obtained by contacting the Center for Disease Control (CDC), Bureau of Laboratories, Atlanta, Georgia 30333.

Table 11.1 Commercially Available Antigens for Complement Fixation Testing

Antigen	Source: Flow Laboratories	Source: Microbiological Associates
Parainfluenza type 1 (HA-2)	X	X (HA2-CB9)
Parainfluenza type 1 (Sendai)	X	
Parainfluenza type 2 (Greek)	X	X (CA)
Parainfluenza type 3 (HA-1)	X	X
Herpes hominis type 1 (MacIntyre)	X	X
Cytomegalovirus (AD-169)	X	X
Adenovirus (group R1-67)	X	X (Strain 6)
Respiratory syncytial (Long)	X	X
Measles (Rubeola)(Edmonston)	X	X (Philadelphia)
Mumps (Enders)	X	X
Varicella-zoster (VZ-10)	X	X (Ellen)
Rubella (Therren)	X	X (Gilchrist)
Coxsackie		
B1 (Conns)		X
B2 (Ohio)		X
B3 (Nancy)		X
B4 (Berchoten)		X
B5 (Faulkner)		X
B6 (Schmitt)		X
Influenza		
A (A/HK/8/68)		X
B (B/Mass/3/66)		X
C (C/Taylor)		X
Poliovirus		
1 (Mahoney)		X
2 (MEF-1)		X
3 (Sankett)		X
Reovirus (Long)		X
Lymphocytic Choriomeningitis (Fortner)		X

that proper antigens may be included in CF testing and results interpreted correctly.

11.2.6 Performing the Complement Fixation Test

Understanding the basic principles and the need for absolute accuracy in all facets of the test are prerequisites to attempting the CF test procedure. The CDC routinely conducts training courses in the performance of the CF test and it is advisable that the novice participate in such a course unless an experienced individual is available in-house.

The LBCF microtiter procedure described below, with slight modifications, is the one used in the authors' laboratory.[2]

[2] The reader is encouraged to obtain the manual *A Guide to the Performance of the Standardized Diagnostic Complement Fixation Method and Adaptation to Micro Test*, which is available from the U. S. Department of Health and Human Services, CDC, Atlanta, Georgia 30333. The 1981 revised addition is Public Health Service Publication, Number 1228.

11.3 MATERIALS

11.3.1 Equipment

Freezer −20°C
Refrigerator 4°C
Incubator 37°C
Water baths 37°C and 56°C
Thermometers for the above
Centrifuge with tachometer
Centrifuge plate carriers
pH meter
Balance
Suction apparatus
Mechanical vibrator
Timer
Test tube racks
Test tubes—13 × 100 mm, 15 × 125 mm
Mechanical diluter (optional)
Microtiter plates, U type
Microtiter droppers or mechanical pipettes with disposable tips 0.025 ml and 0.05 ml
Serologic pipettes—1, 2, 5, and 10 ml
Glassware—Erlenmeyer and volumetric flasks, graduated cylinders, and centrifuge tubes, variously sized beakers, etc.
Mirrored reading mechanism (optional)

11.3.2 Reagents

Antigens
Complement
Gelatin-water solution
Hemolysin (glycerinized)
Stock $MgCl_2$–$CaCl_2$ solution
Stock and working barbital-buffered diluent
Sheep erythrocytes—2.8% suspension
Sera—test serum, positive and negative controls

11.4 REAGENT PREPARATION

11.4.1 Gelatin–Water Solution

1. Add 1 g of gelatin to a 1-L Erlenmeyer flask containing 200 ml distilled water
2. Place over heat and bring to a boil to dissolve gelatin
3. Remove from heat and allow to cool to room temperature
4. Add 600 ml distilled water and cover mouth of flask with screw cap or heavy aluminum foil
5. Label flask with name and date prepared
6. Store at 4°C for no longer than 1 week

11.4.2 Stock $MgCl_2$–$CaCl_2$ Solution

1. Place 70 ml of distilled water in a 100 ml volumetric flask
2. Add 20.3 g $MgCl_2 \cdot 6H_2O$ and 4.4 g $CaCl_2 \cdot 2H_2O$ to the flask
3. Mix thoroughly by swirling
4. Add distilled water to the 100-ml mark and mix as before
5. Label appropriately with name and date prepared and store at 4°C

11.4.3 Stock Buffer Solution

1. To a 2-L screw-capped volumetric flask containing 1500 ml distilled water, add 83.0 g of NaCl
2. Add 10.19 g of Na-5,5-diethyl barbiturate
3. *Completely* dissolve chemicals by swirling the flask
4. Place 36 ml of distilled water in a 100-ml flask and add 4 ml of 10N HCl; mix by swirling; add 36 ml of the 1N HCl to the volumetric flask and mix by swirling
5. Add 5.0 ml stock $MgCl_2$–$CaCl_2$ solution
6. Add distilled water to the 2-L mark, cap tightly, and mix by inverting the flask several times
7. Check the pH of a 1:5 dilution of the buffer
8. If the pH is not 7.3 to 7.4, discard buffer and prepare fresh
9. Label appropriately and store stock buffer at 4°C

11.4.4 Veronal Buffer Diluent

1. On the day of the test, place 200 ml of stock buffer solution in a 1-L screw-capped volumetric flask
2. Add gelatin water to the 1-L mark, cap tightly, and mix thoroughly by inverting the flask several times
3. Check the pH of the solution; if it is not between 7.3 and 7.4, discard and prepare fresh
4. Label flask and hold at 4°C until used in the test; discard unused portion

11.4.5 Washing Sheep Erythrocytes

1. Place 15–20 ml of the preserved whole blood in a 250-ml centrifuge bottle
2. Add approximately 100 ml of cold VBD and mix gently with a pipette
3. Centrifuge at 600 × *g* at 4°C for 10 minutes
4. Aspirate the supernatant fluid with a water faucet aspirator or suction apparatus
5. Gently resuspend the cells in 100 ml cold VBD
6. Centrifuge at 600 × *g* at 4°C for 10 minutes
7. Carefully aspirate the supernatant fluid and white cell layer
8. Add 100 ml cold VBD to the centrifuge bottle and gently resuspend the cells
9. Centrifuge at 600 × *g* at 4°C for 10 min
10. Aspirate the supernatant fluid
11. Resuspend the cells in 80 ml cold VBD and transfer to two 40 ml conical, graduated, centrifuge tubes
12. Centrifuge at 600 × *g* 10 min at 4°C to pack the cells
13. Aspirate the colorless supernatant fluid without disturbing the cells. If supernatant is even slightly pink, cells may be too fragile and should be discarded and new cells obtained and washed as described.
14. Note volume of packed cells and proceed with standardization of cell suspension

11.5 STANDARDIZING THE CELL SUSPENSION

11.5.1 Packed Cell Method

1. Multiply the volume of packed cells obtained after the final centrifugation by 34.7 to determine the volume of VBD required for a 2.8% suspension
2. Place the calculated volume of VBD into an appropriately sized Erlenmeyer flask, using a portion of the total to resuspend the cells in the centrifuge tube and then decant into the Erlenmeyer flask
3. Swirl the flask gently to obtain a homogeneous suspension
4. To determine the accuracy of the dilution, pipette 7.0 ml of the suspension into a 10-ml graduated centrifuge tube having an accuracy of ±0.025 ml in the 0–1 ml range
5. Centrifuge at 600 × *g* for 10 minutes at 4°C
6. If the volume of the packed cells is 0.2 ml, the suspension is accurate and the 7 ml sample may be returned to the original flask and the cells used in the test; if not, a corrected volume must be determined as follows: Corrected volume equals the actual reading of the centrifuge tube divided by 0.2 ml, times the original volume of the cell suspension, (7.0 ml)
7. If the corrected volume is *less* than the existing volume, a portion of the VBD must be removed by centrifuging an aliquot of the cell suspension at 600 × *g* at 4°C for 5 minutes, then pipetting off and discarding the calculated volume of excess buffer. The remainder of the aliquot is returned to the cell suspension and the accuracy of the dilution again determined (step 4)
8. If the corrected volume is *greater* than

the existing volume, additional VBD must be added to the suspension. To determine the amount to be added simply subtract the actual volume from the corrected volume. After adding this to the cell suspension, gently swirl the flask, and again determine the accuracy of the dilution (step 4)

11.6 HEMOLYSIN TITRATION

A hemolysin titration should be done when a new lot of hemolysin or sheep erythrocytes is used.

11.6.1 Preparation of a 1:100 Stock Hemolysin Dilution

1. Place 4.0 ml of a 5% phenol/normal saline solution (V/V) in a 125 ml Erlenmeyer flask
2. Add 94.0 ml of cold VBD to the flask and mix by swirling
3. Add 2.0 ml of glycerinized hemolysin to the flask and mix by swirling
4. Label with name and date prepared and store at 4°C

11.6.2 Preparation of a 1:1000 Hemolysin Dilution

1. Place 9.0 ml of cold VDB in a 15 × 125 mm tube labeled 1:1000 hemolysin
2. Add 1.0 ml of the 1:100 dilution of hemolysin and mix well

11.6.3 Preparation of Additional Hemolysin Dilutions

1. Label six 15 × 125 mm test tubes with the dilutions shown in Table 11.2 (column three)
2. Using a 5.0-ml pipette, place the designated volumes of VBD into the appropriately labeled tubes as indicated in Table 11.2 (column one)
3. Add 1.0 ml of the 1:1000 hemolysin dilution to each of the tubes using a 5.0-ml pipette

11.6.4 Preparation of 1:400 Dilution of Complement

1. Place undiluted complement in an ice bath
2. Place 99.75 ml of cold VBD in a 125 ml flask
3. For accuracy, at least 0.4 ml of complement should be drawn up into a pipette and the tip wiped before 0.25 ml of complement is delivered dropwise into cold VBD and the excess complement discarded; gently swirl the flask to mix
4. Place the 1:400 dilution of complement at 4°C for at least 20 minutes before using; do not use after 2 hours

11.6.5 Preparation of Sensitized Cells for Hemolysin Titration

1. Label two sets of seven 13 × 100 mm test tubes with the hemolysin dilutions starting with 1:1000 and continuing as shown in Table 11.2 (column three); place each set of labeled tubes starting with the 1:1000 dilution and ending with the 1:8000 dilution in separate racks; put one rack aside to be used later in the hemolysin titration
2. Place 1.0 ml of the standardized 2.8% sheep erythrocyte suspension in each of the seven tubes
3. Thoroughly mix the 1:1000 hemolysin dilution previously prepared and add 1.0 ml slowly, *with constant stirring*, to the tube labeled 1:1000; to each of the labeled tubes add 1.0 ml of its corresponding hemolysin dilution in a like manner (i.e., 1.0 ml of the 1:1500 hemolysin dilution into the tube labeled 1:1500)
4. Shake the rack and place it in a 37°C water bath for 15 minutes

Table 11.2 Preparation of Additional Hemolysin Dilutions

VBD (ml)	1:1000 Hemolysin Dilution (ml)	Final Dilution
0.5	1.0	1:1500
1.0	1.0	1:2000
1.5	1.0	1:2500
2.0	1.0	1:3000
3.0	1.0	1:4000
7.0	1.0	1:8000

11.6.6 Preparing the Hemolysin Titration

1. Place 0.4 ml of cold VBD in each of the tubes in the extra rack of tubes prepared (step 1, 11.6.5)
2. To each of the seven tubes, add 0.4 ml of the 1:400 complement dilution that had been held at 4°C for at least 20 minutes
3. Shake the rack to mix
4. Cells sensitized with a hemolysin dilution (step 4, 11.6.5) are added in 0.20 ml amounts to the tube comparably labeled (i.e., cells sensitized with the 1:1000 hemolysin dilution are added to the tube labeled 1:1000)
5. Place the rack in a 37°C water bath for 1 hour; shake the rack after 30 minutes of incubation

11.7 PREPARATION OF COLOR STANDARDS

11.7.1 Hemoglobin Solution

1. Pipette 1.0 ml of the thoroughly mixed 2.8% cell suspension into a 15 × 125 mm test tube
2. Add 7.0 ml distilled water and shake the tube until all the cells are lysed
3. Add 2.0 ml *stock* buffer (11.4.3) to the test tube
4. Mix thoroughly and set aside until needed

11.7.2 0.28 Percent Cell Suspension

1. Remove the 2.8% cell suspension from the refrigerator and swirl to obtain a homogeneous suspension
2. Pipette 1.0 ml of the suspension into a 15 × 125 mm tube
3. Add 9.0 ml cold VBD to the test tube
4. Mix well and set aside until needed

11.7.3 Color Standards

1. Label 11 serologic tubes (13 × 100 mm or the size to be used in the test) with the percentage of hemolysis in increments of 10% starting with 0%; label this tube with the date and time of preparation; place tubes in a rack in ascending order
2. Using a 2-ml pipette, deliver the volume of hemoglobin solution shown in Table 11.3 to the appropriately labeled test tubes
3. Using a 2-ml pipette, deliver the volume of 0.28% cell suspension to each of the appropriately labeled tubes as shown in Table 11.3
4. Shake the rack vigorously to mix well
5. Centrifuge the tubes at 600 × *g* for 5 minutes
6. Gently remove the tubes from the centrifuge and place them in their rack without disturbing the pellet and store at 4°C until needed

Table 11.3. Preparation of Color Standards

Reagents	Percent Hemolysis										
	0	10	20	30	40	50	60	70	80	90	100
Hemoglobin solution	0.0	0.1	0.2	0.3	0.4	0.5	0.6	0.7	0.8	0.9	1.0
0.28% Cell suspension	1.0	0.9	0.8	0.7	0.6	0.5	0.4	0.3	0.2	0.1	0.0

11.7.4 Determining Hemolysin Dilution Needed for Sensitization of 2.8 Percent Cell Suspension

1. Remove the hemolysin titration tubes from the water bath (11.6.6) and centrifuge the tubes at 600 × *g* for 5 minutes

Figure 11.1. Determination of the optimal hemolysin dilution.
The location for each of the hemolysin dilutions on the X axis is determined by dividing the lowest dilution by the next higher dilution times the number of blocks allotted on the X axis.

Example:
1000/1500 × blocks = 6.7 (hemolysin dilution, 1:1500)
1000/2000 × 10 blocks = 5.0 (hemolysin dilution, 1:2000)
1000/2500 × 10 blocks = 4.0 (hemolysin dilution, 1:2500)

The closed squares show an acceptable titration and the open squares, an unacceptable titration. The optimal dilution in this titration is 1:2000 (i.e., the second dilution in the plateau).

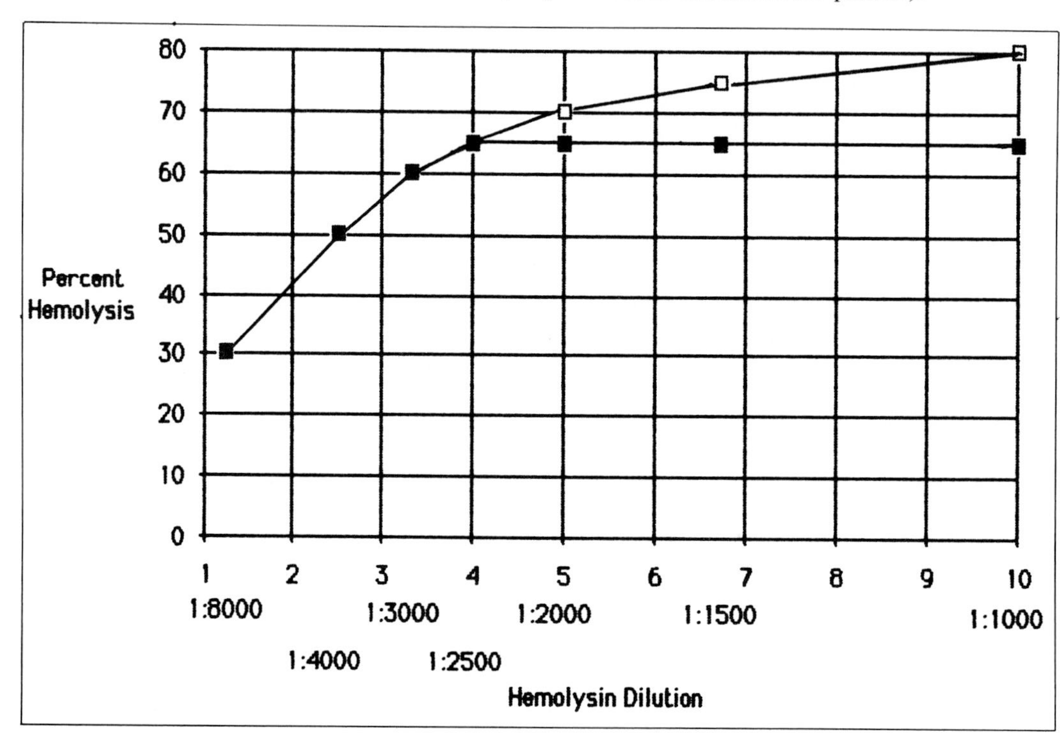

2. Compare each tube with the prepared color standards; if the tube matches a color standard, read and record the percent hemolysis; if not, interpolate to the nearest 5% and record (commercial complement diluted 1:400 will usually yield 30% to 80% hemolysis for optimally sensitized cells as shown in Figure 11.1—to obtain the correct percentage hemolysis with a less active complement it may be necessary to use a 1:300 dilution, or if the complement is particularly potent, it may be possible to use a 1:500 dilution)
3. Plot on linear graph paper the amount of hemolysis obtained with each dilution of the hemolysin (Figure 11.1)
4. Draw a line through the plotted points and determine the area where the line levels off, or plateaus; this is the level at which additional hemolysin produces no marked increase in percent hemolysis (Figure 11.1)
5. The *second* dilution on the plateau is the optimal hemolysin dilution to be used for sensitizing cells until a new lot of hemolysin or red cells is used (e.g., 1:2000, in Figure 11.1)

11.8 COMPLEMENT TITRATION

11.8.1 Sensitizing Cells

1. Place 1 ml of the standardized 2.8% cell suspension in a 13 × 100 mm tube
2. Using the 1:100 stock hemolysin, prepare the optimal dilution of hemolysin as determined in the hemolysin titration
3. Add 1 ml of this hemolysin dilution to the cells, swirl the tube rapidly
4. Place the tube in a 37°C water bath for 15 minutes

11.8.2 Setting Up the Complement Titration

1. Label two sets of four 13 × 100 mm serologic test tubes with the numbers 1 to 4 and place in a rack in order, one set behind the other; the titration is run in duplicate
2. Place the designated volumes of VBD and complement in the two sets of appropriately labeled test tubes as shown in Table 11.4 and shake the rack to mix the reagents
3. Add 0.2 ml sensitized cells to each test tube
4. Shake the rack and place it in a 37°C water bath to incubate for a total of 30 minutes; after 15 minutes, remove the rack, shake it to resuspend cells that have not lysed, and return the rack to the water bath

11.8.3 Reading Percent Hemolysis

1. Remove the rack from the water bath after 30 minutes incubation, and centrifuge the test tubes at 600 × *g* for 5 minutes
2. Compare each tube in the first set with the color standards; if the tube matches a standard, read and record the percent hemolysis; if not, interpolate to the nearest 5% and record the reading
3. Do the same with the duplicate set of test tubes
4. Calculate the average percentage hemolysis for each pair of test tubes

11.8.4 Determining Volume of Complement Producing 50 Percent Hemolysis

1. Using the conversion factors shown in Table 11.5, obtain the X/100-X ratio value for each pair of test tubes; repeat the procedure if lysis exceeds 90% or is less than 10%
2. Plot on log–log graph paper the ratio values (X/100-X) obtained (X axis) against the volume of the 1:400 dilution of complement in ml (Y axis)
3. If two values fall to the left and two to the right of the vertical 1.000 line on the

Table 11.4. Complement Titration

	Reagent (ml)		
Tube Number	VBD	1:400 Dilution of Complement	Sensitized Cells
1	0.60	0.20	0.20
2	0.55	0.25	0.20
3	0.50	0.30	0.20
4	0.40	0.40	0.20

X axis, the dilution of complement was correct; if more than two points fall to the right of that line, too much complement was used and the titration should be repeated using a 1:500 dilution; conversely, if more than two points fall to the left of that line, too little complement was used and the titration must be repeated using a lower dilution (i.e., 1:300 dilution of complement)

4. Draw a line between the two plotted points for test tubes 1 and 2 and another line between the two plotted points for test tubes 3 and 4 (Figure 11.2)
5. Determine the midpoint on each of the lines and draw a second line between these two points (Figure 11.2)
6. Determine the slope of the line just drawn by measuring horizontally from a point near the end of the line to a point 10 cm to the right; measure the vertical distance in centimeters from that point to the line joining the two midpoints (first lines drawn) and divided the value by 10 cm to determine the slope of the line; if the slope is within ±10% of 0.20, the titration is acceptable; if not, the titration must be repeated; reproducibility demands that the slope of this line fall within ±10% of 0.20

Table 11.5. Conversion Values for Percent of Lysis

X[a]	$\frac{X}{100-X}$[b]	X	$\frac{X}{100-X}$	X	$\frac{X}{100-X}$
10	0.111	40	0.67	70	2.33
15	0.176	45	0.82	75	3.00
20	0.250	50	1.00	80	4.00
25	0.330	55	1.22	85	5.70
30	0.430	60	1.50	90	9.00
35	0.540	65	1.86		

[a] Percent of lysed cells.
[b] Percent of lysed cells/percent of nonlysed cells ratio.

11.8.5 Determination of Complement Dilution Needed for Antigen Titration and Diagnostic Test

1. Find the intersect of the vertical 1.000 line and the line joining the two midpoints; draw a dotted line horizontally from this point to meet the Y axis on the left (Figure 11.2)
2. Read the volume in milliliters of the 1:400 dilution of complement at the point where the dotted line intersects the Y axis. This volume contains one unit of complement capable of hemolyzing 50% (1 CH_{50}) of the cells; to obtain the five units required in the test (5 CH_{50}), multiply this volume by 5
3. Because the 5 CH_{50} must be added to the test in a 0.4-ml volume, the dilution of complement necessary to achieve this must be calculated as follows: The dilution of complement used in the titration is multiplied by the desired volume (0.4 ml) and divided by the volume calculated to contain 5 CH_{50}.

Example:

The volume containing 5 CH_{50} at 1:400 dilution is 1.4 ml (5 × 0.29 ml).

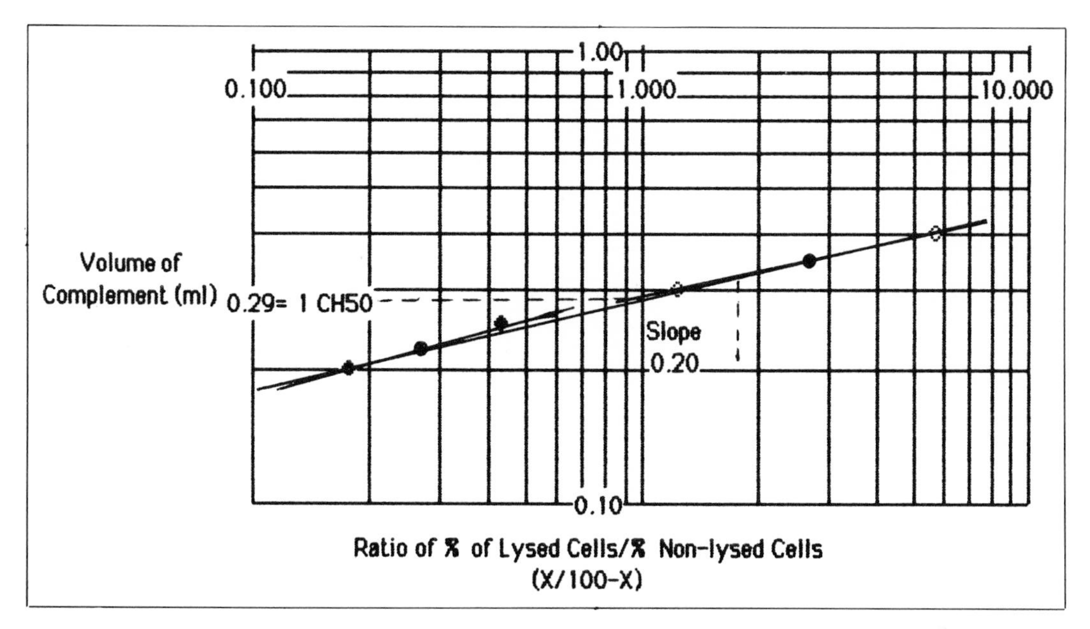

Figure 11.2. Determination of the optimal volume (ml) of complement.

Thus: 400 × 0.4/1.4 = 114
Therefore: 1:114 is the desired dilution of complement to achieve 5 CH_{50} in 0.4 ml

11.9 ANTIGEN TITRATION

11.9.1 Preparation of Initial 1:8 Dilution of Antiserum

1. Place 6.3 ml VBD into a serologic test tube
2. Add 0.9 ml of the antiserum to the test tube and mix
3. Cover the mouth of the test tubes and place in a 56°C water bath for 30 minutes to inactivate the serum

11.9.2 Setting Up and Labeling Test Tubes

1. Label six serologic test tubes for twofold master dilutions of antiserum (i.e., 1:16, 1:32, through 1:512)
2. Label two sets of seven serologic test tubes for twofold antigen dilutions (i.e., 1:2, 1:4, through 1:128—two will be labeled 1:2, two labeled 1:4, through 1:128); one set will be used for complement controls and the other for the antigen titration
3. Label one test tube for the known positive antigen at its optimal dilution
4. Label one test tube for normal tissue antigen dilution, the same dilution as the optimal dilution of the known positive antigen
5. Label and arrange test tubes in racks as outlined in Table 11.6; each square represents a test tube

11.9.3 Preparation of Serum Dilutions

1. Place 3.0 ml cold VBD in each of the six labeled antiserum test tubes
2. From the test tube containing the cooled, inactivated, 1:8 dilution of the antiserum, remove 3.0 ml and add it to the VBD in the tube labeled 1:16 and mix thoroughly; for the sequential dilutions transfer 3.0 ml to the 1:64 dilution tube

Table 11.6. Antigen Titration

	Antigen Dilutions (0.2 ml/Tube)	Antiserum Dilution (0.2 ml/assay tube) 1:8	1:16	1:32	1:64	1:128	1:256	1:512
Assay Tubes	1:2							
	1:4							
	1:8	In addition to antigen and antiserum						
	1:16	add 0.4 ml of complement containing						
	1:32	5 CH_{50} to each assay tube.						
	1:64							
	1:128							
Serum Control	Antiserum VBD 5 CH_{50}	0.2 ml 0.2 ml → 0.4 ml						
Tissue Controls	Antiserum 1:2 Normal Tissue Antigen 5 CH_{50}	0.2 ml 0.2 ml → 0.4 ml						
Known Positive Antigen Control	Antiserum Known Positive Antigen[a] 5 CH_{50}	0.2 ml 0.2 ml → 0.4 ml						
Cell Control	VBD	0.8 ml						

[a] Used at optimal dilution previously determined by titration.

and so on until the 1:512 dilution has been made

11.9.4 Preparation of Antigen Dilutions

1. Place 3.0 ml cold VBD in each of the test antigen dilution test tubes
2. Add 3.0 ml of the test antigen to the VBD in the tube labeled 1:2 and mix well
3. Transfer 3.0 ml from the 1:2 dilution tube to the tube labeled 1:4 and mix well; continue this transfer procedure until all dilutions are made
4. Prepare similar dilutions of a normal tissue antigen
5. Prepare the optimal dilution of the known positive antigen

11.9.5 Setting Up the Antigen Titration

Before setting up the antigen titration, the technician should become thoroughly familiar with Table 11.6. In a sense, two tests are being conducted simultaneously and it is imperative that the correct reagents and volumes are added to the correct tubes

1. Place 0.2 ml of the 1:8 serum dilution in each of the complement–serum control tubes

Table 11.6. (*continued*)

	Components	Complement Controls 5-Unit CH_{50}	2.5-Unit CH_{50}	1.25-Unit CH_{50}
Complement-Antigen Controls	1:2 Antigen	0.2 ml	0.2 ml	0.2 ml
	VBD	0.2 ml	0.4 ml	0.5 ml
	C	0.4 ml	0.2 ml	0.1 ml
	1:4 Antigen	0.2 ml	0.2 ml	0.2 ml
	VBD	0.2 ml	0.4 ml	0.5 ml
	C	0.4 ml	0.2 ml	0.1 ml
	1:8 Antigen	0.2 ml	0.2 ml	0.2 ml
	VBD	0.2 ml	0.4 ml	0.5 ml
	C	0.4 ml	0.2 ml	0.1 ml
	1:16 Antigen	0.2 ml	0.2 ml	0.2 ml
	VBD	0.2 ml	0.4 ml	0.5 ml
	C	0.4 ml	0.2 ml	0.1 ml
	1:32 Antigen	0.2 ml	0.2 ml	0.2 ml
	VBD	0.2 ml	0.4 ml	0.5 ml
	C	0.4 ml	0.2 ml	0.1 ml
	1:64 Antigen	0.2 ml	0.2 ml	0.2 ml
	VBD	0.2 ml	0.4 ml	0.5 ml
	C	0.4 ml	0.2 ml	0.1 ml
	1:128 Antigen	0.2 ml	0.2 ml	0.2 ml
	VBD	0.2 ml	0.4 ml	0.5 ml
	C	0.2 ml	0.2 ml	0.1 ml
Controls	1:8 Antiserum	0.2 ml	0.2 ml	0.2 ml
	VBD	0.2 ml	0.4 ml	0.5 ml
	C	0.4 ml	0.2 ml	0.1 ml
C-Tissue Controls	1:2 Normal Tissue Antigen	0.2 ml	0.2 ml	0.2 ml
	VBD	0.2 ml	0.4 ml	0.5 ml
	C	0.4 ml	0.2 ml	0.1 ml
C-Known Antigen Controls	Known Positive Antigen[a]	0.2 ml	0.2 ml	0.2 ml
	VBD	0.2 ml	0.4 ml	0.5 ml
	C	0.4 ml	0.2 ml	0.1 ml
C-VBD Controls	VBD	0.4 ml	0.6 ml	0.7 ml
	C	0.4 ml	0.2 ml	0.1 ml

2. Place 0.2 ml of the 1:8 serum dilution in all but the cell control tube in the first row; place 0.2 ml of the 1:16 serum dilution in each of the tubes in the next row, etc., until the various dilutions have been added to the appropriate tubes
3. Add 0.2 ml of the test antigen dilutions to the appropriate assay tubes and to the complement–antigen control tubes
4. Add 0.2 ml of 1:2 normal tissue antigen to the tissue control and the complement–tissue control tubes
5. Add 0.2 ml of optimal dilution known positive antigen to the known positive antigen control tubes and the complement-known antigen control tubes
6. Add 0.2 ml cold VBD to the seven serum control tubes
7. Add the volumes of VBD designated in Table 11.6 to the complement-control tubes and the cell-control tube
8. Shake the rack vigorously to mix thoroughly and let it incubate for 15 minutes at room temperature (22°–25°C)

11.9.6 Preparing Complement for Antigen Titration

1. Prepare at least 40 ml of the optimal dilution of complement as determined in the complement titration; to determine the volume of undiluted complement needed to prepare 40 ml of the desired complement concentration, divide the desired volume by the reciprocal of the dilution of complement determined in the complement titration (For example, 40 ml of a 1:114 dilution of complement are needed, therefore, 40 ml/114 = 0.3 ml of undiluted complement must be added to 39.7 ml cold VBD)
2. Swirl the flask gently to mix, but avoid foaming
3. Place the diluted complement in an ice bath and hold at 4°C for at least 20 minutes before using
4. Remove the ice bath with the diluted complement from the refrigerator and add 0.4 ml of the cold diluted complement to each of the antigen titration tubes, with the exception of the cell control tube, and to each of the complement control tubes requiring 5 CH_{50} units; add 0.2 ml to each of the complement control tubes requiring 2.5 CH_{50} units and 0.1 ml to the remainder of the complement control tubes (1.25 CH_{50})
5. Shake each rack to adequately mix reactants and store at 4°C for 15 to 18 hours, usually overnight

11.9.7 Preparing Sensitized Cells

1. Place 12 ml of the standardized 2.8% cell suspension in a 50-ml Erlenmeyer flask
2. Place 12 ml of the optimal hemolysin dilution in another 50-ml Erlenmeyer flask
3. Decant the hemolysin dilution into the cell suspension while rapidly swirling the cells in the flask
4. Place the flask containing the cells in a 37°C water bath for 15 minutes
5. While the cells are incubating, remove the test racks from the refrigerator and allow to warm to room temperature
6. Remove the color standards, prepared within the past 24 hours, from the refrigerator and allow them to warm to room temperature
7. When the 15-minute cell incubation period is over, remove the flask from the water bath and add 0.2 ml of the cell suspension to all tubes in the antigen titration, including the complement controls; mix well by shaking the racks
8. Incubate the tubes by placing the racks in the 37°C water bath for 30 minutes
9. Prepare record sheets for the titration (Table 11.6)

11.9.8 Determining Optimal Antigen Dilution

1. Examine each tube for lysis. Centrifuge at 600 × *g* for 5 minutes any tubes showing incomplete lysis

Table 11.7. Acceptable Percent Hemolysis in Control Test Reactions[a]

	Number of CH_{50} Units		
Control Testing	5	2.50	1.25
Antigen	100	85–100	0–75
VBD	100	90–100	40–75
Antisera	100	90–100	0–75
Tissue	100	85–100	0–75

[a] Values less than those shown are considered to be anticomplementary.

2. Read and record the percentage hemolysis of the complement controls using the color standards; interpolate to the nearest 5%
3. Compare the complement control readings with those in Table 11.7 to determine if they are acceptable; if not acceptable, repeat test
4. Read and record titration results
5. Selection of the optimal antigen dilution requires careful evaluation of the titration results based on the following criteria
   - Approximately 30% hemolysis end points represent the optimal antigen dilutions; draw a line enclosing these, interpolating as necessary (Table 11.8.)
   - All dilutions showing anticomplementary activity plus the next highest dilution must be excluded from the curve
   - Use the lower of two dilutions giving identical fixation reactions
   - If the identical fixation reactions are within the optimal dilution curve, select the dilution that shows the greater fixation in the next higher dilution
   - The known reference antigen titer must be within the twofold dilution of the test antigen dilution selected

11.10 DIAGNOSTIC TEST: MICRO METHOD

11.10.1 Preparing 1:8 Dilutions of Sera

1. Appropriately label serologic test tubes 15 × 85 mm, one for each of the 1:8 dilutions of the unknown, known positive, and known negative sera, and place all tubes in a rack
2. Place 0.7 ml cold, fresh (less than 24 hours old) VBD into each tube
3. Add 0.1 ml of unknown serum to its labeled dilution tube and mix with the pipette; repeat the procedure with the next unknown serum and continue until 1:8 dilutions of all the unknown sera are made; repeat the procedure with the known positive and known negative sera
4. Place the rack in the 56°C water bath for 30 minutes to inactivate the sera; remove and allow to cool to room temperature

11.10.2 Labeling Plates

1. Label the rows of wells on plates for the diagnostic test according to Table 11.9; each square represents a well

11.10.3 Preparing Antigen Dilutions

1. Determine the volume of test antigens required by multiplying the number of wells receiving test antigen by 0.025 ml; add 0.5 ml to allow for pipetting; do the same for the normal tissue antigen
2. Using cold VBD, prepare the test antigen at the optimal dilution as determined in the antigen titration
3. Prepare normal tissue antigen at the same dilution as the test antigen
4. Store antigen dilutions in the refrigerator (4°C) until needed

11.10.4 Preparing Twofold Dilutions of Sera

1. Using a 0.025-ml dropper pipette, place 0.025 ml of VBD into each well, which

Table 11.8. Determination of the Optimal Antigen Dilution

Component	Antigen Dilution	Percent Hemolysis									
		Reference Antiserum Dilutions							Complement–Antigen Controls		
		1:8	1:16	1:32	1:64	1:128	1:256	1:512	5 CH_{50}	2.5 CH_{50}	1.25 CH_{50}
Test antigen	1:2	0	0	0	20	40	60	70	40	0	0
	1:4	0	0	0	30	60	70	80	70	50	0
	1:8	0	0	0	0	30	45	80	100	100	35
	1:16	0	0	0	0	30	85	90	100	100	40
	1:32	0	0	0	0	30	90	100	100	100	50
	1:64	0	50	75	100	100	100	100	100	100	50
	1:128	75	90	100	100	100	100	100	100	100	40
Serum control	None	100	100	100	100	100	100	100	100	100	40
Tissue control	1:2	100	100	100	100	100	100	100	100	100	30
Known reference antigen	Optimal	100	100	100	100	100	100	100	100	100	40
									Complement–VBD Controls		
									100	100	55

Optimal dilution is 1:16. Were it not for the anticomplementary activity noted in the first two dilutions, 1:8 would have been the optimal dilution. This may not be used because of the need to have at least one dilution between the optimal dilution and the last dilution exhibiting anticomplementary activity. Even though dilutions 1:16 and 1:32 are equally acceptable, the lower dilution is the one of choice.

is to contain a serum dilution greater than 1:8

2. Using a 0.2- or 0.5-ml pipette, add 0.05 ml of the cooled 1:8 dilutions of unknown, known negative, and known positive sera to their appropriately labeled wells
3. Using the 0.025-ml loops (microdiluters), pick up 0.025 ml of the 1:8 dilutions and transfer to the wells labeled 1:16 dilution; twirl the loops rapidly for 4 seconds to mix; repeat the process through the 1:256 dilution of each serum

11.10.5 Setting Up the Diagnostic Test

1. Using a 0.025-ml dropper pipette, add 0.025 ml of the optimal dilution of test antigen to the wells in the test rows and to the wells for complement–antigen control
2. In the same manner, add 0.025 ml of normal tissue antigen dilution to the rows of tissue control wells and complement–tissue control wells and cold VBD to the rows of serum control wells
3. Using an appropriate dropper pipette, add the designated volumes of cold VBD to the complement control wells
4. Turn on the mechanical vibrator (Syntron Jogger, Model J-1A, Syntron Co., Homer City, PA, Thomas shaking apparatus, A. H. Thomas, Catalog No. 8927, or similar apparatus)
5. Place the plates on the vibrator to mix for 20 to 30 seconds; remove the plates *before* turning off the vibrator

11.10.6 Preparing Diluted Complement

1. Determine the volume of diluted complement required for the test by multiplying the number of wells in the test by 0.05 ml; add an additional 1 ml to account for loss of fluid when pipetting
2. Calculate the volumes of VBD and complement needed to prepare the required volume of complement at the dilution containing 5 CH_{50} as determined in the complement titration in tubes

   X = V/C where:
   X is the volume of undiluted complement needed
   V is the required volume
   C is the reciprocal of the dilution of complement determined in the complement titration

   Example:
   40 ml of a 1:114 dilution of complement are needed
   X = 40 ml (V)/110 (C) = 0.4 ml
   0.4 ml C + 39.6 ml diluent
3. Place the calculated volume of VBD diluent into a 100-ml flask
4. Add the calculated volume of complement dropwise to the VBD and swirl the flask gently to mix well; avoid foaming
5. Place the flask in an ice bath in the refrigerator for at least 20 minutes
6. Determine the volume of 1:2 dilution of the diluted complement containing 5 CH_{50} in 0.05 ml needed in the test by multiplying the number of wells for the 2.5 unit controls in the test by 0.05 ml; add 0.4 ml excess; calculate the volume of VBD and complement needed and prepare the dilution (step 2)
7. Prepare the same volume of a 1:4 dilution for the 1.25 unit controls

11.10.7 Adding Complement to the Wells

1. Using a 0.05 ml dropper pipette add 0.05 ml of diluted complement containing 5 CH_{50} to each assay and to each five unit complement control wells; shake each plate immediately after all the complement has been added
2. Add 0.05 ml of the 1:2 and 1:4 dilutions to the appropriate complement control

Table 11.9. Micro Method Diagnostic Test

		Components (ml)	Serum Dilution					
			1:8	1:16	1:32	1:64	1:128	1:256
Unknown Serum Assay	Test	Unknown Serum	0.025					
		Test Antigen	0.025	→				
		5 CH_{50}	0.050					
	Serum Controls	Unknown Serum	0.025					
		VBD	0.025	→				
		5 CH_{50}	0.050					
	Tissue Controls	Unknown Serum	0.025					
		Normal Tissue Antigen*	0.025	→				
		5 CH_{50}	0.050					
Known Positive Serum Assay	Test	Known Positive Serum	0.025					
		Test Antigen	0.025	→				
		5 CH_{50}	0.050					
	Serum Controls	Known Positive Serum	0.025					
		VBD	0.025	→				
		5 CH_{50}	0.050					
	Tissue Controls	Known Positive Serum	0.025					
		Normal Tissue Antigen[a]	0.025	→				
		5 CH_{50}	0.050					
Known Negative Serum Assay	Test	Known Negative Serum	0.025					
		Test Antigen	0.025	→				
		5 CH_{50}	0.050					
	Serum Controls	Known Negative Serum	0.025					
		VBD	0.025	→				
		5 CH_{50}	0.050					
	Tissue Controls	Known Negative Serum	0.025					
		Normal Tissue Antigen[a]	0.025	→				
		5 CH_{50}	0.050					
	Cell Control	VBD	0.1					

[a] Normal tissue antigen used at same dilution as test antigen.
[b] The 2.5-unit and 1:25-unit are a 1:2 and a 1:4 dilution of the 5 CH_{50} in 0.05 ml of complement, respectively.

wells (i.e., 2.5 CH_{50} and 1.25 CH_{50}, respectively)

3. Stack the plates one on top of the other with an empty plate on the top to serve as a cover
4. Incubate the plates overnight in the refrigerator for 15 to 18 hours at 4°C

11.10.8 Preparing Sensitized Cells

1. Multiply the number of wells in the test by 0.025 ml to determine the volume of sensitized cells needed; add at least 1.0 ml for pipetting
2. After removing the cell suspension from the refrigerator and swirling it to obtain

Table 11.9. (*continued*)

Complement Controls for Test

Complement Controls	Components (ml)	5-Unit CH_{50}	2.5-Unit[b] CH_{50}	1-Unit[b] CH_{50}
Antigen	Test Antigen	0.025	0.025	0.025
	VBD	0.025	0.025	0.025
	Complement	0.050	0.050	0.050
Tissue[a]	Normal Tissue Antigen	0.025	0.025	0.025
	VBD	0.025	0.025	0.025
	Complement	0.050	0.050	0.050
VBD	VBD	0.050	0.050	0.050
	Complement	0.050	0.050	0.050

a good suspension, place in an appropriately sized flask a volume of the cell suspension equal to half the volume of sensitized cells needed

3. Add an equal volume of optimal hemolysin dilution to the cells in the flask while swirling it rapidly
4. Place the flask in a 37°C water bath for 15 minutes

11.10.9 Adding Sensitized Cells to Wells

1. Remove test plates from the refrigerator and allow to warm to room temperature
2. Add 0.025 ml of the sensitized cells to each well and gently tap the side of the plate to initiate mixing
3. Turn on the mechanical vibrator
4. Place the test plates on the vibrator and remove when the cells are suspended; remove the plates *before* turning off the vibrator
5. Cover each plate with 3-inch transparent plastic or cellophane tape
6. Place the plates in a 37°C incubator for 30 minutes; do *not* stack the plates

11.10.10 Reading and Recording Test Results

1. Make certain that the less than 24-hour-old color standards (0%, 30%, 50%, 70%, 90%, and 100%) for hemolysis evaluation are available
2. Prepare record sheets for the test (Table 11.9)
3. When the 30-minute incubation period is over, remove the plates
4. Using a 0.025-ml dropper pipette add 5 drops of the 30% color standard to a well on each plate used in the test; add 5 drops of *each* of the standards to individual wells on the plate containing the complement controls
5. Centrifuge the plates for 5 minutes at 300 × *g*; if this is not possible, place the plates in the refrigerator for 2 or 3 hours until the cells settle
6. Read and record the complement con-

Table 11.10. Acceptable Limits of Hemolysis in Complement Controls

	CH_{50} Units		
Type of Control	5	2.5	1.25
Antigen	100	85–100	0–75
VBD	100	90–100	40–75
Unknown serum	75–100		
Tissue	100	85–100	0–75
Positive serum	100	90–100	0–75

trols by comparison with the color standards
7. Compare the complement control readings with those in Table 11.10 to determine their acceptability
8. If all of the complement controls are acceptable, read and record the percentage hemolysis in each well in the test; numerical values may be assigned to the percentage of hemolysis (i.e., 0% = 4, 30% = 3, 50% = 2, 70% = 1, 90% = ±, and 100% = −)

REFERENCES

Ada, G.L., and Perry, B.T. 1954. Studies on the soluble complement-fixing antigens of influenza virus III: The nature of the antigens. Australian J. Exp. Biol. Med. Sci. 32:177–186.

Bankowski, R.A., Wichmann, R.W., and Kummer, M. 1953. A complement-fixation test for identification of immunological types of the virus of vesicular exanthema of swine. Am. J. Vet. Res. 14:145–149.

Beveridge, W.I.B., and Burnet, F.M. 1946. The cultivation of viruses and rickettsiae in the chick embryo. Medical Research Council, Special Report Series No. 256. London: His Majesty's Statistics Office, pp. 1–92.

Black, F.L., and Melnick, J.L. 1954. The specificity of the complement fixation test in poliomyelitis. Yale J. Biol. Med. 26:385–393.

Black, F.L., and Melnick, J.L. 1955. Appearance of soluble and cross-reactive complement-fixing antigens on treatment of poliovirus with formalin. Proc. Soc. Exp. Biol. Med. 89:353–355.

Casals, J., and Palacios, R. 1941. The complement fixation test in the diagnosis of virus infections of the central nervous system. J. Exp. Med. 99:429–449.

Casals, J., and Palacios, R. 1949. Acetone–ether extracted antigens for complement fixation with certain neurotropic viruses. Proc. Soc. Exp. Biol. Med. 70:339–343.

Casey, H.L. 1965. Adaptation of LBCF method of microtechnique. In Standard Diagnostic Complement-Fixation Method and Adaptation to Microtest. Public Health Monograph No. 74. Public Health Service Publication No. 1228. Washington, D.C.: U.S. Government Printing Office.

Chambers, L.A., Cohen, S.S., and Clawson, J.R. 1950. Studies on commercial typhus vaccines II. The antigenic fractions of disrupted epidemic typhus rickettsiae. J. Immunol. 65:459–463.

Churcher, G.M., Sheffield, F.W., and Smith, W. 1959. Poliomyelitis virus flocculation: The reactivity of unconcentrated cell-culture fluids. Br. J. Exp. Pathol. 40:87–95.

Ciuca, A. 1929. The reaction of complement-fixation in foot-and-mouth disease as a means of identifying the different types of virus. J. Hyg. 28:325–339.

Clark, D.H., and Casals, J. 1958. Techniques for hemagglutination and hemagglutination-inhibition with arthropod-borne viruses. Am. J. Trop. Med. Hyg. 7:561–573.

Cohen, G.H., and Wilcox, W.C. 1966. Soluble antigens of vaccinia infected mammalian cells. I. Separation of virus-induced soluble antigens into two classes on the basis of physical characteristics. J. Bacteriol. 92:676–686.

Craigie, J., and Wishart, F.O. 1936a. The complement fixation reaction in variola. Canad. Pub. Health J. 27:371–379.

Craigie, J., and Wishart, F. O. 1936b. Studies on the soluble precipitable substances of vaccinia. II. The soluble precipitable substances of dermal vaccine. J. Exp. Med. 64:819–830.

DeBoer, C.J., and Cox, H.R. 1947. Specific complement-fixing diagnostic antigens for neurotropic virus diseases. J. Immunol. 55:193–204.

Dulbecco, R. 1952. Production of plaques in monolayer tissue cultures by single particles of an animal virus. Proc. Natl. Acad. Sci. 38:747–752.

Ende, M., Van den Polson, A., and Turner, G.S. 1957. Experiments with the soluble antigen of rabies in suckling mouse brain. J. Hyg. 55:361–373.

Enders, J.F., Weller, T.H., and Robbins, F.C. 1949. Cultivation of the Lansing strain of poliomyelitis virus in cultures of various human embryonic tissues. Science 109:85–87.

Enders, J.F., and Peebles, T.C. 1954. Propagation in tissue cultures of cytopathogenic agents from patients with measles. Proc. Soc. Exp. Biol. Med. 80:277–286.

España, C., and Hammon, W.McD. 1948. An improved benzene-extracted complement-fixing antigen applied to the diagnosis of the arthropod-borne virus encephalitides. J. Immunol. 59:31–44.

French, E.L. 1952. Murray Valley encephalitis. Isolation and characterization of the aetiological agent. Med. J. Australia 1:100–103.

Girardi, A.J., Warren, J., Goldman, C., and Jeffries, B. 1958. Growth and CF antigenicity of measles virus in cells deriving from human heart. Proc. Soc. Exp. Biol. Med. 98:18–22.

Goodpasture, E.W., Woodruff, A.M., and Buddingh, G.J. 1931. The cultivation of vaccine and other viruses in the chorio-allantoic membrane of chick embryos. Science 74:371–372.

Goodpasture, E.W., Woodruff, A.M., and Buddingh, G.J. 1933. Use of embryo chick in investigation of certain pathological problems. South. Med. J. 26:418–420.

Halonen, P.E., Casey, H.L., Stewart, J.A., and Hall, A.D. 1967. Rubella complement fixing antigen prepared by alkaline extraction of virus grown in suspension culture of BHK-21 cells. Proc. Soc. Exp. Biol. Med. 125:167–172.

Hammon, W.McD., and Sather, G.E. 1969. Arboviruses. In E.H. Lennette and N.J. Schmidt (eds.), Diagnostic Procedures for Viral and Rickettsial Infections, 4th ed. American Public Health Association, pp. 227–280.

Havens, W.P., Jr., Watson, D.W., Green, R.H., Lavin, G.I., and Smadel, J.E. 1943. Complement fixation with neurotropic viruses. J. Exp. Med. 77:139–153.

Hawkes, R. 1979. General principles underlying laboratory diagnoses of viral infections. In E.H. Lennette and N.J. Schmidt (eds.), Diagnostic Procedures for Viral, Rickettsial and Chlamydial Infections, 5th ed. American Public Health Association, pp. 3–48.

Henle, G., Harris, S., and Henle, W., 1948. The reactivity of various human sera with mumps complement fixation antigens. J. Exp. Med. 88:133–147.

Henle, G., Harris, S., Henle, W., and Deinhardt, F. 1955. Propagation and primary isolation of mumps virus in tissue culture. Pro. Soc. Exper. Biol. Med. 89:556–560.

Hilleman, M.R., and Werner, J.H. 1954. Recovery of new agent from patients with acute respiratory illness. Proc. Soc. Exp. Biol. Med. 85:183–188.

Howitt, B.F. 1937. The complement fixation reaction in experimental equine encephalomyelitis, lymphocytic choriomeningitis and St. Louis type of encephalitis. J. Immunol. 33:235–250.

Hoyle, L. 1952. Structure of the influenza virus. The relation between biological activity and chemical structure of virus fractions. J. Hyg. 50:229–245.

Huebner, R.J., Rowe, W.P., Ward, T.G., Parrott, R.H., and Bell, J.A. 1954. Adenoidal–pharyngeal conjunctival agents. A newly recognized group of common viruses of the respiratory system. N. Engl. J. Med. 251:1077–1086.

Kidd, J.G., and Friedewald, W.F. 1942. A natural antibody that reacts *in vitro* with a sedimentable constituent of normal tissue cells. I. Demonstration of the phenomenon. J. Exp. Med. 76:543–556.

LeBouvier, G.L., Laurence, G.D., Parfitt, E.M., Jennens, M.G., and Goffe, A. 1954. Typing of poliomyelitis virus by complement fixation. Lancet ii:531–532.

Lennette, E.H., Wiener, A., Neff, B.J., and Hoffman, M.N. 1956a. A chick embryo-derived complement-fixing antigen for western equine encephalomyelitis. Proc. Soc. Exp. Biol. Med. 92:575–577.

Lennette, E.H., Wiener, A., Ota, M.I., Fujimoto, F.Y., and Hoffman, M.N. 1956b. Rapid identification of isolates of Western equine encephalomyelitis virus by the complement fixation technique. Am. J. Hyg. 64:270–275.

Lennette, E.H. 1969. General principles underlying laboratory diagnosis of viral and rickettsial infections. In E.H. Lennette and N.J. Schmidt (eds.), Diagnostic Procedures for Viral and Rickettsial Diseases, 4th ed. New

York: American Public Health Association, pp. 78–176.

Lichter, A.G. 1953. Observations on anticomplementary reactions. A.M.A. Arch. Dermat. Syph. 67:362–368.

Lief, F.S., and Henle, W. 1956a. Studies on the soluble antigen of influenza virus. I. The release of S antigen from elementary bodies by treatment with ether. Virology 2:753–771.

Lief, F.S., and Henle, W. 1956b. Studies on the soluble antigen of influenza virus. III. The decreased incorporation of S antigen into elementary bodies of increasing incompleteness. Virology 2:753–771.

Maltaner, F. 1946. Significance of thromboplastic activity of antigens used in complement fixation tests. Proc. Soc. Exp. Biol. Med. 62:302–304.

Paucker, K., Lief, F.S., and Henle, W. 1956. Studies on the soluble antigen of influenza virus. IV. Fractionation of elementary bodies labeled with radio-active phosphorus. Virology 2:798–810.

Rice, C.E. 1948a. Inhibitory effects of certain avian and mammalian antisera in specific complement fixation systems. J. Immunol. 59:365–378.

Rice, C.E. 1948b. Some factors influencing the selection of a complement-fixation method. II. Parallel use of the direct and indirect techniques. J. Immunol. 60:11–21.

Robbins, F.C., Enders, J.F., and Weller, T.H. 1950. Cytopathogenic effect of poliomyelitis viruses in vitro on human embryonic tissues. Proc. Soc. Exp. Biol. Med. 75:370–374.

Robbins, F.C., Enders, J.F., Weller, T.H., and Florentino, G.L. 1951. Studies on the cultivation of poliomyelitis viruses in tissue culture. V. The direct isolation and serologic identification of virus strains in tissue culture from patients with non-paralytic and paralytic poliomyelitis. Am. J. Hyg. 54:286–293.

Rowe, W.P., Huebner, R.J., Gilmore, L.K., Parrott, R.H., and Ward, T.G. 1953. Isolation of a cytopathogenic agent from human adenoids undergoing spontaneous degeneration in tissue culture. Proc. Soc. Exp. Biol. Med. 84:570–573.

Ruckle, G., and Rogers, K.D. 1957. Studies with measles virus. II. Isolation of virus and immunological studies in persons who have had the natural disease. J. Immunol. 78:341–355.

Schafer, W., and Zillig, W. 1954. Über den Aufbau des Virus-Elementarteilchens der klassischen Geflügelpest. I. Gewinnung, physikalisch-chemische und biologische Eigenschaften einiger Spaltprodukte. Ztschr. Naturforsch 9b:779–788.

Schmidt, N.J., and Lennette, E.H. 1956. Modification of the homotypic specificity of poliomyelitis complement fixing antigens by heat. J. Exp. Med. 104:99–120.

Schmidt, N.J., Lennette, E.H., Doleman, J.H., and Hagens, S.J. 1957. Factors influencing the potency of poliomyelitis complement-fixing antigens produced in tissue culture systems. Am. J. Hyg. 66:1–19.

Schmidt, N.J. 1957. An inquiry into the use of the complement fixation test for the typing of poliomyelitis viruses. Am. J. Hyg. 66:119–130.

Schmidt, N.J. 1969. Tissue culture methods and procedures for diagnostic virology. In E.H. Lennette and N.J. Schmidt (eds.), Diagnostic Procedures for Viral and Rickettsial Diseases, 4th ed. New York: American Public Health Association, pp. 79–178.

Sever, J.L. 1962. Application of a microtechnique to viral serological investigations. J. Immunol. 88:320–329.

Smadel, J.E., Baird, R.D., and Wall, M.J. 1939. Complement fixation in infections with the virus of lymphocytic choriomeningitis. Proc. Soc. Exp. Biol. Med. 40:71–73.

Sosa-Martinez, J., and Lennette, E.H. 1955. Studies on a complement fixation test for herpes simplex. J. Bacteriol. 70:205–215.

Stevenson, W.D.H., and Butler, M.B.E. 1933. Dermal strains of vaccinia virus grown on the chorio-allantoic membrane of chick embryos. Lancet Vol. i, No. 225:228–230.

Suggs, M.R., Jr., Casey, H.L., Sligh, D.D., Fodor, A.R., and McLimmans, W.F. 1961. A batch-type concentration and purification procedure for poliovirus complement fixing antigen. J. Bacteriol. 82:789–791.

Svedmyr, A., Enders, J.R., and Holloway, A. 1952. Complement fixation with Brunhilde and Lansing poliomyelitis viruses propagated in tissue culture. Proc. Soc. Exp. Biol. Med. 79:296–309.

Svedmyr, A., Enders, J.R., and Halloway, A. 1953. Complement fixation with the three types of poliomyelitis viruses propagated in tissue culture. Am. J. Hyg. 57:60–70.

Takatsy, G. 1950. A new method for the preparation of serial dilutions in a quick and accurate way. Kiserletes Orvostudomany 2:293–296.

Taylor-Robinson, D., and Downie, A.W. 1959. Chickenpox and herpes zoster. I. Complement fixation studies. Br. J. Exp. Pathol. 40:398–409.

Weller, T.H. and Witton, H.M. 1958. The etiologic agents of varicella and herpes zoster. Serologic studies with the viruses as propagated in vitro. J. Exp. Med. 108:869–890.

Westwood, J.C.N., Appleyard, G., Taylor-Robinson, D., and Zwartouw, H.T. 1960. The production of high titre poliovirus in concentrated suspensions of tissue culture cells. Br. J. Exp. Pathol. 41:105–111.

Westwood, J.C.N., Zwartouw, H.T., Appleyard, G., and Titmuss, D.H.J. 1965. Comparison of the soluble antigens and virus particle antigens of vaccinia virus. J. Gen. Microbiol. 38:47–53.

Whitney, E., Kraft, L.M., Lawson, W.B., and Gordon, I. 1953. Noninfectious complement-fixing antigen from embryonated hens' eggs infected with lymphocytic choriomeningitis virus. Proc. Soc. Am. Bact. p. 50.

Woodruff, A.M., and Goodpasture, E.W. 1931. The susceptibility of the chorioallantoic membrane of chick embryos to infection with the fowl-pox virus. Am. J. Pathol. 7:209–222.

Neutralization

Harold C. Ballew

12.1 INTRODUCTION

Neutralization of a virus is defined as the loss of infectivity through reaction of the virus with specific antibody. To perform the virus neutralization test, virus and serum are mixed under appropriate conditions, incubated, and then used to inoculate a susceptible living host to detect unneutralized virus. The presence of unneutralized virus may be detected by reactions such as cytopathic effect (CPE), plaque formation, and metabolic inhibition in cell cultures; death of or paralysis in animals; and pock formation on the chorioallantoic membrane (CAM) of embryonated hens' eggs.

The neutralization test has been used in virology longer than any other serologic procedure. Despite its relative antiquity, neutralization is one of the most specific and widely used serologic procedures in diagnostic virology. Neutralization techniques can be used in virology to identify a virus isolate or to measure the antibody response of an individual to a virus. Because of its high immunologic specificity, the neutralization test is often the standard against which the specificity of other serologic procedures is evaluated.

One of the first principles formulated in virus serologic testing is the so-called "percentage law" (Andrewes and Elford, 1933). This law states that when virus is added to excess antibody, the percentage of virus not neutralized is the same regardless of the amount of virus added. With the advent of plaquing procedures for viruses, the neutralization reaction could be accurately evaluated (Dulbecco et al., 1956). In their studies on the kinetics of the reaction, they found that there is a linear interdependence between the rate of neutralization and the concentration of antibody, and there is always a fraction of virus that is not inactivated.

Other investigators who have helped establish the basic properties of the neutralization test are Tyrrell and Horsfall (1953), Salk et al. (1954), and Mandel (1960). The contributions of these and other investigators to the neutralization test are not described in this chapter but are adequately reviewed in Horsfall and Tamm (1960) and Maramorosch and Koprowski (1967).

12.2 STANDARDIZATION OF TEST MATERIALS

Before the neutralization test is performed, the known components that are to be used must be standardized. To identify a virus isolate, a known pretitered antiserum or standardized serum pool is used. Conversely, to measure the antibody response of an individual to a virus, a known pretitered virus is employed.

12.2.1 Virus

The known virus consists of extracts from infected tissues, or fluids from cell cultures and embryonated hens' eggs. This virus can be purchased commercially or prepared by inoculating a susceptible host system with a stock virus and harvesting it at the optimal time. This known virus should always be titrated in the host system to be used for the neutralization test, whether purchased commercially or prepared by host system inoculation.

To titrate a known virus or virus isolate, prepare serial tenfold dilutions in a maintenance medium and inoculate a susceptible host system with fixed volumes of each dilution. Observe the host for signs of infection, which indicates virus that is not neutralized. The virus endpoint titer is the reciprocal of the highest dilution of virus that infects 50% of the host systems. This endpoint dilution contains one 50% tissue culture infecting dose ($TCID_{50}$) of virus per unit volume, or the amount of virus that will infect 50% of the cell cultures inoculated. In animals, if death is the criterion employed, then this endpoint dilution is called the 50% lethal dose (LD_{50}). The concentration of virus generally used in the neutralization test is 100 TCD_{50} or 100 LD_{50} per unit volume (Figure 12.1).

12.2.2 Serum

A specific immune serum can be purchased commercially or prepared by immunizing susceptible animals and harvesting the serum at the optimal time. The antiserum should always be titrated in the neutralization test against its homologous virus, whether it is purchased commercially or prepared by immunization.

To titrate a specific immune serum or test serum, prepare serial twofold dilutions of the serum and mix each dilution with an equal volume of standardized virus (usually 100 TCD_{50}). The virus and serum mixtures are usually incubated for 1 hour at room temperature or at 37°C. The time and temperature for incubating the virus and serum mixtures varies with different viruses. Inoculate a susceptible host system with each virus–serum mixture. The serum antibody titer is the reciprocal of the highest dilution of the antiserum protecting against the virus. The endpoint dilution contains one antibody (Ab) unit per unit volume. The standardized concentration of antiserum generally used in the neutralization test is 20 antibody units per unit volume (Figure 12.2). In Figure 12.2, 1 Ab unit/0.1 ml is used as an example to demonstrate how antibody units are calculated.

Earlier studies have shown that the neutralization titer of a serum is reduced in certain circumstances by heat, dilution, repeated freezing–thawing, and prolonged storage at 4°C. Part of the neutralization titer may be restored by adding "accessory factors," undefined substances that often are present in freshly collected serum or serum that has been maintained in a frozen state, to the virus–serum mixtures when they are prepared. However, it is usually possible to detect seroconversion in the absence of accessory factors. In addition, these factors may be eliminated by heating the sera to be tested at 56°C for 30 minutes. Many laboratories do nothing about these factors. The sera may be used without inactivation or without the addition of fresh serum. There is no general agreement on the importance of accessory factors, but accessory factors and nonspecific neutralizing substances have been discussed in detail by Ginsberg and Horsfall (1949), Sabin (1950), Tyrrell and Horsfall (1953), and Mandel (1960).

12.2.3 Host System

One of the basic requirements for all neutralization tests is a living host system to demonstrate unneutralized virus. In the

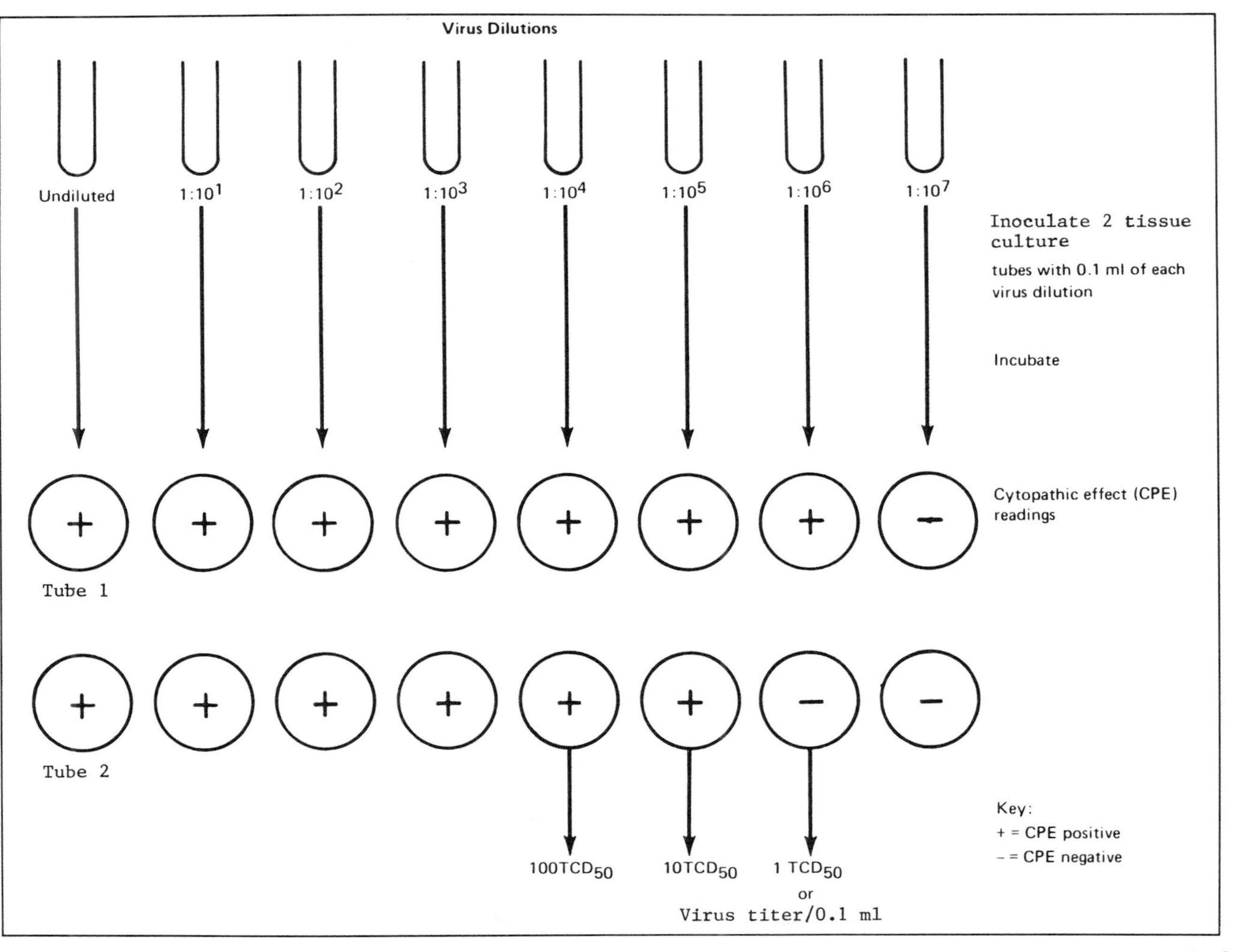

Figure 12.1 This virus titration demonstrates that the endpoint titer is the highest dilution of the virus that infects 50% of the cell cultures tubes inoculated.

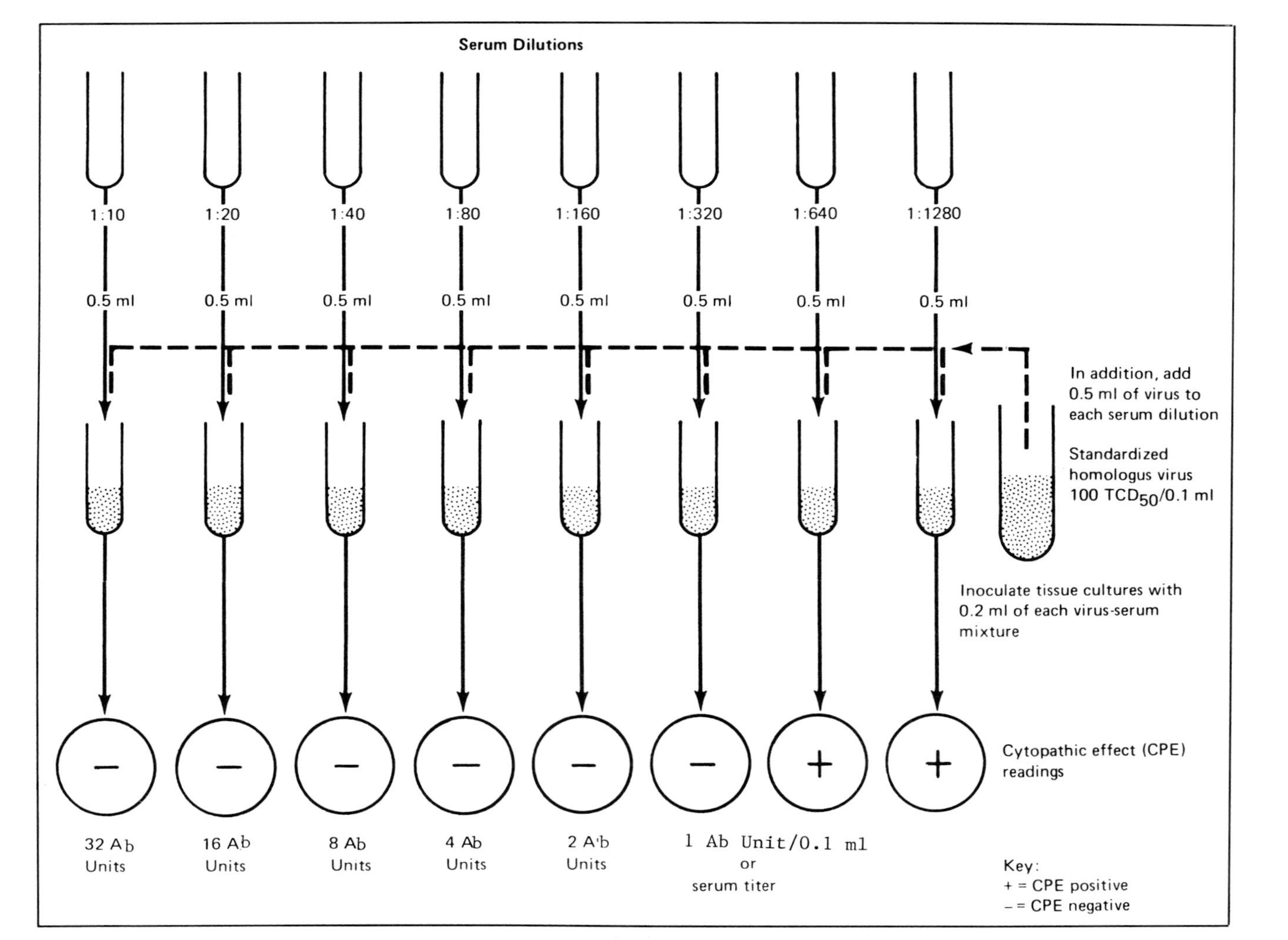

Figure 12.2. This serum titration demonstrates that the antibody titer is the highest dilution of the antiserum protecting against the standardized virus.

neutralization procedure, virus and serum are mixed, incubated, and then injected into or exposed to a susceptible host system in which the presence of surviving virus may be detected. The host used is primarily determined by the infectious and lethal capacity of the virus, as well as by host availability. The three host systems commonly used for the neutralization test are cell cultures, embryonated hens' eggs, and mice.

12.2.3.a Cell Cultures

Where feasible cell cultures are the preferred host. Cell cultures are the most important hosts for performing the neutralization test because they are susceptible to a wide range of viruses, are readily available, and have no immune system to influence the test. The two types of cell cultures used in the neutralization test are suspension and monolayer cultures. In the suspension culture, the virus-infected cells grow floating in the medium. After the virus replicates in the cells, progeny are released into the medium and the unneutralized virus may be detected by reactions such as CPE, change in pH of the medium, or by hemadsorption. In monolayer cell cultures the virus replicates in cell cultures that have been overlaid with agar (Dulbecco et al., 1956). This "agar cover," prevents the virus from establishing secondary foci of infection and keeps the initial infection localized, giving rise to plaques. One plaque is produced by each infectious or plaque forming unit (PFU) in the original virus suspension. The counting of plaques is a very accurate method for quantitating virus. The prevention of PFU with specific antiserum in the neutralization test is called "plaque reduction."

12.2.3.b Embryonated Hens' Eggs

The inoculation of various tissues or cavities of the developing embryo for attempted virus isolation and identification is standard procedure in many laboratories. The route of inoculation depends on the virus to be isolated. For example, the amniotic cavity is used for isolating influenza viruses and mumps virus, and the CAM is often used for isolating variola virus, vaccinia virus, and herpes simplex virus (HSV). Unneutralized virus will replicate in the embryonated hens' egg and can be detected by the death of the embryo, by pock formation on the CAM, or by the agglutination of erythrocytes using amniotic or allantoic fluids containing hemagglutinating virus.

12.2.3.c Mice

Adult and suckling mice are the animals most frequently used for virus isolation and identification. The white Swiss mouse is used extensively for isolating and determining unneutralized arboviruses, certain enteroviruses, and rabiesvirus. The age of the mouse has a great influence on its susceptibility to disease. Thus, suckling mice may be more susceptible to viruses than adult mice. Unneutralized virus is usually determined by death of or paralysis in infected mice.

12.3 TEST PROCEDURES

Cell culture is the most frequently used host system for performing the neutralization test. In the following types of neutralization tests, suspension cell cultures primarily will be used as the host system.

12.3.1 Constant Virus, Varying Serum

Mix a selected dilution of virus (usually 100 TCD_{50}, as determined from previous titration) with varying dilutions of acute- and convalescent-phase sera. Incubate the virus–serum mixture and then inoculate a susceptible host system with the mixture. The reciprocal of the highest dilution of acute- and convalescent-phase sera protecting the host against the virus is the serum titer.

12.3.1.a Materials

1. Susceptible host system such as cell cultures, embryonated hens' eggs, or mice
2. Maintenance medium
3. Known positive virus (pretitered and standardized to contain 100 TCD_{50} or LD_{50}/0.1 ml)
4. Acute- and convalescent-phase sera (usually inactivated at 56°C for 30 minutes)

12.3.1.b Procedures

1. Prepare serial twofold dilutions of the acute- and convalescent-phase sera (1:10 through 1:5120) in 0.5-ml volumes
2. Mix 0.5 ml of standardized known positive virus with each serum dilution
3. Dilute the known standardized virus (1:10 through 1:1000)
4. Incubate the virus–serum mixtures at about 37°C for 1 hour
5. After incubation, inoculate each of three cell culture tubes with 0.2 ml of each virus–serum mixture (when inoculating mice, use 0.02–0.04 ml of mixture)
6. Inoculate three cell cultures with 0.1 ml of known positive virus dilutions (undiluted through 1:1000); these dilutions are used as a back titration to confirm the test potency of 100 TCD_{50} (Figure 12.1)
7. Include three uninoculated cell cultures for controls
8. Incubate the cell cultures at 33°–35°C in a slanted position
9. Observe the cell cultures daily for CPE

12.3.1.c Interpretation

A fourfold rise in antibody titer between the acute- and convalescent-phase sera is considered to be diagnostically significant. In this test, CPE in cell cultures was used to detect unneutralized virus. This unneutralized virus may be detected in embryonated hens' eggs by pocks on the CAM and in mice by death or paralysis.

12.3.2 Constant Antiserum, Varying Virus

For virus identification varying dilutions of the virus are mixed with a constant antiserum dilution. These dilutions are incubated to allow the virus and antiserum to react. Each virus and serum mixture is then inoculated into a susceptible host system. The dilution of virus that infects 50% of the host systems is considered the endpoint dilution.

12.3.2.a Materials

1. Susceptible host system such as cell cultures, embryonated hens' eggs, or mice
2. Maintenance medium, such as Eagle's modified essential medium (EMEM)
3. Known positive antiserum (inactivated at 56°C for 30 minutes and standardized to contain 20 Ab units/0.1 ml)
4. Known negative serum (inactivated at 56°C for 30 minutes)
5. Virus isolate

12.3.2.b Procedures

1. Prepare serial tenfold dilutions (1:10 through $1:10^8$) of the virus isolate in maintenance medium
2. Mix 0.5 ml of each isolate dilution (undiluted through $1:10^8$) with 0.5 ml of known standardized antiserum
3. Mix 0.5 ml of each isolate dilution (undiluted through $1:10^8$) of the virus with 0.5 ml of known negative serum
4. Incubate the virus–serum mixtures at 37°C for 1 hour
5. Inoculate each of three cell culture tubes with 0.2 ml of each virus–serum mixture (when inoculating mice, use 0.02–0.04 ml of the mixture)
6. Include three uninoculated cell cultures for controls
7. Incubate the cell cultures at 33°–35°C in a slanted position
8. Observe the cell cultures daily for CPE

12.3.2.c Interpretation

The dilution of virus that infects 50% of the host system is considered the endpoint dilution. A difference of at least 2 logs or two tubes must be demonstrated between the normal and the immune antiserum to show significant neutralization. Virus dilutions are generally made using tenfold dilutions. The standardized antiserum dilution in this test is selected based on its ability to neutralize 100 TCD_{50} or 100 LD_{50} of virus. The neutralization of 100 TCD_{50} (LD_{50}) of virus gives the greatest amount of sensitivity and specificity for the test.

12.3.3 Constant Virus, Constant Antiserum

A selected dilution of virus (usually 100 TCD_{50} as determined from prior titration) is mixed with a selected dilution of known antiserum (usually 20 Ab units/0.1 ml). The mixture is incubated and then injected into a susceptible host system for observation of unneutralized virus. The virus is identified if the antiserum neutralizes the infectivity of the virus.

12.3.3.a Material

1. Susceptible host system such as cell cultures, embryonated hens' egg, or mice
2. Maintenance medium
3. Known positive antiserum (inactivated at 56°C for 30 minutes and standardized to contain 20 Ab units/0.1 ml)
4. Unknown virus isolated (pretitered and standardized to contain 100 TCD_{50}/0.1 ml)

12.3.3.b Procedures

1. Mix 0.5 ml of the standardized unknown virus isolate with 0.5 ml of the known standardized positive antiserum
2. Dilute the standardized unknown virus isolate (1:10 through 1:1000)
3. Incubate the virus–serum mixture at about 37°C for 1 hour
4. Inoculate each of three susceptible cell culture tubes with 0.2 ml of the virus–serum mixture (when inoculating mice, use 0.02–0.04 ml of the mixture)
5. Inoculate three susceptible cell culture tubes with 0.1 ml of the standardized unknown virus dilutions (undiluted through 1:1000). These dilutions are used as a back titration to confirm the test potency of 100 TCD_{50}
6. Include three uninoculated cell culture tubes for controls
7. Incubate the cell cultures at 33°–35°C in a slanted position
8. Observe the cell cultures daily for CPE

12.3.3.c Interpretation

The absence of CPE in the cell cultures inoculated with the virus–serum mixtures identifies the virus isolate. Antibody in serum can be determined also by running a single dilution of a test serum against a standard dose of known virus. Neutralization indicates the presence of specific antibody.

12.3.4 Varying Virus, Varying Antiserum

This type of neutralization test should be used with caution. The titers of the unknown virus and the known antiserum have not been predetermined. Varying dilutions of the virus and antiserum are made, combined, incubated, and then inoculated into a susceptible host system in which the presence of unneutralized virus may be detected. The test can actually give maximum information about both the virus and the antiserum in relation to one another. Although this test design can be used for neutralization results, it is mainly used as a routine procedure in the block titration of antigen in the complement fixation test.

12.4 CALCULATIONS OF 50 PERCENT ENDPOINTS

The 50% endpoint can be used in several different reactions. A 50% mortality ratio is expressed as LD_{50} and a 50% infective dose, as ID_{50}; in cell cultures, a CPE in 50% of the cultures is expressed as TCD_{50}.

12.4.1 Reed–Muench Method

See Tables 12.1 and 12.2.

12.4.2 Karber Method for Calculating 50 Percent Mortality

Karber Formula: Negative logarithm of the 50% end-point titer = negative logarithm of the highest virus concentration used

$$\left[\frac{(\text{Sum of \% Mortality at each dilution} - 0.5)}{100} \times (\text{logarithm of dilution})\right]$$

Example:

Virus Dilution	*Mortality Ratio*	*% Mortality*
10^{-1} (1:10^1)	8/8	100
10^{-2} (1:10^2)	8/8	100
10^{-3} (1:10^3)	8/8	100
10^{-4} (1:10^4)	4/8	50
10^{-5} (1:10^5)	0/8	0

$$= -1.0 - \left[\left(\frac{(100 + 100 + -100 + 50)}{100} - 0.5\right) \times (\log_{10})\right]$$

$$= -1.0 - \left[\left(\frac{350}{100} - 0.5\right) \times (\log_{10})\right]$$

$$= 1.0\ [(3.5 - 0.5) \times 1]$$

$$= -1.0 - 3.0$$

$$= 4.0$$

Table 12.1. Calculation of 50 Percent Mortality (Virus Titration)

Virus Dilution	Deaths per Number Inoculated	Cumulative Deaths	Cumulative Survivors	Mortality Ratio	Percent Mortality
10^{-4} (1:10^4)	5/5	10	0	10/10	100%
10^{-5} (1:10^5)	4/5	5	1	5/6	83%
10^{-6} (1:10^6)	1/5	1	5	1 6	17%
10^{-7} (1:10^7)	0/5	0	10	0/10	0%

Interpolation Formula:

$$\frac{\%\ \text{Mortality Greater than 50\%–50\%}}{\%\ \text{Mortality Greater than 50\%–\% Mortality less than 50\%}}$$

Substituting: $\dfrac{83 - 50}{83 - 17} = \dfrac{33}{66} = 0.5$

a. Multiply the interpolative value times the negative $\log_{10}$ of the dilution ratio.
   Negative $\log_{10}$ of the dilution ration 10 = -1
   Interpolative value = $\times$ 0.5
   Corrected interpolative value = -0.5

b. The endpoint dilution associated with 50% mortality is located between the 10^{-5} and 10^{-6} dilution.

c. The $\log_{10}$ of the 50% endpoint dilution is estimated by adding the corrected interpolative value to the $\log_{10}$ of the dilution above 50%

   $-5 + (-0.5) = -5.5$

d. The 50% endpoint dilution is estimated at $10^{-5.5}$.

e. The 50% − titer is estimated at $10^{5.5}$.

Table 12.2. Calculation of 50 Percent Endpoint Dilution of the Virus Plus Constant Immune Serum

Virus Dilution	+	Constant Serum	Deaths per Number Inoculated	Cumulative Deaths	Cumulative Survivors	Mortality Ratio	Percent Mortality
10^{-2} (1:10^2)	+	IS	5/5	10	0	10/10	100
10^{-3} (1:10^3)	+	IS	3/5	5	2	5/7	71
10^{-4} (1:10^4)	+	IS	2/5	2	5	2/7	29
10^{-5} (1:10^5)	+	IS	0/5	0	10	0/10	0

Interpolation Formula:

$$\frac{\%\text{ Mortality Greater than }50\%-50\%}{\%\text{ Mortality Greater than }50\%-\%\text{ Mortality less than }50\%}$$

Substituting: $\frac{71 - 50}{71 - 29} = \frac{21}{42} = 0.5$

a. Multiply the interpolative value times the negative $\log_{10}$ of the dilution ratio.
   Negative $\log_{10}$ of the dilution ration 10 $= -1$
   Interpolative value $= \times\ 0.5$
   Corrected interpolative value $= -0.5$
b. The endpoint dilution associated with 50% mortality is located between the 10^{-3} and 10^{-4} dilutions.
c. The $\log_{10}$ of the 50% endpoint dilution is estimated by adding the corrected interpolative value to the $\log_{10}$ of the dilution above 50%

$$-3 + (-0.5) = -3.5$$

d. The 50% endpoint dilution is estimated at $10^{-3.5}$.
e. The 50% – titer is estimated at $10^{3.5}$.

The logarithmic difference between the 50% titers of the virus titration in procedure 1 and the virus plus immune serum titration in procedure 2 is 2.0 [5.5 – 3.5]. Generally, a reduction of at least 2 in $\log_{10}$ (titer) of the virus must be demonstrated by a test serum to show significant neutralization.

The 50% endpoint dilution $= 10^{-4.0}$.
The 50% endpoint titer $= 10^{4.0}$.

12.5 PREPARATION OF SERUM POOLS

To avoid the time and expense required for typing each enterovirus isolate using individual antiserum, the use of serum pools is recommended (Lim and Benyesh-Melnick, 1960; Schmidt et al., 1961). In one type of serum pool, several antisera that react against different enterovirus serotypes are combined into a pool. A number of pools containing different antisera are set up. Neutralization of the isolate by one of the serum pools indicates that the isolate is one of the enteroviruses whose antiserum is included in the pool. For final identification, the isolate is run in a neutralization test against each antiserum in the pool. In the second type of serum pool, each antiserum is present in two different intersecting pools. For example, antibody to poliovirus type 1 (PV-1) is the only antiserum that is included in both pools C and F, and PV-1 is the only virus that will be neutralized by both of these pools. The method of incorporating antisera into intersecting pool schemes is not included in the chapter but is adequately reviewed by Lennette and Schmidt (1979) and Melnick et al. (1977).

Serum pools should be prepared with only known pretitered antisera. The procedure for preparing enterovirus serum pools follows.

> High-titered antiserum must be used. The titers of each serum must be high enough so that their combination in a pool will result in a total serum concentration not greater than 10%. Titer each serum individually against 100 TCD_{50}/0.1 ml of its prototype virus (Figure 12.2).

1. Prepare serial twofold dilutions (1:10 through 1:1280) of the serum and mix each dilution with an equal volume containing 100 TCD_{50} of the virus
2. After incubation at 37°C for 1 hour, inoculate three cell culture tubes (animals may be used) with 0.2 ml of each virus–serum mixture
3. Determine the titer of each serum by examining the cell cultures for CPE; this endpoint dilution or antibody titer is the last dilution of serum demonstrating complete neutralization (no CPE); this endpoint dilution contains 1 Ab unit/0.1 ml
4. 20 Ab units/0.1 ml are required in the pool; to calculate 20 Ab units/0.1 ml, divide 20 into the denominator of the dilution containing 1 Ab unit/0.1 ml

Example: $\frac{320}{20} = 16$ (thus, a 1/16 dilution contains 20 Ab units/0.1 ml)

With the sera titered and the dilution containing 20 Ab units of serum/0.1 ml calculated, incorporate the sera into the pool at the dilution that will give 20 Ab units/0.1 ml.

Example: To make a pool of polio 1, 2, and 3, immune serum (IS) have the following titers:

Polio 1 IS, 1:5120
Polio 2 IS, 1:20,480
Polio 3 IS, 1:10,240

Divide the titers by 20 to get 20 Ab units/0.1 ml.

Dilution containing 20 Ab units/0.1 ml:

$$\text{Polio 1 IS } \frac{5120}{20} = 256$$

$$\text{Polio 2 IS } \frac{20{,}480}{20} = 1024$$

$$\text{Polio 3 IS } \frac{10{,}240}{20} = 512$$

Decide on the volume of the pool you want to make. To make a 1000 ml polio pool using the figures in the example above, divide the denominator of the dilution containing 20 Ab units/0.1 ml into the volume of the pool desired to obtain the amount of each undilute serum to be added to the pool.

Example: $\text{Polio 1 IS } \frac{1000}{256} = 3.90 \text{ ml}$

$$\text{Polio 2 IS } \frac{1000}{1024} = 0.98 \text{ ml}$$

$$\text{Polio 3 IS } \frac{1000}{512} = 1.95 \text{ ml}$$

Total the volumes of sera to be added:

3.90 ml
0.98 ml
+1.95 ml
6.83 ml

Subtract this from the total volume of the pool to determine the amount of dilutent to be used:

1000.00 ml
− 6.83 ml
993.17 ml

Combine the diluent (maintenance medium) with the immune sera in the following volumes:

Maintenance medium	993.17 ml
Polio 1 IS	3.90 ml

Polio 2 IS	0.98 ml
Polio 3 IS	1.95 ml

Aliquot the pools and store frozen at −20°C or lower. This results in a pool containing 20 Ab units/0.1 ml of each serum.

12.6 INTERPRETATION OF SEROLOGIC RESULTS

The neutralization test is frequently used to identify a virus isolate and is also used to indicate a recent virus infection by demonstrating a significant rise in antibody titer to a specific virus using paired sera. Usually, a fourfold rise in antibody titer between an acute- and a convalescent-phase serum run in the same test is considered to be diagnostically significant. Recent infections cannot be distinguished from past infections by examining a single serum specimen. A high antibody titer with a single serum specimen is no guarantee of recent infection.

However, a single serum specimen may be of value in determining if an individual has been exposed to a viral agent at some time in the past. In some diseases, the presence of antibody against the causative agent indicates immunity to the disease. Information gained from a single serum specimen may also be of epidemiologic value in determining the number of individuals in a population exposed to certain agents and, where applicable, in assessing the herd immunity.

The serum collection time is extremely important in all serologic tests. The acute serum should be collected as soon as possible after the onset of disease. The convalescent serum is usually collected 2 to 3 weeks later. If the acute serum is drawn too late, antibodies may be approaching maximal levels; therefore, a significant rise in antibody titer may not be demonstrated.

12.6.1 Interpretation of Tests for Viral Antibodies

The following test examples demonstrate the basic principles for interpreting serologic results and do not represent tests for any particular virus.

Key: + = Virus infectivity, − = No virus infectivity.

Test 1

	Undiluted	1:10	1:20	1:40	1:80	1:160	1:320	1:640	1:1280	Serum Controls
S-1 Acute-phase serum	+	+	+	+	+	+	+	+	+	−
S-2 Convalescent-phase serum	+	+	+	+	+	+	+	+	+	−

The results in this test indicate the individual has not been exposed to the virus tested.

Test 2

	Undiluted	1:10	1:20	1:40	1:80	1:160	1:320	1:640	1:1280	Serum Controls
S-1 Acute-phase serum	−	−	+	+	+	+	+	+	+	−
S-2 Convalescent-phase serum	−	−	−	−	−	+	+	+	+	−

The results demonstrate an eightfold rise in antibody titer between the acute and convalescent serum. The results suggest recent infection with the virus tested.

Test 3

	Undiluted	1:10	1:20	1:40	1:80	1:160	1:320	1:640	1:1280	Serum Control
Single Serum	–	–	–	–	–	–	+	+	+	–

The 160 antibody titer on a single serum indicates past infection with the antigen tested at some time in the past. A high antibody titer is no guarantee of a recent infection.

Test 4

	Undiluted	1:10	1:20	1:40	1:80	1:160	1:320	1:640	1:1280	Serum Controls
S-1 Acute-phase serum	–	–	–	–	–	+	+	+	+	–
S-2 Convalescent-phase serum	–	–	–	–	–	+	+	+	+	–

The 80 antibody titers of the paired sera indicate infection with the antigen tested at some time in the past. It is very important to collect the sera at the proper time. If the acute serum is drawn too late, the antibody titer may have already risen to such a point that, when it is compared with the convalescent serum titer, an antibody rise may not be present.

REFERENCES

Andrewes, C.H., and Elford, W.J. 1933. Observation on anti page 1: The percentage law. Br. J. Exp. Pathol. 14:367–374.

Dulbecco, R., Voit, M., and Strickland, A.G.R. 1956. A study on the basic aspects of neutralization of two animal viruses, western equine encephalitis virus and poliomyelitis virus. Virology 2:162–205.

Ginsberg, H.S., and Horsfall, F.L., Jr. 1949. A labile component of normal serum which combines with various viruses. Neutralization of infectivity and inhibition of hemagglutination by the component. J. Exp. Med. 90:475–495.

Horsfall, F.L., Jr., and Tamm, I., eds. 1960. Viral and Rickettsial Infections of Man, 4th ed. Philadelphia: JB Lippincott.

Lennette, E.H., and Schmidt, N.J., eds. 1979. Diagnostic Procedures for Viral, Rickettsial and Chlamydia Infections, 5th ed. Washington, D.C.: American Public Health Association.

Lim, K.A., and Benyesh-Melnick, M. 1960. Typing of viruses by combination of antiserum pools. Application of typing of enteroviruses (coxsackie and ECHO). J. Immunol. 84:309–317.

Mandel, B. 1960. Neutralization of viral infectivity. Characterization of virus–antibody complex, including association, disassociation and host cell interaction. Ann. N.Y. Acad. Sci. 83:515–527.

Maramorosch, K., and Koprowski, H., eds. 1967. Methods in Virology, Vol. III. New York: Academic Press.

Melnick, J.L., Schmidt, N.J., Hampil, B., and Ho, H.H. 1977. Lyophilized combination pools of enterovirus equine antisera: Preparation and test procedures for the identification of field strains of 19 group A coxsackie serotypes. Intervirology 8:172–181.

Sabin, A.B. 1950. The dengue group of viruses and its family relationship. Bact. Rev. 14:225–232.

Salk, J.E., Younger, J.S., and Ward, E.N. 1954. Use of color change of phenol red as the indicator in titrating poliomyelitis virus or its antibody in a tissue-culture system. Am. J. Hyg. 60:214–230.

Schmidt, N.J., Guenther, R.W., and Lennette, E.H. 1961. Typing of ECHO virus isolates of immune serum pools, the interseting serum scheme. J. Immunol. 87:623–626.

Tyrrell, D.A.J., and Horsfall, F.L., Jr. 1953. Neutralization properties. J. Exp. Med. 97:845–862.

BIBLIOGRAPHY

Ballew, H.C., Forrester, F.T., Lyerla, H.C., Velleca, W.M., and Bird, B.R. 1977. Laboratory Diagnosis of Viral Diseases. Atlanta: Centers for Disease Control.

Ballew, H.C., Forrester, F.T., Lyerla, H.C., Velleca, W.M., Bird, B.R., and Roberts, J.D. 1979a. Basic Laboratory Methods in Virology. Atlanta: Centers for Disease Control.

Ballew, H.C., Forrester, F.T., and Lyerla, H.C. 1983. Laboratory Methods for Diagnosing Respiratory Virus Infections. Atlanta: Centers for Disease Control.

Ballew, H.C., Lyerla, H.C., and Forrester, F.T. 1979b. Laboratory Methods for Diagnosing Herpesvirus Infections. Atlanta: Centers for Disease Control.

Habel, K., and Salzman, N.P. 1969. Fundamental techniques in virology. New York: Academic Press.

Hatch, M.H., and Marchetti, G.E. 1971. Isolation of echoviruses with human embryonic lung fibroblast cells. Appl. Microbiol. 22:736–737.

Hsiung, G.D. 1982. Diagnostic Virology, 3rd ed. New Haven: Yale University Press.

Lennette, E.H., Balows, A., Hausler, W.J., Jr., and Truant, J.P., eds. 1980. Manual of Clinical Microbiology, 3rd ed. Washington, D.C.: American Society of Microbiology.

Lennette, D.A., Specter, S., and Thompson, K.D., eds. 1979. Diagnosis of viral infections. Baltimore: University Park Press.

Melnick, J.L., and Hampil, B. 1965. WHO collaborative studies on enterovirus reference antisera. (Bulletin WHO 33) Geneva: World Health Organization.

Wallis, C., and Melnick, J.F. 1967. Virus aggregation as the cause of non-neutralizable persistent fraction. J. Virol. 1:478–488.

13

Hemagglutination Inhibition and Hemadsorption

Leroy C. McLaren

13.1 HEMAGGLUTINATION INHIBITION TEST

A wide variety of viruses possess the property of hemagglutination [i.e., the ability to agglutinate the erythrocytes (RBC) of one or more animal species under certain conditions]. Influenza, parainfluenza, arboviruses, adenoviruses, rubella, and some strains of picornaviruses have this property. Antibodies to these viruses have the ability to react with the virus and prevent viral hemagglutination. This is the principle of the hemagglutination inhibition test (HAI), which is extremely valuable in the serologic diagnosis of infections. The test is most frequently used in the routine clinical virology laboratory for influenza and parainfluenza viruses, but laboratories with special interests also use this test for the serologic diagnosis, or viral identification, of other hemagglutinating viruses of medical importance.

The HAI test is simple to perform and requires relatively inexpensive equipment and reagents. Serial dilutions of patients' sera are allowed to react with a fixed dose of viral hemagglutinin (HA), followed by the addition of agglutinable erythrocytes. In the presence of antibody, the ability of the virus to agglutinate the erythrocytes is inhibited. In some viral systems, the test can be complicated by the presence of nonspecific viral inhibitors (nonantibody) in sera that must, therefore, be treated prior to being used. The presence of such inhibitors can give rise to false-positive results in the HAI test.

The specificity of the HAI test varies with the virus. The reaction can be highly specific for certain viruses (influenza–parainfluenza groups) and less specific for other viruses (arboviruses).

The method presented in this section will concern the influenza virus system with which most routine clinical virology laboratories are concerned. Conditions for conducting the HAI test for other viruses are presented in the Appendix. Specific methods for these viruses can be found in a number of excellent references (Lennette and Schmidt, 1979; Lennette et al., 1979; Howard 1982; McLean, 1982; Washington, 1985).

13.1.1 Materials

1. Blood from a guinea pig or other mammalian or avian species
2. Amniotic, allantoic, or tissue culture fluid containing influenza virus
3. Alsever's solution
4. Phosphate buffered saline (PBS), 0.01 *M*, pH 7.2

5. Disposable microtiter plates, "U" type
6. Calibrated diluting loops, 0.025 and 0.050 ml
7. Dropping pipettes, 0.025 and 0.050 ml per drop
8. Microtiter plate reading mirror
9. 37°C and 56°C water baths
10. Receptor destroying enzyme (RDE), 100 units/ml

See Appendix for sources.

13.1.2 Preparation of Red Blood Cells Suspension

1. Guinea pig, human, or chicken blood is collected in either Alsever's solution or in heparin to prevent clotting; cells are washed three times in PBS by centrifugation at 1800 × *g* for 5 minutes and suspended in PBS as a 10% stock suspension, which should be stored for no more than 1 week at 4°C
2. For use in the following tests the red blood cell (RBC) stock suspension should be diluted to 0.5% in PBS prior to use (1 ml of 10% RBC added to 19 ml of PBS)

13.1.3 Titration of Hemagglutinin

1. Influenza stock suspension of a contemporary strain of each influenza type A and type B strain are diluted twofold, from 1:10 to 1:2560, employing a 0.05-ml volume to each well
2. Include an RBC control well by adding 0.05 ml PBS
3. Add 0.05 ml of the diluted RBC suspension to each mitrotiter well, gently mix, and incubate at room temperature until the RBC settle to the bottom of each well (1–2 hours)
4. Read the HA titer by determining the highest dilution of virus capable of causing agglutination; agglutinated RBC will form a confluent pattern of RBC on the bottom of the well; nonagglutinated RBC will form a discrete button at the bottom of the well (Figure 13.1).
5. The HA titer is the reciprocal of the highest dilution of virus showing agglutination and represents 1 HA unit/0.05 ml of virus
6. For the HAI test, the virus suspension will be diluted to contain 4 HA units/0.025 ml (or 8 HA units/0.05 ml); thus, if the HA titer is 160 then, for the HAI test, the original virus stock will be diluted 1:20

13.1.4 Serum Treatment

Many influenza and parainfluenza virus strains are sensitive to serum factors that may inhibit hemagglutination nonspecifically. In most instances, these inhibitors can be removed from human sera by treat-

Figure 13.1. Determination of viral hemagglutinin titer.

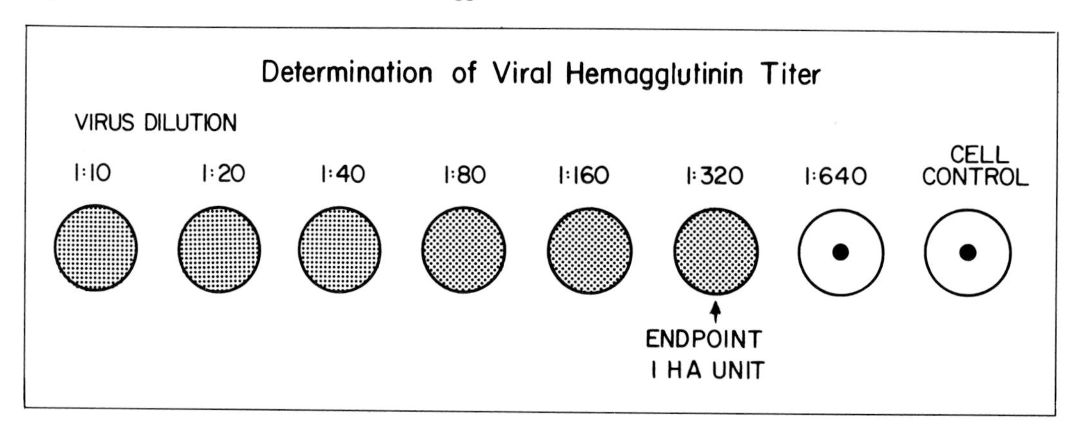

ment with RDE and by heat inactivation, potassium periodate (KIO) treatment, or with kaolin.

13.1.5 RDE Treatment

1. Add 0.4 ml of RDE (100 units/ml), which is commercially available, to 0.1 ml of serum and incubate overnight in a water bath at 37°C
2. Add 0.3 ml of 2.5% sodium citrate to the RDE–serum mixture and incubate in a 56°C water bath for 30 minutes
3. Add 0.2 ml of PBS to each inactivated serum mixture to give a final serum dilution of 1:10

13.1.6 KIO Treatment

1. Add 0.3 ml of 0.01 *M* KIO_4 to 0.1 ml of serum and incubate the serum–KIO_4 mixture for 15 minutes at room temperature
2. Add 0.3 ml of 1% glycerol (in PBS) to neutralize excess KIO_4
3. Add 0.3 ml PBS to each serum mixture to give a final serum dilution of 1:10

13.1.7 Kaolin Treatment

1. Prepare a 1:5 dilution of serum
2. Add an equal volume of acid washed kaolin (25% suspension in PBS) to the diluted serum; mix thoroughly and allow to stand at room temperature for 30 minutes with intermittent shaking
3. Centrifuge at 3000 rpm for 30 minutes
4. Carefully remove the supernatant serum, which has a final dilution of 1:10

13.1.8 Removal of Naturally Occurring Agglutinins for Red Blood Cells

Some sera may contain agglutinins for human O or chicken RBC. In the HAI test that follows, if the serum control wells (serum and RBC, but no virus) show agglutination, the agglutinins can be adsorbed with erythrocytes and the HAI test repeated.

1. Add 0.1 ml of a 50% suspension of the RBC being agglutinated to 1.0 ml of the 1:10 dilution of heat-inactivated (56°C for 30 minutes) serum
2. Incubate for 1 hour at 4°C and remove the RBC by centrifugation at 2000 rpm for 10 minutes

13.1.9 Hemagglutination Inhibition Test for Influenza

Paired sera (acute and convalescent) treated to remove nonspecific inhibitors should be tested for antibodies to one or more contemporary strains of virus. The hemagglutinin titer for each virus employed should be previously titrated for preparation of 4 HA units/0.025 ml in PBS. Reference antiserum for each virus should be included to confirm the identity of viral strains used in the test.

1. Prepare twofold dilutions of each treated serum, from 1:10 to 1:2560
2. Add 0.025 ml serum dilution to wells for each virus employed
3. Add 0.025 ml of virus suspension (containing 4 HA units/0.025 ml) to each serum dilution
4. Include a hemagglutinin antigen control for each virus used (0.025 ml PBS and 0.025 ml hemagglutinin)
5. Include a serum control for each serum being titrated (0.025 ml of 1:10 dilution of serum and 0.025 ml PBS)
6. A back titration of the test hemagglutinin antigen dilution is necessary to insure that the amount of hemagglutinin antigen used in the HAI test contains 4 HA units/0.025 ml
   - Add 0.05 ml PBS to each of five wells
   - Add 0.05 ml diluted test antigen to the first well
   - Make serial twofold dilutions with a 0.05-ml loop
   - Add 0.05 ml RBC to each well
7. Include an RBC control (0.05 ml PBS)
8. Gently shake the microtiter plate after serum dilutions and hemagglutinin have

been added and incubate at room temperature for 30 minutes

9. Add 0.05 ml RBC suspension to each well
10. Gently shake the microtiter plate and incubate the plate at room temperature until the RBC control shows a button at the bottom of the well
11. The serum controls should show the absence of RBC agglutination
12. The hemagglutinin control should show positive agglutination
13. The HAI titer of each serum is defined as the highest dilution of serum that completely inhibits hemagglutination
14. The HAI titer of the acute phase serum is 10, and for the convalescent serum the titer is 160; fourfold or greater rise in HAI titers is interpreted as significant and is indicative of recent influenza virus infection or vaccination (Figure 13.2 illustrates an example of an acute and convalescent serum titrated against a single influenza strain)

13.1.10 Other Viruses for Which the Hemagglutination Inhibition Test is Applicable

The procedure described above is applicable with some modifications for the parainfluenza, rubella, measles, reovirus, adenovirus, enterovirus, and togavirus groups. The appropriate chapter in this volume should be consulted for details. However, examples of modifications that are necessary for these viruses are summarized below.

1. Members of the parainfluenza virus group agglutinate human O, guinea pig, and chicken RBC; however, in addition to RDE heat treatment of sera, adsorption of sera with guinea pig RBC is usually necessary if guinea pig RBC are used in the HAI test.
2. For rubella virus HAI tests, 1-day-old chick or goose RBC are used. The diluent employed is not PBS. HEPES-Saline-albumen-gelatin (HSAG) diluent is employed. In addition, sera must be treated with $MnCl_2$-heparin and adsorbed with chick RBC to remove nonspecific inhibitors and agglutinins (see Appendix).

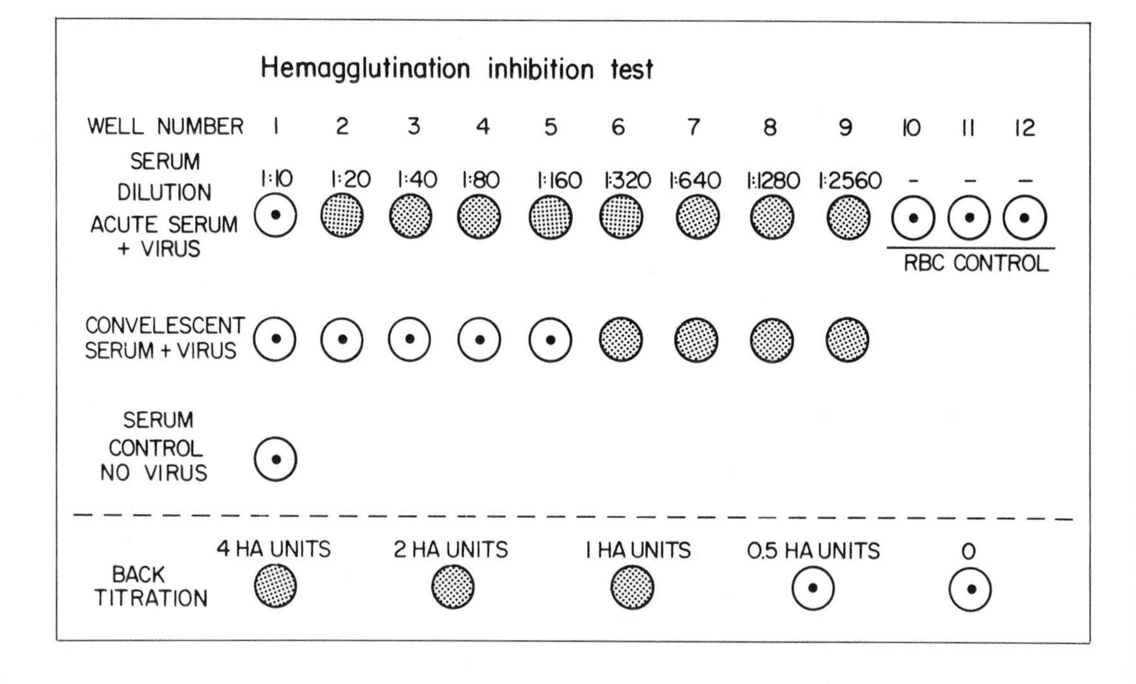

Figure 13.2. Hemagglutination inhibition.

3. For group I adenovirus serotypes, rhesus monkey RBC are usually used, and for adenoviruses in group II, rat RBC. Sera must be heat-inactivated at 56°C for 30 minutes and adsorbed with the type of RBC being used in the HAI test before testing.
4. For reoviruses, human O RBC are used. Sera must be heat inactivated and adsorbed with kaolin.
5. For members of the togavirus group, hemagglutinin occurs optimally at different pH values ranging from pH 6.0 to 7.4. Heat inactivation and kaolin adsorption of sera are usually employed.
6. Some serotypes of enteroviruses agglutinate human O RBC but the incubation temperature of the hemagglutinin and HAI tests is critical. Coxsackie A-20, A-21, A-24, and echovirus types 3, 11, 13, and 19 hemagglutinate at 4°C. Coxsackie B-1, B-3, B-5, and echovirus 6, 7, 12, 20, and 21 hemagglutinate at 37°C. Heat inactivation and kaolin adsorption of sera are employed for the HAI test.

13.2 HEMADSORPTION

Influenza and parainfluenza viruses, as well as mumps and Newcastle's disease virus (NDV), can replicate in a variety of cell culture systems. Commonly employed cell cultures in clinical virology laboratories include primary monkey kidney cells, and continuous cultures of Madin-Darby canine kidney (MDCK) cells and the LLC-MK_2 cell line derived from rhesus monkey kidney. These viruses frequently do not produce extensive CPE; however, infected cells can be detected by the hemadsorption (HAd) technique. These viruses mature by budding from the plasma membrane of infected cells and, because they can react with human O, chicken, or guinea pig RBC, the addition of a RBC suspension to the cell monolayers results in the RBC adsorbing onto the surface of infected cells (Figure 13.3).

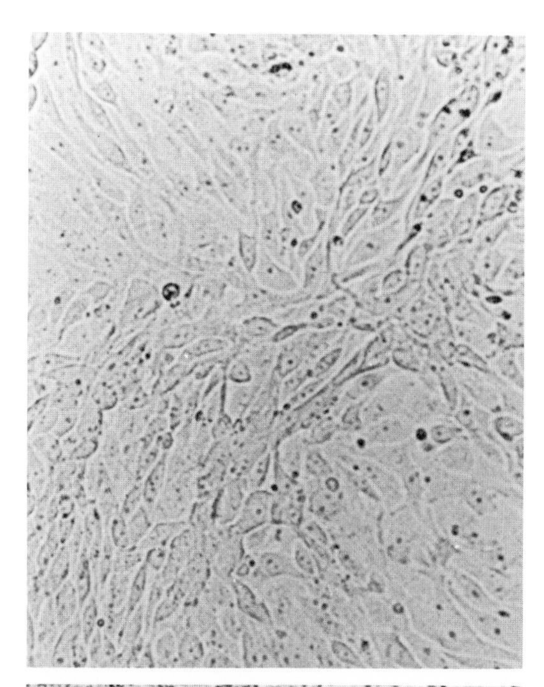

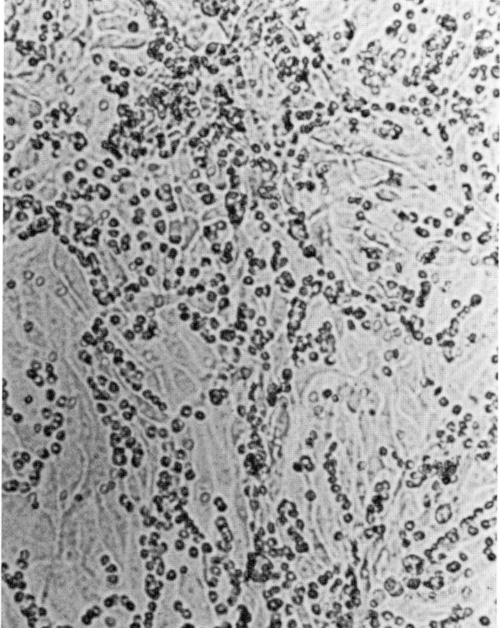

Figure 13.3. Hemadsorption. (A) Uninoculated primary monkey kidney cells. (B) Primary monkey kidney cells infected with influenza A.

13.2.1 Materials

1. Guinea pig blood washed three times in PBS by centrifugation and suspended in Alsever's solution at a concentration of

20%; RBC should not be kept for more than 1 week at 4°C
2. Uninoculated control cell cultures
3. Cell cultures inoculated with clinical specimens
4. Cell cultures inoculated with hemadsorbing strain of virus
5. Cold PBS

See Appendix for sources.

13.2.2 Methods

1. Remove fluids from control and inoculated cell cultures to sterile tubes
2. Add 0.25 ml of 0.5% guinea pig RBC suspension to each cell culture
3. Incubate the cell cultures at either 4°C or room temperature for 30 minutes
4. Carefully remove the unadsorbed RBC by rinsing the cell monolayers with cold PBS
5. Observe the cells microscopically for HAd; the RBC should not adsorb to uninoculated cells (RBC suspensions stored for more than 1 week frequently adsorb to uninoculated cell cultures of kidney origin)
6. Culture fluids from cells showing positive HAd can be subcultured to newly prepared cell cultures for subsequent identification by HAI, or HAd-inhibition employing reference antisera to influenza and parainfluenza viruses.

13.3 HEMADSORPTION INHIBITION TEST

Inhibition of HAd by means of reference antisera can be used for identification of viral isolates. Second passage to replicate cell cultures of viral isolates produces HAd. At least four cell cultures need to be inoculated for identification of influenza type A and B isolates, and if parainfluenza types 1, 2, and 3 are to be included, six additional cultures.

13.3.1 Methods

1. Infected and uninoculated control cell cultures are rinsed with Hanks' balanced salt solution (HBSS)
2. Specific viral antisera that have been RDE heat treated and absorbed with guinea pig RBC are added to one set of inoculated and control cultures with 0.8 ml of each antiserum, which have been diluted to 2.5% concentration in HBSS
3. The antisera are allowed to cover the entire cell monolayer at room temperature for 30 minutes
4. Add 0.2 ml of 0.4% guinea pig RBC to each cell culture and incubate at room temperature for 30 minutes
5. Examine the cell cultures microscopically for inhibition of HAd; isolates are identified by the serum inhibition of HAd

REFERENCES

Lennette, E.H., and Schmidt, N.J., eds. 1979. Diagnostic Procedures for Viral, Rickettsial and Chlamydial Infections, 5th ed. Washington, D.C.: American Public Health Association.

Lennette, D.A., Specter, S., and Thompson, K.D., eds. 1979. Diagnosis of Viral Infections: The Role of the Clinical Laboratory. Baltimore: University Park Press.

Howard, C.R., ed. 1982. New Developments in Practical Virology: New York, Alan R. Liss.

McLean, D.M. 1982. Immunological Investigation of Human Virus Disease. Edinburgh: Churchill Livingstone.

Washington, J.A., ed. 1985. Laboratory Procedures in Clinical Microbiology. 2nd edition New York: Springer–Verlag.

APPENDIX

Reagents

1. *Phosphate buffered saline* (PBS) pH 7.2 is prepared as follows:

   Disodium phosphate (Na_2HPO_4), 1.096 g

Monosodium phosphate (NaH_2PO_4), 0.315 g

Sodium chloride, 8.500 g

Distilled water, q.s.ad, 1.000 l

2. *Alsever's solution* is available from commercial sources or can be prepared as follows:

   Dextrose, 20.500 g

   Sodium citrate ($Na_3C_6H_5O_7.2H_2O$), 8.000 g

   Citric acid ($C_6H_8O_7.H_2O$), 0.550 g

   Sodium chloride, 4.200 g

   Distilled water, q.s.ad, 1.000 L

3. *Erythrocytes* from human and a variety of animal species in Alsever's solution are available commercially from Flow Laboratories, M.A. Bioproducts, and numerous other suppliers of virologic products

4. *Hemagglutinating viral antigens* are prepared by inoculation of 10-day-old embryonated chicken eggs or sensitive cell cultures; for most clinical virology laboratories it is more convenient to obtain contemporary strains of influenza A and B or parainfluenza virus strains from commercial sources (Flow Laboratories or M.A. Bioproducts); other hemagglutinating viral antigens are also available from these sources and include measles, mumps, and rubella viruses

5. *Reference antisera* for most hemagglutinating viruses are available from Flow Laboratories or M.A. Bioproducts

6. *Receptor destroying enzyme* (RDE) is available from Flow Laboratories and M.A. Bioproducts and should be shipped on dry ice; upon receipt, RDE should be stored at -70°C

7. *Kaolin* for serum treatment is available from Flow Laboratories or prepared as a 25% suspension of acid-washed kaolin in PBS

8. *HEPES-saline-albumen-gelatin* (HSAG) diluent is available from Flow Laboratories and M.A. Bioproducts

9. *Heparin–$MnCl_2$* reagent is available from Flow Laboratories or can be prepared by adding equal parts of heparin (5000 units/ml) and 1 *M* $MnCl_2$ (39.6 g $MnCl_2.4H_2O$ in 200 ml distilled water); the heparin–$MnCl_2$ solution should be stored at 4°C and prepared every 1 to 2 weeks

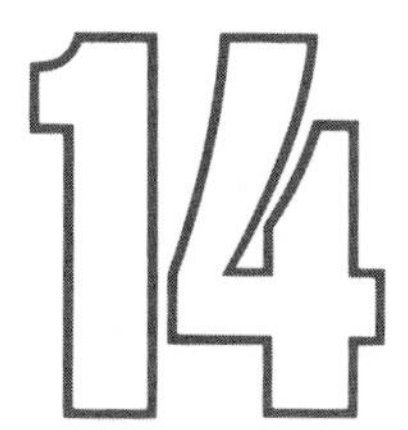

Immune Adherence Hemagglutination

Evelyne T. Lennette and David A. Lennette

14.1 INTRODUCTION

Each of the various serologic test procedures described in this volume has its advantages and disadvantages. Their utilities are relative and depend on the specific requirements of the laboratorian. Clearly, the selection of an appropriate assay requires careful analysis, as no assay is ideal for all situations. The following discussion is oriented to the needs of a clinical diagnostic laboratory. The needs of a clinical laboratory are for test procedures of specificity and sensitivity suitable for both determination of immunity and diagnosis of current infections. The procedure should be convenient and rapid to perform, on a single sample or on a hundred. The demand on equipment and reagents should be within the reach of smaller laboratories with limited resources. The test antigens should be varied and readily available. Of significance to a reference laboratory is the ease with which the assay may be adapted for use with new or additional test antigens.

Some years ago, when we reviewed many of the serologic procedures applicable for use in virology, it was clear that two procedures came close to meeting most of the requirements just outlined. One was the complement fixation (CF) test, already in wide use, and the other was immune adherence hemagglutination (IAHA) assay, a test virtually unknown in the U.S. Although much progress has been made in the development of potentially useful serologic procedures during the intervening years, the situation has not changed greatly. CF remains the mainstay in most clinical virology laboratories, despite its low sensitivity compared with various other techniques: radioimmunoassay (RIA), enzyme-linked immunosorbent assay (ELISA), hemagglutination inhibition (HAI), and immunofluorescence assay (IFA). Its continued use is probably due to the variety of antigens that are commercially available for the test, and the large amount of experience with the procedure for clinical diagnosis, as published in the medical literature. In contrast, the limited inventory of antigens that are commercially available for ELISA tests, the reluctance to deal with the hazards and waste disposal problems of RIA procedures, and the unsuitability of IFA tests for examining large numbers of specimens, have all limited the adoption by laboratorians of these assays for more widespread use.

When we first decided to evaluate the IAHA assay in 1976, immune adherence procedures were known only to a few researchers working on hepatitis A virus and varicella virus in the United States, and to a number of Japanese workers who were instrumental in developing the procedure for clinical use. After 8 years of extensive use of the IAHA assay in our laboratory, and its continued comparison with other serologic methods, we believe that the IAHA test could and should replace the CF test as a general serodiagnostic method.

14.2 HISTORICAL BACKGROUND

The IAHA assay is based on a phenomenon known as serologic adhesion, observed by Levaditi (1901). After injecting antibody-coated *Vibrio cholera* into guinea pigs, he noted platelet aggregation (or adhesion) to the vibrios. Similar observations were reported about the same time by Laveran and Mesnil using *Trypanosoma lewisi* and immune rats (1901). Kritchewski later showed that this adhesion reaction required complement (Lamanna, 1957). Other French and Russian investigators then improved and applied the adhesion test for the in vitro assay of antibodies to trypanosomes, leishmania, leptospira, and spirochetes (Lamanna, 1957). In 1952, Nelson showed that similar adhesion occurs if human erythrocytes are substituted for platelets. The ag-

Table 14.1. Applications of IAHA

Organism	Antigens	References
Bacterial	*Legionella pneumophila*	Lennette et al., 1979
Chlamydial	*Chlamydia psittaci* *Chlamydia trachomatis*	Lennette and Lennette, 1978
Fungal	Blastomyces *Histoplasma capsulatum* (mycelial phase)	Lennette, unpublished
Rickettsial	Q-Fever	Lennette, unpublished
Viral	Adenoviruses	Ito and Tagaya, 1966
	BK papovavirus	Lennette and Shah, unpublished
	Cytomegalovirus	Dienstag et al., 1976
	Dengue	Inouye et al., 1980
	Enteroviruses	Ito and Tagaya, 1966
	Epstein–Barr virus	Lennette et al., 1982
	Herpes simplex	Ito and Tagaya, 1966
	Hepatitis A	Miller et al., 1975
	Hepatitis B surface	Okochi et al., 1970
	Hepatitis B core	Tsuda et al., 1975
	Japanese encephalitis	Inouye et al., 1981
	Mammary tumor virus	Nagayoshi et al., 1981
	Measles (rubeola)	Lennette and Lennette, 1978
	Norwalk agent	Kapikian et al., 1978
	Rabies	Budzko et al., 1983
	Rotavirus	Kapikian et al., 1981; Nagayoshi et al., 1980
	SV40	Lennette and Shah, unpublished
	Varicella-zoster	Gershon et al., 1976
Soluble	Paul-Bunnell heterophil	Lennette et al., 1982

glutination was examined under a microscope after mixing antigen, serum, complement, and erythrocytes on a slide. Nelson coined the name immune adherence for the reaction in 1963. The simple procedure was improved when Ito and Tagaya adapted it for use with microtest plates (1966). Although the potential utility of the IAHA procedure was recognized early, it failed to come into widespread use due to one shortcoming: The agglutination reaction was reversible, and was not stable enough for reliable reading of the reactions. This problem was finally overcome when Mayumi et al. introduced the use of dithiothreitol (DTT) as a stabilizer (1971). With that single improvement, the repertory of antigens found suitable for use in the IAHA test quickly grew to include enteroviruses, adenoviruses, and hepatitis B virus (HBV). The Japanese also used IAHA for detection of hepatitis B surface antigen (HBsAg) in serum. In the U.S., IAHA was first used in hepatitis A seroepidemiologic studies. Table 14.1 provides a list of many of the successful applications of IAHA assays, showing that it has worked with viral, bacterial, fungal, and mycoplasmal antigens.

14.3 IAHA MICROTITER PROCEDURE

In the actual IAHA test, antibodies (Ab) and antigens (Ag) are allowed to form complexes in the first incubation. This is followed by the addition of complement (C). The resultant Ab–Ag–C complexes can then react with C3b receptors on human erythrocytes to cause hemagglutination.

The equipment, reagents, and manipulations required for the performance of the IAHA test are similar to those used for the CF test. The following is a step-by-step outline of the procedure. See appendix for composition of reagents.

1. Test sera are diluted in diluting buffer Veronal-buffered saline (VBS) at 1:4 and then heated at 56°C for 30 minutes to inactivate complement in the sera
2. Prior to their use, V-well microtiter plates are rinsed with VBS containing gelatin (GVB); the rinse solution is discarded by inverting the plates over a sink, then rapping the inverted plates against a hard surface to remove residual buffer
3. One drop (25 μl) VBS containing bovine serum albumin (BVB) is added to every well that will be used
4. One drop of each inactivated serum is added to the first and eighth well of a corresponding row on the test plate; microdilutors are then used to prepare serial twofold dilutions of the added sera (two sets of seven and five wells)
5. One drop of positive test antigen (previously titrated to contain one to two units of reactivity) is added to the first seven wells of each row; identically diluted negative (control) antigen is added to wells eight through twelve; antigen concentrations are determined by block titrations with control reference sera
6. The plates are shaken for 10 seconds with a vibrating mixer, then incubated at 37°C for 30 minutes; it is also acceptable to incubate the plates overnight at refrigerator temperature, especially if the antigen is unstable; all plates should be covered during the incubations to minimize evaporation, which may affect results
7. One drop of diluted guinea pig complement (1:100 in BVB) is added to all wells on the test plates. The plates are again mixed and incubated at 37°C for 40 minutes; the exact dilution of complement is determined by a prior block titration (ordinarily, only one unit of complement is added)
8. At the end of the incubation, the reaction is stopped by the addition of one drop of DTT–EDTA to each well, immediately followed by the addition of one drop of 0.4% suspension of human erythrocytes, type O in GVB

9. Hemagglutination should be complete and readable within 1 hour; positive reactions are those showing >50% agglutination; the agglutination pattern is usually stable at room temperature for several days

Either "U" or "V" bottomed microtest plates can be used for IAHA tests, although the hemagglutination patterns are easier to read on the "V" type plates. We have found that polystyrene plastic plates cause nonspecific binding of erythrocytes, which can be eliminated by prewetting the plates with buffer containing carrier protein. If gelatin is used in the prewetting buffer, the specific lot of gelatin should be screened for suitability. Some lots contain heat-labile substances that interfere in the IAHA test; the interference can be destroyed by autoclaving the gelatin stock solution.

Some years ago, we noticed that many "acute phase" sera appear to contain immune complexes that persist even after the diluted serum has been "inactivated" by heating at 56°C. The complexes will react to give positive IAHA reactions, even in the absence of added test antigens. Sera containing these immune complexes are not found as frequently as are sera that are reported as "anticomplementary" in the CF test; nevertheless, the immune complexes may be a complication. We have been able to eliminate the nonspecific reactions due to these immune complexes by diluting test sera in barbital–saline adjusted to pH 3 (essentially not buffered) and then heating the diluted sera at 56°C for the usual 30-minute period. It appears that this modified inactivation procedure dissociates the complexes. If the initial dilution is not greater than 1:5, the dilution is almost neutral and no adjustment of pH is necessary, as serum is adequately buffered.

Commercially available antigens for use with the CF test have been satisfactory for use in the IAHA test, provided the antigen titer was at least 8. Antigen concentrations needed for the IAHA are generally only one-fourth or one-eighth that required for the CF test, which may represent a significant cost savings. Our laboratory has had occasion over the years to prepare antigens and compare them with the commercially available materials. In nearly all cases, antigens prepared from tissue culture material in-house has been of better quality and higher titer than the commercially available antigens. Infected and uninfected cell cultures are suspended separately in saline equal in volume to one-fifth of their original growth medium. The cells are then disrupted by three cycles of freezing and thawing or by sonication. The disrupted suspension is clarified by centrifugation at 10,000 × *g* for 10 minutes. Further purification or concentration usually is not necessary. The preparation can be dispensed in small volumes and frozen at −70°C for indefinite periods of storage before use. With proper conditions for viral replication, antigen titers of 32 or greater should be expected. Inactivation of residual virus infectivity by means of psoralens and longwave ultraviolet irradiation has given us excellent results, using the methods of Hanson et al. (1978). We have also prepared bacterial antigens using the confluent growth obtained on appropriate agar media. The bacteria are scraped off the agar, suspended in saline, and inactivated by heating or the addition of formalin (at 0.1%). The cell suspension is then washed twice by centrifugation at 10,000 × *g* for 20 minutes. The washed sediment is then adjusted to a 10% suspension and sonicated to disrupt the cells. For disruption of 1–2 ml of suspension, we use three cycles of 15 seconds each at 10 W output of a micro-probe type sonicator. The preparation is clarified at 10,000 × *g* and stored at −70°C until use. The antigens can be prepared very efficiently in this manner and the antigen titer achievable is usually between 500 and 2000.

Egg-derived antigens have not been satisfactory in the IAHA test. Such antigens include rickettsiae and chlamydiae grown in yolk sacs. Extraction of lipids from these antigens with Freon 113 enhances the interpretation of antigen–antibody reactions, at the cost of decreased antigen titers. Freon 113 extraction is also needed to obtain a satisfactory rotavirus antigen from stools, as is purification of marmoset-liver derived hepatitis A virus (HAV) by density gradient centrifugation.

The stability of viral and bacterial antigens for the IAHA test is very good, provided they do not become contaminated. Addition of 0.01% thimerosal preservative reduces, but will not eliminate, such contamination. A minor problem encountered with frozen antigen preparations is the precipitation of antigen that has been stored frozen and thawed. This can usually be counteracted by brief sonication of the thawed antigen preparation.

14.4 RED BLOOD CELLS

Primate erythrocytes and non-primate platelets are reported to be suitable indicator cells for the IAHA test (Nelson, 1963). The most widely used cell type is human type O erythrocytes. The nature of the reactive site on the cells is unknown. The presence of C3 receptors on the cells is necessary, but not sufficient to provide IAHA reactivity, as only about one of three donors have suitable erythrocytes (Klopstock et al., 1963). The receptors are sensitive to proteolytic enzymes and to neuraminidase, consistent with a glycoprotein composition (E.T. Lennette, unpublished results). The ability of cells to react in the IAHA procedure, in the presence of immune complexes, is apparently a permanent property. Erythrocytes obtained from one of the authors are known to have been reactive during a span of 8 years. Hence, the suitability of any particular donor needs to be checked only once. It seems that obtaining a supply of reactive erythrocytes is the main obstacle preventing many laboratories from adopting IAHA procedures. There is no commercial supply of pretested blood, so that each laboratory needs to find its own supply. Our laboratory uses two sources of blood: laboratory personnel and local blood banks. Type O blood from a blood bank can be used for 2 to 3 weeks after collection of the sample segments (which are often discarded) from blood bags may supply enough cells for a useful number of IAHA procedures, at a rate of one segment per day for our laboratory. Often there are eight or more segments from each blood bag that are not used by the blood bank, which could be made available to a laboratory for use in the IAHA test. Blood collected in EDTA is not suitable.

In our laboratory, a reference antigen and a set of reference sera are used to screen blood samples for suitable erythrocytes. Only erythrocytes that give acceptable serum titers are selected for use. The segments can be stored for up to 2 weeks at 4°C before use. Although this approach for obtaining erythrocytes is less convenient than having access to a panel of preselected donors, we find it presents the fewest problems. We do not recommend pooling cells from randomly selected donors, as pooled erythrocytes give substantially lowered sensitivity with the IAHA procedure. Limited efforts have been made towards increasing the shelf-life of erythrocytes for use in the IAHA test. Cells fixed with glutaraldehyde or formaldehyde can be used in IAHA procedures, but we find fixed cells less satisfactory than fresh cells. Erythrocytes may be frozen for later use, using the method of Lawrence and Wentworth (1985).

14.5 HEMAGGLUTINATION PATTERNS

The agglutination pattern seen in the IAHA test is uniform and does not vary appreciably with different test antigens. However, one should become familiar with the agglutination patterns produced by individual sera. The patterns in microtest plates are best examined with aid of a magnifying mirror. A specific reaction should appear as uniformly granular agglutination and a negative reaction should appear as a cell button with a smooth outline. While reading test results, one should compare the patterns obtained with test sera to those obtained with the control or reference sera. With some practice, it is usually possible to differentiate specific agglutination from nonspecific reactions. In addition to the agglutination sometimes found with negative control antigens, some sera give uncharacteristic agglutination with the test antigen. Specific reactions give a slightly granular or coarse agglutination, whereas, nonspecific reactions often produce a fine textured or matte agglutination. Even when freshly collected sera are inactivated in low pH diluent, about 2% of the specimens tested have equal titers with positive and control antigens. It is usually necessary to resort to another test system to evaluate such sera.

14.6 DETECTION OF IgM ANTIBODIES

It is well-established that IgM antibodies are not efficiently detected by the CF assay. As both CF and IAHA tests detect complement-activating immune complexes, it is somewhat surprising to discover that IgM antibodies are readily detected by IAHA assay. The first evidence that IgM antibodies react well in IAHA came from the application of IAHA testing to the detection of the Paul–Bunnell heterophil antibodies associated with infectious mononucleosis. The IAHA test was both more sensitive and more specific than the standard differential heterophil agglutination test and ox-cell hemolysis test (Lennette et al., 1978). Paul–Bunnell heterophil antibodies are exclusively of the IgM class. Other evidence that IgM antibodies to varicella-zoster virus (VZV) are reactive in IAHA came from work using sera that were fractionated by sucrose gradient density centrifugation (Gershon et al., 1981). Fractions containing either VZV specific IgM or IgG were both reactive in the IAHA test, provided that the IAHA test was performed promptly after fractionation of the sera. In our experience, sucrose has a detrimental effect on IAHA reactive IgM even if the sera are only briefly stored.

We have observed that the detection of IgM antibodies by IAHA depends on the quality of the test antigens used. Using Paul–Bunnell heterophil antibody containing sera, it was possible to show that antigen lots varied greatly in reactivity. Results obtained by block-titration of different lots of antigen can differ almost tenfold. This observation has been extended to other commercially available viral antigens that are used in CF tests [e.g., cytomegalovirus (CMV) and VZV antigens]. Using IgM containing reference sera, the lot-to-lot variation of an antigen from the same supplier is seen as a shift in the optimal titer of each lot of antigen. Although the optimal antigen titer varies between lots, the reference sera titers do not vary. In contrast, variations in antigen preparations obtained from different suppliers also affect the titer of the test sera. That is, the reference serum titer may vary greatly according to the source of antigen. This variation is seen mainly with IgM containing sera and not appreciably with sera from immune donors, suggesting different antigenic reactivities for IgM and IgG. At present, there is no standardization among suppliers as to the composition of viral test antigens. The differences are noticeable in the CF test, but are more pronounced in the IAHA test, especially when sera containing IgM antibodies are tested. We recommend that serum panels for antigen titration and evaluation should include known IgM-positive sera when possible.

14.7 SENSITIVITY AND SPECIFICITY

IAHA is most often compared with CF, probably due to their technical similarities and their reagent requirements. Both procedures also measure complement activation by immune complexes, the IAHA directly and the CF by complement depletion. However, there are a few differences between the two tests that explain the advantages found with IAHA.

First, the tests differ in the nature of the indicator system used to measure the immune complexes formed in vitro. In the CF test, a measured amount of complement is added to the test wells and any complement not bound in the test reactions is then indicated by the addition of sheep RBC–anti-sheep RBC complex. Remaining complement reacts to lyse the sheep RBC. This procedure is inherently insensitive to small differences in depletion of the excess complement added at the beginning of the test. In the IAHA procedure, only the complement that reacts in the test is measured. The test is not adversely affected by excess complement and small differences in complement concentration. Any agglutination above a very low background is significant. This difference in mechanism between the two tests accounts for the four- to eightfold increase in sensitivity shown by the IAHA method.

The increased sensitivity of the IAHA test has been shown to be adequate for the reliable use of the IAHA test for determination of immunity, even in immunosuppressed patients (Gershon et al., 1976). Agents for which this has been shown include VZV, CMV, EBV, and others (Table 14.1). This increased sensitivity, together with the ability to detect IgM antibodies, allows demonstration of a more rapid and pronounced titer increase during an infection using IAHA than can be seen with the CF test. Using the CF test, a minimum of 2 weeks is advocated for the reliable detection of antibody titer changes between acute- and convalescent-phase sera. With the IAHA test, 8- to 16-fold titer changes can frequently be found with sera collected 3 to 5 days apart, during the acute- or early convalescent-phase of illness.

IAHA is comparable to IFA, rather than CF, in its sensitivity, as shown by parallel serologic testing of numerous sera using a variety of antigens. Although IFA titers are often slightly higher than those obtained with IAHA, the ability of the IFA test to differentiate positive from negative sera is the same as that of the IAHA test. In our laboratory, IFA and IAHA tests can be used almost interchangeably, providing useful and complementary test systems for a wide range of antigens.

According to published reports of serologic studies in VZV patients; IAHA is comparable in reactivity to both fluorescent antibody to membrane antigen (FAMA) and neutralization tests (Gershon et al., 1976). For the determination of postimmunization immunity to rabies, IAHA was found to be as suitable as the fluorescent-focus inhibition (FFI) assay, which is commonly used (Budzko et al., 1983). The FFI assay takes several days to complete and requires the use of live rabies virus.

Only very limited comparisons have been made of IAHA against RIA and ELISA tests, and it appears that IAHA is slightly less sensitive than these methods. The sensitivity difference does not appear to be significant for routine clinical applications.

A property of IAHA that is not yet well studied is its ability to detect antigenic differences. For example, IAHA has been reported to be more type-specific than CF for rotavirus identification. IAHA is also reported to be useful for serotyping dengue virus isolates (Suntharee et al., 1981). In our own laboratory, IAHA has been used to subtype *Legionella pneumophila*, using hyperimmune rabbit sera. The ability to detect antigenic differences, against a background of crossreacting specificities, allows the IAHA to be used in monitoring antigen purification. IAHA has been used for that purpose in purification of rotavirus extracted from stools. In the purification of EBV antigens, IAHA was found to be both sensitive to specific antigens and insensitive to the presence of other antigens (impurities) in the virus antigen preparations.

14.8 ADVANTAGES AND DISADVANTAGES

The principal limitation of the IAHA lies in its inability to differentiate antibodies of different immunoglobulin subclasses. Although the IAHA assay detects IgM antibodies as well as IgG antibodies, the two classes cannot be measured separately, unless they are initially separated by column chromatography or density gradient centrifugation. In addition, IAHA will not react with antibodies that do not fix complement, IgA for example. Another disadvantage of the IAHA is that it is difficult to use with hemagglutinating viruses. In theory, it is possible to obtain antigens free of hemagglutinins. In practice, commercial CF antigens for viruses, such as influenzas and mumps, often do contain hemadsorbing activity and are not suitable for IAHA use.

We have found the IAHA test to be satisfactory in routine use for reasons other than sensitivity and specificity. The procedure is quite economical in the use of reagents and supplies. Due to its increased sensitivity, compared with the CF test, the amounts of antigen and complement used may be reduced about fourfold, on the average. Although no special equipment is required to perform IAHA tests, equipment is sold that would permit partial to full automation of the test procedure. The endpoints are usually very sharp, with 4+ agglutination in one well and no agglutination in the next well. Thus, there is little uncertainty or subjectivity in obtaining an accurate titer, which is a common problem with reading IFA tests, for example. Titers are very reproducible, both within test runs and between runs. This reproducibility makes it easy to detect significant titer changes.

Another advantage of IAHA tests is the simplicity of pretest preparations. Every test component of the CF procedure has to be monitored carefully and titered for each test batch. With the IAHA procedure, every component is added in excess, and needs to be titered only once for each lot of reagent, as long as the reagent is stable during storage. While the CF test usually requires overnight incubation, the IAHA test is completed within 3 to 4 hours.

REFERENCES

Budzko, D.B., Charamella, L.J., Jelinek D., and Anderson, G.R. 1983. Rapid test for detection of rabies antibodies in human serum. J. Clin. Microbiol. 17:481–484.

Dienstag, J.L., Cline, W.L., and Purcell, R.H. 1976. Detection of cytomegalovirus antibody by immune adherence hemagglutination. Proc. Soc. Exp. Biol. Med. 153:543–548.

Gershon, A.A., Kalter, Z.G., and Steinberg, S. 1976. Detection of antibody to varicella-zoster virus by immune adherence hemagglutination. Proc. Soc. Exp. Biol. Med. 151:762–765.

Gershon, A.A., Steinberg, S.P., Borkowsky, W., Lennette, D., and Lennette, E. 1981. IgM to varicella-zoster virus: Demonstration in patients with and without clinical zoster. Ped. Infect. Dis. 1:164–166.

Hanson, C.V., Riggs, J.L., and Lennette, E.H. 1978. Photochemical inactivation of DNA and RNA viruses by psoralen derivatives. J. Gen. Virol. 40:345–358.

Inouye, S., Matsuno, S., Hasegawa, A., Miyamura, K., Kono, R., and Rosen, L. 1980. Serotyping of dengue viruses by an immune adherence hemagglutination test. Am. J. Trop. Med. Hyg. 29:1389–1393.

Inouye, S., Matsuno, S., and Kono, R. 1981. Difference in antibody reactivity between complement fixation and immune adherence hemagglutination tests with virus antigens. J. Clin. Microbiol. 14:241–246.

Ito, M., and Tagaya, I. 1966. Immune adherence hemagglutination test as a new sensitive method for titration of animal virus antigens and antibodies. Japan J. Med. Sci. Biol. 19:109–126.

Kapikian, A.Z., Greenberg, H.B., Cline, W.L., Kalica, A.R., Wyatt, R.G., James, H.D., Jr., Lloyd, N.L., Chanock, R.M., Ryder, R.W., and Kim, H.W. 1978. Prevalence of antibody to the Norwalk agent by a newly developed immune adherence hemagglutination assay. J. Med. Virol. 2:281–294.

Kapikian, A.Z., Cline, W.L., Greenberg, H.B., Wyatt, R.G., Kalica, A.R., Banks, C.E.,

James, H.D., Jr., Flores, J., and Chanock, R.M. 1981. Antigenic characterization of human and animal rotaviruses by immune adherence hemagglutination assay (IAHA): Evidence for distinctness of IAHA and neutralization antigens. Infect. Immun. 33:415–425.

Klopstock, A., Schwartz, J., and Zipkis, N. 1963. Individual differences of the reactivity of human erythrocytes in the immune adherence haemagglutination test. Vox Sang. 8:382–383.

Lamanna, C. 1957. Adhesion of foreign particles to particulate antigens in the presence of antibody and complement (serological adhesion). Bact. Rev. 21:30–45.

Laveran, A., and Mesnil, F. 1901. Recherches morphologiques et experimentales sur le trypanosome des rats (Tr. lewisi Kent.). Ann. Inst. Pasteur 15:673–714.

Lawrence, T.G., and Wentworth, B.B. 1985. Freezing and rejuvenation of human O erythrocytes for use in the immune adherence hemagglutination test. J. Clin. Microbiol. 22:654–655.

Lennette, D.A., Lennette, E.T., Wentworth, B.B., French, M.L.V., and Lattimer, G.L. 1979. Serology of Legionnaires' disease: Comparison of indirect immunofluorescent antibody, immune adherence hemagglutination and indirect hemagglutination tests. J. Clin. Microbiol. 10:876–879.

Lennette, E.T., Henle, G., Henle, W., and Horwitz, C.A. 1978. Heterophil antigen in bovine sera detectable by immune adherence hemagglutination with infectious mononucleosis sera. Infect. Immun. 19:923–927.

Lennette, E.T., and Lennette, D.A. 1978. Immune adherence hemagglutination: Alternative to complement-fixation serology. J. Clin. Microbiol. 7:282–285.

Lennette, E.T., Ward, E., Henle, G., and Henle, W. 1982. Detection of antibodies to Epstein–Barr virus capsid antigen by immune adherence hemagglutination. J. Clin. Microbiol. 15:69–73.

Levaditi, C. 1901. Sur l'etat de la cytose dans la plasma des animaux normaux et des organismes vaccines contre le vibrion chloerique. Ann. Inst. Pasteur 15:894–927.

Mayumi, M.K., Okochi, K., and Nishioka, K. 1971. Detection of Australia antigen by means of immune adherence hemagglutination test. Vox Sang. 20:178–181.

Miller, W.J., Provost, P.J., McAleer, W.J., Ittensohn, O.L., Villarejos, V.M., and Hilleman, M.R. 1975. Specific immune adherence assay for human hepatitis A antibody application to diagnostic and epidemiologic investigations. Proc. Soc. Exp. Biol. Med. 149:254–261.

Nagayoshi, S., Yamaguchi, H., Ichikawa, T., Miyazu, M., Morishima, T., Ozaki, T., Isomura, S., Suzuki, S., and Hoshino, M. 1980. Changes of the rotavirus concentration in faeces during the course of acute gastroenteritis as determined by the immune adherence hemagglutination test. Eur. J. Pediat. 134:99–102.

Nagayoshi, S., Imai, M., Tsutsui, Y., Saga, S., Takahashi, M., and Hoshino, M. 1981. Use of the immune adherence hemagglutination test for titration of breast cancer patients' sera cross-reacting with purified mouse mammary tumor virus. GAN 72:98–103.

Nelson, D.S. 1963. Immune adherence. Adv. Immunol. 3:131–180.

Okochi, K., Mayumi, M., Haguino, Y., and Saito, N. 1970. Evaluation of frequency of Australia antigen in blood donors of Tokyo by means of immune adherence hemagglutination technique. Vox Sang. 19:332–337.

Suntharee, R., Charnchudhi, C., Sompop, A., Kanai, C., Igarashi, A., and Inouyes, S. 1981. Isolation and identification of dengue viruses combined use of C6/36 cells and the immune adherence hemagglutination test. Jpn. J. Med. Sci. Biol. 34:375–9.

Tsuda, F., Takahashi, T., Takahashi, K., Miyakawa, Y., and Mayumi, M. 1975. Determination of antibody to Hepatitis B core antigen by means of immune adherence hemagglutination. J. Immunol. 115:834–838.

APPENDIX

1. *Veronal Buffered Saline* (VBS) 5X stock: Dissolve 43.0 g NaCl and 4.6 g diethylbarbituric acid in 950 ml of warmed (lower than 65°C) deionized water. Adjust the pH with NaOH to 7.4. Add 2.5 ml of $MgCl_2/CaCl_2$ solution and adjust

final volume to 1 L. The $MgCl_2/CaCl_2$ solution should contain 20.33 g $MgCl_2.6H_2O$ and 4.4 g of $CaCl_2$ in 100 ml water.

2. *Serum dilution buffer:* VBS 1X is prepared by diluting above stock solution fivefold with water. Adjust the pH to 3.0 with 2N HCl.
3. *GVB:* 2.5% autoclaved stock gelatin is added to VBS 1X to a final concentration of 0.125%.
4. *BVB:* Bovine serum albumin fraction V is added to 1X VBS to a final concentration of 1 mg/ml.
5. *EDTA–DTT–VBS buffer:* Two parts of 0.1 *M* EDTA, pH 7.5 (disodium ethylenediaminetetraacetic acid) is added to three parts of 1X VBS. DTT is added to a final concentration of 3 mg/ml. EDTA–VBS mixture can be stored indefinitely at 5°C. Once DTT is added, the solution should not be used after 4 weeks at 5°C.

IgM Determinations

Kenneth L. Herrmann

15.1 INTRODUCTION

The presence of specific antibody activity in various serum immunoglobulins (Ig) was reported as early as the 1930s (Heidelberger and Pederson, 1937). Subsequent studies demonstrated that the first immunoglobulins to appear after a primary antigenic stimulus were of the IgM class. These IgM antibodies reportedly disappeared rapidly, usually within a few weeks, and were replaced by IgG antibodies that persisted for a longer period.

The transient nature of the IgM antibody response appears to hold true for most primary viral infections, and the determination of specific antiviral IgM antibodies is now well established as a potentially valuable method for the rapid diagnosis of recent viral infections. Such an approach provides a considerable advantage over classical serology, which required the demonstration of a significant rise in antibody titer between paired acute- and convalescent-phase serum specimens. For this approach to be successful, the IgM antibody response must be specific, must be measurable with adequate reliability and sensitivity, and must be transient (i.e., present only with recent active infection with the specific virus).

15.2 METHODS USED FOR IgM ANTIBODY DETERMINATION

Since the introduction of the first diagnostic applications of IgM determination, a variety of methods have been developed and applied for this purpose (Table 15.1). These methods can generally be separated into four groups: a) those based on comparing titers before and after chemical inactivation of serum IgM proteins, b) those based on physiochemical separation of IgM from the other serum immunoglobulin classes, c) those based on solid phase indirect immunoassays using labeled anti-human IgM, and d) reverse "capture" solid phase IgM assays. Each of these approaches will be briefly discussed.

15.2.1 Methods Based on Chemical Inactivation of IgM

Mercaptans have the capacity to split the IgM molecule into immunologically inactive parts by breaking the disulfide bonds between the polypeptide chains. Some laboratories have attempted to use this simple technique to detect virus-specific IgM antibodies (Banatvala et al., 1967) Briefly, serum antibody titers are measured by standard serologic methods before and after

Table 15.1. Assay Methods for IgM Antibody

Methods based on chemical inactivation of IgM
Alkylation-reduction by mercaptans
Methods based on physiochemical separation of IgM
Chromatographic gel filtration
Sucrose density gradient ultracentrifugation
Ion exchange chromatography
Affinity chromatography
Protein A absorption
Solid phase indirect immunoassays using labeled anti-human IgM
Immunofluorescence assay (IFA)
Radioimmunoassay (RIA)
Enzyme immunoassay (EIA)
Reverse capture solid phase IgM assays
IgM antibody capture enzyme immunoassay
IgM antibody capture radioimmunoassay (MACRIA)
Solid phase immunosorbent technique (SPIT)
Other assays
Radioimmunodiffusion
Radioimmunoprecipitation
Counterimmunoelectrophoresis
Anti-IgM hemagglutination
Latex-IgM agglutination

mercaptan treatment. The presence of specific IgM antibody is indicated by a significant decrease in titer of the treated serum sample. The method is simple, but very insensitive. For this test to be positive (i.e., demonstration of a fourfold or greater decrease in titer between treated and untreated serum), at least 75% of the total virus specific antibody must be of the IgM class. This would be true only during the very early stages of most viral infections and, therefore, its diagnostic value would be very limited. For this reason, this method is not recommended as an acceptable approach for virus-specific IgM determination. On the other hand, treatment with 2-mercaptoethanol (2-ME) or dithiothreitol (DTT) may be a useful control step in conjunction with various physiochemical immunoglobulin separation methods described later in this chapter (Caul et al., 1974; Pattison, 1982).

15.2.2 Methods Based on Physiochemical Separation of IgM

15.2.2.a Gel Filtration

Column chromatographic methods have been used for many years to separate and isolate serum IgM antibodies. Sephacryl S-300 or Sephadex G-200 (Pharmacia, Inc.) are the gels of choice for fractionation of serum immunoglobulins. Sephacryl S-300 is a mixture of sepharose and acrylamide, and offers several advantages over Sephadex G-200, because it does not need to be rehydrated and will allow high flow-rate filtration of serum under pressure without overpacking or deforming the column (Morgan-Capner et al., 1980). Chromatography columns are packed as directed by the gel manufacturer. Serum lipoproteins and nonspecific cell agglutinins will elute from these gel columns in the same fractions as IgM, so must be removed prior to the fractionation if they will interfere with the reliability of the assay or specific antibody activity in the fractions. Pretreated serum is layered on the top of the gel column and eluted through the column with Tris-buffered saline (0.02 *M* Tris in 0.15 *M* NaCl). Discrete fractions are collected for titration of antibody activity. Each new column should be standardized with known specific IgM-positive and -negative sera. With both Sephacryl S-300 and Sephadex G-200 columns, IgM is eluted in the first protein peak and IgG in the second. IgA may be present also in the first peak eluted from the Sephadex G-200 column but not from the Sephacryl S-300 column. Both columns are equally useful for separating IgM from IgG antibodies.

The specificity of the gel filtration IgM test for diagnosis of viral infections is very high, provided a number of factors that can cause false-positive results are recognized (Pattison et al., 1976). Prolonged storage of sera at -20°C and bacterial contamination of the sera may make the serum pretreatment ineffective. Also, if the serum has been preheated at 56°C or higher, IgG will aggregate and will elute in the IgM fractions after gel filtration. To minimize misinterpretation of the gel fractionation test, any presumptive IgM antibody activity in the first peak should be shown to be 2-ME sensitive.

15.2.2.b Sucrose Density Gradient Ultracentrifugation

The most commonly used method of fractionating sera for detecting virus-specific IgM antibodies is high-speed ultracentrifugation of the serum on a sucrose density gradient. Because IgM proteins have a higher sedimentation coefficient (19S) compared with other immunoglobulin proteins (7–11S), IgM antibodies can be separated from other antibodies by rate-zonal centrifugation. Lipoprotein molecules, including most of the nonspecific inhibitors of rubella hemagglutination, are very light and remain close to the top of the gradient during centrifugation. The technique was introduced in 1968 for the rapid diagnosis of recent rubella infection by demonstration of the IgM antibodies (Vesikari and Vaheri, 1968). Since that time, modifications of the method have been published (Forghani et al., 1973; Caul et al., 1976), and the test has been applied to the diagnosis of other viral infections (Al-Nakib, 1980; Hawkes et al., 1980). In my laboratory, sucrose density gradient ultracentrifugation has been found to be the most reliable and specific method for rubella and measles IgM antibody assay. Given the necessary equipment, this separation procedure is relatively easy to perform.

The method as performed in my laboratory is described as follows. A density gradient is prepared by layering 1.4 ml amounts of 37%, 23%, and 10% (w/v) solutions of sucrose in phosphate buffered saline (PBS), pH 7.2, on a 0.2-ml cushion of 50% sucrose in a 5-ml ultracentrifuge tube [Figure 15.1(A)]. The tube is allowed to equilibrate for 4 to 6 hours at 4°C. The test serum is diluted 1:2 in PBS (0.15 ml of serum mixed with 0.15 ml PBS), pretreated (if necessary) to remove nonspecific serum

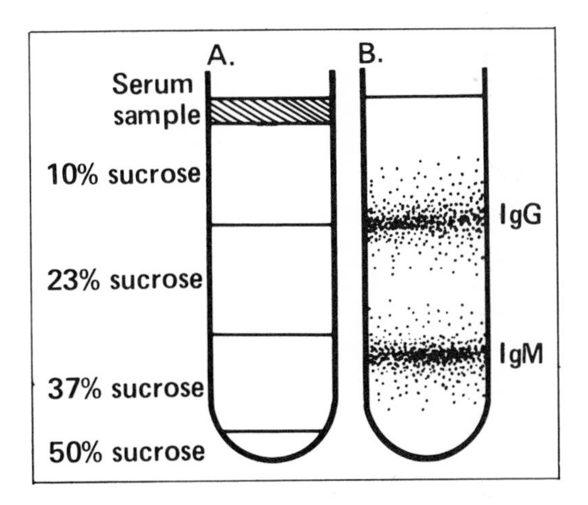

Figure 15.1. Diagram of sucrose density gradient before (A) and after (B) ultracentrifugation.

components that would interfere with the assay of IgM antibody, and 0.2 ml of the diluted serum is carefully layered on top of the gradient. The gradient is then centrifuged at 157,000 × *g* for 16 hours in a swinging-bucket rotor. Ten to 12 fractions (about 0.4 ml each) are collected by puncturing the bottom of the tube. The IgM antibodies concentrate in the bottom three or four fractions, IgG antibodies separate primarily in fractions six to eight, and the nonspecific lipoprotein inhibitors remain in the top layers of the gradient [Figure 15.1(B)].

The isolated IgM fractions can then be tested with any suitable serologic tests for the virus in question. The first four fractions (i.e., those presumed to contain IgM) must be checked for the presence of contaminating human IgG by radial immunodiffusion using low-level IgG agar gel immunodiffusion plates. Suitable IgG assay plates are available commercially. In addition, the specificity of the IgM antibody activity in fractions 1 to 4 may be confirmed by demonstrating its sensitivity to 2-ME treatment.

Studies have shown that sucrose density gradient ultracentrifugation may not be as sensitive as some of the more recently developed indirect immunoassays or IgM capture immunoassays. However, because of its high degree of specificity and overall reliability, this method is generally considered the standard for comparison of other new IgM antibody tests. Sucrose gradient ultracentrifugation is a rather laborious procedure, and the high cost of the necessary equipment places it out of financial reach of most clinical diagnostic laboratories. The more recent introduction of vertical rotors with reorienting gradients, however, now makes it possible to reduce the centrifugation time from 16 hours to only 2 hours, making it feasible to test considerably larger numbers of specimens in a given time.

15.2.2.c Other Physiochemical Separation Methods

Other less frequently utilized methods for physically separating IgM from the other serum immunoglobulins include ion exchange chromatography (Johnson and Libby, 1980), affinity chromatography (Barros and Lebon, 1975), and staphylococcal protein A absorption (SPA-Abs) (Ankerst et al., 1974). Ion exchange chromatography is based on the differential binding of IgM and IgG to anion exchange resins. The commercially available Quik-Sep IgM# Isolation System (Isolab, Inc., Akron, OH) is an example of such a filtration system. Affinity chromatography employs columns of anti-human IgM covalently bound to sepharose beads to isolate the serum IgM for subsequent assay for specific viral antibodies. Neither of these two methods have received much attention for IgM antibody assay. On the other hand, SPA-Abs has attracted considerable interest as a simple and rapid screening method for IgM antibody. SPA, a cell wall protein present in some *Staphylococcus aureus* strains, binds to the Fc receptor of the IgG molecule, and can be used to absorb and remove the IgG component of serum. SPA-Abs does not remove all IgG, however, and up to 5% of the original IgG antibody activity may still remain following absorption. This residual antibody activity must not be mistakenly interpreted as representing IgM antibody. Absorbed and unabsorbed serum samples are run in parallel by a standard doubling dilution assay, so the percentage of residual antibody activity can be approximated. A decrease in titer of less than four doubling dilutions (i.e., residual of 12.5% or more of original titer) is presumptive evidence of specific IgM, and results should be confirmed by a more definitive IgM assay. For example, a serum with an unabsorbed titer of 256 and a titer of 64 after SPA absorption (i.e., a two dilution decrease or 25% of original titer) would be considered presumptive positive for specific IgM, whereas, a serum with titers of 256 and 16 before and after SPA-Abs (i.e., a four dilution decrease or about 6% residual) would be interpreted as negative for specific IgM. Another frequent use of the SPA absorbent reagent is for pretreating serum to remove excess IgG and possible IgG–IgM immune complexes before testing by one of the solid phase indirect immunoassays using labeled anti-human IgM (see below). This potentially increases the sensitivity and specificity of the later assays. Several commercial sources of SPA absorbent reagent are available in the U.S.

15.2.3 Solid Phase Indirect Immunoassays

The availability of class-specific antiglobulins has led to the adaptation of several other serologic techniques, including indirect immunofluorescence assay (IFA), enzyme immunoassay (EIA), and radioimmunoassay (RIA) for detecting virus-specific IgM antibodies. The general principle for these methods is that test serum is incubated with viral antigens that are bound to a solid phase surface, and specific IgM antibodies bound to the antigen, are subsequently detected with anti-human IgM antibody labeled with a suitable marker (Figure 15.2).

Because of the technical simplicity of these methods, commercial indirect immunoassay IgM kits for several viruses, including rubella virus, herpes simplex virus (HSV), cytomegalovirus (CMV), Epstein–Barr virus (EBV), rotavirus, and hepatitis

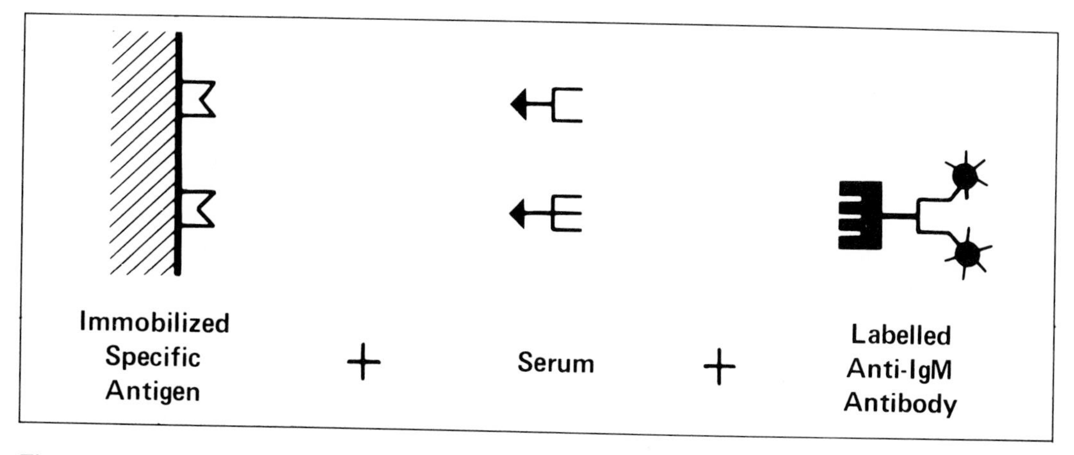

Figure 15.2. Schema of solid phase indirect immunoassay for IgM antibody.

A virus and hepatitis B surface antigen (HBsAg) recently have become available in the U.S. and Europe. However, there are concerns over a number of pitfalls in such methods that may limit the sensitivity and specificity of these assays. These pitfalls can be grouped generally into three categories: a) the quality of available reagents, b) interference by IgM-class rheumatoid factor (RF), and c) competition between specific IgG and IgM antibodies in patient serum specimens for available antibody-binding sites on the solid phase antigen. Each of these factors can play an important role in the reliability of an indirect IgM antibody assay system.

Substantial improvements in the quality of reagents for these assays have occurred during the past few years. More highly purified antigens and highly specific anti-IgM globulin conjugates are now available; however, no standards for the specificity and potency of these reagents have yet been established. False-positive results may occur in these IgM antibody assays if IgM with anti-IgG activity but with no antiviral specificity (i.e., RF) becomes attached to complexes of specific IgG and the bound solid phase antigen (Figure 15.3). On the other hand, failure to detect specific antiviral IgM may occur in sera with high levels of specific antiviral IgG due to the

Figure 15.3. Schema of false-positive IgM due to RF interference in the solid phase indirect immunoassay.

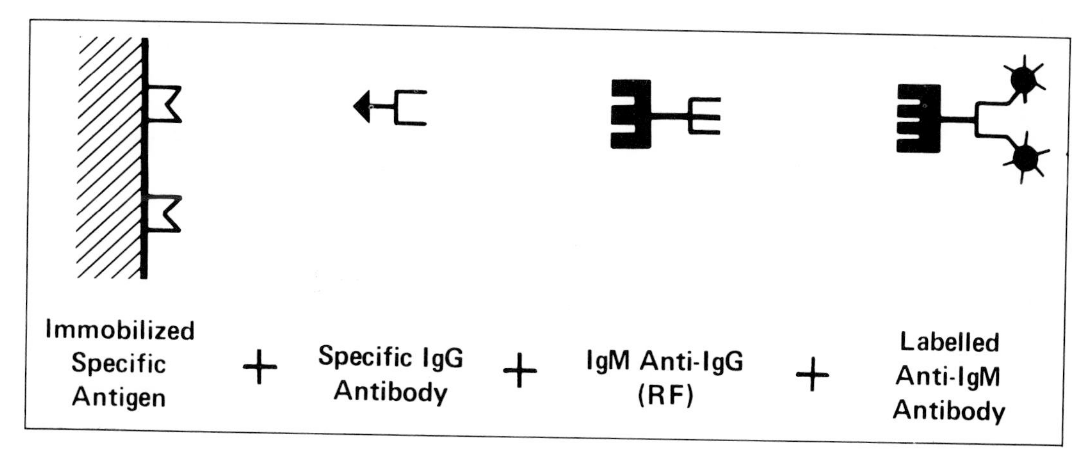

competition for antigen binding sites. Both of the above problems can be minimized by preadsorption of the sera with SPA or protein A-Sepharose. Such preadsorption has been shown to effectively eliminate nonspecific IgM activity from sera with known RF and to significantly increase the sensitivity of the specific IgM assay by removing most competing IgG (Kronvall and Williams, 1969).

15.2.3.a Indirect Immunofluorescence

The first of the indirect assays to be applied for the determination of antiviral IgM antibodies was the IFA (Baublis and Brown, 1968). The antigen in the test usually consists of infected cells fixed on microscope slides. The method is essentially identical to IFA for IgG antibodies, except that fluorescein-labeled antihuman IgM is used. Monoclonal antihuman IgM conjugates have recently become available.

The reading of IFA–IgM tests requires considerable skill and experience. Nonspecific staining may cause false-positive readings; but, on the other hand, an experienced FA microscopist may eliminate false-positive results by differentiating patterns of specific and nonspecific fluorescence, a possibility that does not exist in RIA and EIA tests. In experienced hands and with the use of high quality reagents, the IFA–IgM test can be a sensitive and reliable method. This method should be avoided by those laboratories lacking substantial skill and past experience with IFA.

15.2.3.b Radioimmunoassay

Solid phase RIA has been used to detect viral antibodies since the early 1970s. The use of purified viral antigens bound to a suitable solid phase surface (polyvinylchloride or polystyrene) has eliminated much of the nonspecific background reactivity caused by anticellular or antinuclear antibodies observed in earlier RIA. With purified antigens the preparation of comparable control antigens is often impossible, and the specificity of the assay must be evaluated by comparison with a reference method and with results obtained after blocking by specific hyperimmune serum.

The major advantages of RIA for specific IgM antibody are the high sensitivity of the method and the potential for automation. Competitive inhibition of IgM reactivity by specific IgG antibodies, a common drawback in IFA–IgM assays, has not been a problem in RIA–IgM tests (Knez et al., 1976). The major disadvantage of RIA has been the relatively short shelf-life of the ^{125}I-labeled antihuman immunoglobulin conjugates compared with IFA or EIA reagents.

15.2.3.c Enzyme Immunoassay

Enzyme immunoassay was first reported for the detection of specific antiviral IgM antibodies by Voller and Bidwell (1976). In principle, this method is identical to the RIA. The EIA uses antihuman IgM labeled with enzyme, usually alkaline phosphatase or horseradish peroxidase. The antigen preparations and solid phase supports used in EIA are identical to those used in RIA. The sensitivity and specificity of the two assays are comparable. Competition between specific IgM and IgG, not observed in RIA, has been reported with EIA (Heinz et al., 1981). The enzyme conjugates for EIA tests have a long shelf-life, compared with the iodinated conjugates for RIA, and thus are more practical for commercial development. Several commercial EIA–IgM test kits are now available in the U.S. and in Europe. These tests generally may be read on automated multichannel spectrophotometers, and the results are generated and interpreted by the aid of microcomputers.

15.2.4 Reverse "Capture" Solid Phase IgM Assays

Another approach for avoiding the problems of competitive interference and nonspecific reactivity seen with the traditional

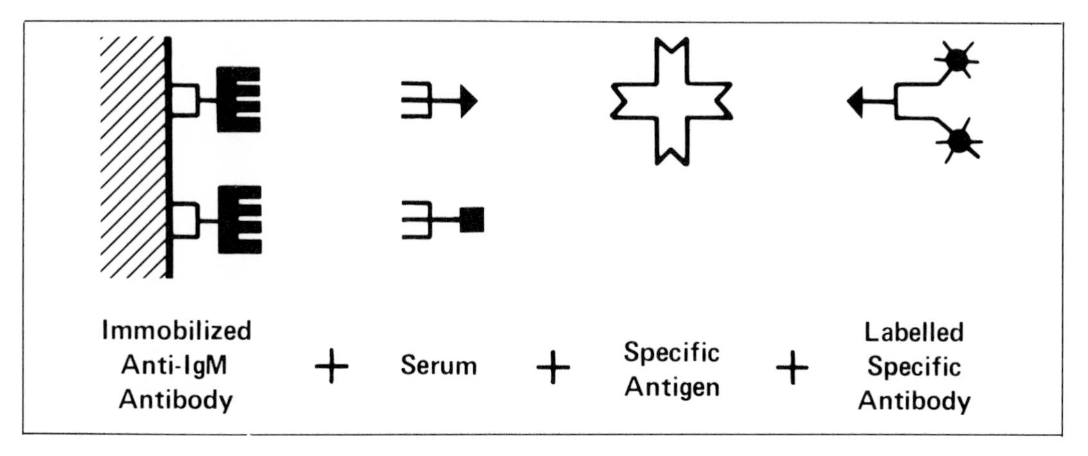

Figure 15.4. Schema of reverse capture solid phase immunoassay for IgM antibody.

indirect immunoassays described in the preceding section is the reverse IgM antibody capture method (Figure 15.4). This approach, first described for detecting antiviral IgM antibodies by Duermeyer and van der Veen (1978), employs a solid phase surface sensitized with an anti-human–IgM antibody to "capture" and bind the IgM antibodies in a serum specimen, after which IgG and any immune complexes in the specimen are washed away. Exposure of the bound IgM antibody to specific viral antigen, followed by the addition of a second, labeled antiviral antibody, completes the test. This approach is attracting considerable support, and its use for detecting IgM antibodies to a number of viruses, including hepatitis A, HBsAg, CMV, HSV, and rubella, has been reported. Alternatively, when the antigen is a hemagglutinating virus, red blood cells can be used as an indicator to produce hemagglutination or hemadsorption (Krech and Wilhelm, 1979).

The reverse capture solid phase IgM assays have proven to be very sensitive and specific (Mortimer et al., 1981). Our experience confirms that the IgM capture assays are potentially more sensitive than our standard assays based on sucrose density gradient fractionation. Because the first step in the IgM capture assay leads to separation of IgM antibodies from other serum components, competition between IgG and IgM does not occur. RF interference, however, potentially exists with these IgM capture assays, as it does with the indirect IgM immunoassays described earlier. IgM–RF may be captured and bound on the solid phase and then, in turn, bind labeled antiviral IgG antibodies. In general, however, interference with RF in these IgM capture assays can be controlled by two simple methods. First, because RF binds to the Fc portion of the IgG molecule, false-positive results can be avoided by using labeled $F(ab')_2$ fragments as the indicator antibody. Second, because the affinity of RF for aggregated IgG is much greater than for native IgG, the use of aggregated IgG in the serum dilution buffer to block the binding sites of the RF is suggested. In general, the reverse capture IgM assays are less susceptible to interference by RF than traditional indirect solid phase IgM assays. Because the reverse capture IgM assay does offer some simple methods to eliminate the RF interference problem, there is no doubt that this approach is potentially more specific than the indirect immunoassays that utilize labeled anti-human IgM conjugates.

15.3 INTERPRETATION OF ASSAY TO DETECT IgM ANTIBODIES

The demonstration of specific antiviral IgM antibodies may be interpreted as indicating a recent or current infection with the virus in question only if the IgM response is specific (i.e., these IgM antibodies are not produced by any other infection or condition). The absence of specific IgM antibodies, on the other hand, rarely can be used as evidence to exclude a recent infection with a given virus. Variations in the temporal appearance of IgM antibodies, including the occurrence of prolonged IgM antibody responses, can result in difficulties in interpreting the significance of the test results in relation to the clinical illness in question.

False-positive IgM antibody results may occur due to crossreactions between closely related viruses. Such crossreactions have been reported for togavirus (Wolff et al. 1981) and in coxsackie B virus infections (Schmidt et al., 1968). In general, the heterologous IgM antibody responses are low compared with homologous titers.

Evidence suggesting the occurrence of true polyclonal IgM production in cases of acute infectious mononucleosis has been reported by Morgan-Capner et al. (1983). Their report suggests that production of various IgM antibodies may result from EBV-induced stimulation of B lymphocytes already committed by prior antigenic stimulation. These results emphasize the importance for careful interpretation of positive virus specific IgM together with the complete clinical picture.

In infections with viruses belonging to groups of closely related strains or serotypes (adenoviruses, enteroviruses, parainfluenza, or togaviruses), serodiagnosis using specific IgM testing may be complicated by the possible absence of a specific IgM response, as well as by possible false-positive reactions to related viruses. Specific IgM antibody responses generally are absent in reinfections or reactivations of latent virus infections and may be very weak or absent in certain immunocompromised patients.

Finally, the expected duration of the specific IgM response must be considered when interpreting the significance of observed specific IgM antibody. Generally, the IgM antibody response following an acute viral infection is of limited duration, usually 1 to 2 months. However, prolonged IgM antibody responses have been observed in complicated infections, chronic infections, congenital infections, and in some immunosuppressed patients. The persistence of specific IgM in these cases appears to be related to the persistence of viral antigen (or even replicating virus) in the patient. Occasionally, prolonged IgM antibody responses have been observed without any apparent reason. Also, as more sensitive methods are developed for the detection of specific antiviral IgM antibodies, the time following an acute infection during which specific IgM is detectable will be extended. For the diagnosis of an acute infection, the ideal maximum duration of specific IgM antibodies should be 2 to 3 months. It may therefore be necessary to limit the sensitivity of some assays to retain the optimal diagnostic usefulness of the methods.

The diagnostic value of specific IgM antibody assay is variable and is dependent on the virus and the infection in question. Generally transient IgM responses are characteristic of acute viral infections caused by viruses that elicit long-lasting immunity. Such responses are seen with rubella, measles, mumps, and hepatitis A viruses. In these infections, a reliable diagnosis can usually be made by specific IgM antibody testing of a single serum specimen taken early in the illness. For other virus infections, such as HSV or CMV, the diagnostic usefulness of such tests is much more limited.

Several sensitive and reliable methods for the determination of specific antiviral IgM antibodies have now been developed. Reagents and kits for the performance of these tests are now available commercially for some virus infections. These methods, when adopted for routine use in clinical laboratories, should bring considerable improvement to viral diagnostic services.[1]

---

[1] Use of trade names is for identification only and does not imply endorsement by the Public Health Service or by the U.S. Department of Health and Human Services.

REFERENCES

Al-Nakib, W. 1980. A modified passive-haemagglutination technique for the detection of cytomegalovirus and herpes simplex virus antibodies: Application in virus-specific IgM diagnosis. J. Med. Virol. 5:287–293.

Ankerst, J., Christensen, P., Kjellen, L., and Kronvall, G. 1974. A routine diagnostic test for IgA and IgM antibodies to rubella virus: Absorption of IgG with *Staphylococcus aureus*. J. Infect. Dis. 130:268–273.

Banatvala, J.E., Best, J.M., Kennedy, E.A., Smith, E.E., and Spence, M.E. 1967. A serological method for demonstrating recent infection by rubella virus. Br. Med. J. 3:285–286.

Barros, M.F., and Lebon, P. 1975. Separation des anticorps IgM anti-rubeole par chromatographie d'affinite. Biomedicine (Express) 23:184–188.

Baublis, J.V., and Brown G.C. 1968. Specific responses of the immunoglobulins to rubella infection. Proc. Soc. Exp. Biol. Med. 128:206–210.

Caul, E.O., Hobbs, S.J., Roberts, P.C., and Clarke, S.K.R. 1976. Evaluation of simplified sucrose gradient method for the detection of rubella-specific IgM in routine diagnostic practice. J. Med. Virol. 2:153–163.

Caul, E.O., Smyth, G.W., and Clarke, S.K.R. 1974. A simplified method for the detection of rubella-specific IgM employing sucrose density fractionation and 2-mercaptoethanol. J. Hyg. (Camb.) 73:329–340.

Duermeyer, W., and van der Veen, J. 1978. Specific detection of IgM antibodies by ELISA, applied in hepatitis A. Lancet ii:684–685.

Forghani, B., Schmidt, N.J., and Lennette, E.H. 1973. Demonstration of rubella IgM antibody by indirect fluorescent antibody staining, sucrose density gradient centrifugation and mecaptoethanol reduction. Intervirology 1:48–59.

Hawkes, R.A., Boughton, C.R., Ferguson, V., and Lehmann, N.I. 1980. Use of immunoglobulin M antibody to hepatitis B core antigen in diagnosis of viral hepatitis. J. Clin. Microbiol. 11:581–583.

Heidelberger, M., and Pedersen, K.O. 1937. The molecular weight of antibodies. J. Exp. Med. 65:393–414.

Heinz, F.X., Roggendorf, M., Hofmann, H., Kunz, C., and Dienhardt, F. 1981. Comparison of two different enzyme immunoassays for detection of immunoglobulin M antibodies against tick-borne encephalitis virus in serum and cerebrospinal fluid. J. Clin. Microbiol. 14:141–146.

Johnson, R.B., Jr., and Libby, R. 1980. Separation of immunoglobulin M (IgM) essentially free of IgG from serum for use in systems requiring assay of IgM-type antibodies without interference from rheumatoid factor. J. Clin. Microbiol. 12:451–454.

Knez, V., Stewart, J.A., and Zeigler, D.W. 1976. Cytomegalovirus-specific IgM and IgG response in humans studied by radioimmunoassay. J. Immunol. 117:2006–2013.

Krech, U., and Wilhelm, J.A. 1979. A solid-phase immunosorbent technique for the rapid detection of rubella IgM by hemagglutination inhibition. J. Gen. Virol. 44:281–286.

Kronvall, G., and Williams, R.C., Jr. 1969. Differences in anti-protein A activity among IgG subgroups. J. Immunol. 103:828–833.

Morgan-Capner, P., Davies, E., and Pattison, J.R. 1980. Rubella-specific IgM detection using Sephacryl S-300 gel filtration. J. Clin. Pathol. 33:1082–1085.

Morgan-Capner, P., Tedder, R.S., and Mace, J.E. 1983. Rubella-specific IgM reactivity in sera from cases of infectious mononucleosis. J. Hyg. Camb. 90:407–413.

Mortimer, P.P., Tedder, R.S., Hambling, M.H., Shafi, M.S., Burkhardt, F., and Schilt, U. 1981. Antibody capture radioimmunoassay for anti-rubella IgM. J. Hyg. (London) 86:139–153.

Pattison, J.R. 1982. Laboratory Investigation of Rubella. Public Health Laboratory Service Monograph Series No. 16. London: Her Majesty's Stationery Office.

Pattison, J.R., Mace, J.E., and Dane, D.S. 1976. The detection and avoidance of false-positive reactions in tests for rubella-specific IgM. J. Med. Microbiol. 9:355–357.

Schmidt, N.J., Lennette, E.H., and Dennis, J. 1968. Characterization of antibodies produced in natural and experimental coxsackievirus infections. J. Immunol. 100:99–106.

Vesikari, T., and Vaheri, A. 1968. Rubella: A method for rapid diagnosis of a recent infection by demonstration of the IgM antibodies. Br. Med. J. 1:221–223.

Voller, A., and Bidwell, D.E. 1976. Enzyme-immunoassays for antibodies in measles, cytomegalovirus infections and after rubella vaccination. Br. J. Exp. Pathol. 57:243–247.

Wolff, K.L, Muth, D.J., Hudson, B.W., and Trent, D.W. 1981. Evaluation of the solid-phase radioimmunoassay for diagnosis of St. Louis encephalitis infection in humans. J. Clin. Microbiol. 14:135–140.

16

Staphylococcus Protein A—Antibody Conjugates

Gerald J. Lancz and Steven Specter

16.1 PROPERTIES AND ACTIVITIES OF PROTEIN A

The analysis of the antigenic components of a bacterium represented an important means of cataloging bacteria for the purpose of classification, as well as of potential benefit for the identification of bacteria associated with clinical diseases. Thus, in 1940, Verwey provided the description of a protein antigen which is present in *Staphylococcus aureus* but not present in coagulase negative nonpathogenic staphylococci, that is, a type-specific protein derived from *Staphylococcus*. Interest in this protein lay dormant until the investigations of the immunochemical and antigenic nature of this material by Jensen et al. (1961) and Lofkvist and Sjoquist (1963). Subsequently, this material was designated as protein A of *S. aureus* (SPA) to differentiate it from immunogenic polysaccharides (Grov et al., 1964).

Protein A has been detected in approximately 95% of the *S. aureus* strains that have been examined, although the amount of SPA produced by *S. aureus* varies with the particular bacterial strain (Forsgren, 1970; Kronvall et al., 1971; Bind et al., 1978). However, there is no known correlation between the relative amount of SPA produced and the virulence of a particular strain of *S. aureus* (Forsgren, 1972). The Cowan I strain (ATCC 12598) which produces relatively large amounts of SPA, is used as the prototype strain. Protocols describing the growth, processing and inactivation with heat and formalin of *S. aureus* have been presented (Goding, 1978; Kessler, 1981), although inactivated Cowan I *S. aureus* is also commercially available. The SPA is synthesized by the organism and is linked to the peptidoglycan structure forming the bacterial cell wall (Sjoquist et al., 1972). Each bacterial cell has been estimated to contain 80,000 SPA molecules (Kronvall et al., 1970). Thus, the SPA represents a relatively large portion of the bacterial cell mass.

SPA has a molecular weight of 42,000. The protein is relatively stable to reducing agents, extremes of temperature, and pH (Sjoholm, 1975). The C-terminus of the molecule is bound to the bacterial cell wall and the N-terminus of the protein contains four binding sites (Sjodahl, 1977). Each of these sites has the ability to bind to γ-globulins of most mammalian species (Richman et al., 1982b). In addition, SPA contains tyrosine, which can be iodinated with ^{125}I (Langone, 1978; Cleveland et al., 1979) or the protein may be tagged with a fluorescent dye or an enzyme (Yolken and Leister, 1981). It is the globulin binding capability of SPA, together with its ability to react with appropriate tracer molecules, that have stimulated interest and use of SPA in clinical, diagnostic, and research laboratories. Highly purified SPA, as well as SPA that is radioiodinated or linked to enzymes, is also commercially available.

It is the Fc portion of the immunoglobulins that binds the SPA. However, this union does not interfere with the antigen binding capacity of the Fab portion of immunoglobulin molecules (Forsgren and Sjoquist, 1966; Langone et al., 1978). Although SPA will bind immunoglobulins of most mammalian species, there is considerable difference in the binding affinity that is observed (Langone, 1978; Richman et al., 1982b). It is principally the IgG molecule that binds SPA, although some human IgM proteins also react with SPA (Harboe and Folling, 1974; Lind et al., 1975). Because SPA bound human IgG molecules so efficiently, it was thought that absorption of human serum with *S. aureus* that contained SPA could be used to remove IgG from serum. Thus, any antibody activity that remained would be IgM-specific antibodies. However, it was shown that SPA binds human IgG_1, IgG_2, and IgG_4 very well, whereas, IgG_3 does not bind (Kronvall and Williams, 1969; Hjelm, 1975). Further, absorption of serum with *S. aureus* containing SPA does not remove IgA, IgE, or substantial amounts of the IgM (Saltvedt and Harboe, 1976; Lind et al., 1975). Thus, the initial promise of utilizing a simple *S. aureus* absorption protocol for the detection of IgM-specific antibodies to rubella or other parasitic agents has not been fulfilled. The realization that not all IgA or IgG bind SPA, and that some IgM proteins also bind SPA, stresses the need for caution in employing *S. aureus* absorption as a means of detecting IgM specific rubella virus antibodies, as occasionally false-positive results have been observed (Leinikki et al., 1978; Crovari et al., 1979; Field et al., 1980).

Because SPA binds to immunoglobulins of most mammalian species, it can be utilized effectively in detecting antigen–antibody interactions in indirect assays that employ immunoglobulins obtained from different animal species (Richman et al., 1982b). Furthermore, SPA does not bind to Fc receptors that are induced in cells following viral infection (Yasuda and Milgrom, 1968; Costa et al., 1978) and its nonspecific binding to laboratory materials and plastics does not appear to be problematic (Cleveland et al., 1979). A more detailed review of the biochemical and biological properties and the developmental investigations leading to the use of SPA in the diagnostic and research laboratory has been published recently (Richman, 1983).

SPA, either in purified form or attached to the intact bacterium, can be employed to great advantage in the clinical laboratory for the detection and identification of viral antigens in clinical specimens. The next two sections of this chapter will each deal with a specific utilization of SPA in protocols that are directly utilizable by the clinical laboratory. In the first protocol, a procedure employing intact *S. aureus* containing SPA will be described and in the second protocol, utilization of purified SPA to detect viral antigens will be described.

16.2. VIRUS DETECTION BY *S. AUREUS* CONTAINING PROTEIN A

The protocol described below has been used for the identification of herpes simplex virus (HSV) following its primary isolation in tissue culture. The procedure makes use of *S. aureus* that has been preadsorbed with anti-HSV immunoglobulins (Mogensen and Dishon, 1981; Lancz and Specter, 1982). A number of reports have described the detection of virus antigens by initially incubating an antiserum with the virus infected cells and subsequently allowing *S. aureus* to adsorb to antibody that has reacted with viral antigens (Huang and Okorie, 1978, 1979; Mogensen and Dishon, 1981).

It is advantageous to use a high titer antiserum that is specific for the particular viral agent in question. This serum must be devoid of antibodies that are directed against *S. aureus*. We have used New Zealand rabbits for immunization, employing purified HSV as the immunogen. The virus is replicated in rabbit cells that are grown in tissue culture so that nonviral membrane components in the virion would be species compatible with the host in which the antiserum would be raised. The HSV is purified by differential centrifugation and then by a rate zonal sedimentation in sucrose gradients (Lancz, 1980). This purified HSV is dialyzed against phosphate buffered saline (PBS; pH 7.4) and is injected as described (Smith et al., 1971; Lancz and Bradstreet, 1976). The serum is obtained by standard techniques and is monitored for neutralizing activity directed against the immunizing strain of HSV-1 or -2.

A volume of a 10% (v/v) formalinized *S. aureus* suspension is washed three times with PBS with intermittent pelleting by centrifugation. Following the third wash, the pelleted bacteria are resuspended in a volume of undiluted rabbit anti-HSV serum that is equal to the initial volume of the bacterial suspension (Figure 16.1). Adsorption is allowed to proceed for 1 hour at room temperature, the bacteria are washed three times with PBS as described above, and fi-

Figure 16.1. Flowchart depicting protocol used to prepare *S. aureus* containing antibody (Sa-AB) and normal rabbit serum (Sa-NRS).

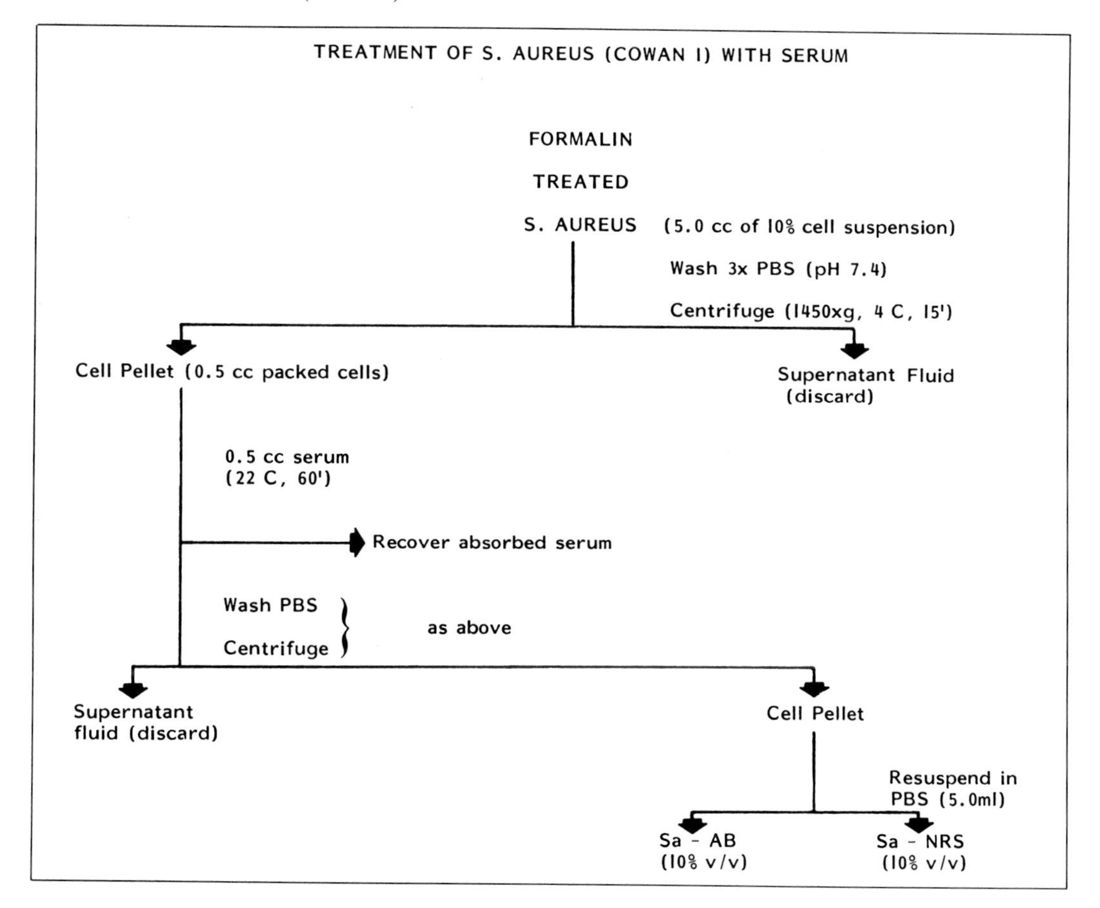

nally suspended in PBS to a final concentration of 10%. A control preparation is obtained by substituting normal rabbit serum for the rabbit anti-HSV serum in the above protocol. We have used these preparations, stored at 4°C, for 6 months without any significant loss of biological activity.

Cell cultures (25-cm^2 culture flasks) are inoculated with the clinical specimen and examined for the development of virus associated cytopathology. This cytopathology is readily evident in a variety of host cells that are commercially available, including MRC-5, RhMK, Vero, and others. The verification of the isolation of HSV from the clinical specimen is performed with the *S. aureus* adsorbed with anti-HSV serum using the following protocol: The culture fluid is decanted from the cell monolayer and is saved as a source of virus for any subsequent inoculation that might be required. The cell sheet is gently washed twice with PBS and the culture vessel is then stood on end to enable thorough drainage of excess fluids. These are carefully removed by aspiration. After placing the cell culture on a horizontal surface, an infected cell area receives a 10–20 μl drop containing *S. aureus* preabsorbed with HSV antibody and a second, separate area receives a drop containing *S. aureus* that was preabsorbed with normal serum. The areas of the sheet covered by the drops of *S. aureus* are outlined with a marker and are sufficiently distant from one another to avoid intermingling of the two preparations. The cell culture is incubated at 37°C for 15 minutes and is then briefly washed three times with 5 ml PBS. The saline is slowly and gently rotated over the cell sheet to remove the unreacted mass of bacteria. Following the third wash, the cell culture is again thoroughly drained and the remaining fluid is aspirated. A 1-ml volume of 1% methylene blue is added and the cell sheet is incubated at room temperature for 10 minutes. The methylene blue is removed by aspiration and the cell sheet may be briefly washed once with PBS. The cells are then viewed microscopically and the cellular area displaying cytopathology is examined for the adherence of blue-stained bacteria to the cell surface (Figure 16.2). The adherence of *S. aureus* containing HSV antibody to infected cell areas showing cytopathology coupled with the absence of nonspecific cellular attachment of *S. aureus* absorbed with normal serum is indicative of the isolation and identification of HSV from the clinical specimen. In most instances, 18 to 24 hours of growth in culture are required before confirmation of the isolation of HSV can be performed. On occasion, it has been possible to detect and identify HSV by 12 to 15 hours postinoculation. It is interesting to note that a recent study by Mogensen and Dishon (1983) reported success with the detection of HSV and varicella-zoster virus (VZV) infected cells by the direct examination of cells in the clinical specimen with antiserum adsorbed to *S. aureus*. This approach has the advantage of reducing the time required to make a positive diagnosis, which is especially important in those cases when rapid detection of HSV is of paramount importance for patient management (e.g., encephalitis or vaginal infection in near-term pregnant women). As with any immunoassay, the reliability of the test is dependent on the specificity and relative potency of the antiserum employed. Using a single antiserum preparation, the detection of HSV antigens in cell culture employing *S. aureus* adsorbed with HSV antiserum and by indirect immunofluorescence, it was apparent that these two procedures are of relative equal sensitivity (Lancz and Specter, 1982).

The protocol described above is a test that facilitates the detection, identification, and verification of virus-specific antigens concomitant with the primary isolation of the virus in tissue culture. Therefore, the protocol as described could be adapted for use to detect any virus or virus-specific antigen in cell culture. Although there is a potential applicability for the use of *S. aureus* to which specific viral antibodies are adsorbed, there are as yet no commercially

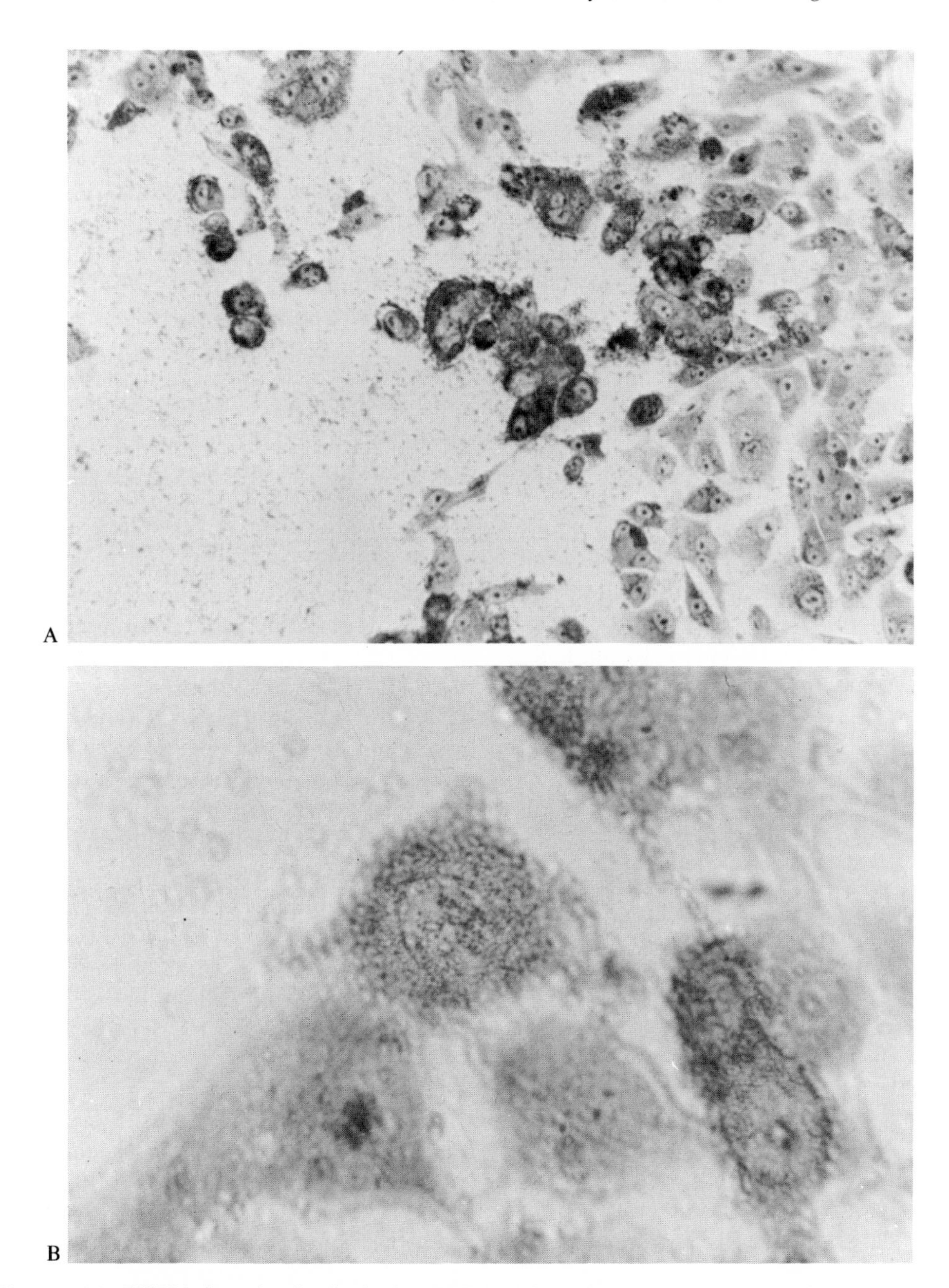

Figure 16.2. HSV infected cells displaying CPE and Sa-AB adhering to their surface. (A) × 100; (B) × 400 (By kind permission of Alan R. Liss, Inc.)

available test kits for this purpose. It should be noted that antiserum that is raised following injection with whole virus will neutralize the homologous HSV type and will also neutralize the heterologous HSV type, to a lesser degree. This is due to the antigenic crossreactivity between HSV-1 and -2 (Nahmias and Dowdle, 1968; Herrmann and Stewart, 1980). Thus, the production of a single antiserum to HSV can be used to detect the primary isolation in cell culture of most strains of both types of HSV. A disadvantage is that HSV is identified but not typed by this procedure. The substitution of a type-specific HSV antibody that is commercially available would enable the simultaneous serologic identification and typing of HSV following its primary isolation in cell culture. A principle advantage of the protocol as described is the simplicity of the procedure, itself. There is no requirement for any specialized equipment other than an inverted light microscope. Radioactive molecules and enzyme reactions are not used. The test is exceedingly simple to perform and requires no specialized training of personnel. The most limiting aspects of antigen detection by *S. aureus* that contains SPA are the requirements for an antiserum devoid of anti-*S. aureus* activity and for the primary isolation of the virus in cell culture. This may be obviated by the cytospin technique described by Mogensen and Dishon (1983).

16.3 UTILIZATION OF PURIFIED PROTEIN A IN THE DIAGNOSIS OF VIRAL INFECTIONS

A variety of protocols have been described that utilize purified SPA in the detection of immunoglobulins directed against viral antigens or in the immunologic confirmation of a virus isolated from a clinical specimen. Each system exploits the interaction between SPA and immunoglobulin molecules most often in a radioimmune assay or an enzyme immunoassay. What follows is a synoptic description of a protocol developed by Richman et al. (1982a) that permits the identification of HSV, cytomegalovirus (CMV), and VZV subsequent to their isolation in vitro.

Cultures of human cells are inoculated with a clinical specimen and examined daily for the development of CPE. Following a freeze–thaw cycle, a 100 μl volume of the cell suspension is passed through a glass fiber filter. The filter pad traps the virus infected cellular material and serves as the matrix for the subsequent reactions. In this particular system, the filters are placed in a plate that has 96 wells, each well has a small hole. The plate fits into a manifold, such that the filtering activity is activated when a vacuum is applied (Cleveland et al., 1982). In this way, debris that contains viral antigens is trapped on the filters, and the filters can be repeatedly washed by the application of vacuum to the manifold. Specific virus antiserum is then added to the wells and allowed to react with the antigens. The filters are washed free of excess serum and incubated with a conjugate of SPA and horseradish peroxidase (SPA–HRP). If the appropriate viral antigens are present, the Fab portions of the immunoglobulin react with the specific antigenic determinants and the Fc portion of the immunoglobulin binds the SPA–HRP. The filters are washed free of unbound SPA–HRP, the substrate is added and the development of the colored product is monitored visually or the reaction fluid can be transferred and evaluated spectrophotometrically (Richman et al., 1982a).

The protocol as described is technically simple and rapid, requiring 2 to 3 hours to complete, but is dependent on the filter plate and the vacuum manifold that is commercially available. The sensitivity and specificity of the system is incumbent on having access to type-specific or high quality antiviral serum. Each antiserum must be block titrated against both the specific virus used to raise the antiserum and related viruses. In this way, should there be a degree of antigenic crossreactivity between viruses, the serum used in the test can be diluted to eliminate the immunoglobulins that react to the virus that is antigenically crossreactive, while retaining sufficient IgG for the antigens of the specific virus in question (e.g., the low level of crossreactivity seen between HSV and VZV). Because of the greater degree of antigenic crossreactivity between HSV-1 and -2, a type-specific serum is more difficult to obtain and requires absorption of the serum with the HSV type that is heterologous to that used to raise the antiserum. Alternatively, type-specific monoclonal antibodies can be used in these assays. The use of a pool that contains a few type specific monoclonal antibodies would reduce the possibility that a clinical isolate might not react with a single monoclonal antibody which would result in a false-negative test. Finally, the SPA–HRP conjugate has been found to be somewhat less sensitive than an antiimmunoglobulin HRP conjugate, both of which are commercially available (Richman et al., 1982a). However, because SPA binds to immunoglobulins from a vast array of mammals, only a single SPA–HRP conjugate is needed to detect antiviral antibodies that have been prepared in different animal species. Thus, in this regard, the SPA–HRP conjugate is more versatile than the antiimmunoglobulin HRP conjugate.

The protocol described above is referred to as an immunofiltration technique. However, the use of SPA–HRP in the diagnosis of virus infections is readily adaptable in slightly modified protocols. For example, titration of a serum sample for antiviral antibodies is accomplished by incubating the dilutions of the serum in antigen coated wells of a microtiter plate. Following wash cycles, SPA–HRP conjugate is added and subsequently detected as described above. Alternatively, the detection of specific virus antigen is accomplished by an antigen trapping protocol using antiviral serum coated microtiter plates. The trapped antigen is reacted with antibody a second time, which is then incubated with the SPA–HRP conjugate. (The details of these protocols appear elsewhere in this volume and, therefore, are not described here. See Chapter 10). Thus, SPA, either in a purified form or attached to the *S. aureus*, has been and will continue to be exploited in the development of assays for the diagnosis of viral infections that are both technically simple and rapid.

REFERENCES

Bind, J.L., Chiron, J.P., and Denis, F. 1978. Protein A research in 1200 strains of staphylococcus (using a rapid detection technic by passive hemagglutination with glutaraldehyde-treated erythrocytes). CR Soc. Biol. (Paris) 172:212–215.

Cleveland, P.H., Richman, D.D., Oxman, M.N., Wickham, M.G., Binder, P.S., and Worthen, D.M. 1979. Immobilization of viral antigens on filter paper for a (^{125}I) staphylococcal protein A immunoassay: A rapid and sensitive technique for detection of herpes simplex virus antigens and antiviral antibodies. J. Immunol. Meth. 29:369–386.

Cleveland, P.H., Richman, D.D., Oxman, M.N., and Worthen, D.M. 1982. A rapid serologic technique for typing herpes simplex viruses. J. Clin. Microbiol. 15:402–407.

Costa, J., Yee, C., Nakamura, Y., and Rabson, A. 1978. Characteristics of the Fc receptor induced by herpes simplex virus. Intervirology 10:32–39.

Crovari, P., Gasparini, R., Bono, A., Lusian, F., and Tassi, G.C. 1979. Use of staphylococcal protein A for the detection of rubella virus IgM antibodies. Boll. Ist. Sieroter. Milan 58:371–381.

Field, P.R., Shanker, S., and Murphy, A.M. 1980. The use of protein A-sepharose affinity chromatography for separation and detection of specific IgM antibody in acquired rubella infection: A comparison with absorption by staphylococci containing protein A and density gradient ultracentrifugation. J. Immunol. Meth. 32:59–70.

Forsgren, A. 1970. Significance of protein A production by staphylococci. Infect. Immun. 2:672–673.

Forsgren, A. 1972 Pathogenicity of *Staphylococcus aureus* mutants in general and local infections. Acta Pathol. Microbiol. Scand. (B) 80:564–570.

Forsgren, A., and Sjoquist, J. 1966. Protein A from *S. aureus*. I. Pseudoimmune reaction with human γ-globulin. J. Immunol. 97:822–827.

Goding, J.W. 1978. Use of staphylococcal protein A as an immunological reagent. J. Immunol. Meth. 20:241–253.

Grov, A., Myklestad, B., and Oeding, P. 1964. Immunochemical studies on antigen preparations from *Staphylococcus aureus*. Acta Pathol. Microbiol. Scand. (C) 61:588–596.

Harboe, M., and Folling, I. 1974. Recognition of two distinct groups of human IgM and IgA based on different binding to staphylococci. Scand. J. Immunol. 3:471–482.

Herrmann, K., and Stewart J. 1980. Diagnosis of herpes simplex virus type 1 and 2 infections. In A.M. Nahmias, W.R. Dowdle, and R.F. Schinazi, (eds.), The Human Herpes Viruses: An Interdisciplinary Perspective. New York: Elsevier.

Hjelm, H. 1975. Isolation of IgG_3 from normal human sera and from a patient with multiple myeloma by using protein A-Sepharose 4B. Scand. J. Immunol. 4:633–640.

Huang, A., and Okorie, T. 1978. Rapid diagnosis using surface analysis by bacterial adherence. Lancet ii:1146.

Huang, A., and Okorie, T. 1979. Surface analysis by bacterial adherence to virus infected cells. J. Infect. Dis. 140:147–151.

Jensen, K., Neter, E., Gorzynski, E.A., and Anzai, H. 1961. Studies on toxic products of *Staphylococcus*. Acta Pathol. Microbiol. Scand. (C) 53:191–200.

Kessler, S.W. 1981. Use of protein A-bearing staphylococci for the immunoprecipitation and isolation of antigens from cells. Methods Enzymol. 73:442–458.

Kronvall, G., and Williams, R.C., Jr. 1969. Differences in anti-protein A activity among IgG subgroups. J. Immunol. 103:828–833.

Kronvall, G., Quie, P.G., and Williams, R.C., Jr. 1970. Quantitation of staphylococcal protein A: Determination of equilibrium constant and number of protein A residues on bacteria. J. Immunol. 104:273–278.

Kronvall, G., Dossett, J.H., Quie, P.G., and Williams, R.C., Jr. 1971. Occurrence of protein A in staphylococcal strains; quantitative aspects and correlation to antigenic and bacteriophage types. Infect. Immun. 3:10–15.

Lancz, G.J. 1980. Physical integrity of herpes simplex virus following thermal inactivation. Arch. Virol. 64:375–381.

Lancz, G., and Bradstreet, J. 1976. pH mediated inhibition of the cell to cell spread of herpes simplex virus infection. Arch. Virol. 52:37–46.

Lancz, G., and Specter, S. 1982. A simple and rapid test for the identification of clinical herpes simplex virus isolates. J. Med. Virol. 10:11–15.

Langone, J.J. 1978. (^{125}I) protein A: A tracer for general use in immunoassay. J. Immunol. Meth. 24:269–285.

Langone, J.J., Boyle, M.D.P., and Borsos, T. 1978. Studies on the interaction between protein A and immunoglobulin G. I. Effect of protein A on the functional activity of IgG. J. Immunol. 121:327–332.

Leinikki, P.O., Shekarchi, I., Dorsett, P., and Sever, J.L. 1978. Determination of virus-specific IgM antibodies by using ELISA: Elimination of false-positive results with protein A-sepharose absorption and subsequent IgM antibody assay. J. Lab. Clin. Med. 92:849–857.

Lind, I., Harboe, M., and Folling, I. 1975. Protein A reactivity of two distinct groups of human monoclonal IgM. Scand. J. Immunol. 4:843–848.

Lofkvist, T., and Sjoquist, J. 1963. Purification of staphylococcal antigens. Intl. Arch. Allergy 23:289–305.

Mogensen, S., and Dishon, T. 1981. The use of *Staphylococcus aureus* rich in protein A in the detection of herpes simplex virus antigens. Acta Path. Microbiol. Scand. Sect. B 89:427–432.

Mogensen, S., and Dishon, T. 1983. Rapid detection of herpes simplex virus and varicella-zoster virus in clinical specimens by the use of *Staphylococcus aureus* rich in protein A. Acta Path. Microbiol. Immunol. Scand. Sect B 91:83–88.

Nahmias, A., and Dowdle, W. 1968. Antigenic and biological differences in herpes virus hominis. Prog. Med. Virol. 10:110–159.

Richman, D.D. 1983. The use of staphylococcal protein A in diagnostic virology. Curr. Top. Microbiol. Immunol. 104:159–176.

Richman, D.D., Cleveland, P.H., and Oxman, M.N. 1982a. A rapid enzyme immunofiltration technique using monoclonal antibodies to serotype herpes simplex virus. J. Med. Virol. 9:299–305.

Richman, D.D., Cleveland, P.H., Oxman, M.N., and Johnson, K.M. 1982b. The binding of staphylococcal protein A by the sera of different animal species. J. Immunol. 128:2300–2305.

Saltvedt, E., and Harboe, M. 1976. Binding of IgA to protein A-containing staphylococci: Relationship to subclasses. Scand. J. Immunol. 5:1103–1108.

Sjodahl, J. 1977. Repetitive sequences in protein A from *Staphylococcus aureus*. Arrangement of five regions within the protein, four being highly homologous and Fc-binding. Eur. J. Biochem. 73:343–351.

Sjoholm, I. 1975. Protein A from *Staphylococcus aureus*. Spectropolarimetric and spectrophotometric studies. Eur. J. Biochem. 51:55–61.

Sjoquist, J., Movitz, J., Johansson, I.B., and Hjelm, H. 1972. Localization of protein A in the bacteria. Eur. J. Biochem. 30:190–194.

Smith, J., Rodriguez, J., and McKee, A. 1971. Biological characteristics of cloned populations of herpes simplex virus types 1 and 2. Appl. Microbiol. 21:350–357.

Verwey, W.F. 1940. A type specific antigenic protein derived from the Staphylococcus. J. Exp. Med. 71:635–644.

Yasuda, J., and Milgrom, F. 1968. Hemadsorption by herpes simplex-infected cell cultures. Intl. Arch. Allergy. Appl. Immunol. 33:151–170.

Yolken, R.H., and Leister, F.J. 1981. Staphylococcal protein A-enzyme immunoglobulin conjugates: Versatile tools for enzyme immunoassays. J. Immunol. Meth. 43:209–218.

17

Considerations for the Automation of Virology Testing

Bryan L. Kiehl

17.1 INTRODUCTION

Although most clinical virology laboratories are not automated today, the development of effective antiviral drug therapy, preventive health care (screening), and/or laboratory consolidation may change this. Those tests that are requested in large numbers and do not require immediate results are the ideal candidates for automation.

The performance characteristics of a diagnostic test, reagent cost, capital expenses, technician requirements, turnaround time, and reliability are the principal concerns of the laboratory. Once a test is identified as a likely candidate for automation, there no doubt will be several competing manufacturers. This will necessitate some form of comparison prior to adopting a system. Selection of the specific kit(s) to be utilized in the laboratory should be evaluated within the context of the laboratory's requirements.

17.2 PRELIMINARY EVALUATION

For most laboratories, the choice will be difficult. If test volume is sufficient to warrant batch processing and "on demand" testing is unusual, gathering information about test kits on the market is probably appropriate. The cost per test and the anticipated capital expenses should be determined for currently available systems. Make sure to read a package insert so that FDA reviewed claims, rather than advertisements, can be used to make the next series of decisions. Advertising claims can be difficult to interpret. The following are examples of data or information needed to select among available kits based on your laboratory's anticipated activity:

1. What is the average number of patient samples per week?
2. What is the number of runs required to provide appropriate patient management? How many runs per day or week are necessary?
3. How often will on demand (stat) testing be required?
4. Rank in order of priority the following three catagories: cost, turnaround time, accuracy.
5. Could technicians with less training be substituted?
6. What is the clinical purpose for testing?
7. Determine the types of populations to be tested.
8. Determine current cost for testing; sep-

arate labor, overhead (including instrument depreciation and contracts), and reagent costs.

All tests require at least one negative and one positive control; many kits require more. At times, the automated laboratory must increase the number of controls to assure the reliability of the results. The best way to compare various kits is to decide how many patient specimens and controls will be incorporated into an average run. The cost per week, month, or year can then be determined. If stat tests are required and an alternative method is not available, the estimated expenses associated with these determinations also should be included. However, a test that has been designed for automation in the laboratory may not be the test of choice for stats. Whenever possible, a second procedure designed for single specimen testing is preferable.

A range of acceptable limits for reagent cost, labor cost, turnaround time, and accuracy must be chosen. One need not assign restrictive limits initially. If more than one kit is technically acceptable, cost and turnaround time should be compared before a final selection is made. These calculations, however, will be based on and reflect individual and particular laboratory situations. For example, automation will only be efficient if the total cost (reagent, labor, and capital costs) is less than or equal to the current cost to run the test. In addition, the automated test must be accomplished in the same time as the current method used or offer significant advantages worth the premium. After consultation with the clinical staff, the primary use for the test (screening for negatives or definitive diagnosis of positives) is assigned.

Current virus testing procedures generally are labor intensive. Although kits designed for automation usually have higher reagent cost, they are often designed to be labor saving. Estimates of required technician time by the minute or determining the percentage of actual time per day or week a technician is dedicated to a test are commonly used for calculating labor costs. If lesser paid employees can perform the test, an estimate of two labor costs should be made.

Most manufacturers price a kit so that final cost will not immediately eliminate their test from consideration. Because of this, price should be considered during this early phase but more carefully scrutinized later in the decision making process. Estimation of labor costs based on package insert protocols should be used. During the final experimental evaluation, confirmation of the labor demands for each individual laboratory should be made.

The "expected results" section of the package insert should provide a description of the trial population(s) tested by the manufacturer and the results of this evaluation when compared with other accepted procedures. It is not uncommon for serologic tests to demonstrate comparable performance for similiar populations. Alternately, comparison of an isolation test or a direct antigen test may yield significantly different results when tested in a variety of populations. If the tested population markedly differs from the population for whom this test will be employed, then all performance claims must be viewed suspiciously. The rate of positives (prevalence), specimen collection method, means of transportation to the laboratory, and cell culture protocols all can affect the test performance. Experimentation using the anticipated patient population and laboratory technical staff will be necessary to make a final judgement.

The "specific performance" section of the package insert contains the sensitivity, specificity, and reproducibility measures of the test. The sensitivity and specificity is generally dependent on the field trial design and, therefore, on the tested populations. Again, these results should be considered only approximations.

If only one kit meets the needs of the laboratory and the test has been validated by laboratories of comparable clinical experience, little comparison testing is required. However, two or more tests with similar claims and costs often are acceptable. The next step will be to conduct experiments to validate the claims and labor estimates for these tests. By carefully defining the requirements of an acceptable test, extensive patient specimen trial comparisons can be eliminated. The purpose is not to repeat the extensive testing of the manufacturer, but only to prove that the kit performs to expectations. A screening test to select negative patients will be evaluated differently from a test designed to provide a definitive diagnosis. If the test is intended for both purposes, two comparisons are necessary.

If the purpose for offering the test cannot be easily resolved, consultation with the clinical staff may be necessary. It must be made clear that only true positives and true negatives be identified; however, most tests misidentify some specimens. *If the purpose is to identify a negative patient, the test with the highest specificity is required. If the purpose is to identify positive patients, a high sensitivity is necessary.* During premarket development, the kit manufacturer endeavors to determine the relative importance to customers of sensitivity and specificity. The cutoff between negative and positive results is set accordingly. If the laboratory's intended purpose differs from the manufacturer's target market, the cutoff may be set inappropriately.

If the test volume is sufficiently large to warrant automation, it may be assumed that many negative patients are being screened for the viral agent or antibody. In these cases, a test with a high negative predictive value is best. When necessary, definitive diagnosis of positive results using a confirmation protocol or different test may be the best compromise.

Before beginning experimental comparison testing, prediction of the population distribution of positives will aid in the experimental design. For example, most laboratories report a 20–30% isolation rate (prevalence) for herpes simplex virus (HSV). However, a subset of this population is at risk, pregnant women who are routinely screened for HSV. The prevalence of positives in this population may be as low as 1–2%. The following case study illustrates a selection process of three hypothetical diagnostic systems for HSV detection.

17.3 CASE STUDY

The following are the performance characteristic claims of three kits a diagnostic laboratory wishes to consider.

Criteria	*Kit A*	*Kit B*	*Kit C*
Sensitivity	93.0	85.2	97.8
Specificity	98.6	99.8	93.1

More than one prevalence rate is anticipated in the population, so three test cases are generated to illustrate the effects on the predictive values of the three kits. An assumption that the antigen titer distribution is the same at each prevalence will be made. Using the reported sensitivities and specificities, the predictive values are calculated based on a 500-patient population. The calculated performance values at various prevalence rates are shown in Tables 17.1, 17.2, and 17.3.

All three systems adequately predict negative patients at any anticipated prevalence rate. Because prediction of negatives

Table 17.1. Expected Values for Kits Tested in a Population Demonstrating 2% Prevalence

Kit	TN	TP	FN	FP	PPV	NPV
A	483	9	1	7	56	99.8
B	489	9	1	1	90	99.8
C	489	10	0	34	23	100.0

Abbreviations: TN, true negative; TP, true positive; FN, false-negative; FP, false-positive; PPV, positive predictive value; NPV, negative predictive value.

Sensitivity = TP/(TP + FN)
Specificity = TN/(TN + FP)
PPV = TP/(TP + FP)
NPV = TN/(TN + FN)

Table 17.2. Expected Values for Kits Tested in a Population Demonstrating 10% Prevalence

Kit	TN	TP	FN	FP	PPV	NPV
A	444	47	3	6	89	99.3
B	449	43	7	1	98	98.4
C	419	49	1	31	61	99.8

See Table 17.1 for abbreviations.

is probably a primary concern of the laboratory running a large number of HSV tests, any one of these systems would be appropriate for this purpose. If HSV diagnosis is also an intended use, only kit B would be useful in the low prevalence population (Table 17.1). Even though the kit has the lowest sensitivity, the high specificity of the assay eliminates false-positives from consideration. *When a positive is rare, the kit with the highest specificity is usually the test of choice*. In higher prevalence populations, kits A and C may be judged to be appropriate. If the price of kits A and C were significantly less than kit B, the laboratory might consider a confirmation test using kit B for all positive results reported by kits A or C as a cost-effective HSV system in the laboratory.

The purpose for the above exercise is not to universally detail how to choose a kit for any particular laboratory, but to offer an example. There are far too many variables to describe. Each laboratory will need to set their own priorities (cost, turnaround time, accuracy, etc.). By using the information provided by the manufacturer and clearly defining the laboratory's needs, expensive experimentation with kits that will not meet the laboratory's needs can be avoided. The "what if" time spent by the director can be very productive.

Table 17.3. Expected Values for Kits Tested in a Population Demonstrating 30% Prevalence

Kit	TN	TP	FN	FP	PPV	NPV
A	345	140	10	5	97	97.2
B	349	128	22	1	99	94.1
C	326	147	3	24	86	99.1

See Table 17.1 for abbreviations.

17.4 COMPARISON TESTING

The previous example assumes the sensitivity and specificity reported in the package insert is accurate. Comparison testing must confirm the performance characteristics. Before starting experimentation, be certain that the purpose of the experiment is clearly stated. An experiment designed to demonstrate equality between two systems usually requires fewer specimens than an experiment designed to demonstrate a difference between methods. It is always best to consult with a biostatistician before starting. The investigator does not wish to spend time and money collecting and testing unnecessary specimens. It is even more disappointing during data review to discover that no satisfactory conclusion is possible because too few samples were tested. Table 17.4 shows the required number of specimens at various sensitivities and prevalence rates to demonstrate two methods are equally effective.

If the anticipated sensitivity of kit to be tested is 85% and the prevalence of positive results is approximately 15%, at least 1307 specimens must be tested to demonstrate equal performance to the reference method. This table is not intended as a substitute for statistical advice, but demonstrates how one might predict sample requirements. Because a series of assumptions have been made to generate Table 17.4, the information may not be appropriate for all experimental designs.

In a low-prevalence population, validation of the sensitivity may be an expensive undertaking. If all positive results will be retested by a second procedure to confirm the diagnosis, validating the sensitivity of the test is unwarranted. An assumption that the average titer in the low-prevalence population is similar should be made unless

Table 17.4. Required Number of Specimens for a Given Test Sensitivity and Population Prevalence Rate

	Sensitivity (percentage)			
Prevalence (percentage)	80	85	90	95
5	4920	3920	2760	1460
10	2460	1960	1380	730
15	1640	1307	920	487
20	1203	980	690	365
25	984	784	552	292

reported to the contrary. With this assumption, determining the sensitivity in a high prevalence population and assuming the same sensitivity in lower prevalence populations is usually most expeditious.

17.4.1 Determining True Positive and True Negative Results

Currently employed cell culture or electron microscopy procedures are commonly used today, but new tests may replace them if they predict the clinical condition more accurately. Selection of the true positive and true negative population determine all other kit performance characteristics, yet this selection process, at best, remains an inaccurate art.

All specimens yielding discrepant results must be evaluated. Whenever possible, a second reference method is run concurrently. Caution must be used when comparing the test results with the clinical interpretation. Because the physician collecting the specimen has usually decided to collect based on a clinical impression, the test results and the collection process are not independent events. On the other hand, an independent detailed clinical history of *all* patients is informative.

17.4.2 Immune Status Testing

Evaluation of newly developed serologic tests that are used to screen immune status is fraught with additional problems. Today, only when testing a vaccine under development can one determine protective levels of antibody. Most newer technologies will attempt to correlate results to an established immunity level established using other procedures. When validating tests of immune status in the laboratory, a large number of samples within 2 to 3 standard deviations of the cutoff must be tested. Repeat testing with another reference method will be necessary to resolve discrepancies between the method under evaluation and the established reference method being used.

Immune status cannot be determined for virus infections for which an immune level has not been established. These serologic tests measure the presence or absence of antibody. Validation of the cutoff is similar to the validation used for antigen testing.

17.4.3 Validating the Cutoff of an Antigen Test

All kits will report a cutoff value that is an appropriate value used by most laboratories. After completing a comparison of the kits, validation of the cutoff for the tested population is appropriate. First, determine the mean of all true negatives. Calculate the standard deviation. It is safe to assume that no true negative occurs above three standard deviations (greater than 99% confidence limit). To correct daily test variation, most manufacturers have set the cutoff relative to the mean negative value of the test run. It is beyond the scope of this chapter to describe how a final cutoff is determined. Nevertheless, to validate the selected cutoff, determine each run cutoff according to the package insert. If the cutoff is below 3 S.D. above the negative mean, occurrence of false-positive results should be suspected.

After analysis of the experimental data, a cutoff more appropriate for the laboratory may be derived. For example, it might be noted that Kit A, from the earlier case study, set the cutoff at 2 S.D. above the mean and Kit B set the cutoff at 3.6 S.D. above the mean. It is possible that both kits are similar and the apparent performance differences can be explained by the manufacturer's cutoff selection. A detailed analysis may permit selection of a similar cutoff in Kit A and, consequently, achieve the desired results. The laboratory comparison study will rarely generate sufficient data to allow such changes. Additional testing to assure the validity is required. A laboratory should not attempt to alter the package insert information without extensive expertise in assay development. Warranties and performance support from the manufacturer are quickly withdrawn if the kit is not used as intended or directed. However, some specialty laboratories may find this situation acceptable.

Automation in the clinical virology laboratory will not occur until physician demand for viral diagnostic testing increases. Introduction of effective drug therapy, recognition of cost efficient screening programs, and/or consolidation of testing in large reference laboratories will be the forces moving virologists in this direction. As this occurs, manufacturers will provide reagents and systems designed for these purposes and the selection of the best combination of these will be required. Ultimately, overall cost (labor, reagents, impact of patient stay and treatment) will be the primary concern. Only kits that are cost-effective and provide acceptable results will remain. It will be the responsibility of the clinical virologist to make this selection.

18

Susceptibility Testing for Antiviral Agents

M. Nixon Ellis

18.1 HISTORY

The recent development of effective therapy of certain viral infections has emphasized the need for rapid methods to determine the sensitivity of viruses to these new compounds. Although the plaque reduction assay (McLaren et al., 1983) yields reproducible results and has been the standard technique to determine the susceptibility of viruses to antivirals for several decades, it is cumbersome to perform, costly in materials and personnel time, and not readily amenable to sensitivity testing of large numbers of isolates. The dye-uptake (DU) method of determining viral sensitivity to inhibition by antiviral agents has been described (McLaren et al., 1983). This technique is an adaption of a system developed for the assay of interferon (Finter, 1969). The preferential uptake of a vital dye (neutral red) by viable cells over damaged cells forms the basis of this method. The relative extent of viral cytopathic effect (CPE) in different cultures may be determined by the relative amounts of dye bound. The dye taken up by living cells may be eluted into acid–alcohol and quantitated colorimetrically. This technique may be adapted to automated and computerized systems to allow for mass screening of isolates.

This work was supported by the Veterans Administration and by contract AI-22681 from the National Institutes of Health.

18.2 MATERIALS

1. Eagle's minimal essential medium (EMEM) containing 5% fetal calf serum (FCS) 0.075% sodium bicarbonate, 75 U/ml pencillin G, 75 μg/ml of streptomycin, 2 m*M* L-glutamine and buffered with HEPES to pH 6.5–7.0.
2. Vero cells, continuous line of African green monkey kidney cells (Flow)
3. Culture plates, 96-well flat-bottom (Costar no. 3596, Cambridge, MA)
4. Small, disposable, 12 × 72 mm sterile disposable test tubes with caps
5. Pipettes, 1.0, 5.0 and 10.0 ml sterile
6. Disposable, sterile 50-μl dropping pipettes (Dynatech, Alexandria, VA)
7. Sterile sealing tape (Dynatech)
8. Sterile blotter papers
9. Cornwall syringe, 1.0 and 2.0 ml, with 8-channel manifold
10. Neutral red dye (Sigma Chemical Co., St. Louis, MO)

11. Phosphate buffered saline (PBS) (0.1 *M*), pH 5.8–6.0
12. Test tube racks
13. Citrate methanol buffer [molecular weights: citric acid, 210.14 (0.1 *M* = 21 g/L = 4.2 g/200 ml); sodium citrate, 294.1 (0.1 *M* = 29 g/L = 2.94 g/100 ml): 0.1 M citric acid, 157.5 ml; 0.1 *M* sodium citrate, 92.5 ml; deionized H_2O, 250.0 ml; methanol and/or ETOH, 500.0 ml]
14. Ice bucket
15. Water bath, 37°C
16. Mini-Mash apparatus (MA Bioproducts, Walkersville, MD)
17. Multichannel spectrophotometer for 96-well plates (Titertek Multiskan, Flow Laboratories, McLean, VA)
18. Autodiluter II (Dynatech)
19. Frozen virus specimens, including reference laboratory strains of herpes simplex virus types 1 and 2 (HSV-1, -2)
20. Hemacytometer
21. Filters (0.45 μ*M*)
22. Ultrasonic cleaning bath (Sonicor Instrument Corp.)

18.3 DYE-UPTAKE ASSAY

18.3.1 Virus Infectivity Assays

1. Prepare suitable volumes (e.g., 1.8 ml) of complete EMEM with 5% FCS in sterile tube with metal closures; keep on ice
2. Rapidly thaw virus sample; briefly sonicate (≃30 seconds) in ultrasonic cleaning bath to disrupt any virus aggregates; keep on ice
3. Prepare tenfold dilution series of test virus in tubes containing EMEM: assuming 1.8 ml volumes of media using a 1-ml pipette measure 0.2 ml of the original virus suspension into the first tube; make further serial dilutions up to 10^{-6} and hold on ice
4. With the Dynatech dispenser fitted with fresh manifold, dispense 100 μl EMEM 5% FCS into rows 1 and 2, 50 μl medium into rows 3 through 8, and 300 μl into row 9
5. Using the Dynatech dispenser, dispense 200 μl of Vero cell suspension (1×10^5 cells/ml) into all wells of the first eight vertical rows of a flat-bottomed, 96-well plate
6. Using sterile 50-μl disposable dropper add 50 μl of the 10^{-6} dilution of virus to all wells of row 8 of the plate; add 50 μl of the 10^{-5} dilution of virus to all wells of row 7; add 50 μl of the appropriate dilution of virus to the other rows until the final virus dilution (10^{-1}) goes into all wells of row 3
7. The plate, therefore, should have the following arrangement (see Table 18.1): Each test well should contain 300 μl of liquid
8. Each plate is sealed with a sheet of sterile sealing film and the lid replaced
9. Incubate for 72 hours at 37°C in a 5% CO_2 incubator
10. After 72-hour incubation, examine the plate for gross contamination and ex-

Table 18.1. Virus Infectivity Assay Plate Arrangement

	Row Numbers									
Contents	1	2	3	4	5	6	7	8	9	10–12
Medium μl	100	100	50	50	50	50	50	50	300	Empty
Cells 200 μm	+	+	+	+	+	+	+	+		
Virus 50 μl			−1	−2	−3	−4	−5	−6		

treme pH changes of medium; check control wells in rows 1 and 2 for cell confluence; examine some wells of rows 3 and 4 for virus CPE; if CPE is absent from these rows do not continue

11. If CPE is present in row 3, then carefully remove the sealing film while under a safety-hood to avoid producing an aerosol-containing infectious virus
12. Using the Dynatech dispenser and the dye manifold add 50 μl of a 0.15% solution of neutral red in PBS (pH 6.0) to each well
13. Incubate the plate for 45 minutes at 37°C in CO_2 incubator
14. After incubation, briefly check several wells for the presence of neutral red crystals; if extensive crystallization has occurred, then high background readings may be obtained
15. Use the Mini-Mash, to rinse out excess dye with PBS pH 6.0.
    - Aspirate dye off
    - Fill wells to the top with PBS
    - Aspirate PBS off
    - With manifold resting on plate, alternatively rinse and aspirate from each well
16. Using Dynatech dispenser and buffer manifold add 100 μl of buffer (citrate methanol, pH 4.2) to each well; gently rock the plate to ensure even elution of dye into buffer
17. The optical density of the solution is determined at 540 nm using multichannel spectrophotometer designed for 96-well plates; The mean optical density (OD) of the cell control wells is assigned a value of 100%, the control (blank) wells a value of 0% and the dilution of virus producing a 50% OD reading (i.e., 50% inhibition of cell growth is determined from a computer-programmed linear regression analysis; for additional information on computer program, contact R. Harvey, Burroughs Wellcome Co.); the titer of each virus pool is expressed as a 50% dye uptake (DU_{50}) value, i.e. the reciprocal of the dilution of virus producing a 50% reduction in neutral red dye-uptake by the cells

18.3.2 Virus Inhibition Assay

1. Prepare initial drug solution in EMEM; Because 50 μl of drug is mixed in the well with 250 μl (1:6 dilution) of other solutions, the initial drug solution should be six times more concentrated than the initial concentration to be tested
2. Using a 50-μl sterile disposable dropper, add 50 μl of the initial drug solutions to all wells of rows 2, 3 and 4
3. Use Dynatech diluter to serially dilute the drug while dispensing 50 μl of complete EMEM (5% FCS); use sterile Cornwall syringe (1 ml) and manifold (8-channel), flame and prewet diluters before use; fill the blot and rinse trays
    - Set dispensers for rows 1, 2, 4, to 12
    - Set diluter for rows 4 to 11
    - Set blot cycle on
    - Place labeled microtiter plate in position
    - Press the RUN button
4. Using the Dynatech dispenser add an additional 50 μl EMEM to all wells of row 1
5. Prepare suspension of Vero cells (1×10^5 cells/ml) in EMEM (5% FCS); using Dynatech dispenser with sterile Cornwall syringe (2.0 ml) add 200 μl of the cell suspension into each well of the plate
6. Make up suspension of virus to be tested to contain 30 DU_{50} virus/50 μl; using sterile disposable dropper add 50 μl of virus suspension to all wells of rows 3 to 12
7. The plate should have the following format (see Table 18.2):

   row 1 = cell control

   row 2 = drug control

   row 3–11 = drug dilution serials

   row 12 = virus control

Table 18.2. Virus Inhibition Assay Plate Arrangement

	Row Numbers											
Contents	1	2	3	4	5	6	7	8	9	10	11	12
Medium μl	100	50		50	50	50	50	50	50	50	50	50
Drug 50 μl		+	+	+								
Cells 200 μl	+	+	+	+	+	+	+	+	+	+	+	+
Virus 50 μl			+	+	+	+	+	+	+	+	+	+

The number of replicates in each row may be eight or less (it is recommended that eight are used); in each group of experiments include a standard laboratory strain as an internal control

8. Seal the plate with sterile sealing tape; incubate the plate at 37°C in CO_2 incubator
9. Prepare four serial tenfold dilutions of the test virus suspension used to infect the drug-treated cells; using sterile 50-μl dropper, add 50 μl of the 10^{-4} dilution of virus to all wells of row 7; add 50 μl of preceding dilution in sequence to rows 4 through 6; 50 μl of test virus suspension added to all wells of row 3; the back titration plate should have the following arrangement (see Table 18.3):
10. After 72-hour incubation, neutral red dye is added to both drug plates and back titration plates as described in virus infectivity assay; plates are read with drug plates first, followed by back titration plate
11. Linear regression analysis of the data was used to determine the concentration of drug producing a 50% reduction in viral CPE in relation to cell controls (0%) and virus controls (100%); this concentration of drug is the 50% inhibitory dose value (ID_{50})
12. The exact dose of the challenge virus is determined by reading the back titration plate as described in virus infectivity assays

18.4 OTHER VIRUSES AND DRUGS

The DU antiviral assay initially was used for the determination of the sensitivity of herpes simplex clinical isolates to inhibition by acyclovir; however, the technique can also be used for screening other compounds for activity against HSV or for screening other viruses. Other drugs that have been screened against HSV in this assay include: adenine arabinoside, 5-iodo-2′-deoxyuridine and phosphonoacetic acid. The major requirements for the successful use of this system are a cytopathogenic virus and cells that will readily form monolayers in the microtiter plates; however, it must be noted that sensitivity testing systems for viruses

Table 18.3. Back Titration Plate Arrangement

	Row Numbers							
Contents	1	2	3	4	5	6	7	8
Medium μl	100	100	50	50	50	50	50	300
Cells 200 μl	+	+	+	+	+	+	+	
Virus 50 μl			undiluted	−1	−2	−3	−4	

are not well standardized, so that considerable preliminary work must be performed in establishing any virus–cell system. For example, even within the same assay, different results may be obtained if different cell lines are used (Harmenberg et al., 1980; Field et al., 1980).

18.5 TECHNICAL PROBLEMS WITH THE DYE-UPTAKE ASSAY

18.5.1 Monolayer Peeling and Dye Crystallization

Two of the major problems that we encountered with the DU assay concern peeling of the cell monolayer prior to elution of neutral red dye or crystallization of the neutral red. Cell peeling has been attributed to overgrowth of the cells ($>1 \times 10^5$ cells/ml), mycoplasmal contamination, or overly vigorous aspiration of dye from the cell monolayers. One way to avoid damaging cells during the rinsing process, particularly if the Mini-Mash apparatus is not functioning properly, is to perform the following. After incubation of plates with dye, gently flick stained plates into a sink, then using rubber tubing attached to a PBS reservoir, gently flood the plate with PBS and then flick the plate into a sink.

The problem of neutral red dye crystallization can be handled in one of several ways. First, one must be very careful in formulating the neutral red solution because the pH is critical. The neutral red must be made up in 0.1 *M* PBS with the pH of the PBS at 5.8–6.0. If the pH is not within this range, crystallization may occur, which will lead to extremely high optical density values. It is also very important to filter the neutral red once it is made up. Finally, if problems still occur, one other change can be made in the assay protocol. After 72 hours of incubation of the plates, the cell medium is removed prior to the addition of dye. This will prevent the neutral red from being affected by the pH of the spent medium.

18.6 ADVANTAGES OF THE DYE-UPTAKE ASSAY

The DU assay has several advantages over the standard plaque reduction method. The availability of automated or semiautomated equipment makes this test ideal for the large laboratory that has many samples to process. In our laboratory we can easily assay 45 to 60 viruses per week. In addition to reducing labor time, there is also a considerable reduction in the amount of cells and reagents consumed in testing.

Perhaps the most significant difference between the DU assay and the plaque reduction method is the fact that it is done with a liquid overlay. This overlay does not inhibit the lateral spread of released extracellular virus which is restricted by agarose or immune serum globulin in other assays. It has been suggested (McLaren et al., 1983) that small populations of drug-resistant virus present in clinical isolates may have a chance to replicate and, thus, produce viral CPE. In clinical isolates of HSV, both drug-sensitive and -resistant populations exist (Parris and Harrington, 1982). Presumably, due to the difference in the techniques, the DU assay has been shown to be slightly more sensitive for the detection of small populations of resistant virus than the plaque reduction test; however, reconstruction experiments have indicated that small fractions ($<25\%$) of drug-resistant virus are below the levels of sensitivity of this assay (McLaren et al., 1983).

The reproducibility of the assay is good and as a control a standard laboratory strain of virus should be included in all assays. The assay should be repeated if the ID_{50} for the control virus is greater than 2 standard deviations from the mean determined in previous testing.

18.7 DISADVANTAGES OF THE DYE-UPTAKE ASSAY

The cost of the automated equipment may be a drawback of the test for the small research laboratory with few viruses to test. Additionally, mechanical problems with the automated equipment could cause severe delays.

The DU assay has been criticized for being less sensitive (ID_{50} values tenfold greater) than the plaque reduction assay (McLaren et al., 1983). This increased DU ID_{50} value may be due to the greater challenge dose used because 30 DU_{50} per well equal approximately 500 plaque forming units (PFU) per well. (McLaren et al., 1983). This points to a more general limitation of all sensitivity tests, namely the ID_{50} values can be significantly altered by the amount of challenge virus. For example, if the challenge dose of a drug-sensitive virus is extremely high, the ID_{50} value of the isolate may be within the resistant range. More importantly, however, it should be noted that since there has been no established correlation between in vitro sensitivity and in vivo response to therapy in humans, differences in absolute in vitro ID_{50} values between different assay systems have little relevance, and the definition of "sensitive" and "resistant" must be unique for each assay system.

Finally, another important limitation of any of these drug sensitivity tests involves the type of cell used in the test. Even when the same assay is used, quite different results can be obtained if different types of cells are used. This is particularly true if the drug being screened is dependent on cellular components for conversion to an active form. As noted earlier, the standardization of any of these drug sensitivity testing systems is poor. One cannot compare plaque reduction assay results with those obtained in the DU assay. These assay systems are still not used widely enough to make general statements about results. At this time, comparisons can only be made if the same assay, virus, and cell type are used in the tests.

REFERENCES

Field, H.J., Darby, G., and Wildy, P. 1980. Isolation and characterization of acyclovir-resistant mutants of herpes simplex virus. J. Gen. Virol. 49:115–124.

Finter, N.B. 1969. Dye uptake methods of assessing viral cytopathogenicity and their application to interferon assays. J. Gen. Virol. 5:419–425.

Harmenberg, J., Wahren, B., and Oberg, B. 1980. Influence of cells and virus multiplicity on the inhibition of herpesviruses with acycloguanosine. Intervirology 14:239–244.

McLaren, C., Ellis, M.N., and Hunter, G.A. 1983. A colormetric assay for the measurement of the sensitivity of herpes simplex viruses to antiviral agents. Antiviral Res. 3:223–234.

Parris, D.S., and Harrington, J.E. 1982. Herpes simplex virus variants resistant to high concentrations of acyclovir exist in clinical isolates. Antimicrob. Agents Chemother. 22:71–77.

19

Nucleic Acid Hybridization

Douglas D. Richman

19.1 INTRODUCTION

The foundation of diagnostic virology is virus isolation. Despite the many advantages of virus isolation, this approach requires technical experience and sophistication, yields results within days to weeks after the collection of the clinical specimen, and is not capable of detecting certain agents. Those agents not amenable to diagnosis by virus isolation include the viruses responsible for hepatitis and gastroenteritis. The alternative approach for the identification of a virus in a clinical specimen involves the detection of viral components. These viral components include structural proteins (which usually are identified on the basis of their antigenicity), virus-induced enzymes, and viral nucleic acids. The detection of viral antigens forms the basis for the standard rapid viral diagnostic techniques: immunofluorescence microscopy, immunoperoxidase microscopy, radioimmunoassays, and enzyme immunoassays. The detection of virus-specific enzymes, such as the neuraminidase of influenza virus, the thymidine kinase of herpes simplex virus (HSV) and varicella-zoster virus (VZV), the DNA polymerase of hepatitis B virus (HBV), the reverse transcriptase of retroviruses, and protease of picornaviruses, have not been put to practical use in the diagnostic lab and remain at the investigational stage. The detection of viral nucleic acids in clinical specimens, as with virus-induced enzymes, has not yet become a practical approach for rapid viral diagnosis.

There are several features of nucleic acid hybridization that have prompted investigation of this technology for the detection of viruses in clinical specimens. First, even short nucleotide sequences can be identified as unique to the genome of specific agents and, therefore, can constitute probes of very high specificity. Second, the high avidity of complementary nucleic acids strands for each other should permit highly sensitive hybridization assays. Finally, molecular cloning should make possible the availability of virtually unlimited amounts of standardized reagents for the detection of any organism. Many laboratories recently have applied the techniques of cloning and hybridization of nucleic acids for the detection of viral nucleic acids in clinical specimens (Table 19.1). Many of the studies listed in Table 19.1 were intended to develop techniques for viral diagnosis; others were intended to investigate the pathoge-

This work was supported by the Veterans Administration and by Contract AI-22681 from the National Institutes of Health.

Table 19.1. Examples of Detection of Viral Nucleic Acid in Clinical Specimens by Hybridization

Virus	Reference
Hepatitis B	Berninger et al., 1982
	Weller et al., 1982
	Scotto et al., 1983a, b
	Lie-Injo et al., 1983
Herpes simplex virus	Stalhandski and Petterson, 1982
	Redfield et al., 1983a, b
Varicella-zoster virus	Seidlin et al., 1984
Cytomegalovirus	Chou and Merigan, 1982
	Martin et al., 1984
	Spector et al., 1984
Epstein–Barr virus	Andiman et al., 1983
	Sixbey et al., 1984
Adenovirus	Brigati et al., 1983
	Virtanen et al., 1983
	Takiff and Straus, unpublished
	Ranki et al., 1983
Papillomavirus	Zachow et al., 1982
	Durst et al., 1983
	Gissmann et al., 1983
	Lancaster et al., 1983
Papovavirus	Grinnell et al., 1983
Rotavirus	Flores et al., 1983
Influenza A virus	Richman et al., unpublished
Enterovirus	Rotbart et al., 1984
Human T-cell leukemia virus	Wong-Staal et al., 1983

netic role of certain viruses, such as HBV in hepatocellular carcinoma and papillomavirus in carcinoma of the cervix.

19.2 TECHNIQUES

Because the technology of nucleic acid cloning and hybridization is evolving so rapidly, any description of methods in the form of a laboratory manual would be obsolete by the time it was published. A number of excellent publications on DNA biochemistry are available for the beginner (and the practitioner) (Kornberg, 1980; Minson and Darby, 1982; Watson et al., 1983). *Molecular Cloning: A Laboratory Manual* (Maniatis et al., 1982) has become a standard in the field.

19.2.1 Hybridization

Nucleic acid is detected in solid phase by one of two general approaches. In situ hybridization detects, as target nucleic acid, the viral DNA or RNA in cytologic preparations or tissue sections. The second approach detects target viral nucleic acid fixed to a nylon or nitrocellulose filter (Meinkoth and Wahl, 1984). The detection of nucleic acid on filters is conceptually simple. DNA is an extremely stable molecule. It resists exposure to such harsh conditions as 0.3 *M* NaOH at 60°C. DNA in clinical specimens is freed from proteins, RNA, and membranes that are degraded under these conditions. The resulting denatured single-stranded DNA can then be applied to nylon or nitrocellulose filters by blotting or filtration. When the filter is dehydrated by baking, the nucleic acid forms bonds with the filter and resists subsequent washing even at high temperature. The result is a solid phase containing immobilized nucleic acid derived from a clinical specimen. Probes of labeled virus specific nucleic acid can then be hybridized to the target nucleic acid that is bound to the filter, for the detection and identification of any viral nucleic acid that is present.

19.2.2 Preparation of Nucleic Acid Probes

Nucleic acid probes are prepared from cloned subgenomic restriction endonuclease fragments of viral genomes. The fragments that are used are selected for their reactivity with a broad range of strains of the virus of interest and for their lack of reactivity with other viruses or with host cell DNA. The nucleic acid probe can be labeled by any of a number of techniques. The most common labeling method is nick-translation in which DNase and DNA polymerase are used to nick and repair an agent-specific DNA molecule and, in the process, to incorporate into it ^{32}P labeled deoxynucleoside triphosphates (Rigby et al., 1977). After hybridization of such a probe with filters containing DNA from the clinical specimen, bound radiolabel is detected by autoradiography. Reexamination of such filters with different probes is possible because the probe can be dissociated and removed under conditions that do not remove the specimen-derived DNA from the filter. Such DNA-containing filters also can be stored for months.

19.3 DIAGNOSIS USING NUCLEIC ACID HYBRIDIZATION

Studies performed in our laboratory (Redfield et al., 1983a, b) have utilized several cloned fragments of HSV-specific DNA[1]. These restriction endonuclease fragments are inserted into the plasmid vector pBR322, which can be replicated to a high number of copies and easily purified from broth cultures of *Escherichia coli*. An assay in which a ^{32}P-labeled DNA probe is used can detect 1 pg of purified HSV DNA, 10^4 plaque forming units of cell-free HSV or as few as four HSV-infected cells (Figure 19.1). When a probe prepared from HSV-1 DNA is used, the assay is 78% as sensitive as viral culture with swabs of HSV-2 infected genital lesions and 90% as sensitive with HSV-1 infected eye lesions in the diagnosis of infection (Figure 19.2). No false-positive results are obtained with culture negative specimens. Furthermore, no hybridization is observed with uninfected, VZV-infected, or cytomegalovirus (CMV)-infected cells, and specimens from herpes zoster lesions give uniformly negative results (Redfield et al., 1983a).

The specificity of a DNA probe may vary greatly with the portion of the genome comprising the probe. Several subgenomic fragments of both HSV and CMV have been shown to hybridize with mammalian DNA (Peden et al., 1982; Puga et al., 1982). Whether this crossreactivity proves to be due to true sequence homology or to guanosine–cytosine rich sequences will not alter the fact that such fragments are undesirable as diagnostic probes for viral nucleic acid in the presence of human cells. The virus specificity of a probe can also depend on the fragment selected. The 3′ end of enterovirus genomes can be used to detect most enteroviruses (Hyypia et al., 1984; Rotbart et al., 1984), whereas, a poliovirus probe consisting only of the first 220 5′ nucleotides reacts only with poliovirus RNA (Rotbart et al., 1984).

The Bam H1 A fragment of HSV-1 is only two- to fourfold less sensitive for the detection of HSV-2 DNA in comparison with HSV-1 DNA (Figure 19.3). Fragments from the junction regions of HSV-1 (601) and HSV-2 (802) yield absolutely type-specific probes under the conditions employed (Figure 19.3). These probes can then be

[1] The HSV-specific DNA was generously provided by Dr. Bernard Roizman of the University of Chicago.

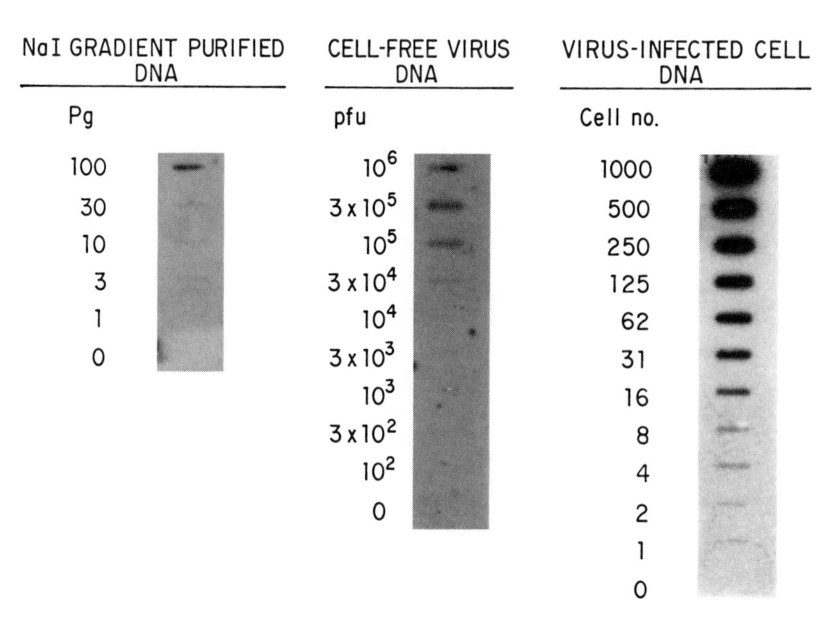

Figure 19.1. (above). HSV-1 DNA from NaI gradient purified HSV DNA, from cell-free virus and from virus-infected cells, was extracted, base treated, and applied to nitrocellulose filters. Autoradiographs are shown following hybridization of these filters with probe prepared from the cloned Bam H1 A fragment of HSV-1, radiolabeled with ^{32}P by nick-translation. (Redfield et al., 1983a).

Figure 19.2 (right). Autoradiographs of the hybridization of ^{32}P-labeled HSV and control probes to DNA from swabs of eyes with keratoconjunctivitis that were culture positive for HSV-1. DNA was extracted and applied to nitrocellulose filters in duplicate. The filters were then hybridized with HSV or control probes radiolabeled with ^{32}P by nick-translation. Culture negative specimens gave uniformly negative results. (Redfield et al., 1983a and b).

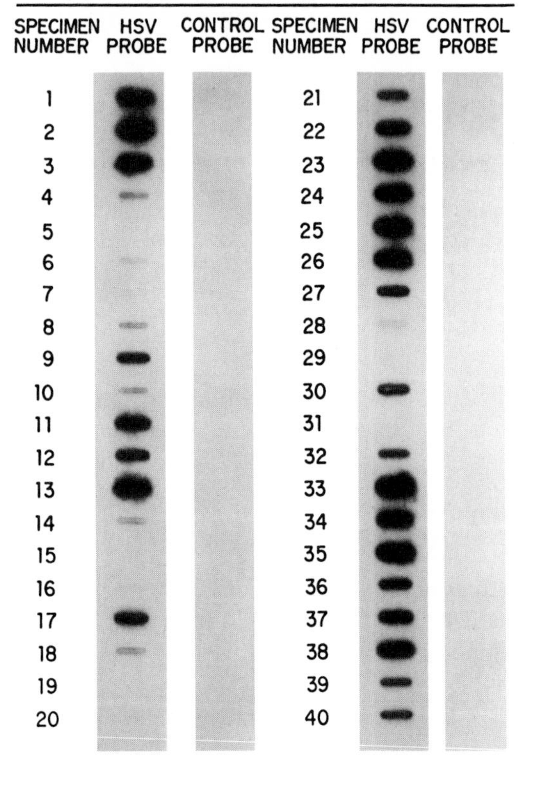

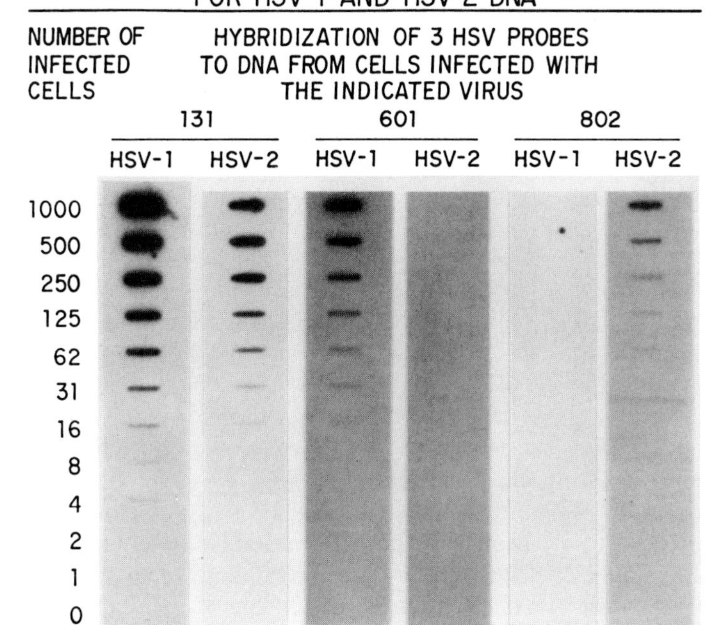

Figure 19.3. Hybridization of three different cloned probes with DNA from dilutions of human fibroblast cells infected with HSV-1 or -2. Probe 131 was prepared from the Bam H1 A fragment of HSV-1, 601 from the junction fragment of HSV-1, and 802 from the junction fragment of HSV-2. (Reproduced with permission of University of Chicago Press from Richman et al., 1984).

SPECIFICITY OF TYPE SPECIFIC PROBES FOR HSV DNA FROM UNRELATED CLINICAL ISOLATES

SPECIMEN NUMBER	HSV TYPE	PROBE 601	PROBE 802
1	1		
2	1		
3	1		
4	1		
5	1		
6	1		
7	1		
8	1		
9	1		
10	1		
11	2		
12	2		
13	2		
14	2		
15	2		
16	2		
17	2		
18	2		
19	2		
20	2		

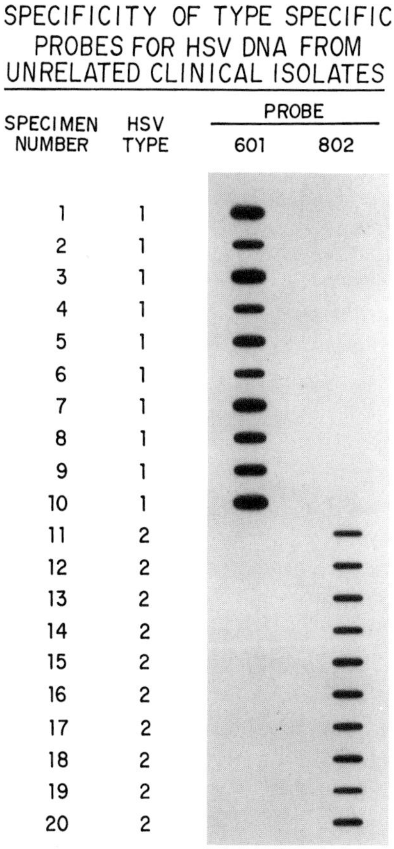

Figure 19.4. Specificity of type-specific probes for DNA in tests with 20 epidemiologically unrelated clinical isolates of HSV. The isolates, which had been characterized with type-specific monoclonal antibodies were examined with type-specific probes from the junction regions of HSV-1. (Reproduced with permission of the University of Chicago Press from Richman et al., 1984).

used for the detection, identification, and typing of HSV DNA in specimens from a number of epidemiologically unrelated patients (Figure 19.4).

19.4 ADVANTAGES

The use of techniques to detect viral components, such as DNA, is especially useful for agents that are slow or difficult to grow. For example, Spector et al. (1984), using the technique just described with cloned restriction endonuclease fragments of CMV, succeeded in detecting CMV DNA in urine and buffy-coat specimens from a number of patients, including neonates and bone marrow transplant recipients. Hybridization yielded positive results quickly and often with specimens from a number of immunosuppressed patients, including some whose viral culture results were negative but who were subsequently shown to have cytomegaloviral disease. The availability of this technology for an agent like CMV, for example, portends a number of clinically useful applications. With the advent of chemotherapeutically promising nucleotide analogs for CMV and EBV (Tocci et al., 1984), the identification of patients at high risk for lethal CMV infections could provide an important application of this new diagnostic approach. This approach merits investigation for the identification of blood and blood products at risk to transmit CMV, a complication well documented both in neonates and in bone marrow transplant recipients (Yeager et al., 1981; Hersman et al., 1982). In addition, this DNA hybridization protocol should yield a method to quantitate antiviral effects both in vitro and in vivo, which is slow and difficult to quantitate with cell culture techniques (Gadler, 1983).

19.5 LIMITATIONS

Although these results are encouraging and clearly demonstrate the feasibility of this approach for rapid viral diagnosis, there are a number of major limitations that must be overcome before nucleic acid hybridization becomes generally useful. First, the extraction of nucleic acid from clinical materials can be quite cumbersome. The preparation of DNA from eye swabs or from herpetic vesicles and ulcers is a simple process (Redfield et al., 1983a); however, DNA present in buffy coats or swabs of materials that contain large amounts of mucus (e.g., respiratory secretions or cervical swabs), can be more difficult to prepare. The preparation of DNA from these materials often requires phenol extraction to eliminate the protein that competes with nucleic acid for binding sites on filters.

The adaptation of hybridization techniques for the detection of single stranded RNA in clinical specimens represents an even more difficult technical problem. The NaOH extraction methods that are used for DNA hydrolyzes single-stranded RNA. In addition, clinical materials in which we want to detect ssRNA viruses (respiratory secretions and stools) contain high levels of ribonuclease activity that degrade RNA as soon as it is extracted from virions or host cells. Utilizing ribonuclease inhibitors, we have modified a number of RNA extraction techniques (Thomas, 1980; White and Bancroft, 1982) to permit the extraction of influenza A viral RNA from respiratory secretions. This RNA was applied to nitrocellulose filters and was detected with probes prepared from plasmids containing clones of DNA complementary to the NP (nucleoprotein) and M (matrix) genes of influenza A/Udorn/1972 (H3N2) virus. The initial results detecting influenza RNA appear quite encouraging. Similarly, other investigators have obtained encouraging preliminary results in efforts to detect enteroviruses (Rotbart et al., 1984).

A second aspect of this approach requiring improvement is sensitivity of detection. For example, in the HSV DNA detection assay already described, 10% of culture positive specimens were not detected by the hybridization assay (Redfield et al., 1983a, b). The intensity of the signal obtained with the hybridization assay roughly correlates with the amount of infectious virus in the specimens (Figure 19.5). Specimens with the lowest titers of viral infectivity are the ones missed by the nucleic acid hybridization assay. This may or may not prove to be a problem with CMV, which is more difficult to culture than HSV. Nevertheless, increased sensitivity for any agent would be desirable because an assay of greater sensitivity permits shorter times for hybridization and development, resulting in a more rapid diagnostic technique.

Several approaches for improving sensitivity are under investigation. One ap-

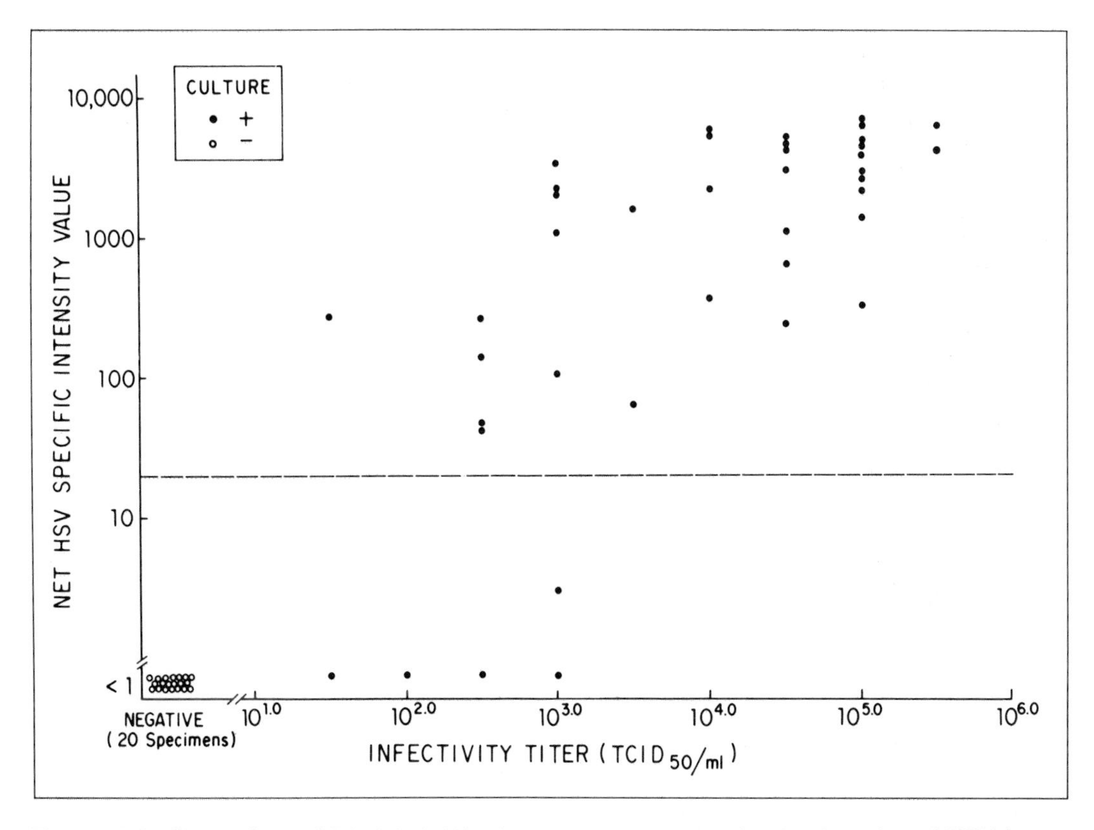

Figure 19.5. Comparison of DNA hybridization and viral isolation for the detection of HSV in eye swab specimens. Forty culture positive and 20 culture negative eye swabs were divided into aliquots for titration of viral infectivity and DNA detection. Viral infectivity is plotted against the intensity of the signal detected on the autoradiograph of the DNA detection assay. The signal intensity represents the area under the curve of the densitometer tracing of the autoradiograph. (Reproduced with permission of the University of Chicago Press from Richman et. al., 1984).

proach involves the use of RNA transcripts as probes rather than the nick-translated DNA (Green et al., 1983). Such transcripts can be labeled to much higher specific activity, and much lower background signals can be obtained with the use of ribonuclease to hydrolyze RNA probes that are nonspecifically bound to the filter and not annealed specifically to viral nucleic acid. Another approach to improve sensitivity would be the development of a nonradiolabeled probe that contained a reporter molecule conferring greater sensitivity. It is hoped that such a reporter molecule could be bound to the nucleic acid probe to such high specific activity that a large number of reporter molecules per unit of target cell DNA could be attained, or that some sort of amplification system was available to render detection of the reporter molecule extremely sensitive.

19.6 ALTERNATIVE METHOD FOR HYBRIDIZATION

There are a number of reasons to encourage conversion to nonradiolabeled probes. One advantage of nucleic acid probes radiolabeled with ^{32}P-nucleotides has been that of extreme sensitivity due to the high level of specific activity in the probe (greater than 10^8 disintegrations per minute per microgram DNA). However, these probes are usable for less than 2 weeks. Handling and disposal of ^{32}P is potentially hazardous and expensive. Langer et al. (1981) recently introduced a biotin labeled nucleotide that offered the first practical alternative to radiolabeled nucleic acid probes. This biotinylated deoxyuridine triphosphate (dUTP) can be incorporated in nick-translation reactions as efficiently as ^{32}P-labeled dUTP (Langer et al., 1981). The resulting biotin labeled DNA probe can be used both in the detection of nucleic acid immobilized to filters and for in situ hybridization (Brigati et al., 1983). Biotin labeled probes have a long shelf-life. In our laboratory a biotinylated HSV probe is still fully active after 1 year. Extremely high probe concentrations (as high as 1 μg/ml) can be utilized, and hybridization time, thus, can be reduced to as little as 1 hour (Leary et al., 1983). Biotin labeled polynucleotides also reduce assay time by eliminating the need for autoradiographic detection of the signal. At the completion of the hybridization reaction, biotinylated probe bound to the specimen can be detected within 1 to 2 hours by incubation of the filter first with biotin–avidin–enzyme complexes and then with a substrate yielding a colored product that can be detected visually (Leary et al., 1983).

The biotinylated dUTP derivatives provide the first useful alternative to radiolabeled probes. This innovation eliminates the problems of risk, disposal, and short shelf-life that are associated with radioactive probes. With biotinylated probes for EBV nucleic acid, Sixbey et al. (1984) demonstrated with in situ hybridization the replication of EBV in shed oropharyngeal epithelial cells during infectious mononucleosis. Nevertheless, much investigation is being directed to develop better nonradioactive probes. The stimuli for this approach include the need for more sensitive detection systems, wider availability of reagents at a reasonable cost, and simpler and more rapid procedures.

19.7 CONCLUSIONS

Nucleic acid hybridization is an area of intensive and rapidly advancing investigation. This approach is inherently specific and sensitive. It will clearly play an important role in studies of the pathogenesis of viral infections. The ultimate role of the detection of viral nucleic acid for rapid viral diagnosis will depend on the results of the ongoing investigations to develop methods that are rapid and simple.

ACKNOWLEDGMENTS

The contributions of Sara Albanil, David Redfield, Michael Oxman, and Geoffrey Wahl are greatly appreciated.

REFERENCES

Andiman, W., Gradoville, L., Heston, L., Neydorff, R., Savage, M.E., Kitchingman, G., Shedd, D., and Miller, G. 1983. Use of cloned probes to detect Epstein-Barr viral DNA in tissues of patients with neoplastic and lymphoproliferative diseases. J. Infect. Dis. 148:967–977.

Berninger, M., Hammer, M., Hoyer, B., and Gerin, J. L. 1982. An assay for the detection of the DNA genome of hepatitis B virus in serum. J. Med. Virol. 9:57–68.

Brigati, D.J., Myerson, D., Leary, J.J., Spalholz, B., Travis, S.Z., Fong, C.K.Y., Hsiung, G.D., and Ward, D.C. 1983. Detection of viral genomes in cultured cells and paraffin-embedded tissue sections using biotin-labeled hybridization probes. Virology 126:32–50.

Chou, S., and Merigan, T.C. 1982. Rapid detection and quantitation of human cytomegalovirus in urine through DNA hybridization. N. Engl. J. Med. 308:921–925.

Durst, M., Gissmann, L., Ikenberg, H., and zur Hausen, H. 1983. A papillomavirus DNA from a cervical carcinoma and its prevalence in cancer biopsy samples from different geographic regions. Proc. Natl. Acad. Sci. 80:3812–3815.

Flores, J., Purcell, R.H., Perez, I., Wyatt, R.G., Boeggeman, E., Sereno, M., White, L., Chanock, R.M., and Kapikian, A. 1983. A dot hybridization assay for detection of rotavirus. Lancet i:555–558.

Gadler, H. 1983. Nucleic acid hybridization for measurement of effects of antiviral compounds on human cytomegalovirus DNA replication. Antimicrob. Agents Chemother. 24:370–374.

Gissmann, L., Wolnik, L., Ikenberg, H., Koldovsky, U., Schnurch, H.G., and zur Hausen, H. 1983. Human papillomavirus types 6 and 11 DNA sequences in genital and laryngeal papillomas and in some cervical cancers. Proc. Natl. Acad. Sci. 80:560–653.

Green, M., Maniatis, T., and Melton, D.A. 1983. Human β-globin pre-mRNA synthesized in vitro is accurately spliced in Xenopus oocyte nuclei. Cell 32:681–694.

Grinnell, B.W., Padgett, B.L., and Walker, D.L. 1983. Distribution of nonintegrated DNA from JC papovavirus in organs of patients with progressive multifocal leukoencephalopathy. J. Infect. Dis. 147:669–675.

Hersman, J., Meyers, J.D., Thomas, E.D., Buckner, C.D., and Clift R. 1982. The effect of granulocyte transfusions on the incidence of cytomegalovirus infection after allogeneic marrow transplantation. Ann. Intern. Med. 96:149–152.

Hyypia, T., Stalhandske, P., Vainionpaa, R., and Pettersson, U. 1984. Detection of enteroviruses by spot hybridization. J. Clin. Microbiol. 19:436–438.

Kornberg, A. 1980. DNA Replication. San Francisco: W. H. Freeman and Co.

Lancaster, W.D., Kurman, R.J., Sanz, L.E., Perry, S., and Jenson, A.B. 1983. Human papillomavirus: Detection of viral DNA sequences and evidence for molecular heterogeneity in metaplasias and dysplasias of the uterine cervix. Intervirology 20:203–212.

Langer, P.R., Waldrop, A.A., and Ward, D.C. 1981. Enzymatic synthesis of biotin-labeled polynucleotides: Novel nucleic acid affinity probes. Proc. Natl. Acad. Sci. 78:6633–6637.

Leary, J.J., Brigati, D.J., and Ward, D.C. 1983. Rapid and sensitive colorimetric method for visualizing biotin-labeled DNA probes hybridized to DNA or RNA immobilized on nitrocellulose: Bio-blots. Proc. Natl. Acad. Sci. 80:4045–4049.

Lie-Injo, L.E., Balasegaram, M., Lopez, C.G., and Herrera, A.R. 1983. Hepatitis B virus DNA in liver and white blood cells of patients with hepatoma. DNA 2:301–308.

Maniatis, T., Fritsch, E.F., and Sambrook, J., eds. 1982. Molecular Cloning: A Laboratory Manual. New York: Cold Spring Harbor Laboratory.

Martin, D.C., Katzenstein, D.A., Yu, G.S.M., and Jordan, M.C. 1984. Cytomegalovirus viremia detected by molecular hybridization and electron microscopy. Ann. Intern. Med. 100:222–225.

Meinkoth, J., Wahl, G. 1984. Hybridization of nucleic acids immobilized on solid supports. Anal. Biochem. 138:267–284.

Minson, A.C., and Darby, G. 1982. Hybridization techniques. In C. Howard (ed.), New Developments in Practical Virology. New York: Alan R. Liss.

Peden, K., Mounts, P., and Hayward, G.S. 1982. Homology between mammalian cell DNA sequences and human herpesvirus genomes detected by a hybridization procedure with high-complexity probe. Cell 31:71–80.

Puga, A., Cantin, E.M., and Notkins, A.L. 1982. Homology between murine and human cellular DNA sequences and the terminal repetition of the S component of herpes simplex virus type 1 DNA. Cell 31:81–87.

Ranki, M., Palva, A., Virtanen, M., Laaksonen, M., and Soderlund H. 1983. Sandwich hybridization as a convenient method for the detection of nucleic acids in crude samples. Gene 21:77–85.

Redfield, D.C., Richman, D.D., Albanil, S., Oxman, M.N., and Wahl, G.M. 1983a. Detection of herpes simplex virus in clinical specimens by DNA hybridization. Diag. Microbiol. Infect. Dis. 1:117–128.

Redfield, D.C., Richman, D.D., Cleveland, P.H., Albanil, S., Oxman, M.N., and Wahl, G.M. 1983b. Detection and typing of herpes simplex virus by DNA hybridization (abstract no. 741). In: Program and abstracts of the 23rd Interscience Conference on Antimicrobial Agents and Chemotherapy. Washington, D.C.: American Society for Microbiology.

Richman, D.D., Cleveland, P.H., and Oxman,

M.N. 1982. A rapid enzyme immunofiltration technique using monoclonal antibodies to serotype herpes simplex virus. J. Med. Virol. 9:299–305.

Richman, D.D., Cleveland, P.H., Redfield, D.C., Oxman, M.N., and Wahl, G.M. 1984. Rapid Viral Diagnosis. J. Infect. Dis. 149:298–310.

Rigby, P.W., Dieckmann, M., Rhodes, C., and Berg, P. 1977. Labeling deoxyribonucleic acid to high specific activity in vitro by nick translation with DNA polymerase. J. Mol. Biol. 113:237–251.

Rotbart, H.A., Levin, M.J., and Villarreal, L.P. 1984a. Use of subgenomic poliovirus DNA hybridization probes to detect the major subgroups of enteroviruses. J. Clin. Microbiol. 20:1105–1108.

Rotbart, H.A., Levin, M.J., and Villarreal, L.P. 1984b. Detection of enteroviruses by DNA–RNA dot hybridization—Progress toward a rapid diagnosis (abstract) Pediat. Res. 18:285A.

Scotto, J., Hadchouel, M., Hery, C., Alvarez, F., Yvart, J., Tiollais, P., Bernard, O., and Brechot, C. 1983a. Hepatitis B virus DNA in children's liver diseases: Detection by blot hybridisation in liver and serum. Gut 24:618–624.

Scotto, J., Hadchouel, M., Hery, C., Yvart, J., Tiollais, P., and Brechot, C. 1983b. Detection of hepatitis B virus DNA in serum by a simple spot hybridization technique: Comparison with results for other viral markers. Hepatology 3:279–284.

Seidlin, M., Takiff, H.E., Smith, H.A., Hay, J., and Straus, S.E. 1984. Detection of varicella-zoster virus by dot-blot hybridization using a molecularly cloned viral DNA probe. J. Med. Virol. 13:53–61.

Sixbey, J.W., Nedrud, J.G., Raab-Traub, N., Hanes, R.A., and Pagano, J.S. 1984. Epstein–Barr virus replication in oropharyngeal epithelial cells. N. Engl. J. Med. 310:1225–1230.

Spector, D.H., Hock, L., and Tamashiro, C. 1982. Cleavage maps for human cytomegalovirus DNA strain AD169 for restriction endonucleases EcoRI, BglII, and HindIII. J. Virol. 42:558–582.

Spector, S., Rau, J.A., Spector, D.H., and McMillan, R. 1984. Detection of human cytomegalovirus in clinical specimens by DNA–DNA hybridization. J. Infect. Dis. 150:121–126.

Stalhandski, P., and Pettersson, U. 1982. Identification of DNA viruses by membrane filter hybridization. J. Clin. Microbiol. 15:744–747.

Thomas, P.S. 1980. Hybridization of denatured RNA and small DNA fragments transferred to nitrocellulose. Proc. Natl. Acad. Sci. 77:5201–5205.

Tocci, M.J., Livelli, T.J., Perry, H.C., Crumpacker, C.S., and Field, A.K. 1984. Effects of the nucleoside analog 2′-nor-2′-deoxyguanosine on human cytomegalovirus replication. Antimicrob. Agents Chemother. 25:247–252.

Virtanen, M., Laaksonen, M., Soderlund, H., Palva, A., Halonen, P., and Ranki, M. 1983. Novel test for rapid viral diagnosis: Detection of adenovirus in nasopharyngeal mucus aspirates by means of nucleic-acid sandwich hybridization. Lancet i:381–383.

Watson, J.D., Tooze, J., and Kurtz, D.T., eds. 1983. Recombinant DNA: A Short Course. New York: W. H. Freeman and Co.

Weller, I.V.D., Fowler, M.J.F., Monjardino, J., and Thomas H.C. 1982. The detection of HBV–DNA in serum by molecular hybridization: A more sensitive method for the detection of complete HBV particles. J. Med. Virol. 9:273–280.

White, B.A., and Bancroft, F.C. 1982. Cytoplasmic dot hybridization. Simple analysis of relative mRNA levels in multiple small cell or tissue samples. J. Biol. Chem. 257:8569–8572.

Wong-Staal, F., Hahn, B., Manzari, V., Colombini, S., Franchini, G., Gelmann, E., and Gallo, R.C. 1983. A survey of human leukaemias for sequences of a human retrovirus. Nature 302:626–628.

Yeager, A.S., Grumet, F.C., Hafleigh, E.B., Arvin, A.M., Bradley, J.S., and Prober, C.G. 1981. Prevention of transfusion-acquired cytomegalovirus infections in newborn infants. J. Pediat. 98:281–287.

Zachow, K.R., Ostrow, R.S., Bender, M., Watts, S., Okagaki, T., Pass, F., and Fara, A.J. 1982. Detection of human papillomavirus DNA in anogenital neoplasias. Nature 300:771–773.

Section 2

Viral Pathogens

20

Respiratory Tract Infections

Paul E. Palumbo and R. Gordon Douglas, Jr.

20.1 INTRODUCTION

Respiratory viral illnesses including common colds, sore throats, and influenza constitute man's most common affliction. Often considered nuisance illnesses, they result in more time lost from work and school than any other diseases. In addition, influenza and pneumonia/bronchiolitis in infants, the elderly, and immunocompromised individuals produce substantial mortality.

Respiratory viral illnesses are characterized by multiple etiology. There are eight distinct clinical syndromes produced by viral infection of the respiratory tract and more than 200 serologically distinct viruses that have been implicated as etiologic agents. Thus, the major challenges for the clinician are to determine when to obtain specimens for viral diagnosis, and for the virologist what viruses to test. Because of the frequency of these illnesses and their relatively mild nature, it is not cost-effective to obtain specimens in most instances. Viral diagnosis is used to identify epidemics of viral disease so that clinical and epidemiologic criteria can be made in subsequent cases. In addition, viral diagnosis is critical when antiviral therapy is contemplated. There are two approved drugs in the U.S., amantadine, which is effective against Influenza A virus, and ribavirin, approved for use against RSV. A few drugs may be approved in the future, such as rimantadine and ribavirin for indications other than RSV.

20.1.1 Viruses

As shown in Table 20.1, members of several distinct virus families and genera constitute the respiratory viruses. Although differing in size, nucleic acid composition, presence or absence of a lipid envelope, and other physical and chemical properties, these viruses share in common the ability to infect the human respiratory tract.

20.1.2 Syndromes

In contrast with the diversity of virus types that infect the respiratory tract there are a limited number of clinical responses to infection. Namely, rhinitis (common cold), pharyngitis, laryngitis, tracheobronchitis, laryngotracheobronchitis (croup), bronchiolitis, and pneumonia. In addition, predominantly systemic illnesses with only minimal signs and symptoms in the respiratory tract can occur as a result of infections with respiratory viruses. In many instances, signs and symptoms referable to more than one

Table 20.1. Viruses Causing Respiratory Tract Infections in Humans

Family	Genus	Types and Groups
Orthomyxoviridae	Influenzavirus	Influenza virus types A, B, and C
Paramyxoviridae	*Paramyxovirus*	Parainfluenza virus types 1–4
	Pneumovirus	Respiratory syncytial virus
Picornaviridae	Rhinovirus	111 serotypes
	Enterovirus	Group A, coxsackievirus, 23 types
		Group B, coxsackievirus, 6 types
		Echovirus, 31 types
		Enterovirus, 4 types
Coronaviridae	Coronavirus	3 types
Adenoviridae	Mastadenovirus	36 serotypes

Modified from Douglas, R. G., Jr., (1984), with permission.

anatomic site may be exhibited. For example, rhinitis and pharyngitis or pharyngitis and laryngitis commonly occur together.

In this chapter we describe the major groups of viruses, the clinical and pathogenic features of the illnesses produced by them, and the method of their diagnosis.

20.2 INFLUENZA

Influenza A virus was the first of the respiratory viruses to be isolated. It was recovered following the inoculation of fertilized hens' eggs by Smith et al. (1933). Influenza B was isolated by Francis (1941) and influenza C by Taylor (1951). Other hallmarks in the history of influenza include discovery of the phenomenon of hemagglutination by Hirst (1943), development of effective inactivated vaccines beginning in the 1950s and the FDA approval for human use of the antiviral agent, amantadine, in 1967.

The influenza viruses are medium sized (80–100 nm, mean diameter) helical viruses enveloped in a host cell-derived lipid membrane. The envelope surface contains two types of glycoprotein spikes, the neuraminidase and the hemagglutinin. Other proteins include the matrix, which is associated with the membrane, the nucleoprotein, three polymerases (PB_2, PB_1, PA), and a nonstructural protein. The single-stranded RNA consists of eight segments, each coding for a separate protein. Type specificity (A,B,C) is conferred by the internal proteins, matrix, and nucleoprotein. Influenza A virus and, to a lesser extent, influenza B virus exhibit antigenic variation. The frequently occurring mild to moderate changes in antigenicity are referred to as antigenic drift, and result from point mutations in the hemagglutinin or the neuraminidase. Such mutations yield changes of one amino acid in the primary structure of these glycoproteins, and may lead to changes in secondary or tertiary structure. More pronounced antigenic variation results from reassortment of gene segments, a phenomenon called antigenic shift. Influenza A subtypes are named according to this antigenic structure: subtype designation as H1, H2, H3 for hemagglutinin and N1, N2 for neuraminidase and strain designation by year and site of isolation. Thus, A/Bangkok/79 H3N2 refers to an influenza A virus belonging to the H3N2 subtype possessing the antigenicity of virus isolated in 1979—prototype Bangkok strain.

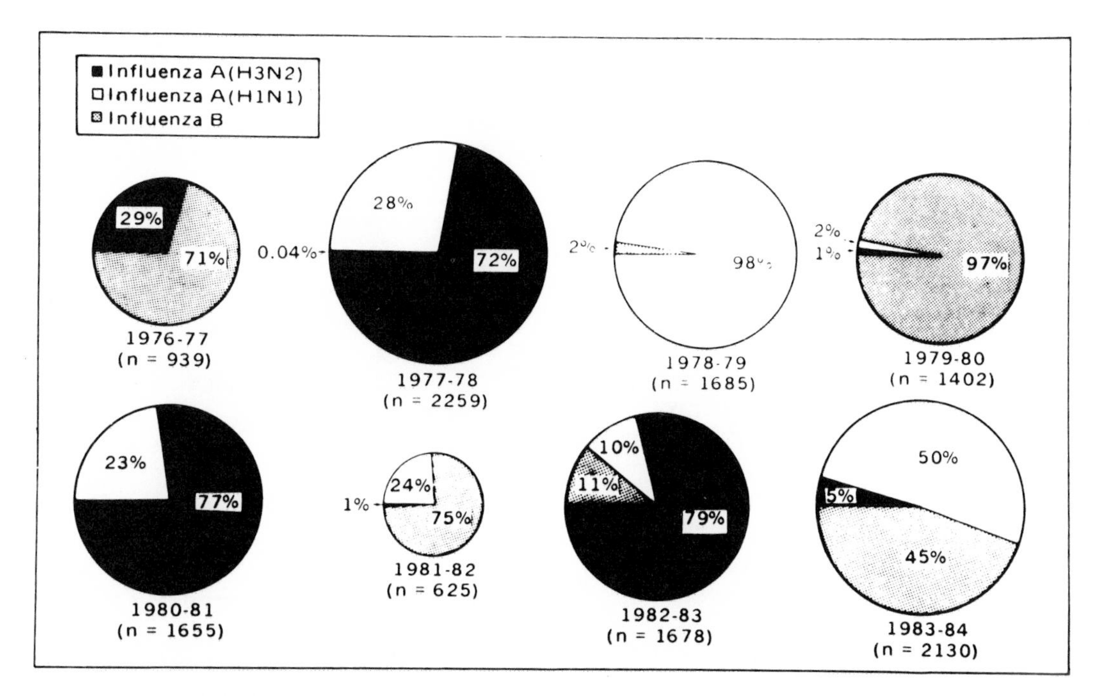

Figure 20.1. Isolations of influenza viruses reported to CDC by collaborating civilian and military laboratories—U.S., 1976–1984. (Morbidity and Mortality, 1984, reproduced by permission.)

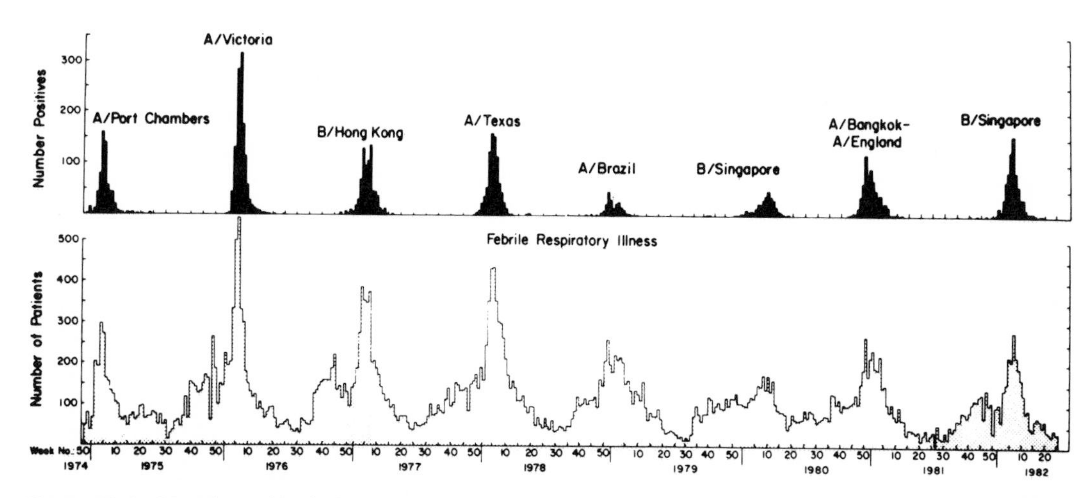

Figure 20.2. Number of isolations of major respiratory viruses from patients with febrile respiratory illnesses who attended sentinel primary care clinics and from persons observed in the Houston Family Study, Houston, Texas, 1974–1980. (Glezen et al., 1984, reproduced by permission.)

Influenza is an epidemic or pandemic disease. Nearly every winter outbreaks occur with viruses often exhibiting antigenic drift from those observed the previous winter (Glezen and Couch, 1978). At intervals of 10 to 30 or more years, pandemic influenza occurs due to antigenic shift. The nature of epidemic influenza in the U.S. is shown in Figure 20.1 and the sharp 5 to 7 week pattern of epidemic influenza in a community is shown in Figure 20.2.

In normal volunteers who possess little or no serum antibody, illness begins abruptly 36 to 48 hours following intranasal

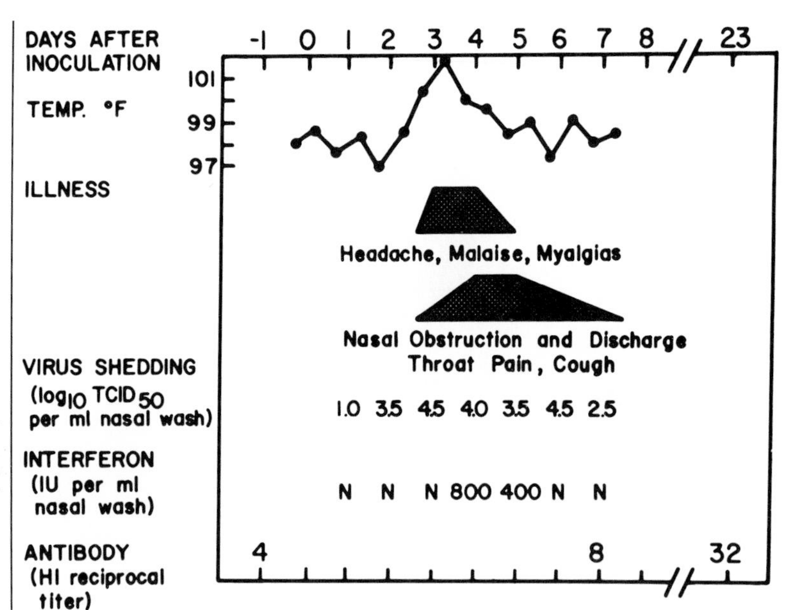

Figure 20.3. Case report of a volunteer inoculated by nose drops with 1000 $TCID_{50}$ influenza A/Udorn/72 (H3N2) virus. N, negative. (Douglas, 1975, reproduced by permission.)

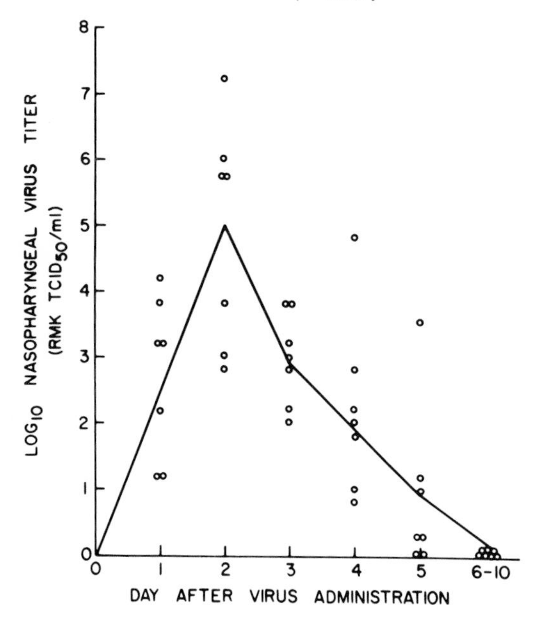

Figure 20.4. Quantities of influenza A (H3N2) recovered from nasal wash (NW) specimens obtained daily after inoculation from seven infected volunteers. RMK, rhesus monkey kidney. (Murphy et al., 1973, reproduced by permission.)

virus inoculation. The predominant symptoms are fever, myalgias, headache, and malaise (Figure 20.3) (Douglas, 1975). Cough, sore throat, and nasal discharge are commonly present. The febrile period lasts 2 to 3 days, and as the systemic symptoms subside, respiratory symptoms such as sore throat, nasal discharge, obstruction of nasal passages, and cough become more prevalent. Generally, most symptoms subside within 7 days, although cough and lassitude may persist 2 to 3 additional weeks.

Virus shedding parallels illness directly, and the virus titer is reflected by the severity of illness (Figure 20.4). An interferon response to virus infection usually is detectable in nasal secretions and in serum. It peaks 1 day later than virus shedding and correlates with recovery from illness (Figure 20.3) (Murphy et al., 1973). Antibody is detectable in serum and in nasal secretions 10 to 14 days after inoculation and reaches peak titers by 21 to 28 days postinfection (Couch and Kasel, 1983).

Virus isolation and identification is the standard laboratory method of approach to diagnose influenza virus infections, although many methods to directly examine clinical specimens for virus antigen have been described and are becoming available. Frequently collected specimens are nasal and throat washes, or swabs of nasopharynx and throat (see Chapter 2 for specimen collection, handling and transport). Optimal virus isolation is obtained from a nasopharyngeal wash but a combined nose–

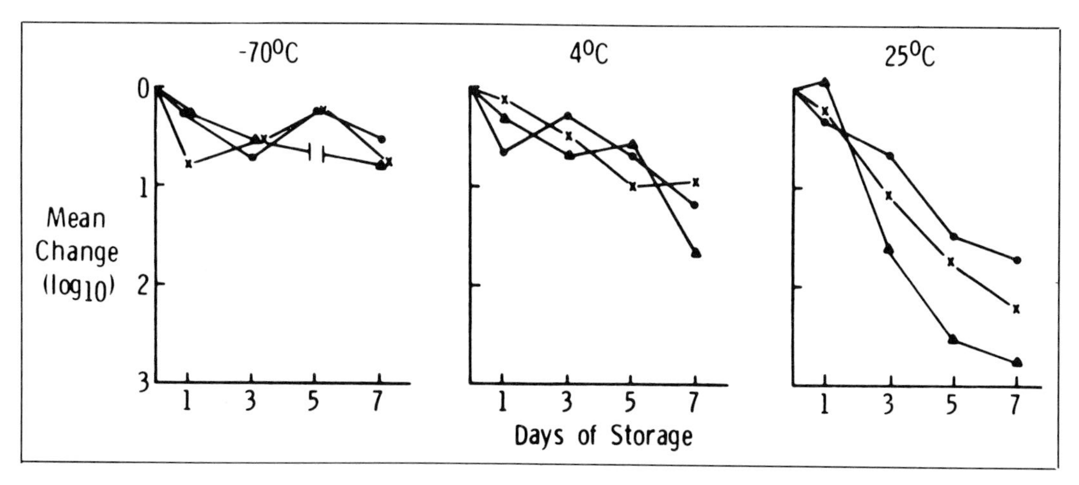

Figure 20.5. Infectivity decay curves for type A/Port Chalmers-like (H3N2) virus at different storage temperatures in (●) veal infusion broth, (x) Hanks balanced salt solution, and (▲) charcoal viral transport media. Concentration of virus at time zero was 3.7 $\log_{10}$/ml (Baxter et al., 1977, reproduced by permission.)

throat swab is acceptable (Douglas, 1975). Maintenance at 4°C on ice, or freezing at −70°C if inoculation into cell culture is to be delayed 5 days, is essential (Figure 20.5) (Baxter et al., 1977). However, a single freeze–thaw cycle may reduce the number of specimens positive for virus by as much as 60% (Smith and Reichrath, 1974).

More recently, rhesus and cynomolgus monkey kidney cells, and continuous cell lines such as Madin–Darby canine kidney (MDCK) and rhesus kidney (LLC-MK2) are used to isolate influenza virus. No single cell line or culture method has completely supplanted the others due to differing sensitivities of influenza types, subtypes, and strains. Classically, embryonated eggs and primary monkey kidney cells have been the methods of choice. Embryonated eggs are inoculated in the allantoic and amniotic cavities and the fluids collected from those cavities are tested for viral hemagglutinin after a 3-day incubation period. Many laboratories find cell culture preferable to egg inoculation and utilize primary monkey kidney cells, either rhesus or cynomolgus, and/or a continuous cell line such as MDCK and LLC-MK2. Because the latter two cell lines have been found to be equally sensitive (Table 20.2), they have become the cells of

Table 20.2. Reisolation of Influenza Viruses from Stored Clinical Specimens

	Number Positive/Number Tested (percentage) in:			
	RMK	CYNO	MDCK	LLC
Influenza A/Texas	35/33 (66)	67/91 (74)	30/38 (79)	39/52 (75)
Influenza A/USSR	36/77 (47)	61/91 (67)	39/64 (61)	66/78 (46)
Influenza B/HK	52/82 (63)	79/94 (84)	45/62 (73)	48/77 (62)
Parainfluenza type 1	35/56 (63)	61/92 (66)	0/36 (0)	36/49 (73)
Parainfluenza type 2	12/45 (21)	33/53 (62)	0/8 (0)	8/22 (36)
Parainfluenza type 3	28/57 (49)	52/92 (56)	10/35 (29)	30/49 (61)

Modified from Frank et al. (1979), with permission.

choice because of their uniform sensitivity and availability (Frank et al., 1979; Orstavik, 1981). Inoculated cell cultures are incubated in the absence of serum (which may have viral inhibitory factors) and in the presence of trypsin (Frank et al., 1979; Orstavik, 1981) (see Parainfluenza Viruses). Most influenza isolates produce little or no cytopathology. Vacuolation and cell lysis may be seen with influenza B by 3 to 5 days, but is rare with influenza A. Thus, detection of influenza isolates in cell culture is performed by hemadsorption using guinea pig, human type O or chicken erythrocytes (RBC). Influenza types A and B hemadsorb at both 4° and 22°C, whereas, type C hemadsorbs chicken RBC only at 4°C due to the presence of a receptor destroying enzyme active at 22°C (Kendal, 1975; Glezen, 1980). Methods for further typing and subtyping influenza isolates include hemagglutination inhibition, hemadsorption inhibition, complement fixation, and neutralization, tests usually performed at state and reference laboratories. These tests are described elsewhere in this volume.

Direct examination of clinical specimens using a direct immunofluorescent technique was first described by Liu (1956). Subsequently, an indirect immunoflourescent technique was described (McQuillen et al., 1970). An experienced laboratory can detect and identify influenza virus at rates comparable to viral isolation in cell culture. In addition, radioisotope and enzyme immunoassays have been developed, which can identify the virus and virus-specific antigens in nasopharyngeal secretions (Yolken et al., 1980; Sarkkinen et al., 1981) but they are not generally available (Figure 20.6).

Diagnosis of infection also can be made by detecting a serologic response when sera obtained during the acute and convalescent phase of infection are compared. Complement fixation tests are useful to detect the influenza virus type (A, B, or C), because this reaction depends primarily on the nucleoprotein antigen. Strain-specific diagnosis is accomplished by hemagglutination inhibition, but this test requires having the correct antigen. Although not often used,

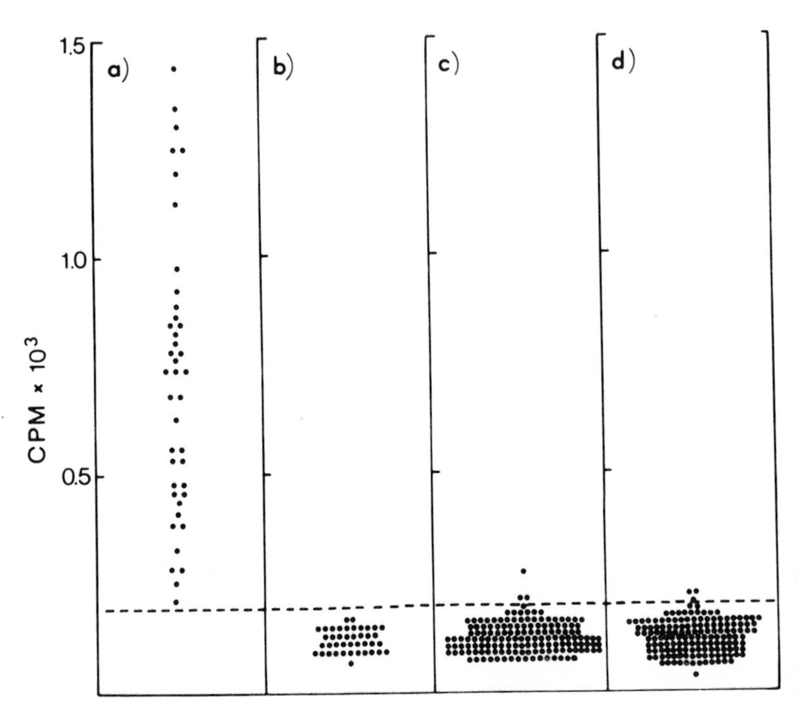

Figure 20.6. Radioimmunoassay results expressed as counts per minute values for 41 influenza A virus (IF) positive nasopharyngeal specimens tested in influenza A RIA (a) and in influenza B RIA (b) and for 150 influenza A IF negative specimens tested correspondingly in influenza A (c) and B (d) RIA. The dotted line represents the cutoff line at 200 cpm. (Sarkkinen et al., 1981, reproduced by permission.)

neutralization is probably the best predictor of susceptibility to infection. All serologic methods are of limited clinical usefulness (e.g., epidemiology) because of the requirement for convalescent sera.

20.3 PARAINFLUENZA VIRUSES

The paramyxoviruses are large, enveloped RNA containing viruses belonging to the *Paramyxoviridae* family, which also contains respiratory syncytial, measles, and mumps virus. There are four serologic types of parainfluenza viruses, three of which are important human disease pathogens. The first report of these paramyxoviruses was by Chanock (1956). They cause significant respiratory disease in children in both severity and numbers, but little disease other than an occasional common cold or case of laryngitis in adults. In children, types 1 and 2 cause fall outbreaks of respiratory disease, which includes colds, pharyngitis, tracheobronchitis and, most importantly, croup (Figure 20.2). Epidemiologically, types 1 and 2 occur in the fall and often occur biannually. Parainfluenza virus type 3 infection often occurs endemically year round, but has been reported to cause spring epidemics following influenza outbreaks (Glezen et al., 1984). Although parainfluenza type 3 may produce croup, more typically, it produces a spectrum of infection that more closely mimics that produced by respiratory syncytial virus, including bronchiolitis and pneumonia.

Croup is an illness characterized by respiratory obstruction on inspiration due to swelling and edema of the larynx and related structures. A child, most often aged 2 to 4, experiencing an afebrile upper respiratory infection develops sudden onset of respiratory stridor, which may remit and exacerbate over several days. The etiology is viral and may be caused by parainfluenza types 1 and 2 most frequently, but also respiratory syncytial virus, parainfluenza type 3, influenza A and influenza B. Occasionally, patients are hospitalized and may require tracheostomy or intubation.

These viruses are present in the posterior pharynx and may be shed for a period of several days to 1 week (Chanock et al., 1961, 1963). Specimens for virus isolation are collected by means of a swab or wash of the posterior pharynx or nasopharynx. Clinical specimens should be stored at 4°C for periods up to 24 hours, or at −60°C or below for more extended periods, as the virus is rapidly inactivated at ambient temperatures (Parkinson et al., 1982). Even at temperatures of −70°C, significant viral infectivity may be lost (Parkinson et al., 1980).

Prompt inoculation of specimens into cell cultures is critical for successful virus isolation (Parkinson et al., 1980). The "classical" cells for isolation of parainfluenza viruses have been primary monkey kidney, either rhesus or cynomolgus. These cells have the disadvantage of being intermittently infected with simian virus (SV) type 5, which has properties similar to parainfluenza viruses. The cell cultures should be maintained in the presence of SV-5 antisera or SV-5 must be ruled out during identification procedures for viral isolates (Hermann and Hable, 1970).

Due to the high cost, limited availability, and problem of intermittent SV-5 infection of primary monkey kidney cells, alternatives have been sought (Frank et al., 1979; Orstavik, 1981). A continuous cell line of rhesus kidney cells, LLC-MK2, was demonstrated to be equally sensitive to parainfluenza viruses, but required the presence of 0.6–2.0 μg/ml trypsin (Table 20.2). They are clearly superior to MDCK cells. It has been shown that proteases are necessary for parainfluenza viral activation, and it is hypothesized that proteases present in primary cell cultures are absent in continuous cell lines; thus, the requirement for trypsin (Homma and Ohuchi, 1973; Silver et al., 1978; Frank et al., 1979; Orstavik, 1981). Human embryonic kidney cells also are sensitive to parainfluenza viruses. Cell cultures are incubated in the absence of serum that may be inhibitory, at 33°–36°C, and for best results, a roller apparatus is employed.

The majority of parainfluenza viruses do not produce cytopathogenic effects in cell culture. In monkey kidney cells, about 51% of type 2, 27% of type 4, and a much lower percentage of types 1 and 3 produce CPE, which consists of a syncytial cell pattern (Hermann and Hable, 1970). The customary screening procedure to detect virus is hemadsorption, using guinea pig or human type O RBC. This is performed after inoculated cells have been incubated for 5 and 10 days. Suspensions of 0.1% RBC are added and incubated at 4°C (30 minutes); if hemadsorption is not observed cultures are further incubated at 25°C for 30 minutes. Identification of virus in hemadsorption-positive cell cultures is performed by hemadsorption inhibition, hemagglutination inhibition, or neutralization tests using commercially available antisera specific for each of the four parainfluenza viruses. An enzyme-linked immunosorbent assay (ELISA) for parainfluenza serotypes that can be used to detect and identify virus irrespective of its ability to replicate has been described recently (Parkinson et al., 1982).

By 10 days after inoculation, 75–85% of cell cultures infected with parainfluenza types 1, 2, and 3 are positive, while type 4 isolates often require more than 15 days for primary isolation (Hermann and Hable, 1970). At the Mayo Clinic, 22% of positive isolates were reported by 5 days and 60% by 10 days (Hermann and Hable, 1970). Blind subpassage is usually not required, and when using primary monkey kidney cells, subpassages may lead to simian virus activation.

Rapid diagnosis of parainfluenza virus infection from cell smears of clinical specimens, utilizing indirect immunoflourescence can be performed (Gardner et al., 1971; Wong et al., 1982). This test can be highly sensitive and specific, but is dependent on the quality of the antisera utilized.

Serology can be useful for retrospective diagnoses, with complement fixation, hemagglutination inhibition, and neutralization tests being most frequently employed. Problems with heterotypic responses are common as are heterologous responses between mumps and parainfluenza viruses, reflecting shared antigenic components (Lennette and Jensen, 1963).

20.4 RESPIRATORY SYNCYTIAL VIRUS

Respiratory syncytial virus (RSV) is a large paramyxovirus (120–200 nm) that is sufficiently different from the parainfluenza viruses to be classified in a separate genus, pneumovirus. The differences include a smaller ribonucleoprotein helix, and the absence of hemagglutinin, neuraminidase, and hemolysin activities associated with the envelope glycoproteins.

Respiratory syncytial virus is the major cause of bronchiolitis and pneumonia of infancy, although these also may result from infection with parainfluenza type 3, influenza A virus, influenza B virus, and occasionally other respiratory viruses. Respiratory syncytial virus infection occurs in annual winter epidemics which last 2 to 5 months (Figure 20.2). Recent reports suggest that there are at least two strains of RSV (Hendry et al., 1984; Anderson et al., 1984).

Characteristically, illness occurs in a male child aged 1 to 6 months, but it may occur at any time up to 2 years of age. After 3 to 4 days of afebrile upper respiratory infection, the child develops respiratory difficulty manifested by hyperventilation and retractions. Examination of the chest reveals expiratory prolongation, wheezes, and rhonchi, and chest radiograph reveals hyperinflation. There may or may not be a pulmonary infiltrate. In more severe cases, anoxia is present and the child must be hospitalized. Virus is shed from the nasopharynx in high titer for prolonged periods of

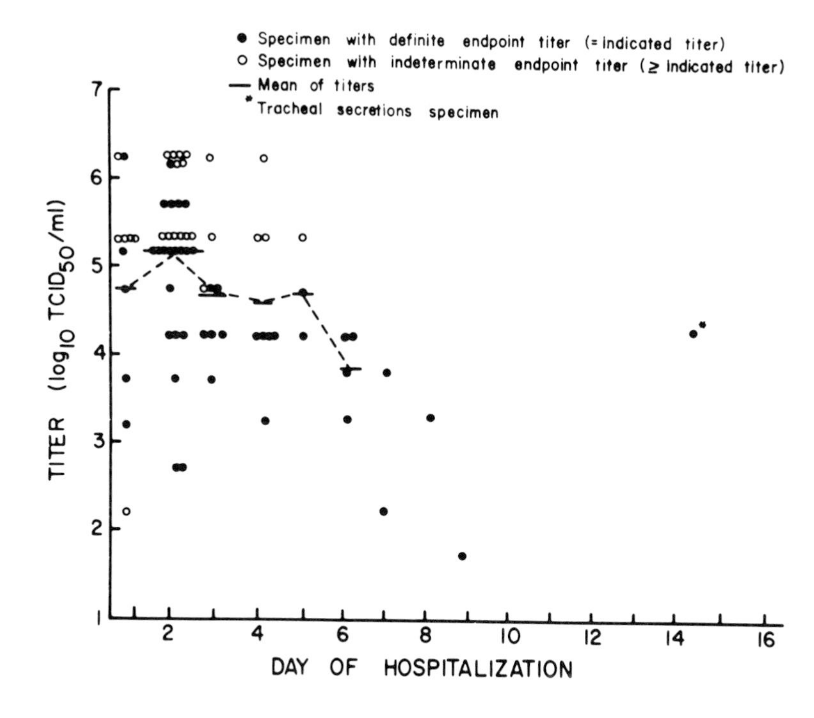

Figure 20.7. Distribution of nasal wash titers of respiratory syncytial virus in infants according to the day of hospitalization. (Hall et al., 1975, reproduced with permission.)

time (Figure 20.7) (Hall and Douglas, 1975).

Both viral isolation in cell culture as well as direct examination of respiratory secretions are widely used and commercially available for diagnosis of RSV. Nasal washes yield the best specimens for virus isolation. A 60% decrease in virus isolation rate was noted when the specimen tested was a nasopharyngeal swab (Hall et al., 1975). Specimens should be transported and stored at 4°C because viral infectivity decreases rapidly at room temperature or after freeze–thawing (Hambling, 1964). Best results are obtained when the specimen is inoculated onto cell culture as soon after collection as possible. Bedside inoculation of specimens on a pediatric ward has been reported with a 77% isolation rate among patients with bronchiolitis during an RSV epidemic (Bromberg et al., 1983).

HEp-2 and HeLa cells are the most sensitive to RSV infection and are widely employed for its isolation (Jordan, 1962). Typical CPE consists of the formation of large multinucleated syncytia, which is detectable in an average of 4 to 5 days (Hall et al. 1975). Formation of large syncytia in these cell lines is strong presumptive evidence for RSV isolation. Definitive identification can be performed by immunofluorescence, complement fixation or virus neutralization.

Many methods for direct identification of respiratory syncytial virus in clinical specimens are available. Both direct and indirect immunofluorescence (IF) have been described and are most commonly utilized (Gardner and McQuillin, 1968; Gray et al., 1968; Minnich and Ray, 1982; Lauer, 1982). The IF tests are highly sensitive and specific but test results are dependent on the quality of antisera utilized and the experience of the observer. Reverse passive hemagglutination (RPH) also has been described for detection of RSV in nasopharyngeal secretions (Cranage et al., 1981). Of 25 specimens positive by cell culture and/or IF, 24 were positive by RPH. Finally, ELISA kits recently have been developed for the rapid detection of RSV (Hendry and McIntosh, 1982; Hornsleth et al., 1982; Lauer et al., 1984). Both polyclonal as well as monoclonal antibodies have been employed in the capture phase of the assay.

These assays have been reported to be highly sensitive and equally specific when compared with results from IF and cell culture (McIntosh et al., 1982; Lauer et al., 1984). In addition, they detect infectious virus, as well as nonviable virus antigen, an advantage when dealing with a labile virus or a previously frozen sample.

A variety of serologic tests can be used to confirm past RSV infection. These include complement fixation, neutralization, or ELISA assays. In addition, indirect IF has been described for the study of the specific immune response to primary and subsequent RSV infections (Welliver and Ogra, 1983).

20.5 RHINOVIRUS

Rhinoviruses, discovered in 1956 nearly simultaneously by Pelon et al. (1957) and by Price (1956), are best isolated in cells grown in roller culture incubated at 33°C. The rhinoviruses are the most important cause of common colds in adults, and are an important cause of such illnesses in children.

Rhinovirus is the largest subgroup of the *Picornaviridae* family, consisting of 80 established serotypes, and a large but unknown number of untyped or untypeable strains (Kapikian et al., 1967, 1971.) They are small RNA containing naked viruses that differ from other picornaviruses in that they are acid (pH 3–5) labile and heat sensitive (replication is inhibited at 37°C). They have no common antigen. Although heterologous antibody responses may occur, they are of little diagnostic or clinical significance. There is a question of antigenic change or shift within this group because of the discovery of new rhinovirus strains in successive epidemics and slight antigenic changes within the same type over time (Mogabgab et al., 1975).

Rhinovirus infections occur year round although a peak of infection occurs each year in the early fall (Figure 20.8) and a second peak often occurs in the spring.

In normal, antibody-free volunteers inoculated intranasally with rhinovirus, the incubation period is 2 to 4 days. Predominant symptoms are nasal, such as airway obstruction, catarrah, and sneezing, but sore throat, malaise, headache, and cough often occur. Fever is unusual; occasionally

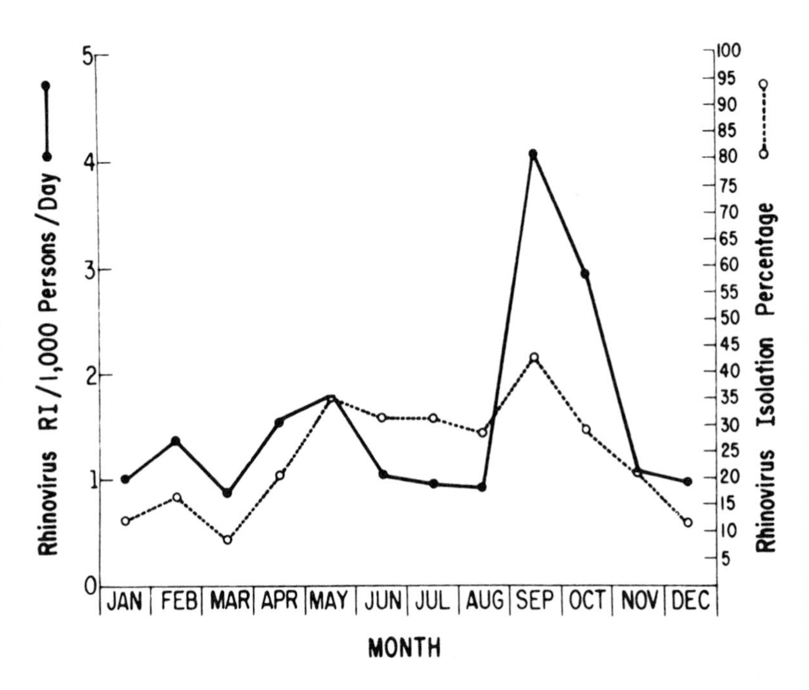

Figure 20.8. Combined data for the 3-year period, March 1963 to March 1966. Depicting the seasonal variation in the percentage of sampled respiratory illnesses yielding rhinoviruses and in the rate of rhinovirus illness derived by application of this percentage to the total rate of respiratory illness. (Gwaltney et al., 1966, with permission.)

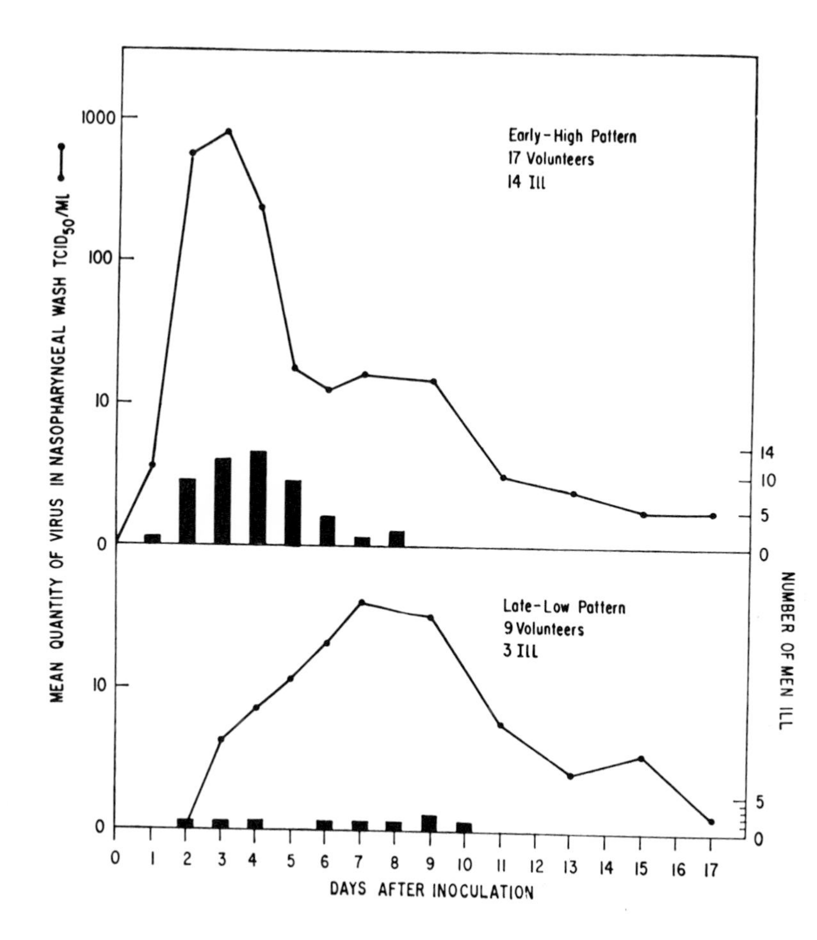

Figure 20.9. Quantitative virus shedding in nasopharyngeal wash specimens. Men in the early-high group began to shed virus on day 1 or 2; men in late-low group began to shed virus on day 3 or later. Bars represent number of men ill each day. (Douglas et al., 1966, reproduced by permission.)

patients exhibit tracheobronchitis. Severity of illness is clearly associated with quantities of virus shed (Figure 20.9).

Rhinovirus is most consistently isolated from nasopharyngeal washes. Isolation rates of virus from nasal and pharyngeal swabs are lower (Figure 20.10) (Cate et al., 1964; Gwaltney et al., 1966; Hendley et al., 1969). In a study of children positive for rhinovirus, 50% had positive cultures from nasal secretions, whereas, their throat swabs were negative (Hendley et al., 1969.) In addition, 176 specimens from individuals with respiratory illness revealed an isolation rate of 86% for nasal secretions, 60% for throat swabs, and 49% for salivary secretions (Gwaltney et al., 1966.) Thirty-five of the positive specimens were confirmed as rhinovirus infection. Best results are obtained with rapid inoculation into cell cultures. Storing the specimen on wet ice for 0.5 to 3.5 hours results in little loss of infectivity (Higgins et al., 1966). If the specimen must be stored for longer periods (>24 hours), the specimen should be frozen at −70°C.

The best cell lines for isolation of rhinovirus from clinical specimens are human embryonic kidney, human diploid cell lines (especially WI-26 and WI-38), and HeLa M cells (Brown and Tyrrell, 1964; Phillips et al., 1965; Higgins, 1966; Lewis and Kennett, 1976). Primary monkey kidney cells, although used in the initial isolation of several strains, are not a host that will consistently yield the broad range of rhinovirus serotypes. Also, different lots of normally sensitive cell lines frequently show variability in isolation rates for rhinoviruses (Phillips et al., 1965; Higgins, 1966; Ham-

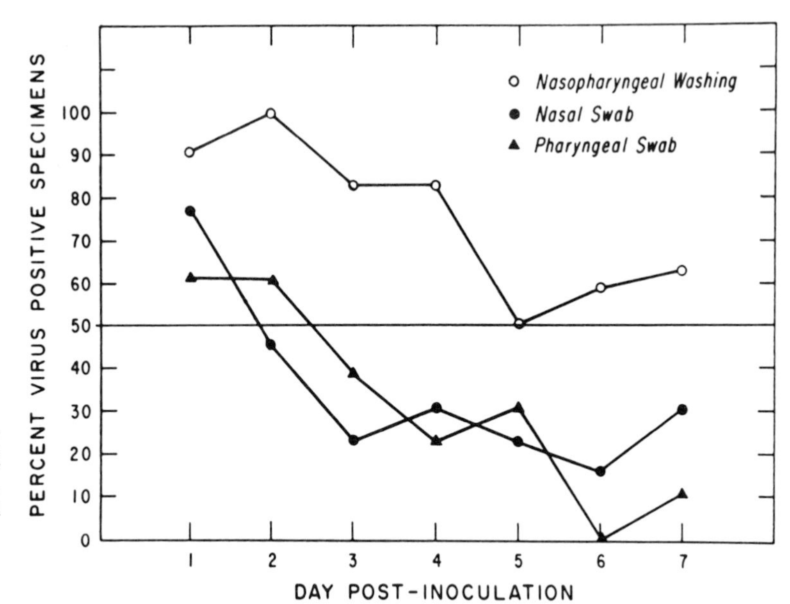

Figure 20.10. Comparison of rhinovirus yield by source of specimen. (Cate et al., 1964, reproduced by permission.)

parian, 1979). Optimal conditions for rhinovirus isolation include incubation of cell cultures at 33°C, the use of roller cell cultures and medium maintained at neutral pH. Cytopathic effects (CPE) consisting of cell rounding and refractility is usually observed in the first week and often by 48 hours. Specimens are generally held a maximum of 14 days.

Differentiation of rhinovirus from other viruses which may cause CPE in these cell lines, most notably enteroviruses, is difficult although rapidity of progression of CPE with rhinoviruses is slower than with enteroviruses. Other methods of differentiation include demonstration of lability to pH 3 (enteroviruses are acid stable) and a resistance to inactivation at 56°C for 30 minutes in the presence of $MgCl_2$. Due to the abundance of rhinovirus serotypes, identification of type for clinical isolates and diagnostic serology are difficult and time consuming. Standard neutralization or microneutralization tests can be performed utilizing panels of single sera or a series of pooled sera (Douglas et al., 1968; Monto and Bryan, 1974). These tests are not performed routinely in most virology laboratories. Unfortunately, no rapid detection systems for clinical specimens or direct smears utilizing fluorescence, enzyme-linked immunoassay (ELISA), or radioimmunoassay (RIA) are currently available.

20.6 CORONAVIRUSES

The coronaviruses were first detected in the mid-1960s (Tyrrell and Bynoe, 1965; Hamre and Procknow, 1966). They are medium sized (60–220 nm) enveloped viruses, containing single-stranded RNA. They are covered with large club or petal shaped projections. Because they are more difficult to isolate, they have been studied less than other respiratory viruses. Coronaviruses are responsible for common cold illnesses in adults and children, especially in winter months.

The central issue in current knowledge and past studies concerning coronaviruses has been the extraordinary difficulty in isolating strains from clinical specimens in the laboratory, opposed to animal coronaviruses, which are isolated with relative ease. The original human coronavirus was isolated in human embryonic kidney cells (strain 229 E), whereas, all additional strains have been isolated in organ cultures of human origin (trachea, lung) (McIntosh, 1974). Serologic studies performed on coronavirus isolates have shown extensive crossreaction between different isolates, but some antigenic subgrouping has been possible. Strains 229 E and LP are related and distinct from organ culture (OC) strains 38, 43, and 44, and from strain B 814 (McIntosh et al., 1969; McIntosh, 1974).

In practice, coronavirus infections are diagnosed by serologic methods. Strain 229 E can be isolated in human embryonic kidney or intestinal cells, or WI-38 cells (Kapikian et al., 1969). Specimens consist of washings or swabs from the nose or throat. Initial CPE, after several days incubation, appears as cellular elongation followed by cell rounding (Kapikian et al., 1969). For primary isolation of all other strains, the use of organ cultures of human embryonic trachea is necessary (Tyrrell and Bynoe, 1965; McIntosh, 1974), and even when such cultures are available, identification of isolates requires the use of neutralization, hemagglutination inhibition (OC-38), indirect IF, or electron microscopy.

The diagnostic method of choice for strain 229 E and the OC-38 group is serologic. Antigens for these strains can be prepared and utilized for complement fixation, ELISA (Kraaijeveld et al., 1980), IF (McIntosh et al., 1978) and hemagglutination inhibition assays (Kaye and Dowdle, 1975).

20.7 ADENOVIRUSES

The adenoviruses were first isolated by cocultivating established cell lines with adenoid and tonsillar tissues from asymptomatic children undergoing tonsillectomy and adenoidectomy (Rowe et al., 1953). The adenoviruses are infrequent contributors to the spectrum of respiratory viral illnesses in adults outside military situations, although they are somewhat more common in children.

Adenoviruses are medium sized (60–90 nm) naked DNA containing viruses. The DNA is enclosed in an icosahedron consisting of 252 capsomeres (240 hexons, 12 pentons). Fiber antigens are attached to each of the pentons. Hexons contain group specific complement fixation antigens, whereas, most type specific antigens reside in the fiber. Despite their relative lack of importance in respiratory illness in civilian populations, adenoviruses have been extensively studied. There are 36 distinct serotypes that can be classified into three major groups (Table 20.3), based on differential agglutination of rat and rhesus RBC (Rosen, 1960) and further subdivided into ten subgroups by differential RBC agglutination (Hierholzer, 1973). Groups II and IV in Table 20.3 have similar properties with re-

Table 20.3. Classification of Human Adenoviruses by DNA Content and by Agglutination of Erythrocytes

Subgroup	Types[a]	DNA Content[b]	Oncogenic Potential[c]	Agglutination of RBC	
				Rhesus	Rat
I	3, 7, 11, 14, 16, 21	49–52	Weak	+	0
II	8–10, 13, 15, 17, 19, 20, 22–30	57–61	None	+ or 0	+
III	1, 2, 4–6	57–59	None	0	Partial
IV	12, 18, 31	48–49	High	0	Partial

[a] Types 32–35 have not been sufficiently characterized for inclusion in this table.
[b] Percentage guanine + cytosine.
[c] Highly oncogenic adenoviruses induce tumors in newborn hamsters within 2 months after inoculation; weakly oncogenic viruses induce tumors in fewer animals in 4 to 18 months. Even those adenoviruses that are not oncogenic transform nonpermissive rodent cells in vitro.
Modified from Davis et al. (1980), with permission.

spect to agglutination of erythrocytes and, thus, are considered a single agglutination group. In addition, these groups share a common antigen. Adenoviruses also can be classified into groups based on the guanosine and cytosine content of their DNA and their oncogenic potential for laboratory animals. These correspond closely to the serologic groups except that oncogenic groups III and IV (Table 20.3) have similar agglutination properties.

The viruses most commonly found in the human respiratory tract belong to group III. Several of those in group I cause severe respiratory disease. Group II viruses rarely cause human disease, except for types 8 and 19, the major causes of epidemic keratoconjunctivitis.

In children, the low numbered serotypes cause a small proportion of most of the respiratory syndromes including pneumonia (Table 20.4) (Baum, 1985). Adenovirus types 3 and 7 (occasionally 1, 4, 14) produce pharyngoconjunctival fever, often transmitted in swimming pools in summer months. On occasion, parents of young children may be infected with the low-numbered adenovirus serotypes. In civilian adults, however, adenovirus disease is uncommon. In military recruits, adenovirus types 4, 7 and occasionally, 3, 11, 14, and 21 produce acute respiratory disease (ARD). Most often ARD presents as a febrile pharyngitis with or without exudative tonsillitis. Febrile tracheobronchitis with or without pneumonia may also occur.

Adenovirus types 34, 35, and 39 may produce a severe bilateral pneumonia in immunosuppressed hosts, such as renal transplant recipients (Stalder et al., 1977; Keller et al., 1979; de Jong et al., 1983). In this setting, the differential diagnosis among respiratory viruses includes cytomegalovirus (CMV) and herpes simplex virus (HSV), rather than the more common respiratory viruses discussed in this chapter.

Occasionally adenovirus respiratory infection with residual lung disease may be seen in normal hosts (Pearson et al., 1980). The virus is relatively easy to isolate using human epithelial cell lines as the host. Specimens may be obtained by throat or nasopharyngeal wash or by swabs of throat, nasopharynx, conjunctivae, or rectum. Specimens are transported and/or collected in balanced salt solution (BSS) or veal infusion broth (VIB) with 0.5% gelatin or bovine serum albumin (BSA) added. The virus is quite stable at 4°C or can be maintained frozen (between −20° and −70°C) without loss of infectivity (Grayston et al., 1958; Kasel, 1979).

Table 20.4. Respiratory Diseases Caused by Adenoviruses

Group Affected	Syndromes	Common Causal Adenovirus Serotypes
Infants	Coryza, pharyngitis (most asymptomatic)	1, 2, 5
Children	Upper respiratory disease	1, 2, 4–6
	Pharyngoconjunctival fever	3, 7
Young Adults	Acute respiratory disease and pneumonia	3, 4, 7
Immunocompromised	Pneumonia with dissemination	34, 35, 39
	CNS disease including encephalitis	7, 12, 32

Modified from Baum (1985), with permission.

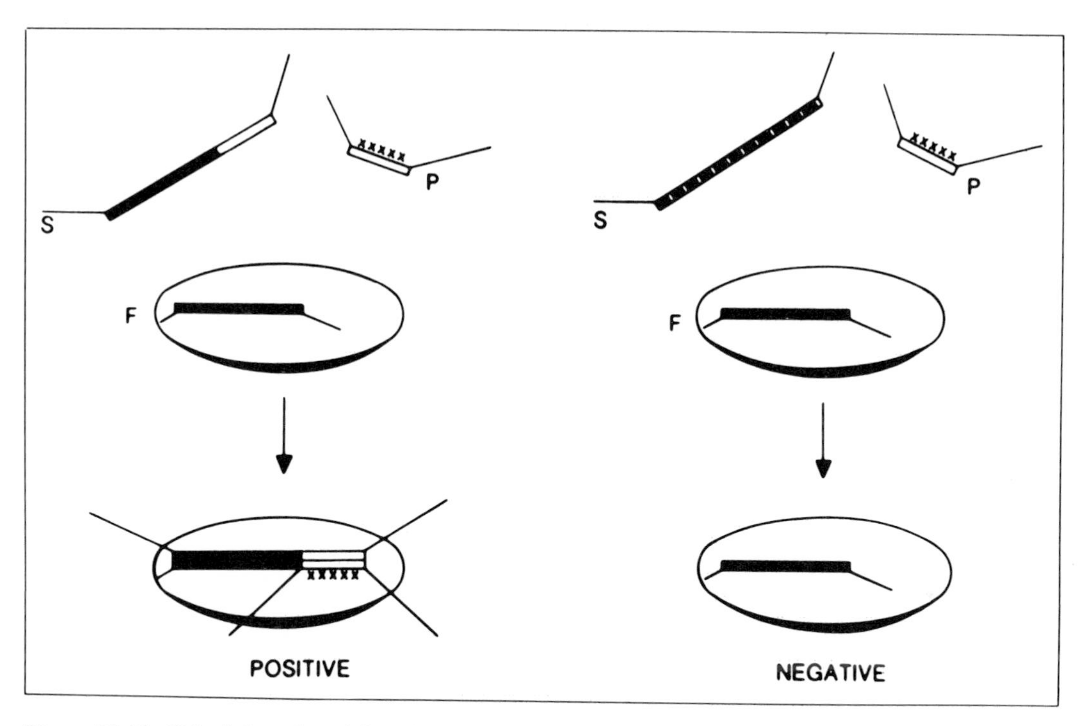

Figure 20.11. Principles of nucleic acid sandwich hybridization. S, sample; P, probe reagent; F, filter reagent. (Virtanen et al., 1983, reproduced by permission.)

Virus isolation is achieved by inoculating specimens onto cell cultures of human epithelial origin, such as human embryonic kidney (HEK), HEp-2, and HeLa. Higher viral titers in HeLa cells are obtained when either laboratory strains or primary isolates are incubated at 37°C, compared with 33°C (Fuchs and Wigand, 1975). Three cell lines were compared for their sensitivity to adenovirus and other viruses by Rutala et al.; among 72 adenovirus specimens, 100% were detected in HEK, 7% in baboon kidney cells, and only 11% in WI-38 fibroblasts. The HEK cells were also routinely superior for displaying CPE earlier (Rutala et al., 1977). Cytopathogenic effect often can be visualized in the first week of incubation and is generally apparent after 14 days. The time required for the development of CPE is usually group and serotype dependent, with some strains requiring blind subpassage and up to 30 days incubation before CPE is evident. Grayston et al. reported that 85% of 377 positive samples were isolated by 14 days, whereas, the last 15% required up to 33 days (following a blind subpassage at 14 days) (Grayston et al., 1958). Viral CPE in cell culture is evidenced by increased acidity of the medium, and development of enlarged, rounded cells in grape-like clusters.

Identification of an isolate as adenovirus can be achieved by using antiserum to the common group antigen (i.e., antisera to hexon antigen) in either a complement fixation or indirect IF test. Such antisera can be used for more rapid identification (after only 2 to 4 days) of virus in tissue culture by means of indirect IF or ELISA (Harmon et al., 1979). Subgrouping is accomplished with hemagglutination using rat and rhesus RBC (Table 20.3), whereas, serotyping is achieved with type-specific antisera in a neutralization, hemagglutination inhibition, or counter immunoelectrophoresis assay (Hierholzer and Barme, 1974). One of the most recent methods described for rapid adenovirus detection, which bypasses tissue culture, is nucleic acid hybridization (Figure 20.11). DNA sequences from adenovirus types 2 and 3 have been cloned and are used as both capture molecules and as ^{125}I-labeled probes. The test can be performed on respiratory mucus, requires about 20 hours, is specific, and can identify subgroups of adenovirus. With the advent of enzyme-linked DNA probes, this test should compare favorably with direct detection of viral antigen using enzyme-linked specific antibodies.

Most of the tests mentioned for identification of adenoviruses in cell culture are also useful for serologic diagnosis of infection. Hexon antigens (not type-specific) are used in a complement fixation assay, whereas, type-specific serology can be obtained with hemagglutination-inhibition and neutralization assays. More recently, the development of RIA and ELISA for the detection of group-common antibodies has been described (Scott et al., 1975; Roggendorf et al., 1982).

REFERENCES

Anderson, L.J., Tsou, C., Stone, V., et al. 1984. Different Antigenic Sites on Respiratory Syncytial Virus Proteins. (abstract) ICAAC #923, p. 252.

Baum, S.G. 1985. Adenovirus. In G.L. Mandell, R.G. Douglas, Jr., and J.E. Bennett (eds.), Principles and Practice of Infectious Diseases, 2nd ed. New York: John Wiley and Sons, pp. 988–993.

Baxter, B.D., Couch, R.B., Greenberg, S.B., and Kasel, J.A. 1977. Maintenance of viability and comparison of identification methods for influenza and other respiratory viruses in humans. J. Clin. Micro. 6:19–22.

Bromberg, K., Clarke, L.M., Clarke, L.M., Curran, E.D., and Sierra, M.F. 1983. Inoculation of HEp-2 cells for respiratory syncytial virus isolation. Clin. Pediat. 22:62–63.

Brown, P.K., and Tyrrell, D.A.J. 1964. Experiments on the sensitivity of strains of human fibroblasts to infection with rhinoviruses. Br. J. Exp. Path. 45:571–578.

Cate, T.R., Couch, R.B., and Johnson, K.M. 1964. Studies of rhinovirus in volunteers: Production of illness, effect of naturally acquired antibody, and demonstration of a protective effect not associated with serum antibody. J. Clin. Invest. 43:56–67.

Chanock, R.M. 1956. Association of a new type of cytopathogenic myxovirus with infantile croup. J. Exp. Med. 104:555–576.

Chanock, R.M., Bell, J.A., and Parrott, R.H. 1961. Natural history of parainfluenza infection. Perspect. Virol. II:126–138.

Chanock, R.M., Parrott, R.H., Johnson, K.M., Kapikian, A.Z., and Bell, J.A. 1963. Myxoviruses: Parainfluenza. Am. Rev. Resp. Dis. 88:152–166.

Couch, R.B., and Kasel, J.A. 1983. Immunity to influenza in man. Ann. Rev. Microbiol. 37:529–549.

Cranage, M.P., Stott, E.J., Nagington, J., and Coombs, R.R.A. 1981. A reverse passive hemagglutination test for the detection of respiratory syncytial virus in nasal secretions from infants. J. Med. Virol. 8:153–160.

Davis, B.D., Dulbecco, R., Eisen, H.N., and Ginsberg, H.S., eds. 1980. Microbiology, 3rd ed. New York: Harper and Row pp. 1049–1050.

de Jong, P.J., Valderrama, G., Spigland, I., and Horowitz, M.S. 1983. Adenovirus isolates from the urines of patients with the acquired immunodeficiency syndrome. Lancet i:1293.

Douglas, R.G., Jr. 1975. Influenza in man. In E.D. Kilbourne (ed.), The Influenza Virus and Influenza. New York: Academic Press.

Douglas, R.G., Jr. 1984. Respiratory diseases. In: G.J. Galasso et al. (eds.), Antiviral Agents and Viral Diseases of Man, 2nd ed. New York: Raven Press.

Douglas, R.G., Jr., Cate, T.R., Gerone, P.J., and Couch, R.B. 1966. Quantitative rhinovirus shedding patterns in volunteers. Am. Rev. Resp. Dis. 94:159–167.

Douglas, R.G., Jr., Fleet, W.F., Cate, T.R., and Couch, R.B. 1968. Antibody to rhinovirus in human sera. I. Standardization of a neutralization test. Proc. Soc. Exp. Biol. Med. 127:497–502.

Francis, T., Jr. 1940. A new type of virus from epidemic influenza. Science 92:405–408.

Frank, A.L., Couch, R.B., Griffis, C.A. and Baxter, B.D. 1979. Comparisons of different tissue cultures for isolation and quantification of influenza and parainfluenza viruses. J. Clin. Microbiol. 10:32–36.

Fuchs, N., and Wigand, R. 1975. Virus isolation and titration at 33° and 37°C. Med. Microbiol. Immun. 161:123–126.

Gardner, P.S., and McQuillin, J. 1968. Application of immunofluorescent antibody technique in rapid diagnosis of respiratory syncytial virus infection. Br. Med. J. 3:330–343.

Gardner, P.S., McQuillin, J., McGuckin, R., and Ditchburn, R.K. 1971. Observations on clinical and immunofluorescent diagnosis of parainfluenza virus infections. Br. Med. J. 2:7–12.

Glezen, W.P., 1980. Influenza C virus infection. Arch. Intern. Med. 140:1278.

Glezen, W.P., and Couch R.B. 1978. Interpandemic influenza in the Houston area 1974–1976. N. Engl. J. Med. 298:587–592.

Glezen, W.P., Frank, A.L., Taber, L.H., and Kasel, J.A. 1984. Parainfluenza virus Type 3: Seasonality and risk of infection and reinfection in young children. J. Infect. Dis. 150:851–857.

Gray, K.G., MacFarlane, D.E., and Sommerville, R.G. 1968. Direct immunofluorescent identification of respiratory syncytial virus in throat swabs from children with respiratory illness. Lancet i:446–448.

Grayston, J.T., Loosli, C.G., Smith, M., McCarthy, M.A., and Johnston, P.B. 1958. Adenoviruses I. The effect of total incubation time in HeLa cell cultures on the isolation rate. J. Infect. Dis. 103:75–101.

Gwaltney, J.M., Hendley, J.O., Simon, G., and Jordan, W.S. 1966. Rhinovirus in an industrial population. I. The occurrence of illness. N. Engl. J. Med. 275:1261–1268.

Hall, C.B., and Douglas, R.G., Jr. 1975. Clinically useful method for the isolation of respiratory syncytial virus. J. Infect. Dis. 131:1–5.

Hall, C.B., Douglas, R.G., Jr., and Geiman, J. 1975. Quantitative shedding patterns of respiratory syncytial virus. J. Infect. Dis. 131:151–156.

Hambling, M.H. 1964. Survival of the respiratory syncytial virus during storage under various conditions. Br. J. Exp. Path. 45:647–655.

Hamparian, V.V. 1979. Rhinoviruses. In: E.H. Lennette and N.J. Schmidt (eds.), Diagnostic Procedures for Viral, Rickettsial and Chlamydial Infections, 5th ed. New York: American Public Health Association.

Hamre, D., and Procknow, J.J. 1966. A new virus isolated from the human respiratory tract. Proc. Soc. Exp. Biol. Med. 112:190–193.

Harmon, M.W., Drake, S., and Kasel, J.A. 1979. Detection of adenovirus by enzyme-linked immunosorbent assay. J. Clin. Micro. 9:342–346.

Hendley, J.O., Gwaltney, J.M., and Jordan, W.S. 1969. Rhinovirus infections in an industrial population. Am. J. Epid. 89:184–196.

Hendry, R.M., Godfrey, E., Talis, A.L., et al. 1984. Evidence for Concurrent Circulation of Several Antigenically Distinct Strains of Respiratory Syncytial Virus. (abstract) ICAAC #922, p. 251.

Hendry, R.M., and McIntosh, K. 1982. Enzyme-linked immunosorbent assay for the detection of respiratory syncytial virus infection: development and description. J. Clin. Microbiol. 16:324–328.

Hermann, E.C., and Hable, K.A. 1970. Experiences in laboratory diagnosis of parainfluenza viruses in routine medical practice. Mayo Clin. Proc. 45:177–188.

Hierholzer, J.C. 1973. Further subgrouping of the human adenoviruses by differential hemagglutination. J. Infect. Dis. 128:541–550.

Hierholzer, J.C., and Barme, M. 1974. Counterimmunoelectrophoresis with adenovirus type-specific anti-hemagglutinin sera as a rapid diagnostic method. J. Immun. 112:987–995.

Higgins, P.G. 1966. The isolation of viruses from acute respiratory infections. IV. A comparative study of the use of cultures of human embryo kidney and human embryo diploid fibroblasts (WI-38) Mon. Bull. Min. Health-Pub. Health Serv. Lab. 25:223–229.

Higgins, P.G., Ellis, E.N., and Boston, D.G. 1966. The isolation of viruses from acute respiratory infections. III some factors influencing the isolation of viruses from cases studied during 1962–1964. Mon. Bull. Min. Health-Pub. Health Serv. Lab. 25:5–17.

Hirst, G.K. 1941. The agglutination of red cells by allantoic fluid of chick embryos infected with influenza virus. Science 94:22.

Homma, M., and Ohuchi M. 1973. Trypsin action on the growth of sendai virus in tissue culture cells. J. Vir. 12:1457–1465.

Hornsleth, A., Friis, B., Anderson, P., and Brene, E. 1982. Detection of RSV in nasopharyngeal secretions by ELISA: Comparison with fluroescent antibody technique. J. Med. Vir. 10:273–281.

Jordan, W.S. 1962. Growth characteristics of respiratory syncytial virus. J. Immunol. 88:581–590.

Kapikian, A.Z., Conant, R.M., Hamparian, V.V., Chanock, R.M., Chapple, P.J., Dick, E.C., Fenters, J.D., Gwaltney, J.M., Jr., Hamre, D., Holper, J.C., Jordon, W.S., Jr., Lenette, E.H., Melnick, J.L., Mogabgab, W.J., Mufson, M.A., Phillips, C.A., Schielole, J.H., and Tyrrell, D.A.J. 1967. Rhinoviruses: a numbering system. Nature 213:761.

Kapikian, A.Z., Conant, R.M., Hamparian, V.V., Chanock, R.M., Dick, E.C., Gwaltney, J.M., Jr., Hamre, D., Jordon, W.S., Jr., Kenney, G.E., Lennette, E.H., Melnick, J.L., Mogabgag, W.J., Phillips, C.A., Schieble, J.H., Stott, E.J., and Tyrrell, D.A.J. 1971. A collaborative report: rhinoviruses-extension of the numbering system. Virology 43:524.

Kapikian, A.Z., James, H.D., Kelly, S.J., Dees, J.H., Turner, H.C., Macintosh, K., Kim, H.W., Parott, R.H., Vincent, M.M., and Chanock, R.M. 1969. Isolation from man of "avian infectious bronchitis virus like" viruses (coronaviruses) similar to 229E virus, with some epidemiological observations. J. Infect. Dis. 119:282–290.

Kasel, J.A. 1979. Adenoviruses. In: E.H. Lennette and N.J. Schmidt (eds.), Diagnostic Procedures for Viral, Rickettsial and Chlamydial Infections. New York: American Public Health Association.

Kaye, H.S., and Dowdle, W.R. 1975. Seroepidemiologic survey of coronavirus (Strain 229 E) infections in a population of children. Am. J. Epid. 101:238–244.

Keller, E.W., Rubin, R.H., Black, P.H., and Hirsch, M.S. 1979. Isolation of adenovirus type 34 from a renal transplant recipient with interstitial pneumonia. Transplantation 23:188.

Kendal, A.P. 1975. A comparison of "influenza c" with prototype myxoviruses: receptor-destroying activity (neuraminidase) and structural polypeptides. Virology 65:87–99.

Kraaijeveld, C.A., Reed, S.A., and MacNaughton, M.R. 1980. Enzyme-linked immunosorbent assay for detection of antibody in volunteers experimentally infected with human coronavirus strain 229 E. J. Clin. Micro. 12:493–497.

Lauer, B.A. 1982. Comparison of virus culturing and immunofluorescence for rapid detection of respiratory syncytial virus in nasopharyngeal secretions: sensitivity and specificity. J. Clin. Microbiol. 16:411–412.

Lauer, B., Masters, H., et al., 1984. Rapid Diagnosis of RSV in Nasal Secretions by ELISA. (abstract) ICAAC #921.

Lennette, E.H., Jensen, F.W., Guenther, R.W., and Magoffin, R.L. 1963. Serologic responses to parainfluenza viruses in patients with mumps virus infection. J. Lab. Clin. Med. 61:780–788.

Lewis, F.A. and Kennett, M.L. Comparison of rhinovirus-sensitive HeLa cells and human embryo fibroblasts for isolation of rhinoviruses from patients with respiratory disease. J. Clin. Micro. 3:528–532.

Liu, C. 1956. Rapid diagnosis of human influenza infection from nasal smears by means of fluorescein-labelled antibody. Proc. Soc. Exp. Biol. Med. 92:883–887.

McIntosh, K. 1974. Coronavirus. A comparative review. Curr. Top. Micro. Immunol. 63:85–129.

McIntosh, K., Hendry, R.M., Fahnestock, M.L., and Pierik, L.T. 1982. Enzyme-linked immunosorbent assay for detection of RSV Infection: Application to clinical samples. J. Clin. Microbiol. 16:329–333.

McIntosh, K., Kapikian, A.Z., Hardison, K.A., Hartley, J.W., and Chanock, R.M. 1969. Antigenic relationship among coronaviruses of man and between human and animal coronaviruses. J. Immun. 102:1109–1118.

McIntosh, K., McQuillan, J., Reed, S.R., and Gardner, P.S. 1978. Diagnosis of human coronavirus infection by immunofluorescence. J. Med. Vir. 2:341–346.

McQuillen, J., Gardner, P.S., and McGuckin, R. 1970. Rapid diagnosis of influenza by immunofluorescent techniques. Lancet 2:690–695.

Minnich, L.L. and Ray, C.G. 1982. Comparison of direct and indirect immunofluorescence staining of clinical specimens for detection of respiratory syncytial virus antigen. J. Clin. Microbiol. 15:969–970.

Mogabgab, W.J., Holmes, B.J. and Pollack, B. 1975. Antigenic relationships of common rhinovirus types from disabling upper respiratory illnesses. Dev. Biol. Stand. 28:400–411.

Monto, A.S., and Bryan, E.R. 1974. Microneutralization test for detection of rhinovirus antibodies. Proc. Soc. Exp. Biol. Med. 145:690–694.

Morbidity and Mortality Weekly Report, July 27, 1984, Volume 33 #29 pp. 417–432.

Murphy, B.R., Baron, S., Chalhub, E.G., Uhlendorf, C.P., and Chanock, R.M. 1973. Temperature sensitive mutants of influenza. IV. Induction of interferon in the nasopharynx by wild-type and a temperature-sensitive recombinant virus. J. Infect. Dis. 128:488–493.

Orstavik, I. 1981. Susceptibility of continuous lines of monkey kidney cells to influenza and parainfluenza viruses in the presence of trypsin. Actr. Path. Microbiol. Scand. Section B, 89:179–183.

Parkinson, A.J., Muchmore, H.G., Scott, L.V., and Miles, J.A.R. 1980. Parainfluenza virus isolation enhancement utilizing portable cell culture system in the field. J. Clin. Micro. 11:535–536.

Parkinson, A.J., Scott, E.N., and Muchmore, H.G. 1982. Identification of parainfluenza virus serotypes by indirect solid-phase enzyme immunoassay. J. Clin. Micro. 15:538–541.

Pearson, R.D., Hall, W.J., Menegus, M.A., and Douglas, R.G., Jr. 1980. Diffuse pneumonitis due to adenovirus type 21 in a civilian. Chest 78:107–109.

Pelon, W., Mogabgag, W.J., Phillips, I.A., and Pierce, W.E. 1957. A cytopathogenic agent isolated from naval recruits with mild respiratory illness. Proc. Soc. Exp. Biol. Med. 94:262–267.

Phillips, C.A., Melnick, J.L., and Grim, C.A. 1965. Human aorta cells for isolation and propagation of rhinoviruses. Proc. Soc. Exp. Biol. Med. 119:843–845.

Price, W.H. 1956. The isolation of a new virus associated with respiratory clinical disease in humans. Proc. Natl. Acad. Sci. U.S.A. 42:892–896.

Roggendorf, M., Wigand, R., et al. 1982. Enzyme-linked immunosorbent assay for acute adenovirus infection J. Vir. Meth. 4:27–35.

Rosen, L. 1960. A hemagglutination-inhibition technique for typing adenoviruses. Am. J. Hyg. 71:120–128.

Rowe, W.P., Huebner, R.J., Gilmore, L.K., Parrott, R.H., and Ward, T.G. 1953. Isolation of a cytopathogenic agent from human adenoids undergoing spontaneous degeneration in tissue culture. Proc. Soc. Exp. Med. 84:570–573.

Rutala, W.A., Shelton, D.F., and Arbiter, D. 1977. Comparative sensitivities of viruses to cell cultures and transport media. J. Clin. Path. 67:397–400.

Sarkkinen, H.K., Halonen, P.E., and Salmi, A.A. 1981. Detection of influenza A virus by radioimmunoassay and enzyme-immunoassay from nasopharyngeal specimens. J. Med. Vir. 7:213–220.

Scott, J.V., Dressman, G.R., Spira, G., et al. 1975. Radioimmunoassay of human serum antibody specific for adenovirus type 5 purified fiber. J. Immun. 115:124–128.

Silver, S.M., Scheid, A., and Choppin, P.W. 1978. Loss on serial passage of rhesus monkey kidney cells of proteolytic activity required for sendai virus activation. Inf. Immun. 20:235–241.

Smith, T.F., and Reichrath, L. 1974. Comparative recovery of 1972–1973 influenza virus isolates in embryonated eggs and primary rhesus monkey kidney cell cultures after one freeze-thaw cycle. Am. J. Clin. Path. 61:579–584.

Smith, W., Andrewes, C.H., and Laidlaw, P.P. 1933. A virus obtained from influenza patients. Lancet 2:66–68.

Stalder, H., Hierholzer, J.C., and Oxman, M.N. 1977. New human adenovirus (candidate adenovirus type 35) causing fatal disseminated infection in a renal transplant recipient. J. Clin. Microbiol. 6:257.

Taylor, R.M. 1951. A further note on 1233 ("influenza c") virus. Arch. Gesamte. Virusforsch. 4:485–500.

Tyrrell, D.A.J., and Bynoe, M.L. 1965. Cultivation of a novel type of common-cold virus in organ cultures. Br. Med. J. 1:1467–1470.

Virtanen, M., Laaksonen, M. Soderlund, H., Palva, A., Halonen, P., and Ranki, M. 1983. Novel test for rapid viral diagnosis: detection of adenovirus in nasopharyngeal mucus aspirates by means of nucleic acid sandwich hybridisation. Lancet 1:381–383.

Welliver, R.C., and Ogra, P.L. 1983. Use of immunofluorescence in the study of the pathogenesis of RSV infection. Ann. N.Y. Acad. Sci. 420:369–375.

Wong, D.T., Welliver, R.C., Riddlesberger, K.R., Sun, M.S., and Ogra, P.L. 1982. Rapid diagnosis of parainfluenza virus infection in children. J. Clin. Micro. 16:164–167.

Yolken, R.H., Torsch, V.M., Berg, R., Murphy, B.R., and Lee, Y.C. 1980. Fluorometric assay for measurement of viral neuraminidase-application to the rapid detection of influenza virus in nasal wash specimens. J. Infect. Dis. 142:516–523.

Enteroviruses

Heinz Zeichhardt

21.1 INTRODUCTION

Enteroviruses comprise one genus in the *Picornaviridae* family. This family additionally contains the following genera: cardioviruses, rhinoviruses, and aphthoviruses (for classification and nomenclature of the viruses cf. Matthews, 1982). The members of the enterovirus group that infect humans are polioviruses, coxsackieviruses group A and group B, echoviruses, hepatitis A virus (HAV), and other enteroviruses (Table 21.1). These are grouped together because of similar physicochemical properties. All enteroviruses inhabit the human alimentary tract and most of them, except HAV are able to infect the central nervous system (CNS). In addition, these viruses induce a variety of clinical syndromes. The reviews of Melnick et al. (1979), Ginsberg (1980), Melnick (1982), Moore and Morens (1984), and Rueckert (1985) are recommended for extensive additional readings on enteroviruses.

21.2 HISTORY OF VIRUS DISCOVERY

Crippling paralytic disease was recorded in ancient times (Melnick, 1982), however, a characterization of poliomyelitis was not reported until the turn of the 19th century. Poliomyelitis was established as a viral disease in 1909 when Landsteiner and Popper (1909) transmitted paralytic disease to monkeys by inoculating them with filtered stool from a patient with paralytic disease. During the next 40 years, animal inoculation was the method of choice for virus inoculation and study. Thus, in 1948, in Coxsackie, New York, a virus was isolated in suckling mice that were inoculated with a cell-free filtrate of stools obtained from two children suffering from paralysis (Dalldorf and Sickles, 1948). This virus, which could not be neutralized by antiserum against any of the three polioviruses, became the first member of the group A coxsackieviruses. The first of the group B coxsackieviruses was isolated in 1949 (Melnick et al., 1949). The major breakthrough for diagnosing and controlling poliomyelitis was the observation that poliovirus could be propagated in human embryonic tissues in culture (Enders et al., 1949). These tissue cultures allowed easy isolation of the viruses and were prerequisite for the development of vaccines, including both the inactivated vaccine of Salk and the attenuated (oral) vaccine of Sabin. After the introduction of tissue culture, the isolation of

Table 21.1. Serotypes of the Human Enteroviruses

Virus	Serotypes
Poliovirus	1–3
Coxsackievirus group A	1–24[a]
Coxsackievirus group B	1–6
Echovirus	1–33[b]
Hepatitis A virus	[c]
Other enteroviruses	68–71

[a] Coxsackievirus type A 23 is the same virus as echovirus type 9.

[b] Echovirus type 10 has been reclassified as reovirus type 1 and echovirus 28 as rhinovirus type 1A. Echovirus type 8 has been deleted because of identity with echovirus type 1.

[c] Formerly classified as enterovirus type 72.

many other enteroviruses was possible. The echoviruses (enteric, cytopathic, human, orphan) were discovered in 1951 by Robbins et al. (1951). These viruses often were isolated from stools of healthy children and, therefore, could not be related to a disease; hence, these "enteric viruses" were called "orphan viruses."

As early as the times of Hippocrates, a disease called infectious jaundice was described (for reviews, see Frösner, 1984; Purcell et al., 1984). However, the HAV was not isolated until the 1960s (Krugman et al., 1967). Initial studies suggested that only marmoset monkeys could be infected and it took until 1979 to cultivate hepatitis A virus in cell cultures (Purcell et al., 1984).

21.3 STRUCTURAL, BIOPHYSICAL, BIOCHEMICAL, AND BIOLOGICAL CHARACTERISTICS OF ENTEROVIRUSES

21.3.1 Structure

Enteroviruses are small, spherical, naked viruses (Figure 21.1). The virus particles have icosahedral symmetry, a diameter of 27–30 nm, a buoyant density in CsCl of 1.34 g/ml, a molecular weight of 8.25×10^6 daltons, and a sedimentation coefficient of 156–160 S (for reviews, see Ginsberg, 1980; Rueckert, 1985; Koch and Koch, 1985).

Figure 21.1. Electron micrograph of poliovirus type 1 particles, negatively stained with 0.5% uranyl acetate. Bar represents 100 nm.

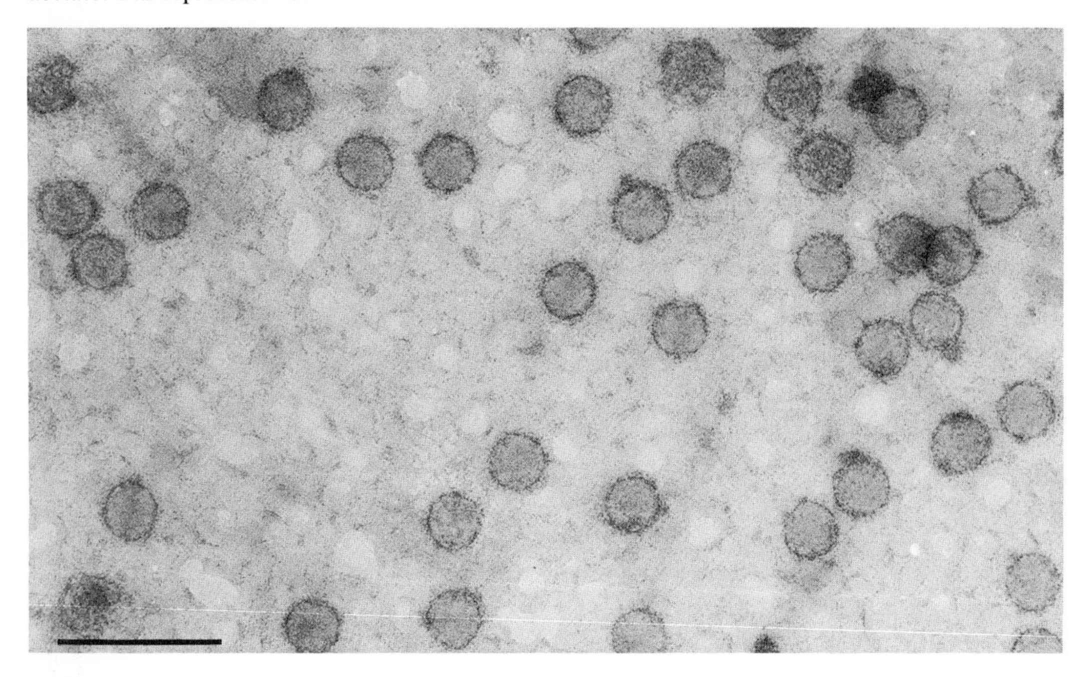

Each virion contains one molecule of single-stranded RNA of about 7500 bases (molecular weight, 2.6 × 10^6 daltons), which serves as genetic information as well as viral messenger RNA ("positive" sense RNA). The genome of poliovirus type 1 has been completely sequenced and that of HAV partially sequenced (Kitamura et al., 1981; Linemeyer et al., 1985). The polygenic RNA is monocystronic and translated into a large precursor protein. This precursor is proteolytically cleaved into the four structural viral capsid proteins, VP1, VP2, VP3, VP4 (the molecular weights determined for poliovirus type 1 are probably typical for all enteroviruses: VP1—33,500 daltons, VP2—30,000 daltons, VP3—26,400 daltons, VP4—7400 daltons), into VPg (molecular weight, 2400 daltons), and the functional proteins, protease and RNA polymerase. VPg is covalently linked to the 5′ end of the RNA. The 3′ end of the RNA contains a poly-A-sequence. A virion contains 60 copies of each VP1, VP2, VP3, and VP4, and one copy of VPg. A virion may also contain a small number of copies of VPO, which is the uncleaved precursor of VP2 and VP4. For poliovirus type 1, VP1, VP3 and, to a lesser extent, VP2 comprise most of the surface of the virus capsid (Wetz and Habermehl, 1979). VP1 and VP3 are in close proximity to each other.

21.3.2 Antigenicity

Preparations of poliovirus contain two antigens that can be detected in complement fixation and precipitin tests: Infective or "native" virus, called D (or N) antigen, and noninfective virus called C antigen or occasionally H (heated) antigen. It seems to be true for all enteroviruses that antigenic sites at the capsid surface determine the type-specific antigenicity that is best investigated in neutralization tests. For poliovirus wild type and attenuated strains, the capsid protein VP1 with one or more antigenic site(s) plays the major role in the interaction with neutralizing antibodies (Emini et al., 1983; Minor et al., 1983). VP2 and VP3 also interact with neutralizing antibodies, however, to lesser extents than VP1 (for review, see Dimmock, 1984). Recent data for HAV show that VP1 also is involved in the neutralization of this virus (Hughes et al., 1984). In contrast, VP2 of coxsackievirus type B3 was reported to comprise the major antigenic site(s) for neutralization (Beatrice et al., 1980).

Several antigenic relationships between enteroviruses have been observed. In neutralization tests poliovirus types 1 and 2 partially crossreact, coxsackievirus types A3 and A8, A11 and A15, A13 and A18; and echovirus types 1 and 8, 6 and 30, and 12 and 29 are antigenically related. The crossreactivity between several enteroviruses observed in the complement fixation test may be due to common antigenic sites of the virus proteins that are located in the interior of the capsid. These sites are accessible only when a soluble antigen is used. Such immunologic crossreactivity was recently confirmed by the immunoblot technique (Mertens et al., 1983). In addition, these antigenic relationships are reflected by limited genomic homologies among the RNA of the different enteroviruses as studied by RNA hybridization (Young, 1973). The following levels of genomic homology were found: 5% or more among all enteroviruses; less than 20% between groups (e.g., polioviruses and coxsackieviruses of group A); 30–50% within groups.

21.3.3 Reactivity of Enteroviruses to Chemical and Physical Agents and Virus Storage

All enteroviruses are resistant to low pH (pH 3) and several proteolytic enzymes, which is the prerequisite for virus passage through the stomach and duodenum. The viruses are resistant to several disinfectants, such as 70% alcohol, 5% lysol, or 1% quaternary ammonium compounds, to ether, deoxycholate, and various other detergents that destroy lipid containing viruses. In general, enteroviruses are inactivated by the following chemicals (see Melnick, 1982; Frösner, 1984; Moore and Morens, 1984): formaldehyde (0.3%), HCl (0.1 N), free residual chlorine (0.3–0.5 ppm) and other halogens (free residual bromine or iodine, $\simeq$ 0.5 ppm $\times$ 10 minutes contact time). Presence of organic matter with the virus may result in protection against inactivation. For this reason 3% formaldehyde is recommended for disinfection. The following physical conditions are inactivating: drying, heat (50°C for 1 hour in the absence of magnesium chloride), light (in the presence of vital dyes, such as neutral red, acridine orange, and proflavine). Hepatitis A virus has a higher stability than the other enteroviruses. Temperatures above 60°C are necessary to destroy its infectivity within a short time (e.g., 85°C for 1 minute).

Enteroviruses are stable for years when stored at −70°C. Storage at −20°C results in some loss in titer over months while enteroviruses in suspension stored at 4°C usually will stay viable for weeks.

21.3.4 Replication of Enteroviruses in Cell Culture

The replication of all enteroviruses so far studied is a mechanism closely associated with cellular membranes. A replication cycle comprises the following steps (for reviews, see Sangar, 1979; Crowell and Landau, 1983):

Virus entry into the cell: Poliovirus, studied most thoroughly of all enteroviruses, enters its host cell by a phagocytosis mechanism (Dales, 1973). As recently shown, the entry is via receptor-mediated endocytosis (Zeichhardt et al., 1985). The virus specifically adsorbs at the cell surface and is internalized into the cell via coated pits, coated vesicles, and endosomes. The site of uncoating (i.e., the release of the viral genome from the virus capsid) has been reported to occur at the cell surface membrane, in the cytoplasm and the endosome and/or lysosome. For poliovirus type 1, uncoating is most likely a pH-dependent process in endosomes and/or lysosomes (Habermehl et al., 1973; Zeichhardt et al., 1985).

Viral protein and RNA synthesis: Viral proteins are synthesized at the rough endoplasmic reticulum after release of VPg from the viral genome. After translation of a large precursor protein that is cleaved (most probably autocatalytically) into viral RNA polymerase, protease and precursor of the capsid proteins, viral RNA synthesis takes place in a replicative complex at the smooth endoplasmic reticulum.

Morphogenesis of virus: The assembly of the virus takes place at the cytoplasmic membranes, however, the association with rough or smooth membranes is still not known. There is likely a membrane associated morphopoietic factor that facilitates the morphogenesis of the mature particle (Putnak and Phillips, 1981).

Virus release: The release of newly synthesized virus from the cell is not clearly understood. It has been shown for poliovirus that progeny virus is found in the supernatant prior to cytolysis. Further, only a small portion of the newly synthesized virus particles are mature infective virions. Ratios of infective virus to total virus particles of $1:10^1–10^3$ have been observed.

The replication cycle takes 6 to 7 hours for poliovirus but is some hours longer for other enteroviruses. An exception is HAV for which replication times of more than 4 weeks were reported when using fecal specimens for primary inoculation. This was reduced to several days, or up to 1 week for virus inocula obtained after several passages (Purcell et al., 1984).

Shortly after infection (1 to 3 hours postinfection) most enteroviruses induce a pronounced inhibition of cellular RNA protein and thereby cellular DNA synthesis (Diefenthal et al., 1973). Most enteroviruses are strongly cytolytic, that is, they induce a cytopathic effect (CPE), resulting in destruction of the cell by lysis. A typical example for such a CPE is shown in Figure 21.2 for poliovirus type 1 infecting a monolayer of HEp-2 cells (Zeichhardt et al., 1982). Most of the infected cells detach from the surface of the culture vessel. The remaining cells withdraw from adjacent cells, round up, and are attached to the substratum by long filopodia (Diefenthal and Habermehl, 1967). The microvilli at the cell surface merge and

Figure 21.2. Scanning electron micrograph of HEp-2 cells infected with poliovirus type 1 (A) compared with mock-infected control cells (B). Bar represents 10 μm. (Zeichhardt, Schlehofer, Wetz, Hampl, Habermehl).

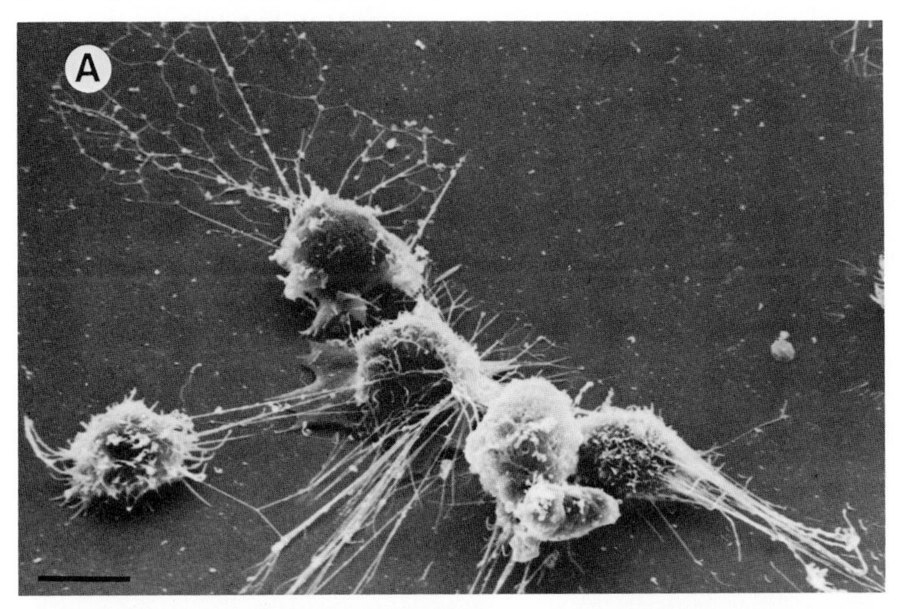

disappear. Ultrathin sections of poliovirus infected cells show drastic changes in the interior of the cell, such as vesicles arranged in clusters in the cytoplasm and a lobed nucleus with irregular distribution of condensed chromatin (Dales et al., 1965). Poliovirus induces characteristic mitotic changes and chromosomal aberrations. The early stage of replication induces enhancement of mitosis, later stages result in an arrest of mitosis in the metaphase (colchicine-like effect). Chromosomal damage is characterized by single chromatid breaks and pulverization (Habermehl et al., 1966; Bartsch et al., 1969).

21.4 EPIDEMIOLOGY

2.4.1 Mode of Transmission

Human enteroviruses have their reservoir only in humans (for growth and pathogenicity in animals, see Table 21.2). Enteroviruses can be isolated from the lower and/or upper alimentary tract and, therefore, can be spread both by the fecal–oral and respiratory routes (Melnick, 1982). In areas with poor sanitary conditions fecal–oral transmission is predominant. Transmission by respiratory routes can occur early in infection because the virus replicates in the upper respiratory tract. Sexual transmission of enteroviruses has not been reported, and blood transfusions and insect bites seem not to be responsible for virus transmission. Enteroviruses can be isolated from sewage, therefore, a fecal–water–oral route of transmission is possible. For HAV, transmission in clams and oysters from fecally contaminated waters is a common source of infection. Food borne acquisition of other enteroviruses has been noted. Nosocomial transmission of enteroviruses typically takes place in newborn nurseries and has been reported for several coxsackieviruses of group A and group B, echoviruses, and HAV.

21.4.2 Geographic, Seasonal, Socioeconomic, Sex, and Age Factors

Enteroviruses are found worldwide. Enterovirus infections characteristically take place during the summer time in areas of

Table 21.2. Growth of Enteroviruses in Human and Monkey Kidney Cell Lines and Pathogenicity for Animals

Viruses	Human Cells	Monkey Kidney Cells	Pathogenicity	
			Mice	Monkeys
Polioviruses	+	+	[a]	+
Coxsackieviruses A	some[b]	some[c]	+[c]	some[d]
Coxsackieviruses B	+	+	+[c]	some
Echoviruses	some[b]	+[e]	some[f]	some
Enteroviruses 68–71	+	+/−	+/−	some
Hepatitis A virus	+[g]	+[g]		+

[a] Some strains of each type have been adapted to mice.

[b] Some strains grow preferentially in, or have been adapted to, human cell cultures. Coxsackievirus types A 11, 13, 15, 16, 18, 20, and 21 may be isolated directly in human cells.

[c] Coxsackievirus types A 7, A 9, and B strains grow readily in monkey kidney cells; some strains grow poorly in mice and fail to produce disease in these animals.

[d] Especially coxsackievirus type A 7.

[e] Echovirus type 21 is cytopathogenic for human epithelial cells but not for monkey kidney cells.

[f] Whereas the prototype and other strains of echovirus 9 are not pathogenic for mice, a number of other strains, especially after passage in monkey kidney cells, produce paralysis in mice (severe coxsackievirus type of myositis).

[g] Not used for routine isolation; propagation is difficult due to long replication times (several weeks) and low virus yield. Some adapted strains of hepatitis A virus have replication times of 3 days (see Purcell et al., 1984).

Modified according to Melnick et al., 1979.

the north temperate zone. In tropical and semitropical areas enterovirus infections occur throughout the whole year. Persons of low socioeconomic status living in urban areas receive a greater exposure to enteroviruses and have a higher incidence of subclinical infections than persons of higher socioeconomic status. This led to the paradox that poliomyelitis in the prevaccine era was a disease of "development" (Melnick, 1982), in other words, improvement in the hygienic and socioeconomic conditions was associated with lower subclinical exposure and an increase of incidence of severe illnesses. The improved hygiene might also have led to a decrease in mixed infections with other enteroviruses of patients infected with polioviruses. Mixed infections with more than one enterovirus at the same time can result in interference, leading to a suppression of replication of one of the viruses. Due to extensive vaccination poliovirus infections are not currently a major problem in developed countries (Figure 21.3, see 21.8.), however, frequent endemics are observed in enclaves, usually religious groups who refuse vaccination.

Diseases due to enteroviruses occur more frequently in males than in females (male/female ratio of 1.5–2.5 : 1) (Moore and Morens, 1984). Generally, young children are the main transmitters of enteroviruses. Echovirus type 9 was found in 50–70% of children compared with 17–33% of adults. The age of initial infection decreases with bad hygienic conditions and low socioeconomic status. It was reported for HAV that in Africa and Asia 90% of the children will be infected within the first 8 years of age, however, in Scandinavia only 30% of the population will be infected by 30 to 40 years of age. Severity of disease is also related to age. Poliovirus infection in adults is more likely to lead to paralysis than in children, and infections with coxsackieviruses of group A and echoviruses are usually milder in children than in adults. In contrast coxsackieviruses of group B, can induce fulminant "viral sepsis," myocarditis, encephalitis, and death more likely in new-

Figure 21.3. Paralytic poliomyelitis attack rates, United States, 1951–1981. From CDC Poliomyelitis Surveillance Summary 1980–1981. Issued December 1982, p. 9. U.S. Dept. H.H.S., P.H.S.

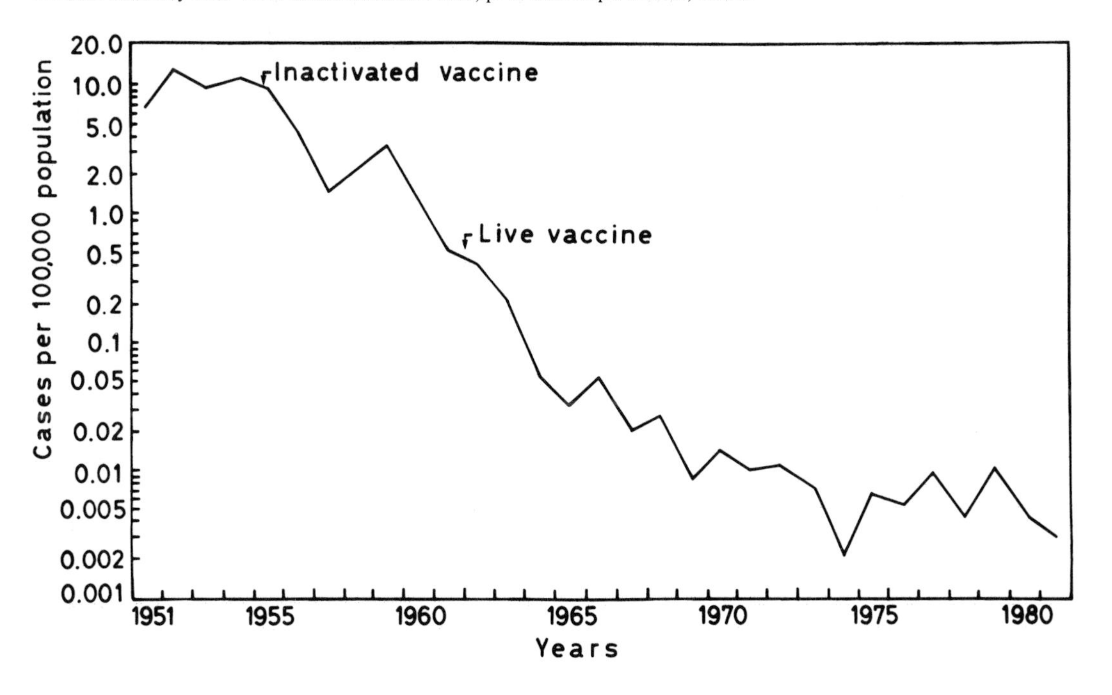

borns than in older children and adults. A reason for the different susceptibility between children and adults for enterovirus infections might be found in a changing pattern of virus-specific receptors during development and aging.

21.4.3 Asymptomatic Infections

Infections with enteroviruses are very common, however, it should be reemphasized that the most common forms of infection are silent, mild, or subclinical. It has been reported that 90–95% of poliovirus infections are asymptomatic, whereas, only 0.1–1.0% of infections cause paralytic poliomyelitis. Asymptomatic infections are most common for polioviruses followed by echoviruses and coxsackieviruses (50%) (reviewed in Moore and Morens, 1984). The high incidence of inapparent infections with enteroviruses may be due to the virus passage of the gut. The cells of the epithelia of the gut normally have a high rate of turnover. Although 10^3 infective virus particles can be reproduced in one infected cell of the gut and consequently 10^6–10^9 viruses per gram can be detected in the stool of enterovirus infected persons, the virus-induced lysis of the cells might be without clinical consequence. Clinical symptoms occur only after massive infection of the gut epithelia.

21.5 PATHOGENESIS AND CLINICAL SYNDROMES

The mechanism of pathogenicity of enterovirus infections is the lytic infection of host cells resulting in severe cytopathic effects (see 21.3.4). Enteroviruses can lead to cyclic infections in their hosts with a viremia and subsequent transport of the virus to the target organs (spinal cord and brain, meninges, myocardium, skin, liver, etc.) (for reviews, see Melnick, 1982; Moore and Morens, 1984).

21.5.1 Polioviruses

The course of infection of polioviruses is best understood of all enteroviruses and, therefore, presented as a typical example (Figure 21.4). The portal of entry of polioviruses is the alimentary tract via the mouth. During the incubation period (6 to

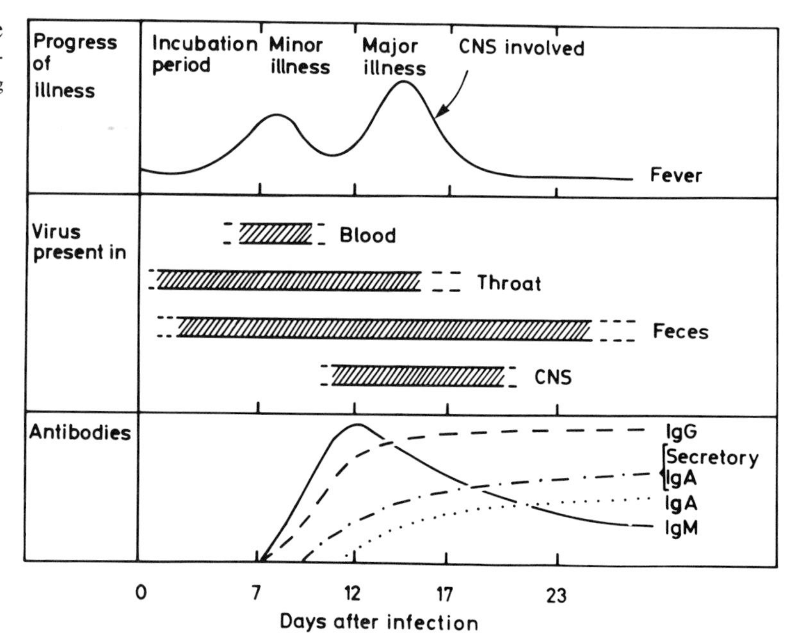

Figure 21.4. The course of infection with poliovirus (idealized). According to Habermehl.

20 days) poliovirus multiplies in the mucosal tissues of the pharynx, the lymphoid tissue (tonsils and Peyer's patches), and/or the gut. For this reason virus is spread via oral and fecal routes beginning shortly after infection. In most cases (90–95%) the virus infection will be apparent (i.e., the patient will be asymptomatic). However, virus can be spread to the draining lymph nodes, leading to a viremia characterized by a recovery of virus from the blood stream for a few days (day 6 to 9 after infection). During this time the first nonspecific clinical symptoms (e.g., fever, malaise, sore throat, sometimes headache and vomiting), will be observed. In about 4–8% of poliovirus infections, the illness will not proceed and only take the form of a minor illness ("abortive poliomyelitis"). If poliovirus infects its target cells in the CNS, nonparalytic poliomyelitis (1–2%) and paralytic poliomyelitis (0.1–1.0%) will occur. In nonparalytic poliomyelitis patients have the same prodromal illness as those with minor illness followed after 3 to 7 days by an illness similar to aseptic meningitis commonly accompanied by high fever, back pain, and muscle spasm. Paralytic poliomyelitis additionally comprises flaccid paralysis (involvement of the whole muscle) or paresis (involvement of only some muscle groups), which is due to spinal and/or bulbar damage. The bulbar poliomyelitis (ascendent infection) is less common than the spinal form and has a poor prognosis due to damage of cerebral nerve or vegetative centers. In the case of the spinal illness, recovery of motor function may occur to some degree after some months, however, remaining paralysis is permanent. The source of these severe clinical symptoms is the very specific extraintestinal target cell range of polioviruses; especially the anterior horn cells of the spinal cord, but also dorsal root ganglia, certain brainstem centers, cerebellum, spinal sensory columns, and occasionally the cerebral motor cortex. Histologic changes first observed are vascular engorgement, accompanied by perivascular infiltration with lymphocytes and also polymorphonuclear neutrophils, plasma cells, and microglia.

Certain factors increase the severity of disease, including very young and very old age, male sex, chronic undernutrition, corticosteroid treatment, physical exertion, hypoxia, cold, irradiation, tonsillectomy, pregnancy, adrenal-related endocrine changes, and possibly hypercholesterolemia (reviewed by Moore and Morens, 1984).

21.5.2 Coxsackieviruses and Echoviruses

Coxsackie- and echoviruses have a less specific extraintestinal target organ range than poliovirus and, therefore, can lead to a wider range of illnesses (Table 21.3). As described for polioviruses, coxsackie- and echoviruses multiply primarily in the pharynx and small intestine and are shed in the feces for up to 1 month and in respiratory secretions for several days. Generally speaking, coxsackieviruses and echoviruses, besides the alimentary tract, can infect the meninges, CNS, myocardium and pericardium, striated muscles, respiratory tract, and skin. Paralysis by coxsackie- and echoviruses are less common and less severe than paralysis induced by polioviruses. Coxsackieviruses usually are more pathogenic than echoviruses. Herpangina (vesicular pharyngitis) is only induced by coxsackieviruses of group A. Common colds and pneumonitis of infants are attributed to several serotypes of coxsackieviruses of group A and enterovirus 68. Epidemic myalgia (Bornholm disease) and pleurodynia are associated with most serotypes of coxsackieviruses of group B. Exanthemata with accompanying fever and pharyngitis are observed with several types of coxsackieviruses of group A and B and echoviruses. Aseptic meningitis very often accompanied by rashes are also induced by several coxsackie- and echoviruses. Meningoencephalitis (especially in children) is

Table 21.3. Enterovirus Infections and Their Clinical Syndromes

Viruses	Types	Clinical Syndromes
Polioviruses types 1–3	1–3	Paralysis (slight to complete muscle weakness), aseptic meningitis, and undifferentiated febrile illness, particularly during the summer
Coxsackieviruses, group A, types 1–24	2, 3, 4, 5, 6, 8, 10	Vesicular pharyngitis (herpangina)
	10	Acute lymphatic or nodular pharyngitis
	2, 4, 7, 9, 10, 23	Aseptic meningitis
	7, 9	Paralysis (infrequently)
	2, 4, 9, 16, 23	Exanthem (macular rash)
	4, 5, 6, 9, 10, 16	Exanthem (vesicular rash; infrequently)
	5, 10, 16	A "hand-foot-and-mouth" disease
	9, 16	Pneumonitis of infants
	21, 24	"Common cold"
	4, 9	Hepatitis
	18, 20, 21, 22, 24	Infantile diarrhea
	24	Acute hemorrhagic conjunctivitis
Coxsackieviruses, group B, types 1–6	1, 2, 3, 4, 5	Pleurodynia
	1, 2, 3, 4, 5, 6	Aseptic meningitis
	2, 3, 4, 5	Paralysis (infrequently)
	1, 2, 3, 4, 5	Severe systemic infection in infants, meningoencephalitis, and myocarditis
	1, 2, 3, 4, 5	Epidemic myalgia (Bornholm disease)
	1, 2, 3, 4, 5	Pericarditis, myocarditis
	4, 5	Upper respiratory illness and pneumonia
	1, 3, 5	Exanthem (macular rash; infrequently)
	5	Hepatitis
	1, 2, 3, 4, 5, 6	Undifferentiated febrile illness
	1, 2, 4	Pancreatitis
	4	Diabetes
Echoviruses, types 1–34	all serotypes, except 12, 24, 26, 29, 32, 33, 34	Aseptic meningitis
	2, 4, 6, 9, 11, 30; possibly 1, 7, 13, 14, 16, 18, 31	Paralysis (infrequently)
	2, 6, 9, 19; possibly 3, 4, 7, 11, 14, 18, 22	Encephalitis, ataxia, or Guillain–Barré syndrome

Table 21.3. (*continued*)

Viruses	Types	Clinical Syndromes
	2, 4, 6, 9, 11, 16, 18; possibly 1, 3, 5, 7, 12, 14, 19, 20	Exanthem
	4, 9, 11, 20, 25; possibly 1, 2, 3, 6, 7, 8, 16, 19, 22	Respiratory disease
	different types have been recovered	Diarrhea (a consistent association has not been established)
	1, 6, 9	Epidemic myalgia (infrequently)
	1, 6, 9, 19	Pericarditis and myocarditis (infrequently)
	4, 9	Hepatic disturbances
Enteroviruses, types 68–71	68	Pneumonia and bronchiolitis
	70	Acute hemorrhagic conjunctivitis
	71	Aseptic meningitis
	71	Meningoencephalitis
	71	Hand-foot-and-mouth disease
Hepatitis A virus		Hepatitis

Modified according to Melnick et al., 1979

seen in infections with several coxsackieviruses of group B and with enterovirus 71 and encephalitis with several echoviruses. Myocarditis and pericarditis are more commonly caused by coxsackieviruses of group B in comparison to coxsackieviruses of group A and echoviruses. Acute hemorrhagic conjunctivitis is associated with enterovirus type 70. Coxsackieviruses of group B seem to be responsible for acute pancreatitis and coxsackievirus type B4 for diabetes. Several echoviruses and coxsackieviruses of group A have been recovered from feces mainly of children during epidemics of gastroenteritis, however, their strict role in epidemic diarrhea is still not completely solved. Some echo- and coxsackieviruses of group B are associated with hepatic disturbances or hepatitis, but HAV is the major source of enterovirus-induced hepatitis.

21.5.3 Hepatitis A Virus

Hepatitis A virus seems to comprise a different mode of pathogenesis than the other enteroviruses (for reviews, see Frösner, 1984; Purcell et al., 1984). Hepatitis A virus infections have a long incubation period (4 weeks) relative to other enteroviruses. So far, there is no evidence that HAV multiplies initially in the pharynx and/or the gut after uptake (fecal–oral route). This might be the reason that in contrast with other enteroviruses HAV cannot be recovered from the pharynx or gut shortly after infection. Virus can only be isolated from the feces of infected patients beginning 2 weeks after infection, which is about 2 weeks before the onset of clinical manifestations (jaundice). Virus recovery decreases shortly before the onset of illness and is mostly negative 2 to 3 weeks after onset. It is postulated that HAV is distributed in a viremic phase to its target organ (i.e. the liver). Virus multiplication takes place in hepatocytes and Kupffer cells. Hepatitis A virus gains access to the stool via the biliary system.

21.5.4 Chronic Infections

Chronic infections are uncommon for enteroviruses, however, persistent infections with some serotypes of coxsackieviruses of group A and B have been observed.

21.5.5 Pregnancy

Pregnancy and infection with enteroviruses are an unsolved problem. In spite of several negative reports, maternal infections during the first trimester of pregnancy are suspected to result in the following anomalies in the infected fetus: coxsackievirus types B2 or B4, urogenital anomalies; coxsackievirus types B3 or B4, cardiovascular anomalies; and coxsackievirus type A9, digestive system malformations (cited in Moore and Morens, 1984). Because the teratogenic risk of intrauterine infections with enteroviruses is lower by several orders of magnitude in comparison with infections with rubella virus, a recommendation of abortion is not generally accepted for women with proven enterovirus infection in the first trimester of pregnancy.

21.6 INCUBATION TIMES

All polioviruses, coxsackieviruses of group A and B, and echoviruses have incubation times ranging from 1 to 35 days, with an average of 1 to 2 weeks (reviewed by Moore and Morens, 1984). The shortest incubation period, 12 to 30 hours, has been reported for local infections of the eye by echovirus type 70. Hepatitis A virus has the longest incubation time of all enteroviruses. Usually, the incubation time is about 4 weeks with a range of 14 to 40 days.

21.7 IMMUNE RESPONSE

Humoral and secretory antibodies play the major role in the immunity to enterovirus infections (Figure 21.4). The role of cellular immunity in these infections is not well defined. Humoral immunity is mediated by type-specific neutralizing IgG, IgM, and IgA, which prevent hematogenous spread of virus to the target organs. IgM appears first after infection (7 to 10 days after infection with polioviruses, coxsackieviruses, and echoviruses, and 4 weeks after infection with HAV).

Virus-specific IgM persists for 4 weeks in 90% of infections. Virus-specific IgG and IgA appear a few days after IgM. IgG persists for years and, therefore, mediates the acquired humoral immunity. Production of antibodies in the CNS after poliovirus infection has been reported. Serum antibodies also may reach the CNS by crossing the blood–brain barrier due to breakdown of the integrity of the meninges. Secretory IgA is induced after 2 to 4 weeks after poliovirus infection and located mainly in nasopharyngeal and gut tissues. Secretory IgA prevents or limits the excretion of polioviruses in the alimentary tract.

21.8 VACCINATION

Of all enteroviruses only infections with polioviruses can presently be prevented by vaccination. Prerequisite for developing anti-poliomyelitis vaccines was the introduction of tissue culture for the production of virus in large quantity (Enders et al., 1949). Two vaccines have led to a dramatic decrease of poliomyelitis (from approximately 14 paralytic cases in 1952 to less than 0.01 cases in 1969 per 100,000 population (Figure 21.3). The first vaccine was a formalin-inactivated vaccine developed by Salk and coworkers and first licensed in 1954 in the U.S. (Salk vaccine). This vaccine, administered intramuscularly, prevents poliovirus infections by inducing humoral neutralizing antibodies. At the same time, live attenuated vaccines were developed by Sabin, Koprowski, Melnick, Cox, and others. This led to introduction of a live vaccine that is orally administered and consists of all three serotypes of poliovirus (Sabin vaccine). The viruses multiply in the gastrointestinal tract and give rise to a subclinical infection. The Sabin vaccine induces humoral IgG as does the Salk vaccine and additionally elicits secretory IgA in the gut. Therefore, Sabin vaccination prevents not only virus spread through the blood stream to the CNS, but also blocks primary virus multiplication in the gut. There is a possibility that one of the types of the trivalent vaccine will not multiply due to interference from one of the other types or from other enteroviruses during mixed infections. For this reason the vaccine is administered three times several weeks apart, in the north temperate zone, preferably not during the summer months when there is high frequency of other enterovirus infections (see 21.4.2). The live Sabin vaccine bears a small risk of developing clinical poliomyelitis including paralysis. Based on data from the U.S., there is a risk of one case of poliomyelitis in vaccines per 8 million doses of the live vaccine administered. A critical evaluation of both vaccines is given by Melnick (1982) and Moore and Morens (1984).

Presently, several laboratories are investigating alternative safe and efficacious vaccines against poliovirus. These vaccines consist of peptides of the immunodominant viral capsid protein VP1 (see 21.3.1. and 21.3.2), either expressed in bacteria by genetic engineering techniques or chemically synthesized (for reviews and references see Brown, 1983; Newmark, 1983). Such vaccines may be expected in a few years.

Prophylaxis with immune serum globulin is usually recommended when the epidemiologic conditions are well known. Immune globulins against polioviruses or HAV from sera of convalescent patients are effective in preventing infections, however, they have to be administered within the first days after exposure.

So far enterovirus infection can be prevented only by prophylactic active or passive immunization and by interruption of the routes of virus transmission. Viral chemotherapeutic agents potent in cell culture systems cannot be utilized for treatment of humans (for review, see Eggers, 1982).

21.9 LABORATORY DIAGNOSIS

Methods used for laboratory diagnosis of enterovirus infections have been summarized by Melnick et al. (1979).

21.9.1 Virus Isolation and Identification

Specimens for virus isolation are usually stools and rectal swabs, throat swabs and washings, and cerebrospinal fluid (CSF). The successful isolation of virus from clinical specimens depends on the time postinfection, which relates to the pathogenesis of enterovirus infections. For example, virus is most readily isolated from the throat shortly after infection up to 15 days or more, from stools and rectal swabs up to 4 weeks after infection, and from the CSF during the time of manifestation of symptoms involving the CNS, usually 2 to 3 weeks after infection (Figure 21.4). Isolation from stools is most promising as virus concentrations in feces are higher than in other specimens (up to 10^6–10^9 virus particles per gram of feces). Viruses inducing vesicular rashes like coxsackievirus types A4, A5, A9, A10, A16, and enterovirus type 71 can be isolated also from the lesions. Virus isolation from blood is successful during viremia (days 6 to 9 after infection), however, due to the short period of this phase virus isolation is generally not attempted from this source. Hepatitis A virus can only be isolated from stools from 2 weeks after infection until the beginning of jaundice (4 weeks after infection). In addition all specimens from target organs generally yield virus if a biopsy or autopsy specimen is taken during the clinical manifestation of disease.

Originally, pathologic lesions produced in mice were used for distinction between coxsackieviruses of groups A and B. More recently cell cultures are most commonly used for isolating most of the enteroviruses (Melnick et al., 1979; Moore and Morens, 1984) (Table 21.2). Cells usually used for growth of polioviruses, coxsackievirus types A7, A9, and A16, coxsackieviruses of group B and echoviruses are human embryonic fibroblasts of the skin or lung, permanent human amnion cells, transformed human cell lines, such as HeLa and HEp-2 cells, and primary monkey cell lines, such as rhesus and African green monkey kidney cells. An exception is the growth of several types of coxsackieviruses of group A. Some of these viruses only replicate in a cell line derived from a human rhabdomyosarcoma cell line or only in newborn mice. Coxsackievirus types A1, A19, and A22 remain as types that require newborn mice for their cultivation (Melnick, 1982).

The appearance of the typical cytopathic effect in cell cultures (see 21.3.4) is proof of the presence of virus in a specimen. Neutralization tests are used for identification of the virus isolates. Most commonly, pools of internationally standardized hyperimmune equine antisera are used for typing the isolates. Typing is performed according to a pattern proposed by Lim and Benyesh-Melnick (1960). Neutralization tests are most favorable for this purpose as neutralizing antibodies are type-specific. In contrast, radioimmunoassays, complement fixation tests, and hemagglutination inhibition tests for viruses with hemagglutinins (see 21.9.2) do not allow type specific identification due to immunologic crossreactions (see 21.3.2).

Identification of HAV is performed in radio- or enzyme immunoassays. Due to the limited propagation of hepatitis A virus in cell culture and its protracted replication cycle (several days to 4 weeks) cell culture techniques are unfavorable for routine virus detection (for review, see Purcell et al., 1984).

Direct virus visualization by electron microscopy is performed in specialized laboratories. Virus can be detected in preparations from stools due to high concentration of virus in feces. This is achieved by using the technique of negative staining. Immune electron microscopy allows a direct virus identification. Specimens containing virus are incubated with virus-specific antiserum and the resulting virus–antibody complexes are visualized. Immunofluorescence may be useful for directly typing enteroviruses in specimens obtained at biopsy or autopsy. However, this technique has not been used extensively.

21.9.2 Serologic Diagnosis

Serologic diagnosis in combination with virus identification is most favorable for confirming an infection with enteroviruses. Documentation of a virus specific serologic response (fourfold or greater rise in antibody titer or a single high titer of virus-specific IgM) proves a recent enterovirus infection. Ideally, serum of a patient should be tested in the beginning of the illness and 7 to 10 days later (Figure 21.4). Neutralization tests using tissue culture are most commonly used for serologic diagnosis of infections with polioviruses, coxsackieviruses, and echoviruses, as neutralizing antibodies react with the virus in a type-specific manner (see 21.3.2 and 21.9.1). For those coxsackieviruses of group A not growing in cell culture, neutralization tests in mice can be performed.

The complement fixation test is another serologic procedure used to diagnose enteroviruses. It is of limited value because it allows only detection of group-specific antibodies. Theoretically, hemagglutination inhibition tests can be performed with some hemagglutinating strains of coxsackieviruses of groups A and B, some echoviruses, and enterovirus type 68.

Detection of antibodies against HAV is performed in radio- or enzyme immunoassays. By applying antibodies against the different human immunoglobulins (IgG or IgM), acute hepatitis A infection can be differentiated from a previous infection. Enzyme immunoassays for coxsackieviruses have been reported (Katze and Crowell, 1980), however, these tests have not been introduced into the routine laboratory so far.

ACKNOWLEDGMENT

Prof. Dr. K.-O. Habermehl, Director, Institut für Klinische und Experimentelle Virologie, Freie Universität Berlin, is gratefully acknowledged for his comments and discussion, as well as the help of R. Joncker in preparing the manuscipt.

REFERENCES

Bartsch, H.D., Habermehl, K.O., and Diefenthal, W. 1969. Correlation between poliomyelitisvirus-reproduction-cycle, chromosomal alterations and lysosomal enzymes. Arch. Ges. Virusforsch. 27:115–127.

Beatrice, S.T., Katze, M.G., Zajac, B.A., and Crowell, R.L. 1980. Induction of neutralizing antibodies by the coxsackievirus B3 virion polypeptide, VP2. Virology 104:426–438.

Brown, F. 1983. Neutralizing site of poliovirus. Nature (London) 304:395–396.

Crowell, R.L., and Landau, B.J. 1983. Receptors in the initiation of picornavirus infections. Comp. Virol. 18:1–42.

Dales, S. 1973. Early events in cell-animal virus interactions. Bact. Rev. 37:103–135.

Dales, S., Eggers, H.J., Tamm, I., and Palade, G.E. 1965. Electron microscopic study of the formation of poliovirus. Virology 26:379–389.

Dalldorf, G., and Sickles, G.M. 1948. An unidentified, filtrable agent isolated from the feces of children with paralysis. Science 108:61–63.

Diefenthal, W., and Habermehl, K.O. 1967. Die Bedeutung mikrokinematographischer Methoden in der Virologie. Res. Film 6:22–30.

Diefenthal, W., Habermehl, K.O., Lorenz, P.R., and Beneke, T. 1973. Virus-induced inhibition of host cell synthesis. Adv. Biosciences 11:127–148.

Dimmock, N.J. 1984. Mechanisms of neutralization of animal viruses. J. Gen. Virol. 65:1015–1022.

Eggers, H.J. 1982. Benzimidazoles. Selective inhibitors of picornavirus replication in cell culture and in the organism. In P.E. Came, and L.A. Caliguiri (eds.), Handbook of Experimental Pharmacology 61. Berlin: Springer Verlag, pp. 377–417.

Emini, E.A., Kao, S.Y., Lewis, A.J., Crainic, R., and Wimmer, E. 1983. The functional basis of poliovirus neutralization determined with monospecific neutralizing antibodies. J. Virol. 46:466–474.

Enders, J.F., Weller, T.H., and Robbins, F.C. 1949. Cultivation of the Lansing strain of poliomyelitis virus in cultures of various human embryonic tissue. Science 109:85–87.

Frösner, G. 1984. Hepatitis A virus. In R.B. Belshe (ed.), Textbook of Human Virology. Littleton, MA: PSG Publishing Co., pp. 707–727.

Ginsberg, H.S. 1980. Picornaviruses. In R. Dulbecco and H.S. Ginsberg (eds.), Microbiology. Hayerstown, MD: Harper & Row, pp. 1095–1117.

Habermehl, K.O., Diefenthal, R., and Diefenthal, W. 1966. Der Einfluβ von Virusinfektionen auf den Ablauf der Zellteilung. Zbl. Bakt. Hyg., I. Orig. 199:273–314.

Habermehl, K.O., Diefenthal, W., and Buchholz, M. 1973. Distribution of parental viral constituents in the course of polioinfection. Adv. Biosciences 11:41–64.

Hughes, J.V., Stanton, L.W., Tomassini, J.E., Long, W.J., and Scolnick, E.M. 1984. Neutralizing monoclonal antibodies to hepatitis A virus: Partial localization of neutralizing antigenic site. J. Virol. 52:465–473.

Katze, M.G., and Crowell, R.L. 1980. Immunological studies of the group B coxsackieviruses by the sandwich enzyme-linked immunosorbent assay (ELISA) and immunoprecipitation. J. Gen. Virol. 50:357–367.

Kitamura, N., Semler, B.L., Rothberg, P.G., Larson, G.L., Adler, C.J., Dorner, A.J., Emini, E.A., Hanecak, R., Lee, J.J., van der Werf, S., Anderson, C.W., and Wimmer, E. 1981. Primary structure, gene organization and polypeptide expression of poliovirus RNA. Nature (London) 291:547–553.

Koch, F., and Koch, G. 1985. The molecular biology of poliovirus. New York: Springer Verlag.

Krugman, S., Giles, J.P., and Hammond, J. 1967. Infectious hepatitis: Evidence for two distinctive clinical, epidemiological and immunological types of infection. J. Am. Med. Assoc. 200:365–373.

Landsteiner, K., and Popper, E. 1909. Übertragung der Poliomyelitis acuta auf Affen. Z. Immunitaetsforsch. Orig. 2:377–390.

Lim, K.A., and Benyesh-Melnick, M. 1960. Typing of viruses by combinations of antiserum pools: Application to typing of enteroviruses (Coxsackie and echo). J. Immunol. 84:309–317.

Linemeyer, D.L., Menke, J.G., Martin-Gallardo, A., Hughes, J.V., Young, A., and Mitra, S.W. 1985. Molecular cloning and partial sequencing of hepatitis A viral cDNA. J. Virol. 54:247–255.

Matthews, R.E.F. 1982. Classification and Nomenclature of Viruses. Basel: Karger Medical Scientific Publishers, pp. 129–132.

Melnick, J.L. 1982. Enteroviruses. In A.S. Evans (ed.), Viral Infections of Humans, Epidemiology and Control. New York; Plenum Medical Books, pp. 187–251.

Melnick, J.L., Shaw, E.W., and Curnen, E.C. 1949. A virus isolated from patients diagnosed as nonparalytic poliomyelitis or aseptic meningitis. Proc. Soc. Exp. Biol. Med. 71:344–349.

Melnick, J.L., Wenner, H.A., and Phillips, C.A. 1979. Enteroviruses. In E.H. Lennette and N.J. Schmidt (eds.), Diagnostic Procedures for Viral, Rickettsial and Chlamydial Infections. Washington, D.C.: American Public Health Association, pp. 471–534.

Mertens, T., Pika, U., and Eggers, H.J. 1983. Cross antigenicity among enteroviruses as revealed by immunoblot technique. Virology 129:431–442.

Minor, P.D., Schild, G.C., Bootman, J., Evans, D.M.A., Ferguson, M., Reeve, P., Spitz, M., Stanway, G., Cann, A.J., Hauptmann, R., Clarke, L.D., Mountford, R.C., and Almond, J.W. 1983. Location and primary structure of the antigenic site for poliovirus neutralization. Nature (London) 301:674–679.

Moore, M., and Morens, D. 1984. Enteroviruses, including polioviruses. In R.B. Belshe (ed.), Textbook of Human Virology. Littleton, MA: PSG Publishing Company, pp. 407–483.

Newmark, P. 1983. Will peptides make vaccines? Nature (London) 305:9.

Purcell, R.H., Feinstone, S.M., Ticehurst, J.R., Daemer, R.J., and Baroudy, B.M. 1984. Hepatitis A virus. In G.N. Vyas, J.L. Dienstag, and H.J. Hoofnagle (eds.), Viral Hepatitis and Liver Disease. Orlando, FL: Grune & Stratton, pp. 9–22.

Putnak, J.R., and Phillips, B.A. 1981. Picornaviral structure and assembly. Microbiol. Rev. 45:287–315.

Robbins, F.C., Enders, J.F., Weller, T.H., and Florentino, G.L. 1951. Studies on the cultivation of poliomyelitis viruses in tissue culture. V. The direct isolation and serologic identification of virus strains in tissue culture from patients with nonparalytic and paralytic poliomyelitis. Am. J. Hyg. 54:286–293.

Rueckert, R.R. 1985. Picornaviruses and their replication. In B.N. Fields (ed.), Virology. New York, Raven Press, pp. 705–738.

Sangar, D.V. 1979. The replication of picornaviruses. J. Gen. Virol. 45:1–13.

Wetz, K., and Habermehl, K.O. 1979. Topographical studies on poliovirus capsid proteins by chemical modification and crosslinking with bifunctional reagents. J. Gen. Virol. 44:525–534.

Young, N.A. 1973. Polioviruses, coxsackieviruses, and echoviruses: Comparison of the genomes by RNA hybridization. J. Virol. 11:832–839.

Zeichhardt, H., Schlehofer, J.R., Wetz, K., Hampl, H., and Habermehl, K.O. 1982. Mouse Elberfeld (ME) virus determines the cell surface alterations when mixedly infecting poliovirus-infected cells. J. Gen. Virol. 58:417–428.

Zeichhardt, H., Wetz, K., Willingmann, P., and Habermehl, K.O. 1985. Entry of poliovirus type 1 and Mouse Elberfeld (ME) virus into HEp-2 cells: Receptor-mediated endocytosis and endosomal or lysosomal uncoating. J. Gen. Virol. 66:483–492.

22

Rotaviruses and Norwalk Viruses

Larry K. Pickering and Bruce H. Keswick

22.1 INTRODUCTION

Diarrheal diseases are among the leading causes of morbidity and mortality worldwide. Rotaviruses and Norwalk viruses are important etiologic agents of human gastroenteritis. They can produce illness in both endemic and epidemic forms and affect all age groups. Establishment of the roles of other viruses, including enteric adenoviruses, caliciviruses, astroviruses, and coronaviruses in human disease will require further research.

22.2 ROTAVIRUSES

22.2.1 Description of Characteristics

22.2.1.a Overview

Rotaviruses belong to a separate genus of the family *Reoviridae* (Joklik, 1981), and possess a genome consisting of 11 segments of double-stranded RNA. The virus is 70 nm in diameter and has an inner and outer capsid but no envelope. The name of the virus is derived from the Latin word "rota" meaning wheel, which it resembles in appearance. Human rotavirus initially was detected by Bishop et al. in 1973 in Australia by thin section electron microscopic (EM) examination of duodenal biopsies from children with acute diarrhea. Subsequently, the virus has been observed by EM in diarrheal stool specimens from people living in various parts of the world. Initially the virus was referred to by several names, including duovirus, infantile gastroenteritis virus, orbivirus, and reovirus-like agent. Rotaviruses from humans are morphologically similar to and share certain common antigens with animal rotaviruses. Rotaviruses are a major cause of disease in humans and in other mammalian and avian hosts.

22.2.1.b Nomenclature

A uniform nomenclature of human rotaviruses has been proposed (Kapikian et al., 1981), now that human rotavirus serotyping has become possible due to the ability to successfully cultivate most human rotaviruses. Rotaviruses are placed into subgroups due to the presence of key capsid antigens and are designated by Roman numerals; rotavirus serotypes are designated by Arabic numerals (Table 22.1). It has been proposed that the term serotype be reserved to identify the antigen that reacts with neutralizing antibodies, as is customary for other viruses, and that the term subgroup

Supported by grant HD-13021 from The National Institutes of Health.

Table 22.1. Suggested Nomenclature of Rotaviruses

Group	Representative Human Rotavirus Strains	Serotype	Subgroup
A (human)	DS-1, S2 or KUN	2	I
	WA, D, K8 or KU	1	II
	M, P, ITO NEMOTO or YO	3	II
	St. Thomas 4, HOCHI, or HOSOKAWA	4	II

be used to refer to type specificity previously established by complement fixation, immunoelectron microscopy (IEM), enzyme immunoassay (EIA), and immune adherence hemagglutination (Kapikian, 1981). Serotyping is currently performed in cell culture. Subgroup specificity does not indicate neutralization specificity, which is a function of another gene product(s) (Kalica et al., 1981; Greenberg et al., 1983), although subgroup specificity is a measure of antigenic diversity. A variety of agents morphologically identical to but antigenically distinct from the typical (group A) rotaviruses have been implicated as a cause of diarrhea in animals, including humans (Eiden et al., 1985; Vonderfecht et al., 1985).

22.2.3 Pathogenesis

22.2.3.a Incidence

The incidence of rotavirus infection has been determined in hospital and community based studies, generally in children. Rotavirus infection accounts for about 15–50% of acute diarrhea cases presenting to hospitals in tropical countries and about 35–60% in temperate areas (Kapikian et al., 1976; Rodriguez et al., 1977; Brandt et al., 1983). The wide ranges reflect differences in age groups studied, methods of detection, geographic locations, and time. In community based studies, rotavirus rates are 5–10% of diarrheal episodes during the first year of life. The higher prevalence in hospital based studies may reflect more severe disease caused by rotavirus than by other agents.

22.2.3.b Pathophysiology

The mechanisms of disease production in patients with diarrhea due to rotavirus include decreased absorption of salt and water related to selective infection of the absorptive intestinal villus cells. This results in net fluid secretion (Cukor et al., 1984). An additional mechanism appears to be poor glucose transport and diminished dissacharidase activity resulting in carbohydrate malabsorption with osmotic diarrhea due to the presence of undigested carbohydrates in the gastrointestinal tract (Davidson et al., 1977; Hyams et al., 1981). The metabolic acidosis that occurs is caused by bacterial interaction on these malabsorbed carbohydrates (Sack et al., 1982).

22.2.3.c Clinical

General. Rotavirus has a mean incubation period of 2 days with a range of 1 to 3 days in children, as well as in experimentally infected adults (Davidson et al., 1975; Kapikian et al., 1983). Excretion of rotavirus in stool can precede the onset of illness and can continue after symptoms of illness have abated. The usual duration of fecal excretion is 8 to 10 days with the quantity of virus per gram of stool being highest in titer shortly after the onset of illness and gradually declining until day 9 or 10 (Konno et al., 1977; Nagayoshi et al., 1980; Vesikari et al., 1981). The disease produced by rotavirus is usually self-limited but symptomatic relapses have been reported (Tallett et al., 1977). The mean duration of illness is 5 to 7 days (Flewett et al., 1975; Konno et al., 1977). Chronic infections with rotavirus can occur in immunodeficient children (Saulsbury et al., 1980).

Rotavirus diarrhea occurs in all age-groups. Symptoms of illness produced by rotavirus vary according to age and may be divergent within an age-group (Table 22.2). Newborn infants most often are asymptomatic and, when diarrhea occurs, it tends to be mild (Chrystie et al., 1978; Bryden et al., 1982; Truant et al., 1982; Rodriguez et al., 1982; Bishop et al., 1983). Severe illnesses have been reported in newborns, particularly premature infants (Dearlove et al., 1983). Findings in Australia indicate that neonatal (first month of life) rotavirus infections may protect infants and children from rotavirus gastroenteritis after the neonatal period (Bishop et al., 1983). The rate of postneonatal rotavirus infection was the same in infants who did and did not experience neonatal rotavirus infection, but symptoms were significantly less frequent and less severe in infants who had had neonatal infection.

In infants and young children rotavirus disease has an abrupt onset characterized by explosive, watery diarrhea and is often associated with vomiting either before or after onset of diarrheal disease. In most countries the peak incidence of rotavirus in-

Table 22.2. Clinical Characteristics of 150 Children Hospitalized with Acute Gastroenteritis

	Percentage Having Each Clinical Finding	
Clinical Finding	Rotavirus Infection Detected (72 patients)	Rotavirus Infection Not Detected (78 patients)
Vomiting	96	58
Fever (°C)		
37.9–39.0	46	29
39	31	33
Total	77	61
Dehydration	83	40
Hypertonic	5	17
Isotonic	95	77
Hypotonic	0	6
Irritability	47	40
Lethargy	36	27
Pharyngeal erythema	49	32
Tonsillar exudate	3	3
Rhinitis	26	22
Red tympanic membrane with loss of landmarks	19	9
Rhonchi or wheezing	8	8
Palpable cervical lymph nodes	18	9

Rodriguez WJ, et al. J. Pediatr 91:188, 1977.

fection occurs in children 6 to 24 months of age (Kapikian et al., 1976; Brandt et al., 1983). Other common clinical features include isotonic dehydration and compensated metabolic acidosis. Dehydration occurs in 40–80% of patients and is usually less than 5%. However, severe dehydration and death have been reported in children (Rodriguez et al., 1977; Hieber et al., 1978; Carlson et al., 1978) and adults (Marrie et al., 1982; Echeverria et al., 1983). Adults typically tend to have asymptomatic or mild disease (Wenman et al., 1979). Rotavirus infection often occurs in adults who are in close contact with young children (Kapikian et al., 1976; Kim et al., 1977; Pickering et al., 1981; Grimwood et al., 1983), and also has been reported in adult travelers, military personnel, elderly who are in institutions, and hospitalized adults (Von-Bonsdorff et al., 1978; Vollet et al., 1979; Halvorsrud et al., 1980; Holzel et al., 1980; Ryder et al., 1981; Marrie et al., 1982). In 18 adult volunteers challenged orally with human rotavirus (HRV), four developed illness and 12 had no illness but developed serologic evidence of infection (Kapikian et al., 1983). In a prospective family study (Wenman et al., 1979) the attack rate in children was 32% and in adults it was 17%. Seventy percent of infected children and 40% of infected adults were symptomatic.

A study by Champsaur et al. (1984) showed that rotavirus shedding was found in 43 (36%) of 119 children with diarrhea and 40 (24%) of 164 children without diarrhea. Virus shedding that was not associated with diarrhea was observed in 71% of neonates, in 50% of 1- to 6-month-old children and in 26% of 7- to 24-month-old children. Shedding of rotavirus by children beyond the neonatal period has been reported to range 0–50% (Konno et al., 1977; Pickering et al., 1978; Brandt et al., 1979; Gurwith et al., 1981; Soenarto et al., 1981; Engleberg et al., 1982; Walther et al., 1983).

Stools are usually watery in patients with rotavirus diarrhea, and do not usually contain blood or fecal leukocytes; mucous is occasionally seen (Pickering et al., 1977; Rodgriguez et al., 1977; Hieber et al., 1978). The finding of pale, fat-containing stools has been reported (Konno et al., 1977; Thomas et al., 1981) suggesting that rotavirus infection can impair the digestion of fat and alter the color of feces. Concurrent respiratory tract symptoms, pharyngitis, and otitis media are frequently associated, but may not be characteristic of rotavirus infection (Rodriguez et al., 1977; Lewis et al., 1979; Goldwater et al., 1979). Rotavirus has not been demonstrated to infect the upper respiratory tract (Lewis et al., 1979; Kapikian et al., 1983), but has been demonstrated in respiratory tract secretions obtained from four patients who were hospitalized with pneumonia (Santosham et al., 1983). The association of rotavirus with intussusception has been made (Konno et al., 1978) but not validated in prospective studies (Mulcahy et al., 1982; Nicolas et al., 1982). Rotavirus has been associated with but not proven to be the cause of several diseases or conditions including Reyes syndrome and encephalitis (Salmi et al., 1978), aseptic meningitis (Wong et al., 1984), sudden infant death syndrome (Yolken et al., 1982), inflammatory bowel disease (Blacklow et al., 1982), neonatal necrotizing enterocolitis (Rotbart et al., 1983), and Kawasaki's syndrome (Matsuno et al., 1983). Comparison of symptoms in children with rotavirus diarrhea with non-rotavirus diarrhea showed greater frequency of watery diarrhea, nausea, vomiting, abdominal pain, dehydration, and loss of appetite in the rotavirus group (Rodriguez et al., 1977), but such a great deal of overlap occurs that symptoms alone cannot be used to differentiate the cause of an episode of diarrhea.

Fluid replacement with oral glucose–electrolyte solutions or by parenteral fluid in severe cases is an important supportive measure (Black et al., 1981; Santosham et al., 1982). Use of γ globulin-containing antibodies to rotavirus has been found to delay excretion of rotavirus and reduce the duration and quantity of virus excretion in low birth-weight infants (Barnes et al., 1982).

Season. Many reports describe a higher occurrence of rotavirus diarrhea in winter in temperate climates (Kapikian et al., 1976; Brandt et al., 1983; Gurwith et al., 1983; Konno et al., 1983) and a lack of seasonality in tropical climates (de Torres et al., 1978; Suzuki et al., 1981). For unknown reasons, however, seasonal discrepancies in the occurrence of rotavirus diarrhea occur in different geographic locations. Low indoor relative humidity has been proposed as important in enhancement of transmission of rotavirus (Brandt et al., 1982). Human rotavirus detection has been reported to increase in several countries during the dry season (Maiya et al., 1977; Hieber et al., 1978; Black et al., 1982; Paul et al., 1982), but relative humidity was found not to be important in Japan (Konno et al., 1983).

Nutrition. Several investigators have reported an increased severity of diarrhea, including diarrhea due to rotavirus in undernourished children. There is no agreement on the impact of poor nutritional status on the occurrence of subsequent diarrhea episodes due to rotavirus.

Mixed infection. Simultaneous infection of children by rotavirus and other viral, bacterial, and parasitic enteropathogens has been documented, particularly in developing countries (Rodriguez et al., 1977; Pickering et al., 1978; Black et al., 1980; Brandt et al., 1983). These mixed infections have not been of greater duration than infections due to rotavirus alone.

Transmission. Rotaviruses are most commonly transmitted from person to person by the fecal–oral route. Several studies have reported a rapid person to person spread of rotavirus infection, which in the absence of a common source (e.g., contaminated water supply) permits speculation that a respiratory route of spread may have occurred in these cases (Foster et al., 1980; Gurwith et al., 1981). Transmission within families, day care centers, and in the hospital environment has been reported (Pickering et al., 1981; Rodriguez et al., 1982; Grimwood et al., 1983). Nosocomial spread of rotavirus may be facilitated in day care centers by contact with the contaminated hands of uninfected attendants (Samadi et al., 1983) or by asymptomatic or mildly ill infected hospital staff, or parents (Flewett, 1983). Nosocomial outbreaks in hospitalized patients may be controlled by closing the affected ward and restricting staff movement for up to 10 days (Flewett, 1983). Increased family size is reportedly a significant risk factor for acquiring rotavirus infection (Gurwith et al., 1983). Increased indoor crowding may facilitate an airborne mode of transmission, which may not be only respiratory but also due to environmental contamination by fecally shed rotavirus (Gurwith et al., 1983; Keswick et al., 1983). Water borne outbreaks also have been reported (Lycke et al., 1978; Sutmoller et al., 1982; Hopkins et al., 1984; Tao et al., 1984) but common source outbreaks of infection do not occur frequently. Importance of the following factors in transmission of rotavirus is unknown: temperature, relative humidity, domestic animals, food, water supply, families, and sanitary conditions. Excretion of rotaviruses by asymptomatic persons may be important in the transmission of this pathogen.

Sequential Infection. Sequential postneonatal rotavirus infection of individual children by different subgroups or serotypes occurs; however, the actual frequency is unknown (Fonteyne et al., 1978; Rodriguez et al., 1978; Yolken et al., 1978; Bishop et al., 1983). Information is needed regarding the severity of symptoms in primary HRV infections and reinfections by both homotypic and heterotypic serotypes.

22.2.4 Isolation Sites

Rotaviruses may be shed in stools in high numbers, up to 10^{11} particles per gram of feces. Thus, stools are the most important specimens to collect for detection of the virus. Stools collected on days 3 to 5 following the onset of illness are most likely to contain virus, whereas, stools collected 8 days after the onset of illness rarely contain virus (Cukor et al., 1984; Vesikari et al., 1981). Stool specimens should be examined immediately or frozen at −70°C until examined. For some methods of identification, such as ELISA, rectal swabs may be useful; however, for direct isolation in cell culture, a rectal swab sample may not be useful due to insufficient material (Wyatt et al., 1983). Rotavirus has been detected in respiratory tract secretions from children with pneumonia (Santosham et al., 1983) but rotavirus has not been found in respiratory tract secretions of children with diarrhea.

22.2.5 Stability

The stability of rotaviruses has been studied using simian SA-11 rotavirus as a model of HRV. The SA-11 virus is stable over the pH range of 3.5–10.0, but infectivity may be lost by treatment with low concentrations (5mM) of EDTA (ethylenediamine tetraacetic acid) or by heating at 50°C for 15 minutes in 2 *M* $MgCl_2$ (Estes et al., 1979). Rotavirus survival on environmental surfaces is enhanced by the presence of fecal material (Keswick et al., 1983). The presence of fecal material also protects the viruses from the action of disinfectants. An ethanol containing disinfectant (Desderman) was reported effective (Brade et al., 1981) against SA-11 rotavirus, whereas, other skin disinfectants, such as chlorhexidine gluconate and providone-iodine were found to be ineffective (Tan et al., 1981). Lysol, formalin, and an iodophor preparation also may be effective. Chlorine has been shown to rapidly inactivate SA-11 in water. There is evidence that human strains of rotavirus may be more resistant to the action of chlorine than animal strains (Butler et al., 1983). Rotaviruses have been found in sewage and treated drinking water where the presence of solids (fecal material) may enhance their survival (Smith et al., 1982; Deetz et al., 1984; Keswick et al., 1984). Likewise rotavirus has been detected on surfaces in day care centers (Keswick et al., 1983) and undoubtedly will be found in the hospital environment.

22.2.6 Isolation Methods

Due to their fastidious nature, HRVs were difficult to cultivate from clinical specimens until 1980, when a single strain of HRV, Wa, was successfully adapted to grow on serial passage in primary African green monkey kidney (AGMK) cell culture after 11 prior serial passages in newborn gnotobiotic piglets (Wyatt et al., 1980). Circumvention of the fastidious nature of HRV was accomplished by genetic reassortment during mixed infection with a temperature-sensitive mutant of a cultivatable bovine rotavirus. The HRV gene responsible for growth restriction in vitro (gene 4, in 88K outer capsid protein) was replaced by the corresponding gene from the tissue culture adapted, bovine rotavirus (Greenberg et al., 1981; Greenberg et al., 1983). Subsequently, direct isolation of numerous HRV without serial passage in animals has been successful (Hasegawa et al., 1982; Kutsuzawa et al., 1982; Birch et al., 1983). Isolates were recovered in roller tube cultures of MA104 cells by treating stool specimens with trypsin before inoculation and by incorporating a small amount of trypsin in the maintenance medium. Successful cultivation of six of six HRV strains in cynomolgus monkey kidney (CMK) was reported by Hasegawa et al. in 1982. Growth in the CMK cells was detected with immune adherence hemagglutination on second passage, whereas, only one of these six strains grew to limited extent in MA104 cells. Five of the six strains were adapted to efficient growth in CMK cells and none grew serially in MA104 cells. In a study of various cells, AGMK cells were more sensitive than

CMK and MA104 cells for supporting growth of HRV isolated from diarrheal stools of infants and young children (Wyatt et al., 1983). Although MA104 cells have been used successfully for isolation of many strains of HRV (Wyatt et al., 1983), MA104 and primary CMK cells appear to be less sensitive than primary AGMK cells (Naguib et al., 1984) (Table 22.3). Preliminary characterization of naturally occurring temperature-sensitive bovine rotavirus mutants generated via gene reassortment, indicated that the product of genome segment 10 can influence virus absorption to MA104 cells (Sabara et al., 1984).

Successful isolation of rotaviruses from clinical samples requires that virus containing samples be pretreated with trypsin and inoculated into roller tube cultures, which are maintained in trypsin supplemented medium (Urasawa et al., 1982; Wyatt et al., 1983). A variety of cell lines have been studied for this purpose including MDBK, PK-15, BSC-1, LLC-MK2, BGM, CV-1, and MA104. In addition, primary cells from a variety of species are widely used. Primary AGMK cells and the MA104 cell line have proven to be the most useful. In particular, MA-104 cells have been used for rotavirus isolation, passage, and plaque formation (Estes et al., 1983).

An approximately 10% suspension of feces in phosphate buffered saline (PBS) or fluid in which a rectal swab had been vigorously agitated are centrifuged at 300 × *g* for 30 minutes and pretreated with 10 μg trypsin before inoculation of cell cultures. Roller tubes receive 0.1 ml of the fecal suspension and are incubated for 60 minutes at 37°C. The cells are then washed once and fed with Eagles minimum essential medium (EMEM) containing 0.5 μg trypsin/ml (Wyatt et al., 1983). DEAE dextran (100 μg/ml) can be added to enhance infectivity in MA104 cells, but may prove toxic to primary cells (Wyatt et al., 1983). Cells are incubated for up to 10 days at 37°C and observed daily for cytopathic effect (CPE). Cultures should then be frozen for further identification by fluorescence assay or enzyme immunoassay. Wyatt et al. (1983) found that 39 of 73 rotavirus-positive fecal specimens could be cultured in MA104 or primary AGMK cells using this method. It should be noted, however, that none of 23 rectal swab samples collected and tested yielded a cultivatable rotavirus, thus, stool specimens should be collected for rotavirus isolation.

The recent development of a plaque assay for HRV has provided an opportunity to study immunity to specific serotypes, viral replication, and genetic variation (Urasawa et al., 1982; Wyatt et al., 1983; Estes et al., 1983). Virus samples are incubated with an equal volume of trypsin solution (20 μg/ml) at 38°C for up to 90 minutes. Virus dilutions then are made in L-15 medium

Table 22.3. Cultivation in Different Cell Types and Subgroup Determination by Enzyme Immunoassay of Rotavirus Recovered from Children with Diarrhea

Cell type	Number of Rotavirus-Positive Specimens Inoculated	Number of Specimens Positive at Indicated Passage Level (percentage)		Results of Subgroup (SG) Assay Determination		
		First	Second	SG 1	SG 2	Undetermined
MA104	12	11 (92)	4 (33)[a]	4	3	8
Primary CMK	11	9 (82)	5 (45)[b]	2	10	2
Primary AGMK	12	11 (92)	10 (83)[a,b]	14	38	5

[a] Number of cultivatable versus noncultivatable strains in AGMK compared with MA104 (Fisher exact test, $p < 0.05$).
[b] Number of cultivatable versus noncultivatable strains in AGMK compared with CMK (Fisher exact test, $p < 0.05$).
Adapted from Naguib et al. J. Clin. Microbiol. 19:210–212, 1984.

containing 0.5% gelatin, glutamine, and antibiotics (L-15/gel). One milliliter of each sample is inoculated onto cell monolayers, which have been washed three times with L-15/gel. The virus is allowed to adsorb for 1 hour at 37°C, then the monolayer is washed once with L-15/gel. The overlay medium consists of EMEM with Earle's salts, 0.9% agarose, glutamine, antibiotics, and trypsin (0.5 μg/ml). Cultures are incubated at 37°C. A second overlay is added 4 to 5 days later, which includes EMEM, agarose, and neutral red (0.067 mg/ml). Plaques are counted 1 to 2 days later and the plates may be fixed with formalin and stained with crystal violet (Wyatt et al., 1982).

A radioimmunofocus assay for quantitation of cell culture-adapted HRV has been developed (Liu et al., 1984). The method appears to be more sensitive than the plaque assay and useful to detect and quantify strains of rotavirus that do not produce plaques, which may be up to 34% of isolates (Wyatt et al., 1983). This technique may be a useful means of serotyping cell culture adapted strains of rotavirus.

22.2.7 Identification

22.2.7.a General

The ability to study the epidemiology of rotaviruses and to identify characteristics of individual strains depends on availability of rapid, practical detection assays. ELISA testing for rotavirus antigen in stool, using standardized reagents, has proved useful and applicable in field settings. Testing for rotavirus serotypes and for serum neutralizing antibody require more sophisticated laboratory techniques. Several methods of identification of rotavirus or their antigens in fecal extracts have been developed and include EM (Flewett et al., 1973); IEM (Kapikian et al., 1974); immunoassays using antibodies labeled with radioisotopes (Kalica et al., 1977), enzymes (Yolken et al., 1977), lectins (Prevot et al., 1981); agglutination techniques, reversed passive hemagglutination (Sanekata et al., 1979), immune adherence hemagglutination (Matsuno et al., 1978; Kapikian et al., 1981), staphylococcal coagglutination (Skaug et al., 1983), latex agglutination (Hughes et al., 1984), agglutination of antibody coated erythrocytes (Bradburne et al., 1979; Sanekata et al., 1983), counter immunoelectrophoresis (CIE) (Middleton et al., 1976; Tufvesson et al., 1976), complement fixation (Kapikian et al., 1975), immunofluorescence in cell culture (Banatvala et al., 1975), fluorescent virus precipitation (Peterson et al., 1976), polyacrylamide gels to detect rotaviral nucleic acid (Herring et al., 1982).

The CIE, complement fixation, immune adherence hemagglutination, and other agglutination techniques (Haikala et al., 1983; Sanekata et al., 1983; Skaug et al., 1983) generally are less sensitive than EIA. A direct monoclonal antibody radioimmunoassay for detection of rotavirus antigen has been demonstrated to be rapid, sensitive, and specific. There was concordance with the Rotazyme test in 96% of the 177 specimens tested (Cukor et al., 1984). Each assay detection method will be discussed in the following sections.

22.2.7.b Electron Microscopy

Electron microscopy was the original diagnostic technique for identification of rotavirus; it is now the reference method for identification of rotaviruses. It can be used to detect rotaviruses in stool specimens because of their characteristic 70-nm size and morphology (Figure 22.1). Because rotaviruses from different species and serotypes cannot be distinguished visually, the technique is limited. It is useful because other viruses (astro, calici, corona, and adeno) also may be detected in stool specimens and dual virus infections can be detected.

22.2.7.c Immunoelectron Microscopy

Immunoelectron microscopy has been employed for detection and identification of ro-

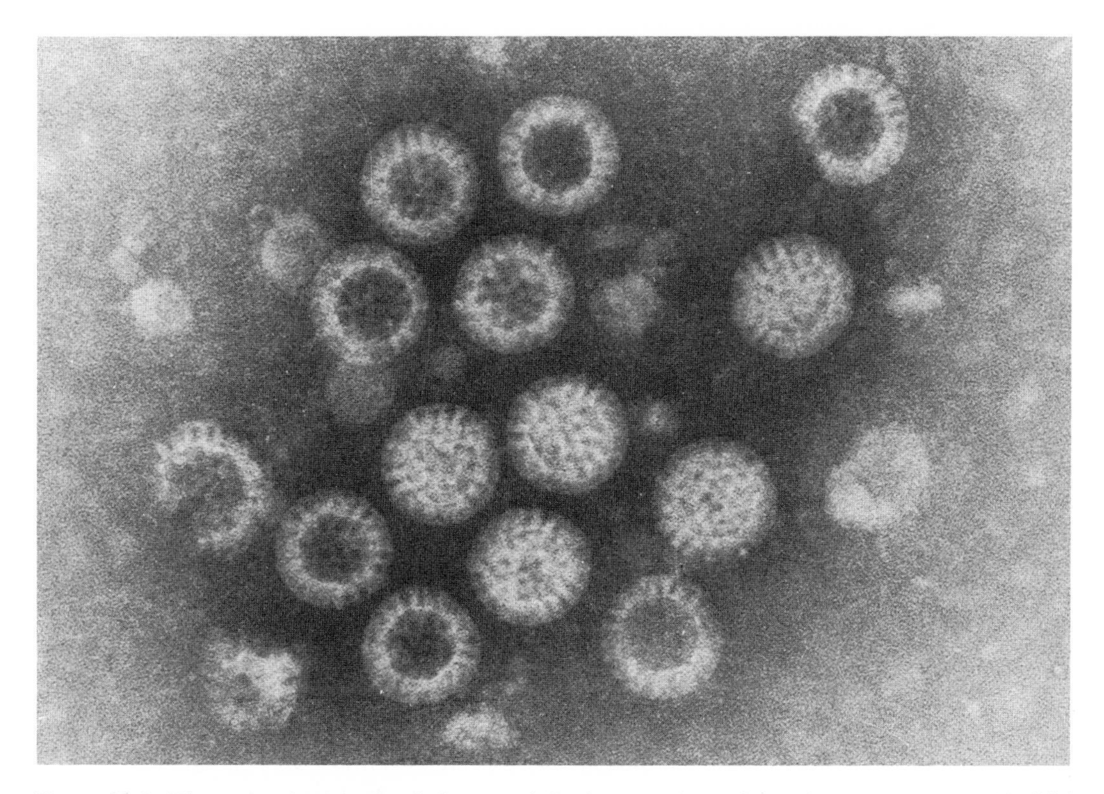

Figure 22.1. Electron micrograph of virus particles in a stool specimen from a two-year-old child with diarrhea (magnified × 238,000).

taviruses and other viruses in feces (Brandt et al., 1979); it is more sensitive than EM, and in some cases may be more useful because some viruses are more readily identifiable by this technique. In this technique a known antiserum is heat inactivated at 56°C for 30 minutes and clarified by ultracentrifugation at 100,000 × *g* for 60 minutes. A 0.1-ml portion of serum is added to 0.5 ml of fecal suspension and incubated at 37°C for 1 to 2 hours or overnight at 4°C. The volume is then made up to 5 ml and the sample is pelleted by ultracentrifugation at 35,000 × *g* for 90 minutes. The supernatant is discarded and the pellet is transferred to coated grids, negatively stained, and examined by conventional techniques. Newer EM assays employ a special ultracentrifuge rotor for grid preparation (Hammond et al., 1981). Solid phase IEM appears to be a rapid and simple technique for typing clinical HRV strains and is a promising test for study of the epidemiology of HRV serotypes (Gerna et al., 1984). Typing results by solid phase IEM appear to be in complete agreement with those obtained by neutralization assay. Rotavirus strains have been subgrouped by ELISA using monoclonal antibodies developed against the 42,000-dalton protein of two rotavirus strains (White et al., 1984).

22.2.7.d ELISA Techniques

A number of enzyme immunoassays have been developed to detect rotavirus in clinical specimens and have been demonstrated to be easy, reliable, and more sensitive than EM (Keswick et al., 1982; Morinet et al., 1984). Currently, several commercially prepared kits are available (Table 22.4). In each of the tests a solid phase bead or well of a microtiter plate is coated with an anti-rotavirus antibody (capture antibody). The

Table 22.4. Commercially Available Diagnostic Kits for Detection of Rotavirus

Name	Source	Solid Phase	Confirmatory Step Included
Rotazyme®	Abbott Laboratories, North Chicago, IL.	Bead-tube	No
Bio-EnzaBead®	Litton Bionetics, Charleston, S.C.	Bead-microtiter plate	Yes
Enzygnost-Rotavirus®	Calbiochem-Behring, LaJolla, CA.	Microtiter plate	No
Pathfinder®	Kallestad Laboratories, Austin, TX.	Microtiter plate	Yes
Rotalex®	Medical Technology Corporation, Somerset, N.J.	Latex agglutination	Yes
Rotavirus immunoassay	International Diagnostic Laboratories, Chesterfield, MO.	Microtiter plate	Yes

fecal suspension is added, incubated, and washed. In most kit forms of the assay a "direct" test is used in which a second antibody directed against rotavirus that has been conjugated to an enzyme (e.g., horseradish peroxidase or alkaline phosphatase) is added. The commercially available Rotazyme® test is an example of a direct assay. It is rapid and depends on the use of anti-simian rotavirus serum. However, it appears to be less sensitive in detecting HRV than most EM or indirect EIA assays. Rotazyme® has been reported to produce many false-positive results in stool specimens collected from neonates (Krause et al., 1983).

A refinement of the test includes using a bead or well for each sample, which has been coated with a preimmune antirotavirus serum to serve as a control and to identify false-positive reactions. The optical density value of the specimen reacted with the preimmune serum (N, negative) is then subtracted from the OD of the sample reacted with the hyperimmune serum (P, positive), yielding a value that can be compared with an established cutoff value for the test, or the sample is often considered positive if the P/N ratio is greater than 2 (Blacklow et al., 1980). Instructions are included with the commercial kits. The bead system offers the advantage of convenience if a few specimens are to be read; a plate offers easier handling and the ease of reading the assay automatically. Some kits combine the features of both. A double determinant assay described by Yolken and Leister (1982) reduces the total time of the ELISA to about 40 minutes. The Rotazyme® and Enzygnost® kits provide sensitive methods of detecting rotavirus; however, the Rotazyme® kit is expensive and Enzygnost® is convenient only when large numbers of specimens must be assayed. Compared with EM, the Rotazyme® assay was reported in two separate studies to have a sensitivity of 98% and 93% and a specificity of 92% and 95%, respectively (Yolken et al., 1981; Rubenstein et al., 1982). Monoclonal antibody based EIA may be more sensitive and predictive than other rotavirus detection systems using polyclonal antibodies (Cukor et al., 1984; Knisley et al., 1986).

The ELISA assay also is used in research applications for serotyping human strains and for detection of antibodies.

These are not currently available to the clinical laboratory in kit form (Zissis and Lambert, 1980).

Solid phase radioimmunoassay (RIA) has been comparable with EM for detection of rotavirus in stool specimens (Kalica et al., 1977; Cukor et al., 1978). The procedure is useful when large numbers of specimens need to be tested and where equipment for counting radioisotopes is available.

22.2.7.e Agglutination Techniques

Several agglutination techniques have been used to detect rotavirus antigen in stool specimens. These tests include reverse passive hemagglutination, immune adherence hemagglutination, staphylococcal coagglutination, latex agglutination, and agglutination of antibody coated erythrocytes.

Immune adherence hemagglutination (IAHA) has been used to type human isolates and may be of use in the clinical laboratory. In this test rotavirus is grown in cell culture and concentrated for use as antigen. Twenty-five μl of sera or virus antigen are added to wells in a round bottom microtiter plate and appropriate dilutions made (Kapikian et al., 1981). After incubation, dithiothreitol solution, guinea pig complement, and human type O red blood cells are added in sequence and incubated. Hemagglutination patterns are scored 0 to 4^+. (Kapikian et al., 1981). The availability of a commercial kit form of this assay will enhance its applicability because it is easy to perform and does not require expensive plate readers. There is good correlation between this test and EM in detecting virus. There is strong evidence (Kapikian et al., 1981; Matsuno et al., 1982) that the IAHA and neutralization typing antigens on rotaviruses are coded for by different genes, therefore IAHA can provide information on the subgroup of the virus, but not the serotype.

A simple latex agglutination (LA) test (Rotalex®), is commercially available. Latex particles are precoated with rabbit anti-Nebraska calf diarrhea virus (NCDV) antibodies and by means of crossreactivity of this antibody with HRV, agglutination becomes macroscopically evident within 1 minute. A 10% or 20% (w/v) fecal extract in PBS or EMEM, clarified by low-speed centrifugation, is prepared. Use of this extract avoids nonspecific agglutination due to solid debris present in stools. Specificity of Rotalex® is excellent; no false-positive reactions were observed in a study by Morinet et al. (1984). Positive tests should be confirmed by neutralization. Although the reproducibility appears good, more studies are needed to draw a definitive conclusion. Rotalex® is rapid and does not require complicated or expensive equipment. Rotalex® appears to be more sensitive than EM (Sanekata et al., 1981; Haikala et al., 1983; Morinet et al., 1984). An LA test using latex beads sensitized with anti-simian–SA-11 IgG detected as few as 9×10^5 rotavirus particles versus 4.5×10^5 particles by the Rotazyme® (Hughes et al., 1984). This LA test does not require expensive equipment, is simple to perform, but is not as sensitive as the Rotazyme® test. It could be used to screen large numbers of stool specimens or for routine diagnostic requirements for laboratories in which only a few specimens are assayed weekly.

A solid phase aggregation-coupled erythrocyte method for virus quantitation is available (Bradburne et al., 1979). Sheep erythrocytes sensitized with antibodies against HRV strain Wa and antibodies against calf rotavirus strain NCDV can detect HRV in stool by hemagglutination of the indicator cells (Sanekata and Okada, 1983). Cells sensitized with antibodies against human strain Wa were more sensitive than cells sensitized with NCDV. This method appears useful for specimens with high background contamination.

22.2.7.f Counterimmunoelectrophoresis

One of the earliest techniques for detection of rotavirus antigen was CIE (Middleton et al., 1976; Tufvesson et al., 1976). Counterimmunoelectrophoresis is less sensitive compared with EIA and EM for rotavirus antigen detection. Using NCDV, Wa, and SA-11 rotavirus antisera, the sensitivity is in the 60–70% range (Hammond et al., 1984). It appears to be specific except for rotavirus detection in newborn infants, a finding similar to that for the Rotazyme® test (Krause et al., 1983; Hammond et al., 1984). The test has the advantage that large numbers of specimens can be examined rapidly.

22.2.7.g Complement Fixation

The complement fixation (CF) test for detection of rotavirus has problems including sensitivity and anticomplementary activity of stool suspensions (Tufvesson et al., 1976; Middleton et al., 1976), although Zissis and Lambert (1978) have had success with a modified test. This modified CF test was as sensitive as EM and IEM for detecting virus. Zissis and Lambert (1978) used the CF test to serotype human rotavirus and found this method to be as sensitive as most ELISA systems, but it appears that this test measures subgroups and not serotypes (Kapikian et al., 1981).

22.2.7.h Immunofluorescence

Immunofluorescent (IF) staining of cell cultures inoculated with fecal preparations permits identification of rotavirus (Barnett et al., 1975; Birch et al., 1979). The sensitivity of the test can be increased by centrifuging inocula at 3000 $\times$ g to enhance virus–cell attachment (Banatvala et al., 1975) or by the use of microtiter plates using LLC-MK2 cells (Bryden et al., 1977). The IF assay appears to be as sensitive as EM but less sensitive than IEM (Morinet et al., 1984). In addition to the EM and IEM methods, IF requires special and costly equipment and is not easy to adapt to the analysis of large numbers of specimens. Rotavirus antigen also can be detected in infected cells or histologic tissue sections using an indirect immunoperoxidase test or a peroxidase–antiperoxidase test (Graham and Estes, 1979). These methods also can be applied to IEM (Altenburg et al., 1979; Petrie et al., 1982).

22.2.7.i Fluorescent Virus Precipitin Test

Virus in fecal specimens may be covered with antibody and detected using a fluorescent virus precipitin test (FVPT). After the viral particles are coated with antibody the resulting virus–antibody particles aggregate and can be detected by immunofluorescence. This method has been found to be as sensitive as EM for detection of bovine and HRV (Foster, 1975; Peterson et al., 1976; Yolken et al., 1977).

22.2.7.j Electropherotyping

The migration pattern of the 11 genome segments of dsRNA of rotavirus following electrophoresis of the viral RNA in polyacrylamide gels is called RNA electropherotyping. The RNA of rotaviruses display a characteristic pattern upon electrophoresis in acrylamide gels (Estes et al., 1984) (Figure 22.2). A recent improvement in the staining technique has increased greatly the sensitivity of this assay (Herring et al., 1982). Although the relationship of electropherotype to serotype and protection is unknown, this technique may be useful in epidemiologic studies particularly in hospital or institutional outbreak investigations (Chanock et al., 1983), because the similarity of isolates can be readily determined.

Electrophoretic separation of the segmented genomes of HRV has been established to be a valuable means of studying the epidemiology of infection caused by rotavirus (Espejo et al., 1980; Laurenco et al., 1981; Rodger et al., 1981; Schnagl et al., 1981; Albert et al., 1983; Konno et al., 1984). Studies have demonstrated extensive

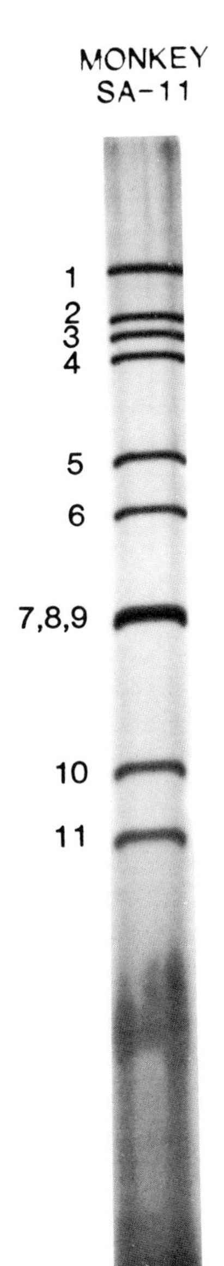

Figure 22.2. RNA electrophoretic patterns of simian (SA-11) rotavirus. Prepared by A. Joanne Bednarz-Prashad, Ph.D.

diversity in the genomic patterns of rotavirus (Rodger et al., 1981; Schnagl et al., 1981), the sequential appearance of rotavirus electropherotypes with a limited number of different RNA patterns at any given time (Rodger et al., 1981), cocirculation of several rotavirus strains with different RNA segment profiles in local outbreaks of infection (Rodriguez et al., 1983), and a change in the prevalence of rotavirus genotypes over several years (Espejo et al., 1980). Differences in electropherotypic patterns, mobility patterns of viral RNA segments, do not necessarily reflect differences in antigenic or biologic properties, and strains showing identical electropherotypes may not have identical RNA segments (Clark et al., 1982). In addition, rotaviruses with similar serologic characteristics may differ in the mobility of their RNA segments. It does appear that mobility patterns of viral RNA segments usually correlate with immunologic properties of the virus and strains with different genomic RNA segments appear to be distinct, a property permitting the procedure to be used in epidemiologic studies. Human rotaviruses show extensive diversity in their segmental genomic patterns.

The RNA of strains of rotaviruses obtained from patients hospitalized with diarrhea during two winter epidemics of rotavirus infection revealed a single but different dominant electropherotype during the first 2 or 3 months of each epidemic. Additional electropherotypes appeared during the later periods of the same outbreaks (Konno et al., 1984). Nicolas et al. (1984) studied three different environments during the winter of 1981–1982 and found the presence of novel RNA patterns, the coexistence of multiple electrophoretic patterns in rotaviruses isolated from specimens from a given ward and that spread of a given electropherotype was limited. These changes in rotavirus genotypes may represent recombination or reassortment of RNA subunits in nature.

22.3 NORWALK VIRUSES

22.3.1 Description of Characteristics

22.3.1.a Overview

The discovery of the first recognized human Norwalk gastroenteritis virus of medical importance was reported in 1972 when Kapikian et al., using IEM, observed increased aggregation of virus-like particles with convalescent serum when compared with acute serum, in stools of volunteers who developed gastroenteritis after oral challenge with a stool filtrate from a patient involved in an outbreak of diarrhea in Norwalk, Ohio in 1968. The responsible agent was named Norwalk virus. Since then, IEM has been used to strengthen the association between viral particles and gastroenteritis in many outbreaks of diarrhea (Kaplan et al., 1982).

22.3.1.b Nomenclature

The Norwalk virus particle is 27 nm in size. Norwalk virus is the best studied member of a group of morphologically similar but antigenically unrelated small round viruses, which are often seen in stool specimens by EM. These small round viruses are 25 to 35 nm and can be categorized as a) those with distinctive morphology, such as the astroviruses and calicivirus-like particles, and b) those without distinctive morphology, referred to as Norwalk-like or parvovirus-like viruses.

The Norwalk-like viruses have been linked to epidemics or family outbreaks of gastroenteritis. They have been named for the geographic location of the outbreak or the presumed source of infection and include Hawaii, W-Ditchling, Cockle, Parramatta, Marin County, and Snow Mountain (Cukor et al., 1984). Based on IEM testing, there is evidence to suggest that there are at least four or five antigenically distinct agents. The Norwalk, Hawaii, and W-Ditchling agents are serologically distinct. The Marin County and Snow Mountain agents possess antigenic differences from each other and from Hawaii agent and Norwalk virus. The Norwalk, Hawaii, W-Ditchling, and Snow Mountain agents have been shown in human volunteers to be transmissible and pathogenic (Wyatt et al., 1974). The significance and etiologic roles of these Norwalk-like viruses remains to be established.

22.3.2 Pathogenesis

22.3.2.a Incidence

Serologic studies have demonstrated that there is a higher prevalence of antibodies to Norwalk virus among children in developing countries than among children in the U.S. and other developed countries (Greenberg et al., 1979; Blacklow et al., 1979; Cukor et al., 1980), and that a high incidence of seroconversion to Norwalk virus occurs in the second and third years of life (Black et al., 1982). Serologic studies of families (Pickering et al., 1982) have demonstrated that Norwalk virus causes family outbreaks of diarrhea. Norwalk virus has a worldwide distribution and in the U.S. 50% of adults have developed antibodies by the time they are 40 years of age (Greenberg et al., 1979).

22.3.2.b Pathophysiology

The Norwalk virus is a 27-nm diameter, nonenveloped, round particle. It has not been possible to delineate its structure with any degree of clarity. Inoculation of Norwalk virus into chimpanzees resulted in the shedding of virus in feces and in the production of a Norwalk specific immune response (Wyatt et al., 1978). Administration of the virus to other laboratory animals including a number of primate species has not resulted in production of illness (World Health Organization, 1980). When a fecal filtrate containing the Norwalk virus was administered orally to human volunteers, approximately 50% developed illness (Blacklow et al., 1972).

Pathologic findings of the mucosa of the proximal small bowel in adult volunteers infected with Norwalk virus showed abnormal histologic findings including mucosal inflammation, absorptive cell abnormalities, villous shortening, crypt hypertrophy, and increased epithelial cell mitosis. The gastric and colonic mucosa remain histologically normal during infection with Norwalk viruses (Agus et al., 1973; Meeroff et al., 1980). Abnormal findings persisted for at least 4 days after clinical symptoms ceased (Schreiber et al., 1973). The virus has not been detected within involved mucosal cells.

22.3.2.c Clinical

Much of what is known about the clinical aspects and immunologic properties of Norwalk virus has been the result of volunteer studies or of large water or food borne outbreaks. In volunteers fed Norwalk virus, diarrhea and/or vomiting occur. No prolonged illness or long-term side effects have been observed in these volunteer subjects (Blacklow et al., 1972). The illness is generally mild and characterized by nausea, vomiting, diarrhea, and abdominal cramps (Table 22.5). Vomiting is the predominant symptom among children, whereas, diarrhea is more common among adults (Kaplan et al., 1982). The vomiting may be caused by a decrease in gastric emptying, which has been demonstrated in adult volunteers (Meeroff et al., 1980). The illness is generally mild but may necessitate hospitalization or result in death in elderly, debilitated persons (Kaplan et al., 1982). Stools generally do not contain blood or mucus.

Seroconversion for antibody to Norwalk virus was most frequent during the cool, dry period in Bangladesh; however, transmission appears to have occurred all year (Black et al., 1982). In the U.S. outbreaks of Norwalk virus gastroenteritis occur during all months of the year and in various locations including elementary schools, colleges, camps, recreational areas, nursing homes, restaurants, cruise ships, and from municipal water supplies (Kaplan et al., 1982). Outbreaks of gastroenteritis due to Norwalk virus may involve persons of all ages, occur during all seasons and generally end in 1 week. Longer outbreaks occur when new groups of susceptibles are introduced (Kaplan et al., 1982). The incubation period of illness has a mean of 24 to 48 hours with a range of 4 to 77 hours.

Transmission of Norwalk virus occurs as a result of person to person spread and by both water (drinking and swimming) and food. Studies of the shedding patterns of Norwalk virus in individuals demonstrated that maximal shedding occurs at the onset of illness and shortly thereafter (Thornhill et al., 1975). Excretion may occur in low

Table 22.5. Presence of Symptoms in 38 Outbreaks of Norwalk Virus Infection

Symptom	Number of Outbreaks Positive for Symptom	Percentage of Patients with Symptoms (range)
Nausea	30	79 (51–100)
Vomiting	34	69 (25–100)
Diarrhea	34	66 (21–100)
Abdominal cramps	30	71 (17–90)
Headache	22	50 (17–80)
Fever	29	37 (13–71)
Chills	14	32 (5–74)
Myalgias	14	26 (11–73)
Sore throat	7	18 (7–32)

Adapted from Kaplan JE, et al. Ann. Intern. Med. 96:756, 1982.

concentrations for weeks after the illness. Viral particles also have been demonstrated in vomitus (Greenberg et al., 1979).

Infection with Norwalk-like viruses occurs as sporadic cases and family clusters, but more commonly as large outbreaks, which are water borne (Taylor et al., 1981; Baron et al., 1982; Kaplan et al., 1982) or food borne (Kaplan et al., 1982; Kuritsky et al., 1984; Riordan et al., 1984; Morse et al., 1986). Norwalk-like virus does not appear to be a zoonosis and, therefore, the sources of water borne outbreaks probably are human excreters. Some of the food borne outbreaks have been associated with shellfish, such as oysters (Grohmann et al., 1980; Gunn et al., 1982; Gill et al., 1983; Morse et al., 1986). It was assumed that the oysters were contaminated in the sea by human sewage containing the virus. Food borne outbreaks not involving shellfish have occurred in which direct contamination of foods such as salads and cake frosting by food handlers excreting virus have been reported (Griffin et al., 1982; Pether et al., 1983; Riordan et al., 1984; Kuritsky et al., 1984).

22.3.3 Isolation Sites

Norwalk virus has been isolated from stools and from vomitus (Greenberg et al., 1978; 1979). Virus shedding in stools reaches a maximum close to the onset of illness and rarely occurs after 72 hours following onset of illness (Blacklow and Cukor, 1980). Virus has not been detected from all individuals with Norwalk virus illness. Stools and paired sera should be collected and frozen at −70°C until they can be processed to confirm infection by Norwalk virus.

22.3.4 Stability

Limited information is available on the stability of Norwalk virus. The virus is stable upon storage at −70°C, to acid treatment (pH 2.7 for 3 hours) and to ether, properties similar to other diarrheal agents (Dolin et al., 1972). The virus also may be resistant to heat treatment at 60°C for 30 minutes (Cukor et al., 1984). Recent information from our laboratory suggests that Norwalk virus is also resistant to chlorine used to treat drinking water.

22.3.5 Isolation and Identification

Because none of the Norwalk virus agents grow in cell culture or in a readily available laboratory animal model, the isolation of Norwalk virus is currently not possible. Identification of Norwalk-like agents is accomplished using IEM, IAHA, or RIA techniques. It should be realized that all tests for viruses of the Norwalk group depend on reagents in critically short supply. These have to be obtained from volunteers. Therefore, diagnosis of Norwalk virus infection is currently available only on a research basis.

Immunoelectron microscopy has been used successfully (Kapikian et al., 1982) to visualize Norwalk virus and related agents in diarrheal stools. Due to the small size and amorphous surface of Norwalk virus, IEM, rather than EM, is required to identify the agent in stool specimens. Immunoelectron microscopy, using acute and convalescent sera, can help to establish the etiology of an illness by comparing the degree to which particles are coated with antibody. Specimens in this type of study should be read under code to establish the accuracy of a specific response (Kapikian et al., 1982).

The radioimmunoassay for Norwalk virus developed by Greenberg et al. (1978) is more sensitive than IEM and is able to detect both soluble and particulate antigens. The assay is dependent on the availability of acute and convalescent sera from volunteers and purified virus. Wells of a microtiter plate are coated with either preinfection serum or convalescent serum in a manner analogous to that used for the rotavirus ELISA. After washing, the stool specimen to be examined is added and the plate is again incubated and washed prior to the addition of an iodinated antibody to

Norwalk virus. The P/N ratio is calculated from the counts of wells coated with convalescent serum to counts of wells coated with preinfection serum. A ratio of 2 or greater is considered positive for Norwalk virus. This test also may be modified to detect antibodies to Norwalk virus by employing a blocking assay (Greenberg et al., 1978; Kapikian et al., 1982). In the blocking test a 50% reduction in activity is taken as the endpoint.

One further method for Norwalk virus serology is the IAHA (Kapikian et al., 1978). Although useful for large scale serologic studies, this assay is not practical for detecting virus in stool specimens. Furthermore, this assay requires purified Norwalk virus and screening of donor red blood cells to ensure an effective assay. Thus, assays for Norwalk virus will remain the domain of a few research laboratories until reagents become more readily available. In turn, this is dependent on the development of an animal model or the propagation of the virus in vitro.

22.4 ENTERIC ADENOVIRUSES AND OTHER GASTROENTERITIS VIRUSES

Adenoviruses of new serotypes 40–41 have been found in stools of patients with gastroenteritis (Yolken et al., 1982; Chiba et al., 1983; Brown et al., 1984; Uhnoo et al., 1984; Brandt et al., 1985). These viruses have been uncultivable by methods routinely applied to detection of the well characterized 39 conventional serotypes of adenoviruses. It has been found that an early event is blocked which prevents replication of enteric adenoviruses in cell culture (Takiff et al., 1981; Takiff and Strauss, 1982). This defect may be overcome by using adenovirus transformed cells designated 293 cells. In addition some success has been achieved with Chang cells (Wigand et al., 1983). Because adenoviruses share a group antigen, the presence of adenoviruses in stools can be detected by immunochemical assays (Yolken et al., 1982). Yolken et al. (1982) also have developed an adenovirus (enteric type) assay. The group specific test can utilize commercially available reagents; however, the type-specific ELISA requires known isolates and specially prepared antisera. Adenoviruses also may be detected in stool samples by EM as described earlier. In fact, enteric-type adenoviruses were first observed in stools that contained adenovirus-like particles, but failed to yield virus upon cultivation in vitro. At this time, there are no commercially available diagnostic kits for enteric adenoviruses. Enteric-type adenoviruses have become increasingly recognized as an important cause of gastroenteritis (Yolken et al., 1982; Uhnoo et al., 1984), which may lead to the development of commercially available diagnostic kits.

The other types of viruses associated with gastroenteritis, astrovirus, calicivirus, and coronavirus, are identified by laboratory methodologies employing reagents that are not commercially available to the clinical laboratory

REFERENCES

Agus, S.G., Dolin, R., Wyatt, R.G., Tousimis, A.J., and Northrop, R.S. 1973. Acute infectious nonbacterial gastroenteritis: Intestinal histopathology: Histologic and enzymatic alterations during illness produced by the Norwalk agent in man. Ann. Intern. Med. 7:18–25.

Albert, M.J., Bishop, R.F., and Shann, F.A. 1983. Epidemiology of rotavirus diarrhea in the highlands of Papua, New Guinea, in 1979, as revealed by electrophoresis of genome RNA. J. Clin. Microbiol. 17:162–164.

Altenburg, B.C., Graham D.Y., and Estes, M.K. 1979. Ultrastructural immunocytochemistry of rotavirus-infected cells. In

Proceedings of the 37th Annual Meeting of the Electron Microscopy Society of America, pp. 40–41.

Banatvala, J.E., Totterdell, B., Chrystie, I.L., and Woode, G.N. 1975. In vitro detection of human rotaviruses. Lancet ii:821.

Baron, R.C., Murphy, F.D., and Greenberg, H.B. et al. 1982. Norwalk/related illness: An outbreak associated with swimming in a recreational lake and secondary person to person transmission. Am. J. Epidemiol 115:163–172.

Barnes, G.L., Hewson, P.H., McLellan, J.A., Doyle, L.W., Knoches, A.M.L., Kitchen, W.H., and Bishop, R.F. 1982. A randomized trial of oral gammaglobulin in low-birth-weight infants infected with rotavirus. Lancet ii:1371–1373.

Barnett, B.B., Spendlove, R.S., Peterson, M.W., Hsu, L.Y., LaSalle, V.A., and Egbert, L.N. 1975. Immunofluorescent cell assay of neonatal calf diarrhoea virus. Canad. J. Comp. Med 39:462–465.

Birch, C.J., Lehmann, N.I., Hawker, A.J., Marshall, J.A., and Gust, I.D. 1979. Comparison of electron microscopy, enzyme-linked immunosorbent assay, solid-phase radioimmunoassay, and indirect immunofluorescence for detection of human rotavirus antigen in faeces. J. Clin. Pathol. 32:700–705.

Birch, C.J., Rodger, S.M., Marshall, J.A., and Gust, I.D. 1983. Replication of human rotavirus in cell culture. J. Med. Virol. 11:241–250.

Bishop, R.F., Davidson, G.P., Holmes, I.H., and Ruck, B.J. 1973. Virus particles in epithelial cells of duodenal mucosa from children with acute nonbacterial gastroenteritis. Lancet ii:1281–1283.

Bishop, R.F., Barnes, G.L., Cipriani, E., and Lund, J.S. 1983. Clinical immunity after neonatal rotavirus infection. A prospective longitudinal study in young children. N. Engl. J. Med. 309:72–76.

Black, R.E., Merson, M.H., Taylor, P.R., Yolken, R.H., Yunus, M., Alim, A.R.M.A., and Sack, D.A. 1981. Glucose versus sucrose in oral rehydration solutions for infants and young children with rotavirus associated diarrhea. Pediatrics 67:79–83.

Black, R.E., Greenberg, H.B., Kapikian, A.Z., Brown, K.H., and Becker, S. 1982. Acquisition of serum antibody to Norwalk virus and rotavirus and relation to diarrhea in a longitudinal study of young children in rural Bangladesh. J. Infect. Dis. 145:483–489.

Blacklow, N.R., Dolin, R., Fedson, D.S., DuPont, H., Northrop, R.S., Hornick, R.B., and Chanock, R.M. 1972. Acute infectious nonbacterial gastroenteritis: Etiology and pathogenesis. Ann. Intern. Med. 76:993–1008.

Blacklow, N.R., Cukor, G., Bedigian, M.K., Echeverria, P., Greenberg, H.B., Schreiber, D.S., and Trier, J.S. 1979. Immune response and prevalence of antibody to Norwalk enteritis virus as determined by radioimmunoassay. J. Clin. Microbiol. 10:903–909.

Blacklow, N.R., and Cukor, G. 1982. Viruses and gastrointestinal disease. In D.A.J. Tyrrell and A.Z. Kapikian (eds.), Virus Infections of the Gastrointestinal Tract. New York: Marcel Dekker, pp. 75–87.

Bradburne, A.F., Almeida, J.D., Gardner, P.S., Moosai, R.B., Nash, A.A., and Coombs, R.R.A. 1979. A solidphase system (SPACE) for the detection and quantification of rotavirus in faeces. J. Gen. Virol. 44:615–623.

Brade, L., Schmidt, W.A.K., and Gattert, I. 1981. Zur relativen Wirksamkeit von Disinfektionmitteln gegenuber rotaviren. Zentralb. Bakteriol. (Orig. B.) 174:151–159.

Brandt, C.D., Kim, H.W., Yolken, R.H., Kapikian, A.Z., Arrobio, J.O., Rodriguez, W.J., Wyatt, R.G., Chanock, R.M., and Parrott R.H. 1979. Comparative epidemiology of two rotavirus serotypes and other viral agents associated with pediatric gastroenteritis. Am. J. Epidemiol. 110:243–254.

Brandt, C.D., Kim, H.W., Rodriguez, W.J., Thomas, L., Yolken, R.H., Arrobio, J.O., Kapikian, A.Z., Parrott, R.H., and Chanock, R.M. 1981. Comparison of direct electron microscopy, immune electron microscopy, and rotavirus enzyme-linked immunosorbent assay for detection of gastroenteritis viruses in children. J. Clin. Microbiol. 13:976–981.

Brandt, C.D., Kim, H.W., Rodriguez, W.J., Arrobio, J.O., Jeffries, B.C., Stallings, E.P.,

Lewis, C., Miles, A.J., Chanock, R.M., Kapikian, A.Z., and Parrott, R.H. 1983. Pediatric viral gastroenteritis during eight years of study. J. Clin. Microbiol. 18:71–78.

Brandt, C.D., Kim, H.W., Rodriguez, W.J., Arrobio, J.O., Jeffries, B.C., Stallings, E.P., Lewis, C., Miles, A.J., Gardner, M.K., and Parrott, R.H. 1985. Adenovirus and pediatric gastroenteritis. J. Infect. Dis. 151:437–443.

Brown, M., Petric, M., and Middleton, P.J. 1984. Diagnosis of fastidious enteric adenovirus 40 and 41 in stool specimens. J. Clin. Microbiol. 20:334–338.

Bryden, A.S., Davies, H.A., Thouless, M.E., and Flewett, T.H. 1977. Diagnosis of rotavirus infection by cell culture. J. Med. Microbiol. 10:121–125.

Bryden, A.S., Thouless, M.E., Hall, C.J., Flewett, T.H., Wharton, B.A., Mathew, P.M., and Craig, I. 1982. Rotavirus infections in a special-care baby unit. J. Infection 4:43–48.

Butler, M., and Harakeh, M.S. 1983. Inactivation of rotavirus in wastewater effluents by chemical disinfection. In M. Butler, A.R. Medlen, and R. Morris (eds.), Viruses and Disinfection of Water and Wastewater. Guildford, U.K.: University of Surrey Print Unit, pp. 282–289.

Carlson, J.A.K., Middleton, P.J., Szymanski, M.T., Huber, J., and Petric, M. 1978. Fatal rotavirus gastroenteritis: An analysis of 21 cases. Am. J. Dis. Child. 132:477–479.

Champsaur, H., Questiaux, E., Prevot, J., Henry-Amar, M., Goldszmidt, D., Bourjouane, M., and Bach, C. 1984a. Rotavirus carriage, asymptomatic infection, and disease in the first two years of life. I. Virus shedding. J. Infect. Dis. 149:667–674.

Champsaur, H., Henry-Amar, M., Goldszmidt, D., Prevot, J., Bourjouane, M., Questiaux, E., and Bach C. 1984b. Rotavirus carriage, asymptomatic infection, and disease in the first two years of life. II. Serological response. J. Infect. Dis. 149:675–682.

Chanock, S.J., Wenske, E.A., and Fields, B.N. 1983. Human rotaviruses and genome RNA. J. Infect. Dis. 148:49–50.

Chiba, S., Nakamura, I., Urasawa, S., Nakata. S., Taniguchi, K., Funinaga, K., and Nakao, T. 1983. Outbreak of infantile gastroenteritis due to type 40 adenovirus. Lancet i:954–957.

Chrystie, I.L., Totterdell, B.M., and Banatvala, J.E. 1978. Asymptomatic endemic rotavirus infections in the newborn. Lancet i:1176–1178.

Clark, I.N., McCrae, M.A. 1982. Structural analysis of electrophoretic variation in the genome profiles of rotavirus field isolates. Infect. Immun. 36:492–497.

Cukor, G., Berry, M.K., and Blacklow, N.R. 1978. Simplified immunoassay for detection of human rotavirus in stools. J. Infect. Dis. 138:906–910.

Cukor, G., Blacklow, N.R., Echeverria, P., Bedigian, M.K., Puruggan, H., and Basaca-Sevilla, V. 1980. Comparative study of the acquisition of antibody to Norwalk virus in pediatric populations. Infect. Immun. 29:822–823.

Cukor, G., Perron, D.M., Hudson, R., and Blacklow, N.R. 1984. Detection of rotavirus in human stools by using monoclonal antibody. J. Clin. Microbiol. 19:888–892.

Cukor, G., and Blacklow, N.R. 1984. Human viral gastroenteritis. Microbiol. Rev. 48:157–179.

Davidson, G.P., Bishop, R.F., Townely, R.R., Holmes, I.H., and Ruck, B.J. 1975. Importance of a new virus in acute sporadic enteritis in children. Lancet i:242–246.

Davidson, G.P., Gall, D.G., Petric, M., Butler, D.G., and Hamilton, J.R. 1977. Human rotavirus enteritis induced in conventional piglets. J. Clin. Invest. 60:1402–1409.

Dearlove, J., Latham, P., Dearlove, B., Pearl, K., Thomson, A., and Lewis, I.G. 1983. Clinical range of neonatal rotavirus gastroenteritis. Br. Med. J. 286:1473–1475.

Deetz, T.R., Smith, E.M., Goyal, S.M., Gerba, C.P., Vollet, J.J., Tsai, L., DuPont, H.L., and Keswick, B.H. 1984. Occurrence of rota- and enteroviruses in drinking water in a developing nation. Water Res. 18:567–571.

deTorres, B.V., DeIlja, R.M., and Esparza, J. 1978. Epidemiological aspects of rotavirus infection in hospitalized Venezuelan children with gastroenteritis. Am. J. Trop. Med. Hyg. 27:567–572.

Dolin, R., Blacklow, N.R., DuPont, H.L., Buscho, R.F., Wyatt, R.G., Kasel, J.A., Hornick, R., and Chanock, R.M. 1972. Biological properties of Norwalk agent of acute infectious nonbacterial gastroenteritis. Proc. Soc. Exp. Biol. Med. 140:578–583.

Echeverria, P., Blacklow, N.R., Cukor, G., Vibulbandhitkit, S., Changchawalit, S., and Boonthai, P. 1983. Rotavirus as a cause of severe gastroenteritis in adults. J. Clin. Microbiol. 18:663–667.

Eiden, J., Vonderfecht, S., and Yolken, R.N. 1985. Evidence that a novel rotavirus-like agent of rats can cause gastroenteritis in man. Lancet ii:8–11.

Engleberg, N.C., Holburt, E.N., Barrett, T.J., Gary, G.W., Jr., Trujillo, M.H., Feldman, R.A., and Hughes, J.M. 1982. Epidemiology of diarrhea due to rotavirus in an indian reservation: Risk factors in the home environment. J. Infect. Dis. 145:894–898.

Espejo, R.T., Munoz, O., Serafin, F., and Romero, P. 1980. Shift in the prevalent human rotavirus detected by ribonucleic acid segment differences. Infect. Immun. 27:351–354.

Estes, M.K., Graham, D.Y., Mason, B.B., Smith, E.M., and Gerba, C.P. 1979. Rotavirus stability and inactivation. J. Gen. Virol. 43:403–409.

Estes, M.K., Palmer, E.L., and Obijeski, J.F., 1983. Rotaviruses: A review. Microbiol. Immunol. 105:123–184.

Estes, M.K., Graham, D.Y., and Dimitrov, D.H., 1984. The molecular epidemiology of rotavirus gastroenteritis. Prog. Med. Virol. 29:1–22.

Flewett, T.H., Bryden, A.S., and Davies, H. 1973. Virus particles in gastroenteritis. Lancet ii:1497.

Flewett, T.H., Bryden, A.S., Davies, H., and Morris, C.A. 1975. Epidemic viral enteritis in a long-stay children's ward. Lancet i:4–5.

Flewett, T.H. 1983. Rotavirus in the home and hospital nursery. Br. Med. J. 287:568–569.

Fonteyne, J., Zissis, G., and Lambert, J.P. 1978. Recurrent rotavirus gastroenteritis. Lancet i:983.

Foster, L.G., Peterson, M.W., and Spendlove, R.S. 1975. Fluorescent virus precipitin test. Proc. Soc. Exp. Biol. Med. 150:155–160.

Foster, S.O., Palmer, E.L., Gary, G.W., Jr., Maron, M.L., Hermann, K.L., Beasley, P., and Sampson, J. 1980. Gastroenteritis due to rotavirus in an isolated Pacific Island Group: An epidemic of 3,439 cases. J. Infect. Dis. 141:32–39.

Gerna, G., Passarani, N., Battaglia, M., and Percivalle, E. 1984. Rapid serotyping of human rotavirus strains by solid phase immune electron microscopy. J. Clin. Microbiol. 19:273–278.

Gill, O.N., Cubitt, W.D., McSwiggan, D.A., Watney, B.M., and Bartlett, C.L.R. 1983. Epidemic of gastroenteritis caused by oysters contaminated with small round structured viruses. Br. Med. J. 287:1532–1534.

Goldwater, P.N., Chrystie, I.L., and Banatvala, J.E. 1979. Rotaviruses and the respiratory tract. Br. Med. J. 2:1551.

Graham, D.Y., and Estes, M.K. 1979. Comparison of methods of immunocytochemical detection of rotavirus infections. Infect. Immun. 26:686–689.

Greenberg, H.B., Wyatt, R.G., Valdesu, J., Kalic, A.R., London, W.T., Chanock, R.M., and Kapikian, A.Z. 1978. Solid-phase microtiter radioimmunoassay for detection of the Norwalk strain of acute nonbacterial, epidemic gastroenteritis virus and its antibodies. J. Med. Virol 2:97–108.

Greenberg, H.B., Wyatt, R.G., and Kapikian, A.Z. 1979a. Norwalk/virus in vomitus. Lancet i:55.

Greenberg, H.B., Valdesuso, J., Kapikian, A.Z., Chanock, R.M., Wyatt, R.G., Szmuness, W., Larrick, J., Kaplan, J., Gilman, R.H., and Sack, D.A. 1979b. Prevalence of antibody to the Norwalk virus in various countries. Infect. Immun. 26:270–273.

Greenberg, H.B., Kalica, A.R., Wyatt, R.G., Jones, R.W., Kapikian, A.Z., and Chanock, R.M. 1981. Rescue of noncultivatable human rotavirus by gene reassortment during mixed infection with ts mutants of a cultivatable bovine rotavirus. Proc. Natl. Acad. Sci. USA 78:420–424.

Greenberg, H., McAuliffe, V., Valdesuso, J., Wyatt, R., Flores, J., Kalica, A., Hoshino, Y., and Singh, N. 1983. Serological analysis of the subgroup protein of rotavirus, using monoclonal antibodies. Infect. Immun. 39:91–99.

Griffin, M.R., Sorowiec, J.J., and McCloskey, D.I., et al. 1982. Foodborne Norwalk/virus. Am. J. Epidemiol. 115:178–184.

Grimwood, K., Abbott, G.D., Fergusson, D.M., Jennings, L.C., and Allan, J.M. 1983. Spread of rotavirus within families: A community based study. Br. Med. J. 287:575–577.

Grohmann, G.S., Greenberg, H.B., Welch, B.M., and Murphy, A.M. 1980. Oyster-associated gastroenteritis in Australia. The detection of Norwalk virus and its antibody by immune electron microscopy and radioimmunoassay. J. Med. Virol 6:11–19.

Gunn, R.A., Janowski, H.T., Lieb, S., Prather, E.C., and Greenberg, H.B. 1982. Norwalk virus gastroenteritis following raw oyster consumption. Am. J. Epidemiol. 115:348–51.

Gurwith, M., Wenman, W., Hinde, D., Feltham, S., and Greenberg, H. 1981. A prospective study of rotavirus infection in infants and young children. J. Infect. Dis. 144:218–224.

Gurwith, M., Wenman, W., Gurwith, D., Brunton, J., Feltham, S., and Greenberg, H. 1983. Diarrhea among infants and young children in Canada: A longitudinal study in three northern communities. J. Infect. Dis. 147:685–692.

Haikala, O.J., Kokkonen, J.O., Leinonen, M.K., Nurmi, T., Mantyjarvi, R., and Sarkkinen, H.K. 1983. Rapid detection of rotavirus in stool by latex agglutination: Comparison with radioimmunoassay and electron microscopy and clinical evaluation of the test. J. Med. Virol. 11:91–97.

Halvorsrud, J., and Orstavik, I. 1980. An epidemic of rotavirus-associated gastroenteritis in a nursing home for the elderly. Scand. J. Infect. Dis. 12:161–164.

Hammond, G.W., Hazelton, P.R., Chuang, I., and Klisko, B. 1981. Improved detection of viruses by electron microscopy after direct ultracentrifuge preparation of specimens. J. Clin. Microbiol. 14:210–221.

Hammond, G.W., Ahluwalia, G.S., Klisko, B., and Hazelton, P.R. 1984. Human rotavirus detection by counterimmunoelectrophoresis versus enzyme immunoassay and electron microscopy after direct ultracentrifugation. J. Clin. Microbiol. 19:439–441.

Hasegawa, A., Matsuno, S., Inouye, S., Kono, R., Tsurukubo, Y., Mukoyama A., and Saito, Y. 1982. Isolation of human rotaviruses in primary cultures of monkey kidney cells. 1982. J. Clin. Microbiol. 16:387–390.

Herring, A.J., Inglis, N.F., Ojeh, C.K., Snodgrass, D.R., and Menzies, J.D. 1982. Rapid diagnosis of rotavirus infection by direct detection of viral nucleic acid in silver-stained polyacrylamide gels. J. Clin. Microbiol. 16:473–477.

Hieber, J.P., Shelton, S., Nelson, J.D., Leon, H., and Mohs, E. 1978. Comparison of human rotavirus disease in tropical and temperate settings. Am. J. Dis. Child. 132:853–858.

Holzel, H.D., Cubitt, W., McSwiggan, D.A., Sanderson, P.J., and Church, J. 1980. An outbreak of rotavirus infection among adults in a cardiology ward. J. Infection 2:33–37.

Hopkins, R.S., Gaspard, G.B., Williams, F.P., Karlin, R.J., Cukor, G., and Blacklow, N.R. 1984. A community waterborne gastroenteritis outbreak: Evidence for rotavirus as the agent. Am. J. Pub. Health. 74:263–265.

Hughes, J.H., Tuomari, A.V., Mann, D.R., and Hamparian, V.V. 1984. Latex immunoassay for rapid detection of rotavirus. J. Clin. Microbiol. 20:441–447.

Hyams, J.S., Krause, P.J., and Gleason, P.A. 1981. Lactose malabsorption following rotavirus infection in young children. J.Pediat. 99:916–918.

Joklik, W.F. 1981. Structure and function of the reovirus genome. Microbiol. Rev. 45:483–501.

Kalica, A.R., Purcell, R.H., Sereno, M.M., Wyatt, R.G., Kim, H.W., Chanock, R.M., and Kapikian, A.Z. 1977. A microtiter solid-phase radioimmunoassay for detection of the human reovirus-like agent in stools. J. Immunol. 118:1275–1279.

Kalica, A.R., Greenberg, H.B., Wyatt, R.G., Flores, J., Sereno, M.M., Kapikian, A.Z., and Chanock, R.M. 1981. Genes of human (strain Wa) and bovine (strain UK) rotavirus that code for neutralization and subgroup antigens. Virology 11:385–390.

Kapikian, A.Z., Wyatt, R.G., Dolin, R., Thornhill, T.S., Kalica, A.R., and Chanock, R.M. 1972. Visualization by immune electron microscopy of a 27-nm particle associated with acute infectious nonbacterial gastroenteritis. J. Virol. 10:1075–1081.

Kapikian, A.Z., Kim, H.W., Wyatt, R.G., Rodriguez, W.J., Ross, S., Cline, W.L., and Parrott, R.H. 1974. Reovirus like agents in stools: Association with infantile diarrhoea and development of serologic tests. Science 185:1049–1053.

Kapikian, A.Z., Cline, W.L., Mebus, C.A., Wyatt, R.G., Kalica, A.R., James, H.D., Van Kirk, D., Chanock, R.M., and Kim, H.W. 1975. New complement-fixation test for the human reovirus-like agent of infantile gastroenteritis. Lancet i:1056–1061.

Kapikian, A.Z., Kim, H.W., Wyatt, R.G., Cline, W.L., Arrobio, J.O., Brandt, C.D., Rodriguez, W.J., Sack, S.A., Chanock, R.M., and Parrott, R.H. 1976. Human reovirus-like agent as the major pathogen associated with winter gastroenteritis in hospitalized infants and young children. N. Engl. J. Med. 294:965–972.

Kapikian, A.Z., Greenberg, H.B., Cline, W.L., Kalica, A.R., Wyatt, R.G., James, W.D., Lloyd, N.L., Chanock, R.M., Ryder, R.W., and Kim, H.W. 1978. Prevalence of antibody to the Norwalk agent by a newly developed immune adherence hemagglutination assay. J. Med. Virol. 2:281–294.

Kapikian, A.Z., Cline, W.L., Greenberg, H.B., Wyatt, R.G., Kalica, A.R., Banks, C.E., James, H.D., Flores, J., and Chanock, R.M. 1981. Antigenic characterization of human and animal rotaviruses by immune adherence hemagglutination assay (IAHA): Evidence for distinctness of IAHA and neutralization antigens. Infect. Immun. 33:415–425.

Kapikian, A.Z., Greenberg, H.B., Wyatt, R.G., Kalica, A.R., and Chanock, R.M. 1982. The Norwalk group of viruses—Agents associated with epidemic viral gastroenteritis. In D.A. Tyrrell and A.Z. Kapikian (eds.), Virus Infections of the Gastrointestinal Tract. New York: Marcel Dekker, pp. 147–177.

Kapikian, A.Z., Wyatt, R.G., Levine, M.M., Yolken, R.H., Vankirk, D.H., Dolin, R., Greenberg, H.B., and Chanock, R.M. 1983. Oral administration of human rotavirus to volunteers: Induction of illness and correlates of resistance J. Infect. Dis. 147:95–106.

Kaplan, J.E., Gary, G.W., Baron, R.C., Singh, N., Schonberger, L.B., Feldman, R., and Greenberg, H.B. 1982a. Epidemiology of Norwalk gastroenteritis and the role of Norwalk virus in outbreaks of acute nonbacterial gastroenteritis. Ann. Intern. Med. 96:756–761.

Kaplan, J.E., Goodman, R.A., Schonberger, L.B., et al. 1982b. Gastroenteritis due to Norwalk virus: An outbreak associated with a municipal water supply. J. Infect. Dis. 146:190–197.

Keswick, B.H., Hejkal, T.W., DuPont, H.L., and Pickering, L.K. 1982. Evaluation of a commercial EIA kit for rotavirus detection. Diag. Microbiol. Infect. Dis. 1:111–115.

Keswick, B.H., Pickering, L.K., DuPont, H.L., and Woodward, W.E. 1983a. Prevalence of rotavirus in children in day care centers. J. Pediat. 103:85–86.

Keswick, B.H., Pickering, L.K., DuPont, H.L., and Woodward, W.E. 1983b. Survival and detection of rotaviruses on environmental surfaces in day care centers. Appl. Environ. Microbiol. 46:813–816.

Keswick, B.H., Gerba, C.P., DuPont, H.L., and Rose, J.B. 1984. Detection of enteric viruses in treated drinking water. Appl. Environ. Microbiol. 47:1290–1294.

Kim, H.W., Brandt, C.D., Kapikian, A.Z., Wyatt, R.G., Arrobio, J.P., Rodriguez, W.J., Chanock, R.M., and Parrott, R.H. 1977. Human reovirus-like agent (HRVLA) infection: Occurrence in adult contacts of pediatric patients with gastroenteritis. J. Am. Med. Assoc. 238:404–407.

Knisley, C.V., Bednarz-Prashad, J., and Pickering, L.K. 1986. Detection of rotavirus in stool specimens using monoclonal and polyclonal antibody based assay systems. J. Clin. Microbiol. 23:897–900.

Konno, T., Suzuki, H., Imai, A., and Ishida, N. 1977. Reovirus-like agent in acute epidemic gastroenteritis in Japanese infants: Fecal shedding and serologic response. J. Infect. Dis. 135:259–266.

Konno, T., Suzuki, H., Kutsuzawa, T., Imai, A., Katsushima, N., Sakamoto, M., Kitaoka, S., Tsuboi, R., and Adachi, M. 1978.

Human rotavirus infection in infants and young children with intussusception. J. Med. Virol. 2:265–269.

Konno, T., Suzuki, H., Katsushima, N., Imai, A., Tazawa, F., Kutsuzawa, T., Kitaoka, S., Sakamoto, M., Yazaki, N., and Ishida, N. 1983. Influence of temperature and relative humidity on human rotavirus infection in Japan. J. Infect. Dis. 147:125–128.

Konno, T., Sato, T., Suzuki, H., Kitaoka, S., Katsushima, N., Sakamoto, M., Yazaki, N., and Ishida, N. 1984. Changing RNA patterns in rotaviruses of human origin: Demonstration of a single dominant pattern at the start of an epidemic and various patterns thereafter. J. Infect. Dis. 149:683–687.

Krause, P.J., Hyams, J.S., Middleton, P.J., Herson, V.C., and Flores, J. 1983. Unreliability of rotazyme ELISA test in neonates. J. Pediat. 10:259–262.

Kuritsky, J.N., Osterholm, M.T., Greenberg H.B., Korlath, J.A., Godes, J.R., Hedberg, C.W., Forfang, J.C., and Kapikian, A.Z. 1984. Norwalk gastroenteritis: A community outbreak associated with bakery product consumption. Ann. Intern. Med. 100:519–521.

Kutsuzawa, T., Konno, T., Suzuki, H., Kapikian, A.Z., Ebina, T., and Ishida, N. 1982. Isolation of human rotavirus subgroups 1 and 2 in cell culture. J. Clin. Microbiol. 16:727–730.

Laurenco, M.H., Nicolas, J.C., Cohen, J., Scherrer, R., and Bricout, F. 1981. Studies of human rotavirus genome by electrophoresis: Attempt of classification among strains isolated in France. Ann. Virolog. (Paris) 132E:161–173.

Lewis, H.M., Parry, J.V., Davies, H.A., Parry, R.P., Mott, A., Dourmashkin, R.R., Sanderson, P.J., Tyrrell, D.A.J., and Valman, H.B. 1979. A year's experience of the rotavirus syndrome and its association with respiratory illness. Arch. Dis. Child. 54:339–346.

Liu, S., Birch, C., Coulepis, A., and Gust, I. 1984. Radioimmunofocus assay for detection and quantitation of human rotavirus. J. Clin. Microbiol. 20:347–350.

Lycke, E., Bloomberg, J., Berg, G., Ericksson, A., and Madsen, L. 1978. Epidemic acute diarrhoea in adults associated with infantile gastroenteritis virus. Lancet ii:1056–1057.

Maiya, P.P., Pereira, S.M., Mathan, M., Bhat, P., Albert, M.J., and Baker, S.J. 1977. Aetiology of acute gastroenteritis in infancy and childhood in southern India. Arch. Dis. Child. 52:482–485.

Marrie, T.J., Lee, S.H.S., Faulkner, R.S., Ethier, J., and Young, C.H. 1982. Rotavirus infection in a geriatric population. Arch. Intern. Med. 142:313–316.

Matsuno, S., and Nagayoshi, S. 1978. Quantitative estimation of infantile gastroenteritis virus antigens in stools by immune adherence haemagglutination test. J. Clin. Microbiol. 7:310–311.

Matsuno, S., Inouye, S., Hasegawa, A., and Kono, R. 1982. Assay of human rotavirus antibody by immune adherence hemagglutination with a cultivable human rotavirus as antigen. J. Clin. Microbiol. 15:163–165.

Matsuno, S., Utagawa, E., and Sugiuna, A. 1983. Association of rotavirus infection with Kawasaki syndrome. J. Infect. Dis. 148:177.

Meeroff, J.C., Schreiber, D.S., Trier, J.S., and Blacklow, N.R. 1980. Abnormal gastric motor function in viral gastroenteritis. Ann. Intern. Med. 92:370–373.

Middleton, P.J., Petric, M., Hewitt, C.M., Szymanski, M.T., and Tam, J.S. 1976. Counterimmunoelectroosmophoresis for the detection of infantile gastroenteritis virus (orbi group) antigen and antibody. J. Clin. Pathol. 29:191–197.

Morinet, F., Ferchal, F., Colimon, R., and Perol, Y. 1984. Comparison of six methods for detecting human rotavirus in stools. Eur. J. Clin. Microbiol. 3:136–140.

Morse, D.L., Guzewich, J.J., Hanrahan, J.P., Stricof, R., Shayegani, M., Deibel, R., Grabau, J.C., Nowak, N.A., Herrmann, J.E., Cukor, G., and Blacklow, N.R., 1986. Widespread outbreak of clam and oyster associated gastroenteritis. Role of Norwalk virus. N. Engl. J. Med. 314:678–681.

Mulcahy, D.L., Kamath, K.R., DeSilva, L.M., Hodges, S., Carter, I.W., and Cloonan, M.J. 1982. A two-part study of the aetiological role of rotavirus in intussusception. J. Med. Virol. 9:51–55.

Nagayoshi, S., Yamaguchi, H., Ichikawa, T., Miyazu, M., Morishima, T., Ozaki, T., Isomura, S., Suzuki, S., and Hoshino, M. 1980. Changes of the rotavirus concentration in faeces during the course of acute gastroenteritis as determined by the immune adherence hemagglutination test. Eur. J. Pediat. 134:99–102.

Naguib, T., Wyatt, R.G., Mohieldin, M.S., Zaki, A.M., Imam, I.Z., and DuPont, H.L. 1984. Cultivation and subgroup determination of human rotaviruses from Egyptian infants and young children. J. Clin. Microbiol. 19:210–212.

Nicolas, J.C., Ingrand, D., Fortier, B., and Bricout, F. 1982. A one-year virological survey of acute intussusception in childhood. J. Med. Virol 9:267–271.

Nicolas, J.C., Pothier, P., Cohen, J., Lourenco, M.H., Thompson, R., Guimbaud, P., Chenon, A., Dauvergne, M., and Bricout, F. 1984. Survey of human rotavirus propagation as studied by electrophoresis of genomic RNA. J. Infect. Dis. 149:688–693.

Paul, M.O., and Erinle, E.A. 1982. Influence of humidity on rotavirus prevalence among Nigerian infants and young children with gastroenteritis. J. Clin. Microbiol. 15:212–215.

Peterson, M.W., Spendlove, R.S., and Smart, R.A. 1976. Detection of neonatal calf diarrhea virus, infant reovirus-like diarrhea virus, and a coronavirus using the fluorescent virus precipitin test. J. Clin. Microbiol. 3:376–377.

Pether, J.V.S., and Caul, E.O. 1983. An outbreak of food borne gastroenteritis in two hospitals associated with a Norwalk like virus. J. Hyg. (London) 91:343–350.

Petrie, B.L., Graham, D.Y., Hanssen, H., and Estes, M.K. 1982. Localization of rotavirus antigens in infected cells by ultrastructural immunocytochemistry. J. Gen. Virol. 63:457–467.

Pickering, L.K., DuPont, H.L., Olarte, J., Conklin, R., and Ericsson, C. 1977. Fecal leukocytes in enteric infections. Am. J. Clin. Pathol. 68:562–565.

Pickering, L.K., Evans, D.J., Munoz,, O., DuPont, H.L., Coello, P., Vollet, J.J., Conklin, R.H., Olarte, J., and Kohl, S. 1978. Prospective evaluation of enteropathogens in children with diarrhea in Houston and Mexico. J. Pediat. 93:383–388.

Pickering, L.K., Evans, D.G., DuPont, H.L., Vollet, J.J., and Evans, D.J. 1981. Diarrhea due to shigella, rotavirus and giardia in day care centers: Prospective study. J. Pediat. 99:51–56.

Pickering, L.K., DuPont, H.L., Blacklow, N.R., and Cukor, G. 1982. Diarrhea due to Norwalk virus in families. J. Infect. Dis. 146:116–117.

Pickering, L.K., and Woodward, W.E. 1982. Diarrhea in day care centers. Pediat. Infect. Dis. 1:47–52.

Prevot, J., and Guesdon, J.L. 1981. A lectin immunotest using erythrocytes as marker (ERYTHRO-LIT) for detection and titration of rotavirus antigen. Ann. Virolog. 132 E:529–542.

Riordan, T., Craske, J., and Roberts, J. et al. 1984. Foodborne infection by a Norwalk like virus (small round structured virus). J. Clin. Pathol. 37:817–820.

Rodger, S.M., Bishop, R.F., Birch, C., McLean, B., and Holmes, I.H. 1981. Molecular epidemiology of human rotaviruses in Melbourne, Australia, from 1973 to 1979 as determined by electrophoresis of genome ribonucleic acid. J. Clin. Microbiol. 13:272–278.

Rodriguez, W.J., Kim, H.W., Arrobio, J.O., Brandt, C.D., Chanock, R.M., Kapikian, A.Z., Wyatt, R.G., and Parrott, R.H. 1977. Clinical features of acute gastroenteritis associated with human reovirus-like agent in infants and young children. J. Pediat. 91:188–193

Rodriguez, W.J., Kim, H.W., Brandt, C.D., Yolken, R.H., Arrobio, J.O., Kapikian, A.Z., Chanock, R.M., and Parrott, R.H. 1978. Sequential enteric illness associated wtih different serotypes. Lancet ii:37.

Rodriguez, W.J., Kim, H.W., and Brandt, C.D., et al. 1982. rotavirus: A cause of nosocomial infection in the nursery. J. Pediat. 101:274–277.

Rodriguez, W.J., Kim, H.W., Brandt, C.D., Gardner, M.K., and Parrott, R.H. 1983. Use of electrophoresis of RNA from human rotavirus to establish the identity of strains involved in outbreaks in a tertiary care nursery. J. Infect. Dis. 148:34–40.

Rotbart, H.A., Levin, M.J., Yolken, R.H.,

Manchester, D.K., and Jantzen, J. 1983. An outbreak of rotavirus-associated neonatal necrotizing enterocolitis. J. Pediat. 103:454–459.

Rubenstein, A.S., and Miller, M.F. 1982. Comparison of an enzyme immunoassay with electron microscopic procedures for detecting rotavirus. J. Clin. Microbiol. 15:938–944.

Sabara, M., and Babiuk, L.A. 1984. Identification of a bovine rotavirus gene and gene product influencing cellular attachment. J. Virol. 51:489–496.

Sack, D.A., Rhoads, M., Molla, A., Molla, A.M., and Wahed, M.A. 1982. Carbohydrate malabsorption in infants with rotavirus diarrhea. Am. J. Clin. Nutr. 36:1112–1118.

Salmi, T.T., Arstila, P., and Koivikko, A. 1978. Central nervous system involvement in patients with rotavirus gastroenteritis. Scand. J. Infect. Dis. 10:29–31.

Samadi, A.R., Hug, M.H., and Ahmed, Q.S. 1983. Detection of rotavirus in handwashings of attendants of children with diarrhea. Br. Med. J. 286:188.

Sanekata, T., Yoshida, Y., and Oda, K. 1979. Detection of rotavirus from faeces by reversed passive haemagglutination method. J. Clin. Pathol. 32:963.

Sanekata, T., Yoshida, Y., and Okada, H. 1981. Detection of rotavirus in faeces by latex agglutination. J. Immunol. Meth. 41:377–385.

Sanekata, T., and Okada, H. 1983. Human rotavirus detection by agglutination of antibody coated erythrocytes. J. Clin. Microbiol. 17:1141–1147.

Santosham, M., Daum, R.S., Dillman, L., Rodriguez, J.L., Luque, S., Russell, R., Kourany, M., Ryder, R.W., Bartlett, A.V., Rosenberg, A., Benenson, A.S., and Sack, R.B. 1982. Oral rehydration therapy of infantile diarrhea. N. Engl. J. Med. 306:1070–1076.

Santosham, M., Yolken, R.H., Quiroz, E., Dillman, L., Oro, G., Reeves, W.C., and Sack, R.B. 1983. Detection of rotavirus in respiratory secretions of children with pneumonia. J. Pediat. 103:583–585.

Saulsbury, F.T., Winkelstein, J.A., and Yolken, R.H. 1980. Chronic rotavirus infection in immunodeficiency. J. Pediat. 97:61–65.

Schnagl, R.D., Rodger, S.M., and Holmes, I.H. 1981. Variation in human rotavirus electropherotypes occurring between rotavirus gastroenteritis epidemics in central Australia. Infect. Immun. 33:17–21.

Schreiber, D.S., Blacklow, N.R., and Trier, J.S. 1973. The mucosal lesion of the proximal small intestine in acute infectious non-bacterial gastroenteritis. N. Engl. J. Med. 288:1318–1323.

Skaug, K., Figenschau, K.J., and Ostravik, I. 1983. A rotavirus staphylococcal co-agglutination test. Acta Pathol. Microbiol. Immunol. Scand. Sect. B 91:175–178.

Smith, E.M., and Gerba, C.P. 1982. Development of a method for the detection of human rotavirus in water and sewage. Appl. Environ. Microbiol. 43:1440–1450.

Soenarto, Y., Sebodo, T., Ridho, R., Alrasjid, H., Rohde, J.E., Bugg, H.C., Barnes, G.L., and Bishop, R.F. 1981. Acute diarrhea and rotavirus infection in newborn babies and children in Yogyakarta, Indonesia, from June 1978 to June 1979. J. Clin. Microbiol. 14:123–129.

Sutmoller, F., Azeredo, R.S., Lacerda, M.D., Barth, O.M., Pereira, H.G., Hoffer, E., and Schatzmayr, H.G. 1982. An outbreak of gastroenteritis caused by both rotavirus and *Shigella sonnei* in a private school in Rio de Janeiro. J. Hyg. 88:285–293.

Suzuki, H., Amano, Y., Kinebuchi, H., Vera, E.G., Davila, A., Lopez, J., Gustabo, R., Konno, T., and Ishida, N. 1981. Rotavirus infection in children with acute gastroenteritis in Ecuador. Am. J. Trop. Med. Hyg. 30:293–294.

Takiff, H.E., Strauss, S.E., and Garon, C.F. 1981. Propagation and in vitro studies of previously non-cultivable enteral adenoviruses in 293 cells. Lancet ii:832–834.

Takiff, H.E., and Strauss, S.E. 1982. Early replicative block prevents the efficient growth of fastidious diarrhea associated adenovirus in cell culture. J. Med. Virol. 9:93–100.

Tallett, S., MacKenzie, C., Middleton, P., Kerzner, B., and Hamilton, R. 1977. Clinical, laboratory and epidemiological features of viral gastroenteritis in infants and children. Pediatrics 60:217–222.

Tan, J.A., and Schnagl, R.D. 1981. Inactivation

of a rotavirus by disinfectants. Med. J. Aust 1:19–23.

Tao, H., Guangmu, C., Changan, W., Aenli, Y., Zhaoying, F., Tungxin, C., Zinyi, C., Weiwe, Y., Xuejian, C., Shuasen, D., Xiaoguang, L., and Weicheng, C. 1984. Waterborne outbreak of rotavirus diarrhoea in adults in China caused by a novel rotavirus. Lancet i:1139–1142.

Taylor, J.W., Gary, G.W., and Greenberg, H.B. 1981. Norwalk related gastroenteritis due to contaminated drinking water. Am. J. Epidemiol. 114:584–92.

Thomas, M.E.M., Luton, P., and Matimer, J.Y. 1981. Virus diarrhoea associated with pale fatty faeces. J. Hyg. 87:313–319.

Thornhill, T.S., Kalica, A.R., Wyatt, R.G., Kapikian, A.Z., and Chanock, R.M. 1975. Pattern of shedding of the Norwalk particle in stools during experimentally induced gastroenteritis in volunteers as determined by immune electron microscopy. J. Infect. Dis. 132:28–34.

Truant, A.L., and Chonmaitree, T. 1982. Incidence of rotavirus infection in different age groups of pediatric patients with gastroenteritis. J. Clin. Microbiol. 16:568–569.

Tufvesson, B., and Johnsson, T. 1976. Immunoelectroosmophoresis for detection of reolike virus: Methodology and comparison with electron microscopy. Acta Pathol. Microbiol. Scand. Sect. B 84:225–228.

Uhnoo, I., Wadell, G., Svensson, L., and Johansson, M.E. 1984. Importance of enteric adenoviruses 40 and 41 in acute gastroenteritis in infants and children. J. Clin. Microbiol. 20:365–372.

Urasawa, S., Urasawa, T., and Taniguchi, K. 1982. Three human rotavirus serotypes demonstrated by plaque neutralization of isolated strains. Infect. Immun. 38:781–784.

Vesikari, T., Sarkkinen, H.K., and Maki, M. 1981. Quantitative aspects of rotavirus excretion in childhood diarrhoea. Acta Paediat. Scand. 70:717–721.

Vollet, J.J., Ericsson, C.D., Gibson, G., Pickering, L.K., DuPont, H.L., Kohl, S., and Conklin, R.H. 1979. Human rotavirus in an adult population with travelers' diarrhea and its relationship to the location of food consumption. J. Med. Virol 4:81–87.

Von-Bonsdorff, C.H., Hovi, T., Makela, P., Morttinen, A. 1978. Rotavirus infections in adults in association with acute gastroenteritis. J. Med. Virol. 2:21–28.

Vonderfecht, S.L., Miskuff, R.L., Eiden, J.J., and Yolken, R.H. 1985. Enzyme immunoassay inhibition assay for the detection of rat rotavirus-like agent in intestinal and fecal specimens obtained from diarrheic rats and humans. J. Clin. Microbiol. 22:726–730.

Walther, F.J., Bruggeman, C., Daniels-Bosman, M.S.M., Pourier, S., Grauls, G., Stals, F., and Bogaard, A.V.D. 1983. Symptomatic and asymptomatic rotavirus infections in hospitalized children. Acta Paediat. Scand. 72:659–663.

Wenman, W.M., Hinde, D., Feltham, S., and Gurwith, M. 1979. Rotavirus infection in adults: Result of a prospective family study. N. Engl. J. Med. 301:303–306.

White, L., Perez, I., Perez, M.A., Urbina, G., Greenberg, H., Kapikian, A., and Flores, J. 1984. Relative frequency of rotavirus subgroups 1 and 2 in Venezuelan children with gastroenteritis as assayed with monoclonal antibodies. J. Clin. Microbiol. 19:516–520.

Wigand R., Baumeister, H.G., Maass, G., Kuhn, J., and Hammer, H.J. 1983. Isolation and identification of enteric adenoviruses. J. Med. Virol. 11:233–240.

Wong, C.J., Price, Z., Bruckner, D.A. 1984. Aseptic meningitis in an infant with rotavirus gastroenteritis. Pediat. Infect. Dis. 3:244–246.

World Health Organization Subgroup of the Scientific Working Group on Epidemiology and Etiology. 1980. Rotavirus and other viral diarrhoeas, Vol. 58. Geneva: World Health Organization, pp. 183–198.

Wyatt, R.G., Dolin, R., Blacklow, N.R., DuPont, H., Buscho, R., Thornhill, T.S., Kapikian, A.Z., and Chanock, R.M. 1974. Comparison of three agents of acute infectious nonbacterial gastroenteritis by cross-challenge in volunteers. J. Infect. Dis. 129:709–714.

Wyatt, R.G., Greenberg, H.B., Dalgard, D.W., Allen, W.P., Sly, D.L. Thornhill, T.S., Chanock, R.M., and Kapikian, A.Z. 1978. Experimental infection of chimpanzees with the Norwalk agent of epidemic viral gastroenteritis. J. Med. Virol. 2:89–96.

Wyatt, R.G., James, W.D., Bohl, E.H., Theil, K.W., Saif, L.J., Kalica, A.R., Greenberg, H.B., Kapikian, A.Z., and Chanock, R.M. 1980. Human rotavirus type 2: Cultivation in vitro. Science 207:189–191.

Wyatt, R.G., Greenberg, H.B., James, W.D., Pittman, A.L., Kalica, A.R., Flores, J., Chanock, R.M., and Kapikian, A.Z. 1982. Definition of human rotavirus serotypes by plaque reduction assay. Infect. Immun. 37:110–115.

Wyatt, R.G., James, H.D., Pittman, A.L., Hoshino, Y., Greenberg, H.B., Kalica, A.R., Flores, J., and Kapikian, A.Z. 1983. Direct isolation in cell culture of human rotaviruses and their characterization into four serotypes. J. Clin. Microbiol. 18:310–317.

Yolken, R.H., Kim, H.W., Clem, T., Wyatt, R.G., Kalica, A.R., Chanock, R.M., and Kapikian, A.Z. 1977a. Enzyme-linked immunosorbent assay (ELISA) for detection of human reovirus-like agent of infantile gastroenteritis. Lancet ii:263–267.

Yolken, R.H., Wyatt, R.G., Kalica, A.R., Kim, H.W., Brandt, C.D., Parrott, R.H., Kapikian, A.Z., and Chanock, R.M. 1977b. Use of a free viral immunofluorescence assay to detect human reovirus-like agent in human stools. Infect. Immun. 16:467–470.

Yolken, R.H., Wyatt, R.G., Zissis, G., Brandt, C.D., Rodriguez, W.J., Kim, H.W., Parrott, R.H., Urrutia, J. J., Mata, L., Greenberg, H.B., Kapikian, A.Z., and Chanock, R.M. 1978. Epidemiology of human rotavirus types 1 and 2 as studied by enzyme-linked immunosorbent assay. N. Engl. J. Med. 299:1156–1161.

Yolken, R.H., and Leister, F.J. 1981. Evaluation of enzyme immunoassays for the detection of human rotavirus. J. Infect. Dis. 144:379.

Yolken, R.H., Lawrence, F., Leister, F., Takiff, H.E., and Strauss, S.E. 1982. Gastroenteritis associated with enteric type adenovirus in hospitalized infants. J. Pediat. 101:21–26.

Yolken, R.H., and Leister, F. 1982. Rapid multiple-determinant enzyme immunoassay for the detection of human rotavirus. J. Infect. Dis. 146:43–46.

Yolken, R.H., and Murphy, M. 1982. Sudden infant death syndrome associated with rotavirus infection. 1982. J. Med. Virol. 10:291–296.

Zissis, G., and Lambert, J.P. 1978. Different serotypes of human rotavirus. Lancet i:38–39.

Zissis, G., and Lambert, J.P. 1980. Enzyme-linked immunosorbent assays adapted for serotyping of human rotavirus strains. J. Clin. Microbiol. 11:1–5.

Chronic Viral Hepatitis

Mario R. Escobar

23.1 INTRODUCTION

23.1.1 Nature and Types of Hepatitis

Hepatitis is an inflammatory disease that may result in necrosis of the liver. Hepatitis can result from noninfectious causes including biliary obstruction, primary biliary cirrhosis, Wilson's disease, drug toxicity and drug hypersensitivity reactions, or from an infectious process associated with certain viruses. Occasionally, it may occur also as a complication of leptospirosis, syphilis, tuberculosis, toxoplasmosis, or amebiasis (Jawetz et al. 1984).

Viral hepatitis refers more specifically to a primary infection of the liver by one of four etiologically associated but different hepatotropic viruses. These four types of hepatitis include: type A (infectious hepatitis, epidemic jaundice, short incubation hepatitis); type B (serum or transfusion hepatitis, homologous serum jaundice, long incubation hepatitis), non-A, non-B (NANB) hepatitis; and type D (Δ) hepatitis. With rare exceptions the clinical manifestations of acute hepatitis caused by these four virus types are virtually identical. Other viruses that can cause sporadic hepatitis include yellow fever virus, human cytomegalovirus (CMV), Epstein–Barr virus (EBV), rubella virus, herpes simplex virus (HSV), varicella-zoster virus (VZV) and some enteroviruses. Histopathologic lesions resulting from these viruses have many common features and the etiologic agent cannot be distinguished on this basis.

Improved immunoassays for the detection of serologic markers of viral hepatitis and the increase of our knowledge and understanding of the immunopathogenesis of viral infections have advanced the field of viral hepatitis very rapidly during the last several years. This chapter presents information regarding the immunodiagnosis of viral hepatitis and a discussion of the clinical and pathologic sequelae in those individuals who did not fully recover from acute hepatitis.

23.1.2 Brief Historical Review of the Agents of Viral Hepatitis

Certain clinical conditions characterized by jaundice were probably grouped together as hepatitis since antiquity. It was not until the seventeenth and eighteenth centuries that epidemics of what would now be considered type A hepatitis were well documented. Many of these epidemics occurred during wars between nations when large concentrations of military personnel provided the epidemiologic setting for spread of the disease. In 1947, the terms hepatitis A and hepatitis B were proposed to designate "infectious" and "serum" hepatitis, respectively. A large series of epidemiologic studies and experiments employing animals and human volunteers of all ages spanning from the mid-1940s to the early 1960s provided clear evidence that these two types of viral hepatitis were microbiologically and epidemiologically distinct (Krugman et al., 1962; 1967). Despite the numerous in vivo and in vitro studies, which resulted in a better understanding of the nature of viral hepatitis, none of the intensive efforts to identify the etiologic agents succeeded until later. The first indication of type B hepatitis came from Bremen, Germany in 1885 involving shipyard workers who had been vaccinated against smallpox a few months earlier with glycerinated vesicular fluid of other vaccinated humans. Similar reports of parenterally transmitted viral hepatitis were published in subsequent years. Although these cases occurred in different settings (e.g., diabetes, tuberculosis, and venereal disease clinics, as well as immunization and blood transfusion services), all had in common the use of contaminated materials and instruments (Zuckerman, 1976).

23.1.2.a Hepatitis A

The series of events that took place following the discovery of the Australia antigen by Blumberg (discussed in further detail below) were instrumental in elucidating the nature and identity of the etiologic agent of hepatitis A. Holmes et al. (1969) conclusively demonstrated that marmoset monkeys were susceptible to infection with hepatitis A virus (HAV) and reported on the use of the marmoset model to perform a neutralization test as a measure of serum antibody (Holmes et al., 1973). Hepatitis A virus particles were observed by Feinstone et al. (1973) in the feces of infected humans and by Provost et al. (1975) in the livers of infected marmosets. Until the isolation of HAV in tissue culture by Provost et al. (1979), the knowledge of this virus had evolved from immunologic identification of virus antigens. Many techniques have been developed for the detection of HAV in clinical specimens. Additionally, serologic tests have been devised to demonstrate the presence of antibodies to HAV. More recently, in vitro studies of HAV in a variety of cells in culture (Purcell et al., 1984), the availability of animal models (Dienstag et al., 1975; Mao et al., 1981), and the molecular cloning of the genome (Ticehurst et al., 1983) have improved greatly the prospects for the development of a live attenuated vaccine.

23.1.2.b Hepatitis B

In 1961, Blumberg et al. (1964) performed a systematic analysis of the sera of multitransfused patients for isoprecipitins employing an agar gel double diffusion procedure. This study led to the discovery in 1962 (Blumberg, 1964) of a human lipoprotein polymorphism (AG system) demonstrating inherited antigenic difference among low density (β) lipoproteins. It was in 1963 when, serendipitously, a precipitating antibody was detected in the serum of a multitransfused patient with hemophilia, which reacted with only a single member of the 24 sera in the test panel, thus, revealing an antigen clearly different from those lipoprotein antigens previously identified. Because this unique antigen was detected in the serum obtained from an Australian aborigine, it was called the Australian antigen and designated by the symbol Au[1] (Blumberg, 1964). The epidemiologic significance, as well as the genetics and physical characteristics of the newly discovered antigen, was rapidly investigated during the course of several years following this important discovery. However, confirmation of its association with acute viral hepatitis was not published until 1968 by investigators in Italy, Japan (Okochi and Murakami, 1968), and the United States (Blumberg et al., 1968) working independently. One of the best controlled clinical studies was carried out at Willowbrook (Krugman et al., 1979) where the new antigen was isolated from the sera of patients who suffered from "long-incubation-period serum hepatitis." Within a few years from this momentous discovery, the hepatitis B virus (HBV) was structurally characterized, its serologic determinants defined, and its etiologic role in acute and chronic hepatitis recognized. Although many questions concerning the host–virus interaction still remain unanswered, the accumulation of enough basic scientific information led to the successful development of an efficacious HBV vaccine within slightly over a decade following the identification of the virus (Szmuness et al., 1980). Moreover, the recently discovered viruses that are structurally and immunologically related to the human HBV but are pathogenic for animals, will provide important experimental models of hepatitis and hepatocellular carcinoma (Ganem, 1984).

23.1.2.c Hepatitis non-A, non-B

Non-A, non-B hepatitis causes over 90% of cases of posttransfusion hepatitis and 20% of sporadic acute viral hepatitis. It has been established on the basis of clinical, epidemiologic, serologic, and electron microscopic observations that the agents of NANB hepatitis are distinct from the known viruses responsible for acute and chronic hepatitis in humans. Infection with the agents of hepatitis A and B, as well as with other viruses also associated with hepatitis (e.g., CMV and EBV), has been excluded in a number of patients with acute hepatitis. This exclusion was possible as a result of the development of sensitive serologic tests to diagnose infections by these other agents. It was also known that hepatitis could occur in individuals who had previously recovered from hepatitis B (Hoofnagle et al., 1977; Mosley et al., 1977). Studies in chimpanzees confirmed these findings and showed conversely that hepatitis B could be transmitted to animals that had recovered from NANB hepatitis (Tabor et al., 1978; Gerety et al., 1984). In contrast to an unconfirmed series of reports from the same laboratory (Hantz et al., 1980; Trepo et al., 1981), no HBV-like particles or serologic crossreactivity with HBV have been observed in NANB hepatitis (Hoofnagle et al., 1977). Analogous experiments were carried out to differentiate hepatitis A from NANB (Hoofnagle et al., 1977; Tabor et al., 1978; Khuroo, 1980). Studies have eliminated CMV, EBV, and some other viruses as etiologic agents of NANB hepatitis (Tabor et al., 1978). Diagnosis of NANB hepatitis remains a diagnosis of exclusion because all direct means for identification of specific antigens or antibodies have been unsuccessful. In other words, our knowledge of this type of hepatitis is at a stage that is comparable with our knowledge of hepatitis A and B 25 years ago, because of the absence of defined serologic markers. A review of the literature is misleading in this respect because there are more than 30 reports of detection systems for NANB; yet there is general agreement among experts in the field that none of these serologic systems is sufficiently specific, sensitive, or will reproducibly detect NANB hepatitis (Alter and Hoofnagel, 1984).

23.1.2.d Hepatitis D

The immunofluorescent detection of the Δ/anti-Δ antigen system in liver cell nuclei and in serum of HBV carriers in southern Italy was first reported by Rizzetto et al. (1977). The Δ agent, or hepatitis D virus (HDV), is unique in that it consists of an RNA genome smaller than the genome of conventional viruses but larger than that of viroids of plants. It requires HBV infection for its replication. Since its discovery, many reports have been published dealing with all aspects of HDV infection. Although many of these aspects need further study, our basic knowledge of this type of hepatitis is greater than that of NANB hepatitis. In fact, the recent availability of commercial reagent kits for its diagnosis by the clinical laboratory, as well as the application of a probe for the direct measurement of Δ RNA in clinical samples has opened new possibilities for a better understanding of the host–

HDV and HBV–HDV interactions in both acute and chronic hepatitis.

23.2 BRIEF DESCRIPTION OF THE CHARACTERISTICS OF THE VIRUSES

23.2.1 Hepatitis A Virus

It has been proposed recently that HAV should be classified as enterovirus 72. Hepatitis A virus is a 27–32 nm spherical naked particle with cubic symmetry. The viral capsid consists of 32 capsomeres arranged in an icosahedral conformation. The unit structure of the capsid antigen consists of four polypeptides: viral proteins 1 through 4 (VP1–VP4). The capsid surrounds a linear molecule of single-stranded RNA made up of about 8100 nucleotides with a molecular weight of approximately 2.25×10^6 daltons (Ticehurst et al., 1983). It has been suggested that the RNA has positive polarity; that the 3′ end of the RNA is polyadenylated and, that the 5′ end has a small protein, the so-called "viral protein, genomic" (VPg), which may aid the virus in attaching to ribosomes, as is the case with other enteroviruses (Ganem et al., 1984). Hepatitis A virus is stable to treatment with ether (20%), acid (pH 2.4), and heat (60°C/hr). It is destroyed by autoclaving (120°C/20 min), by boiling in water for 5 minutes, by dry heat (180°C/hr), by ultraviolet irradiation (1 minute at 1.1 watts), and by treatment with formalin (1:4000 for 3 days at 37°C). Its infectivity can be preserved for at least 1 month after lyophilization and storage at 25°C and 42% relative humidity or for years at −20°C.

23.2.2 Hepatitis B Virus

Hepatitis B virus, also referred to as the "Dane particle," belongs to a new class called "Hepadna viruses." The virion is a complex 42-nm double shelled particle. The outer surface, or envelope, contains hepatitis B surface antigen (HBsAg) and surrounds a 27-nm inner core that contains hepatitis B core antigen (HBcAg). Inside the core is the genome of HBV, a single molecule of circular DNA. One strand is complete, containing all of the genetic information for the production of HBsAg and HBcAg but is "nicked" (i.e., its 3′ and 5′ ends are not joined) (Hoofnagle, 1981). The complementary strand contains 50–90% of the complementary sequences, thus, leaving a single-stranded or gap region. The DNA molecule consists of 3200 base pairs and has a molecular weight in excess of 2×10^6 daltons. It is the variable length of the gap region that results in genetically heterogeneous particles with a wide range of buoyant densities. It has been proposed that this unique structure of the viral core may be biologically relevant for integration of the DNA genome into the chromosome of the hepatocyte and, thus, would represent an important step in the eventual development of hepatocellular carcinoma (Hoofnagle, 1981). In addition, the HBV core also contains DNA-dependent DNA polymerase (which acts to complete the single-stranded region of the DNA), as well as hepatitis B e antigen (HBeAg). The latter can be detected in the serum of HBV infected individuals either as a single protein with a molecular weight of 19,000 daltons or complexed with serum immunoglobulin with a molecular weight of about 300,000 daltons. Its major clinical and epidemiologic significance relates to the finding that it appears to be a reliable serologic marker for the presence of high levels of HBV and, thus, a high degree of infectivity. Other markers with the same connotation include the Dane particle, DNA polymerase, or specific HBV-DNA. However, the procedures employed to demonstrate the presence of these markers are tedious and complicated in contrast to the use of a single enzyme immunoassay for detection of HBeAg.

HBsAg is the viral component found in the highest concentration in the serum of infected individuals (up to 10^{13} particles per milliliter). Its stability does not always coincide with that of the DNA particle. Nevertheless, both are stable at −20°C for more than 20 years and stable after repeated freezing and thawing. Infectivity is not destroyed by incubation at 37°C for 60 minutes and the virus remains viable after dessication and storage at 25°C for 1 week. The Dane particle (but not HBsAg) is inactivated at higher temperatures (100°C for 1 minute) or when incubated for longer periods (60°C for 10 hours) depending on the amount of virus present in the sample. HBsAg is stable at pH 2.4 for up to 6 hours with loss of viral infectivity. Sodium hypochlorite, 0.5% (e.g., 1:10 household bleach) destroys antigenicity within 3 minutes at low protein concentrations, but undiluted serum may require a higher bleach concentration (5%). HBsAg is not destroyed by UV irradiation of plasma or other blood products and viral infectivity may be retained after such treatment. Hepatitis B virus is unevenly distributed during Cohn ethanol fractionation of plasma. Most of the virus is contained in fraction I (fibrinogen, factor VIII) or III (prothrombin complex), whereas, HBsAg is found in fractions II (γ globulin) and IV (plasma protein).

23.2.3 Hepatitis non-A, non-B Viruses

Clinical and epidemiologic studies and cross-challenge experiments in chimpanzees have shown that there are at least two NANB hepatitis agents (Shimizu et al., 1979; Bradley et al., 1980). Virus-like particles have been identified by immunoelectron microscopy (IEM) in human serum and liver tissue. The serum of patients with chronic NANB hepatitis has revealed 37-nm particles, whereas, the liver of patients with acute NANB has well defined 27-nm particles inside the hepatocyte nucleus. The viability of these agents appears to be destroyed by heating at 60°C/10 hours (Gerety et al., 1984), and may be inactivated by chloroform and formalin (Bradley et al., 1983; Feinstone et al., 1983).

23.2.4 Hepatitis D Virus

Hepatitis D virus is a 35–37 nm particle with a buoyant density of 1.24–1.25 g/ml. It contains HBsAg on the surface and Δ antigen and a small RNA genome (with a molecular weight of 5.5×10^5) in the interior (Rizzetto et al., 1980). The particle consists of an organized structure that prevents hydrolysis of the internal RNA, thus, protecting the infectivity of the virus (Hoyer et al., 1983). The Δ antigen is distinct from the known antigenic determinants of HBV. It is localized to hepatocyte nuclei that have no HBcAg present. It has a buoyant density in cesium chloride of 1.28 g/ml and a molecular weight of 68,000 daltons (Jawetz et al., 1984). No homology with the HBV genome has been found using hybridization techniques. The Δ agent is believed to be a defective virus that replicates only in HBV-infected cells. It appears to be resistant to treatment with EDTA, detergents, ether, nuclease, glycosidases, and acid; but partial to complete inactivation follows treatment with alkali, thiocyanate, guanidine hydrochloride, trichloroacetic acid, and proteolytic enzymes.

23.3 PATHOGENESIS OF VIRAL HEPATITIS

During its clinical course, viral hepatitis can lead to any one of four events: a) typical, icteric hepatitis; b) subclinical and anicteric hepatitis; c) fulminant hepatitis; or d) chronic hepatitis. The clinical manifestations of acute hepatitis for each of the four etiologic types of viral hepatitis are very similar. For every clinically apparent case of typical acute (icteric) hepatitis, there are several subclinical cases that can be documented only by the detection of antibody to one of the hepatitis viruses when a history of hepatitis or jaundice is denied. Fulminant hepatitis is a rare but dramatic outcome of viral hepatitis, which presents as an acute disease characterized by hepatic failure and symptoms of hepatic encephalopathy. Chronic hepatitis is a rare aftermath of HAV infection, but it does occur in about 10% of type B and 40–60% of NANB hepatitis cases. Chronic Δ hepatitis generally can be seen in an individual with stable or inactive chronic HBV infection.

23.3.1 Hepatitis A

The average incubation period for HAV infection is about 1 month. Hepatitis A virus infection rates are higher in close contact settings and antibody prevalence rates increase with age. Up to 50% of cases may be anicteric. Unlike HBV infection, a chronic carrier state does not occur and fatalities are very rare; infected patients recover fully and are immune to reinfection.

The pathogenesis of HAV infection is not fully understood. It appears that HAV enters the portal blood from the intestine and is transported to the liver. The specific cells involved in the primary site of infection have not been identified. In experimentally infected chimpanzees and marmosets, hepatitis A antigen has not been detected in intestinal cells during the acute illness (Mathiesen et al., 1977, 1978). However, this does not rule out the possibility of an earlier intestinal phase of replication. Viral antigen can be detected by immunofluorescence (IF) in the cytoplasm of chimpanzee and marmoset hepatocytes beginning approximately 2 weeks after inoculation and continuing for 3 to 4 weeks thereafter. This viral antigen is usually found about 1 week before liver enzyme elevations or histopathologic evidence of hepatitis. The ability of HAV to persist in humans is unresolved, however, viral antigen has been shown to persist in the hepatocytes of some marmosets for 3 to 4 months after inoculation with HAV (Mathiesen et al., 1978). Virus can no longer be detected in the feces of these animals after the appearance of serum antibodies. It cannot be ascertained, however, whether or not infectious virus continues to be shed at a low level during the extended period when antigen is detected in the liver.

23.3.2 Hepatitis B

The incubation period of HBV is usually about 3 months, with a range between 45 and 180 days. Although HBV is transmitted predominantly by the percutaneous route, it is also recognized that it can be transmitted by other means, because HBsAg has been detected in virtually every type of body fluid. Spread of HBV by oral and/or genital contact has been demonstrated. Although HBV preferentially replicates in hepatocytes, it has been reported to replicate in nonhepatocytes and in lymphoid elements of hemopoietic tissue (Blum et al., 1984).

The natural history of the different forms of HBV infection, as well as the pathogenetic mechanisms responsible for the clinical course of hepatitis type B, are not clear. A diagram outlining the potentially immunopathologic sequelae of HBV infection is shown in Figure 23.1. HBV infection may result in a variety of syndromes, including subclinical infection with or without hepatitis B surface antigenemia (<50%), acute hepatitis with resolution of illness (30–40%), fulminant hepatitis or subacute hepatic necrosis with possible death within 3 months (1–3%), chronic active hepatitis frequently resulting in cirrhosis, chronic persistent hepatitis, or a silent carrier state with minimal (or absent) liver damage or even primary hepatocellular carcinoma (5–10% of adults after either symptomatic or subclinical acute hepatitis B).

No single immunologic alteration, as identified in laboratory testing, adequately explains the differences in outcomes among those infected with HBV. Differences in outcome may be related to immunomodulatory activities such as competition between cytolytic T-cells and circulating antibody to HBcAg, serum and liver-derived immunoregulatory molecules and immunoregulatory T-cell influences (Dienstag, 1984).

23.3.2.a The Host

Certain host factors have been considered as potential pathogenic determinants of viral hepatitis. These include age, genetic background (Blumberg et al., 1970), sex, physiologic state, and the immune response

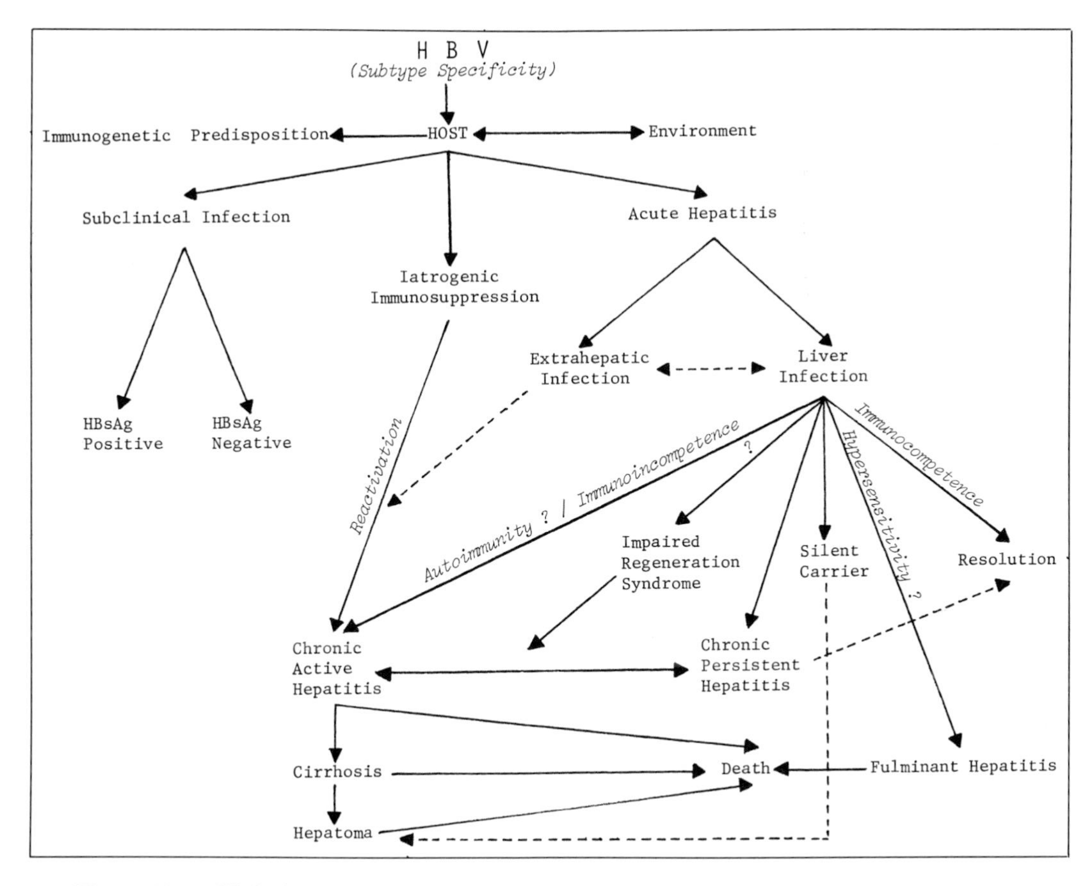

Figure 23.1. Clinical course and immunopathologic sequelae of hepatitis B virus infection.

of the host to viral or autoantigens (Dudley et al., 1971).

Recent work suggests a link between a variety of human diseases and alleles within the major histocompatibility gene complex (HLA). Despite a weak association between HBV infection and some HLA phenotypes, the most convincing evidence for a genetic role in this disease is its very high prevalence in Down's syndrome and in certain Asian and African populations. However, environmental factors could just as easily explain these findings (Levy and Chisari, 1981). The clinical course that occurs in an individual following infection with HBV may be determined—at least in part—by three factors: a) immunogenetic predisposition as related to the nature of the "match between the antigens of the virus and those of the host"; b) immunocompetence as regards the humoral and cellular immune responses of the infected individual; and c) iatrogenic immunosuppression occurring as a result of immunosuppressive drug therapy, which may lead to HBV reactivation.

23.3.2.b The Virus

The persistence of virus replication in silent carriers (Dudley et al., 1972), the exacerbation of biochemical abnormalities (e.g., liver enzymes, etc.), which follow withdrawal of immunosuppressive therapy in patients with chronic active hepatitis B and the occurrence of often severe acute hepatitis in HBsAg-positive cancer patients following withdrawal of cytotoxic chemotherapy (Dienstag, 1984), suggest that the immune response to HBV, and not the virus itself, is cytopathic for hepatocytes. Nonetheless, viral persistence is often associated with chronic hepatitis (Melnick et al., 1977) suggesting that the virus may play the role of a trigger if not a causative factor in the pathogenesis of this disease. The possibility that virus strain, dose and route of inoculation, or cofactors such as drugs, play a role in determining severity and chronicity has not been discounted and there are even suggestions that these factors are important (Barker and Murray, 1972; Schaefer et al., 1984).

23.3.3 Hepatitis non-A, non-B

Non-A, non-B hepatitis has an incubation period ranging from 2 to 12 weeks postexposure. Two incubation periods are found within this range, a short incubation period of 2 to 4 weeks or a longer time interval of 8 to 12 weeks. This is consistent with the existence of at least two etiologic agents.

As long as the agents of NANB hepatitis remain unidentified, it will be virtually impossible to consider the immunopathogenic mechanisms responsible for this type of hepatitis. Nevertheless, three main features that are related to its pathogenesis are: a) The characteristic pattern of enzyme elevations with marked spontaneous fluctuations in serum alanine aminotransferase levels. This enzyme pattern is rare in hepatitis B, except during periods of reactivation (Hoofnagle and Alter, 1984). b) The histopathologic pattern characterized by fatty metamorphosis. It has been suggested that this feature is more typical of a cytopathic, rather than an immunologically, mediated injury (Dienes et al., 1982). c) The high incidence of chronicity with at least 50% of the patients developing chronic liver disease following acute NANB hepatitis (Dienstag, 1983). Chronic NANB hepatitis tends to be a mildly symptomatic or a silent disease that nevertheless progresses to cirrhosis in about 20% of cases.

23.3.4 Hepatitis D

The incubation period of hepatitis D varies from 2 to 12 weeks, with shorter incubation periods present in HBV carriers who are superinfected with the agent than in susceptible individuals who are simultaneously infected with both HBV and HDV. That is, the mechanism and outcome of the infection are different in the normal individual and in the individual who is already infected with HBV at the time of exposure to HDV. In normal individuals exposed to HDV (and simultaneously to HBV), the HBsAg necessary to rescue the infective pathogen is induced acutely by the HBV component of the infectious inoculum. In this context, a limiting factor for the expression of HDV is the short duration of the hepatitis B viremia (Smedile et al., 1981; Caredda et al., 1983). The clinical course of the HDV-associated disease depends on whether the carrier was asymptomatic or suffered from chronic active hepatitis B prior to infection with HDV. The silent carrier may present with an acute HBsAg-positive hepatitis with all the features of classic hepatitis B, except for the possible lack of anti-core IgM antibody (Farci et al., 1983). The disease is frequently severe, perhaps due to the explosive replication of HDV in the patient. This form accounts for a significant number of fulminant HBsAg-positive hepatitis cases in Europe and the U.S. (Smedile et al., 1982; Govindarajan et al., 1984). The acute hepatitis D has a tendency to become chronic, which creates a diagnostic problem if the past HBsAg reactivity of the patient is unknown, as transition to chronicity with HDV could resemble progressive hepatitis B (Farci et al., 1983). Conversely in patients with prior HBV-associated hepatitis, the primary infection with HDV may actually coincide with an apparent flare up of the chronic hepatitis. Such exacerbations have been fatal in some cases (Raimondo et al., 1983). The liver damage in hepatitis D has been associated with a virus mediated cytopathic mechanism, thus, explaining the failure of conventional immunosuppressive therapy to alter the course of the disease (Rizzetto et al., 1983). The activity of the chronic hepatitis correlates with the presence but not with the amount of Δ antigen in the liver. The immunologic features of the

HDV infection are still under investigation. However, it has been observed that children, who typically develop mild cases of hepatitis B, often suffer severe and even fatal cases of hepatitis D (Popper et al., 1983). The same occurs in patients on renal dialysis (Smedile et al., 1982), suggesting that immunologic factors may be less important in the pathogenesis of hepatitis due to HDV compared with HBV.

23.4 PATHOLOGY OF VIRAL HEPATITIS

23.4.1 Hepatitis A

Ordinarily, acute HAV and HBV human infections cannot be distinguished by histopathologic criteria. Hepatitis A virus infection has no chronic stage, and the lesions found in chronic viral hepatitis do not apply to HAV infection. In the chimpanzee, certain differences between the hepatic lesions of HAV and HBV have been reported (Dienstag et al., 1975). In HAV infections, the changes in the hepatocytes were primarily in the periportal areas and parenchymal histopathology was less severe than portal inflammation. In contrast, HBV infection tended to involve the entire lobular parenchyma but was predominantly centrilobular. The portal areas were less affected in HBV than in HAV infection. Although these experimental findings may mimic those in human hepatitis, the small number of biopsies taken from hepatitis A patients cannot support this assumption.

23.4.2 Hepatitis B

As described for hepatitis A, the histopathologic changes in the human liver during acute HBV infection have not been distinguished from those of acute HAV infection, except by special stains. Hepatitis B virus can be demonstrated in liver tissue by detection of HBsAg and HBcAg in infected hepatocytes (Phillips and Poucell, 1981). HBsAg has been detected by histochemical stains, such as aldehyde fuchsin (Shikata et al., 1974), aldehyde thionine (Shikata et al, 1974), modified orcein (Shikata et al., 1974), modified trichrome (Gubetta et al., 1977) Victorian blue, (Tanaka et al., 1981), and resorcin fuchsin (Senba, 1982). Immunohistochemical procedures, such as IF and immunoperoxidase (Huang, 1975), as well as electron and IEM methods using immunoferritin and immunoperoxidase (Huang and Neurath, 1979) have been employed successfully for the histologic detection of HBV infection.

The histologic picture in HBV infection varies according to the severity of the disease, ranging from loss of scattered individual hepatocytes and replacement by inflammatory cells, to massive necrosis of whole lobules of the liver (Peters, 1975). Characteristic lesions include acidophilic bodies (the dehydrated remnants of individual hepatocytes, sometimes with pycnotic nuclei), hepatocytolysis, "ballooned" hepatocytes that appear swollen and pale, excess lipofuscin pigment (resulting from breakdown of liver cells), and scattered or portal inflammation consisting of accumulations of Kupffer cells, lymphocytes, plasma cells, eosinophils, and fibroblasts. Chronic hepatitis may contain some of these elements and, in addition, varying degrees of fibrosis. Viral inclusions have not been reported in either acute or chronic viral hepatitis.

23.4.3 Hepatitis non-A, non-B

Non-A, non-B hepatitis can progress from the acute phase through the severe stages as those described for HBV infection, but with a considerably higher tendency to become chronic. Although NANB hepatitis was described only a few years following the discovery of HBV, the etiologic agents of this disease have yet to be identified. Nonetheless, more than 13 different particulate structures have been associated with NANB hepatitis (Alter and Hoofnagle, 1984). These have shown great diversity in size and morphologic characteristics and none has been independently confirmed or shown to be aggregated by an antibody proven to be specific for the NANB agent. It is probable that most of these particles represent subcellular organelles, lipoprotein, or adventitious agents, but no definitive statement can be made until more reliable reagents become available for IEM (Alter and Hoofnagle, 1984). The histologic alterations found in human liver biopsy specimens, for the most part, are similar to those seen in cases of HBV infection, except that the lesions in NANB hepatitis are the result of virus mediated cytotoxicity rather than immune-mediated cytotoxicity as is characteristic of hepatitis B on light microscopy, with the former usually being less severe (Popper, 1984). In this regard, there is a certain similarity between the lesion produced in hepatitis NANB and D (Popper, 1984). In addition, there have been cytoplasmic tubular structures observed in thin-section electron micrographs of chimpanzee liver infected with human NANB agents. These cytoplasmic tubules consist of double-unit membranes with an electron dense center contiguous with the smooth endoplasmic reticulum. Also observed in chimpanzee liver were 20- to 27-nm intranuclear particles. The intranuclear and cytoplasmic ultrastructures were originally thought to be mutually exclusive and perhaps to represent the morphologic footprint of two NANB agents (Shimizu et al., 1979). Later studies, however, not only disproved their mutual exclusivity but, as reported recently (Schaff et al., 1984), the nuclear particles are neither associated with a second NANB agent nor are they specific for NANB infection (Alter and Hoofnagle, 1984). Nevertheless, the cytoplasmic tubular structures constitute a valuable histologic marker of NANB infection even though only a single recent report noted their detection in liver biopsies of humans with NANB hepatitis (Watanabe et al., 1984).

23.4.4 Hepatitis D

The morphologic changes seen in NANB and Δ hepatitis appear to be very similar with predominant cytotoxic rather than lymphocytotoxic effect (Popper, 1984).

Biopsy specimens of hepatitis B carriers who had HDV infection demonstrated eosinophilic clumping of the cytoplasm of the hepatocyte, sometimes progressing to acidophilic bodies and only rarely fine droplet steatosis; lobular inflammatory cells were lymphocytes, mainly in the sinusoidal lumen, as well as macrophages with periodic acid–Schiff (PAS)-positive nonglycogenic granules in the perisinusoidal spaces, whereas, some parenchymal areas with significant hepatocellular alterations were free of inflammatory cells. By contrast, portal tracts were expanded and infiltrated by a large number of mononuclear cells (Rizzetto et al., 1983). These lesions in the acute stage were in distinct contrast to lesions of hepatitis B in chimpanzees. In HBV hepatocellular degeneration, including acidophilic bodies, was associated with many lymphocytes, often in close contact with normal and abnormal hepatocytes (Bianchi and Gudat, 1979). These cytotoxic lesions characteristic of HDV also have been reported in studies of hepatitis D in young South American Indians, chronic carriers of HBV, during an outbreak in Venezuela (Hadler et al., 1984). It was found that in the initial stage, within a few days after onset of symptoms, the liver exhibited numerous small fat droplets in hepatocytes that underwent focal necrosis, preceded by eosinophilic alterations and acidophilic bodies. Most of the inflammatory cells within the lobular parenchyma were PAS-positive macrophages, while the expanded portal tracts contained a large number of lymphocytes. Delta (Δ) antigen was detected immunochemically in the nuclei of scattered hepatocytes. Subsequent stages of massive necrosis and postnecrotic collapse were not associated with detectable Δ antigen, whereas, autopsy specimens, obtained several months after onset of symptoms where there was a transition to cirrhosis underway, contained large amounts of Δ antigen. Finally, there is a similarity in the histopathologic features of hepatitis D and Labrea or black fever, the etiology of which remains unknown (Andrade et al., 1983).

23.5 LABORATORY DIAGNOSIS OF VIRAL HEPATITIS

23.5.1 Histologic, Biochemical, and Hematologic Tests

Liver biopsy, when indicated, provides tissue for a histologic diagnosis of hepatitis. Tests for abnormal liver function, such as alanine aminotransferase (ALT; formerly SGPT) and bilirubin, supplement the clinical, pathologic, and epidemiologic data. Alanine amino transferase levels in acute hepatitis range between 500 and 2000 units, and are almost never below 100 units; they are usually higher than those of serum aspartate transaminase (AST: formerly SGOT). A sharp elevation of ALT with a short duration (3 to 19 days) is more suggestive of HAV infection, and a gradual rise with a longer duration (35 to 200 days) is more consistent with HBV infection. Fluctuating levels of ALT are more characteristic of NANB hepatitis. Serum albumin levels are decreased and total serum globulin is increased. In many patients with HAV infection, there is an abnormally high level of total IgM 3 to 4 days after ALT begins to rise. Hepatitis B patients have normal to slightly elevated IgM levels. Leukopenia is typical in the preicteric phase and may be followed by a relative lymphocytosis. Large atypical lymphocytes similar to those found in infectious mononucleosis may occasionally be present on smears prepared for differential white blood cell counts, but these generally do not exceed 10% of the total lymphocyte population.

23.5.2 Cell Culture for Virus Isolation

Of all the viruses associated with primary viral hepatitis only HAV has been isolated in cell culture, in 1979 (Provost et al.). Its propagation was quickly confirmed by others and its host range in vitro was extended to other cells of primate origin, mainly primary, secondary, and continuous kidney cells (Provost et al., 1981; Binn et al., 1984) and diploid fibroblasts (Gauss-Müeller et al., 1981). Initially, virus grew poorly in cell culture with no cytopathic effect and was characterized by a long replicative cycle measured in weeks and low yield of virus and viral antigen. Staining of cultures with fluorescein-labeled anti-HAV revealed the presence of discrete fluorescent foci that slowly enlarged with time, suggesting that the virus was mostly cell-associated and not readily released into the tissue culture medium. It has been reported that the incubation period has been shortened to 3 weeks and the infectivity titers have been increased to 10^7 to 10^9 tissue culture 50% infectious doses per milliliter of lysed cells (Binn et al., 1984; Purcell et al., 1984). In addition, a rapidly replicating strain of HAV has been developed in Buffalo green monkey kidney cells, which can be passaged every 3 days. This rapidly replicating virus can be used to test for neutralizing anti-HAV and is as sensitive as commercially available radioimmunoassays (RIA) for anti-HAV antibody detection (Purcell et al., 1984), a finding consistent with that of Lemon and Binn (1983).

23.5.3 Serologic Markers and Immunodiagnosis in Viral Hepatitis

23.5.3.a Hepatitis A

The presence of HAV in stool and the humoral immune responses of the host from time of exposure up to 12 weeks postinfection vis a vis the presence of clinical symptoms and liver enzyme elevation are illustrated in Figure 23.2. Initial evaluation of acute viral hepatitis should include tests for HAV and HBV. Assays used 10 years ago for anti-HAV included complement fixation, IEM, and immune adherence hemagglutination (Chernesky et al., 1984). Currently, the most widely used immunoassay for IgM class anti-HAV employs anti-human IgM as a solid phase capture reagent, wherein reaction with a dilution of the patient's test serum results in the bind-

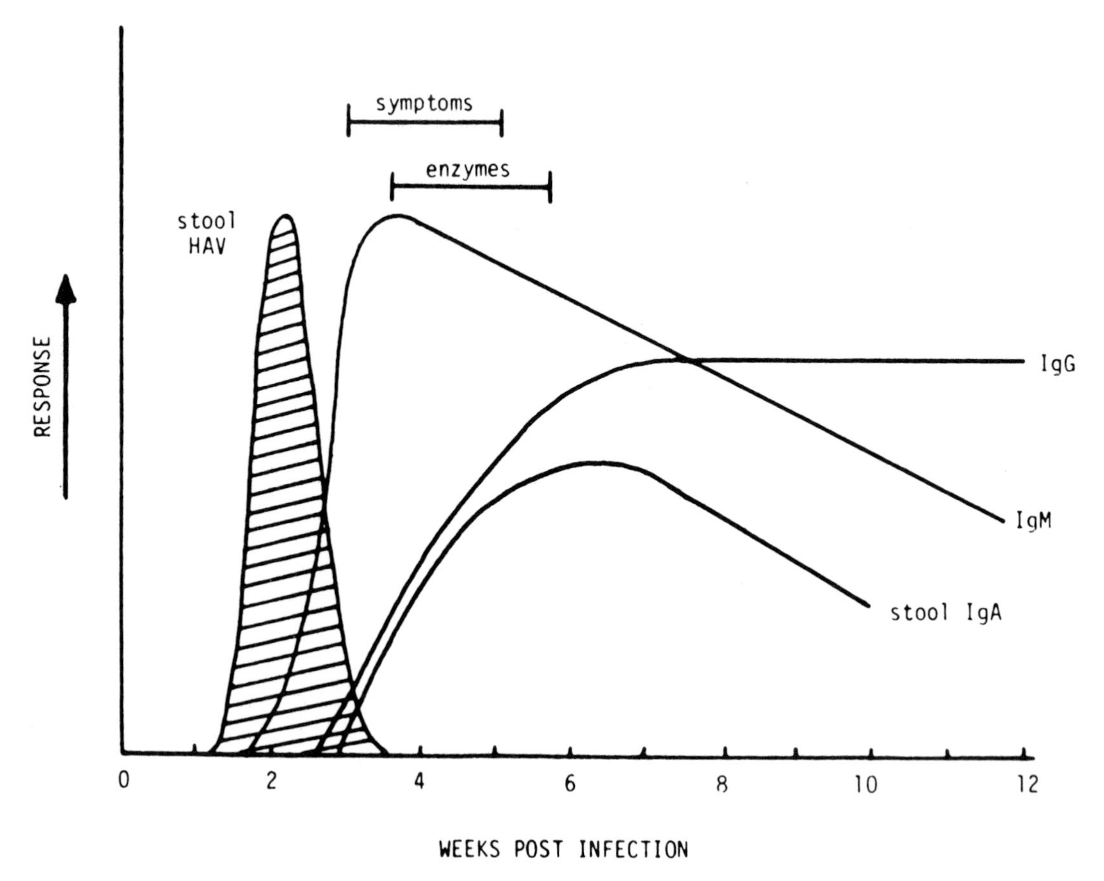

Figure 23.2. Virus detection and humoral immune responses in hepatitis A. Reproduced from Chernesky et al., 1984.

ing of a portion of all IgM molecules present in the specimen. In second and third reaction sequences, HAV and labeled anti-HAV probe are used to detect the presence of IgM anti-HAV, and either radioactive- or enzyme-conjugated anti-HAV probes may be utilized. These tests are sensitive, specific, and practical for diagnosing hepatitis A. The key for specificity of the test is that the solid phase must not bind human IgG, because even minimum binding will lead to interpreting high levels of IgG anti-HAV as low levels of IgM (Decker et al., 1984). Anti-HAV IgM can also be detected with tests that measure total anti-HAV immunoglobulin after serum treatment with staphylococcal protein A (Bradley et al., 1979), but caution must be exercised because this treatment does not remove all subclasses of IgG. Positive results may be confirmed by treatment of the serum with 2-mercaptoethanol. Of course, tests for the detection of total anti-HAV immunoglobulins also may be used to assess past immunity in a single serum or for diagnostic purposes when paired sera show a fourfold or higher titer rise. Solid phase IgM-specific immunoassays are best because they are extremely accurate in measuring anti-HAV IgM responses up to 4 to 6 months (Chernesky et al., 1984). An alternate procedure for anti-HAV IgM detection using murine-derived monoclonal anti-HAV has been reported recently (Decker et al., 1984). It is more convenient than the conventional solid phase immunoassay described earlier and appears to have certain advantages for diagnostic effectiveness.

In most cases, it is impractical to test fecal specimens for HAV particles or antigens because most patients have few, if any, particles in their feces during the clinical phase of the illness, and what is present may be heavily coated by antibody preventing serologic detection. Nonetheless, examination of feces from contacts of patients with hepatitis A, during an epidemic outbreak, may be useful epidemiologically (Chernesky et al., 1984).

23.5.3.b Hepatitis B

The present unavailability of cell cultures for isolation of HBV limits the laboratory diagnosis of hepatitis B to the detection of serologic markers by sensitive immunoassays. Third generation procedures including solid phase immunoassays using enzyme- or radioactive-labeled conjugates, latex agglutination, and hemagglutination tests are commercially available for the detection of at least six serologic markers of HBV infection. These markers include HBsAg, anti-HBs, anti-HBc IgG, anti-HBc IgM, HBeAg, and anti-HBe. Other serologic markers of HBV infection include viral DNA polymerase (Kaplan et al., 1973), HBcAg (Rizzetto et al., 1981), and HBV DNA (Berninger et al., 1982). However, tests for the presence of these latter three markers are impractical for routine use and do not add significantly to the information provided by the immunoassays for HBeAg detection. This relatively complicated system of serologic markers of HBV infection may help establish the stage of the disease, degree of infectivity, prognosis, and immune status of the patient. The various phases of HBV infection as defined by the reactions obtained using these serologic markers are shown in Table 23.1. A typical serologic profile of HBV-associated markers after onset of acute hepatitis B is illustrated in Figure 23.3. A recent review deals

Table 23.1 Serodiagnostic Profiles of HBV Infection

	Serological markers					
Interpretation	HBsAg	HBeAg	IgM anti-HBc	Total anti-HBc	Anti-HBe	Anti-HBs
Acute infection						
Incubation period	+	+	−	−	−	−
Acute phase	+	+	+	+	−	−[b]
Early convalescent phase	+	−[a]	+	+	+	−
Convalescent phase	−	−	+	+	+	−
Late convalescent phase	−	−	−[c]	+	+	+
Long past infection	−	−	−	+[d]	+/−	+[d]
Chronic infection						
Chronic active hepatitis	+[e]	+/−[f]	+/−	+	+/−[f]	−[g]
Chronic persistent hepatitis	+	+/−	+/−	+	+/−	−
Chronic HBV carrier state	+	+/−	+/−	+	+/−	−
HBsAg immunization	−	−	−	−	−	+

[a] HBeAg "rarely" may persist for weeks to months after disappearance of HBsAg.

[b] Anti-HBs occasionally appears before the clearance of HBsAg in acute infection.

[c] IgM anti-HBc may persist for over 1 year after onset of acute infection when very sensitive assays are employed.

[d] Total anti-HBc and anti-HBs may be detected together or separately long after acute infection.

[e] HBsAg-negative chronic infection may occur where anti-HBc is the only detectable serological marker.

[f] Chronic infection is usually associated with liver disease more likely in the presence of HBeAg than of anti-HBe.

[g] Anti-HBs may be present in chronic infection more likely due to previous infection with a different HBsAg subtype than to antigen-antibody complexes.

Modified from Chernesky et al., 1984.

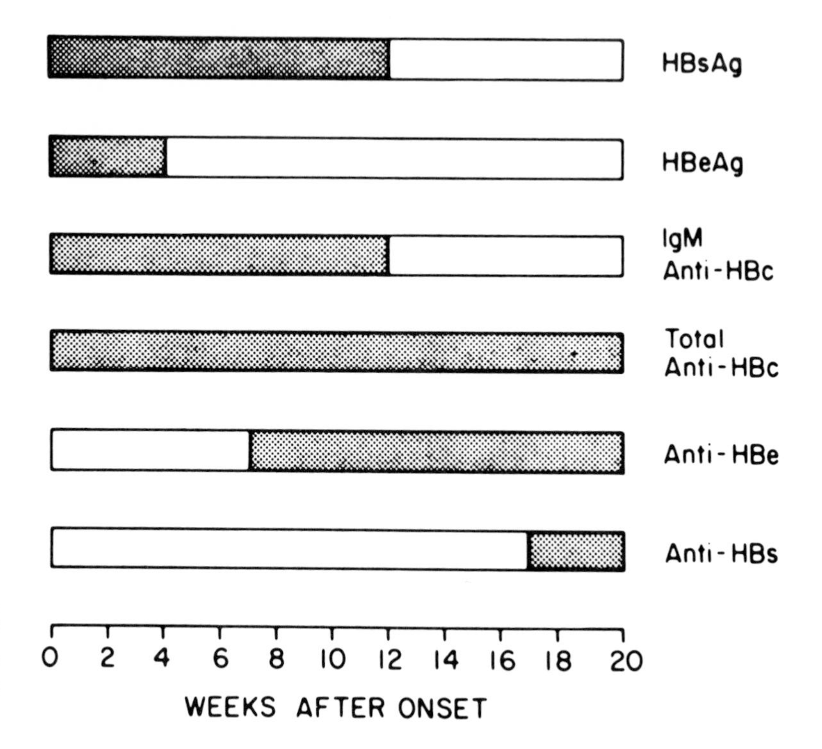

Figure 23.3. Serological profile of HBV markers in acute hepatitis B. Reproduced from Chernesky et al., 1984.

with the application and interpretation of tests for the detection of these serologic markers (Chernesky et al., 1984). This source of information is recommended for further details. Briefly, HBsAg appears prior to anti-HBc; and, therefore, the presence of HBsAg with a negative test for anti-HBc indicates very early infection in non-vaccinated individuals. Testing for HBeAg and anti-HBe provides information about the relative infectivity of the patient. The presence of HBeAg correlates with abundant circulating Dane particles and/or elevated viral DNA polymerase activity. It has been shown also to have a strong correlation with increased risk of transmitting infection upon accidental needlestick, and the transmission of HBV from mother to baby during the perinatal period or following exposure in household settings. Conversely, a positive result for anti-HBe indicates a low number of Dane particles and reduced risk of transmitting infection. Furthermore, serial HBeAg testing may be useful in assessing recovery from acute infection. Persistence of HBeAg by RIA at 10 weeks or more after the onset of symptoms has the same prognostic value as persistence of HBsAg at 4 to 6 months, and both may be useful in predicting development of the carrier state. Detection of anti-HBe indicates that recovery from infection is underway and complete resolution of infection is likely. Periodic testing (e.g., every 4 to 6 weeks) for HBsAg is always indicated when managing a patient with acute HBV hepatitis. When HBsAg has disappeared, anti-HBs testing is highly desirable to confirm immunologic recovery from infection and the presence of long-lasting immunity to reinfection. An enzyme immunoassay for the detection of anti-HBc IgM has recently become available. This marker is very useful in the differential diagnosis between: a) acute hepatitis B and a chronic HBsAg carrier state when HBsAg is undetectable; b) acute hepatitis B with development of chronic HBsAg carrier state; and c) non-B hepatitis in a healthy HBsAg carrier. In the latter, the anti-HBc IgM titer is usually much lower (Frösner and Franco, 1984). High anti-HBc titers may be useful in the early diagnosis of fulminant hepatic necrosis (Gitlin, 1984). It has been suggested that the presence of anti-HBc among HBsAg-positive carriers places them in a higher risk group for the development of primary hepatocellular carcinoma (Beasley and Hwang, 1984).

23.5.3.c Hepatitis non-A, non-B

Specific diagnosis of NANB hepatitis has been hampered by the absence of defined serologic markers. A review of the literature is misleading in this respect because none of the serologic markers reported to date has been confirmed as reproducible and specific for NANB (Alter and Hoofnagle, 1984). Thus, at the present time, there is no specific test for NANB and it is probable that there is so little circulating NANB antigen that no existing assay is sufficiently sensitive to allow for its detection in serum. Avidin–biotin amplification schemes hold some promise for detecting an NANB antigen in liver tissue. Until a specific test becomes available diagnosis of NANB hepatitis by exclusion of other causes is the accepted practice.

23.5.3.d Hepatitis D

Initially, the detection of the Δ antigen in the nuclei (and occasionally in the cytoplasm) of hepatocytes was achieved by direct IF using either frozen unfixed or fixed embedded liver biopsy specimens (Rizzetto et al., 1977). Later, the Δ antigen was visualized in liver biopsy specimens by indirect immunoperoxidase staining (Govindarajan et al., 1984). Two reports on the use of the enzyme immunoassay (Crivelli, et al., 1984; Shattock and Morgan, 1984) for the detection of Δ antigen and anti-Δ led the way for the routine use of this assay in the clinical laboratory. The very recent commercial availability of a licensed RIA kit for the detection of antibodies to the Δ agent, thus, will reduce the need for liver biopsy in many cases. A commercial enzyme immunoassay kit, which is presently under clinical evaluation, also has been developed by the same manufacturer. Although detecting antigen in serum is possible, it is not a practical approach at this time.

Acute HBV and acute HDV coinfection is indicated when HBsAg, anti-HBc IgM, HBeAg, and anti-HD IgM are positive. A patient with this pattern will recover from HDV infection as he is able to resolve the hepatitis B disease, because HDV depends on HBV for replication. Otherwise the patient would also be very likely to become a chronic HDV carrier. Chronic HBV infection with superimposed HDV infection is indicated when HBsAg, anti-HBc IgG, and anti-HD IgM are positive. The clinical type of the HDV infection in the presence of clinical symptoms of hepatitis can be further determined by titering the anti-HD antibodies. That is, an anti-HD titer of less than 10^2 is consistent with an acute infection and a titer of more than 10^3 with chronic infection. The same pattern above with the loss of anti-HD IgM in the absence of clinical symptoms is consistent with the recovery from HDV infection or onset of chronicity, again, depending on the anti-HD titer. An anti-HD titer of less than 10^2 should indicate recovery and of more than 10^3 the onset of chronic asymptomatic HDV infection. The disappearance of HBsAg and persistence of anti-HBc IgG and anti-HD IgG antibodies are compatible with the probable recovery from HBV and HDV infections. This can be further confirmed by demonstrating that anti-HBs antibodies are also present.

REFERENCES

Alter, H.J., and Hoofnagle, J.H. 1984. Non-A, non-B: Observations on the first decade. In G.N. Vyas, J.D. Dienstag, and J.H. Hoofnagle (eds.), Viral Hepatitis and Liver Disease. Orlando, FL: Grune & Stratton.

Andrade, Z.A., Santos, J.B., and Prata, A. 1983. Histopatologia de hepatite de labrea. Rev. Soc. Bras. Med. Trop. 16:31–40.

Barker, L.F., and Murray, R. 1972. Relationship of virus dose to incubation time of clinical hepatitis and time of appearance of hepatitis associated antigen. Am. J. Med. Sci. 263:27–33.

Beasley, R.P., and Hwang, L.Y. 1984. Epidemiology of hepatocellular carcinoma. In G.N. Vyas, J.L. Dienstag, and J.H. Hoofnagle (eds.), Viral Hepatitis and Liver Disease. Orlando, FL: Grune & Stratton.

Berninger, M., Hammer, M., Hoyer, B., and Gerin, J.L. 1982. An assay for the detection of the DNA genome of hepatitis B virus in serum. J. Med. Virol. 9:57–68.

Bianchi, L., and Gudat, F. 1979. Immunopathology of hepatitis B. In H. Popper and F. Schaffner (eds.), Progress of Liver Diseases, Vol. VI. New York: Grune & Stratton.

Binn, L.N., Lemon, S.M., Marchwicki, R.H., Redfield, R.R., Gates, N.L., and Bancroft, W.H. 1984. Primary isolation and serial passage of hepatitis A virus strains in primate cell cultures. J. Clin. Microbiol. 20:28–33.

Blum, H.E., Haase, A.T., and Vyas, G.N. 1984. Origin of replication of hepatitis B virus DNA in human liver. (Abst. 3B.32). In G.N. Vyas, J.L. Dienstag, and J.H. Hoofnagle (eds.), Viral Hepatitis and Liver Disease. Orlando, FL: Grune & Stratton, p. 633.

Blumberg, B.S. 1964. Polymorphisms of the serum proteins and the development of isoprecipitins in transfused patients. Bull. N.Y. Acad. Med. 40:377–386.

Blumberg, B.S., Sutnick, A.I., and London, W.T. 1968. Hepatitis and leukemia. Their relation to Australia antigen. Bull. N.Y. Acad. Med. 44:1566–1586.

Blumberg, B.S., Sutnick, A.I., and London, W.T. 1970. Australia antigen as a hepatitis virus: Variation in host response. Am. J. Med. 48:1–8.

Bradley, D.W., Fields, H.A., McCaustland, K.A., Maynard, J.E., Decker, R.H., Whittington, R., and Overby, L.R. 1979. Serodiagnosis of viral hepatitis A by a modified competitive binding radioimmunoassay for immunoglobulin M anti-hepatitis A virus. J. Clin. Microbiol. 9:120–127.

Bradley, D.W., Maynard, J.E., Cook, E.H., Ebert, J.W., Gravelle, C.R., Tisiquaye, K.N., Kessler, H., Zuckerman, A.J., Miller, M.F., Ling, C., and Overby, L.R. 1980. Non-A, non-B hepatitis in experimentally infected chimpanzees: Cross-challenge and electronmicroscopic studies. J. Med. Virol. 6:185–201.

Bradley, D.W., Maynard, J.E., Popper, H., Cook, E.H., Ebert, J.W., McCaustland, K.A., Schable, C.A., and Fields, H.A. 1983. Posttransfusion non-A, non-B hepatitis: Physicochemical properties of two distinct agents. J. Infect. Dis. 148:254–265.

Caredda, F., d'Arminio Monforte, A., Rossi, E., Farci, P., Smedile, A., Tappero, G., and Moroni, M. 1983. Prospective study of epidemic delta infection in drug addicts. In G. Verme, F. Bonino, and M. Rizzetto (eds.), Viral Hepatitis and Delta Infection. New York: Alan R. Liss.

Chernesky, M.A., Escobar, M.R., Swenson, P.D., and Specter, S.C. 1984. Laboratory diagnosis of hepatitis viruses. Washington, D.C.: American Society of Microbiologists, Cumitech 18, pp. 1–12.

Crivelli, O., Rizzetto, M., Lavarini, C., Smedile, A., and Gerin, J.L. 1981. Enzyme-linked immunosorbent assay for detection of antibody to the HBsAg-associated delta antigen. J. Clin. Microbiol. 14:173–177.

Decker, R.H., Dawson, G.J., and Mushahwar, I.K. 1984. Monoclonal antibodies to hepatitis A virus: Applications to the serologic detection of IgM anti-HAV. In G.N. Vyas, J.L. Dienstag, and J.H. Hoofnagle (eds.), Viral Hepatitis and Liver Disease. Orlando, FL: Grune & Stratton.

Dienes, H.P., Popper, H., Arnold, W., and Lobeck, H. 1982. Histologic observations in human hepatitis non-A, non-B. Hepatology 2:562–571.

Dienstag, J.L., Feinstone, S.M., Purcell, R.H., Hoofnagle, J.H., Barker, L.F., London, W.T., Popper, H., Peterson, J.M., and Kapikian, A.Z. 1975. Experimental infection of chimpanzees with hepatitis A virus. J. Infect. Dis. 132:532–535.

Dienstag, J.L. 1983. Non-A, non-B hepatitis. Gastroenterology 85:439–462, 743–768.

Dienstag, J.L. 1984. Immunologic mechanisms in chronic viral hepatitis. In G.N. Vyas, J.L. Dienstag, and J.H. Hoofnagle (eds.), Viral Hepatitis and Liver Disease. Orlando, FL: Grune & Stratton.

Dudley, F.J., Fox, R.A., and Sherlock, S. 1971. Relationship of hepatitis associated antigen (H.A.A.) to acute and chronic liver injury. Lancet i:12.

Dudley, F.J., Fox, R.A., and Sherlock, S. 1972. Cellular immunity and hepatitis associated Australia antigen liver disease. Lancet i:723–726.

Farci, P., Smedile, A., Lavarini, C., Piantino, P., Crivelli, O., Caporaso, N., Toti, M., Bonino, F., and Rizzetto, M. 1983. Delta hepatitis in inapparent carriers of hepatitis B surface antigen. Gastroenterology 85:669–673.

Feinstone, S.N., Kapikian, A.Z., and Purcell, R.H. 1973. Hepatitis A: Detection by immune electron microscopy of a virus-like antigen associated with acute illness. Science 182:1026–1028.

Feinstone, S.M., Mihalik, K.B., Kamimura, T., Alter, H.J., London, W.T., and Purcell, R.H. 1983. Inactivation of hepatitis B virus and non-A, non-B hepatitis by chloroform. Infect. Immun. 41:816–821.

Frösner, G.G., and Franco, E. 1984. New developments and open questions in the serology of hepatitis A and B. In G.N. Vyas, J.L. Dienstag, and J.H. Hoofnagle (eds.), Viral Hepatitis and Liver Disease. Orlando, FL: Grune & Stratton.

Ganem, D. 1984. Animal models in hepatitis research. In G.N. Vyas, J.L. Dienstag, and J.H. Hoofnagle (eds.), Viral Hepatitis and Liver Disease. Orlando, FL: Grune & Stratton.

Gauss-Müeller, V., Frösner, G.G., and Deinhardt, F. 1981. Propagation of hepatitis A virus in human embryo fibroblasts. J. Med. Virol. 7:233–239.

Gerety, R.J., Tabor, E., Schaff, Z., Seto, B., and Coleman, W.G., Jr. 1984. Non-A, non-B hepatitis agents. In G.N. Vyas, J.L. Dienstag, and J.H. Hoofnagle (eds.), Viral Hepatitis and Liver Disease. Orlando, FL: Grune & Stratton.

Gitlin, N. 1984. The treatment of viral hepatitis: Uncomplicated acute viral hepatitis and fulminant viral hepatic necrosis. In G.N. Vyas, J.L. Dienstag, and J.H. Hoofnagle (eds.), Viral Hepatitis and Liver Disease. Orlando, FL: Grune & Stratton.

Govindarajan, S., Chin, K.P., Redecker, A.G., and Peters, R.L. 1984. Fulminant B. Viral hepatitis: Role of delta agent. Gastroenterology 86:1717–1720.

Govindarajan, S., Lim, B., and Peters, R.L. 1984. Immunohistochemical localization of the delta antigen associated with hepatitis B virus liver biopsy sections embedded in araldite. Histopathology 8:63–67.

Gubetta, L.L., Rizetto, M., Crivelli, O., Verme, G., and Arico, S. 1977. A trichrome stain for the intrahepatic localization of the hepatitis B surface antigen (HBsAg). Histopathology 1:227–288.

Hadler, S.C., De Monzon, M., Ponzetto, A., Anzola, E., Rivero, D., Mondolfi, A., Bacho, A., Francis, D.P., Gerber, M.A., and Thung, M.A. 1984. An epidemic of severe hepatitis due to delta virus infection in Yupca Indians of Venezuela. Ann. Intern. Med. 100:339–344.

Hantz, O., Vitvitski, L., and Trepo, C. 1980. Non-A, non-B hepatitis: Identification of hepatitis B-like virus particles in serum and liver. J. Med. Virol. 5:73–86.

Holmes, A.W., Wolfe, L., Rosenblate, H., and Deinhardt, F. 1969. Hepatitis in marmosets: Induction of disease with coded specimens from a human volunteer study. Science 165:816–817.

Holmes, A.W., Deinhardt, F., Wolfe, L., Frösner, G.G., Peterson, D., Casto, B., and Conrad, M. 1973. Specific neutralization of human hepatitis type A in marmoset monkeys. Nature (London) 243:419–420.

Hoofnagle, J.H., Gerety, R.J., Smallwood, L.A., and Barker, L.F., 1977. Subtyping hepatitis B. Antigen and antibody by radioimmunoassay. Gastroenterology 72:290–296.

Hoofnagle, J.H. 1981. Serological markers of hepatitis B virus infection. Ann. Rev. Med. 32:1–11.

Hoofnagle, J.H., and Alter, H.J. 1984. Chronic viral hepatitis. In G.N. Vyas, J.L. Dienstag, and J.H. Hoofnagle (eds.) Viral Hepatitis and Liver Disease. Orlando, FL: Grune & Stratton.

Hoyer, B., Bonino, F., and Ponzetto, A. 1983. Properties of delta-associated ribonucleic acid. In G. Verme, F. Bonino, and M. Rizzetto (eds.), Viral Hepatitis and Delta Infection. New York: Alan R. Liss.

Huang, S.N. 1975. Immunohistochemical demonstration of hepatitis B core and surface antigens in paraffin sections. Lab. Invest. 33:88–95.

Huang, S.N., and Neurath, A.R. 1979. Immunohistologic demonstration of hepatitis B viral antigens in liver with reference to its significance in liver injury. Lab. Invest. 40:1–17.

Jawetz, E., Melnick, J.L., and Adelberg, E.A. 1984. Review of Medical Microbiology,: Lange Medical Publishers, Chap 32. Los Altos, CA, pp. 416–428.

Kaplan, P.M., Greenman, R.L., Gerin, J.L., Purcell, R.H., and Robinson, W.S. 1973. DNA polymerase associated with human hepatitis B antigen. J. Virol. 12:995–1005.

Khuroo, M.S. 1980. Study of an epidemic of non-A, non-B hepatitis: Possibility of another human hepatitis virus distinct from posttransfusion non-A, non-B type. Am. J. Med. 68:818–824.

Krugman, S., Ward, R., and Giles, J.P. 1962. The natural history of infectious hepatitis. Am. J. Med. 32:717–728.

Krugman, S., Giles, J.P., and Hammond, J. 1967. Infectious hepatitis: Evidence for two distinct clinical, epidemiological and immunological types of infection. J. Am. Med. Assoc. 200:365–373.

Krugman, S., Overby, L.R., Mushahwar, I.K., Ling, C., Frösner, G., and Deinhardt, F. 1979. Viral hepatitis type B. Studies on natural history and prevention re-examined. N. Engl. J. Med. 300:101–107.

Lemon, S.M., and Binn, L.N. 1983. Serum neutralizing antibody response to hepatitis A virus. J. Infect. Dis. 148:1033–1039.

Levy, G.A., and Chisari, F.V. 1981. The immunopathogenesis of chronic HBV induced liver disease. In H.C. Thomas (ed.), Immunopathology of Hepatitis B Virus Infection. New York: Springer International.

Mao, J.S., Go, Y.Y., Huang, H.Y., Yu, P.H., Huang, B.Z., Ding, Z.S., Chen, N.L., Yu, J.H., and Xie, R.H. 1981. Susceptibility of monkeys to human hepatitis A virus. J. Infect. Dis. 144:55–60.

Mathiesen, L.R., Feinstone, S.M., Purcell, R.H., and Wagner, J.A. 1977. Detection of hepatitis A antigen by immunofluorescence. Infect. Immun. 18:524–530.

Mathiesen, L.R., Feinstone, S.M., Wong, D.C., Skinhoej, P., and Purcell, R.H. 1978. Enzyme-linked immunosorbent assay for detection of hepatitis A antigen in stool and antibody to hepatitis A antigen in sera: Comparison with solid phase radioimmunoassay, immune electron microscopy and immune adherence hemagglutination assay. J. Clin. Microbiol. 7:184–193.

Melnick, J.L., Dreesman, G.R., and Hollinger, F.B. 1977. Viral hepatitis. Sci. Am. 44:237.

Mosley, J.W., Redecker, A.G., Feinstone, S.M., and Purcell, R.H. 1977. Multiple hepatitis viruses in multiple attacks of acute viral hepatitis. N. Engl. J. Med. 296:75–78.

Okochi, K., and Murakami, S. 1968. Observations on Australia antigen in Japanese. Vox Sang. 15:376–385.

Peters, R.L. 1975. Viral hepatitis: A pathologic spectrum. Am. J. Med. Sci. 270:17–31.

Phillips, M.J., and Poucell, S. 1981. Modern aspects of the morphology of viral hepatitis. Hum. Pathol. 12:1060–1084.

Popper, H., Thung, S.N., Gerber, M.A., Hadler, S.C., De Monzon, M., Ponzetto, A., Anzola, E., Rivera, D., Mondolfi, A., Bracho, A., Francis, D.P., Gerin, J.L., Maynard, J.E., and Purcell, R.H. 1983. Histologic studies of severe delta agent infection in Venezuelan indians. Hepatology 3:906–912.

Popper, H. 1984. Summary of workshop on pathology. In G.N. Vyas, J.L. Dienstag, and J.H. Hoofnagle (eds.), Viral Hepatitis and Liver Disease. Orlando, FL: Grune & Stratton.

Provost, P.J., Ittensohn, Q.L., Villarejos, V.M., and Hilleman, M.R. 1975. A specific complement-fixation test for human hepatitis A employing CR326 virus antigen. Diagnosis and epidemiology. Proc. Soc. Exp. Biol. Med. 148:962–969.

Provost, P.J., and Hilleman, M.R. 1979. Propagation of human hepatitis A virus in cell cultures *in vitro*. Proc. Soc. Exp. Biol. Med. 160:213–221.

Provost, P.J., Giesa, P.A., McAleer, W.J., and Hilleman, M.R. 1981. Isolation of hepatitis A virus *in vitro* in cell cultures directly from human specimens. Proc. Soc. Exp. Biol. Med. 167:201–206.

Purcell, R.N., Feinstone, S.M., Daemer, R.J.,

Ticehurst, J.R., and Baroudy, B.M. 1984. Approaches to the development of hepatitis A vaccines: The old and new. In R.M. Chanock and R.A. Lerner (eds.), Cold Spring Harbor Symposium Modern Approaches to Vaccines: Molecular and Chemical Basis of Virus Virulence and Immunogenicity, pp. 59–63.

Raimondo, G., Longo, G., and Squadrito, G. 1983. Exacerbation of chronic liver disease due to hepatitis B surface antigen after delta infection. Br. Med. J. 286:845.

Rizzetto, M., Canese, M.G., Airco, S., Crivelli, O., Bonino, F., Trepo, C., and Verme, G. 1977. Immunofluorescence detection of a new antigen–antibody system (δ/Anti-δ) associated to the hepatitis B virus in the liver and in the serum of HBsAg carriers. Gut 18:997–1003.

Rizzetto, M., Hoyer, B., and Canese, M.G. 1980. Delta antigen. The association of delta antigen with hepatitis B surface antigen and ribonucleic acid in the serum of delta-infected chimpanzees. Proc. Soc. Natl. Acad. Sci. USA 77:6124–6128.

Rizzetto, M., Shih, J.W.K., Verme, G., and Gerin, J.L. 1981. A radioimmunoassay for HBcAg in the sera of HBsAg carriers: Serum HBcAg, serum DNA polymerase activity, and liver HBcAg immunofluorescence as markers of chronic liver disease. Gastroenterology 80:1420–1427.

Rizzetto, M., Verme, G., Recchia, S., Bonino, F., Farci, P., Arico, S., Calzia, R., Picciotto, A., Columbo, M., and Popper, H. 1983. Chronic HBsAg hepatitis with intrahepatic expression of delta antigen. An active and progressive disease unresponsive to immunosuppressive treatment. Ann. Intern. Med. 98:437–441.

Schaefer, R.L., Wolfe, R.L., Steinfeld, C.M., Kawas, E.E., Martin, D.A., Hinds, M.W., Johnson, R.H., Hayes, J.R., Jacobs, J., Pendleton, J., Jr., Redeker, A.G., Roberto, R.R., and Chin, J. 1984. Fulminant hepatitis B among parenteral drug abusers—Kentucky, California. Morb. Mortal. Wkly. Rep. 33:70–77.

Schaff, Z., Tabor, E., Jackson, D.R., and Gerety, R.J. 1984. Ultrastructural alterations in serial liver biopsy specimens from chimpanzees experimentally infected with a human non-A, non-B hepatitis agent. Virchows Arch. B Cell. Path. 45:301–312.

Senba, M. 1982. Staining methods for hepatitis B surface antigen (HBsAg) and its mechanism. Am. J. Clin. Pathol. 77:312–315.

Shattock, A.G., and Morgan, B.M. 1983. Sensitive enzyme-immunoassay for the detection of delta antigen and anti-delta using serum as a delta antigen source. J. Med. Virol. 12:73–82.

Shikata, G., Uzawa, T., Yoshiwara, N., Akatsuka, T., and Yamazaki, S. 1974. Staining methods of Australia antigen in paraffin section. Jpn. J. Exp. Med. 44:25–36.

Shimizu, Y.K., Feinstone, S.N., Purcell, R.H., Alter, H.J., and London, W.T. 1979. Non-A, non-B hepatitis: Ultrastructural evidence for two agents in experimentally infected chimpanzees. Science 205:197–200.

Smedile, A., Dentico, P., Zanetti, A., Sagnelli, E., Nordenfelt, E., Actis, G., and Rizzetto, M. 1981. Infection with HBV-associated delta (δ) agent in HBsAg carriers. Gastroenterology 81:992–997.

Smedile, A., Farci, P., Verme, G., Czredda, F., Cargnel, A., Caporaso, N., Dentico, P., Trepo, C., Opolon, P., Gimson, A., Vergani, D., Williams, R., and Rissetto, M. 1982. Influence of delta infection on severity of hepatitis B. Lancet ii:945–947.

Szmuness, W., Stevens, C.E., Harley, E.J., Zang, E.A., Oleszko, W.R., William, D.C., Sadovsky, R., Morrison, J.M., and Kellner, A. 1980. Hepatitis B vaccine: Demonstration of efficacy in a control clinical trial in a high-risk population in the United States. N. Engl. J. Med. 303:833–841.

Tabor, E., Drucker, J.A., Hoofnagle, J.H., April, M., Gerety, R.J., Seeff, L.B., Jackson, D.R., Barker, L.F., and Pineda-Tamondong, G. 1978. Transmission of non-A, non-B hepatitis from man to chimpanzee. Lancet i:463–466.

Tanaka, K., Mori, W., and Suwa, K. 1981. Victoria blue-nuclear fast red stain for HBs antigen detection in paraffin section. Acta Pathol. Jpn. 31:93–98.

Ticehurst, J.R., Racaniello, V.R., Baroudy, B.M., Baltimore, D., Purcell, R.H., and Feinstone, S.M. 1983. Molecular cloning and characterization of hepatitis A virus cDNA. Proc. Natl. Acad. Sci. U.S.A. 80:5885–5889.

Trepo, C., Vitvitski, L., and Hantz, O. 1981. Non-A, non-B hepatitis virus: Identification of a core antigen–antibody system that cross reacts with hepatitis B core antigen and antibody. J. Med. Virol. 8:31–47.

Watanabe, S., Reddy, K.R., Jeffers, L.J., Dickinson, G.M., O'Connell, M., and Schiff, E.R. 1984. Electron microscopic evidence for non-A, non-B hepatitis markers and virus-like particles (Abst. 3A.29). In Immunocompromised Humans. In G.N. Vyas, J.L. Dienstag, and J.H. Hoofnagle (eds.), Viral Hepatitis and Liver Disease. Orlando, FL: Grune & Stratton, p. 620.

Zuckerman, A.J. 1976. Twenty-five centuries of viral hepatitis. Rush-Presbyt. St. Luke's Med. Bull. 15:57–82.

Rabies

George R. Anderson and George H. Burgoyne

24.1 HISTORY

Diseases similar or identical to rabies were described by the Greeks and Romans between the years three to five hundred B.C. The first isolation of the virus is credited to Zinke, who in 1804 used virus-infected saliva to transmit the disease to a normal dog (Zinke, 1804). In 1903, Negri described the presence of intracytoplasmic inclusion bodies in the neurons of rabid animals. Negri's findings provided the basis upon which the diagnosis of rabies rested for almost three quarters of a century.

In the latter part of the nineteenth century, Louis Pasteur and a few of his scientific contemporaries, working with crude techniques and limited means of communication, generated much of the information that has formed the basis of our knowledge about rabies.

Dr. George Newman (1904) stated, "Although rabies was mentioned by Aristotle and has been studied by a large number of workers, the additions of Pasteur have been greater than all other additions to our knowledge of the disease put together." This same author refers to the comments of Dr. Rose Bradford, who stated that "Pasteur has established, a) that the virus was not only in the saliva but also in the central and peripheral nervous system—yet absent from the blood, b) that the disease was most readily inoculated in the nervous system, c) that by suitable means the virus could be attenuated, and d) that by means of an attenuated virus preventive and even curative methods might be adopted."

Webster and Clow (1936) and Kanazawa (1936) reported the first isolations of rabies virus in cell culture. That same year, Galloway and Elford determined the size of rabies virus to be in the range of 100–150 nm (Galloway and Elford, 1936).

Rabies research accelerated following World War II. During the 35 years between 1945 and 1980, significant contributions were made toward the diagnosis and control of this disease. At least three of these contributions deserve mention as part of the historical perspective.

The first was the adaptation of the Flury strain of rabies virus to the developing chick embryo (Koprowski and Cox, 1948). Through the process of adaptation (40 to 50 passages), the virus lost its virulence for dogs but not its ability to infect dogs or induce protection against challenge with street virus. This work set the stage for the development of attenuated rabies vaccines suitable for use in dogs. The attenuated rabies vaccines have proven to be highly efficacious, with the result that the domestic dog is no longer a significant reservoir for the dissemination of rabies to humans in developed countries.

The second major contribution of this era was the development and specific application by Goldwasser and Kissling (1958) of fluorescent antibody procedures for use in the diagnosis of rabies. This procedure not only facilitated diagnosis but opened new vistas for studying the location and the distribution of the virus in tissues and cell cultures.

The third major contribution was made by Wiktor et al. (1964;1966), who provided critical information regarding the adaptation and cultivation of rabies virus in cell culture. This work provided the essential groundwork for the development of tissue culture-derived vaccines and for further elucidation of the structure, replication, and composition of the virion.

Since the time of Pasteur, two basic types of rabies virus have been recognized: the wild or naturally occurring viruses referred to as street viruses and laboratory or passaged viruses that are referred to as fixed viruses. Fixed viruses are derived from street virus. Fixed viruses have been passed through animals a sufficient number of times to modify their behavior from that of street virus.

The onset of rabies in animals following the inoculation of fixed virus is rapid and usually occurs within a predictable time frame. This is not the case with street virus, where the incubation period is longer and the time of onset of the disease somewhat unpredictable. Fixed viruses more commonly induce the paralytic form of the disease, whereas street viruses tend to induce the aggressive or furious form of the disease. The pathogenesis of these viruses differ somewhat in that street viruses usually spread more readily from peripheral sites to the central nervous system.

24.2 CHARACTERISTICS OF THE VIRUS

The rabies virus is a bullet shaped RNA virus that matures in the plasma membranes of the host cell (Murphy, 1975). The virus consists of a central helical nucleocapsid and an outer membrane or envelope. Small filamentous projections or spikes arise from the surface of the envelope membrane. The nucleocapsid is composed of nucleoprotein (NP) and single strand of RNA. The envelope membrane contains two non-glycosylated proteins (m_1 and m_2), a glycoprotein (G) and lipid. Rabies virus can be disrupted by a combination of chemical and physical treatments (Sokol, 1975). Sodium deoxycholate solubilizes the lipid component of the membrane coat and breaks the virus particle into nucleocapsid and the envelope components. The important glycoprotein surface projections of the envelope can be released by treatment of the virion with a nonionic detergent such as Nonidet P-40 (NP-40). Further separation of the individual components can be accomplished by density gradient centrifugation and column chromatography. The protective antigen, a hemagglutinin, and at least one and possibly two soluble antigens are associated with the virion. The protective antigen is derived primarily from the envelope glycoprotein, the soluble antigens are derived from the nucleocapsid, and the hemagglutinin from envelope components of the intact virus particle. The glycoprotein (G) component is responsible for the induction of neutralizing antibody (Wiktor et al., 1973). Dietzschold, et al. (1983) recently described the isolation of a soluble rabies glycoprotein antigen (G_s). The G_s and G antigens were found to have identical antibody binding characteristics but the G_s antigen, unlike the G antigen, is a weak immunogen. Certain fixed virus strains contain two classes of glycoprotein: one class with three glycosylated tryptic peptides; the other class with only one glycosylated tryptic peptide. The hemagglutinin is apparently composed of aggregates or reassociated aggregates of glycoprotein. The reaggregated glycoprotein confers a high level of protection against a rabies challenge.

The rabies virus is classified as a member of the Lyssa subgroup of Rhabdovi-

Table 24.1. Lyssa Subgroup Viruses

Virus	Principal Species Involved	Disease Occurrence
Rabies	Bats, foxes, dogs, cats, skunks, wolves, raccoons, cattle	Worldwide
Lagos bat	Bats	Nigeria, Central Africa
Mokola	Man, shrews	Nigeria
Duvenhage	Man	South Africa
Kotonkan	Mosquito	Nigeria
Obodhiang	Mosquito	Sudan

ruses. Presently five other viruses are included in this subgroup (Table 24.1). The degree of relatedness of the six viruses in the Lyssa subgroup was determined by cross neutralization, cross protection, and complement fixation tests. Rabies virus is most closely related to the Lagos bat and Mokola viruses. The latter two viruses would appear to serve as bridge viruses between rabies and the Kotonkan and Obodhiang viruses.

The work of Shope on the viruses of the Lyssa serogroup (Tignor and Shope, 1972; Shope, 1975) and that of Wiktor et al. (1980) with nucleocapsid antigens from a variety of rabies virus strains suggests that some degree of heterogeneity exists among rabies virus strains. Whether this heterogeneity is significant in terms of diagnosis or the ability of the vaccine to induce cross protection remains to be determined. Rabies vaccines for animal and human use have been made from a wide variety of strains with evidence of cross protection against heterologous challenge strains.

Rabies virus is inactivated by heat, sunlight, tissue enzymes, ultraviolet irradiation, and a number of chemical agents including formalin, quaternary ammonium compounds, strong acids, 70% ethanol, and lipid solvents. Virus infectivity may be retained for one or two weeks at room temperature (20°C) and is preserved for extended periods of time by low temperature storage (−70°C or below), by freeze drying, or by storage of infected tissue in 50% glycerol.

Beta propiolactone (BPL) and the aziridines or their derivatives inactivate rabies virus with minimum alteration of the envelope protein (Larghi and Nebel, 1980). BPL has been used extensively for the inactivation of rabies virus in the preparation of vaccines and virus antigens for use as diagnostic reagents. Gamble et al. (1980) demonstrated the inactivation of rabies virus in brain tissue in situ by subjecting mice to 1.26 megarads of γ radiation. This procedure reduces the risk associated with the handling of brains or brain tissues from animals harboring rabies virus.

24.3 PATHOGENESIS

Rabies is one of the few infectious diseases in which the exact time and site of exposure (infection) is usually known. It would seem that having this specific information would facilitate our understanding of the pathogenesis of this disease. Unfortunately, this is not the case. A detailed review of the literature suggests that there is still confusion regarding the mechanism involved in the spread of the virus from point of entry—usually muscle, connective tissue, or abraded skin—to the central nervous system. The early studies in animals indicated that the virus gained access to the central nervous system via peripheral nerve pathways and not through hematogenous spread. These early investigations have been supported by the work of Dean et al. (1963), who demonstrated that the progress of rabies virus from a leg or foot wound could be blocked by amputation or the use of local anesthetics.

Murphy (1975) demonstrated virus budding from the plasma membranes of myocytes, supporting the concept that the virus replicates locally prior to its migration to the central nervous system. The virus was then observed to progress to neuromuscular and neurotendinal spindels. Baer (1975) reported that removal of the perineurium, epineurium, and perineurial epithelium in the legs of rats prior to virus inoculation did not prevent the development of clinical rabies, though in some cases the appearance of symptoms was delayed. When the nerve fasciculus was removed from the leg infected with rabies virus, the development of rabies was prevented. The progression of the virus was not inhibited following demyelination or the disintegration of axons. Baer postulated that the virus moves passively through the tissue spaces between Schwann's cells or the interstitial nerve spaces within the nerve bundle.

The studies of Watson et al. (1981) further clarify events at the site of inoculation. Sequential histochemical and immunofluorescent staining procedures of the same tissue sections identified the axon terminal of the neuromuscular junction (NMJ) and the presence of virus at the NMJ. A portion of the nerve at the NMJ is not protected by perineurial sheath. Watson postulated that the virus moves from the point of entry in the body into the extracellular space and then gains access to the unsheathed terminal of the motor axon. The virus then traverses the motor axon to the spinal cord. Viral antigen was detected consistently at the NMJ within one hour postinoculation; however, by 72 and 95 hours viral antigen was detected only rarely at this site.

The incubation period from exposure to development of overt clinical rabies is extremely variable. In humans the incubation period may be as short as 13 days or as long as 2 years. However, the vast majority of human rabies cases develop within 20 to 60 days following exposure. The incubation period of rabies in animals in their natural habitat is less well defined but appears to proximate that described for humans. The incubation period in experimental animals is influenced by a number of factors, including the route of injection, the species and strain of animal, the quantity of virus injected, the strain of virus, and the vaccination history of the challenged animal. All warm blooded animals and some cold blooded animals are susceptible to rabies. There are, however, only a relatively few animals that serve as a reservoir for rabies and are of true concern to the clinical virologist.

Rabies is found in the United States and Canada most commonly in bats, skunks, foxes, and raccoons and less commonly in dogs, cats, cattle, horses, sheep, and humans. In South and Central America, rabies is found commonly in vampire bats, cattle, dogs, and cats. In Western Europe, the fox has become the dominant reservoir of rabies. It is rare to find rabies in wild rodents or wild rabbits, in fact, sufficiently rare that many laboratories no longer examine rodents or rabbits for evidence of rabies except in extenuating circumstances, e.g., where there is an unprovoked attack. All of the standard laboratory animal species are susceptible to rabies, with the hamster showing the highest degree of susceptibility. There is variability in resistance or susceptibility between species, between strains within a species, and between individual animals within the same strain. Lodmell (1983) demonstrated that resistance to challenge with street virus within inbred strains of mice was dominant and not controlled solely by the H-2 locus. The degree of susceptibility or resistance of individual mice may profoundly influence the course of the clinical disease. Resistant mice may show few if any signs of clinical rabies even though the challenge virus is found in their spinal cord and brain 5 to 7 days after peripheral inoculation. In the resistant animals, clinical symptoms are rare as the virus does not appear to induce malfunction of the motor nerves. The incubation period in susceptible mice is relatively short. These animals become paralyzed and prostrate, and death ensues rapidly.

Rabies manifests itself in many forms. Domestic and wild animals infected with street (wild) virus most often become agitated and aggressive and will attempt to bite inanimate objects or other animals. The symptomatology may pass eventually to that more typical of the paralytic form of the disease. Laboratory animals infected with street virus may show some agitation, e.g., jumping and aggressiveness, but more often develop the paralytic or "dumb" form of the disease. In this form of the disease, the animal begins to tremor and becomes unsteady with the development of a slowly ascending paralysis. Prior to death the animal is usually prostrate and nonresponsive. Fixed virus tends to induce the paralytic form of the disease though there are exceptions to this rule.

Laboratory animals infected with fixed virus strains generally show symptoms within 5 to 7 days following intracerebral challenge and usually die within 10 days. After peripheral challenge with fixed virus, virus can be demonstrated in the cord in about 72 hours and in the brain in about 96 hours. When street virus is injected intramuscularly infection progresses more slowly in laboratory animals than it does with fixed virus, with virus appearing in the brain 6 to 9 days following experimental infection. Clinical symptoms appear about 10 to 15 days after challenge. French authors, early in this century, coined the term "septinévrite" to characterize the general dissemination of rabies virus throughout the nervous system of an infected animal (VanRooyen and Rhodes, 1948). Septinévrite confirms the affinity of the virus for all of the nerve cells in the body and reflects the centrifugal spread of the virus throughout the body, including the salivary glands and respiratory tract mucosa. Whatever the mechanism for the transfer of virus from nerve cell to saliva, the fact remains that the involvement of the salivary gland is the essence of rabies pathogenesis, for without salivary gland involvement the disease would gradually disappear. Among certain wildlife species rabies may be transmitted on occasion by inhalation or ingestion, but these routes of transmission play a minor role in the overall disease picture.

Vaughan et al. (1963;1965) investigated the excretion of wild rabies virus in dogs and cats. Twenty-three of 26 infected cats showing clinical symptoms excreted virus in the saliva. The three cats that did not excrete the virus developed the paralytic form of the disease. The earliest that rabies virus was detected in the saliva of cats was 1 day before the onset of clinical symptoms. With one exception, salivary excretion of virus continued until death. In contrast, rabies virus was found in the saliva of only 24 of 54 infected dogs. In dogs, virus excretion began as early as 3 days before onset of illness. Once detected, salivary excretion of the virus in the dog continued until death.

The studies of Vaughan et al. are important to the clinical virologist because they confirm three assumptions that have been paramount in determining the risk of infection following exposure:

1. If a domestic animal does not develop clinical rabies within 5 to 7 days after biting a human or another animal, the chances of virus excretion at the time of bite are negligible
2. Not all animals that develop clinical rabies excrete virus in their saliva.
3. The bite of a cat is potentially more dangerous than that of a dog considering that 90% of the rabid cats were virus excretors as opposed to only 45% of the rabid dogs.

24.4 VIRUS IDENTIFICATION

In 1903, Negri described the appearance of eosinophilic staining inclusion bodies in the cytoplasm of nerve cells from animals infected with rabies virus. These inclusion bodies are commonly referred to as Negri bodies. Negri bodies occur most frequently in the hippocampus (Ammon's horn), the cerebellum, and the base of the brain. These bodies may also be found in other parts of the brain, the spinal cord, and the ganglion cells of the central nervous system. For most of this century the diagnosis of rabies was based on the identification of Negri bodies in impression smears from sections of brain tissue. Negri body identification is still the most economical and rapid means of diagnosis of rabies in animals; however, inclusion body identification is the least sensitive and is generally unsuitable for use in animals infected with the fixed or adapted strains of rabies virus.

The histopathology of rabies infection is neither peculiar to nor characteristic of the disease except for the presence of Negri bodies. Very little histopathology is noted in the brain during the course of the disease. At autopsy, leukocytic infiltration, perivascular cuffing, and mononuclear cuffing of nerve cells are the most common neuropathological features of rabies; however, these may merely reflect generalized degenerative changes rather than virus-induced pathology. Specific virus-induced pathologic changes are observed in other infections of the central nervous system, e.g., the equine encephalitides.

Laboratories in some parts of the world still use the conventional Negri body techniques for the diagnosis of rabies. Eventually, these conventional techniques will undoubtedly be replaced by more accurate and sensitive immunoassay procedures that do not require extensive training of personnel or a sizeable equipment expenditure.

The fluorescent antibody (FA) test is now the universal test of choice for the identification of rabies antigen in infected tissues (Gardner and McQuillin, 1980). This test has the advantage of being easy to perform, with the end result obtainable in 2 to 4 hours. The FA test is based on the principle that high titered rabies antibody previously conjugated with a fluorescent dye will bind to rabies antigen present in animal tissues or infected cell cultures.

There are some points that should be taken into consideration with the use of the FA test for the diagnosis of rabies:

1. The individual performing the test should have normal color vision.
2. It is essential to use positive and negative antigen controls to make certain that the FA conjugate is working properly.
3. The FA conjugated antibody or the FA stained impression smear should receive minimal exposure to natural light.
4. Several areas of the brain and/or salivary glands should be tested for the presence of rabies antigen.
5. If rabies antigen is not detected in a brain impression smear by FA after careful examination, it can be assumed that the saliva of the suspect rabid animal was noninfectious.
6. The test will not distinguish between wild and fixed strains of rabies virus.
7. Any suspect animal can be euthanized immediately for FA examination (with the Negri body test it is desirable to allow the animal to develop symptoms or die before examining the brain).
8. Decomposed brain is unsuitable for FA examination.
9. Stained smears should not be left under ultraviolet light for an extended period before examination, because this causes a decrease in fluorescence.

Impression smears for the FA diagnosis of rabies may be made from a number of sites in the brain, from salivary gland tissue, and even from the surface of the cornea. The corneal smear is a noninvasive procedure that has potential application in establishing a diagnosis in humans.

It is not uncommon for a virology laboratory to receive severely damaged or decomposed brains for examination for rabies. Decomposed brains should never be examined by either the FA or Negri body techniques. The chance for error is too great. Rarely, a decomposed brain may be inoculated into mice in lieu of in vitro examination. Even this approach is fraught with uncertainty, for the virus may have been destroyed by the action of proteolytic enzymes. Fresh or nondecomposed damaged brain (damaged by gunshot) may be examined by FA with the provision that a negative finding with brain material from a highly suspect animal must be interpreted with caution and may require the use of the mouse inoculation test.

24.5 THE ISOLATION AND CULTIVATION OF RABIES VIRUS

Rabies virus may be isolated from the brain, spinal cord, or salivary glands of infected animals. The brain of a suspect animal is usually the tissue of choice for the isolation and/or demonstration of rabies virus or virus antigens. The hippocampus (Ammon's horn), the cerebellum, and the thalamus are the sites of the brain usually selected for the isolation of the virus. The salivary glands are often overlooked as a source of virus even though the old adage "that an animal bites with its mouth not its brain" still applies. Unfortunately, the salivary glands are more difficult to locate and identify, particularly in very small animals. The isolation of rabies virus from infected animals is best accomplished by direct intracerebral inoculation of laboratory animals with brain or salivary gland tissue suspension.

The adult mouse is used most commonly for the primary isolation of rabies virus because it is susceptible, available, and easy to handle. Hamsters and neonatal mice are highly susceptible to rabies virus and, when available, may be used in addition to or in lieu of adult mice. These animals (hamsters and neonatal mice) are valuable for detecting attenuated rabies viruses or confirming that inactivated rabies vaccines are innocuous.

Fluorescent antibody methodology has supplanted animal inoculation in the public health diagnostic laboratory for the diagnosis of rabies. The only situation for which animal inoculation is still useful is for those tissues where the cell structure is sufficiently disturbed (damaged) to make it difficult to recognize and properly characterize fluorescent foci.

Rabies virus will propagate in a variety of cell cultures, including primary hamster kidney cells, human diploid cells, rhesus monkey diploid cells, chick embryo fibroblasts, mouse embryo brain cells, and many others. Although cell cultures have not been practical where isolation of the virus is required for the definitive diagnosis of natural infection, Smith et al. (1978) reported the isolation of street virus strains in a chick embryo-related (CER) cell line with an efficiency rate that parallels mouse inoculation. If this work can be confirmed in a variety of laboratory situations, the CER cell line may offer a useful alternative to animal inoculation. Wild or street virus strains of rabies virus need to be adapted to tissue culture before consistent high titer virus production occurs. Adaptation can be accomplished by rapid serial passage; by the harvest, transfer, and subculture of infected cells; or by the mixing and co-cultivation of infected cells with noninfected cells.

The BHK 21 continuous cell line, derived from baby hamster kidney, has become a standard cell line for virology laboratories working with rabies virus (Wiktor and Clark, 1975). Rabies virus adapts easily to this cell line and, once adaptation has occurred, this cell line will support consistently high levels of virus replication. The BHK 21 cell has been used for antigen production, growth studies, testing sera for neutralizing antibody, and the production of virus for physical, chemical, and morphological studies.

Virus-infected cells are harvested by treatment of monolayers with trypsin or the use of rubber scraping devices. Release of infectious virus and/or the viral protective antigen can be obtained by rapid cycles of freezing and thawing. Ultrasound has also been used for this purpose but appears to offer little or no advantage over the freeze–thaw method.

Viruses adapted for growth in primary hamster kidney cells or in rhesus diploid cells reach maximum titers in 4 to 5 days postinoculation. However, protective antigen is released and accumulates beyond the time of maximum virus growth. The optimal time for antigen harvest is at 8 to 10 days postinoculation.

Fernandes et al. (1964) were able to establish chronic rabies infection in rabbit endothelial cell cultures. These cells, observed over an extended period, displayed no demonstrable alteration in cell metabolism or cell replication in the continued persistence of virus infection.

Interferon production has been demonstrated in tissue culture cells infected with high multiplicities of rabies virus. Interferon has been shown to inhibit rabies virus infection and its early induction may be critical to protect individuals who are infected prior to receipt of any rabies vaccine.

24.5.1 Detection of Rabies Antibody

Unlike so many other viral diseases, the detection and quantitation of antibody has not played an important role in the diagnosis, treatment, or surveillance of rabies infection in animal or man. This is now changing. In the last few years, several relatively simple, sensitive, low-cost testing procedures have made rabies antibody testing practical. In addition, the development of a number of promising innovative new rabies vaccines for use in humans has resulted in a significant need for qualitative and quantitative rabies antibody determinations.

Detectable rabies antibody appears 14 to 18 days after the onset of clinical symptoms and may rise to very high levels prior to death. The presence of circulating serum antibody may be the best and only criterion to confirm the diagnosis of rabies in nonvaccinated humans prior to death.

Antibody analysis following rabies vaccination is the simplest and most efficient way of determining vaccine efficacy and the immune status of an individual. Currently, the most widely used test for detecting rabies antibody is the rapid fluorescent focus inhibition test (RFFIT), developed by Smith et al. (1973). The RFFIT is a virus neutralization test performed in cell culture. There is a high degree of correlation between the RFFIT and the rabies antibody neutralization test performed in mice. The RFFIT test is sensitive and specific and the results of this test can be reported within 48 hours, whereas the results of the neutralization test in mice require 10 to 14 days for completion and reporting.

Two other in vitro rabies antibody tests appear to be worthy of mention. Budzko et al. (1983) reported the adaptation of immunoadherence hemagglutination (IAHA) for use in the detection and measurement of rabies antibody. This procedure is simple to perform, low-cost, and offers a quick turnaround time (under 4 hours). IAHA detects complement binding antibody. The data obtained by IAHA correlate well with RFFIT, although the IAHA test is not as sensitive as RFFIT for detecting low levels of rabies antibody.

Atanasiu et al. (1977) and Nicholson and Prestage (1982) quantitated rabies antibody in human sera by application of enzyme-linked immunosorbent assay (ELISA). The antibody values obtained by the ELISA test appeared to correlate well with those obtained by an in vitro neutralization test. The results of the ELISA test are available within 48 hours. The fact that the ELISA test is rapidly becoming a universal tool in the clinical laboratory could make it highly attractive to the clinical virologist for use in the detection and quantitation of rabies antibody.

24.6 CONSIDERATIONS FOR THE CLINICAL VIROLOGIST

There is no disease that produces any greater degree of hysteria among the uninformed, and sometimes the informed, than rabies. There are many misconceptions and myths surrounding this disease. Therefore, it is very important for the clinical virologist to be fully informed about rabies so that facts and knowledge supersede misconceptions and myths.

Rabies is not transmitted through intact skin and rarely through intact mucosa. Rabies is not acquired by humans by touching animal skin, animal excreta, or a variety of fomites, as long as there is no break in the skin.

Most animals that bite are not rabid. The bite of a rabid animal does not always result in overt clinical rabies nor does antirabies treatment after exposure guarantee protection against clinical rabies.

The bite of a rabid carnivore is more serious than that of a herbivore. This is probably accounted for by the fact that carnivore saliva contains hyaluronidase and other spreading factors. Rabies has rarely been transmitted to man from a herbivore (cow, horse, sheep) in the U.S. (VanRooyen and Rhodes, 1948). However, in the Middle East countries, donkey and camel bites are an important mode of rabies transmission to man.

The location of a bite is of paramount consideration in determing risk and whether to institute treatment. Facial or finger bites pose greater risk than bites incurred on the trunk or other parts of the extremities. Multiple bites increase risk over a single bite. The degree of mortality risk was classified by Webster according to the location and the severity of the bite (VanRooyen and Rhodes, 1948).

Clinical virologists, laboratorians, veterinarians, animal control officers, and others with a higher risk of contact with rabid animals, the tissues of rabid animals, or infected tissue cultures should be vaccinated against rabies before exposure occurs. A sufficient number of injections of rabies vaccine should be given to prime the immune mechanism to recognize and respond to rabies protective antigen. Priming is confirmed by the appearance in the serum of rabies neutralizing antibody of the IgG class (Turner, 1978). Usually, two injections of a potent rabies vaccine of tissue culture origin spaced four weeks apart are needed to sensitize the immune system to rabies antigen. A single booster at six months following the primary vaccine series will reinforce that sensitization so that an additional booster will not be needed for at least 1 year. Once the primary series is completed, the requirement for additional boosters will depend, in part, on the degree of risk of exposure to rabies. For example, persons who work with rabies virus in a laboratory should have the antibody titer of their serum determined every 6 months and, if necessary, given a booster to maintain adequate antibody levels. Others, such as animal control officers, should have their antibody titer determined every 2 years and receive a booster to maintain adequate antibody levels. If the means to determine antibody titers is not available, then these workers should be given a booster after the indicated interval.

In humans, in domestic animals, and in wild animals, rabies may mimic other diseases. Even though human rabies is a truly rare disease, and although rabies is found less frequently in domestic animals, in Western countries one should never dismiss this disease from clinical consideration. Cardinal clinical signs of rabies, such as dysphagia, oculomotor paralysis, excessive salivation, and hyperesthesia or hypalgesia, may be overlooked. Many a veterinarian has sought a foreign body in the mouth of a rabid animal.

Currently there is only one type of rabies vaccine that is being distributed nationwide for use in humans in the U.S. This is an inactivated vaccine derived from rabies virus-infected human diploid cells. This vaccine produces an excellent antibody response following a preexposure primary series of two or three injections (Anderson et al., 1980). It also appears to protect against clinical rabies when administered *intramuscularly* in a primary five-dose regimen following exposure (Nicholson, 1982). Treatment failures and some inadequate antibody responses have been reported from Africa following the *intradermal* inoculation of this vaccine (Centers for Disease Control, 1983). The reason for these treatment and antibody response failures following intradermal use of the vaccine has not been fully ascertained, but there is some suggestion that the problems may be related to the use of the vaccine in individuals in an immunosuppressed state as a result of disease or drug treatment.

Recently there was also a report that human diploid vaccine triggers severe serum sickness type reactions in a small percentage of individuals following a booster injection (Centers for Disease Control, 1984). This problem is now under investigation.

An inactivated rabies vaccine derived from rhesus diploid cells has been tested extensively in clinical trials in humans in the U.S. (Berlin et al., 1982; 1983). This vaccine has been shown to induce high levels of serum antibody when administered intramuscularly in a primary series of two or three doses. Following adequate primary stimulation, a booster injection of this vaccine stimulates antibody recall in practically all recipients. So far, adverse reactions following both primary and booster injection of the rhesus diploid cell vaccine have been minimal. This vaccine is now in the process of being licensed for use in the U.S.

Rabies immune globulin (RIG), human origin, is available for use on a postexposure basis. RIG should be used in conjunction with vaccine and should be administered as quickly as possible following exposure (Rubin et al., 1973).

Possibly the most effective and probably the most overlooked method of treatment for persons bitten by an animal or cut by a contaminated instrument or bone fragment is local treatment at the site of the wound. The cleansing of a bite wound can be accomplished with a soap solution, a detergent solution, or 50–70% ethyl alcohol. If rabies antiglobulin or antiserum is available, it should be infused in and around the wound site. Local treatment must be administered within 1 hour of the time of injury in order to remove or neutralize virus before it disseminates from the site of the wound.

24.6.1 Unique Features of the Disease

The "Early Death Syndrome" and the failure of high levels of rabies serum antibody to guarantee protection in previously vaccinated individuals following postexposure treatment are two unusual features of this disease.

A number of investigators have reported the occurrence of early deaths in experimental animals who received their first rabies vaccine or vaccine series immediately following virus challenge. Challenged vaccinated animals who do not survive often die 3 to 4 days earlier than the nonvaccinated challenge controls. The early death phenomenom has been reported in mice, guinea pigs, rhesus monkeys, and humans (Blancou et al., 1980; Porterfield, 1981; Prabhakar and Nathanson, 1981). Various hypotheses have been presented to explain this unusual feature of rabies, e.g., that vaccine virus may initially inhibit natural defense mechanisms; that rabies vaccine antigen may combine with small quantities of performed rabies antibody thereby making the antibody unavailable to act against wild virus; or that following antigen stimulation, blocking antibodies may prevail over neutralizing antibody.

VanRooyen and Rhodes in their 1948 edition, *Virus Diseases of Man*, referred to a statement from the work of Proca and Bobes: "that antirabies treatment may actually bring about death from rabies after a shorter incubation period than in untreated persons." A hypothesis was offered that fixed vaccine virus inhibits defense mechanisms and accelerates the passage of street virus. So the "Early Death Syndrome" in rabies is really nothing new. It has just taken us 40 years to catch up with what our predecessors already knew.

The second unique feature of rabies worthy of mention is the failure of antibody to protect a certain percentage of animals who have been exposed to rabies virus and then vaccinated following exposure.

In our laboratories we have observed clinical disease and death in monkeys and guinea pigs whose serum contained high levels of rabies antibody prior to the appearance of clinical symptoms. These animals were vaccinated following virus exposure in experiments that were designed to simulate what happens most commonly with humans, i.e., one must attempt to generate immunity de novo in the face of an exposure.

Wiktor was unable to protect a group of mice when vaccine was administered 24 hours after infection even though these animals had a mean antibody titer of 1000 seven days after infection (Wiktor, 1978). None of these animals had demonstrable antirabies cell mediated immunity at seven days postinfection. In animals vaccinated at the time of exposure or prior to exposure, a cell mediated immunity was demonstrable at 7 days after infection and the majority of these animals survived.

These studies indicate that rabies-neutralizing antibody alone is not enough to guarantee protection in postexposure treatment situations. It is probable that any vaccine intended for postexposure use should be able to stimulate cell mediated immunity, humoral immunity, and interferon in order to ensure the development of adequate protection.

Rabies itself is a unique disease. It presents to us, in a single entity, a microcosm of unsolved disease processes. The virus can establish a symbiotic relationship with certain hosts. Cell mediated immunity may be as important in rabies as it is in cancer. Practically all species of warm blooded animals are susceptible to this disease. Some host cells are modified sufficiently by the virus to no longer be recognized as "self."

One hundred years after Louis Pasteur, the excitement that he faced is still there. Rabies remains of interest and is a challenge for the clinical virologist.

REFERENCES

Anderson, L., Winkler, G., Hafkin, B., Keelyside, R., D'Angels, L., and Deitch, M. 1980. Clinical experience with a human diploid cell rabies vaccine. J.A.M.A. 244:781–784.

Atanasiu, P., Savy, V., and Perrin, P. 1977. Epreuve immunoenzymatique pour la detection rapide des anticorps antirabique. Ann. Microbiol. 128A:489–498.

Baer, G. 1975. Pathogenesis to the central nervous system. In G. Baer (ed.), The Natural History of Rabies. New York: Academic Press, 181–198.

Berlin, B., Mitchell, J., Burgoyne, G., Oleson, D., Brown, W., Goswick, C., and McCullough, N. 1982. Rhesus diploid rabies vaccine (adsorbed) A new rabies vaccine I. Results of initial clinical studies of pre-exposure vaccination. J.A.M.A. 247:1726–1728.

Berlin, B., Mitchell, J., Burgoyne, G., Oleson, D., Brown, W., Goswick, C., and McCullough, N. 1983. Rhesus diploid vaccine (adsorbed). A new rabies vaccine II. Results of clinical studies simulating prophylactic therapy for rabies exposure. J.A.M.A. 249:2663–2665.

Blancou, J., Andral, B., and Andral, L. 1980. A model in mice for the study of the early death phenomenon after vaccination and challenge with rabies virus. J. Gen. Virol. 50:433–435.

Bishop, D. 1979. Rhabdoviruses. Vol. 1, Part 3. West Palm Beach, FL: CRC Press, 6–9.

Budzko, D., Charamella, L., Jelinek, D. and Anderson, G. 1983. Rapid detection of rabies antibodies in human serum. J. Clin. Microbiol. 17:481–484.

Centers for Disease Control. 1983. Human rabies—Kenya. Morbidity and Mortality Weekly Report. 32:494–495.

Centers for Disease Control. 1984. Systemic allergic reactions following immunization with human diploid cell rabies vaccine. Morbidity and Mortality Weekly Report. 33:185–187.

Cox, J., Dietzschold, B., Weiland, F., and Schneider, L. 1980. Preparation and characterization of rabies virus hemagglutinin. Infect. Immun. 30:572–577.

Dean, D., Baer, G., and Thompson, W. 1963. Studies on the local treatment of rabies infected wounds. Bull. W.H.O. 28:477–486.

Dietzshold, B., Wiktor, T., Wunner, W., and

Varrichio, A. 1983. Chemical and immunological analysis of the rabies soluble glycoprotein. Virology 124:330–337.

Fernandes, M., Wiktor, T., and Kiprowski, H. 1964. Endosymbiotic relationship between animal viruses and host cells. A study of rabies virus in tissue culture. J. Exp. Med. 1220:1099–1116.

Galloway, I., and Elford W. 1936. Size of virus of rabies ("fixed" strain) by ultrafiltration analysis. J. Hyg. (Lond.) 36:532–535.

Gamble, W., Chappell, W., and George, E., 1980. Inactivation of rabies diagnostic reagents by gamma radiation. J. Clin. Microbiol. 12:676–678.

Gardner, P., and McQuillin, J. 1980. Rapid virus diagnosis application of immunofluorescence. In Rabies. London: Butterworths, 174–184.

Goldwasser, R., and Kissling, R. 1958. Fluorescent antibody staining of street and fixed virus rabies antigens. Proc. Soc. Exp. Biol. Med. 98:219–223.

Kanazawa, K. 1936. Sur la culture in vitro du virus de la rage. Jpn. J. Exp. Med. 14:519–522.

Koprowski, H., and Cox, H. 1948. Studies on chick-embryo adapted rabies virus I—cultural characteristics and pathogenicity. J. Immunol. 60:533–554.

Larghi, O., and Nebel, A. 1980. Rabies virus inactivation by binary ethylenimine: New method for inactivated vaccine production. J. Clin. Microbiol. 11:120–122.

Lodmell, D. 1983. Genetic control of resistance to street rabies virus in mice. J. Exp. Med. 157:451–460.

Murphy, F. 1975. Morphology and morphogenesis. In G. Baer (ed.), The Natural History of Rabies. New York: Academic Press, 33–61.

Negri, A. 1903. Beitrag zum Studium der Aetiologie der Tollwuth. Ztschr. fur Hyg. und Intekhonskr. 43:507–528.

Newman, G. 1904. Pasteur's treatment for rabies. Bacteriology and Public Health. London: John Murray, 420.

Nicholson, K. 1982. Human diploid-cell-strain rabies vaccine IM 3. 53–61.

Nicholson, K., and Prestage, H. 1982. Enzyme linked immunosorbent assay: A rapid reproducible test for the measurement of rabies antibody. J. Med. Virol. 9:43–49.

Porterfield, J. 1981. Antibody-mediated enhancement of rabies virus. Nature 290:542.

Prabhakar, B., and Nathanson, N. 1981. Acute rabies death mediated by antibody. Nature 290:590–591.

Rubin, R., Sikes, K., and Gregg, M. 1973. Human rabies immune globulin clinical trials and effects of serum antigamma globulins. J.A.M.A. 224:871–874.

Shope, R. 1975. Rabies virus antigenic relationships. In G. Baer (ed.), The Natural History of Rabies. New York: Academic Press, 141–152.

Smith, A., Tignor, G., Emmons, R., and Woodie, J. 1978. Isolation of field rabies virus strains in CER and murine neuroblastoma cell cultures. Intervirology 9:359–361.

Smith, J., Yager, R., and Baer, G. 1973. A rapid tissue culture test for determining rabies neutralizing antibody. M. Kaplan, and H. Koprowski (eds.). Laboratory Techniques in Rabies, 3rd ed. Geneva, Switzerland: WHO Publications, 354–357.

Sokol, F. 1975. Chemical composition and structure of rabies virus. In G. Baer (ed.), The Natural History of Rabies. New York: Academic Press, 79–113.

Tignor, G., and Shope, R. 1972. Vaccination and challenge of mice with viruses of the rabies serogroup. J. Infect. Dis. 125:322–327.

Turner, G., 1978. Immunoglobulin (IgG) and (IgM) antibody responses to rabies vaccine. J. Gen. Virol. 40:595–604.

VanRooyen, C., and Rhodes, A. 1948. The Rabies Virus. London: Thomas Nelson, 811–833.

Vaughan, J., Gerhardt, P., and Newell, K. 1965. Excretion of street rabies virus in the saliva of dogs. J.A.M.A. 193:363–368.

Vaughan, J., Gerhardt, P., and Paterson, J. 1963. Excretion of street rabies virus in saliva of cats. J.A.M.A. 184:705–708.

Watson, H., Tignor, G., and Smith, A. 1981. Entry of rabies virus into the peripheral nerves of mice. J. Gen. Virol. 56:371–382.

Webster, L., and Clow, A. 1936. Propagation of rabies virus in tissue culture and the successful use of culture as an antirabic vaccine. Science 84:487–488.

Wiktor, T. 1966. Dynamics of rabies virus infection in tissue culture. Symposia Series in Immunobiological Standardization. New York: Karger, 1, 65–80.

Wiktor, T. 1978. Cell mediated immunity and post exposure protection from rabies by inactivated vaccines of tissue culture origin. Dev. Biol. Stand. 40:255–264.

Wiktor, T., and Clark, H. 1975. Growth of rabies virus in cell culture. In G. Baer (ed.), The Natural History of Rabies. New York: Academic Press, 155–179.

Wiktor, T. Fernandes, M., and Koprowski, H. 1964. Cultivation of rabies virus in human diploid cell strain WI-28. J. Immunol. 93:353–366.

Wiktor, T. Flamand, A., and Koprowski, H. 1980. Use of monoclonal antibodies in diagnosis of rabies virus infection and differentiation of rabies and rabies related viruses. J. Immunol. Methods 1:43–66.

Wiktor, T., Gyorgy, H., Schlumberger, H., Sokol, F., and Koprowski, H. 1973. Antigenic properties of rabies virus components, J. Immunol. 110:269–276.

Zinke, G. 1804. Neue Ansichten der Hundswuth, ihrer Ursachen und Folgen, nebst einher sichern Behandlungsart der von Follen Thieren gebissenen. Menschen Gabler, Jena 16, 212.

Arboviruses

Robert E. Shope

25.1 HISTORY

The early white settlers in both Africa and the America were familiar with diseases we now know were caused by arboviruses. Yellow fever virus was responsible for a clearly identified epidemic in the Yucatan in 1648 and induced a much feared illness along the major rivers and in the seaports of the New World throughout the days of sailing ships. *Aedes aegypti* mosquitoes were the urban vectors of yellow fever. This mosquito also transmitted dengue virus, the cause of dengue fever and dengue hemorrhagic fever and shock syndrome. Dengue fever was rampant in the southern United States until the 1920s, when populations of the vector mosquito were controlled. Both dengue and yellow fever continue to occur in tropical America and Africa, although yellow fever can be prevented by vaccination. Today, dengue hemorrhagic fever and shock syndrome (Halstead, 1980) is a major, lethal, epidemic disease of children in Southeast Asia, and appeared in Cuba in 1981 for the first time in the New World.

In modern times the primary clinical manifestation of life-threatening arboviral disease in North America has been encephalitis. Three mosquito borne viruses that cause human encephalitis were discovered during the 1930s. Western equine encephalitis (WEE) virus was isolated in 1930 from horses (Meyer et al., 1931) and in 1938 was associated with encephalitis in humans in California; since then it has been isolated regularly in the plains and plateaus of the western U.S. and Canada. Eastern equine encephalitis (EEE) was isolated in 1933 (TenBroeck and Merrill, 1933) and was associated with encephalitis in horses; it was subsequently recovered in 1938 from people along the Atlantic coast and more recently along the Gulf coast and in upper New York State and Michigan. In 1933, St. Louis encephalitis (SLE) virus caused an epidemic of encephalitis in St. Louis and surrounding communities with 1095 reported cases (Cumming et al., 1935). Endemic (rural) SLE is prevalent each year in much of the western U.S., and urban epidemics occur every 7 to 10 years in widely scattered loci in cities such as Chicago, Philadelphia, and Houston. The last major epidemic, in 1975, was responsible for 1815 reported cases of SLE.

California encephalitis virus was isolated in 1943 from *Aedes* mosquitoes in California and was later associated serologically with three pediatric encephalitis cases in California (Hammon and Reeves, 1952). Not until 1964, however, was the full significance of the California group viruses realized. In that year a virus closely related to California encephalitis virus was isolated from the stored brain of a child who had died in 1960 in LaCrosse, Wisconsin (Thompson et al., 1965). Starting in the early 1960s, the LaCrosse (LAC) subtype has been associated in the United States with about 30 to 140 cases per year of California group encephalitis. Two other closely related California group viruses, snowshoe hare (SSH) and Jamestown Canyon (JC), have been etiologically associated with a small number of encephalitis cases in the U.S. and Canada since 1980 (Fauvel et al., 1980; Grimstad et al., 1982). Another mosquito borne virus, Venezuelan equine encephalitis (VEE), induced encephalitis in Florida, where three endemic cases were recorded (Ehrenkranz et al., 1970). Similarly, according to Centers for Disease Control (CDC) reports, the tick transmitted Powassan (POW) virus has caused only ten reported cases of human encephalitis in the U.S. and Canada since it was first isolated from the brain of a child in Powassan, Ontario in 1958.

Colorado tick fever (CTF) is another arboviral syndrome, prevalent in the western mountain region of the U.S. The disease was known as mountain fever as early as 1855, but the etiologic virus isolated from the blood of febrile patients was not described until 1944 (Florio et al., 1944). Colorado tick fever is a tick borne virus that causes diphasic fever, muscle aches, malaise and, occasionally, hemorrhagic or central nervous system (CNS) complications in

Table 25.1. The Distribution and Natural Cycles of Arboviral Diseases Found in North America

Disease	Disease Distribution	Virus	Vector	Vertebrate	Annual Number of U.S. Cases Reported to CDC
Eastern encephalitis	Atlantic and Gulf coasts Upper New York State, Michigan	Eastern encephalitis	Mosquito	Birds, horses, pheasants	0–14
Western encephalitis	Western United States and Canada	Western encephalitis	Mosquito	Birds	0–133
Venezuelan encephalitis	Florida	Venezuelan encephalitis	Mosquito	Forest & swamp rodents	0–1
St. Louis encephalitis	North America	St. Louis encephalitis	Mosquito	Birds	15–1815
Powassan encephalitis	U.S. and Canada	Powassan	Tick	Woodchuck, skunk, other small mammals	0–1
LaCrosse encephalitis[a]	North Central and Northeast U.S.	LaCrosse	Mosquito	Chipmunk, squirrel	30–160
Colorado tick fever	Rocky Mountain states	Colorado tick fever	Tick	Ground squirrel	At least 200[b]
Dengue fever	Caribbean, in travelers to U.S.	Dengue, types 1–4	Mosquito	Humans	45 confirmed in 1982[b]

[a] California encephalitis, snowshoe hare encephalitis, and Jamestown Canyon encephalitis are caused by viruses closely related to LaCrosse virus; the diseases are known collectively as California group encephalitis.

[b] Not reportable diseases.

children. It is most common in campers, hikers, and other persons coming in contact with *Dermacentor andersoni* ticks.

Detailed reviews are recommended for SLE (Monath, 1980), for California group encephalitis (Calisher and Thompson, 1983), for alphaviruses and flaviviruses (Schlesinger, 1980), and for bunyaviruses (Bishop and Shope, 1979).

25.2 BIOCHEMICAL, SEROLOGIC, AND EPIDEMIOLOGIC CHARACTERISTICS OF ARBOVIRUSES

Over 400 different serotypes of arboviruses are known, of which at least 90 have been incriminated in human disease. Arboviruses encountered in patients in North America belong to one of four genera: *Alphavirus* (family Togaviridae), *Flavivirus* (family Flaviviridae), *Bunyavirus* (family Bunyaviridae), and *Orbivirus* (family Reoviridae). Table 25.1 lists those human disease arboviruses indigenous to or apt to be encountered in travellers to North America.

Eastern, western, and Venezuelan encephalitis viruses are alphaviruses. These are single-stranded, spherical RNA viruses, 40–70 nm in diameter. The RNA genome has positive (sense) polarity. The viruses have a lipid envelope into which are inserted virus specified glycoprotein spikes that endow the particle with the property of hemagglutination. The virus multiplies in the cytoplasm of vertebrate and invertebrate cells and matures by budding through the plasma membrane. There are several complexes of closely related viruses within the *Alphavirus* genus. Eastern equine, WEE, and VEE viruses are serologically related but each belongs to a different complex and thus is easily distinguished. The hemagglutination inhibition (HI) test and the enzyme-linked immunosorbent assay (ELISA) demonstrate the crossreactions; the neutralization test is relatively specific.

The clinical virologist must understand the epidemiology in order to entertain a rational index of suspicion when considering a diagnosis. Eastern equine encephalitis virus is maintained in a transmission cycle between birds and *Culiseta melanura* mosquitoes. These are swamp breeding mosquitoes found along the Gulf and Atlantic coasts, in upper New York State, and in Michigan. The distribution of the vector presumably accounts for the distribution of the disease. It is not known how the virus survives the winter, but virus appears each year in the vicinity of swamps first infecting wild birds. This leads to an amplification of the transmission cycle. *Culiseta melanura* feeds almost exclusively on birds. It is thus necessary to implicate other mosquitoes such as *Aedes* and *Coquillettidia*. These mosquitoes transmit virus from birds to horses and pheasants, which develop encephalitis. About 2 weeks later, encephalitis appears in people. There are commercial vaccines for horses for EEE, WEE, and VEE viruses; if these vaccines have been used, the disease in horses will not be available as an indicator to alert the physician of impending illness in people. Cases of EEE occur early in the summer along the Gulf coast, but not until late June and lasting into October in more northern climates.

Western equine encephalitis virus is also maintained in a cycle of birds and mosquitoes, the vector being *Culex tarsalis*. This mosquito occurs in the western part of the U.S. It breeds in ground pools, especially in irrigated areas and river floodplains. Sporadic cases of WEE are detected between June and October each year and range as far north as Canada. Epidemics are usually associated with heavy irrigation, excess runoff from snowpack, or flood conditions in river valleys.

Endemic VEE virus in the U.S. is maintained in a cycle of tropical *Culex* (*Melanoconion*) spp. mosquitoes in the south Florida swamps. The vertebrate hosts are swamp rodents such as the cotton rat. Recognized encephalitis is rare. When it occurs it is usually in older persons who have entered the swamp to fish or for other recreation. In 1971, a major epizootic of VEE in horses moved from Mexico into Texas where it caused encephalitis in horses and people. The virus differed antigenically from the endemic VEE of Florida. All transmission was eliminated after a campaign to vaccinate horses and to kill mosquitoes by aerial insecticide spraying. Venezuelan equine encephalitis infection has not been detected in Texas since.

St. Louis encephalitis and Powassan viruses are flaviviruses. These are spherical, positive sense single-stranded RNA viruses that are 40–50 nm in diameter, somewhat smaller than alphaviruses. They have a lipid envelope into which are inserted virus specified glycoprotein spikes, which are responsible for the property of hemagglutination. The virus multiplies in the cytoplasm of vertebrate and invertebrate cells but, unlike alphaviruses, particles do not bud from the plasma membrane; they mature in association with endoplasmic reticulum. There are several complexes of closely related flaviviruses. St. Louis encephalitis and POW viruses are serologically related but each belongs to a different complex and, thus, is easily distinguished. The HI test and the ELISA demonstrate the crossreactions; the neutralization test is quite specific.

The epidemiology of SLE virus in rural North America is similar to that of WEE. Transmission is maintained in a bird–*Culex tarsalis* cycle with sporadic cases occurring each summer, somewhat later than cases of WEE because SLE virus requires warmer temperatures to replicate efficiently in the mosquito. Unlike WEE, SLE virus has alternate vectors involved in urban outbreaks. St. Louis encephalitis epidemics follow unusually warm and dry periods in urban centers. The vectors are *Culex pipiens* complex mosquitoes, which breed in water with high organic content such as city storm sewers and poorly draining sewage. This mosquito takes blood-meals from birds and humans. Cases of SLE are prevalent from mid-summer until frost. In Florida, SLE virus is transmitted by *Culex nigripalpus*.

LaCrosse, California encephalitis (CE), snowshoe hare, and Jamestown Canyon viruses are bunyaviruses. These are spherical, negative polarity single-stranded RNA viruses. The RNA genome consists of three segments. The viruses replicate in the cytoplasm and mature by budding into smooth surfaced vesicles in association with the Golgi apparatus. They have a lipid envelope into which are inserted spikes made up of two glycoproteins. The spikes confer the ability to agglutinate red blood cells. The particles are 90–100 nm in diameter. California encephalitis is the type virus of the California serogroup. LaCrosse and snowshoe hare viruses are varieties of California encephalitis virus and crossreact extensively in serologic tests; Jamestown Canyon is a subtype of California encephalitis virus and is easily distinguished serologically (Bishop and Shope, 1979).

Encephalitis caused by LaCrosse virus may occur somewhat earlier in the summer than other arbovirus encephalitides because this virus is transmitted transovarially in *Aedes triseriatus*, its mosquito vector. Because the virus passes via the egg, the mosquito can infect vertebrate hosts as soon as the adult emerges and takes its first bloodmeal in the early summer. The insect continues to transmit until frost. This mosquito lays its eggs in water-containing tree holes in the deciduous hardwood forests of Wisconsin, Minnesota, Iowa, Michigan, Ohio, Indiana, Illinois, New York and, to a lesser extent, neighboring states. In recent years it has also adapted to using tires and other peridomestic water containers. Children presenting with encephalitis give a history of living near the woods, of having tire swings or discarded tires in the yard, or of camping or hiking in forests. The mosquito feeds on tree squirrels and chipmunks that become viremic and may serve as amplifying hosts of the virus. The disease is focal and endemic because of the ecology of the vector.

Colorado tick fever virus is an orbivirus. It is a spherical particle of 65–80 nm diameter. The genome consists of 12 segments of double-stranded RNA. Replication is in the cytoplasm. The particles lack an envelope. Colorado tick fever virus has no close serologic relatives in North America.

Colorado tick fever infects *Dermacentor andersoni* ticks inhabiting the western mountain states, including Oregon, Washington, California, Idaho, Nevada, Montana, Wyoming, Utah, Colorado, New Mexico, and South Dakota, as well as western Canada. The immature tick feeds on ground squirrels and other small rodents with a CTF viremia and they serve as amplifying hosts. The tick remains infected transtadially and the adult tick transmits to people. Disease is most common during April and May at lower altitudes and June and July at higher elevations, because the adult tick is most active during these months.

25.3 PATHOGENESIS

Arboviruses gain entry through the skin by the bite of an infected arthropod. Knowledge of the initial events of infection is superficial. The mosquito saliva enters the dermis and at times enters the small capillaries directly when the mosquito's proboscis threads the vessel. It is presumed that the virus replicates initially in the dermal tissues, including the capillary endothelium, although it is also possible that virus is transported directly in the blood to primary target organs. Replication also occurs in the regional lymph nodes and from there the blood is seeded, inducing a secondary viremia, which in turn carries virus to infect muscle and connective tissue cells. This viremia is often of very high titer and is accompanied by fever, leukopenia, and malaise. It is during this viremic phase that an arthropod may feed and become infected. The period between infection and viremia (intrinsic incubation period) is usually short, from 1 to 3 days. Viremia may last 2 to 5 days. Colorado tick fever viremia is of much longer duration because immature red blood cells are infected and virus remains in the blood cells for 2 to 6 weeks.

The vast majority of human arboviral infections are either asymptomatic or self-limited febrile illnesses. Antibody is produced and it complexes with and neutralizes circulating virus. The process is accompanied by complete recovery and leads to the presence of life-long antibody. Occasionally, however, an infected person develops encephalitis. The mechanism of entry of virus into the central nervous system is not completely understood. Nor is it understood why one person develops encephalitis and another apparently similar individual does not. Virus may reach the brain by seeding of cerebral capillaries during viremia, then by direct invasion of the brain parenchyma through the capillary walls. Alternatively, certain neural cells such as the olfactory neurons are exposed directly to circulating blood; viremia may seed these nerve endings and the virus may pass directly to the olfactory lobe of the brain. Regardless of the mechanism, it is important to note that the process of seeding the brain and productive infection of brain cells takes time. By the time the patient presents with encephalitis, serum antibody is usually detectable as is antibody in the cerebrospinal fluid (CSF). At this stage of infection, viremia has ceased and diagnosis is made by serologic assay.

The clinical laboratory findings and histopathology of arboviral encephalitis are often not helpful in arriving at an etiologic diagnosis. A definitive diagnosis can be made only in the virus diagnostic laboratory. The histopathology is characterized by perivascular cuffing, neuronal chromatolysis, cell shrinkage, and neuronophagia. Eastern equine encephalitis brain lesions are unusually necrotizing and are associated with high lymphocyte counts and modestly elevated protein levels in the CSF.

Central nervous system infections with other arboviruses have CSF cell counts < 500 and normally slightly elevated CSF protein.

25.4 CLINICAL DESCRIPTION

The arboviral encephalitides are not readily distinguished on clinical grounds, but the age of the patient has predictive value (Table 25.2). Western equine, Powassan, and LaCrosse encephalitis have the highest attack rates and are also more severe in the pediatric age-group. St. Louis encephalitis is a more common and severe disease in people over 50 years. Fever, malaise, and often severe headache usually precede neurologic signs. LaCrosse encephalitis patients often present with seizures, which are sometimes uncontrollable with antiseizure medication (Chun, 1983). Abnormal reflexes, paresis, and paralysis are common, but the prognosis with conservative treatment is good. Eastern encephalitis patients develop severe damage to the CNS, manifest by rapid progression to drowsiness and coma, convulsions, spasticity, periorbital or facial edema, high fever, and death as early as 3 to 5 days after onset (Feemster, 1957). Western encephalitis is characterized by seizures (90% of infants), lethargy or coma, restlessness, stiff neck, and fever (Kokernot et al., 1953). Remission may be sudden. St. Louis encephalitis usually has a benign course with fever, headache, and uneventful recovery. Individuals over 55 years of age, however, have severe disease with convulsions, paralysis, abnormal reflexes, confusion, and stiff neck. Again, remission may be sudden after an illness of 3 to 10 days. Urinary tract symptoms including frequency, urgency, and incontinence are found in about 25% of patients, sometimes preceding the onset of encephalitic symptoms. The syndrome of inappropriate secretion of antidiuretic hormone has also been described as a common complication of SLE (White et al., 1969).

The differential diagnosis of arboviral encephalitis includes cerebrovascular accidents, neoplasia, and a variety of parasitic, bacterial, mycotic, and spirochetal diseases. Among the viral causes of encephalitis are enteric viruses, mumps, measles, lymphocytic choriomeningitis, rabies, and herpes simplex viruses. Herpes simplex is important to differentiate, because its onset with temporal lobe seizures may closely mimic arboviral encephalitis (es-

Table 25.2. Clinical Features of Some Arboviral Encephalitides of North America

Disease	Age	Onset	Seizures	Case Fatality	Sequelae
Eastern encephalitis	Severe in children and older adults	Abrupt	75%	About 50%	Severe in about 30%
Western encephalitis	Severe in children	Preceded by fever	Common in infants	5–10% under 1 year	Common and severe in children
St. Louis encephalitis	Severe & higher attack rate >50 years	Often abrupt	About 50%	3–20%	Uncommon
Powassan encephalitis	Severe in children	Gradual		About 10%	Frequent and severe
LaCrosse encephalitis	Children	Abrupt	>50%	0.5%	Seizures later in life in 6–13%
Jamestown Canyon encephalitis	Children & adults	Preceded by fever	About 10%	Low	Uncommon

pecially EEE and LaCrosse types). The diagnosis of herpes encephalitis requires brain biopsy and, for best results, treatment with vidarabine or similar drugs should be started early. In Summer–Fall cases (especially in children) however tests to detect antibody to arbovirus should be conducted prior to biopsy.

Colorado tick fever is a self-limited febrile disease of persons exposed to ticks in the Rocky Mountain area of the U.S. In addition to fever, patients have malaise and muscle aches and pains (Goodpasture et al., 1978). The fever may be diphasic and is occasionally accompanied by a maculopapular or petechial rash. Leukopenia is consistently present on days 2 to 6 of fever (Anderson et al., 1985). Patients recover by the ninth or tenth day with weakness but without long-term effects of the disease. Rare complications are encephalitis, myocarditis, or a hemorrhagic syndrome. The disease is usually mild in children; however, adults can be quite ill.

25.5 LABORATORY PROCEDURES

25.5.1 Collection of Specimens

Details of laboratory procedures for arboviruses are described by Shope and Sather (1979). Specimens should be collected aseptically and refrigerated at 4°C for 24 hours or less until tested, or frozen at −60°C or colder if they are to be stored for longer periods. Virus is rarely found in the serum or plasma of encephalitis cases, but virus is readily isolated during the first 3 or 4 days of systemic illnesses such as dengue fever; in CTF the virus is in the erythrocytes for as long as 6 weeks. Acute phase blood and CSF should also be taken for antibody detection. If a diagnosis is not made acutely, then a second serum should be collected 1 week or later. At necropsy, virus is present in the brain in cases of encephalitis, and may be recovered from the spleen, liver, and kidney in systemic illnesses.

25.5.2 Isolation of Virus

Intracerebral inoculation of mice 1 to 3 days of age is the isolation method of choice for those arboviruses causing human encephalitis and for CTF. Alternate isolation systems include Vero, BHK-21, and primary chicken or duck embryo cells. The C6/36 clone of *Aedes albopictus* mosquito cells is highly susceptible to the encephalitis viruses and to dengue and yellow fever viruses, but these cells do not always develop cytopathic effect and the virus usually must be detected by immunofluorescence or subculture into another host.

Colorado tick fever is readily isolated from patients' red blood cells for 2 weeks and sometimes for as long as 6 weeks after onset of fever. For cases of arbovirus encephalitis, the viremic period has passed before the patient is seen by the physician. Virus may still be isolated from brain tissue at biopsy or autopsy. At autopsy, tissue from multiple sites in the brain should be sampled, because arboviruses may replicate better in one part of the brain than another. It is rarely possible to isolate virus from the CSF.

Isolation of virus takes time. Mice inoculated with EEE, WEE, or VEE viruses die in less than 48 hours; mice inoculated with LaCrosse and other California group viruses die within 48 to 72 hours; and mice inoculated with SLE and Powassan viruses die in 4 to 7 days. The time for appearance of plaques or cytopathic effect in cell culture is usually shorter than the incubation period in mice.

Arboviruses are stable at −70°C, or lyophilized and stored at −20°C. Lyophilized viruses may be transported for short periods at room temperature without loss of titer.

Virus is identified by complement fixation, HI, immunofluorescence, neutralization, or ELISA.[1] An antigen is made by

[1] Reference hyperimmune mouse ascitic fluids are available from the CDC, Vector-Borne Viral Diseases Laboratory, Box 2087, Fort Collins, CO 80522, or from the Yale Arbovirus Research Unit, Box 3333, New Haven, CT 06510.

suspending infected mouse brain in veronal buffer (for complement fixation) or in borate saline buffer pH 9.0 (for HI). Alternatively, the brain is weighed and suspended 5% w/v in 8.5% sucrose and extracted twice with cold acetone. The resulting precipitate is dried, rehydrated and used as antigen in the complement fixation, HI, or ELISA methods. For confirmation of the identification, mice are immunized with infected mouse brain of the unknown virus and the resulting antibody is used in reciprocal tests with the antigen of the unknown, and the antigen of and antibodies to reference viruses known to be in the geographic area of exposure. Type-specific monoclonal antibodies reactive in the immunofluorescence test are available from reference centers for the four types of dengue virus.

25.5.3 Detection of Antigen and Antibody

Antigen detection offers a more rapid diagnosis than virus isolation and is the method of choice for CTF virus. Colorado tick fever antigen is readily demonstrated by immunofluorescence in red blood cells taken within the first 2 weeks of infection, and even as late as 6 weeks (Emmons et al., 1969). The immunofluorescence test is also used for the rapid diagnosis of arbovirus encephalitis when brain biopsy or autopsy material is available. Either frozen sections or impression smears may be used. Antigen detection by ELISA may also be employed, although this technique is not in general use.

IgM antibody detection is the method of choice for diagnosis of arbovirus encephalitis. In most cases of encephalitis, onset of CNS signs and symptoms coincides with or follows development of antibody. The viremic phase of the infection has terminated and it is usually no longer possible to isolate virus except in fatal cases or by brain biopsy. With the IgM antibody capture ELISA, a reliable presumptive diagnosis can be made within the first 5 days after onset of CNS signs in most patients. In a study of 29 LaCrosse encephalitis patients, 83% were diagnosed on the day of hospital admission using the IgM capture ELISA with acute phase sera (Jamnback et al., 1982).

The IgM antibody capture ELISA utilizes anti-μ chain antibody to coat the solid phase. This antibody captures the patient's serum IgM or CSF IgM. Anti-μ chain antibody captures total IgM (i.e., arboviral infection-specific as well as other IgM). Other IgM is found in serum, but usually not in substantial quantities in CSF. The color intensity of the IgM antibody capture ELISA depends on the proportion of arboviral-specific IgM, not total IgM. Therefore ELISA using CSF is more sensitive than that of serum. The ELISA indicator system utilizes the arboviral antigen and a conjugated mouse or rabbit specific antiarboviral serum. An analogous method using either ELISA or radioimmune assay has been applied to flaviviral encephalitis (Heinz et al., 1981; Burke et al., 1982). The antibody capture IgM ELISA has not been thoroughly tested with all arboviruses, but it has proved reliable in experience to date.

Many public health and hospital laboratories still rely on demonstration of a fourfold or greater rise in antibody titer using paired acute and convalescent phase sera to diagnose arboviral encephalitis. This classic procedure is not completely satisfactory. Not only is the diagnosis delayed, but a study of LaCrosse encephalitis cases showed that the complement fixation test alone diagnosed only 50%, the HI test alone 79%, and the neutralization test alone 85% (Calisher and Bailey, 1981). The neutralization test, although the most reliable, is expensive and not widely used.

The IgM antibody capture ELISA is rapid. When the differential diagnosis includes herpes encephalitis, the ELISA for arboviral encephalitis should be performed before the brain is biopsied, to preclude an unnecessary invasive procedure.

25.6 TRENDS

Summertime acute encephalitis in the U.S. continues to be an enigma. More than 50% of the reported encephalitis cases are still undiagnosed. Their Summer–Fall seasonality is reason to believe that they may be mosquito borne. The clinical virologist is urged to sample these cases. Already, in the 1980s, snowshoe hare and Jamestown Canyon viruses have emerged as causes of encephalitis. Additional viruses will almost certainly be linked in the future. Table 25.1 shows the annual numbers of reported cases of arboviral encephalitis in the past decade. It is probable that there is marked underreporting. The clinical virologist can help obtain a truer picture.

Imported dengue cases are also of public health concern. In 1982, the CDC recorded 45 serologically confirmed cases; eight of these were in states that had *Aedes aegypti* (Gubler, 1985). So far, dengue virus transmission has not become established in the southern U.S., but to alert authorities to such an event, it is critical that clinical virologists continue surveillance, including laboratory testing of specimens for virus isolation and for serologic confirmation where dengue is suspected.

REFERENCES

Anderson, R.D., Entringer, M.S., and Robinson, W.A. 1985. Virus-induced leukopenia: Colorado tick fever as a human model. J. Infect. Dis. 151:449–453.

Bishop, D.H.L., and Shope, R.E. 1979. Bunyaviridae. In H. Fraenkel-Conrat and R.R. Wagner (eds.), Comprehensive Virology. Vol. 14. New York: Plenum, 1–156.

Burke, D.S., Nisalak, A., and Ussery, M.A. 1982. Antibody capture immunoassay detection of Japanese encephalitis virus immunoglobulin M and G antibodies in cerebrospinal fluid. J. Clin. Microbiol. 16:1034–1042.

Calisher, C.H., and Thompson, W.H. (eds.). 1983. California Serogroup Viruses. New York: Alan R. Liss.

Calisher, C.H., and Bailey, R.E. 1981. Serodiagnosis of LaCrosse virus infections in humans. J. Clin. Microbiol. 13:344–350.

Chun, R.W.M. 1983. Clinical aspects of LaCrosse encephalitis: neurological and psychological sequelae. In C.H. Calisher and W.H. Thompson (eds.), California Serogroup Viruses. New York: Alan R. Liss.

Cumming, H.S. et al. 1935. Report on the St. Louis Outbreak of Encephalitis. Public Health Bulletin No. 214. Washington, D.C.: U.S. Treasury Department, Public Health Service.

Ehrenkranz, N.J., Sinclair, M.C., Buff, E., and Lyman, D.O. 1970. The natural occurrence of Venezuelan equine encephalitis in the United States. N. Engl. J. Med. 282:298–302.

Emmons, R.W., Dondero, D.V., Devlin, V., and Lennette, E.H. 1969. Serologic diagnosis of Colorado tick fever. A comparison of complement-fixation, immunofluorescence, and plaque-reduction methods. Am. J. Trop. Med. Hyg. 18:796–802.

Fauvel, L.M., Artsob, H., Calisher, C.H., Davignon, L., Chagnon, A., Skvorc-Ranko, R., and Belloncik, S. 1980. California group virus encephalitis in three children from Quebec: clinical and serological findings. Can. Med. Assoc. J. 122:60–64.

Feemster, R.F. 1957. Equine encephalitis in Massachusetts. N. Engl. J. Med. 257:701–704.

Florio, L., Stewart, M.O., and Mugrage, E.R. 1944. The experimental transmission of Colorado tick fever. J. Exp. Med. 80:165–188.

Goodpasture, H.C., Poland, J.D., Francy, D.B., Bowen, G.S., and Horn, K.A. 1978. Colorado tick fever: Clinical, epidemiologic, and laboratory aspects of 228 cases in Colorado in 1973–1974. Ann. Intern. Med. 88:303–310.

Grimstad, P.R., Shabino, C.L., Calisher, C.H., and Waldman, R.J. 1982. A case of encephalitis in a human associated with a serologic rise to Jamestown Canyon virus. Am. J. Trop. Med. Hyg. 31:1238–1244.

Gubler, D.J. 1985. Dengue in the United States, 1982. M.M.W.R. 33:9ss–13ss.

Halstead, S.B. 1980. Immunological parameters of togavirus disease syndromes. In R.W. Schlesinger (ed.), The Togaviruses. New York: Academic Press, 107–173.

Hammon, WF.McD., and Reeves, W.C. 1952. California encephalitis virus. A newly described agent. Calif. Med. 77:303–309.

Heinz, F.X., Roggendorf, M., Hormann, H., Kunz, C., and Deinhardt, F. 1981. Comparison of two different enzyme immunoassays for detection of immunoglobulin M antibodies against tick-borne encephalitis virus in serum and cerebrospinal fluid. J. Clin. Microbiol. 14:141–146.

Jamnback, T.L., Beaty, B.J., Hildreth, S.W., Brown, K.L., and Gundersen, C. 1982. Capture immunoglobulin M system for rapid diagnosis of LaCrosse (California encephalitis) virus infections. J. Clin. Microbiol. 16:577–580.

Kokernot, R.H., Shinefield, H.R., and Longshore, W.A. Jr. 1953. The 1952 outbreak of encephalitis in California: Differential diagnosis. Calif. Med. 79:73–77.

Meyer, K.F., Haring, C.M., and Howitt, B. 1931. The etiology of epizootic encephalomyelitis in horses in the San Joaquin Valley. Science 74:227–228.

Monath, T.P. (ed.). 1980. St. Louis Encephalitis. Washington, D.C.: American Public Health Association.

Schlesinger, R.W. (ed.). 1980. The Togaviruses. New York: Academic Press.

Shope, R.E., and Sather, G. 1979. Arboviruses. In E.H. Lennette and N.J. Schmidt (eds.), Diagnostic Procedures for Viral, Rickettsial and Chlamydial Infections. Washington, D.C.: American Public Health Association, 767–814.

TenBroeck, C., and Merrill, M. 1933. A serological difference between eastern and western equine encephalomyelitis virus. Proc. Exp. Biol. Med. 31:217–220.

Thompson, W.H., Kalfayan, B., and Anslow, R.O. 1965. Isolation of California encephalitis group virus from a fatal human illness. Am. J. Epidemiol. 81:245–253.

White, M.G., Carter, N.W., Rector, F.C., and Seldin, D.W. 1969. Pathophysiology of epidemic St. Louis encephalitis. I. Inappropriate secretion of antidiuretic hormone. Ann. Intern. Med. 71:691–702.

26

Papovaviruses

Keerti V. Shah

26.1 INTRODUCTION

Papovaviruses are small, naked, icosahedral viruses that have a double-stranded DNA genome and that multiply in the nucleus. The family *Papovaviridae* consists of two genera, papillomavirus (wart virus) and polyomavirus. Viruses of both genera are widely distributed in nature (Table 26.1). Humans are hosts to at least 25 different human papillomaviruses (HPV types 1 to 25) and to polyomaviruses BK virus and JC virus. Viruses of each genus have a common evolutionary origin. All members of a genus are immunologically related and share some nucleotide sequences, but there is no evidence of an intergeneric relationship between papillomaviruses and polyomaviruses.

Papillomaviruses are larger in size than polyomaviruses (virion diameter of 55 nm compared with 45 nm) and have a larger genome (8×10^3 base pairs compared with 5×10^3 base pairs). All of the genetic information in papillomaviruses is located on one strand. In contrast, the genetic information in polyomaviruses is about equally divided between the two strands. Viruses of the two genera also differ biologically. Papillomaviruses infect surface epithelia and produce warts at the site of multiplication on the skin or the mucous membrane. On the other hand, polyomaviruses, after initial multiplication at the site of entry in the respiratory or the gastrointestinal tract, reach internal organs (kidney, lung, brain) following viremia. Although viruses of both genera can transform cells and produce tumors experimentally, only papillomaviruses are associated with naturally occurring tumors.

26.2 HUMAN PAPILLOMAVIRUSES

The infectious nature of human warts has been suspected for many centuries (Rowson and Mahy, 1967). Its viral etiology was established in 1907 by experimental transmission of warts from person to person by inoculation of a cell-free extract of wart tissue. The virus was visualized in the 1950s soon after electron microscopy came into general use and, on the basis of morphologic similarities and nuclear site of multiplication, was grouped with polyomavirus of mice and vacuolation agent (SV40) of rhesus monkey to form the papova (*pa*pilloma, *po*lyoma, and *va*cuolating agent) group (Melnick, 1962). Warts, because of their characteristic histopathologic features,

This work was supported by National Institutes of Health grants CA-13478 and PO1 AI-16959.

Table 26.1. Natural Hosts of Papovaviruses

Host	Papillomavirus	Polyomavirus
Man	Human papillomaviruses types 1–25	BK virus and JC virus
Monkeys	None characterized	Simian virus 40 of macaques; Simian agent 12 of baboons; lymphotropic papovavirus of African green monkeys
Cattle	Bovine papillomaviruses types 1–6	Bovine polyomavirus
Rabbit	Cottontail rabbit papillomavirus	Rabbit kidney vacuolating virus
Rodents	*Mastomys natalensis* papillomavirus	Polyoma and K viruses of mice; hamster papova virus
Birds	Chaffinch papillomavirus	Budgerigar fledgling disease virus
Other	Horse, dog, sheep, European elk, deer	—

have been recognized at many different sites in humans (skin, genital tract, respiratory tract, oral cavity) and in many mammalian species. However, papillomaviruses still cannot be grown in culture. The existence of a large number of distinct human papillomaviruses became evident only after the development of recombinant DNA technology, which permitted the cloning of viral genomes from different sites and the comparison of these genomes (Orth et al., 1977b; zur Hausen, 1980). Different genotypes of HPV were associated with lesions of specific morphology and at specific anatomic sites (Table 26.2).

26.2.1 Characteristics of the Virus

The virion is nonenveloped and has a diameter of 55 nm, icosahedral symmetry, and 72 capsomers. The viral genome is a double-stranded, circular DNA molecule with 8×10^3 base pairs and a molecular weight of 5.2×10^6 daltons. Complete nucleotide sequences are known for HPV-1, HPV-6, and HPV-11. All of the open reading frames in papillomavirus DNA are located on only one of the two strands, indicating that only one strand carries the genetic information. Detailed physical maps have been constructed for most of the HPV genomes. Few functions have been localized on these maps because biological assays for viral multiplication and transformation are not yet available. However, by alignment of HPV with bovine papillomavirus (BPV) (for which transformation assays are available), the genome can be divided into early and late regions. The early region contains eight open reading frames and the late region two open reading frames.

The genomes of all papillomaviruses share some nucleotide sequences. DNA hybridization performed under conditions of low stringency (which allow for the detection of weakly homologous regions), shows extensive crosshybridization between human papillomaviruses and between human and animal papillomaviruses (Heilman et al., 1980). Surprisingly, HPV-1 and HPV-2 do not show a greater degree of relatedness with one another than they do with viruses of bovine or rabbit origin. In DNA hybridization tests performed under conditions of high stringency, when only strongly homologous regions are detected, HPV do not crosshybridize with animal papillomaviruses. Among the HPV, some viruses can be grouped together on the basis of significant crosshybridization under conditions of high stringency (Kremsdorf et al., 1984) (Table 26.3). An HPV genome that displays less than 50% homology with all other known HPV genotypes in hybridization tests performed under stringent conditions is designated a new genotype. Within each genotype, subtypes are identifiable on the basis of variations in restriction enzyme digest patterns.

Table 26.2. Human Papillomaviruses

Virus	Main Site(s) of Virus Recovery
HPV-1	Deep plantar warts
HPV-2	Common warts
HPV-3	Flat skin warts in normal individuals and in patients with epidermodysplasia verruciformis (EV)
HPV-4	Deep plantar warts and common warts
HPV-5	Pityriasis-like macular lesions in EV patients
HPV-6	Genital and laryngeal papillomas; rarely in genital cancers
HPV-7	Common warts in butchers
HPV-8	Macular lesions in EV patients
HPV-9	Macular lesions in EV patients
HPV-10	Flat skin warts
HPV-11	Genital and laryngeal papillomas; rarely in genital cancers
HPV-12	Macular lesions in EV patients
HPV-13	Oral focal epithelial hyperplasia
HPV-14	Macular lesions in EV patients
HPV-15	Macular lesions in EV patients
HPV-16	Genital tract cancers; few genital tract papillomas
HPV-17	Macular lesions in EV patients
HPV-18	Genital tract cancers
HPV-19–25	Macular lesions in EV patients

Table 26.3. Human Papillomaviruses Grouped on the Basis of Significant Cross-Hybridization Under Stringent Conditions

Site	Viruses
Skin, flat warts	HPV-3 and 10
Genital tract	HPV-6 and 11
Skin, warts in epidermodysplasia verruciformis (EV)	HPV-5, 8, 12, 14, 19, 20, 21, 22, 23
Skin, warts in EV	HPV-9, 15, 17

The viral capsid proteins consist of a major polypeptide of approximately 57 k and a number of minor polypeptides with molecular weights between 43 and 53 k. Purified virions contain four histones of host origin. The virion surface displays type-specific antigenic determinants; immune sera prepared against the whole virus particle show virtually no crossreactivity between different viral types. Conversely, some of the genus-specific determinants, shared by all viral types, are probably located internally; immune sera prepared against disrupted virions are broadly crossreactive (Jenson et al., 1980).

26.2.2 Pathogenesis and Disease Potential

Human papillomaviruses infect only epithelia of skin and mucous membranes. The virus probably infects cells of the lower layers of the epithelium which undergo proliferation and form the wart. Histologically, a wart is localized epithelial hyperplasia with a defined boundary and an intact basement membrane. All layers of the normal epithelium are represented in the wart. The prickle cell layer is irregularly thickened, the granular layer contains foci of koilocytotic cells, and the cornified layer displays hyperkeratosis. The viral capsid antigen and viral particles are found only in the nuclei of cells of the differentiated, nondividing, superficial layers of the wart. In the infected cell, the multiple copies of the viral genome are present in an unintegrated state.

Warts vary widely in their appearance, morphology, site of occurrence, and pathogenic potential. The clinical spectrum of warts ranges from transient, barely noticed, self-limiting, benign tumors of the skin to infections that lead to malignancies of the skin or of the genital tract. Many factors determine the clinical significance of papillomavirus infection.

26.2.2.a Location of Lesion

This is best exemplified by laryngeal papilloma. Although the tumors are benign, they may cause life-threatening respiratory obstruction because of their location on the vocal cord.

26.2.2.b Genotype of Virus

There is considerable correlation between genotype of the infecting virus and the morphology and site of the lesion (Tables 26.4 and 26.5). For example, almost all flat warts of the skin yield HPV-3 or HPV-10. Most deep plantar warts are caused by HPV-1 and common warts by HPV-2. Virus types HPV-6 and HPV-11 are recovered from most of the genital warts (condylomas). Oncogenic potential is also correlated with viral genotype. In the genital tract, HPV-16 and HPV-18 are strongly associated with malignancies and HPV-6 and HPV-11 with benign warts. In the rare dermatologic disorder epidermodysplasia verruciformis (EV), lesions caused by HPV-5, HPV-8, and HPV-13 have a greater tendency to convert to malignancy than lesions caused by several other types (Orth et al., 1977a).

Table 26.4. Viral Genotypes Most Frequently Recovered from Skin Warts

Type of Wart	Virus Genotype
Plantar wart	HPV-1, HPV-4
Common wart	HPV-2, HPV-4
Flat wart	HPV-3, HPV-10

Table 26.5. Viral Genotypes Most Frequently Recovered from Patients with Epidermodysplasia Verruciformis

Type of Lesion	Virus Genotype
Flat skin warts	HPV-3, HPV-10
Red-brown (macular) plaques[a]	HPV-5, 8, 9, 12, 14, 15, 17, 19, 20, 21, 22, 23, 24

[a] Lesions with HPV-5, 8, or 14 have a potential for malignant conversion to squamous cell carcinoma.

26.2.2.c Host Factors

Warts tend to increase in size and numbers in conditions associated with immunologic impairment, especially T-lymphocyte deficiency. This immunologic impairment may be subtle, as in pregnancy, or gross, as in organ transplant recipients and patients receiving anticancer therapy.

Papillomavirus infection is acquired in a variety of ways: through skin abrasions (skin warts), by sexual intercourse (genital warts), during passage through an infected birth canal (juvenile-onset laryngeal papilloma), and probably in other ways (e.g., papillomas of the oral cavity by autoinoculation or by oral sex).

26.2.3 Clinical Types of Warts

26.2.3.a Skin Warts

There are many morphologic types of warts and each type may have preferred locations on the skin (Bunney, 1982). Common warts are found on hands, and there are generally multiple lesions. The warts are characteristically dome-shaped, with numerous conical projections (papillomatosis) that give their surface a velvety appearance. Deep plantar warts (on the bottom surface of the foot) generally occur singly, and have a highly thickened corneal layer (hyperkeratosis). Flat warts (with little or no papillomatosis) almost always occur as multiple warts and are found most often on arms and face and around the knees. The thread-like filiform warts occur most often on the face and neck.

Skin warts are transmitted by direct contact with an infected individual or indirectly by contact with contaminated objects. The incubation period is difficult to estimate but may be as short as 1 week or as long as several months. As a rule, warts in an otherwise healthy individual are few in number and small in size, but a large number of warts may develop in immunodeficient individuals or in apparently normal persons. Most warts regress within two years, probably as a result of cell-mediated immune responses. Treatment or excision of one wart often results in regression of the remaining warts. This may result from a "triggering" of an immune response due to immune-competent cells, which come in contact with antigens that are released as a result of treatment.

Warts are most prevalent in children and young adults. At any one time, as many as 10% of school children may have warts at some site. It is not known if the reduced prevalence in the older population represents acquired immunity, reduced exposure, or both. The incidence of warts in the general population is believed to be increasing. Recreational activity in which bare skin may be exposed to virus contaminated objects (for example, swimming in communally used pools) increases the risk of acquiring warts, especially plantar warts (Bunney, 1982). Types of HPV most frequently recovered from skin warts are HPV-1, -2, -3, -4, and -10 (Table 26.4).

28.2.3.b Epidermodysplasia Verruciformis

Epidermodysplasia verruciformis is a rare, life-long disease in which a patient is unable to resolve the wart virus infection (Jablonska et al., 1972). Most patients exhibit defects of cell-mediated immunity. The disease probably has a genetic basis (Lutzner, 1978). Patients frequently give a history of parental consanguinity and, despite the rarity of the disease, multiple cases occur in some families. It is postulated that EV patients have an inherited immunologic defect as a result of homozygosity for a rare recessive autosomal gene. The nature of the presumed genetic defect is not known.

The onset of the disease occurs in infancy or childhood. The patient develops multiple, disseminated, polymorphic wart-like lesions that tend to become confluent. The warts are of two clinical types: flat warts and red or reddish-brown macular plaques resembling pityriasis versicolor. The warts contain abundant amounts of viral particles, viral antigen, and viral DNA. The flat warts of EV patients yield HPV-3 and -10, the same genotypes that are recovered from flat warts of normal individuals. However, a bewildering variety of viral genotypes are recovered from the macular plaques of EV patients (Orth et al., 1980) (Table 26.5). It is unclear how EV patients acquire these infections because these genotypes are seldom encountered in normal populations.

In about 33% of the cases, multiple foci of malignant transformation arise in the reddish-brown plaques, especially in lesions occurring in areas exposed to sunlight. Histologically, the tumors may be in situ (bowenoid) or invasive squamous cell carcinoma. The tumors grow slowly and are generally nonmetastasizing. The malignant cells contain multiple copies of viral DNA (HPV-5, -8, or -14) but no viral particles or viral antigen. Human papillomavirus DNA is also recovered from metastatic tumor cells (Ostrow et al., 1982).

The carcinomas occurring in EV patients illustrate how several factors working in concert result in papillomavirus-induced malignancy. Viruses of specific genotypes infecting an immunologically impaired host produce malignant transformation in lesions that are exposed to sunlight.

26.2.3.c Genital Warts (Anogenital Warts, Condyloma, Genital Papilloma)

Papillomavirus infection of the genital tract occurs predominantly in young adults and in sexually promiscuous populations. It is increasing in incidence over the past 20 years and is now among the most common sexually transmitted diseases. In the U.S., an estimated 946,000 individuals consulted private physicians for genital warts in 1981, compared with an estimated 169,000 in 1966 (Centers for Disease Control, 1983). The number of comparable consultations for genital herpes in 1981 was 295,000, or about 31% of that for condyloma. In the United Kingdom, the annual incidence of genital warts per 100,000 population rose from about 30 in 1971 to 50 in 1978. In sexually transmitted disease clinics, genital warts account for about 4% of patient visits, compared with 24% of visits for gonorrhea; however, in a population-based study in Rochester, Minnesota, the incidence rate for genital warts was about one half that for gonorrhea (Chuang et al., 1984). In the U.S. and Canada, 1.3–1.6% of routinely collected Papanicolaou smears show cytopathologic evidence of HPV infection (Reid et al., 1980; Meisels et al., 1982).

The incubation period for condylomas is estimated to be between 3 weeks and 8 months, with an average of 2.8 months (Oriel, 1971). About 66% of the sexual partners of condyloma patients develop the disease. Condylomas may be papillary (condyloma acuminatum) or flat (condyloma planum). The most frequent sites for papillary (or exophytic) condylomas are the penis, around the anus, and on the perineum in the male and the vaginal introitus, vulva, the perineum, and around the anus in the female. On the cervix, flat condylomas are far more frequent than papillary condylomas (Meisels et al., 1982). The flat lesion on the cervix was not recognized to be due to papillomavirus infection until the late 1970s. It is now known to be the most common manifestation of genital HPV infection in the female. The lesion is generally seen only by a careful colposcopic examination and is confirmed by cytology and histopathology.

In a large number of infected individuals, condylomas occur at more than one site in the genital tract. Condylomas may increase in number and size during pregnancy and regress after delivery. Immunosuppressed populations, for example, patients with acquired immunodeficiency syndrome (AIDS), have a high prevalence of condylomas.

The genital tract is infected with four characterized and one or more as yet uncharacterized papillomaviruses (Table 26.6). The closely related HPV-6 and -11 are responsible for a large majority of the condylomas (Gissmann et al., 1983). Viral genotypes that are primarily associated with skin warts are rarely found in genital warts.

Table 26.6. Prevalence of Papillomavirus Antigen and Genomes in Genital Tract Lesions

	Percent Prevalence			
		Viral Genome[b]		
Tissue	Capsid Antigen[a]	HPV-6 or HPV-11	HPV-16	HPV-18
Condyloma	50–70	90	6	0
Dysplasias	10–45	20–30	20	0
In situ and invasive cervical cancer	<1	5	35–60	15–25
"Normal" genital tract tissue	<5	<5	<1	<1

[a] By immunoperoxidase test with genus-specific antiserum. The genotype of the infecting virus cannot be identified by this test.

[b] By hybridization under conditions of high stringency.

Many genital warts regress with time but some may persist for long periods. They may cause local irritation and itching, become infected, and cause severe physical and psychological difficulties for the patient if they enlarge in size or increase in numbers. They also pose the risk of malignant transformation and, if present during pregnancy, the risk of transmission to the fetus and consequent laryngeal papilloma in the offspring.

26.2.3.d Cervical Dysplasia (Cervical Intraepithelial Neoplasia) and Cancer of the Cervix

Dysplasias are a progressive spectrum of abnormalities of the cervical epithelium preceding the development of cervical cancer. Numerous studies have been performed to determine if papillomavirus infection of the cervix is benign and self-limiting, or if it has the potential to progress to severe dysplasia and invasive cancer. These studies have been prompted by a number of considerations:

1. Cancer of the cervix has the epidemiologic characteristics of a sexually transmitted disease
2. A majority of lesions previously diagnosed as mild cervical dysplasia are now recognized as flat cervical condylomas
3. The cervix is the most common site of papillomavirus infection and many condylomas are located in the transformation zone where cervical cancer originates
4. Condylomas and severe dysplasias are often found side by side in histologic sections

Condylomas, dysplasias, and genital cancers have been examined for evidence of productive papillomavirus infection, as well as for the presence of papillomavirus-related DNA sequences. The frequency of productive virus infection decreases as the lesion becomes progressively more severe; viral capsid antigen is demonstrated in about 50–70% of condylomas, in 10–45% of dysplasias, and in virtually no lesions of carcinoma in situ or invasive cervical cancer (Guillet et al., 1983; Kurman et al., 1983) (Table 26.6). Similarly, HPV-6 and -11 genomes are recovered from a large majority of condylomas but from a small proportion of cervical cancers. In contrast, HPV-16 and -18 related sequences are detected in a majority of cervical cancers but in a very small proportion of condylomas (Durst et al., 1983; Boshart et al., 1984). Virus type HPV-16 is associated with aneuploid dysplasias (Crum et al., 1984). These findings indicate that women infected with HPV-16 and -18 carry a higher risk of developing cancer than those infected with HPV-6 and -11. Several cell lines originating from cervical cancer tissues, including HeLa cells, possess HPV-18 sequences integrated into the cellular DNA (Boshart et al., 1984).

Virus types HPV-16 and -18 have also been recovered from some tissues of vulvar and penile cancers, and HPV-16 appears to be primarily responsible for bowenoid papulosis.

26.2.3.e Laryngeal Papilloma

This is a chronic, rare, and/or a recurrent disease in which benign viral papillomas in the respiratory tract may become life-threatening because of their location. The vocal cords in the larynx are the site most often affected, although the disease may occur at other locations (e.g., trachea) without laryngeal involvement. The most common presenting symptom is hoarseness of voice or change of voice. The papillomas may produce respiratory distress and obstruction, especially in children. The disease tends to recur following surgical removal of the papilloma and patients may require frequent operations, sometimes as often as every 2 to 4 weeks. Surgery may lead to dissemination of disease to other sites, for example, to the lungs. Malignant conversion of papilloma is rare and is usually associated with a history of previous radiation therapy.

The highest risk of onset of laryngeal

papilloma is under the age of 5 years. About 33–50% of the cases occur by that age, and about 33% of the cases have onset of illness in adult life (Mounts and Shah, 1984). The viral types recovered from both juvenile- and adult-onset disease are HPV-6 and -11, the viruses that are responsible for genital warts (Mounts et al., 1982; Gissmann et al., 1983). The transmission of virus in juvenile-onset cases probably occurs during the process of birth, in the course of fetal passage through an infected birth canal (Hajek, 1956). Mothers of patients with laryngeal papilloma frequently give a history of genital warts during pregnancy. The risk of acquiring laryngeal papilloma for children born to mothers with active genital papillomavirus infection is estimated to be between 1:100 and 1:1000. Cesarean delivery prior to rupture of membrane very likely reduces the risk of virus transmission. It is not known if transmission of virus in adult-onset disease occurs intrapartum, with the virus remaining latent for many years, or if the infection is acquired in other ways, for example, by oral contact with infected genitalia.

26.2.3.f Warts at Other Sites

Several morphologic types of warts occur in the oral cavity. They have been described as common warts, flat warts, condylomas, or respiratory papillomas on the basis of their clinical and histologic features, but the genotypes of the viruses in these lesions have not been identified. A clinically well defined entity, focal epithelial hyperplasia, has been described only in the oral mucosa. The condition occurs with a high frequency in American Indians in North and South America but it has also been seen in other races and in many parts of the world. Clinically, there are discrete, multiple, elevated nodules on the oral mucosa (lips, buccal mucosa, tongue), which may persist for many years and have the histologic appearance of warts. A unique virus, HPV-13, has been isolated from several of these cases (Pfister et al., 1983).

Papillomavirus particles have been found in a small proportion of esophageal papillomas but the infecting genotype has not been identified.

26.2.4 Diagnosis

Clinically, a papillary wart is seldom misdiagnosed as something else, but other dermatologic conditions (e.g., molluscum contagiosum, plantar corns, skin tags) may be mistaken for warts. Histologic examination of the tissue generally establishes the diagnosis of a wart, but does not assist in identification of the genotype of the infecting virus. No serologic tests are available for virus identification. Human papillomaviruses cannot be grown in culture and there is no other source that can be used to regularly obtain viral antigens of known genotypes.

26.2.4.a Tests for Viral Antigen

A broadly crossreactive genus-specific antiserum is available, which is capable of recognizing capsid antigen of all human and animal papillomavirus by immunoperoxidase or immunofluorescence tests (Jenson et al., 1980). Tests can be performed on sections of routinely collected, formalin-fixed, paraffin-processed tissues, as well as on exfoliated cells (Gupta et al., 1983). The viral antigen is present in the nuclei of cells of the superficial layers of the epithelium. For detection of virus, an immunologic test for viral capsid antigen is considerably more sensitive than demonstration of viral particles by electron microscopy (Ferenczy et al., 1981). However, the antigen is not detectable in at least 25% of histologically confirmed warts. In antigen-positive tissues the number of cells displaying antigen is variable, ranging from only one or two cells to a large number of cells in the section. Only a proportion of cytologically affected cells exhibit antigen. Warts at different sites differ markedly with respect to their yield of viral particles and patterns of antigen distribution (Braun et al., 1983). Viral particles and antigen are abundant in some plantar and common warts, but are scarce in genital tract and laryngeal papillomas. In the genital tract, the antigen prevalence decreases as the lesion progresses toward malignancy. Genotype-specific antisera for individual viral genotypes are not available. The identification of the viral genotype in a tissue requires DNA hybridization.

26.2.4.b Identification of Viral Genotypes

The DNA hybridization methods employed for viral genotype identification, and their advantages and disadvantages, are listed in Table 26.7. Hybridization by the Southern transfer method utilizing DNA extracted from fresh or fresh-frozen tissue (Gissman et al., 1983) provides the most sensitive method of genotype diagnosis and also permits identification of viral subtypes. Hybridization tests, however, can also be performed in situ on paraffin sections of formalin-fixed tissues (Beckmann et al., 1985; Gupta et al., 1985). These tests, although they are less sensitive than the Southern transfer method, permit retrospective diagnosis of routinely collected material and correlation of viral genotype with pathologic characteristics of the tissues. Hybridization of cells placed on a nitrocellulose filter and denatured in situ provides a simple method suitable for testing of large numbers of specimens collected by a noninvasive technique (Wagner et al., 1984). Almost all

Table 26.7. DNA Hybridization Techniques for Identification of Papillomavirus Genotypes

Hybridization Method	Procedure	Advantages	Disadvantages
Southern transfer	Extract DNA from fresh or fresh-frozen tissue or cells, fractionate on gel, denature and transfer to filter, hybridize	Sensitive; permits identification of subtype and detailed studies of genome in tissue	Requires unfixed, fresh or carefully stored tissue; lengthy protocol; not suitable for large studies
DNA dot blot hybridization	Extract DNA from fresh or fresh-frozen tissue or cells, denature, spot denatured DNA on filter, hybridize	Sensitive; omits gel fractionation; simpler than Southen transfer	Requires unfixed, fresh, or carefully stored tissue; does not permit subtype identification or detailed study of genome in tissue
Cells denatured in situ on filter	Place cells on filter, denature, hybridize	Sensitive; suitable for large studies	Requires unfixed, fresh, or carefully stored cells; not applicable to tissues; does not permit subtype identification or detailed study of genome in tissue
In situ hybridization of paraffin-processed tissues (or cell smears)	Deparaffinize sections, denature cells, hybridize	Permits retrospective study of routinely collected material; allows identification of cell types harboring genome	Less sensitive than above methods; does not permit subtype identification or detailed study of genome in tissue

of the HPV DNA have been cloned in plasmid vectors. The viral DNA probes for the hybridization tests are prepared by nick translation using either radiolabeled or biotin-labeled nucleotides. As a result, ^{32}P-labeled probes have been employed in filter hybridizations and ^{35}S-labeled and biotin-labeled probes for in situ hybridizations of paraffin sections and cells fixed on slides.

All papillomavirus DNA share some conserved nucleotide sequences. In order to make a specific diagnosis, therefore, hybridizations are performed under conditions of high stringency (i.e., at high effective temperature; for example, at Tm −17°C). Under these conditions, regions of weak homology are not detected and crosshybridization will occur only between closely related viruses (Table 26.3). Conversely, hybridization under conditions of low stringency (e.g., at Tm −43°C) permits the detection of weakly homologous regions, and is useful for screening of tissue DNA for papillomavirus sequences.

26.2.5 Treatment

Most skin warts and genital warts regress spontaneously. The patient seeks treatment for cosmetic reasons, pain, discomfort, and disability depending on the location and size of warts. The most difficult problems for therapy are posed by children with recurrent laryngeal papilloma, patients with EV, pregnant women with genital warts, and warts in immunocompromised individuals. There is no "one-time" treatment for all warts (Bunney, 1982). The therapies in use include application of caustic agents such as podophyllin and salicylic acid, cryotherapy, surgical removal, antimetabolites such as 5-fluorouracil applied in a cream or a solution, immunotherapy with "autogenous vaccines," and treatment with interferon. Both laryngeal papillomas and genital warts are reported to respond to interferon therapy (Haglund et al., 1981; Schonfeld et al., 1984), but recurrence after cessation of therapy is not uncommon.

26.3 HUMAN POLYOMAVIRUSES

The first conclusive evidence of human infection with polyomaviruses was obtained in the mid-1960s when polyomavirus particles were consistently demonstrated by electron microscopy in the enlarged nuclei of oligodendrocytes in the affected areas of brains of patients with progressive multifocal leukoencephalopathy (PML) (Zu Rhein, 1969). In 1971, JC virus (JCV), the causative agent of PML, was isolated from a PML brain in primary human fetal glial cell cultures. In the same year, another polyomavirus BK virus (BKV), was isolated in Vero cell cultures from the urine of a renal transplant recipient (Padgett and Walker, 1976; Gardner, 1977). Subsequent studies have shown that in the immunocompetent host, both JCV and BKV persist in the kidney following clinically inapparent primary infection in childhood, and that they are reactivated in a variety of conditions that impair cell-mediated immune responses. Almost all of the pathologic effects of BKV and JCV infections occur in immunodeficient individuals.

Between 1955 and 1961, millions of people were inadvertently exposed to simian virus 40 (SV40), an oncogenic polyomavirus of Asian macaques, which had contaminated inactivated (Salk) poliovirus vaccines and experimental live poliovirus vaccines prepared from virus pools grown in primary rhesus kidney cultures. There is no persuasive evidence of any ill effect attributable to SV40 in these vaccines (Shah and Nathanson, 1976).

26.3.1 Characteristics of the Virus

The virion is nonenveloped and has a diameter of 44 nm, icosahedral symmetry, and 72 capsomers. The viral genome is a double-stranded, circular DNA molecule with 5×10^3 base pairs and a molecular weight of 3.2×10^6 daltons. Each of the two DNA strands carries about 50% of the genetic information. Complete nucleotide sequences, as well as detailed physical and physiologic maps, are known for both JCV and BKV genomes. There is extensive nucleotide sequence homology between BKV and JCV throughout their genomes with the highest conservation in the late region, which codes for the capsid proteins (Howley, 1980).

Both JCV and BKV hemagglutinate human erythrocytes. The capsid consists of three virus-specified proteins (VP1, VP2, and VP3) and three cellular histones (VP4, VP5, VP6). The major capsid protein, VP1, accounts for more than 70% of the virion mass and has a molecular weight of 39–44 k. Cells infected with or transformed by the viruses express T antigens, which are coded by the "early" regions of the genomes and are not part of the viral capsid. T antigens are required for the initiation of viral DNA synthesis. BK virus codes for a large T antigen (86–97 k) and a small t antigen (17 k).

Despite the extensive nucleotide sequence homology between the two genomes, JCV and BKV can be readily distinguished from one another by immunologic and DNA hybridization tests. Antibodies to the two viruses in human sera display minimal or no crossreactivity in neutralization, hemagglutination-inhibition (HI), or enzyme-linked immunosorbent assays (ELISA). Both JCV and BKV share with other polyomaviruses genus-specific immunologic determinant(s) that are physically located on VP1 but are internal to the virion surface (Shah et al., 1977). Antibodies against the genus-specific determinant(s), prepared by immunization with disrupted capsids, react with all human and animal polyomaviruses.

Both JCV and BKV transform cells in tissue culture and are oncogenic in laboratory animals. JC virus transforms hamster brain cells and human amnion cells, whereas BKV transforms cells of hamster, rat, rabbit, monkey, and mouse origin. Both viruses are oncogenic for newborn hamsters. JC virus also produces cerebral neoplasm in owl and squirrel monkeys, and provides the only model of a primate central nervous system (CNS) tumor caused by a virus (London et al., 1983).

26.3.2 Pathogenesis and Disease Potential

26.3.2.a In Immunocompetent Hosts

Primary infections with JCV and BKV occurs in childhood. Infection with JCV is acquired at a later age than BKV infection. In the U.S., 50% of children develop antibodies to JCV by the age of 10 to 14 years and to BKV by the age of 3 to 4 years. Infection, in healthy children, is most often subclinical. Serologic studies suggest that primary BKV infection may be associated with mild upper respiratory disease but BKV has not been isolated from respiratory secretions (Goudsmit et al., 1982). An occasional case of cystitis in an otherwise normal child may occur as a result of primary BKV infection (Padgett and Walker, 1983).

The viruses persist in the kidney following primary infection. Viral genomes can be demonstrated in cadaver kidney tissues (McCance, 1983). It is likely that after multiplication at the site of entry, the viruses reach the kidney by a process of viremia, and infection of the kidney is associated with transient viruria. However, JCV and BKV have been recovered very rarely from blood, urine, or from any other site in healthy, immunocompetent children.

26.3.2.b In Immunocompromised Hosts

Most of the infections in immunocompromised hosts are the result of reactivation of viruses latent in the kidney and are evidenced by viruria. Conditions in which viruses are reactivated include pregnancy, diabetes, organ transplantation, antitumor therapy, and immunodeficiency diseases. Unchecked virus multiplication after primary infection of immunodeficient individuals may lead to pathologic consequences.

26.3.2.c Progressive Multifocal Leukoencephalopathy

This is a rare, fatal, subacute demyelinating disease of the CNS that results from JCV infection of oligodendrocytes in the brain (Johnson, 1982; Walker and Padgett, 1983). It occurs as a complication of a wide variety of conditions associated with T-cell deficiencies. These conditions include lymphoproliferative disorders, such as Hodgkin's disease, chronic lymphocytic leukemia, and lymphosarcoma; chronic diseases, such as sarcoidosis and tuberculosis; primary immunodeficiency diseases; prolonged immunosuppressive therapy as, for example, in renal transplant recipients and patients with rheumatoid arthritis, systemic lupus erythematosus, and myositis; and AIDS. Most cases of PML occur in middle age or later life, but the disease is being increasingly identified in younger patients, e.g., in children with primary immunodeficiency diseases, in renal transplant recipients, and in AIDS patients. Cases of progressive multifocal leukoencephalopathy (PML) in the older patients are most likely the result of reactivation of latent JCV. In the younger patient, it is possible that unchecked primary JCV infection may lead to PML.

Progressive multifocal leukoencephalopathy has unique pathologic features. The affected area of the brain contains foci of demyelination, which have at their edges enlarged oligodendrocytes. The nuclei of the oligodendrocytes are two to three times their normal size, basophilic, and they may contain basophilic or eosinophilic inclusion bodies. The centers of the demyelinating foci contain macrophages and "reactive" astrocytes. Most lesions also have bizarre, giant astrocytes with hyperchromatic pleomorphic nuclei. Inflammation is minimal or absent. Neurons are unaffected. The characteristics of these lesions and the occurrence of PML in immunodeficient individuals led Richardson (1961) to propose that the key event in the pathogenesis of PML was infection of oligodendrocytes with a common virus, which atypically infected these cells when immune defenses were impaired, and that demyelination was a result of destruction of these cells, which are normally responsible for the formation and maintenance of myelin sheaths. This hypothesis proved correct. The nuclei of affected oligodendrocytes contain abundant numbers of JCV particles. Progressive multifocal leukoencephalopathy is caused by an atypical course of JCV in an immunocompromised host.

Clinically, PML has an insidious onset, and may occur at any time in the course of the underlying illness. The signs and symptoms point to a multifocal involvement of the brain. Impaired speech and vision and mental deterioration are common early features of the disease. The patient remains afebrile and headache is uncommon. The cerebrospinal fluid (CSF) remains normal. As a rule, the disease is progressive, resulting in death within 3 to 6 months after onset. Paralysis of limbs, cortical blindness, and sensory abnormalities occur in later stages. A few patients may survive for years with stabilization of the condition and even apparent remission. A longer survival time is thought to be associated with a more marked inflammatory response in the brain.

The diagnosis of PML can be conclusively established only by pathologic examination of a biopsy or at postmortem. Macroscopically, the brain shows foci of demyelination that may vary widely in size and may become confluent and necrotic in the advanced stages of disease. The lesions are most frequent in the subcortical white matter. The cerebrum is almost always affected. Microscopically, the presence of enlarged oligodendrocyte nuclei around the foci of demyelination is diagnostic. These altered nuclei contain abundant amounts of JCV particles, antigen, and DNA. JC virus particles or antigen are not found in normal brains or in nondiseased areas of PML brains. Small amounts of viral DNA may be recovered from extraneural sites such as kidney, liver, lymph node, and spleen (Grinnell et al., 1983).

26.3.2.d Renal and Bone Marrow Transplant Recipients

About 33–50% of these organ transplant recipients excrete one or both of these viruses in urine in the posttransplant period (Gardner, 1977; Arthur et al., 1983). The duration of viruria varies from a few days to several months. BK virus infection is more common than JCV infection. Most infections are due to reactivation of latent viruses. The frequency of infection is higher in recipients with a history of diabetes (Hogan et al., 1980). The risk of BKV infection is increased if a kidney from a seropositive donor is transplanted into a seronegative recipient (Andrews et al., 1983). The infections rarely lead to severe pathologic consequences and it is unclear if they result in loss of renal function or rejection of allografts. Some cases of ureteral obstruction, an uncommon and late complication in renal transplantation, have been ascribed to JCV and BKV infections.

26.3.2.e Primary Immunodeficiency Diseases

BK virus has been isolated from the urine of patients with primary immunodeficiency diseases. A fatal end result of BKV infection has been reported. A 6-year-old boy with hyperimmunoglobulin M deficiency developed massive BKV viruria, tubulo-interstitial nephritis with viral inclusions in the lesions, and irreversible renal failure (Rosen et al., 1983).

26.3.2.f Pregnancy

Both viruses are reactivated in some women during normal pregnancy. In a prospective study, cytopathology in cells obtained from urine sediment suggested JCV and BKV infections in 3.2% of pregnant women. This was most frequently observed in the last trimester of pregnancy (Coleman et al., 1980). In another study, 16% of the women showed an antibody rise to one or the other virus during pregnancy. All the infections were reactivations of latent viruses in antibody-positive individuals (Andrews et al., 1983). It has been reported that fetal sera may have BKV-specific IgM, indicating transplacental transmission of the virus; these observations have not been confirmed (Andrews et al., 1983).

26.3.2.g Role in Human Malignancies

JC virus and BKV are oncogenic for laboratory animals and they transform cultured cells. These viruses as well as SV40, therefore, have been investigated for their roles in human malignancies (Howley, 1983). Tumors of the urinary tract and the CNS have received special attention. In one instance, multifocal gliomas corresponded topographically to lesions of PML demyelination, suggesting that the tumors arose in these lesions. There are sporadic reports of finding viral genome or viral antigen in individual human tumors, especially in meningiomas, but these observations have not been confirmed. A reproducible and consistent association of JCV or BKV with any human malignancy has not been demonstrated.

26.3.3 Diagnosis

Evidence of multifocal brain disease in an immunocompromised individual suggests the possibility of PML, but definitive diagnosis can be established only by examination of affected tissue obtained by biopsy or at autopsy. The unique histopathologic features of PML are seen by light microscopy. Except for two cases in 1971 in which SV40 was isolated from PML brains, all other cases have yielded JCV. The virus can be specifically identified by an immunoperoxidase test of frozen sections or paraffin sections of the affected tissue using a monospecific anti-JCV serum. Alternatively, the viral genome in the lesion can be identified by hybridization of the total DNA extracted from the affected tissue (Grinnell et al., 1983) or by in situ hybridization of paraffin sections with a JCV probe. Serologic studies are not helpful in the diagnosis of PML. JC virus antibodies are present at the onset of the disease and they do not show any marked increase as the disease progresses. Viral antibodies are not detected in the CSF.

In conditions other than PML, cytomorphology of the urinary tract epithelial cells suggestive of virus excretion in the urine is often the first indication of virus infection. Virus-infected cells are enlarged and their nuclei contain single, large, basophilic inclusions that may occupy the whole nucleus (Coleman, 1975; Kahan et al., 1980; Traystman et al., 1980). There are no cytoplasmic inclusions. Differential diagnosis includes cells infected with cytomegalovirus (CMV) and sometimes urothelial cancer cells. Cytomegalovirus-infected cells may have both nuclear and cytoplasmic inclusions; the nuclear inclusions are small, surrounded by a clear peripheral zone (halo), and are either basophilic or eosinophilic. The malignant cell nucleus has rough-textured chromatin in contrast to the structureless inclusion in the virus-infected nucleus.

The cytologic abnormalities are not always present or clear-cut during viruria, and these abnormalities do not distinguish between JCV and BKV infection. JC virus grows best in primary human fetal glial cells and BKV grows best in primary human embryonic kidney or human diploid fibroblast cells. Both viruses also grow in primary urothelial cell cultures collected from infant urine (Beckmann and Shah, 1983). Isolation of virus in tissue culture is inefficient and it may take several weeks before a specific diagnosis is possible. Human fetal glial cells or urinary tract epithelial cells are difficult to obtain. A recently developed enzyme-linked immunosorbent assay for detection of viral antigens in urine and DNA dot blot hybridization assays for identification of viral genomes in cells from the urinary tract offer the prospect of rapid diagnosis of JCV and BKV (Arthur et al., 1983).

26.3.4 Treatment

Attempts to treat PML have not been successful although some remissions have been reported with the use of nucleic acid base analogs, adenine arabinoside, and cytosine arabinoside (Walker and Padgett, 1983). It would be useful, when possible, to reduce or discontinue immunosuppressive therapy. No attempts have been made to treat urinary infections.

REFERENCES

Andrews, C., Daniel, R., and Shah, K. 1983. Serologic studies of papovavirus infections in pregnant women and renal transplant recipients. In: J.L. Sever and D.L. Madden (eds.), Polyomaviruses and Human Neurological Diseases. New York: Alan R. Liss.

Arthur, R., Shah, K., Yolken, R., and Charache, P. 1983. Detection of human papovaviruses BKV and JCV in urines by ELISA. In J.L. Sever and D.L. Madden (eds.), Polyomaviruses and Human Neurological Diseases. New York: Alan R. Liss.

Beckmann, A., and Shah, K. 1983. Propagation and primary isolation of JCV and BKV in urinary epithelial cell cultures. In: J.L. Sever and D.L. Madden (eds.), Polyomaviruses and Human Neurological Diseases, New York: Alan R. Liss.

Beckmann, A.M., Myerson, D., Daling, J.R., Kiviat, N.B., Fenoglio, C.M., and McDougall, J.K. 1985. Detection of human papillomavirus DNA in carcinomas by in situ hybridization with biotinylated probes. J. Med. Virol. 16:265–273.

Boshart, M., Gissmann, L., Ikenberg, H., Kleinheinz, A., Scheurlen, W., and zur Hausen, H. 1984. A new type of papillomavirus DNA, its presence in genital cancer biopsies and in cell lines derived from cervical cancer. EMBO. J. 3:1151–1157.

Braun, L., Farmer, E., and Shah, K. 1983. Immunoperoxidase localization of papillomavirus antigen in cutaneous warts and Bowenoid papulosis. J. Med. Virol. 12:187–193.

Bunney, M. 1982. Viral Warts: Their Biology and Treatment. Oxford: Oxford University Press.

Centers for Disease Control 1983. Condyloma acuminatum—United States, 1966–1981. M.M.W.R. 32:306–308.

Chuang, T.-Y., Perry, H.O., Kurland, L.T., and Ilstrup, D.M. 1984. Condyloma acuminatum in Rochester, Minn., 1950–1978. I. Epidemiology and clinical features. Arch. Dermatol. 120:469–483.

Coleman, D.V. 1975. The cytodiagnosis of human polyomavirus infection. Acta Cytol. 19:93–96.

Coleman, D., Wolfendale, M., Daniel, R., Dhanjal, N., Gardner, S., Gibson, P., and Field, A. 1980. A prospective study of human polyomavirus infection in pregnancy. J. Infect. Dis. 142:1–8.

Crum, C.P., Ikenberg, H., Richart, R.M., and Gissman, L. 1984. Human papillomavirus type 16 and early cervical neoplasia. N. Engl. J. Med. 310:880–883.

Durst, M., Gissmann, L., Ikenberg, H., and zur Hausen, H. 1983. A papilloma-virus DNA from a cervical carcinoma and its prevalence in cancer biopsy samples from different geographic regions. Proc. Natl. Acad. Sci. U.S.A. 80:3812–3815.

Ferenczy, A., Braun, L., and Shah, K.V. 1981. Human papillomavirus (HPV) in condylomatous lesions of cervix. A comparative ultrastructural and immunohistochemical study. Am. J. Surg. Pathol. 5:661–670.

Gardner, S. 1977. The new human papovaviruses: Their nature and significance. In Recent Advances in Clinical Virology. Waterson, A.P. ed. New York: Livingstone.

Gissmann, L., Wolnik, L., Ikenberg, H., Koldovsky, U., Schnurch, H., and zur Hausen, H. 1983. Human papillomavirus types 6 and 11 DNA sequences in genital and laryngeal papillomas and in some cervical cancers. Proc. Natl. Acad. Sci. U.S.A. 80:560–563.

Goudsmit, J., Wertheim-van Dillen, P., van Strein, A. and van der Noordaa, J. 1982. The role of BK virus in acute respiratory tract disease and the presence of BKV DNA in tonsils. J. Med. Virol. 10:91–99.

Grinnell, B., Padgett, B. and Walker, D. 1983. Distribution of nonintegrated DNA from JC papovavirus in organs of patients with progressive multifocal leukoencephalopathy. J. Infect. Dis. 147:669–675.

Guillet, G., Braun, L., Shah, K. and Ferenczy, A. 1983. Papillomavirus in cervical condylomas with and without associated cervical intraepithelial neoplasia. J. Invest. Dermatol. 81:513–516.

Gupta, J.W., Gupta, P.K., Shah, K.V. and Kelly, D.P. 1983. Distribution of human papillomavirus antigen in cervicovaginal smears and cervical tissues. Int. J. Gynecol. Pathol. 2:160–170.

Gupta, J., Gendelman, H.E., Naghashfar, Z., Gupta, P., Rosenshein, N., Sawada, E., Woodruff, J.D., and Shah, K. 1985. Specific identification of human papillomavirus type in cervical smears and paraffin sections by in situ hybridization with radioactive probes: A preliminary communication. Int. J. Gynecol. Pathol. 4:211–218.

Haglund, S., Lundquist, P.G., Cantell, K. et al. 1981. Interferon therapy in juvenile laryngeal papillomatosis. Arch. Otolaryngol. 107:327–332.

Hajek, E. 1956. Contribution to the etiology of laryngeal papilloma in children. J. Laryngol. 70:166–168.

Heilman, C.A., Law, M.F., Israel, M.A., and Howley, P.M. 1980. Cloning of human papillomavirus genomic DNAs and analysis of homologous polynucleotide sequences. J. Virol. 36:395–407.

Hogan, T., Borden, E., McBain, J., Padgett, B., and Walker, D. 1980. Human polyomavirus infections with JC virus and BK virus in renal transplant patients. Ann. Intern. Med. 92:373–378.

Howley, P. 1980. Molecular biology of SV40 and the human polyomaviruses BK and JC. In: G Klein (ed.), Viral Oncology. New York: Raven Press.

Howley, P. 1983. Papovaviruses: search for evidence of possible association with human cancer. In: L.A. Phillips (ed.), Viruses Associated with Human Cancer. New York: Marcel Dekker.

Jablonska, S., Dabrowski, J., and Jakubowicz, K. 1972. Epidermodysplasia verruciformis as a model in studies on the role of papovaviruses in oncogenesis. Cancer Res. 32:583–589.

Jenson, A., Rosenthal, J., Olson, C., Pass, F., Lancaster, W., and Shah, K. 1980. Immunological relatedness of papilloma viruses from different species. J. Natl. Cancer Inst. 64:495–500.

Johnson, R. 1982. Progressive multifocal leukoencephalopathy. In Viral Infections of the Nervous System. New York: Raven Press.

Kahan, A., Coleman, D., and Koss, L. 1980. Activation of human polyomavirus infection—detection by cytologic technics. Am. J. Clin. Pathol. 74:326–332.

Kremsdorf, D., Favre, M., Jablonska, S., Obalek, S., Rueda, L.A., Lutzner, M.A., Blanchet-Bardon, C., Van Voost Vader, P.C., and Orth, G. 1984. Molecular cloning and characterization of the genomes of nine newly recognized human papillomavirus types associated with epidermodysplasia verruciformis. J. Virol. 52:1013–1018.

Kurman, R., Jenson, A., and Lancaster, W. 1983. Papillomavirus infection of the cervix. II. Relationship to intraepithelial neoplasia based on the presence of specific viral structural proteins. Am. J. Surg. Pathol. 7:39–52.

London, W.T., Houff, S.A., McKeever, P.E., Wallen, W.C., Sever, J.L., Padgett, B.L., and Walker, D.L. 1983. Viral-induced astrocytomas in squirrel monkeys. In: J.L. Sever and D.L. Madden (eds.), Polyomavirus and Human Neurological Diseases. New York: Alan R. Liss.

Lutzner, M.A. 1978. Epidermodysplasia verruciformis: Autosomal recessive disease characterized by viral warts and skin cancer: a model for viral oncogenesis. Bull. Cancer. 65:169–182.

McCance, D. 1983. Persistence of animal and human papovaviruses in renal and nervous tissues. In J.L. Sever and D.L. Madden (eds.), Polyomaviruses and Human Neurological Disease. New York: Alan R. Liss.

Meisels, A., Morin, C., and Casas-Cordero, M. 1982. Human papillomavirus infection of the uterine cervix. Int. J. Gynecol. Pathol. 1:75–94.

Melnick, J. 1962. Papova virus group. Science 135:1128–1130.

Mounts, P., and Shah, K. 1984. Respiratory papillomatosis: Etiological relation to genital tract papillomaviruses. Prog. Med. Virol. 29:90–114.

Mounts, P., Shah, K.V., and Kashima, H. 1982. Viral etiology of juvenile- and adult-onset squamous papilloma of the larynx. Proc. Natl. Acad. Sci. U.S.A. 79:5425–5429.

Oriel, J. 1971. Natural history of genital warts. Br. J. Vener. Dis. 47:1–13.

Orth, G., Breitburd, F., Favre, M., and Croissant, O. 1977a. Papilloma viruses: possible role in human cancer. In H.H. Hiatt, J.D. Watson, and J.A. Winsten (eds.), Origins of Human Cancer. Cold Spring Harbor Conferences on Cell Proliferation, Vol. 4. New York: Cold Spring Harbor Laboratory.

Orth, G., Favre, M., and Croissant, O. 1977b. Characterization of a new type of human papillomavirus that causes skin warts. J. Virol. 24:108–120.

Orth, G., Favre, M., Breitburd, F., Croissant, O., Jablonska, S., Obalek, S., Jarzabek-Chorzelska, M., and Rzesa, G. 1980. Epidermodysplasia verruciformis: A model for the role of papilloma viruses in human cancer. In M. Essex, G. Todaro, and H. zur Hausen (eds.), Viruses in Naturally Occurring Cancers. Cold Spring Harbor Conferences on Cell Proliferation, Vol. 7. New York: Cold Spring Harbor Laboratory.

Ostrow, R., Bender, M., Nhmura, M., Seki, T., Kawashima, M., Pass, F., and Faras, A. 1982. Human papillomavirus DNA in cutaneous primary and metastasized squamous cell carcinomas from patients with epidermodysplasia verruciformis. Proc. Natl. Acad. Sci. U.S.A. 79:1634–1638.

Padgett, B., and Walker, D. 1976. New human papovaviruses. Prog. Med. Virol. 22:1–35.

Padgett, B., and Walker, D. 1983. BK virus and nonhemorrhagic cystitis in a child. Lancet i:770.

Pfister, H., Hettich, I., Runne, U., Gissmann, L., and Chilf, G.N. 1983. Characterization of human papillomavirus type 13 from focal epithelial hyperplasia Heck lesions. J. Virol. 47:363–366.

Reid, R., Laverty, C., Coppleson, M., Isarangkul, W., and Hills, E. 1980. Noncondylomatous cervical wart virus infection. Obstet. Gynecol. 55:476–483.

Richardson, E. 1961. Progressive multifocal leukoencephalopathy. N. Engl. J. Med. 265:815–823.

Rosen, S., Harmon, W., Krensky, A., Edelson, P., Padgett, B., Grinnell, B., Rubino, M., and Walker, D. 1983. Tubulo-interstitial nephritis associated with polyomavirus (BK type) infection. N. Engl. J. Med. 308:1192–1196.

Rowson, K.E.K., and Mahy, B.W.J. (1967). Human papova (wart) virus. Bacteriol. Rev. 31:110–131.

Schonfeld, A., Schattner, A., Crespit, M., et al. 1984. Intramuscular human interferon-B injections in treatment of condylomata acuminata. Lancet i:1038–1042.

Shah, K., and Nathanson, N. 1976. Human exposure to SV40: Review and comment. Am. J. Epidemiol. 103:1–12.

Shah, K., Ozer, H., Ghazey, H., and Kelly, T. Jr. 1977. Common structural antigen of papovaviruses of the simian virus 0-polyoma subgroup. J. Virol. 21:179–186.

Traystman, M.D., Gupta, P.K., Shah, K.V., Reissig, M., Cowles, L.T., Hillis, W.D., and Frost, J.K. 1980. Identification of viruses in the urine of renal transplant recipients by cytomorphology. Acta Cytol. 24:501–510.

Wagner, D., Ikenberg, H., Boehm, N., and Gissmann, L. 1984. Identification of human papillomavirus in cervical smears by deoxyribonucleic acid in situ hybridization. Obstet. Gynecol. 65:767–772.

Walker, D., and Padgett, B. 1983. Progressive multifocal leukoencephalopathy. In H. Fraenkel-Conrat and R.R. Wagner (eds.), Comprehensive Virology, Vol. 18. New York: Plenum Press.

Zu Rhein, G. 1969. Association of papova-virions with a human demyelinating disease (progressive multifocal leukoencephalopathy). Prog. Med. Virol. 11:185–247.

zur Hausen, H. 1980. Papilloma viruses. In: J. Tooze (ed.), DNA Tumor Viruses. New York: Cold Spring Harbor Laboratory.

27

Herpes Simplex Viruses

Fred Rapp

27.1 INTRODUCTION

The herpesviruses are a unique group of viruses with a host range that includes invertebrates and vertebrates of all species examined thus far. The word herpes is Greek, meaning "to creep" and it was originally used to define "an animal that goes on all fours" (herpeton) and "the natural study of reptiles" (herpetology). Apparently, the word came into use because of certain clinical manifestations that were seen before the virus was actually identified as the causative agent of many syndromes. Until a few years ago, herpes was a word familiar only to physicians, virologists, and cold sore sufferers. It has become a household word since it was discovered that the virus causes sexually transmitted recurrent disease. Now almost everyone knows that herpes usually refers to herpes simplex virus type 1 (HSV-1) and herpes simplex virus type 2 (HSV-2), although there are other members included in the human herpesvirus group. These include Epstein–Barr virus (EBV), cytomegalovirus (CMV), and varicella-zoster virus (VZV), which will be discussed in the succeeding chapter.

No information existed on the physical properties of HSV until 1921, when Luger and Lauda demonstrated that HSV passed through a filter with pores small enough to retain bacteria. From that time, the accumulation of information on the characteristics of these viruses has been steady.

27.2 CHARACTERISTICS OF HERPES SIMPLEX VIRUSES

Herpesviruses exhibit a unique morphology that allows rapid identification by electron microscopy. However, differentiation among the herpesviruses by visual examination is very difficult because the morphology is typical of all group members and the differences are slight. The herpesviruses are enveloped, ether-sensitive, and of icosahedral symmetry, consisting of 162 capsomeres. The capsid is approximately 100 nm in diameter with an envelope enlarging the complete virion to 150 nm. The envelope is composed of a lipid bilayer and contains glycoproteins. A structure located between the envelope and the capsid is called the tegument and appears by staining techniques as a layer of amorphous material. The virus capsid contains nucleoprotein in a core; the core is a densely staining area measuring approximately 77.5 nm in diameter and containing the virus DNA.

Herpes simplex virus type 1 and HSV-2 DNA are linear, double-stranded molecules of approximately 95×10^6 daltons

(Frenkel and Roizman, 1971) with a base composition of 67 and 69% guanine plus cytosine, respectively (Kieff et al., 1971). The nucleic acids of HSV-1 and -2 share approximately 47–50% base sequence homology (Kieff et al., 1972) and the buoyant density of the DNA in cesium chloride gradients is 1.726 g/cm^3 for HSV-1 and 1.728 g/cm^3 for HSV-2 (Kieff et al., 1971). The density of the HSV virion has been calculated by measuring the position of the virus band at equilibrium in density gradients. Densities ranging from 1.253 to 1.285 g/cm^3 have been reported, but differences in virus strains, choice of gradient, and cells in which the virus is grown may account for these discrepancies. Using cesium chloride gradients, Roizman and Roane (1961) established the densities for HSV-1 and -2 at 1.271 and 1.267 g/cm^3, respectively.

Herpes simplex virus DNA has a very intriguing structure. Specifically, it consists of two covalently linked components that represent the L or long region (82%) and the S or short and right-hand region (18%) of the HSV genome. Each of these two components consists of unique (U) sequences that are bracketed by inverted repeated sequences (Roizman, 1979; Hyman, 1980). At the present time, the significance of these reiterated sequences is unknown. In addition, the HSV DNA can form four isomeric arrangements (Sheldrick and Berthelot, 1975). In short, the L and S components can invert relative to each other; therefore, the DNA isolated from virions consists of four populations that are approximately equimolar in concentration and differing in the orientations of the U_L and U_S components.

In order to begin to understand the multipotentiality of HSV and its ability to cause a wide spectrum of diseases in humans, a basic knowledge of HSV replication (Roizman, 1980) is required. The replicative cycle of HSV is unlike that of all other intranuclear DNA-containing viruses that have distinct, early and late replicative steps that are separated by the synthesis of progeny virus DNA. The replication of HSV involves the manufacture of more than 50 new polypeptides in addition to the nucleic acids and glycoproteins required to produce the fully infectious virion. The events in the HSV growth cycle require 8 to 16 hours from infection to the end of the growth phase and have been well studied. The initial stages appear to be similar to those of other viruses:

1. Attachment to the cell membrane
2. Penetration of the cytoplasm
3. Release of DNA or a nucleoprotein complex
4. Migration to the nucleus where HSV DNA synthesis and transcription of messenger RNA (mRNA) occur

The mRNA then migrates to the cytoplasm and is translated into various HSV-specific proteins, which are then transported back to the nucleus. Herpes simplex virus DNA maturation and encapsidation by appropriate HSV structural proteins occur in the nucleus; this is followed by budding through the nuclear membrane, which adds envelope material to the virus particles. These particles then travel to the cytoplasm and exit the cytoplasm into extracellular spaces and fluids. Budding, as an alternative mechanism of egress, may take place at the cell membrane.

The synthesis of HSV-specific gene products appears to be highly orchestrated and controlled. Honess and Roizman (1974) have revealed the regulated synthesis of three classes of mRNA that correspond to three groups of HSV-specific polypeptides or glycoproteins (for review, see Roizman, 1980).

Productive HSV infection is usually fatal to the cell and is manifested by a series of cytopathic events, including damage to host chromosomes, margination of the chromatin, development of an early basophilic inclusion body, and a late eosinophilic intranuclear inclusion body (type A) devoid of virus material and representing a scar of virus infection. Most cells eventually become rounded and are destroyed. The replicative cycle of HSV-2 is somewhat longer than that of HSV-1, however, this varies depending on the type of cells used for virus replication.

27.3 PATHOGENESIS OF HERPES SIMPLEX VIRUS INFECTION

Herpes simplex virus infects neonates, children, and adults, and is able to produce diseases of the eye, lip, mouth, skin, genitourinary tract, and central nervous system (CNS). The primary (or initial) infection is in the form of a localized lesion, but this is often accompanied by a severe systemic illness that sometimes is fatal. However, an individual suffering from recurrent infection is usually free from disseminated disease. Herpes simplex virus is ubiquitous throughout the population: Serologic studies have demonstrated that although most adults have circulating antibodies to HSV, only 10–15% of all primary infections produce clinical illness.

In 1962, Schneweis (1962) reported two distinct serologic types of HSV, subsequently designated as types 1 and 2. In the past, most children were exposed to HSV-1 within 18 months of birth as determined by antibody development; this appears to be changing in the U.S. with older first exposure being fairly common. With HSV-1 infection, the initial disease is a gingivostomatitis characterized by vesicles in the oral cavity and, on occasion, elevated temperature. These lesions are often referred to as herpes labialis, facialis febrilis, cold sores, and fever blisters. The lesions usually occur at the mucocutaneous junction of the lip and recur at or close to the same site. The incubation period may last from 2 to 20 days and averages approximately 6 days. The lesions produced by HSV-1 most often are manifested as a group of small vesicles that usually last for 7 days. Scab formation and healing without scarring follow. Inside the mouth, the lesions can be quite painful and have the appearance of ulcers. Other HSV-1–induced diseases of the skin include primary herpes dermatitis (a generalized vesicular eruption), eczema herpeticum (usually a manifestation of a primary infection in which skin is the portal of entry), and traumatic herpes (resulting from traumatic breaks due to burns or abrasions in the normal skin of a susceptible child). Herpetic whitlow is an occupational hazard (dentists, hospital personnel, wrestlers) resulting from infection of broken skin (often on fingers) in contact with virus on another individual. In addition, HSV-1 is responsible for approximately 2500 cases of encephalitis in the U.S., and infections of the eye can lead to keratoconjunctivitis which, in its most serious form, can cause blindness. Sometime after primary exposure HSV-1 appears to enter cells of the trigeminal ganglion, where it remains latent, apparently for the lifetime of the host (Bastian et al., 1972; Baringer and Swoveland, 1973).

The genitourinary tract is also the site of primary and recurrent HSV infections. Initial HSV-2 infection usually follows puberty and the onset of sexual activity. After primary infection, the virus enters a state of latency that usually involves the sacral dorsal root (S2-4) ganglia. Genital HSV infection is usually associated with HSV-2 and, in most cases, it is transmitted sexually. In the female, the infection is manifested by vesicles on the mucous membranes of the labia and the vagina. Severe forms of genital infection result in ulcers that cover the entire area surrounding the vulva. Symptoms of primary infection include itching, pain, and lymphadenopathy. In infected males, tiny vesicles appear on the penis and prepuce; urethritis and a watery discharge also may be present. A primary stomatitis may be accompanied by inguinal lymphadenitis. Recurrences of HSV-2 infection may produce pain during sexual intercourse and result in emotional distress.

Diseases of the mucous membranes can be caused by HSV-1 and -2. These include acute herpetic gingivostomatitis, acute herpetic rhinitis, and genital herpes infection. Herpes gingivostomatitis, a serious infection of the gums, tongue, mouth, and pharynx, usually is seen in children from 1 to 3 years of age and is often accompanied by high fever, swollen gums, irritability, and cervical lymphadenopathy. Acute herpetic rhinitis is a primary infection of the nose recognized by the appearance of tiny vesicles in the nostrils. This condition also usually is associated with fever and enlarged cervical lymph nodes.

Several modes of transmission are responsible for HSV infection. Autoinfection of the urogenital tract with HSV-1 may occur and a higher frequency of orogenital sexual practices may be contributing to an increased incidence of HSV-1 infection in the genital tract. Herpes simplex virus type 2 can be transmitted from mother to fetus during passage through an infected birth canal, which can result in a severe and often fatal generalized herpetic infection of the neonate. This means of herpesvirus infection can be prevented by delivery by Cesarean section.

Herpes simplex virus-induced infections of the eyes can seriously damage sight and may be the result of primary or recurrent disease. These diseases, clinically designated herpetic keratoconjunctivitis and keratitis, can exist in superficial or deep forms. In children and adolescents, a primary infection often begins as a unilateral conjunctivitis with chemosis, edema of the lids, and pain. The disease usually ends in a few days, if limited to the conjunctiva. If the cornea is involved, however, the disease may persist in an acute phase for 2 to 3 weeks and keratitis may remain for several more weeks. If deeper layers of the cornea are involved, as in disciform keratitis, corneal scarring may occur.

The CNS is also a target of HSV infection and when this happens the disease is usually severe. The CNS disease most often associated with herpetic infection is meningoencephalitis. This usually results from a generalized primary infection of newborns, children, and young adults. The virus can be isolated from other organs, as well as the CNS. The disease has a rapid onset and symptoms include fever, chills, headache, meningeal irritation, convulsions, and alterations in reflexes. When the disease is fatal, death usually occurs within 8 to 10 days. Adults can contact meningitis when HSV-2 infects the CNS, and immunocompromised adults can develop a severe generalized disease similar to neonatal herpes; this is occasionally responsible for herpetic hepatitis. Table 27.1 details the diseases caused by HSV.

Table 27.1. Diseases Caused by Herpes Simplex Viruses

Stomatitis
Herpes labialis
Genital lesions
Primary herpetic dermatitis
Eczema herpeticum
Traumatic herpes
Acute herpetic rhinitis
Keratoconjunctivitis
Keratitis
Neonatal herpes
Meningitis
Encephalitis
Herpetic hepatitis
Urethritis
Arthritis
Disseminated rash
Autonomic system dysfunction

27.4 HERPES SIMPLEX VIRUS-INDUCED LATENCY

Following a primary HSV infection, the virus often enters a state of latency in its natural host. As stated, the most acknowledged sites for latent HSV-1 and -2 are the trigeminal ganglion and the sacral dorsal root (S2-4) ganglion, respectively. The pathogenesis of HSV latency and reactivation was described as early as 1929 (Goodpasture, 1929). Latent HSV infections can persist in the presence of humoral antibody and sensitized lymphocytes. Based on work conducted in laboratory animals (Stevens and Cook, 1971; Stevens et al., 1972; Baringer and Swoveland, 1974; Walz et al., 1974; Price et al., 1975; Reeves et al., 1976), it appears that latent HSV-1 resides in the trigeminal ganglion. Autopsies of human trigeminal tissue also resulted in the recovery of HSV-1 (Bastian et al., 1972); however, HSV-1 has not been recovered from other areas of the trigeminal nerve network when cocultivated with cells susceptible to virus replication.

At the present time, it is not known why certain individuals experience recurrences and others do not. Apparently, more than 80% of the human population harbors HSV-1 in a latent form; however, only a small portion experience recurrences. People who do reactivate HSV usually do so after nonspecific stimuli that include fever due to bacterial or viral infections, exposure to ultraviolet irradiation (i.e., sunshine), stress, and possibly hormonal irregularities. The stimuli seem to have very little in common, although an increase in temperature may trigger the reactivation. This reasoning is based on results from experiments using an in vitro latency system in which persistence and reactivation of HSV have been regulated by temperature shifts (O'Neill, 1977; Colberg-Poley et al., 1979). There are probably many factors involved in maintaining latency and in activating latent virus but these remain elusive. Perhaps a deficiency in the immune system of the human host is required for reactivation of the virus and recurrence of disease. Studies investigating the role of the immune system in HSV latency and reactivation are underway, but no immune deficiency has yet been associated with reactivation of HSV.

27.5 DIAGNOSIS OF HERPES SIMPLEX VIRUS INFECTIONS

Herpes simplex viruses are responsible for a large number of clinical conditions and, as a result, diagnosis is difficult and sometimes impossible. Herpes simplex virus has a wide host range in vitro and in vivo, and readily infects experimental animals such as hamsters, mice, rats, guinea pigs, rabbits, and embryonated chicken eggs. These viruses can be grown in primary cell cultures as well as in primate and nonprimate cell lines. To date, the most sensitive cell culture systems are rabbit kidney cells and human amnion cells, although human diploid, HeLa, HEp-2, and monkey Vero cell lines are also very susceptible. When cells are infected with HSV there is a characteristic rounding of the cells. If stain is applied to the cell culture, the cells will exhibit eosinophilic intranuclear inclusions. The most rapid results for isolating HSV are obtained with human amnion cells: extensive cellular destruction occurs in 2 to 3 days. Figure 27.1 is a flow chart for herpesvirus identification and isolation.

Herpes simplex virus can be isolated from oral and genital vesicle fluid, throat swabs and washings, the oropharynx and, on occasion, CSF. Specimens should be obtained during the first 5 days of infection and cultures should be inoculated within 1 hour after collection for best results. Otherwise, refrigeration or freezing should be employed to preserve the specimen.

Newborn mice are particularly sensitive to HSV. When inoculated with the virus intracerebrally or intraperitoneally, the animals will develop encephalitis within a short time. Upon examination of sections stained with hematoxylin and eosin, mouse brain tissue will exhibit Cowdry type A inclusion bodies. Types of HSV can be recognized successfully using the chorioallantoic membranes of embryonated eggs. Herpes simplex virus type 1 will produce well defined pinpoint pocks on the membrane within 3 to 4 days postinoculation, whereas, HSV-2 produces easily recognizable, large, clear pocks. More recently, differential sensitivity to bromodeoxyuridine has been used (HSV-1 is sensitive; HSV-2 is relatively insensitive).

Herpes simplex virus can usually be recognized within 48 hours by the typical cytopathology of the virus in culture and its wide range of host infectivity. Neutralization, complement-fixation, or immunoflu-

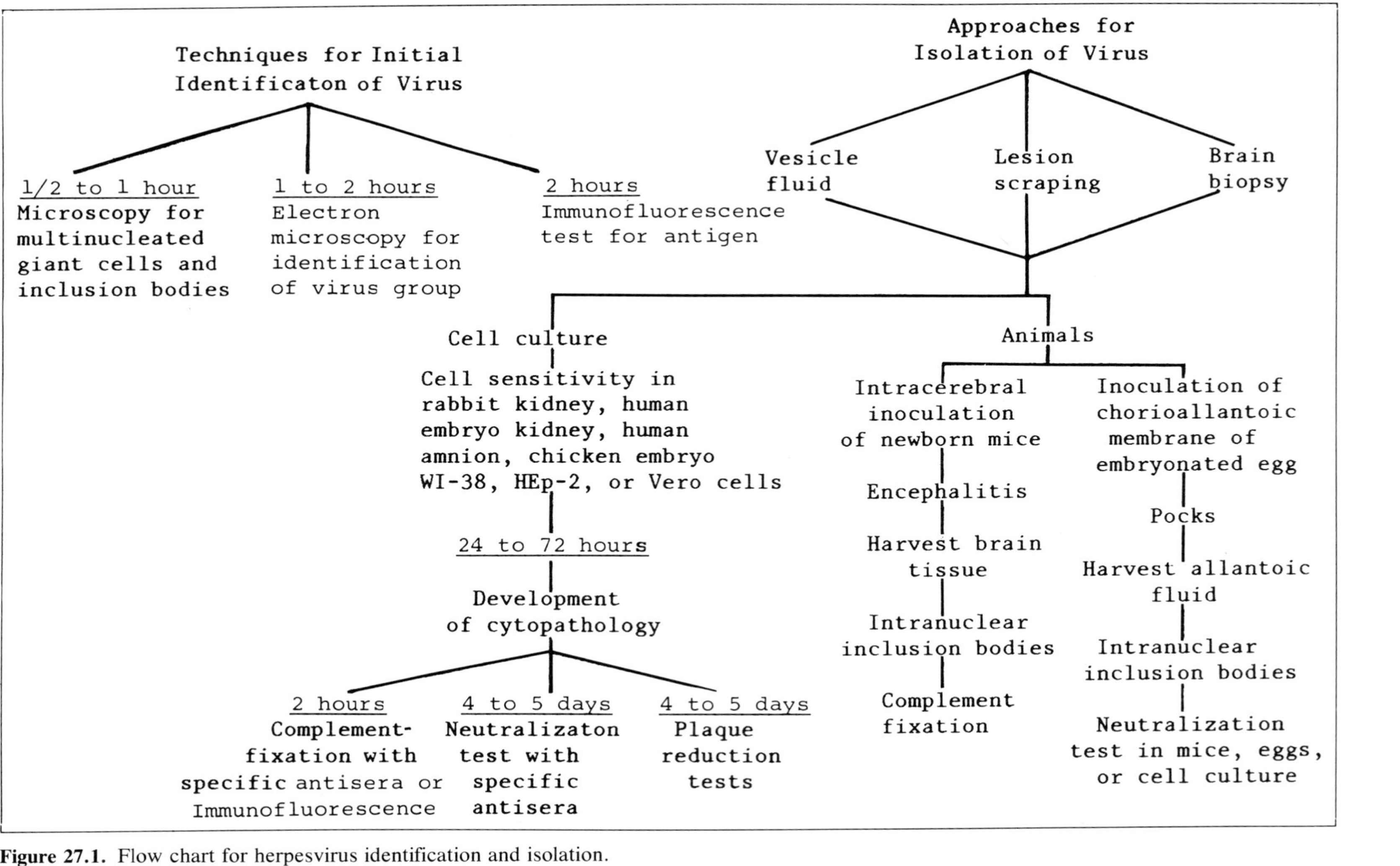

Figure 27.1. Flow chart for herpesvirus identification and isolation.

Table 27.2. Serologic Methods for the Diagnosis of Herpes Simplex Virus Infections

Type of Infection	Method of Detection	Advantages	Disadvantages
Mucocutaneous	FAT	Rapid, specific	—
	CF	Routine procedure	Not sensitive, complex procedure
	IP	Rapid, specific	—
Keratitis	FAT	Rapid, specific	—
	CF	Routine procedure	Not sensitive, complex procedure
Encephalitis	FAT (biopsy specimen)	Rapid, specific	Not completely reliable
	FAT (cells in CSF)	Rapid, specific	Not completely reliable, difficult to interpret, low cell count early in disease
	IP staining of biopsy material	Rapid, specific, uses cell suspensions, permanent slides	—

Abbreviations: FAT = fluorescent antibody-staining technique; CF = complement-fixation; IP = immunoperoxidase.

orescence tests with specific HSV-immune serum can also be performed for final identification. The neutralization test is more specific than the complement fixation test, but a standard fourfold increase in antibody titer is diagnostic for both. The indirect fluorescent antibody-staining technique is the most rapid method for detecting HSV infection and is most often used in diagnosing infections of the eye (Table 27.2).

27.5.1 Diagnosis of Herpes Simplex Viruses in Mucocutaneous Lesions

Immunofluorescence (IF) is the method most widely used to diagnose suspected herpetic mucocutaneous lesions. Scrapings from vesicular lesions of the skin can be examined by IF to detect HSV antigens. The cells are fixed in acetone and then stained using the direct fluorescent antibody technique. A positive fluorescent antibody (FA) test correlates well with the ability to culture HSV in eggs. When using this method, both scrapings and swabs are taken from vesicular lesions. Fluid is obtained from the vesicle by puncture with a sterile needle; it is then swabbed and placed in virus transfer medium for inoculation of cell cultures. Next, the base of the vesicle is scraped and these scrapings are placed into a drop of phosphate–buffered saline (PBS) on slides. After drying, the tissue scrapings are fixed in acetone for 10 minutes at 4°C.

Following this, scrapings can be stained using the indirect FA test, which is the preferred method. Commercially prepared and standardized anti-HSV antiserum can be used. After 30 minutes incubation at 37°C in a damp atmosphere, the slides are washed in PBS and overlaid with a second fluorescein-labeled antibody, incubated for 20 minutes, washed, and mounted in a nonfluorescing mounting medium. The presence of HSV antigens in the scrapings can then be determined with known negative cell cultures and HSV-infected cell cultures as controls.

The medium in which the lesion swab is placed is inoculated onto coverslips in Leighton tubes or onto Petri dishes containing monolayers of appropriate cells. In cultures exhibiting cytopathic effects, the cells are also tested for HSV antigens using the indirect FA test. Most often, a preliminary diagnosis can be made within hours with the indirect FA test; confirmation using cell culture can be made after 1 to 2 days. Available data indicate that this is a very reliable procedure.

When a diagnosis is needed for HSV infections of the cornea, the foregoing procedure can be used after modification. Specifically, the procedure is as follows. The cornea is anesthetized; specimens are obtained by scraping the corneal epithelium with a sterile scalpel blade and then transferring the specimens to sterile PBS on glass slides. The blade is washed in transport medium and a swab of the affected area is placed in the virus transport medium. Standard procedures for the indirect FA test are followed to obtain a preliminary diagnosis while awaiting confirmation from virus culture. The fluorescent antibody test has been found to be the most effective method currently in use to diagnose eye disease due to herpetic infection.

The fluorescent antibody test is also used to diagnose genital HSV infections. The procedures are basically the same as described earlier. The specimen should be taken from the base of the lesion using light curettage and a swab of the lesion should also be taken for inoculation into cell culture.

27.5.2 Diagnosis of Encephalitis due to Herpes Simplex Virus

Mortality rates due to HSV encephalitis can be as high as 70% in untreated cases. Survivors are often scarred for life with serious physical and neurologic abnormalities. As a result, there is an increasing awareness of the need for fast, accurate diagnosis of the agent responsible for the encephalitis. Because chemotherapy is now available to treat encephalitis due to HSV, the need for an excellent diagnostic method is great. Virus culture takes too long and may delay prompt, appropriate treatment if it is the only diagnostic tool employed.

Burr-hole exploration of the temporal lobe can be used to obtain specimens by brain biopsy in order to differentiate between brain tumor, brain abscesses, and viral encephalitis. Using IF tests on specimens removed at brain biopsy, a diagnosis can be made in 3 hours compared with the 24 to 48 hours required for virus isolation from cell culture. The indirect FA test on impression preparations from brain biopsies provides the most rapid and sensitive method of diagnosis. To do this, the surface of the biopsy is pressed several times onto the surface of one or more microscopic slides. When dry, the specimens are fixed in cold acetone for 1 minute. Longer fixation or use of different fixatives will yield suboptimal results. The specimens are then stained using the direct or indirect FA test, although some reports indicate that the indirect FA test allows more flexibility.

It has been demonstrated that HSV antigens can be found in cells from the cerebrospinal fluid of suspected encephalitis patients. Cells in CSF can be separated using a Shandon-Elliott "cytocentrifuge" and then spread onto microscope slides. The cells are fixed in cold acetone and then stained using a conventional, indirect FA test. By this procedure, more than 92% of the cases have been correctly diagnosed. However, this technique is not routinely used because interpretation is difficult and the cell counts in the CSF early in the disease are often too low.

A final method for rapid diagnosis of HSV encephalitis is the indirect immunoperoxidase method. With this method, brain biopsy specimens are teased and separated into a cell suspension, placed on slides, and fixed in acetone. The specimens are treated with anti-HSV antiserum for 45 minutes at 35°C and rinsed with PBS. The specimens are then allowed to react with an anti-species immunoglobulin previously conjugated to peroxidase for 45 additional minutes at 35°C, and washed again thoroughly. The specimens are then stained immediately with Kaplow's reagent, washed, and counterstained with safranin for 30 seconds. The specimens are dehydrated through alcohol to xylene and mounted in Permount. The cells that are positive for

HSV antigens will show blue granules indicating sites of peroxidase activity. This procedure has been described as reliable, rapid, and specific with advantages over the IF technique. Specifically, the advantages are easier interpretation of results, use of cell suspensions rather than frozen sections, permanent preparations, and the use of only a light microscope. In contrast, the complement fixation test has not been shown to be effective in the diagnosis of herpesvirus encephalitis because of the slow rise in antibody titer.

27.6 UNUSUAL FEATURES OF HERPES SIMPLEX VIRUSES

The HSV have several properties that distinguish them from other animal viruses. Perhaps the most unique feature of this complex group of viruses is their low species-specificity for replication. These viruses will replicate in almost any type of cell and in many different animals. Along with nonspecificity for replication and probably because of it, HSV are responsible for a wide spectrum of diseases, from gingivostomatitis to keratoconjunctivitis to encephalitis.

Another property characteristic of HSV is the ability to become latent. They have an extremely high preference for nerve cells and can remain in a quiescent state for many years without causing clinical disease. However, biological as well as external stimuli can trigger the reactivation of HSV and various types of disease can result.

The final unique property of HSV is the ability to transform cells in vitro. Herpes simplex viruses enter a state of latency with no apparent damage to the host. However, it may be possible that host cell damage will occur after many years of virus insult either to the latent cell or to cells exposed to reactivated virus. This damage may render the cell(s) susceptible to transformation. This possibility is reflected by the association of early HSV-2 infection with cervical cancer.

REFERENCES

Baringer, J.R., and Swoveland, P. 1973. Recovery of herpes-simplex virus from human trigeminal ganglions. N. Engl. J. Med. 288:648–650.

Baringer, J.R., and Swoveland, P. 1974. Persistent herpes simplex virus infection in rabbit trigeminal ganglia. Lab. Invest. 30:230–240.

Bastian, F.O., Rabson, A.S., Yee, C.L., and Tralka, T.S. 1972. Herpesvirus hominis: Isolation from human trigeminal ganglion. Science 178:306–307.

Colberg-Poley, A.M., Isom, H., and Rapp, F. 1979. Experimental HSV latency using phosphonoacetic acid. Proc. Soc. Exp. Biol. Med. 162:235–237.

Frenkel, N., and Roizman, B. 1971. Herpes simplex virus: Studies of the genome size and redundancy studied by renaturation kinetics. J. Virol. 8:591–593.

Goodpasture, E.W. 1929. Herpetic infection with especial reference to involvement of the nervous system. Medicine 8:223–243.

Honess, R.W., and Roizman, B. 1974. Regulation of herpesvirus synthesis. I. Cascade regulation of the synthesis of three groups of viral proteins. J. Virol. 14:8–19.

Hyman, R. 1980. Comparison of herpesvirus genome. In F. Rapp (ed.), Oncogenic Herpesviruses. Vol. I. Boca Raton, FL: CRC Press, 1–18.

Kieff, E.D., Bachenheimer, S.L., and Roizman, B. 1971. Size, composition and structure of the deoxyribonucleic acid of herpes simplex virus subtypes 1 and 2. J. Virol. 8:125–132.

Kieff, E.D., Hoyer, B., Bachenheimer, S.L., and Roizman, B. 1972. Genetic relatedness of type 1 and type 2 herpes simplex viruses. J. Virol. 9:738–745.

Luger, A., Lauda, E. 1921. Transmissibility of herpetic keratitis in man to the cornea of rabbit. Wien. Klin. Wochenschr. 34:132.

O'Neill, F.J. 1977. Prolongation of herpes simplex virus latency in cultured human cells by temperature elevation. J. Virol. 24:41–46.

Price, R.W., Katz, B.J., and Notkins, A.L. 1975. Latent infection of the peripheral ANS with herpes simplex virus. Nature 257:686–688.

Reeves, W.C., Di Giacomo, R.G., Alexander, E.R., and Lee, C.K. 1976. Latent herpesvirus hominis from trigeminal and sacral dorsal root ganglia of Cebus monkeys. Proc. Soc. Exp. Biol. Med. 153:258–261.

Roizman, B. 1979. The organization of the herpes simplex virus genomes. 1979. Annu. Rev. Genet. 13:25–57.

Roizman, B. 1980. Structural and functional organization of herpes simplex virus genomes. In F. Rapp (ed.), Oncogenic Herpesviruses. Vol. I. Boca Raton, FL: CRC Press, 19–51.

Roizman, B., and Roane, P.R. Jr. 1961. A physical difference between two strains of herpes simplex virus apparent on sedimentation in cesium chloride. Virology 15:75–79.

Schneweis, K.E. 1962. Zum antigenen aufbau des herpes simplex virus. Z. Immunitaetsforsch. 124:173–196.

Sheldrick, P., and Berthelot, N. 1975. Inverted repetitions in the chromosome of herpes simplex virus. Cold Spring Harbor Symp. Quant. Biol. 39:667–678.

Stevens, J.G., Cook, M.L. 1971. Latent herpes simplex virus in spinal ganglia of mice. Science 173:843–845.

Stevens, J.G., Nesburn, A.B., and Cook, M.L. 1972. Latent herpes simplex virus from trigeminal ganglia of rabbits with recurrent eye infection. Nature (London) New. Biol. 235:216–217.

Walz, M.A., Price, R.W., and Notkins, A.L. 1974. Latent ganglionic infection with herpes simplex virus types 1 and 2: Viral reactivation in vivo after neurectomy. Science 184:1185–1187.

Cytomegalovirus, Varicella-Zoster Virus, and Epstein–Barr Virus

Stephen E. Straus and Holly A. Smith

28.1 INTRODUCTION

This chapter reviews the biological, clinical, and diagnostic features of cytomegalovirus (CMV), varicella-zoster virus (VZU), and Epstein–Barr virus (EBV) infections. Although these viruses are structurally related, they are each quite unique. They possess little or no antigenic or sequence homology, and their genomes are organized somewhat differently, as well. All three agents are extremely common causes of disease in humans. They establish latent infections, have the potential for clinical or subclinical reactivation, and are of special concern when they affect the neonate and immunodeficient hosts. To varying degrees, these viruses have the ability to transform cells. The diagnosis of CMV, VZV, or EBV infection, which still rests largely on clinical grounds, can be supported by specific serologic determinations and confirmed by isolation and identification of the virus.

28.2 CYTOMEGALOVIRUS

28.2.1 Historical Perspectives

Cytomegalovirus was first recognized by the pathoanatomic changes it induces in renal and salivary tissues during the course of congenital infection. In 1881, Ribbert reported enlarged "protozoan-like" cells in the kidney of a stillborn infant (Ribbert, 1904). In 1932, similar changes identified as intranuclear and cytoplasmic inclusions were reported in salivary glands of 12% of infants who died from a variety of causes (Farber and Wolbach, 1932). Three groups of investigators reported in quick succession the propagation in cell culture of the agent, variously termed salivary gland virus, cytomegalic inclusion disease virus and, finally, cytomegalovirus (Rowe et al., 1956; Smith, 1956; Weller et al., 1957). Subsequently, CMV has been shown to be associated with a wide range of clinical syndromes affecting individuals of all ages with normal or impaired immune systems (Ho, 1979).

28.2.2 Biology and Pathogenesis of Infection

Cytomegalovirus is a typical human herpesvirus having an icosahedral nucleocapsid and a glycoprotein-bearing lipid envelope. It has the largest genome complement (150 megadaltons) of any of the herpes viruses. Although epithelial cells, leukocytes, and fibroblasts can be infected with CMV in vivo, only fibroblasts readily support CMV replication in vitro. There are no good animal hosts for human CMV. Compared with herpes simplex virus (HSV) and VZV, replication of CMV proceeds very slowly in vitro. The replicative processes of CMV, like those of all herpesviruses, involve an orderly cascade of biochemical events that include early and late transcription, DNA synthesis, and assembly and release of progeny virions. There is little cell-free virus and infection is spread predominantly to contiguous cells.

The hallmark of CMV infection in vitro or in vivo that distinguishes it from other herpesviruses is the development of perinuclear cytoplasmic inclusions in addition to the typical herpetic intranuclear inclusions (Hanshaw, 1968). It also has a propensity for massive enlargement of the affected cell, hence, the name for the classical cytomegalic inclusion disease.

Cytomegalovirus strains appear to fall into two or three closely related groups (Weller et al., 1960). A uniform serotyping convention has not been agreed upon, and because the strains share sufficient antigenicity, it is practicable to dismiss these differences.

Cytomegaloviruses, like all other herpesviruses, induces a lifelong infection. Monocytes and polymorphonuclear leukocytes serve as reservoirs for CMV, and virus can be shed from the pharynx or in the urine for years after primary infection (Diosi et al., 1969; Hanshaw, 1978). Cytomegalovirus is also a transforming agent and has been associated epidemiologically and biochemically with Kaposi's sarcoma in both the classical (European and African patients) and acquired immunodeficiency syndrome (AIDS) settings (Giraldo et al., 1972).

Immune responses to CMV are complex and involve both humoral and cellular mechanisms. As with other herpesviruses, the cellular mechanisms are most critical. Agammaglobulinemic patients are not at increased risk of suffering serious CMV infection, which is in direct contrast to patients having cellular immune deficiencies, especially organ transplant recipients and AIDS patients who may succumb to progressive visceral CMV disease (Suwansirikul et al., 1977; Macher et al., 1983).

28.2.3 Epidemiology

It appears that CMV is transmissible by several different means, all requiring intimate contact with virus-bearing material. Virus is known to be shed in urine, saliva, semen, breastmilk, and from the cervix, in addition to being carried in circulating white blood cells (Diosi et al., 1969; Lang and Hanshaw, 1969; Hanshaw, 1978; Lang and Kummer, 1972; Montgomery et al., 1972; Stagno et al., 1980). Cytomegalovirus can be spread transplacentally in the perinatal period by direct contact with an infected mother, by needlestick exposure and transfusions, and by transplantation of infected organs or tissues.

In individuals from developing nations (Krech and Jung, 1971) and in promiscuous homosexual men (Drew et al., 1982), the seroprevalence of antibodies to CMV approaches 100%. In individuals in industrialized nations about 50–70% are seropositive by middle age (Krech and Jung, 1971; Steen and Elek, 1965). In about 1% of all live births, the neonates are infected with CMV, but only a small percentage of these infected infants show symptoms of congenital or development problems. The earlier in pregnancy the intrauterine infection occurs, the greater the risk of clinically apparent congenital infection (Hanshaw et al., 1976). The estimated risk of acquiring CMV from transfused blood varies widely, but appears to be around 3% per unit transfused (Armstrong et al., 1976). Fifty to 100% of bone marrow or kidney transplant recipients develop CMV infection (Marker et al., 1980). Most of these infections appear to represent reactivation of the recipient's own virus and are usually mild or subclinical, but primary infection by virus in the donor tissues carries a high risk of morbidity and mortality (Pass et al., 1979). From case studies of patients who acquired CMV after blood transfusion or organ transplantation, it has been possible to determine that the incubation period for CMV infection is generally 4 to 8 weeks.

28.2.4 Clinical Features

Cytomegalovirus infection is truly protean in its clinical expression (Weller, 1971). Most infections are subclinical. Congenital infection is classically associated with hepatosplenomegaly, thrombocytopenia, hemolytic anemia, chorioretinitis, and encephalitis and their sequelae. Typical infections in older individuals are manifested as hepatitis or mononucleosis. Individuals with marked cellular immune impairment, particularly AIDS patients and transplant recipients, can experience aggressive visceral infection associated with pneumonitis, blinding retinitis, hepatitis, esophagitis, or encephalitis.

Management and prevention of CMV infections have been difficult and elusive. Antiviral drugs and interferon can temporarily reduce virus shedding but have had little impact on the progression of congenital cytomegalic disease or infection of the compromised host (Chien et al., 1974). Of the approaches to prevention, three have been somewhat promising. Prophylaxis with high titers of immunoglobulin (Winston et al., 1982) or leukocyte interferon (Hirsch et al., 1983), and a live attenuated CMV vaccine (Starr et al., 1981) may modify the expression of subsequent CMV disease in transplant recipients.

28.2.5 Diagnosis

For diagnostic purposes, the most useful specimens for CMV isolation are throat washings and urine. Virus is shed in the urine, with the highest titers being observed in congenitally infected infants. Shedding may persist from months to years after signs or symptoms of the infection have resolved. Cytomegalovirus has been detected in other body fluids as well, including saliva, breastmilk, cervical secretions, blood, and semen. Biopsy and autopsy materials, particularly of lung, kidney, liver, spleen, brain, and retina are also occasionally processed for CMV isolation.

Urine specimens should be clean-voided, and all other body fluids that are likely to be nonsterile should be transported on ice in culture medium containing antibiotics. For recovery of CMV from blood, 3–5 ml of freshly drawn blood should be immediately placed in a tube containing an anticoagulant, preferably citrate, and then transported to the laboratory on ice. Throat cultures require gargling for 5 to 10 seconds with 5–10 ml of transport medium containing serum and antibiotics.

In the laboratory, aliquots of liquid specimens should be promptly delivered into sensitive cell culture lines. Tissues should be freshly ground with a sterile mortar and pestle, triturated with culture medium, and inoculated into cultures. Blood specimens can be allowed to settle or can be centrifuged at 1000 rpm for 5 to 10 minutes at 4°C. The buffy coat can be removed with the use of a pipette and inoculated into cultures directly or diluted for easier handling in tissue culture medium.

Cytomegalovirus is sensitive to freezing and thawing, and its stability is better assured by the presence of culture additives. If a specimen must be frozen, an equal volume of 70% sorbitol should be added, and a freezer that attains temperatures at or below −70°C must be used.

Cytomegalovirus is extremely sensitive to lipid solvents (Rowe et al., 1956). The virus is thermolabile and infectivity is completely destroyed within 30 minutes by heating at 56°C (Weller et al., 1957). Cytomegalovirus is more stable in distilled water or minimal essential media without sodium bicarbonate (Vonka and Benyesh-Melnick, 1966).

Microscopic examination of CMV-infected cells or tissues can reveal the presence of characteristic large cells with cytoplasmic and intranuclear inclusions. The time of appearance and extent of cytopathic effects induced by CMV depends on the amount of virus present in the specimen. Cytopathic changes appear slowly, occasionally as early as 2 to 3 days after infection, but generally 2 to 3 weeks are required. Some isolates cannot be detected for 4 to 6 weeks. Cytopathic effects induced by CMV cause the cells to take on an irregular elongated shape that is reminiscent of the gladiolus flower (Figure 28.1). The pattern of cytopathic effect is sufficiently typical that other supporting tests are generally not required, but indirect immunofluorescence or immunoperoxidase staining with specific antisera will confirm the diagnosis.

Cytomegalovirus displays strong species and cell line specificity. Human fibroblastic cells best support the growth of CMV. Acceptable fibroblastic culture lines include those prepared from human embryonic tissues, foreskin fibroblasts, or relatively low-passage diploid fetal cell lines, such as WI-38 and MRC-5. All of the foregoing are commercially available.

Because of the time required to grow and identify most CMV isolates, many laboratories have necessarily sought more rapid and convenient means of diagnosis.

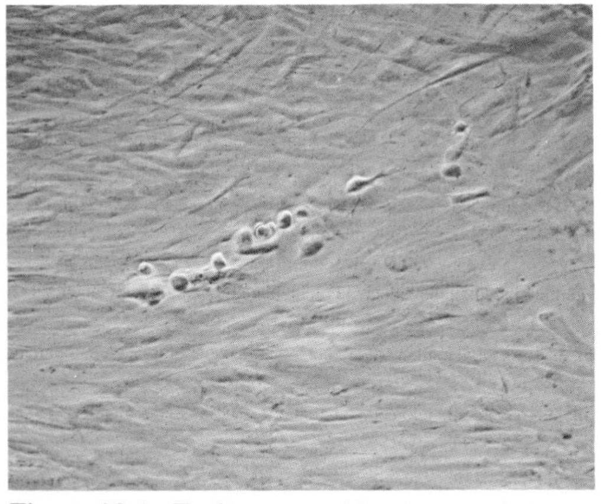

Figure 28.1. Early cytopathic changes induced by CMV in WI-38 cells.

Other more rapid techniques, including cytologic examinations, immunofluorescence microscopy, and electron microscopy have their advantages, but are fraught with either insensitivity, nonspecificity, or the need for elaborate expensive equipment and substantial expertise (Starr and Friedman, 1980; Lee et al., 1978).

Using cytologic studies of urinary sediment, one can make a presumptive diagnosis of CMV infection in 25–50% of the cases of symptomatic congenital infection. These tests are most useful when virus isolation techniques are not available, but they are too insensitive to be of value in most other forms of CMV infections.

A highly specific technique has been developed recently that is at least as sensitive as tissue culture and requires less than 48 hours to perform. It involves dot blot DNA hybridizations using molecularly cloned CMV DNA probes (Chou and Merigan, 1983). This technique is still not appropriate for general laboratories, but modification of the assay to use nonradioactive probes may vastly simplify the technique.

28.3 VARICELLA-ZOSTER VIRUS

28.3.1 Historical Perspectives

Varicella (or chicken pox) is a familiar exanthematous disease that afflicts most humans during childhood or adolescence (Weller, 1983). A unique property of the etiologic agent for varicella is its ability to cause a distinctly different clinical syndrome, zoster. In 1892, after observing cases of varicella in children closely exposed to patients with zoster, von Bokay first suggested that varicella and zoster are related infections (von Bokay, 1909). Classic experiments that confirmed this theory involved direct inoculation of zoster vesicle fluids into children, with the resulting development of varicella (Kundratitz, 1925). Based on these and other observations, Garland (1943) and, later, Hope-Simpson

(1965) postulated that zoster represents reactivation of varicella virus that had persisted in a latent form in the sensory ganglia.

Varicella-zoster virus was first cultivated in monolayer cultures of human cells by Weller (1953). Immunofluorescence and biochemical studies subsequently have proven that the agents recovered from varicella and zoster infections are essentially identical, and represent herpesviruses (Weller and Witton, 1958; Weller et al., 1958; Hyman, 1981).

28.3.2 Biology and Pathogenesis of Infection

Varicella-zoster virus is a typical herpesvirus, consisting of double-stranded linear DNA of approximately 80 megadaltons residing within an icosahedral nucleocapsid and surrounded by a lipid envelope bearing viral-specific glycoproteins (Hyman, 1981). Varicella-zoster virus has a narrow host range, growing best in human fibroblast or epithelial cell lines, moderately well in simian cells, and poorly (if at all) in other mammalian cell lines. The replicative cycle of VZV is similar to that described previously for CMV.

Varicella-zoster virus is thought to be a transforming agent (Gelb et al., 1980) although no human malignancies have been epidemiologically associated with VZV infection. Therefore, it is uncertain whether or not transforming infection occurs in vivo; however, the virus does establish a lifelong latent infection. It is believed that during varicella infection sensory nerve roots become infected either by contiguous spread from cutaneous epithelial cells or by direct viremic spread to the nerve ganglia. The virus persists within selected ganglion cells in a manner presumably identical to that of HSV (Gilden et al., 1983). It is unknown if any virus-specific synthetic processes are active during latency. Through mechanisms that have not been defined as yet, latent VZV may reactivate. Virus released from nerve endings spreads to contiguous epithelial cells in the skin. This infection, zoster, is generally confined to a single cutaneous dermatome (Brunell, 1979). Humoral immunity can be passively transferred with zoster immune globulin, conferring partial protection against primary varicella infection. Otherwise, cellular immunity is of paramount importance in controlling VZV infection.

28.3.3 Epidemiology

Varicella is a highly contagious illness, with about 80–90% of susceptible household contacts acquiring the infection following direct contact with lesions or inhalation of infectious virus in aerosols (Weller, 1979). The incubation period is between 11 and 20 days in nearly all cases. Except in isolated communities, about 90% of individuals contract varicella before adulthood.

Zoster develops in about 10–20% of normal individuals during life, with the risk increasing with advancing age. Patients with cellular immune impairment, especially those with lymphoproliferative malignancies and AIDS, are at a greatly increased risk of developing zoster. Recurrent zoster is infrequent.

28.3.4 Clinical Features

Varicella is an easily recognizable infection in its classic form (Brunell, 1979). Most infections are clinically apparent, with a brief prodrome of fever, chills, and myalgias followed by the development of vesicular lesions surrounded by erythema. The lesions appear first on the head or trunk and are distributed centrifugally. The average case involves some 100 to 500 lesions appearing over 3 to 5 days, with gradual drying, crusting, and healing of lesions over 10 to 14 days. Visceral complications including meningoencephalitis, pneumonitis, and hepatitis are uncommon except in immunodeficient patients, especially leukemic children, for whom the mortality rate from progressive varicella infection is approximately 7%.

Zoster is also easily recognizable, but mild, well localized infections can be mistaken for herpes simplex infection or vice versa. The emergence of lesions is occasionally heralded by neuralgia within the dermatome affected. Grouped vesicular lesions on an erythematous base evolve and coalesce over several days. Few scattered lesions outside of the dermatome are commonly observed, but frank cutaneous or visceral dissemination is a morbid event seen in compromised patients. In the elderly, post zoster neuralgia is a major complication and may continue for weeks to months. Therapy for varicella and zoster infection is largely limited to supportive and symptomatic measures, but vidarabine, acyclovir, and interferon can limit the duration and severity of the illness in the immunodeficient host (Merigan et al., 1978; Peterslund et al., 1981; Whitley and Alford, 1981).

28.3.5 Diagnosis

The best method of isolating VZV begins with a sterile aspiration of fresh vesicles. Clear vesicles generally contain only a few microliters of fluid, which is rich in cell-free virus. By DNA hybridization, we have found that individual lesions often contain 100 pg–10 ng or about 10^6–10^8 genome copies of VZV DNA (Seidlin et al., 1984). The virus is very labile, however, and a large percentage of virions in vesicle fluid appears to be defective; thus, all efforts should be made to inoculate the aspirates as promptly as possible into cell culture. Ideally, this can be done at the bedside, but more practically the aspirates can be delivered promptly to the laboratory in an ice-cold transport medium, such as minimal essential medium or veal infusion broth. On rare occasions, virus has been recovered from buffy coat preparations or throat washes of varicella patients (Gold, 1966; Feldman and Epp, 1976). Tissues obtained at biopsy or autopsy are also examined occasionally for VZV.

Specimens to be tested within 24 hours of collection may be held at 4°C or on ice; for longer intervals, the samples should be frozen at or below −70°C. Nutrient broth, tissue culture media, and skim milk can be used as transport media.

Diploid human cell lines or primary human cell cultures are the most sensitive host systems for isolation of VZV from clinical material. In this laboratory, WI-38 diploid fibroblasts and human embryonic kidney cells have proved to be highly successful for primary isolation of VZV. However, many diploid human cell lines and continuous epithelial lines are suitable. These lines are readily available from commercial sources. Isolation is most practical in roller tube cultures maintained with minimal essential medium supplemented with 10% heat-inactivated fetal bovine serum and antibiotics. Inoculated into each tube are 0.2 ml of transport medium containing the specimen. The remainder of the original sample is stored at −70°C for retesting if necessary.

Inoculated cell cultures are incubated at 36–37°C in roller drums and examined microscopically for evidence of viral cytopathic effects over a period of 30 days. Varicella-zoster virus cytopathic effect is characterized by the emergence of small foci of rounded and swollen refractile cells that typically appear from 5 to 7 days postinoculation, but may be considerably delayed (Figure 28.2). The foci of infected cells expand and eventually involve larger regions of the monolayer. When 50–75% of the monolayer demonstrates cytopathic changes, the culture should be subpassaged or frozen.

Initial identification of VZV infection may be made on the basis of typical cytopathic changes that are more focal and spread far more slowly than those induced by herpes simplex virus. An important characteristic of VZV, which distinguishes it from CMV (which also grows slowly in cell culture), is that VZV grows well in epithelial cells, whereas, CMV produces a cytopathic effect only in fibroblastic cell lines.

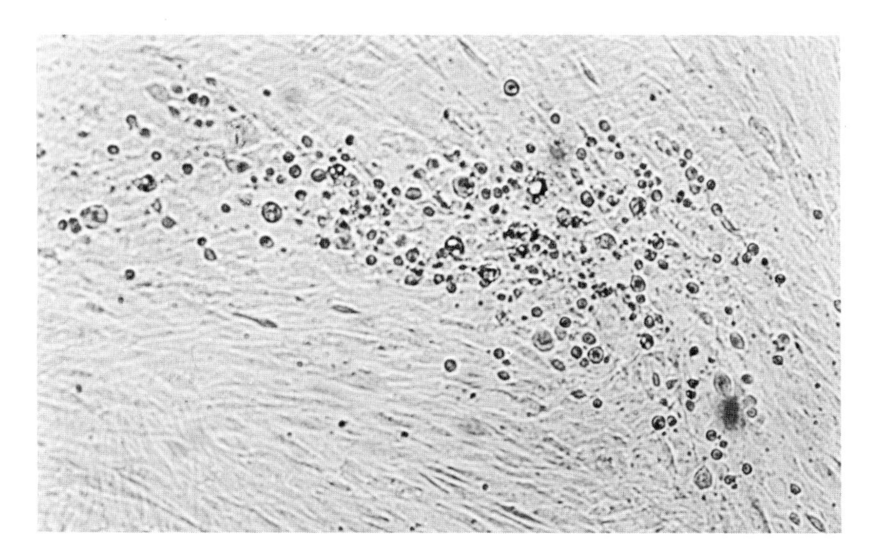

Figure 28.2. Cytopathic changes characteristic of VZV replication in human embryonic fibroblasts (Flow 5000 cell line). (Courtesy of J. Felser.)

Infectious VZV maintains a very close association with the host cell so that the serial propagation of the virus requires passage of virus-infected cells rather than tissue culture fluid. The lack of substantial production of cell-free virus allows for convenient plaquing of VZV in the absence of a gel overlay (Rapp and Benyesh-Melnick, 1963) (Figure 28.3). Viability of frozen infected cells is enhanced with the addition of dimethylsulfoxide or glycerol in the medium and storage at −70°C or in liquid nitrogen (Schmidt, 1980).

Propagation of VZV in small animals or embryonated eggs is usually unsuccessful. Harbour and Caunt (1975), using cell-free virus as the inoculum, demonstrated conclusively that guinea pig cells will support the growth of VZV. Moreover, Edmond et al. (1981) have isolated VZV in guinea pig cells by inoculating them with human vesicle fluid and have further reported that weanling guinea pigs can be infected with virus that had been passaged through guinea pig cell cultures (Myers et al., 1980).

A number of other techniques can be used to support the diagnosis of VZV infection or to confirm the source of cytopathic changes observed in culture. These include electron microscopy of clinical specimens, immunofluorescent staining of specimens or infected cells (Schmidt et al., 1965), a variety of sensitive serologic techniques (Brunell et al., 1971; Williams et al., 1974) and, more recently, the detection of VZV by dot blot DNA hybridization (Seidlin et al., 1984).

28.4 EPSTEIN–BARR VIRUS

28.4.1 Historical Perspectives

The discovery of EBV and its association with infectious mononucleosis was serendipitous. Epidemiologic investigations suggested that mononucleosis had an infectious etiology (Hoagland, 1955), but serious attempts in the 1940s and 1950s to recover an agent or to experimentally transmit the infection were unsuccessful (Evans, 1947). In 1958, however, Burkitt described an unusual lymphoma in African children with a predilection for head and neck involvement. A variety of climatologic and geographic features of the regions in which the

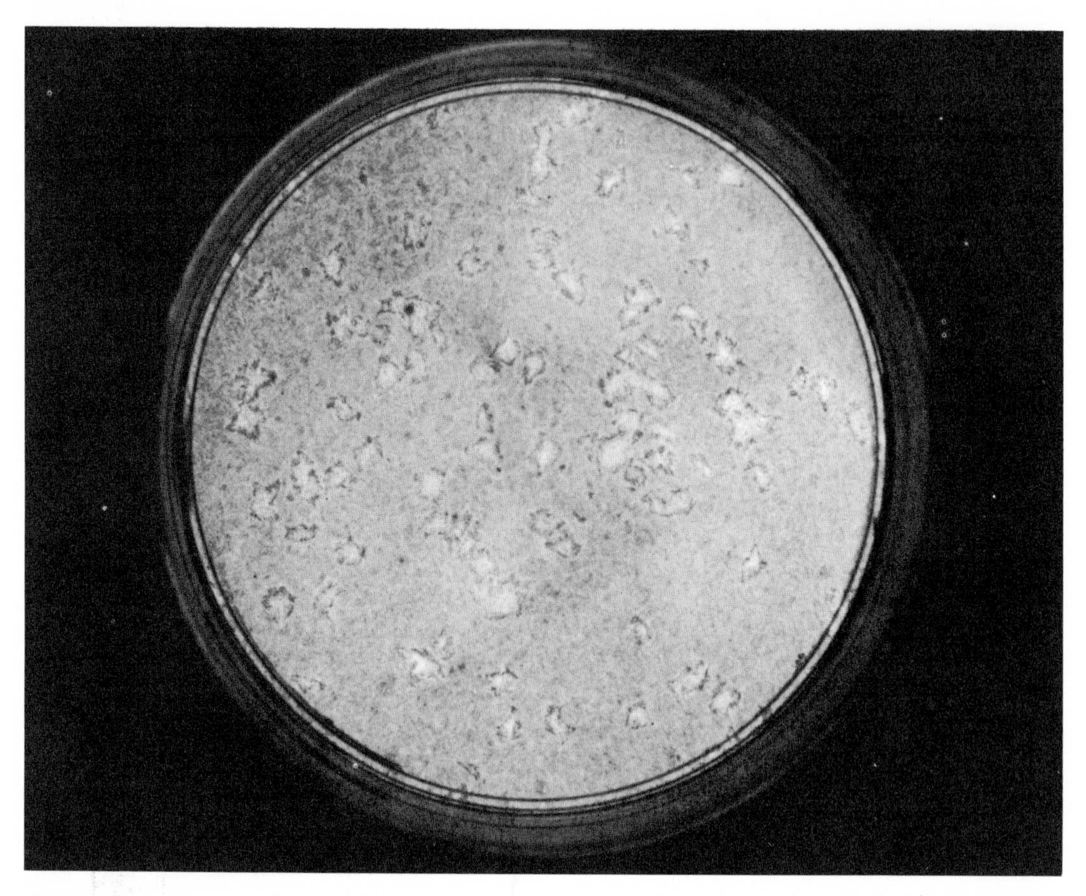

Figure 28.3. Plaques formed by VZV strain Ellen in Flow 5000 cells stained with crystal violet on day 7 of infection. (Courtesy of J. Felser.)

tumor was most prevalent suggested an infectious etiology. Efforts to identify a viral agent led to the discovery, reported by Epstein, Achong, and Barr (1964), of a herpesvirus-like particle in Burkitt tumor biopsy specimens. The Henles (1966), working in Philadelphia, determined that Burkitt's patients have antibodies that react in an indirect immunofluorescence assay with antigens within the lymphomatous tissues. As an appropriate control, they chose nonreacting serum from workers in their laboratory. In 1967, however, one of their technicians developed classical acute infectious mononucleosis, after which her serum reacted positively with Burkitt's tissue antigens (Henle et al., 1968). This occurrence represented the first clue that EBV may be associated with infectious mononucleosis. Through a series of studies in college students, it was determined that heterophile-positive infectious mononucleosis occurred in EBV-seronegative patients and was followed by seroconversion (Niederman et al., 1970). Subsequently, the virus has been recovered from throat washings of acutely infected individuals by propagation in lymphocytes obtained from fetal cord blood (Chang and Golden, 1971). Epstein–Barr virus is now known to be a ubiquitous agent that is associated with acute infectious mononucleosis, but also with a variety of other aggressive lymphoproliferative disorders in congenital or acquired immunodeficiency states (Schooley and Dolin, 1979).

28.4.2 Biology and Pathogenesis

Epstein–Barr virus is a unique human herpesvirus (Epstein and Achong, 1979). Although it exhibits typical herpesvirus-like structural and biochemical properties, its exceedingly narrow host range is remarkable. Two strains of EBV are presently recognized. Isolates similar to HR1 virus primarily induce lytic, productive infections, while isolates similar to the B95-8 strain are predominantly transforming. Most clinical isolates are of the latter type (Hinuma et al., 1967; Miller and Lipman, 1973).

In vitro, EBV can be propagated only in B lymphocytes (Pattengale et al., 1973). Epstein–Barr virus induces a blastogenic transformation and immortalization of B cells. The viral genome remains predominantly in the form of superhelical circular episomes, but integration of linear DNA into the host chromosome also occurs (Adams, 1979). Most cells transformed by EBV express few if any virus-specific proteins (Sugawara et al., 1972). Productive infection with shedding of virus, e.g., in the oropharynx, may be the result of infection of epithelial cells by EBV. Infection of epithelial cells has been induced experimentally by DNA transfection, but its importance for in vivo infection is unknown because only B cells appear to have receptors for EBV (Sixbey et al., 1983).

The EBV-transformed B-cell clones persist for life. The potential for unlimited growth of these lines can be realized in vitro, indicating that in vivo immune mechanisms must be continuously vigilant to prevent unbridled B cell lymphoproliferation. Epstein–Barr virus is known to reactivate intermittently with shedding of virus in pharyngeal secretions and renewed expression of antibodies to selected (early) viral antigens. Current information suggests that most such reactivation events are clinically silent except in severely immunodeficient patients.

28.4.3 Epidemiology

In developing nations, EBV infection is nearly universal during early childhood (Evans, 1974). In industrialized nations, at least 50% of all EBV infections are delayed until late adolescence and adulthood, and it is in this setting that classical infectious mononucleosis is recognized. The infection is most likely to be spread by exchange of infected saliva. Epstein–Barr virus can be recovered frequently from the oropharynx of seropositive patients (Chang et al., 1973). The virus is not easily communicable, however, so that outbreaks within enclosed communities or families are not problematic. Blood transfusions account for a small percentage of EBV infections (Gerber et al., 1969).

28.4.4 Clinical Manifestations

In children, EBV infection is generally asymptomatic. With college-age patients, 33–75% of infections are recognizable and typically take the form of infectious mononucleosis. This is a self-limited illness lasting 2 to 4 weeks, characterized by fever, sore throat, lymphadenopathy, and hepatosplenomegaly (Schooley and Dolin, 1979). There is an increase in the number and proportion of circulating lymphocytes, with the emergence of a substantial number of atypical, antigen-reactive T lymphocytes (Pattengale et al., 1974). Infection of B lymphocytes by EBV induces polyclonal activation, resulting in the transient appearance of heterophile antibodies and autoimmune manifestations.

The wide variety of complications of infectious mononucleosis includes rare (but potentially fatal) acute splenic rupture, hemolytic anemia or thrombocytopenia, hepatitis, and neurologic manifestations, including Guillain–Barré syndrome, Bell's palsy, and encephalitis (Schooley and Dolin, 1979). Infectious mononucleosis is rarely fatal, but such an outcome is predictable in young boys with an X-linked immunodeficiency syndrome (Purtilo et al., 1975).

Epstein-Barr virus has also been associated with Burkitt's lymphoma and nasopharyngeal carcinoma (Zur Hausen et al., 1970). Sera from patients with these malignancies have elevated titers to virus-specific antigens. Tissue specimens contain EBV DNA and express viral antigens as well. Although it remains uncertain whether or not EBV causes these malignancies (or even predisposes the patient to their development), studies of unusual cases in which aggressive EBV-associated polyclonal lymphoproliferation has undergone a monoclonal transformation argue that EBV itself may be capable of inducing malignancy (Hanto et al., 1982).

28.4.5 Diagnosis

Laboratories that routinely perform virus diagnosis generally avoid EBV cultivation. A reliable source of cord blood lymphocytes and the effort, expertise, and time necessary to develop, identify, and test the transformed clones are requirements more readily fulfilled in a research laboratory setting rather than in a routine clinical laboratory setting (Nilsson et al., 1971).

Epstein–Barr virus has been identified in Burkitt's lymphoma tumors, throat washings and saliva of infectious mononucleosis patients, oropharyngeal secretions of renal transplant recipients, patients treated with immunosuppressive drugs, and in malignant epithelial cells of nasopharyngeal carcinoma. The virus is also routinely detected in the saliva of healthy EBV-seropositive individuals.

For virus isolation, throat washings or pooled saliva are collected in RPMI 1640 medium supplemented with 10–20% heat-inactivated fetal bovine serum. Cultures should be placed on ice or immediately frozen at −70°C. Specimens are inoculated into lymphocytes of human fetal cord blood. The suspension cultures are periodically observed and fed biweekly for up to 6 weeks. Epstein–Barr virus infection of susceptible B lymphocytes within the culture results in morphologic transformation and growth. Nontransformed cells will disintegrate, whereas, transformed cells will appear as small but growing clusters of floating cells. Epstein–Barr virus-transformed B lymphocytes proliferate indefinitely under appropriate environmental conditions. To confirm that the growing colonies of lymphocytes are, indeed, EBV-transformed, the cells can be passaged and then affixed to cover slips and subsequently stained with fluorescein-conjugated antisera to the EBV nuclear antigen (Reedman and Klein, 1973).

Another means of documenting EBV infection entails recovery and identification of spontaneously transforming circulating B lymphocytes. All individuals who have had an EBV infection carry latent virus in a small proportion of circulating B lymphocytes. During acute infection as many as 1 of 1000 circulating B cells are EBV-infected and will divide continuously in culture. With remote infection, approximately 1 of 10^6 B lymphocytes will undergo spontaneous outgrowth (Nilsson, 1971).

Because of the difficulty in isolating EBV and in establishing and confirming the identity of presumptively transformed cells, the interested clinician and investigator generally turn to other techniques. In the absence of a cell line that will reliably support productive replication of EBV, viral antigens are detected either by complement-fixation technique, enzyme-linked immunosorbent assay (ELISA), various fluorescent antibody techniques, direct examination by electron microscopy, or DNA hybridization.

By indirect immunofluorescence, the expression of EBV viral capsid antigens can be seen in Burkitt's lymphoma and nasopharyngeal carcinoma tissues in addition to occasional lymphocytes in lymph nodes and liver tissues of patients with very aggressive acute infectious mononucleosis (Nilsson et

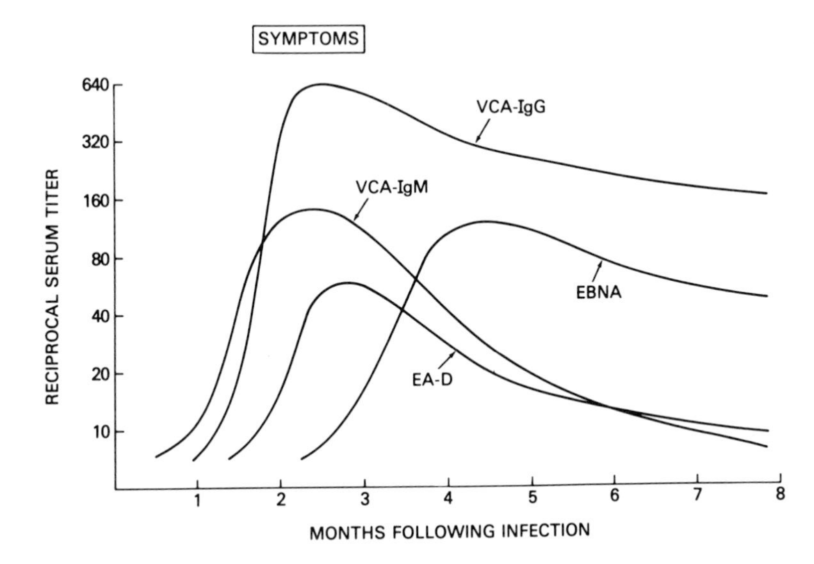

Figure 28.4. Typical kinetics of appearance and persistence of antibodies to EBV-specific antigens after a primary infection. Individuals with recent infection have IgM and IgG antibodies to the viral capsid antigen (IgM-VCA, IgG-VCA). Antibodies to early antigens (EA-D) develop in 70–80% of patients and persist for several months. Several weeks after acute infection, antibodies to EBV nuclear-associated antigen (EBNA) appear and persist for life.

al., 1971). A more sensitive but complicated technique involving anticomplementary immunofluorescence can be used to demonstrate EBV nuclear-associated antigens in a somewhat higher proportion of cells in these tissues (Reedman and Klein, 1973).

Hybridizations of RNA/DNA and DNA/DNA have proven to be specific and sensitive for the detection of EBV nucleic acids in tissues (Lindahl et al., 1976; Saemundsen et al., 1981). These techniques are highly specialized and cumbersome, but for diagnostic purposes may be substantially simplified by using molecularly cloned EBV DNA probes and dot blot or Southern hybridization, as described by Andiman et al. (1983).

Because all of the foregoing techniques remain beyond the capabilities of most diagnostic virology laboratories, the diagnosis of EBV infection typically rests upon clinical observations, the detection of circulating heterophile antibodies in patients with the infectious mononucleosis syndrome, and, most importantly, upon EBV-specific serologic testing. Traditionally, these tests have all been performed with the use of indirect or anticomplementary immunofluorescence microscopy to demonstrate the presence of IgM or IgG antibodies to viral capsid antigens (IgM-VCA, IgG-VCA), diffuse or restricted early antigens (EA-D, EA-R), or EBV nuclear-associated antigen (EBNA). The pattern of EBV-specific antibody titers to these three classes of EBV antigens is an accurate indicator of the patient's status with regard to EBV infection (Figure 28.4) (Henle and Henle, 1966; Henle et al., 1971; Reedman and Klein, 1973). Recently, these techniques have been further simplified by adapting them to ELISA technology. With the availability of specific monoclonal antisera to each of the individual EBV antigen classes, rapid, standardized, and highly reproducible serologic testing should be accessible to all laboratories in the near future (Wallen et al., 1977).

ACKNOWLEDGMENT

We thank J. Felser for photographs of varicella-zoster virus-infected cells and C. Crout and K. Leighty for manuscript preparation and editing.

REFERENCES

Adams, A. 1979. The state of the virus genome in transformed cells and its relationship to host cell DNA. In M.A. Epstein and B.G. Achong (eds.), The Epstein-Barr Virus. Berlin: Springer-Verlag, 155–183.

Andiman, W., Gradoville, L., Heston, L., Neydorff, R., Savage, M.E., Kitchingman, G., Shedd, D., and Miller, G. 1983. Use of cloned probes to detect Epstein-Barr viral DNA in tissues of patients with neoplastic and lymphoproliferative diseases. J. Infect. Dis. 148:967–977.

Armstrong, J.A., Tarr, G.C., Youngblood, L.A., Dowling, J.N., Saslow, A.R., Lucas, J.P., and Ho, M. 1976. Cytomegalovirus infection in children undergoing open-heart surgery. Yale J. Biol. Med. 49:83–91.

Brunell, P.A. 1979. Varicella-zoster virus. In G.L. Mandell, R.G. Douglas Jr., and J.E. Bennett (eds.), Principles and Practices of Infectious Diseases. New York: John Wiley and Sons, 1295–1306.

Brunell, P.A., Cohen, B.H., and Granat, M. 1971. A gel-precipitin test for the diagnosis of varicella. Bull. W.H.O. 44:811–814.

Burkitt, D. 1958. A sarcoma involving the jaws in African children. Br. J. Surg. 46:218–223.

Chang, R.S., and Golden, H.D. 1971. Transformation of human leukocytes from throat washings from infectious mononucleosis patients. Nature 234:359–360.

Chang, R.S., Lewis, J.P., and Abildgaard, C.F. 1973. Prevalence of oropharyngeal excretors of leukocyte-transforming agents among a human population. N. Engl. J. Med. 289:1325–1329.

Chien, L.T., Cannon, N.J., Whitley, R.J., Diethelm, A.G., Dismukes, W.E., Scott, C.W., Buchanan, R.A., and Alford, C.A., Jr. 1974. Effect of adenine arabinoside on cytomegalovirus infections. J. Infect. Dis. 130:32–39.

Chou, S., and Merigan, T.C. 1983. Rapid detection and quantitation of human cytomegalovirus in urine through DNA hybridization. N. Engl. J. Med. 308:921–925.

Diosi, P., Moldovan, E., and Tomescu, N. 1969. Latent cytomegalovirus infection in blood donors. Br. Med. J. 4:660–662.

Drew, W.L., Miner, R.C., Ziegler, J.L., Gullett, J.H., Abrams, J.I., Conant, M.A., Huang, E., Groundwater, J.R., Volberding, P., and Mintz, L. 1982. Cytomegalovirus and Kaposi's sarcoma in young homosexual men. Lancet ii:125–128.

Edmond, B.J., Grose, C., and Brunell, P.A. 1981. Varicella-zoster infection of diploid and chemically transformed guinea pig cells: Factors influencing virus replication. J. Gen. Virol. 54:403–407.

Epstein, M.A., and Achong, B.G. 1979. The Epstein-Barr Virus. Berlin: Springer-Verlag.

Epstein, M.A., Achong, B.G., and Barr, Y.M. 1964. Virus particles in cultured lymphoblasts from Burkitt's lymphoma. Lancet i:702–703.

Evans, A.S. 1947. Experimental attempts to transmit infectious mononucleosis to man. Yale J. Biol. Med. 20:19–26.

Evans, A.S. 1974. New discoveries in infectious mononucleosis. Mod. Med. 1:18–24.

Farber, S., and Wolbach, S.B. 1932. Intranuclear and cytoplasmic inclusions ('protozoan-like bodies') in the salivary glands and other organs of infants. Am. J. Pathol. 8:123–236.

Feldman, S., and Epp, E. 1976. Isolation of varicella-zoster virus from blood. J. Pediatr. 88:265–267.

Garland, J. 1943. Varicella following exposure to herpes zoster. N. Engl. J. Med. 228:336–337.

Gelb, L.D., Huang, J.J., and Wellinghoff, W.J. 1980. Varicella-zoster virus transformation of hamster embryo cells. J. Gen. Virol. 51:171–177.

Gerber, P., Walsh, J.H., Rosenblum, E.N., Purcell, R.H. 1969. Association of EB virus infection with the post-perfusion syndrome. Lancet i:593–596.

Gilden, D.H., Vafai, A., Shtram, Y., Becker, Y., Devlin, M., and Wellish, M. 1983. Varicella-zoster virus DNA in human sensory ganglia. Nature 306:478–480.

Giraldo, G., Beth, E., and Haguenau, F. 1972. Herpes-type virus particles in tissue culture of Kaposi's sarcoma from different geographic regions. J. Natl. Cancer. Inst. 49:1509–1513.

Gold, E. 1966. Serologic and virus-isolation studies of patients with varicella or herpes-zoster infection. N. Engl. J. Med. 274:181–185.

Hanshaw, J.B. 1968. Cytomegaloviruses. In Virology, Monograph No. 3. New York: Springer-Verlag, 1–23.

Hanshaw, J.B. 1978. Congenital cytomegalovirus. In J.S. Remington and J.O. Klein (eds.) Viral Disease of the Fetus and Newborn. Philadelphia: W.B. Saunders, 97–152.

Hanshaw, J.B., Scheiner, A.P., Moxley, A.W., Gaev, L., Abel, V., and Scheiner, B. 1976. School failure and deafness after "silent" congenital cytomegalovirus infections. N. Engl. J. Med. 295:468–470.

Hanto, D.W., Frizzera, G., Gajl-Peczalska, K.J., Sakamoro, K., Purtilo, D.T., Balfour, H.H., Jr., Simmons, R.L., and Najarian, J.S. 1982. Epstein-Barr virus-induced B-cell lymphoma after renal transplantation. Acyclovir therapy and transition from polyclonal to monoclonal B-cell proliferation. N. Engl. J. Med. 306:913–918.

Harbour, D.A., and Caunt, A.E. 1975. Infection of guinea pig embryo cells with varicella-zoster virus. Arch. Virol. 49:39–47.

Henle, G., and Henle, W. 1966. Immunofluorescence in cells derived from Burkitt's lymphoma. J. Bacteriol. 91:1248–1256.

Henle, G., Henle, W., and Diehl, V. 1968. Relation of Burkitt's tumor-associated herpes-type virus to infectious mononucleosis. Proc. Natl. Acad. Sci. U.S.A. 59:94–101.

Henle, G., Henle, W., and Klein, G. 1971. Demonstration of two distinct components in the early antigen complex of Epstein–Barr virus-infected cells. Int. J. Cancer. 8:272–282.

Hinuma, Y., Kohn, M., Yamaguchi, J., Wudarski, D.J., Blakely, J.R., and Grace, J.T., Jr. 1967. Immunofluorescence and herpes type virus particles in the P_3HR-1 Burkitt lymphoma clone. J. Virol. 1:1045–1051.

Hirsch, M.S., Schooley, R.T., Cosimi, A.B., Russell, P.S., Delmonico, F.L., Tolko, F.F., Rubin, N.E., Herrin, J.T., Cantell, K. Farrell, M., Rota, T.R., and Rubin, R.H. 1983. Effects of interferon-alpha on cytomegalovirus activation syndromes in renal-transplant patients. N. Engl. J. Med. 308:1489–1493.

Ho, M. 1979. Cytomegalovirus. In G.L. Mandell, R.G. Douglas, Jr., and J.F. Bennett (eds.), Principles and Practices of Infectious Disease. New York: John Wiley and Sons, 1307–1323.

Hoagland, R.J. 1955. The transmission of infectious mononucleosis. Am. J. Med. Sci. 229:262–272.

Hope-Simpson, R.E. 1965. The nature of herpes zoster: A long-term study and a new hypothesis. Proc. R. Soc. Med. 58:9–20.

Hyman, R.W. 1981. Structure and function of the varicella-zoster virus genome. In A.J. Nahmias, W.R. Dowdle, and R.F. Schinazi (eds.), The Human Herpesviruses: An Interdisciplinary Perspective. New York: Elsevier Press, 63–71.

Krech, U., and Jung, M. 1971. Age distribution of complement-fixing antibodies in Tanzania, 1970. In U. Krech and M. Jung (eds.) Cytomegalovirus Infections of Man. Basel: Karger, 27–28.

Kundratitz, K. 1925. Experimentelle Ubertragung von Herpes Zoster auf den Menschen und die Bezienhungen von Herpes Zoster zu Varicellen. Monatsschr. Kinderheilkd. 29:516–522.

Lang, D., and Kummer, J.F. 1972. Demonstration of cytomegalovirus in semen. N. Engl. J. Med. 287:756–758.

Lang, D.J., and Hanshaw, J.B. 1969. Cytomegalovirus infection and the post-perfusion syndrome: Recognition of primary infection in four patients. N. Engl. J. Med. 280:1148–1149.

Lee, F.K., Nahmias, A.J., and Stagno, S. 1978. Rapid diagnosis of cytomegalovirus in infants by electron microscopy. N. Engl. J. Med. 299:1266–1270.

Lindahl, T., Adams, A., Bjursell, G., Borhkamm, G.W., Kaschka-Dierich, C., and Jehn, U. 1976. Covalently closed circular duplex DNA of Epstein-Barr virus in a human lymphoid cell line. J. Mol. Biol. 102:511–530.

Macher, A.M., Reichert, C.M., Straus, S.E., Longo, D.L., Parrillo, J., Lane, H.C., and

Fauci, A.S. 1983. Death in the AIDS patient: Role of cytomegalovirus. N. Engl. J. Med. 309:1454.

Marker, S.C., Howard, R.J., Simmons, R.L., Kalis, J.M., Connelly, D.P., Najarian, J.S., and Balfour, H.H., Jr. 1980. Cytomegalovirus infection: A quantitative prospective study of 320 consecutive renal transplants. Surgery 89:660–671.

Merigan, T.C., Rand, K.H., Pollard, R.B., Abdallan, P.S., Jordan, G.W., and Fried, R.B. 1978. Human leukocyte interferon for the treatment of herpes zoster in patients with cancer. N. Engl. J. Med. 298:981–987.

Miller, G., and Lipman, M. 1973. Release of infectious Epstein-Barr virus by transformed marmoset leukocytes. Proc. Natl. Acad. Sci. U.S.A. 70:190–194.

Montgomery, R., Youngblood, L., and Medearis, D.N., Jr. 1972. Recovery of cytomegalovirus from the cervix in pregnancy. Pediatrics 49:524–531.

Myers, M.G., Dyer, H.L., and Hausler, C.K. 1980. Experimental infection of guinea pigs with varicella-zoster virus. J. Infect. Dis. 142:414–420.

Niederman, J.C., Evans, A.S., Subrahmanyan, M.S., and McCollum, R.W. 1970. Prevalence, incidence and persistence of EB virus antibody in young adults. N. Engl. J. Med. 282:361–365.

Nilsson, K. 1971. High frequency establishment of immunoglobulin-producing lymphoblastoid cell lines from normal and malignant lymphoid tissue and peripheral blood. Int. J. Cancer 8:432–442.

Nilsson, K., Klein, G., Henle, W., and Henle. G. 1971. The establishment of lymphoblastoid lines from adult and fetal human lymphoid tissue and its dependence on EBV. Int. J. Cancer 8:443–450.

Pass, R.F., Whitley, R.J., Diethelm, A.G., Whelchel, J.D., Reynolds, D.W., and Alford, C.A., Jr. 1979. Outcome of renal transplantation in patients with primary cytomegalovirus infection. Transplant. Proc. 11:1288–1290.

Pattengale, P.K., Smith, R.W., and Gerber, P. 1973. Selective transformation of B lymphocytes by EB virus. Lancet ii:93–94.

Pattengale, P.K., Smith, R.W., and Perlin, E. 1974. Atypical lymphocytes in acute infectious mononucleosis. Identification by multiple T and B markers. N. Engl. J. Med. 291:1145–1148.

Peterslund, N.A., Seyer-Hansen, K., Ipsen, J., Esmann, V., Schonheyder, H., and Juhl, H. 1981. Acyclovir in herpes zoster. Lancet ii:827–830.

Purtilo, D.T., Cassel, C.K., Yang, J.P.S., Stephenson, S.R., Harper, R., Landing, B.H., and Vawter, G.F. 1975. X-linked recessive progressive combined variable immunodeficiency (Duncan's disease). Lancet i:935–940.

Rapp, F., and Benyesh-Melnick, M. 1963. Plaque assay for measurement of cells infected with zoster virus. Science 141:433–434.

Reedman, B.M., and Klein, G. 1973. Cellular localization of an Epstein-Barr virus (EBV)-associated complement-fixing antigen in producer and non-producer lymphoblastoid cell lines. Int. J. Cancer 11:499–520.

Ribbert, H. 1904. Ubeer protozoenartige zellen in der niere eines syphilitischen neugeborenen und in der Parotis von Kindern. Zbl. Allg. Path. u. Pathol. Anat. 15:945–948.

Rowe, W.P., Hartley, J.W., Waterman, S., Turner, H.C., and Huebner, R.J. 1956. Cytopathogenic agent resembling human salivary gland virus recovered from tissue culture of human adenoids. Proc. Soc. Exp. Biol. Med. 92:418–424.

Saemundsen, A.K., Purtilo, D.T., Sakamoto, K., Sullivan , J.L., Synnerholm, A.-C., Hanto, D., Simmons, R., Anvret, M., Collins, R., Klein, G. 1981. Documentation of Epstein-Barr virus infection in immunodeficient patients with life-threatening lymphoproliferative diseases by Epstein-Barr virus complementary RNA/DNA and viral DNA/DNA hybridization. Cancer Res. 41:4237–4242.

Schmidt, N.J. 1980. Manual of Clinical Microbiology, 3rd ed. Washington, D.C.: American Society of Microbiology, 801.

Schmidt, N.J., Lenette, E.H., Woodie, J.D., and Ho, H.H. 1965. Immunofluorescent staining in the laboratory diagnosis of varicella-zoster virus infections. J. Lab. Clin. Med. 66:403–412.

Schooley, R.T., and Dolin, R. 1970. Epstein-Barr virus (infectious mononucleosis). In G.L. Mandell, R.G. Douglas, Jr., and J.E. Bennett (eds.), Principles and Practice of Infectious Diseases. New York: John Wiley and Sons, 1324–1341.

Seidlin, M., Takiff, H.E., Smith, H.A., Hay, J., and Straus, S.E. 1984. Detection of varicella-zoster virus by dot-blot hybridization using a molecularly cloned viral DNA probe. J. Med. Virol. 13:53–61.

Sixbey, J.W., Vesterinen, E.H., Nedrud, J.G., Raab-Traub, N., Walton, L.A., and Pagano, J.S. 1983. Replication of Epstein-Barr virus in human epithelial cells infected in vitro. Nature 306:480–483.

Smith, M.G., 1956. Propagation in tissue cultures of a cytopathogenic virus from human salivary gland virus (SGV) disease. Proc. Soc. Exp. Biol. Med. 92:424–430.

Stagno, S., Reynolds, D.W., Pass, R.F., and Alford, C.A. 1980. Breast milk and the risk of cytomegalovirus infections. N. Engl. J. Med. 302:1073–1076.

Starr, S.E., and Friedman, H.M. 1980. Human cytomegalovirus. In E.H. Lenette, A. Balows, W.J. Hausler, and J.P. Truant (eds.), Manual of Clinical Microbiology, 3rd ed. Washington, D.C.: American Society for Microbiology, 790–797.

Starr, S.E., Glazer, J.P., Friedman, H.M., Farquhar, J.D., and Plotkin, S.A. 1981. Specific cellular and humoral immunity after immunization with live Towne strain cytomegalovirus vaccine. J. Infect. Dis. 143:585–589.

Steen, H., and Elek, S.D. 1965. The incidence of infection with cytomegalovirus in a normal population: A serologic study in greater London. J. Hyg. 63:79–87.

Sugawara, K., Mizuno, R., and Oslo, T. 1972. Epstein-Barr virus associated antigens in non-producing clones of human lymphoblastoid cell lines. Nature New Biol. 239:242–244.

Suwansirikul, S., Rao, N., Dowling, J.N., and Ho, M. 1977. Clinical manifestations of primary and secondary CMV infection after renal transplantation. Arch. Intern. Med. 137:1026–1029.

von Bokay, J. 1909. Ueber den ätiologischen zusammenhaug der varizellen mit gervissen fällen von herpes zoster. Wien. Klin. Wochenschr. 22:1323–1326.

Vonka, V., and Benyesh-Melnick, M. 1966. Thermoinactivation of human cytomegalovirus. J. Bacteriol. 91:221–226.

Wallen, W.C., Mattson, J.M., and Levine, P.H. 1977. Detection of soluble antigen of Epstein-Barr virus by the enzyme-linked immunosorbent assay. J. Infect. Dis. 156:S324–S328.

Weller. T.H. 1953. Serial propagation in vitro of agents producing inclusion bodies derived from varicella and herpes zoster. Proc. Soc. Sys. Br. Med. 83:340–346.

Weller, T.H. 1971. The cytomegalovirus: Ubiquitous agents with protean clinical manifestations. N. Engl. J. Med. 285:203–214, 267–274.

Weller, T.H., 1979. Varicella-herpes zoster virus. In A.S. Evans (ed.), Viral Infection of Humans. Epidemiology and Control. New York: Plenum, 457–480.

Weller, T.H. 1983. Varicella and herpes zoster. Changing concepts of the natural history, control, and importance of a not-so-benign virus. N. Engl. J. Med. 309:1362–1368, 1434–1440.

Weller, T.H., Hanshaw, J.B., and Scott, D.E. 1960. Serologic differentiation of viruses responsible for cytomegalic inclusion disease. Virology 12:130–132.

Weller, T.H., Macauley, J.C., Craig, J.M., and Wirth, P. 1957. Isolation of intranuclear inclusion producing agents from infants with illnesses resembling cytomegalic inclusion disease. Proc. Soc. Exp. Biol. Med. 94:4–12.

Weller, T.H., and Witton, H.M. 1958. The etiologic agents of varicella and herpes zoster: Serologic studies with the viruses as propagated in vitro. J. Exp. Med. 108:869–890.

Weller, T.H., Witton, H.M., and Bell, E.J. 1958. The etiologic agents of varicella and herpes zoster: Isolation, propagation, and cultural characteristics in vitro. J. Exp. Med. 108:843–868.

Whitley, R.J., and Alford, C.A. 1981. Parenteral antiviral chemotherapy of human herpesviruses. In A. Nahmias, W. Dowdle, and R.

Schinazi (eds.) The Human Herpesviruses: An Interdisciplinary Perspective. New York: Elsevier, 478–490.

Williams, V., Gershon, A.A., and Poronell, P.A. 1974. Serologic response to varicella-zoster membrane antigens by indirect immunofluorescence. J. Infect. Dis. 130:669–672.

Winston, D.J., Pollard, R.B., Winston, G.H., Gallagher, J.G., Rasmussen, L.E., Huang, S.N.-Y., Lin, C.-H., Gossett, T.G., Merigan, T.C., and Gale, R.P. 1982. Cytomegalovirus immune plasma in bone marrow transplant recipients. Ann. Intern. Med. 97:11–18.

Zur Hausen, H., Schulte-Holthausen, H., Klein, G., Henle, W., Henle, G., Clifford, P., and Santesson, L. 1970. EBV DNA in biopsies of Burkitt's tumors and anaplastic carcinoma of the nasopharynx. Nature 228:1056–1058.

Poxviruses

James H. Nakano

29.1 DESCRIPTION OF DISEASES

29.1.1 Smallpox (Variola)

Smallpox is believed to have originated sometime after 10,000 B.C. in some agricultural settlement in Asia or Africa (Hopkins, 1983). Supporting evidence that the disease had existed in antiquity was the discovery of the mummy of Ramses V of Egypt, who died at the age of 40 in 1157 B.C. with lesions resembling those of smallpox on the surface of his face, neck, shoulders, and arms (Hopkins, 1983).

Smallpox was endemic in 33 countries in 1967, but because of the intensified effort for global smallpox eradication by the World Health Organization (WHO), the world saw the last case of endemic smallpox, which occurred in Merka, Somalia, in October 1977. Two cases of smallpox, with one death, occurred in Birmingham, England, in August 1978, but these cases were laboratory-associated, rather than the result of endemic infection.

Perhaps the first tangible evidence for the existence of the etiologic agent for smallpox was the description of a structure which is now known as elementary bodies in infections of variola and vaccinia virus reported by Buist in 1887 (cited by Smadel, 1948). This was described also by Paschen in 1906 (cited by Smadel, 1948). The first variola virus isolated was alastrim, a strain also known as variola minor virus, on chorioallantoic membranes of embryonated chicken (Torres and de Castro Teixeira, 1935).

Smallpox is transmitted to humans by person-to-person contact or by fomites. The incubation period for human infection ranges from 7 to 17 days, with an average of 12 days. The virus enters via the upper respiratory tract, and the prodromal period is characterized by fever (38.8 to 39.4°C), chills, headache, backache, vomiting, pain in the limbs, and prostration. The virus invades the lymph glands and is carried through the bloodstream to the internal organs, where the virus reproduces and is shed into the bloodstream. The skin eruption appears on the third or fourth day, as the fever subsides. The eruption develops through the stages of macule, papule, vesicle, and pustule within 5 or 6 days. The distribution of the eruption characteristically involves the face and the limbs. If a person is to recover, the lesions usually begin to dry up at about the tenth day, and scabs will be shed, almost completely, after the third week.

Two types of smallpox identifiable only clinically during smallpox outbreaks are variola major and variola minor. Variola major, which prevailed on the Asian subcontinent, was severe, with case fatality rates of 15–40% (WHO, 1972). Variola minor apparently did not exist until the 19th century (Hopkins, 1983). It is also known as amaas, Kaffir pox, or alastrim (Mardsen, 1948); it first occurred in Southern Africa and the West Indies and spread into Brazil, North American, and Europe (Hopkins, 1983). It was variola minor that existed in Ethiopia and Somalia, where the last case of endemic smallpox was seen in 1977.

29.1.2 Human Monkeypox

Monkeypox was first discovered as a disease entity in captive monkeys in 1958 (von Magnus et al., 1959), but it was not until 1970 in Zaire that it also was discovered to be a disease entity in humans. This occurred 2 years after the last case of smallpox was recorded in the area. It has been called human monkeypox to differentiate it from monkeypox in monkeys. From 1970 (when it was first discovered to infect humans) until April 1986, 345 cases have been verified by laboratory testing. Cases of human monkeypox have occurred in Liberia, Ivory Coast, Sierra Leone, Nigeria, Benin (from Nigeria), Cameroon, and Zaire, with most of the cases occurring in Zaire. Although all the sources of monkeypox virus infection in humans are still unknown, one source may be wild monkeys, because many human patients had direct contact with monkeys before the onset of the disease. Also, a number of species of African monkeys captured in the wild were found to have specific monkeypox antibody determined by a radioimmunoassay adsorption test (Hutchinson et al., 1977). However, because many other individuals with human monkeypox had no contacts with monkeys, other animals such as rodents are suspected as being sources of the disease. In 1985, we isolated monkeypox viruses from a sick squirrel with lesions, identified as Funisciurus anerythrus, captured near a village where a human case of monkeypox was found (Khodakevich et al., 1986).

The disease is clinically similar to smallpox; in fact, it is so similar that the final diagnosis would require laboratory testing if smallpox were still prevalent. The disease begins with 2 to 4 days of prodromal illness with fever and prostration before the eruption. The course of development of lesions is similar to that of smallpox. However, many patients with monkeypox, unlike those with smallpox, show prominent submandibular, cervical, and inguinal lymphadenopathy. Patients usually recover in 2 to 6 weeks. The transmission rate of monkeypox is about 3.3% and the fatality rate is about 15% (compared with 15–40% for variola major, and less than 1% for variola minor) and the overall transmission rate is about 3.3% to known susceptible contacts (Breman et al., 1980). (Compared with 25 to 40% for variola). Undoubtedly, human monkeypox currently is the most important orthopoxvirus disease, because its clinical manifestations are so similar to those of smallpox. Because its transmissibility to susceptible contacts has been much lower than that of smallpox, however, it is not a serious public health problem at the present time. Due to the appearance of two tertiary cases of human monkeypox in 1983 for the first time and the sudden increase in the number of cases since then, the disease is receiving more concern than previously anticipated by WHO.

29.1.3 Vaccinia

Vaccinia is a ''man-made'' disease as long as smallpox vaccination is practiced, because vaccination is the only way in which vaccinia virus is introduced into the human population.

The origin of the virus is unknown, but its existence became known sometime after Jenner's report in 1798 (Jenner, 1798) on smallpox vaccination. Some believe that it evolved from serial passages of cowpox virus on the skin of calves; some claim that it evolved from serial passages of variola virus on the skin of calves; and still others claim that it is a genetic hybrid of cowpox and variola virus. Recently, however, it was suggested that it originated from horsepox virus (Baxby, 1981), which is now apparently extinct. Vaccinia infection in humans can cause:

1. Erythema multiforme, a macular and erythematous rash, which appears 7 to 14 days after smallpox vaccination and is caused by an allergic reaction to vaccine components
2. Generalized vaccinia, a benign disease with multiple lesions appearing on the body of vaccinees whose antibody production was delayed, but adequate
3. Congenital vaccinia, a severe disease of the fetus following primary vaccination of pregnant women
4. Progressive vaccinia (vaccinia gangrenosa or vaccinia necrosum), a very serious infection following a vaccination in individuals apparently with a defective immune system
5. Postvaccinial encephalitis, a serious disease manifested by meningeal signs, ataxia, muscular weakness, paralysis, lethargy, coma, and convulsion
6. Eczema vaccinatum, a serious local or disseminated infection in individuals with eczema or a history of eczema.

Because of the problems accompanying smallpox vaccination, most countries have discontinued vaccination after the declaration of worldwide smallpox eradication in 1980 by WHO. In March 1983, WHO reported that 155 of its 160 member states and associated members had officially discontinued vaccination. In Egypt, revaccination has been discontinued, but primary vaccination continues. In France, primary vaccination has been discontinued, but revaccination continues. In the U.S., vaccination of military personnel continues and does contribute to producing cases of vaccinia, especially in the family contacts of vaccinees. In 1983, the Advisory Committee for Immunization Practice, which meets annually at the Centers for Disease Control (CDC), recommended vaccination only for those individuals who come in contact with orthopoxviruses (including viruses of smallpox, monkeypox, vaccinia, and cowpox) because of their occupation, and that the distribution of smallpox vaccine to vaccinate the general civilian population in the U.S. be discontinued.

In laboratories in which orthopoxviruses are handled, three levels of requirements for smallpox vaccination are recommended according to the viruses involved and with the assumption that the viruses are handled in a class 2-type biosafety cabinet:

1. In a laboratory with smallpox virus, every person entering the special bio-contained laboratory should be vaccinated with smallpox vaccine annually. Only two laboratories currently are in this category; one is at CDC and the other at the Research Institute of Virus Preparations, Moscow, USSR.
2. In a laboratory with monkeypox, vaccinia, and cowpox viruses, the laboratory need not be bio-contained, but entry to the area should be restricted. Every person entering the laboratory should be vaccinated every 3 years. These persons include laboratory workers, service persons, and visitors. Although monkeypox virus is a class 2 security-level agent (class 3 when inoculated into susceptible animals), it does cause systemic infection similar to smallpox in individuals who are not protected by smallpox vaccination, and it has caused a number of secondary and some tertiary transmissions.
3. In a laboratory with vaccinia and cowpox viruses, but without monkeypox virus, only those who are working with the viruses need to be vaccinated every 3 years with smallpox vaccine. Service persons and visitors need not be vaccinated, but an unrestricted entry of these

persons into the laboratory is not recommended during the time when the viruses, especially in high concentration, are being handled.

Laboratories in the third category are most numerous in the U.S. with only a few laboratories in the second category.[1]

29.1.4 Cowpox

Cowpox was known to Edward Jenner, who used this virus in his early smallpox vaccine. It is endemic in Great Britain and in Western Europe, but there is no evidence that it prevailed in the U.S. Reports of cowpox in humans before the 1930s in the U.S. were in patients who were infected by cows suffering with either vaccinia virus infection (previously transmitted from humans) or pseudocowpox (parapox, milker's nodule) virus infection. Of course, the confusion was understandable because the distinction between vaccinia and cowpox viruses was not made until 1939 (Downie et al., 1939). Furthermore, indirect evidence that indicates cowpox virus never existed in the United States is that in recent times in Europe, poxviruses similar to cowpox virus have been isolated from zoo animals such as the giant anteater (Marennikova et al., 1976), the family Felidae, including lions, black panthers, cheetahs, pumas, jaguars, ocelots (Marennikova et al., 1975, Baxby et al., 1979), okapis (Zwart et al., 1971), and also domestic cats (Thomsett et al., 1978). In the U.S., however, not only have we not seen cowpox virus infection in carnivores in zoos and domestic cats, we have not encountered cowpox virus in the poxvirus laboratory which was established at CDC in 1966. Therefore, it is my belief that cowpox virus was never transported from Europe to the United States and it never existed in the U.S.

Cowpox virus infection in humans is transmitted by direct contact with infections on the skin of the udder and teats of cows. The lesions in humans are found on the fingers, with reddening and swelling which develop into papules that become vesicular in 4 to 5 days and heal in 2 to 4 weeks.

Although the disease was always believed to be transmitted to humans by direct contact with infected cows, it is now believed that rats or other rodents (Baxby, 1977) are an important vector for disease transmission. This mode of transmission can explain how zoo animals and domestic cats can be infected and, thus, how domestic cats can transmit the disease to humans.

29.1.5 Buffalopox

Poxvirus disease in buffalo, in the past, has been caused by vaccinia, variola, and cowpox viruses, but a unique poxvirus which is now known as buffalopox virus (Mathew, 1976, Singh and Singh, 1967, Baxby and Hill, 1971) caused outbreaks in buffaloes in India in the 1960s and 1970s. During an outbreak of the disease in buffaloes, it is transmitted to humans by direct contact with an infected buffalo. The lesions are localized on the fingers, hands, and sometimes on the face of humans. No generalized infection has been seen so far and no person-to-person transmission has been reported. The patient recovers from the infection, with the scabs falling off in about 2 weeks.

29.1.6 Whitepox

Six isolates of whitepox virus are known today and were isolated from nonhuman primates and rodents (cited by Nakano, 1979). The first two were isolated from kidney cell cultures of two cynomolgus monkeys in a laboratory in the Netherlands in 1964; the third was isolated from a chim-

[1] Because smallpox vaccine has not been distributed for use with the general civilian population since May 1983, laboratories in categories 2 and 3 can request a supply of vaccinia from the Centers for Disease Control, Division of Host Factors, Clinical Medicine Branch, Atlanta, Georgia 30333, telephone number 404-329-3356.

panzee captured in Zaire in 1971; the fourth from a sala monkey in Zaire in 1973; the fifth from a rodent (Mastomys) in Zaire in 1974; and the sixth from another rodent (Heliosciurus) in Zaire in 1975. The latter four were isolated from wild animals captured during investigations of human monkeypox cases. Whitepox virus cannot be differentiated from variola virus by biological methods or by DNA analysis (Esposito et al., 1978, Dumbell and Archard, 1980). Although we do not know how this virus may affect humans because it was never isolated from humans, we do know from our investigation at CDC that it can cause a smallpox-like or a monkeypox-like disease in African green monkeys (*Cercopethecus aethiops*) (Nakano, 1977). When whitepox virus first became known, investigators theorized that wild animals might be a reservoir for variola virus. However, it has been fairly well established (Dumbell and Kapsenberg, 1983) that at least the first two strains isolated from kidney cell cultures of the two cynomolgus monkeys in 1964 are products of laboratory contamination of a variola virus strain from Vellore, India, which was present in the laboratory at that time. Furthermore, the whitepox viruses from the chimpanzee and the sala monkey were indistinguishable by the biological marker test (Dumbell and Archard, 1980) from Harvey, the international reference strain for variola major virus. Whitepox virus DNA from the sala monkey and DNA from the Harvey strain were indistinguishable in every respect; the virus DNA from the chimpanzee was indistinguishable except for a minor difference (Esposito et al., 1978) when endonuclease-digested DNA was examined by electrophoresis.

29.1.7 Tanapox

Downie et al. (1971) first reported epidemics of Tanapox in humans (1957 and 1962) in the Wapakoma tribe which lives in Kenya along the Tana River. Tanapox was also found in the northeast area of Zaire (Lisala area) in 1975 by a WHO surveillance team that was looking for smallpox and human monkeypox. The diagnosis was verified by the CDC Poxvirus Laboratory. In Zaire, the disease is called the "river smallpox" (WHO investigators M. Szczeniowski and J. Jezek, personal communication).

Tanapox in humans as described by Downie et al. (1971) can begin with a febrile period of 3 to 4 days, and can include backache, severe headache, and prostration, or, as found in many cases in Zaire, the disease may cause very few clinical signs. Lesions seen on uncovered areas of the skin are few, usually one or two. Very few patients examined are found to have more than ten lesions. The lesions appear on the skin of the upper arms, face, neck, or trunk and start as papules similar to those of smallpox and then become vesicles. Fluid, however, is difficult to extract from these vesicles. The lesions become umbilicated without pustulation and usually heal in 2 to 4 weeks. The healing time, however, may extend to 7 weeks. The disease is believed to be transmitted by mosquitoes; the source of the infection is unknown, but monkeys are suspected.

29.1.8 Milker's Nodule

Dr. Jenner called milker's nodule "spurious cowpox," and he knew that it did not provide immunity against cowpox or smallpox; thus he suggested that it is unrelated to cowpox and smallpox. This disease, as first reported in the U.S. in 1940 (Becker, 1940), primarily affects cattle and is known as pseudocowpox. In cows, milker's nodule produces lesions on the skin of the udder and teats, and in calves it produces lesions on the lips and nose. The infection may spread to the head, trunk, and limbs. Milker's nodule is transmitted to humans by direct contact with an infected animal, and the lesion(s) is usually located on the abraded skin area of the fingers and hands. The lesion(s) starts as an erythematous papule 5 to 7 days after exposure and becomes a firm, elastic, bluish-red, and semi-globular nodule, which measures from 1 to 2 cm and has a central depression. The lesion(s) flattens as it heals and disappears in 4 to 6 weeks. The infection sometimes causes the regional lymph nodes to swell.

29.1.9 Bovine Papular Stomatitis

Bovine papular stomatitis (BPS) is a mild infection in calves and is manifested by lesions occurring on the muzzle, margin of the lips, and buccal mucosa. It affects cows and bulls of all ages, but calves seem to be affected most often. Although the calves are infected in the buccal area, cows are not always infected on the teats and the udders as found in pseudocowpox (paravaccinia). Also unlike the pseudocowpox, BPS is found more often in beef animals than in dairy animals. The infection is transmitted to humans, manifested by lesions on the hands and arms, and heals in 3 to 4 weeks. Although there is a great deal of similarity between BPS virus and milker's nodule virus, the question of whether they are the same virus is still unresolved.

29.1.10 Orf

Orf, also known as contagious ecthyma, contagious pustular stomatitis, contagious pustular dermatitis, or sore mouth, is an infection of sheep and goats which is transmitted to humans by direct contact with an infected animal. The infection in sheep was first reported in 1887 and is prevalent worldwide (Tripathy et al., 1981). Infection sites are usually on the fingers, hands, and arms, but are sometimes on the face and neck. Only very occasionally is infection generalized. The lesions develop after an incubation period of 3 to 6 days through maculopapular stages. A red center then develops in the lesions surrounded by a white ring and a red halo. A nodular stage with red and weeping surface follows, often with a central umbilication. The lesions become granulomatous or papillomatous in 3 to 4 weeks, and some become ulcerated and superinfected with bacteria. Healing occurs in 4 to 7 or more weeks.

29.1.11 Molluscum Contagiosum

Molluscum contagiosum (MC) was recognized in 1817 and is found worldwide. Two types of this disease are known to occur (Brown et al., 1981). One, found in childhood, manifests itself with lesions on the face, trunk, and limbs and is transmitted by direct contact from skin to skin or by fomites. This type is common in the tropics. The second, found in young adulthood, manifests itself with lesions located mostly in the lower abdominal area, pubis, inner thighs, and genitalia and is transmitted by sexual contact. The incubation time ranges from 1 week to 6 months. The lesions begin as pimples and become umbilicated papules which are pale pink to white, measuring 2 to 8 mm. The surface area over the lesions often appears tightly stretched with a slight central depression. A semi-solid caseous material can be expressed and used for examination. The disease is self-limiting, but may last from several months to several years.

29.2 DESCRIPTION OF VIRUSES

29.2.1 Physical Characteristics and Classification

Poxviruses that cause human infections are relatively large, brick-shaped, or ovoid virions which possess an external coat containing lipid and tubular or globular protein structures, enclosing two bilateral bodies and a core which contains the genome. The genome is a single molecule of double-stranded DNA with a molecular weight of $130–240 \times 10^6$. The G + C content of orthopoxviruses is 35–40% and of parapoxviruses is about 63%. As shown in Table 29.1, the orthopoxvirus group includes viruses of variola, monkeypox, vaccinia, cowpox, buffalopox, and whitepox; parapoxvirus group includes viruses of milker's

Table 29.1. Taxonomic and Morphologic Classification of Poxviruses of Human Infections

Orthopoxvirus	Unclassified Virus	Parapoxvirus
Brick-Shaped Morphology	Brick-Shaped Morphology[a]	Ovoid Morphology
1. Variola	1. Tanapox	1. Milker's nodule (pseudocowpox, paravaccinia)
2. Monkeypox (human and monkey)	2. Molluscum contagiosum	2. Bovine papular stomatitis
3. Vaccinia		3. Orf (contagious ecthyma, contagious pustular stomatitis, contagious pustular dermatitis, sore mouth)
4. Cowpox		
5. Buffalopox		
6. Whitepox		

[a] Prominent tubules for viruses of tanapox and molluscum contagiosum; envelope for tanapox virus.

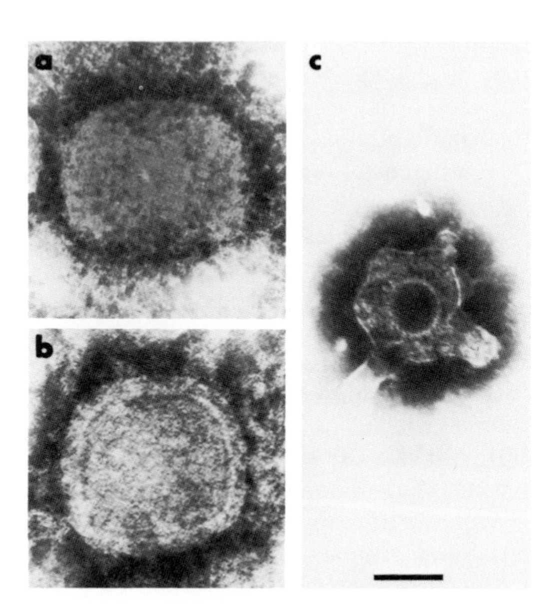

Figure 29.1. A. "M" form of variola virus from Somalia. B. "C" form of variola virus from Somalia. C. a typical varicella virus with envelope. The line represent 100 nm. See the text for the differentiation of "M" and "C" forms.

nodule, BPS, and orf; and the unclassified group includes viruses of tanapox and MC.

There are two morphologic groups: one group as illustrated by Figure 29.1 a and b is brick-shaped and includes the viruses of the orthopoxvirus group and the unclassified group (Table 29.1), and the second group as illustrated by Figure 29.2 a and b is ovoid or elongated and includes the viruses of the parapoxvirus group (Table 29.1). The surface tubules of these viruses are characteristic in that they are uniquely arranged in a parallel and criss-crossing pattern (Figure 29.2). The sizes of the viruses in the orthopoxvirus and the unclassified groups range from 140–230 × 210–380 nm, and the size of the viruses in the parapoxvirus group range from 120–160 × 250–310 nm.

Although cursory comparison of viruses in the *orthopoxvirus group* and the *unclassified group* shows no discernible morphologic difference by electron micros-

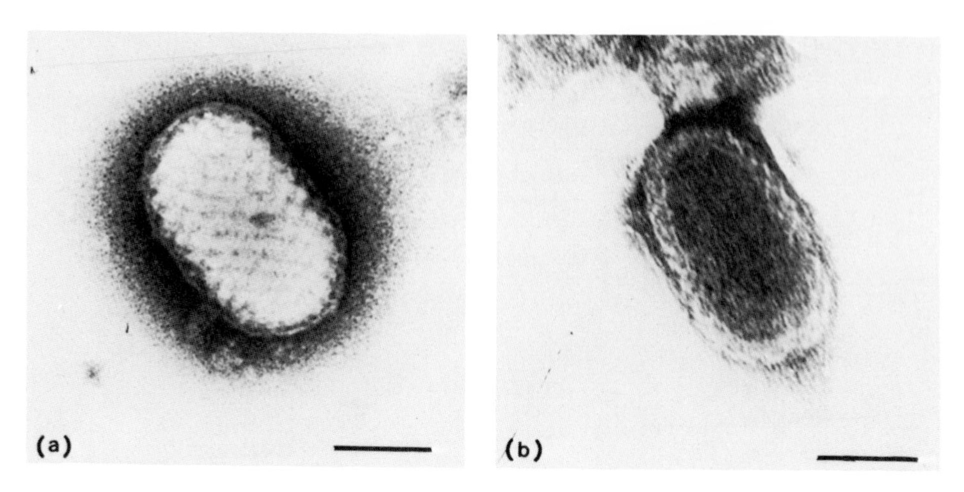

Figure 29.2. Electron microscope photograph of a parapoxvirus. A. "M" form; B. "C" form.

copy, I found that tanapoxvirus and MC virus in field specimens (lesion materials) can, in certain situations, be differentiated from the orthopoxviruses. This is discussed under "electron microscopy" in this chapter.

Note, as shown in Figure 29.1 that an "M" form of poxvirus is a viral particle into which a negative stain had not penetrated; therefore, when examined by electron microscopy, only the surface resembling that of a "mulberry" is seen. A "C" form is a poxvirus particle into which a negative stain had penetrated; therefore, the particle appears to be "capsulated." The "M" form is found more often in vesicular specimens (wet), and the "C" form in scabs (dry).

29.2.2 Description of Antigens

Viruses of the genus Orthopoxvirus are so closely related antigenically that there is no easy, routine serologic method that can differentiate these viruses. However, because of this close antigenic relationship, an antiserum against one virus can be used to identify any virus in this group as an orthopoxvirus. For example, an antivariola virus rabbit serum cannot identify an isolate as variola virus, but can identify the isolate as an orthopoxvirus.

Viruses of milker's nodule, BPS, and orf in the genus Parapoxvirus are not antigenically related to any of the orthopoxviruses and neither of the two groups are related to tanapoxvirus or MC virus. By the complement-fixation test, however, one may be able to see one way cross-reactivity between orthopoxviruses (monkeypox and vaccinia) and tanapoxvirus.

Antigens of orthopoxviruses, parapoxvirus, tanapoxvirus, and MC virus are composed of two types: the structural proteins of virions which are the viral nucleoproteins and the viral surface tubule proteins, and the soluble antigens which are released from infected cells during the course of infection. An additional antigen, a hemagglutinin, which is produced only by orthopoxviruses, is derived from the plasma membrane of the infected cells and is separate from the soluble antigen and the structural proteins. A small percentage of extracellular vaccinia virus particles are found enveloped. The envelope surrounding these particles also contains a hemagglutinin (Payne and Norrby, 1976), which is composed of glycoprotein (Payne, 1979).

29.3 COLLECTION AND HANDLING OF SPECIMENS

For each of the poxvirus infections described in this chapter it is the skin lesions that are the specimen of choice for virus isolation. These poxviruses are relatively stable and therefore remain viable after several weeks of storage even without refrigeration. We have been able to isolate viruses from specimens of smallpox and monkeypox even when the specimens were received at CDC 6 weeks after they were collected in Africa. A sufficient amount of specimen should be submitted to permit effective testing. Because specimens are examined by electron microscopy at CDC they should not be diluted with any "holding" fluid prior to shipment.

Clinical specimens for virus isolation during *orthopoxvirus infections*, should be from skin lesions collected at macular-papular, vesicular-pustular, and crusting stages. During the macular-papular stage, the lesions are scraped onto a slide with a scalpel blade reaching deeply into the lower epithelial layer. During the vesicular-pustular stage, the fluid can be collected by capillary tubes, swabs, or on slides, noting especially to collect the materials from the base of the vesicles where virus is found in higher concentration. During the crusting stage, no fewer than three crusts (preferably ten) are collected and placed in a screw-capped vial and processed. For vaccinia and cowpox, only fragments of crusts may be available for collection.

In *parapoxvirus infections*, vesicular fluid, if available, and crusts should be collected. A biopsy from the periphery of the lesions is occasionally useful when crusts are not available.

For *tanapox,* specimens similar to those for parapox are collected.

For *molluscum contagiosum,* expelled materials from the lesions are collected on swabs or slides.

29.4 METHODS FOR ISOLATION

29.4.1 Chicken Embryo Chorioallantoic Membrane

The procedures for the preparation of the chorioallantoic membrane (CAM) for inoculating viruses and harvesting the membranes are described in Chapter 4 of this book. Fertile chicken eggs must be incubated at 38–39°C for 11 to 13 days. The eggs incubated less than 11 days can be insusceptible to viruses, and those incubated more than 13 days have embryos that are too large to enable use of the CAM.

Although viruses of parapox, tanapox, and MC do not grow on the CAM, orthopoxviruses, especially viruses of smallpox, monkeypox, vaccinia, cowpox, and buffalopox, grow well. The use of the CAM is the best method to isolate these viruses, because not only does it isolate orthopoxviruses, but the viruses can also be identified according to the characteristic morphology of the pocks that each virus produces on the CAM. Because these orthopoxviruses cannot routinely be differentiated serologically, this method of identification is very useful.

29.4.1.a Smallpox Virus

Smallpox virus pocks are about 1 mm in diameter 72 hours postinoculation. They are grayish-white, opaque, convex, raised above the CAM surface, round, regular, with a smooth outer edge, are not hemorrhagic, and all are nearly the same size. The pocks appear like a fried sunny side-up egg with an opaque area with a halo.

29.4.1.b Monkeypox Virus

Monkeypox virus pocks are about the same size as those of smallpox virus at 72 hours but they are flat and ridged along the periphery and not raised above the CAM surface. Many pocks have a crater in the center that appears as a punched-out hole, and many are hemorrhagic when incubated below 35°C. The hemorrhagic appearance in monkeypox virus pocks is caused by the deposition of red blood cells in the surface cell layer of the CAM, not in the pock itself.

29.4.1.c Vaccinia Virus

Vaccinia virus pocks at 72 hours measure 3 to 4 mm in diameter, are flattened with central necrosis and ulceration, and are sometimes hemorrhagic. Certain strains of vaccinia virus produce hemorrhagic pocks. In the pocks with a hemorrhagic appearance, the red blood cells are deposited in the surface cell layer of the CAM, like those found with monkeypox virus pocks.

29.4.1.d Cowpox Virus

Cowpox virus pocks at 72 hours measure 2 to 4 mm in diameter, are flattened and rather round, and have a bright red central area which is caused by the red blood cells deposited in the pock proper, unlike those deposited on the surface cell layer of the CAM for monkeypox virus pocks.

29.4.1.e Buffalopox Virus

Buffalopox virus pocks at 48 hours measure about 1.6 mm in diameter and characteristically produce two types of pocks. One type is white and raised with little or no hemorrhage and the other is flat and gray.

29.4.1.f Whitepox Virus

Whitepox virus pocks at 72 hours are exactly like those of smallpox virus.

29.4.2 Cell Cultures

The use of cell culture for isolating orthopoxviruses is warranted as a back-up method for CAM inoculation, because the CAM can manifest periodic unpredictable "insusceptibility" for these viruses.

29.4.2.a Orthopoxviruses

Orthopoxviruses can be isolated in all human and nonhuman primate cells, e.g. human embryonic diploid cells, LLC-MK_2 (stable rhesus monkey kidney cells), and Vero (stable African green monkey kidney cells). Most of these viruses can grow in cells of other animals such as rabbit, mouse, and hamster. Smallpox virus does require adaptation in rabbit kidney cells before the virus can attain satisfactory growth.

29.4.2.b Smallpox Virus and Whitepox Virus

From clinical specimens, these produce a cytopathic effect (CPE) within 1 to 3 days in LLC-MK_2 or Vero cells with rounding of the cells and the formation of hyperplastic foci seen as small plaques 1 to 3 mm in diameter.

29.4.2.c Monkeypox, Vaccinia, and Cowpox Viruses

These viruses produce CPE in nonprimate cells, and in LLC-MK_2 or Vero cells in 1 to 3 days by fostering cell fusion and thereby forming foci. This is followed by the formation of plaques in 2 to 3 days. The plaques produced by these viruses are much larger, measuring 2 to 6 mm, than those made by smallpox virus. The plaques usually show cytoplasmic bridging. Although these three viruses can be differentiated from smallpox virus by the large plaques they produce, they cannot be differentiated from one another based on their plaque characteristics.

A continuous line of pig embryonic kidney (PEK) cells can be used to differentiate monkeypox virus from smallpox, whitepox, vaccinia, and cowpox viruses because monkeypox virus cannot grow in PEK, but the other four viruses can (Marennikova et al., 1972).

29.4.2.d Buffalopox Virus

Little work has been done to characterize the CPE that buffalopox virus produces in cell cultures, but it grows well in primary monkey kidney cells, primary rabbit kidney cells, hamster kidney cells, and human amnion cells (Mathew, 1976; Baxby and Hill, 1971).

29.4.2.e Orf Virus

Orf virus, a parapoxvirus, grows well in ovine cells such as embryonic ovine kidney and ovine testis. We found that some strains of orf virus can be isolated from clinical specimens in primary rhesus monkey kidney (PRMK) cells. Once isolated, the virus can be passaged in other primate cell lines such as LLC-MK_2. Orf virus from human infections can also be isolated in bovine cells, but that from ovine infections cannot be isolated in bovine cells. The virus produces CPE in 3 to 6 days.

29.4.2.f Milker's Nodule

Milker's nodule virus can be isolated in bovine cells such as bovine embryonic lung and calf testis. Most can also be isolated in ovine cells. Once isolated, the virus can be grown in primate cells such as human diploid cells and LLC-MK_2.

29.4.2.g Tanapoxvirus

Tanapoxvirus was first isolated from humans in human thyroid cells (Downie et al., 1971). Once isolated, the virus grows in WI-38, primary vervet monkey kidney cells, Vero cells, HEP-2, and primary patas monkey kidney cells (Downie et al., 1971). We have been growing the virus in human embryonic lung fibroblast cells, primary African monkey kidney cells, and LLC-MK_2 at the CDC. Of 145 positive cases of tanapox determined using electron microscopy, only 20 isolations have been made by using LLC-MK_2 cells incubated at 33°C instead of at 35–37°C (Nakano, 1982). The virus produces CPE in 6 to 10 days.

29.4.2.h Molluscum Contagiosum Virus

The MC virus has been virtually impossible to grow in any cell culture system.

29.5 METHODS FOR IDENTIFICATION

29.5.1 Electron Microscopy

Electron microscopy (EM) is the best method for the identification of the viruses in smallpox and human monkeypox. Its reliability at CDC is rated at 98.6% compared with 89% for that of viral isolation by CAM (Nakano, 1982). However, all of the specimens for smallpox and human monkeypox received at CDC were sent from Africa or other countries overseas, and the time required for the specimens to arrive in Atlanta was several weeks to 2 months. Therefore, if each specimen were sent in sufficient amount and received fresh, the reliability of EM and CAM could have been very close to 100%.

Electron microscopy is advantageous in that results in most cases can be obtained within 1 hour after receipt of the specimen.

29.5.1.a Orthopoxviruses

Orthopoxviruses cannot be differentiated by EM because they are similar in size and morphology. Use of CAM can, however, prevent mistaking chickenpox for smallpox and human monkeypox. As shown in Figure 29.1, smallpox (Figure 29.1a) and human monkeypox (Figure 29.1b) are similar but the herpesvirus of varicella shown in Figure29.1c is distinctly different from orthopoxviruses.

In our experience, the possibility of finding poxvirus by EM for smallpox and monkeypox was 98.6%; however, for finding herpesvirus in patients with chickenpox it was about 60–70%. Based on this experience, if a patient is suspected of smallpox or monkeypox and no poxvirus or herpesvirus is found by EM, the patient is considered not to have smallpox or monkeypox, but may have chickenpox or another disease.

When a poxvirus (brick-shaped), as stated in Table 29.1, is seen by EM for suspected cases of vaccinia, cowpox, or buffalopox, a tentative diagnosis can be made because of the characteristic clinical picture and the clinical history.

29.5.1.b Parapox Viruses

Parapox viruses are alike in size and morphology and, therefore, they cannot be differentiated from each other (Figure 29.2 a and b). These viruses have tubules on their surface which are arranged in parallel and form a criss-cross pattern. These viruses are easily differentiated from those of orthopoxviruses, tanapoxviruses, and MC viruses by EM.

29.5.1.c Tanapox

Because tanapox lesions are quite characteristic, observing brick-shaped poxvirus by EM can confirm the disease. Although these virions are morphologically similar to those of orthopoxviruses, they can often be differentiated from those of orthopoxviruses. About 90% of the specimens from tanapox patients reveal viral particles that are enveloped when examined by EM (Table 29.1). Furthermore, the surface tubules on tanapoxvirus appear to be more pronounced than those found on viruses of smallpox and monkeypox and other orthopoxviruses when the virions are found in field specimens. Although viruses of smallpox, monkeypox, and vaccinia can show prominent tubules when they are grown in cell cultures or when they are "cleaned" during a virus concentration procedure, generally, prominent tubules are not seen on virus particles in field specimens. Vaccinia virus is occasionally seen with an envelope when it is grown in tissue culture, but we have never seen it enveloped in specimens taken from patients.

29.5.1.d Molluscum Contagiosum

Molluscum contagiosum lesions are characteristic and, therefore, may not require laboratory assistance for diagnosis. Although MC virions cannot be differentiated from virions of orthopoxviruses on the basis of shape and size, most MC virions do have the prominent surface tubules similar to those on the surface of tanapox virions. Therefore, most MC virions can be differentiated from those of orthopoxviruses. Note again that the comparison is made on the basis of virions seen only in field specimens.

Four other methods which can identify the poxviruses mentioned in this chapter are described only briefly, because they cannot be used with the speed and reliability shown by the EM.

29.5.2 Agar Gel Precipitation

Agar gel precipitation (AGP) test has been used extensively for the preliminary identification of smallpox virus antigen. The detailed procedure for AGP has been described (Nakano, 1979). The test identifies orthopoxviruses as a group and, therefore, does not differentiate smallpox, human monkeypox, vaccinia, cowpox, and buffalopox viruses from one another. At CDC, although the efficiency of detecting smallpox by this method has been about 72% (Nakano, 1982), I believe that this percentage could have been much higher if fresh specimens were tested. This is a good substitute if the EM method is not available, because AGP is easily installed for routine use, can yield a positive result in about 2 hours, and can yield a verified negative in 24 hours.

The AGP can be used for the identification of all poxviruses. However, its reliability for the identification of virus groups other than orthopoxvirus group is unknown.

29.5.3 Immunofluorescence

Immunofluorescence (IF) testing by the direct or indirect method has been used to diagnose smallpox, but we found that the method can give false-positive results when a field specimen has been stored unfrozen for more than 7 days. The test identifies orthopoxviruses as a group and cannot differentiate one orthopoxvirus from another. The test has been successfully used also for the antigenic detection of parapoxviruses, tanapoxvirus, and MC virus. However, the reliability of identification of these viruses has not been tested.

29.5.4 Complement Fixation

The complement-fixation (CF) test for the identification of viral antigen is sensitive and was useful in earlier days for the laboratory identification of smallpox virus antigen. Like the other serologic tests, it identifies orthopoxviruses as a group. This test has been used also to detect the presence of parapoxviruses, tanapoxvirus, and MC virus, but unless the lesion materials are "cleaned" by treatment with ether, an anticomplementary reaction often occurs, which makes the results useless.

29.5.5 Stained Smears

Stained smears of lesion materials were often used in the past to look for inclusion bodies present in specimens collected from smallpox patients but not present in specimens from chickenpox patients. The use of EM made this method obsolete, but it can still be used when EM is not available.

Table 29.2 lists seven methods which can identify poxvirus antigens. Although enzyme-linked immunosorbent assay (ELISA) and radioimmunoassay (RIA) are not listed in the table, ELISA has been used for identifying virus antigens of variola, vaccinia, monkeypox, and tanapoxviruses. Except for tanapoxvirus, these tests can only identify each virus as belonging to a group.

Table 29.2. Tests to Identify Poxvirus Antigens

Viruses	EM	CAM	TC	AGP	FA	CF	Stained Smears
Smallpox	+	+	+	+	+	+	+
Monkeypox	+	+	+	+	+	+	+
Vaccinia	+	+	+	+	+	+	+
Cowpox	+	+	+	+	+	+	+
Camelpox	+	+	+	+	+	+	+
Whitepox	+	+	+	+	+	+	
Milker's Nodule	+		+	+[a]	+	+	
BPS	+		+	+[a]	+	+	
Orf	+		+	+[a]	+	+	
Tanapox	+			+[a]	+	+	
MC	+			+[a]	+[a]	+[a]	+

Abbreviations: EM = electron microscopy; CAM = choriallantoic membrane of embryonated chicken, 11 to 13 days old; TC = tissue culture (cell culture); AGP = agar gel precipitation test; FA = fluorescent antibody test; CF = complement-fixation test; BPS = bovine papular stomatitis; MC = molluscum contagiosum.

[a] The test can be used, but its reliability is uncertain.

29.6 SEROLOGIC METHODS FOR ANTIBODY ASSAY

For quick and accurate diagnosis of poxvirus infections, virologic specimens are preferred over serum specimens; however, such a specimen at times cannot be collected because the patient was seen too late in the course of the infection. Thus, a blood specimen for antibody assays must be collected and becomes very important for the diagnosis of an infection. Blood specimens are also important in an epidemiologic survey to determine the extent of a past infection in a population. A list of serologic methods for antibody assays of various poxvirus infections is given in Table 29.3.

29.6.1 Hemagglutination Inhibition Test

As shown in Table 29.3, the hemagglutination inhibition antibody assay can be performed only for smallpox, monkeypox, vaccinia, cowpox, and buffalopox, because only viruses in the orthopoxvirus group produce hemagglutinin.

A standard microtiter method as described in Chapter 14 can be used. Virtually any virus in the orthopoxvirus group can be used as the hemagglutinin (viral antigen) for the HI test; however, because of its easy availability and its fairly high hemagglutinin content, vaccinia virus traditionally has been used. (See Nakano, 1979 for details of preparation.)

A suspension of chicken erythrocytes is used as a standard reagent, but since erythrocytes from only about 50% of the chickens will agglutinate with vaccinia hemagglutinins, erythrocytes must be pretested. (See Nakano, 1979 for details of preparing chicken erythrocyte suspension.)

Sera stored for a long time with inadequate refrigeration and sera from blood samples collected after death may contain nonspecific HI factor. These sera must be treated with periodate (Nakano, 1979). Occasionally sera will contain a nonspecific hemagglutinin which must be removed by absorbing with 50% chicken erythrocytes suspension before the HI test. (See reference by Nakano, 1979, for the procedure.)

Hemagglutination inhibition antibody is detectable within 4 to 7 days after infection in patients with smallpox or human monkeypox. For a patient with a dependable clinical history of either of these dis-

Table 29.3. Serologic Methods in Assaying Poxvirus Antibodies

Poxvirus Infection	HI	NT	IFA	ELISA	RIA	RIAA	CF	AGP
Smallpox	+	+	+	+	+	+	+	+
Monkeypox (human and monkey)	+	+	+	+	+	+	+	+
Vaccinia	+	+	+	+	+	+	+	+
Cowpox	+	+	+	+	+		+	+
Buffalopox	+	+	+	+	+		+	+
Tanapox		+	+	+			+	+
Milker's nodule		+	+	+			+	+
Bovine papular stomatitis		+	+				+	+
Orf		+	+	+			+	+
Molluscum contagiosum		+[a]	+[a]				+[a]	+[a]

Abbreviations: HI = hemagglutination inhibition; NT = neutralization test; IFA = indirect fluorescent antibody; ELISA = enzyme-linked immunosorbent assay; RIA = radioimmunoassay; RIAA = radioimmunoassay-adsorption; CF = complement fixation; AGP = agar gel precipitate.

[a] No routine test can be performed.

eases, an HI titer of greater than 40 in blood specimens collected 4 weeks after onset of the infection may be diagnostically significant, provided that the serum was treated to eliminate nonspecific hemagglutinin and HI factor. The HI test is also suitable to measure antibody response in other orthopoxvirus infection, in addition to smallpox and monkeypox.

Although the HI test is useful in measuring a response to smallpox vaccination in individuals who are vaccinated for the first time or who are being revaccinated after several years (more than 5 years), it is not useful for a large number of individuals who are vaccinated every 3 years and for a greater number of individuals who are vaccinated every year, because the viral replication at the site of inoculation is far less in the latter two groups of individuals. Although HI antibody titer was traditionally believed to be short-lived, we have found HI titers of 10 to 20 after 3 or more years in many patients who had human monkeypox. The HI test identifies orthopoxvirus group antibody and not specific virus antibody.

29.6.2 Neutralization Test

As shown in Table 29.3, an assay for neutralizing antibody assay can be performed for the ten infections listed. For the orthopox group (smallpox, monkeypox, vaccinia, cowpox, and buffalopox), the assay can be done by a method using tissue culture or CAM, but for infections caused by other viruses that do not grow on CAM (tanapox, milker's nodule, BPS, orf, and MC), only the tissue culture method can be used.

Procedures for neutralization test (NT) using tissue culture are found in Chapter 13 and in Nakano (1979) but for procedures using CAM see reference by Boulter (1957).

The live-virus antigen used for orthopoxvirus NT is monkeypox virus, and the neutralizing antibody assay is based on 50% plaque reduction, in which the plaques formed under a liquid medium are counted at 44 to 48 hours after the virus-serum mixture is inoculated on cell cultures. Because of the close antigenic relationship of the viruses within the orthopoxvirus group, as demonstrated by the HI test, NT cannot identify the causative poxvirus from the antibody assay.

In a patient with smallpox, human monkeypox, or vaccinia, and probably with other orthopoxvirus infections, neutralizing antibody is detected in the latter part of the first week or during the second week after infection and the antibody persists for a number of years. Orthopoxvirus NT antibodies can vary from 500 to 2000 and, therefore, a positive control serum with an average titer of 1000 should be included in each test run.

In many cases in which laboratory diagnosis of smallpox or human monkeypox is dependent upon serologic results, only one serum specimen is collected at some time after the onset of rash. If this specimen has a titer of less than 500, a definite diagnosis cannot be made, but if the titer is greater than 1000, the patient probably had the infection.

The neutralizing antibody assay for parapox is the 50% plaque reduction technique using tissue cultures such as human embryonic lung fibroblast cells or LLC-MK_2 ovine or bovine cells. Specific live virus is used to assay its corresponding NT antibody, e.g., orf virus must be used to assay antibody for orf virus, etc. Although the NT for parapoxviruses has been used successfully in the laboratory, its reliability is unknown because its usage has been limited. We do know that a convalescent-phase serum should be collected at 4 to 5 weeks after onset of the infection, however, and any positive result in conjunction with clinical disease confirms the diagnosis. Because the three viruses are antigenically related, the specific etiologic agent cannot be identified unless the animal contact of the patient is identified (e.g., orf virus for sheep, milker's nodule virus for milking cows and their calves, and BPS virus for beef cattle and calves.)

The technique for tanapox is again 50% plaque reduction with tissue culture choices of primary African green monkey kidney cells, human embryonic fibroblast cells, or LLC-MK_2, and with live tanapox virus. Any positive result again confirms the diagnosis of tanapox. Because very few blood specimens collected from cases of human tanapox within 2 weeks after onset of the infection were positive, we recommend that the blood specimens be collected at 4 to 5 weeks after onset of the infection.

Routine NT for MC is not available because the virus does not grow well in any cell system or any known laboratory animal.

29.6.3 Indirect Fluorescent Antibody Test

As shown in Table 29.3, indirect fluorescent antibody (IFA) test can be used for the ten infections listed. Procedure for IFA is described in Chapter 8, but the method used at the CDC differs from the usual in that the infected cells on the test slides are not fixed with acetone or methanol but merely dried thoroughly, leaving the slides with the infected cells overnight in a class II safety cabinet with the air-ventilation switch on.

To detect antibodies to orthopoxviruses, IFA has not been used extensively for smallpox, but has been used for monkeypox. Cells infected with monkeypox virus are the antigen used to detect antibody in the specimen. An IFA titer of greater than 32 in a serum specimen from a case of suspected monkeypox virus is probably diagnostic of the infection if the patient had no previous smallpox vaccination. The monkeypox virus antibody measured by IFA test does not become detectable in many cases until the second week after onset of the infection. The IFA test was not suitable for serologic survey of monkeypox; the IFA titers begin to decrease after 6 months and may not be detectable at 1 or 1 1/2 years.

The IFA test for antibody to parapoxvirus in our experience is more sensitive than the CF test. Although the viruses show some cross-reaction, we have used orf virus for orf antibody assay and milker's nodule virus for milker's nodule and BSP antibody assays. Any IFA positive result is diagnostic of these diseases since these diseases are not commonly found in humans.

The convalescent-phase serum should be collected about 5 weeks after onset of an infection since those collected at 2 to 3 weeks are often negative. Again, as described previously, the etiologic agent can be determined only after the patient's animal contact is known.

The IFA test for antibodies to tanapoxvirus is often negative for serum collected 12 to 15 days after the infection and, therefore, the serum should be collected at 5 to 6 weeks after onset of the infection.

No routine IFA test is presently available for MC since cells infected with MC virus cannot be produced.

29.6.4 Enzyme-Linked Immunosorbent Assay

An indirect ELISA test was adapted at CDC from the indirect ELISA test described by Voller et al. (1976) to detect antibody to orthopoxviruses using monkeypox virus as the antigen. An ELISA test using selective adsorption has been developed by S.S. Marennikova at the Research Institute for Viral Preparations, Moscow, USSR, that can identify specific monkeypox antibody (personal communication).

An ELISA, test used at CDC, has aided in the diagnosis and serologic survey of human monkeypox cases in Africa. An ELISA titer greater than 160 in a serum from a patient with suspected monkeypox who had no previous smallpox vaccination and whose blood had been collected 2 to 6 weeks after onset of the illness is diagnostic for human monkeypox.

The ELISA test has also been adapted at CDC for antibody assays of parapoxviruses (orf and milker's nodule) and tanapoxvirus. For parapoxviruses, the reliability of the test remains to be determined. But for tanapox, the ELISA test, for many patients, is capable of detecting antibody at an earlier time after onset of the disease than the IFA or the NT.

The ELISA test is not available for the assay of MC antibody since the virus virtually cannot be grown in a laboratory.

29.6.5 Radioimmunoassay

The RIA for orthopoxvirus antibody assay was reported by Ziegler et al. (1975) and was used at CDC in diagnosing smallpox, vaccinia, and human monkeypox. Among the serologic tests for poxvirus antibody, RIA is probably the most sensitive in detecting antibodies to orthopoxviruses. However, for the detection of antibodies to smallpox and monkeypox in the very early stage of these illnesses, HI appears to be better than the RIA. Because of its high sensitivity, RIA titers less than 100 are questionable. Radioimmunoassay titers of 3000–20,000 or more can be found in a blood specimen collected within 4 to 6 weeks after onset of human monkeypox.

29.6.6 Radioimmunoassay-Adsorption Test

The radioimmunoassay-adsorption test reported by Hutchinson et al. (1977) has been used at CDC mainly for the identification of specific antibodies to monkeypox virus, but it has also been used for the identification of specific antibody of vaccinia and variola viruses. The RIA test is first used to screen sera with titers greater than 500. After a high-titered serum is adsorbed by normal CAM and vaccinia virus antigen, the residual antibody in the serum is reacted against CAM, vaccinia virus, monkeypox virus, and variola virus. The reaction pattern produced determines whether the serum contains specific antibody to monkeypox, vaccinia, or variola viruses. The test often cannot identify a specific antibody when a blood specimen was collected too early (3 to 4 weeks) after onset of a disease. Also, the test sometimes cannot identify specific antibodies to monkeypox virus if the patient with monkeypox was vaccinated against smallpox at some time before contracting the illness. The test has been useful in identifying cases of monkeypox in a serologic survey and in a situation in which no virologic specimens could be collected.

29.6.7 Complement-Fixation Test and Agar Gel Precipitation for Antibody Assay

The use of these two tests has been discontinued at CDC.

REFERENCES

Baxby, D. 1977. Is cowpox misnamed? A review of ten human cases, 1977. Br. Med. J. 1:1379–1380.

Baxby, D. 1981. Jenner's Smallpox Vaccine. The Riddle of the Origin of Vaccinia Virus. London: Heinemann Educational Books.

Baxby, D., Ashton, D.G., Jones, D., Thomsett, L.R., and Denham, E.M.H. 1979. Cowpox virus infection in unusual hosts. Vet. Rec. 109:175.

Baxby, D., and Hill, B.J. 1971. Characteristics of a new poxvirus isolated from Indian buffaloes. Archiv. ges. Virusforsch. 35:70–79.

Becker, F.T. 1940. Milker's nodules. J.A.M.A. 115:2140–2144.

Boulter, E.A. 1957. The titration of vaccinial neutralizing antibody on chorioallantoic membranes. J. Hyg. 55:50–52.

Breman, J.G., Kalisa-Ruti, Steniowski, M.V., Zanotto, E., Gromyko, A.I., and Arita, I. 1980. Human monkeypox 1970–1979. Bull. WHO 58:165–182.

Brown, S.T., Nalley, J.F., and Kraus, S.J. 1981. Molluscum contagiosum. Sex. Transm. Dis. 8:227–233.

Downie, A.W. 1939. A study of the lesions produced experimentally by cowpox virus. J. Pathol. Bacteriol. 48:361–379.

Downie, A.W., Taylor-Robinson, C.H., Count, A.E., Nelson, G.S., Mason-Bahr, P.E.C., and Matthews, T.C.H. 1971. Tanapox: a new disease caused by a poxvirus. Br. Med. J. 1:363–368.

Dumbell, K.R., and Archard, L.C. 1980. Comparison of whitepox (b) mutants of monkeypox virus with parental monkeypox and with variola-like viruses isolated from animals. Nature 286:29–32.

Dumbell, K.R., and Kapsenberg, J.G. 1983. Laboratory investigation of two "whitepox" viruses and comparison with two variola strains from southern India. Bull. WHO 60:3281–3287.

Esposito, J.J., Obijeski, J.F., and Nakano, J.H. 1978. Orthopoxvirus DNA: strain differentiation by electrophoresis of restriction endonuclease fragmented virion DNA. Virology 89:53–66.

Hopkins, D.R. 1983. Princes and Peasants: Smallpox in History. Chicago: University of Chicago Press.

Hutchinson, H.D., Ziegler, D.W., Wells, D.E., and Nakano, J.H. 1977. Differentiation of variola, monkeypox and vaccinia antisera by radioimmunoassay. Bull. WHO 55:613–623.

Jenner, E. 1798. An inquiry into the causes and effect of the variolae vaccinae, a disease known by the name of the cow pox. Sampson Low No. 7, Soho, London.

Khodakevich, L., Jezek, Z., and Kinzanka, K. 1986. Isolation of monkeypox virus from wild squirrel infected in nature. Letters to the Editor, Lancet 1:98–99. (No. 8472, 11 January 1986).

Marennikova, S.S., Seluhina, E.M., Maltseva, N.N., Cimiskjan K.L., and Macevic, G.R. 1972. Isolation and properties of the causal agent of a new variola-like disease (monkeypox) in man. Bull. WHO 46:599–661.

Marennikova, S.S., Maltseva, N.N., Korneeva, V.I., and Garanina, V.M. 1975. Pox infection in carnivora of the family Felidae. Acta Virol. 19:260.

Marennikova, S.S., Maltseva, N.N., and Korneeva, V.I. 1976. Pox in giant anteater due to agent similar to cowpox virus. Br. Vet. J. 132:182–186.

Mardsen, J.P. 1948. Variola minor, a personal analysis of 13,686 cases. Bull. Hyg. 30:735–746.

Mathew, T. 1976. Comparative studies on the propagation of poxvirus in chick embryo with special reference of buffalopox virus. Kerala J. Vet. Sci. 1:48–56.

Nakano, J.H. 1977. Comparative diagnosis of poxvirus diseases. In E. Kurstak and C. Kurstak (eds.), Comparative Diagnosis of Viral Diseases. Vol. I, Part A. Human and Related Viruses. New York: Academic Press, 289–339.

Nakano, J.H. 1979. Poxviruses. In E.H. Lennette and N.J. Schmidt (eds.), Diagnostic Procedures for Viral, Rickettsial and Chlamydial Infections, 5th Ed. Washington, D.C.: American Public Health Association, 257–308.

Nakano, J.H. 1982. Human poxvirus diseases and laboratory diagnosis. In L.M. de la Maza and E.M. Peterson (eds.), Medical Virology. New York: Elsevier Science Publishing, 125–147.

Payne, L.G., and Norrby, E. 1976. Presence of haemagglutinin in the envelope of extracellular vaccinia virus particles. J. Gen. Virol. 32:63–72.

Payne, L.G. 1979. Identification of the vaccinia hemagglutinin polypeptide from a cell system yielding large amounts of extracellular enveloped virus. J. Virol. 31:147–155.

Singh, I.P., and Singh, S.B. 1967. Isolation and characterization of the aetiologic agent of buffalopox. J. Res. Ludhiana 4:440–448.

Smadel, J.E. 1948. Smallpox and vaccinia. pp. 314–336. In T.M. Rivers (ed.) Viral and Rickettsial Infections of Man. Philadelphia: J.B. Lippincott.

Thomsett, L.R., Baxby, D., Denham, E.M.H. 1978. Cowpox in domestic cats. Vet. Rec. 108:567.

Torres, C.M., and de Castro Teixeira, J. 1935. Culture du virus l'alastrim sus les membranes de l'embryon de poulet. Compt. Rend. Seances Soc. Biol. Filiales 118:1023–1024.

Tripathy, D.N., Hanson, L.E., and Crandell, R.A. 1981. Poxviruses of veterinary importance: Diagnosis of infections. In E. Kurstak and C. Kurstak (eds.), Comparative Diagnosis of Viral Diseases. Vol. III. Vertebrate

Animals and Related Viruses, Part A. DNA Viruses. New York: Academic Press, 268–348.

Voller, A., Bidwell, D., and Bartlett, A. 1976. Microplate enzyme immunoassays for the immunodiagnosis of virus infection. In N.E. Rose and H. Friedman (eds.), Manual of Clinical Immunology. Washington, D.C.: American Society of Microbiology, 506–512.

Von Magnus, P., Andersen, E.K., Petersen, K.B., and Birch-Anderson A. 1959. A pox-like disease in cynomolgus monkeys. Acta Pathol. Microbiol. Scand. 46:156–176.

World Health Organization. 1972. Expert Committee on Smallpox Eradication. WHO Tech. Rep. Ser. No. 493:24–27.

Ziegler, D.W., Hutchinson, H.D., Koplan, J.P., and Nakano, J.H. 1975. Detection by radioimmunoassay of antibodies in human smallpox patients and vaccinees. J. Clin Microbiol. 1:311–317.

Zwart, P., Gispen, R., and Peters, J.C. 1971. Cowpox in okapis: Okapia Johnsloni at Rotterdam Zoo. Br. Vet. J. 127:20–24.

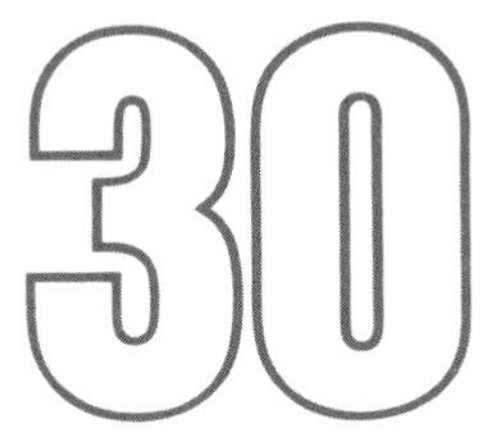

Measles, Mumps, and Rubella

David A. Fuccillo and John L. Sever

30.1 MEASLES VIRUS

Measles, also called rubeola, has been recognized as a disease for almost 3000 years. Man is the only known natural host. Enders and Peebles, in 1954, reported the successful isolation of measles virus in human and rhesus monkey kidney tissue cultures. Vaccines became available in 1963. One of the vaccines was inactivated by formalin and the other was live attenuated vaccine (Schwartz, 1962). The killed vaccine eventually proved less effective and children who received this material were at risk of developing an atypical severe form of the disease when subsequent exposure to live measles virus occurred. A further attenuated variant of the live Edmonston B vaccine was introduced in 1968 and is widely used today. Initially, vaccine was administered to children over 9 months of age but it became apparent that the young infants did not mount an adequate immune response. It is now recommended that vaccination be performed at age 15 months.

30.1.1 Characteristics of Virus

Measles virus is an RNA virus and is classified as a member of the paramyxovirus group. The measles virion is spherical with a diameter ranging from 120–250 nm (Hall and Martin, 1974). It has an envelope composed of glycoproteins and lipids and bears short surface projections. The envelope encloses an elongated helical nucleocapsid in which protein units are spirally arranged around the nucleic acid.

30.1.2 Clinical Aspects

Measles is a highly contagious, acute biphasic disease with a prominent prodrome preceding the exanthemic phase. Measles is spread through direct contact with infected droplets originating from a cough or sneeze or from contaminated fomites. Susceptible persons intimately exposed to a measles patient have a 99% chance of acquiring the disease. Prior to the use of vaccines, more than 90% of the population had measles before 10 years of age. After an incubation period of 9 to 11 days there is an initial 3- to 4-day prodromal period characterized by fever, cough, coryza, and conjunctivitis. The incubation period in adults may last up to 3 weeks. Fever occurs 24 hours or less before other symptoms appear and these increase in severity reaching a peak with the appearance of the rash on the fourth to fifth day.

Bluish-white lesions with a red halo,

Koplik spots, will appear on the buccal or labial mucosa in 50 to 90% of the cases, 2 to 3 days after the onset of the prodrome. These lesions are small, irregular red spots with a bluish-white speck in the center. This lesion, pathognomonic for measles (Koplik, 1896), is on the inner lip or opposite the lower molars. They may be few in number early in the prodrome; however, they increase rapidly to spread over the entire surface of the mucous membranes. A lesion somewhat similar in appearance to Koplik spots has been reported with ECHO-9 (coxsackie A23) and coxsackie A16 and A9 virus infections. The rash is first evident behind the ears or on the forehead. The lesions are red macules, 1–2 mm in diameter, which become maculopapules over the next 3 days. By the end of the second day, the trunk and upper extremities are covered with rash and by the third day the lower extremities are affected. The rash resolves in the same sequence, lasting approximately 6 days. The lesions turn brown and persist for 7 to 10 days and then are followed by a fine desquamation.

The most frequent complication of measles involves infections of the lower respiratory tract. Croupe, bronchitis, bronchiolitis, and, rarely, giant cell interstitial pneumonia may occur. Otitis media is a common bacterial complication of measles. Prior to the advent of antibiotics these complications contributed significantly to a high number of fatalities and significant morbidity. Excluding pneumonia and otitis media, the most frequent serious complication of measles is postinfectious encephalitis. It occurs in 0.1–0.2% of measles patients during any stage of the illness, although it is most common 2 to 7 days after the onset of the exanthem. Mortality is about 30%, with the same proportion of patients showing other permanent residual damage. Other complications include thrombocytopenic purpura, appendicitis, myocarditis, and mesenteric lymphadenitis (Gershon and Krupman, 1979).

Subacute sclerosing panencephalitis (SSPE), also called Dawson's encephalitis, is a late or "slow virus" complication of measles. The incidence of SSPE was approximately 1:100,000 to 1,000,000 cases, but after the advent of the vaccine there was a dramatic decrease in the frequency of the disease. Subacute sclerosing panencephalitis is a progressive, invariably fatal encephalopathy characterized by personality changes, mental deterioration, involuntary movements, muscular rigidity, and death. It usually begins 4 to 17 years after the patient has recovered from measles. Measles virus has been successfully isolated from brain and lymphoid tissues of SSPE patients (Barbosa et al., 1969; 1971).

Transplacental infections have been associated with some fetal effects. There is an apparent increased frequency of abortions and stillbirths. The teratogenic potential of gestational measles has been neither proved nor refuted (Fuccillo and Sever, 1973; South and Alford, 1980).

Atypical measles can occur in children previously vaccinated with killed measles virus vaccines when they become infected with wild measles (Fulginiti et al., 1967). The disease is characterized by fever, a prodromal period, and subsequent rash. During the prodrome, patients may experience malaise, myalgia, headache, nausea, and vomiting. Symptoms usually last for 2 to 3 days and frequently individuals have a sore throat, conjunctivitis, and photophobia along with a nonproductive cough and pneumonia. Chest x-rays often show patchy infiltrates. The rash produced is different from that of typical measles. It can be a mixture of macules, papules, vesicles, and pustules. Frequently, there is a petechial component which begins at the distal extremities and concentrates on the hands, wrists, ankles, and feet and then progresses centrally toward the trunk. Koplik spots have not been reported and the face is rarely involved. Edema often occurs in extremities. The appearance of atypical measles may be confused with Rocky Mountain Spotted Fever.

30.1.3 Laboratory Diagnosis

Diagnosis of a typical case of measles can be made based upon clinical symptoms. However, demonstration of the virus or seroconversion against the virus is necessary to confirm the diagnosis. Best results for isolation of the virus are achieved from specimens taken within the first few days of illness. Measles virus may be isolated from blood, throat, conjunctivae, and urine. Swabs of nasopharyngeal secretions are best for isolation during first 4 to 5 days. Virus may be present in urinary sediment for as long as 1 week. Measles virus grows very slowly in primary human kidney, monkey kidney, and human amnion cell cultures. The virus produces characteristic multinucleate giant cells after 7 to 10 days incubation (Katz and Enders, 1969). Uninoculated, control cultures should always be compared with inoculated cultures. Suspicious cultures can then be tested by hemadsorption with monkey red blood cells. Hemadsorption with monkey red blood cells but not with nonsimian erythrocytes is useful to distinguish the virus from other paramyxoviruses, especially mumps. The most specific and rapid method for identification is the detection of measles antigen present in the multinucleated giant cells using direct immunofluorescence (Nommensen and Dekkers, 1981).

Measles virus is relatively labile to heat. Infectivity is decreased one half when kept at 37°C for 2 hours and complete inactivation can be accomplished at 56°C for 30 minutes (DeJong and Winkler, 1964). It can also be inactivated after exposure to ultraviolet and visible light. The virus is stable from pH 5 to 10.5 with an optimum at pH 7 (Musser and Underwood, 1960). Measles virus is preserved quite readily by storage at −70°C in a suspension medium containing protein. Infectivity may also be maintained at 4°C for several months in a protein containing media or in the presence of stabilizers (McAleer et al., 1980).

The most practical method to make a laboratory diagnosis of measles is to obtain acute and convalescent phase sera and demonstrate a greater than four-fold rise in specific antibody. Hemagglutination inhibition (HI), neutralization, and complement-fixation methods have all been employed with the serodiagnosis of measles. Hemagglutination inhibition tests have been the most useful method in the past, having sensitivity and specificity comparable to neutralization. Recently, newer methods such as enzyme-linked immunosorbent assays (ELISA) and radioimmunoassay (RIA) have been developed which have greater sensitivity than the HI test (Boteler et al., 1983; Rice et al., 1983). Antibody appears within 1 to 2 days after onset of rash and titers peak 10 days to 2 weeks later. The presence of specific IgM antibody can be used to diagnose recent infection.

30.1.4 Control and Prevention

Individuals having an illness compatible with measles should be cared for in such a way that contact with other people or patients is minimal. The communicability of measles virus is extremely high. Therefore, any susceptible individuals who had direct face-to-face contact with the infectious individual should obtain prophylactic treatment. Risk, other than face-to-face, is very low and therefore postexposure prophylaxis is unnecessary. Measles vaccination may prove protective if given within 72 hours of exposure (Centers for Disease Control, 1982a). Immune globulin, given within 6 days of exposure, can prevent or modify measles virus infection. It is indicated for susceptible, close contacts of measles patients, particularly if they are under 1 year of age. If immune globulin is used for a child at this age, measles vaccine should be given about 3 months later but at no less than 15 months of age.

After a further attenuated variant of the Edmonston B vaccine was introduced in 1968, the reported cases of measles took a dramatic downward turn. In 1960, the cumulative total number of cases was 399,852 from week 1 to 35. In 1970 the total was 39,365; in 1981, 2562; in 1982, 1188; and recently in 1983, for the same period, the total number of cases was 1194 (Centers for Disease Control, 1982b). There was hope that 1983 would be the year in which measles was eliminated from the United States but this goal has not been met.

30.2 MUMPS VIRUS

Mumps virus infection was probably first described around the fifth century B.C. by Hippocrates. The name "mumps" is thought to be derived from the mumbling speech of patients afflicted with this disease. The etiologic agent was identified as virus by Johnson and Goodpasture in 1934. The virus was first isolated in the amniotic cavity of chick embryo in 1945 (Habel, 1945). Buynak and Hilleman, in 1966, developed the first successful live attenuated vaccine by passage of the virus in chick embryo cell cultures (Buynak and Hilleman, 1966).

30.2.1 Characteristics of Virus

Mumps virus is a member of the paramyxovirus group. Virus particles range in size from 85–300 nm in diameter (Cantell, 1961). The virus has a single strand of RNA, contains a nucleoprotein core, and has an outer viral envelope. The envelope contains a hemagglutinin, neuraminidase, and a hemolysin.

30.2.2 Clinical Aspects

Man is the only known host and reservoir of the virus. The infection can be either clinically apparent or subclinical. Infection is endemic worldwide, usually affecting the 6 to 10-year-old age group; it occurs predominantly in the spring. The virus usually causes an uncomplicated infection of the salivary glands (e.g., parotid glands) and is manifested by an enlargement of these organs. The parotitis is sudden and may not be preceded by any prodromal symptoms. Swelling of the glands reaches a maximum after 48 hours and they usually remain swollen for a period of 7 to 10 days. There may be little or no increase in body temperature. Approximately 20–30% of postpubertal men acquiring mumps develop epididymo-orchitis between 1 and 2 weeks following the parotitis. Sterility is not a common sequela of infection since only 1–12% of the cases are bilateral.

Another complication of mumps virus infection is meningoencephalitis which has an incidence from 5–10%. Encephalitis is one central nervous system (CNS) complication but mumps virus infection has been linked to other rare CNS complications such as transverse myelitis, cerebellar ataxia, poliomyelitis-like syndrome, and Guillain-Barré syndrome. About 5% of adult females with mumps may develop oophoritis. Other complications such as pancreatitis, thyroiditis, neuritis, inflammation of the eye, and inner ear infection can be encountered. There have been reports of diabetes mellitus being associated with mumps but at present information is inconclusive (Sultz et al., 1975). Other reports that intrauterine mumps can lead to endocardial fibroelastosis have not been confirmed (St. Geme et al., 1966).

Seroepidemiologic surveys have indicated that 80–90% of adults have evidence of prior mumps infection. Mumps is transmitted by saliva containing the virus either by direct transfer, air-suspended droplets, or by recently contaminated fomites. Approximately 85% of susceptible contacts can become infected when first exposed and 25–40% of the infections may be asymptomatic. The virus is thought to multiply in the upper respiratory tract, then invade the bloodstream, and finally affect the salivary glands and other organs. About 18 days elapse between the time of exposure and the first detectable enlargement of the salivary glands. The incubation period may range from 14 to 24 days. The period of communicability can be from 7 days before the salivary gland involvement until 9 days thereafter. The virus is also excreted in the urine for as long as 14 days after onset of illness.

30.2.3 Laboratory Diagnosis

The diagnosis of mumps infection is quite simple when typical parotitis is produced. However, diagnosis by viral isolation or serologic techniques is most useful when the patient presents with an atypical or asymptomatic infection. A viral isolation from the spinal fluid, blood, saliva, and urine confirms the diagnosis of recent mumps infection. Primary monkey kidney cell cultures are the most sensitive hosts for isolating the virus. These may be obtained from rhesus or cynomolgus ceropithecus monkeys. Continuous human cell lines such as HeLa and primary cell cultures of human amnion or human embryonic kidney can also be used for the growth of the virus (Hopps and Parkman, 1979). This virus produces a characteristic cytopathic effect (CPE) with large syncytia. However, some strains may not produce this CPE, and therefore hemadsorption with guinea pig erythrocytes should be performed for identification. Rapid identification of mumps isolates from cell cultures can be accomplished by immunofluorescence staining (Lennette et al., 1975).

The virus is relatively stable and infectivity is changed very little upon storage at 4°C for several days. The virus is stable when stored for a number of weeks at −20°C and for months at −70°C. Stability can be increased with the addition of protein such as 1% bovine albumin, 0.5% gelatin, or 2% serum. Swabs taken for isolation studies can be placed in tubes containing a suitable medium with a protein stabilizer and with antibiotics. Other specimens such as urine and cerebral spinal fluid can be placed on ice where they may be kept for a few hours before inoculation. For longer storage, specimens should be maintained at −70°C (Cantell, 1961).

Serologic diagnosis of mumps infection can be very important, especially in those cases of meningitis or encephalitis that occur in the absence of parotitis. Serologic methods for diagnosis of mumps infections include complement fixation (CF), HI, neutralization, and, more recently, an ELISA test. The CF and HI procedures are of approximately equal sensitivity. The neutralization test is generally more sensitive in detecting antibodies than is the HI test. Recent studies have shown the ELISA test to be more sensitive detecting antibodies in acute phase sera than the CF and more sensitive than HI for the detection of low levels of antibodies (Leinikki et al., 1979; Popow-Kraupp, 1981). Other studies have shown that the use of ELISA to detect IgM is particularly suitable for early diagnosis of mumps infection with one serum specimen (Nicolai-Scholten et al., 1980; Ukkonen et al., 1980).

30.2.4 Control and Prevention

More than 59 million doses of vaccine have been distributed in the United States since licensure in December 1967. Reported cases of mumps in 1967 were 185,691 and in 1982 were 5270. This is a 97% decrease. It appears that incidence rates have declined in all age groups by more than 90%. During this 15-year period, the highest reported rate occurred in 5 to 9 year olds followed by children under 5 years old. These two groups accounted for over 30% of all reported cases. Over the 5 years from 1978 to 1982, this group accounted for about 50% of recorded cases and the risk of infection for 10 to 14 year olds surpassed that for children under 5 years of age. This age-specific change in risk of infection to mumps is similar to those noted for measles and rubella and results from a vaccination policy oriented toward preschool and elementary school children. Individuals who are neither vaccinated nor infected at a young age eventually will be exposed at an older age and subsequently come down with the infection (Centers for Disease Control, 1983c).

Mumps vaccine is considered one of the safest of the childhood immunizing agents but 19 states still do not require proof of mumps infection or immunization as a condition for school entry (Centers for Disease Control, 1982c). Despite this, the 1982–1983 school entrance survey indicated nationwide mumps vaccine coverage of 95%. This recent increase is attributed to the use of combined measles, mumps, and rubella vaccine. This preparation has been the vaccine of choice for the routine immunization of children 15 months of age or older. A benefit cost analysis was performed on a mumps vaccination program in which mumps was given as part of a measles, mumps, rubella combination. It was estimated that costs associated with mumps were reduced by more than 86% with a benefit-cost ratio of 7 to 1 using reported incidence rates. When rates were corrected for underreporting, the benefit-to-cost ratio was 39 to 1. Since outbreaks of mumps are still potentially possible in the unvaccinated population, considerable medical and economic savings can be realized by including mumps in the immunization process as part of the state compulsory school immunization laws (Koplan and Preblud, 1982).

30.3 RUBELLA

Rubella was first recognized in 1815 when Maton described "a rash liable to be mistaken for Scarletina" (Maton, 1815). More than a century later, Gregg noted a high incidence of congenital malformations, mostly cataracts, and reported the association of these abnormalities with rubella contracted in the first trimester of the pregnancy (Gregg, 1941). In 1962, several groups (Parkman et al., 1962; Weller and Neva, 1962) reported the successful isolation of the rubella virus.

Several live attenuated vaccines were developed and the first was licensed in 1969 (Meyer et al., 1966; Stokes et al., 1967). The HPV-77 and Cendehill vaccine strains were introduced first, whereas the presently used vaccine, Wistar RA 27/3, was licensed in 1979 (Plotkin et al., 1967; Ingalls et al., 1970).

30.3.1 Characteristics of Virus

Acquired rubella, also known as German or 3-day measles, is caused by a double-layered, single-stranded RNA virus classified as a togavirus. The virion is a spherical particle with a diameter of 60–70 nm having surface projections 6 nm in length. It has a dense central nucleoid of about 30 nm surrounded by an envelope acquired when virus buds out through the cytoplasmic or plasma membrane of cell (Herrmann, 1979).

30.3.2 Clinical Aspects

The disease is quite contagious and is usually transmitted by respiratory secretions. Before rubella immunization, the disease occurred most commonly in childhood although it affected adults as well. The clinical manifestations of rubella are usually mild; in fact, they may be absent in as much as one third of the individuals infected. Catarrhal symptoms are first to be seen, followed by lymphadenopathy involving the posterior auricular, posterior cervical, and postoccipital lymph nodes, and finally the emergence of a maculopapular rash on the face, then on the neck and trunk. A low grade fever is usually present and some individuals may have transient arthralgia and arthritis, which is usually more severe in adults.

The signs and symptoms of rubella may be difficult to distinguish from other rash-associated diseases. For this reason neither the clinical presentation of apparent rubella nor the absence of classic rubella signs and symptoms can be regarded as definitively diagnostic. Acute rubella lasts from 3 to 5 days and generally requires little treatment. The incubation period for rubella varies from 10 to 21 days, with 12 to 14 days being typical. Infected individuals are usually contagious for 12 to 15 days, beginning 5 to 7 days before the appearance of a rash. Acquired rubella infection almost always confers permanent immunity to the disease.

Acquired rubella in a pregnant woman is no more intrinsically dangerous than it is in any other patient but the potential damage to the fetus can be quite disastrous. The point in the gestation cycle at which maternal rubella occurs greatly influences the severity and risk of the congenital rubella syndrome (CRS). If rubella occurs in the first trimester of pregnancy there is a 25% chance of fetal anomalies; the risk is 50% when infection occurs in the first month. The risk becomes less than 10% by the third month and drops approximately to 6% when infection occurs in the fourth or fifth month. After the fifth month, although fetal infections continue to occur, rubella poses no known threat to the fetus (Sever, 1983). Fetal infection is a result of placental infection during the viremic phase of the disease. One study of mortality associated with CRS showed a 13% death rate in the first 18 months of life (Cooper, 1968). Another study showed a 20% mortality rate in the first 18 months of life (Desmond et al., 1967).

The extent of rubella-produced anomalies varies from one neonate to another. Fetal abnormalities include stillbirths, cataracts, cardiovascular defects, mental retardation, microcephaly, encephalitis, hepatosplenomegaly, bone defects, and growth retardation. Severely affected children are likely to have multiple organ involvment. In neonates with CRS, thrombocytopenia purpura is common, as is low birth weight and failure to thrive.

Chronic infection may persist for months or even years. Rubella immunity develops in most children who have had congenital rubella. However, in late childhood, about a third of these children lose antibody and become susceptible to acquired rubella which, if it occurs, follows a typical benign course (Hardy et al., 1969).

In the past few years late manifestations of CRS have been recognized. These are disabilities that do not make their appearance until years after birth. One of the first disabilities to be found was insulin-dependent diabetes mellitus (Menser et al., 1967; Plotkin and Kaye, 1970). Additional studies revealed a 20% incidence of latent or overt diabetes in a study involving 50 older subjects from the 1964 rubella epidemic (Forrest et al., 1971). All patients had a prenatal history of rubella before the 16th week of gestation. All were deaf and most had other rubella-associated defects of the eyes and heart. In a follow-up study, 40% of these CRS patients had developed evidence of overt or latent diabetes (Menser et al., 1974). Other endocrine disorders have been seen in small numbers of survivors of congenital rubella. They include hypothyroidism (Ziring et al., 1975; Ziring et al., 1977), hyperthyroidism (Floret et al., 1980), hypoadrenalism (Ziring et al., 1977), and a growth hormone deficiency (Preece et al., 1977).

Ocular consequences of CRS are observed during and after the neonatal period. One study describes 13 patients having glaucoma 3 to 22 years after birth. All had cataracts early in life which had either been removed surgically or had resolved spontaneously. Another group of patients were found to have keratic precipitates without other evidence of acute ocular inflammation (Boger, 1980; 1981).

Another disability associated with congenital rubella has been bilateral hearing loss. There has been one report of a CRS child who had numerous audiograms, developed normal speech patterns, and attended to everyday activities until age 10, when signs of progressive deafness began to appear (Desmond et al., 1978).

The last disability associated with CRS is progressive rubella panencephalitis (PRP) (Townsend et al., 1976). This is a "slow virus" manifestation of rubella that is similar to SSPE which is due to measles. Progressive rubella panencephalitis usually appears during the second decade of life. Progressive deterioration of intellectual and motor function occurs with dementia close to the time of death. There is an intense immune response against rubella antigens and high titers of rubella antibody are present in both serum and cerebrospinal fluid (Weller et al., 1964). Virus has been recovered from brain by rescue techniques (Wolinsky, 1978). The pathologic findings are similar to SSPE but without the inclusions and with perivascular deposits (Rosenberg et al., 1981).

Progressive rubella panencephalitis has now been found in two patients with postnatal rubella infection and in nine cases of CRS (Lebon and Lyon, 1974; Weil et al., 1975; Townsend et al., 1975; Wolinsky, 1978). No correlation can be made between the occurrence of PRP and the presence of rubella-associated defects or the severity of neonatal infection.

The pathology produced with congenital rubella appears to result from a chronic viral infection with an inhibition of cell multiplication at critical points in organogenesis. This causes the hypoplastic organ development and other characteristic structural defects seen with this disease (Rawls and Melnick, 1966). The immune response may also contribute to permanent damage in the developing child either by an impaired immunity or by inflicting damage through inflammatory mechanisms (Fuccillo et al., 1974; Rosenberg et al., 1981).

30.3.3 Laboratory Diagnosis

Rubella virus can be readily isolated from a variety of specimens obtained from patients with congenital or acquired infection if care is taken in obtaining and storing specimens. Specimen collection should be as early as possible after the person becomes ill, preferably within 3 to 4 days after symptoms appear. Specimens to be inoculated within 48 hours should be stored at 4°C and transported on ice. If inoculation is delayed, the specimen should be frozen at −70°C immediately after collection. Most body fluids and tissues such as placental or aborted material contain the virus. Nasopharyngeal washings or throat swabs placed in a protein medium containing antibiotics and protein media are particularly useful for the detection of the agent.

Basically, there are two techniques used to detect the isolation of rubella virus—the direct technique and the indirect interference method. Weller and Neva (1962) first introduced the direct method when they described the CPE of rubella virus in primary cultures of human amnion cells (1962). The spectrum of tissue culture systems capable of supporting the growth of rubella with characteristic CPE after adaptation is quite large. These include the continuous rabbit kidney, RK13, LLC-RK-1, vervet monkey kidney, Vero, GMK-AH-1 and BSC-1, rabbit cornea (SIRC), baby hamster kidney, BHK-21, and a number of others. In primary cells, rubella virus very seldom produces CPE and when it is present the changes are slow to appear and difficult to detect (Herrmann, 1979).

The indirect technique is based on the ability of rubella virus to interfere with the replication of enteroviruses such as echo 11 or coxsackie A9 in primary African green monkey kidney. A number of other superinfecting viruses and a variety of cell cultures can also be used to detect rubella propagation (Parkman et al., 1964). Each system has been found to have its advantages and disadvantages, but the interference technique in primary African green monkey kidney with a coxsackie A9 challenge has proven to be a very sensitive technique for detecting the recovery of virus from specimens (Schiff and Sever, 1966).

The variability of the signs and symptoms of rubella and their similarity to those of various other conditions make a clinical history of rubella unreliable and serologic testing essential. Antibody tests are used today to determine immune status and to diagnose rubella infection. These tests have been available for more than 15 years. Hemagglutination inhibition was the first widely used test and is the standard against which other rubella screening and diagnostic tests are measured (Stewart et al., 1967). Other screening assays include neutralization, CF, passive hemagglutination, latex agglutination, and, more recently, ELISA (Vejtorp, 1983). The ELISA test is very sensitive and, in fact, can detect antibody when HI antibody was not found. It was shown that a small group of individuals who initially seroconverted after vaccination and then lost detectable HI antibody still retained antibody when tested by the ELISA method (Buimovici-Klein et al., 1980; Best et al., 1980). Therefore, the presence of any detectable level of rubella antibody or a history of rubella vaccination is accepted presumptive evidence of immunity. To make an accurate diagnosis using serologic testing requires the utilization of tests in the proper relationship to the progress of the disease and, specifically, to the emergence, increase, and reduction of IgG and IgM antibodies. An individual infected with rubella develops both IgG and IgM antibodies along with the appearance of clinical symptoms when they are present. These IgG antibodies increase rapidly for the next 7 to 21 days. IgG levels off, and then remains present and indefinitely protective. Therefore, detection of IgG antibody is useful in indicating immunity but recent rubella infection can only be documented in cases where paired serum specimens, drawn several weeks apart, show a four-fold or greater increase in IgG antibody. Both serum specimens should be run at the same time to eliminate test variation.

IgM antibodies become detectable a few days after onset of symptoms and are at their peak from 8 to 21 days afterward. During the next 4 to 5 weeks, IgM drops off until it is no longer present. The presence of IgM antibody on a single sample thus indicates a recent rubella infection and, in most cases, points to an infection that has occurred within the preceding month (Meurman, 1978; Leinikki et al., 1978).

30.3.4 Control and Prevention

Rubella occurred in epidemic proportions at 6- to 9-year intervals before the widespread use of the rubella vaccine. More than 20,000 cases of congenital rubella syndrome and an unknown number of stillbirths occurred in the U.S. as a result of the 1964 epidemic. This unfortunate event stimulated the quest for an effective rubella vaccine. Three strains of live, attenuated rubella vaccine virus were developed and the first was licensed for use in 1969. Since that time there have been no further rubella epidemics in the United States.

The vaccine presently used is Wistar RA 27/3 (Ingalls et al., 1970) which was licensed in 1979. This vaccine has been effective in inducing immunity in 95% or more of vaccinees. Rash and lymphadenopathy in children and arthralgia in adults are side effects but they are not disabling. Less than 1% of vaccinees report arthritis. There is no indication of increased risk of these reactions in immune patients receiving the vaccine. Rubella vaccine virus has been isolated from products of conception but there has been no demonstration of congenital rubella syndrome (Banatvala et al., 1981). The vaccine is not recommended by Centers for Disease Control (1981; 1983a) to be given to pregnant women. Vaccine is available in monovalent preparations and in combination with measles and mumps. Doctors have been encouraged to use the triple combination for routine vaccination of all children at 15 months of age.

The incidence of reported rubella has fluctuated slightly over the past several years, although there has been a dramatic downward trend for most of the United States. A recent review of the data for the period January 1, 1980 to September 24, 1983 indicates that the final rubella incidence rate for 1983 should be at an all-time low. For 1980 there was a total of 3904 cases of rubella reported, which represents an incidence of 1.7 cases per 100,000 of population. In 1981, the incidence was 0.9 per 100,000, the lowest reported since rubella became a notifiable disease in 1966. In 1982, the incidence was 1.0 per 100,000 and during the first 38 weeks of 1983, 791 cases reported; this is a 61% decrease from the number reported during the same period in 1982. The incidence rate of rubella in children under 15 years of age decreased from 1980 to 1983 while children under 5 years of age still had the highest overall incidence (Centers for Disease Control, 1983b).

According to the National Congenital Rubella Syndrome Registry the incidence rates of confirmed congenital rubella have declined since 1979. Fifty-five cases were reported in 1979, 14 in 1980, and nine in both 1981 and 1982. In 1983 there were three cases reported. The goal of rubella vaccination programs has been to prevent congenital rubella infection. Dramatic results have been accomplished; however, congenital rubella still continues at a very low incidence. The vaccination strategy of doctors in the United States in 1969 was aimed at controlling rubella in preschool and young school-aged children, the known reservoirs for rubella transmission. If it were possible to control disease in this population, exposure of susceptible pregnant females to rubella virus would also be decreased. This approach resulted in dramatic declines as previously mentioned. However, this vaccination strategy had less effect on rubella incidence in persons 15 and older. Approximately 10–20% of this population continues to be susceptible, similar to that during prevaccine years. Now increased efforts are being made to immunize susceptible junior and senior high school students and to enforce rubella immunization requirements for school entry. Military recruits are also receiving rubella vaccine. Also, physicians are encouraged to test all women of childbearing age and to immunize susceptible women when not pregnant. Until such time that the susceptibility rate of postpubertal women is effectively lowered, congenital rubella will continue to occur. It has been estimated that a child with congenital rubella causes an average lifetime expenditure of $221,600 (Centers for Disease Control, 1984).

REFERENCES

Banatvala, J.E., O'Shea, S., Best, J.M., Nicholls, M.V., and Cooper, K. 1981. Transmission of RA27/3 rubella vaccine strain to products of conception (letter). Lancet i:392.

Barbosa, L.H., Fuccillo, D.A., Sever, J.L., and Zeman, W. 1969. Subacute sclerosing panencephalitis: Isolation of measles virus from a brain biopsy. Nature 221:974.

Barbosa, H.L., Hamilton, R., Wiltig, B., Fuccillo, D.A., Sever, J.L., and Vernon, M.L. 1971. Subacute sclerosing panencephalitis: Isolation of suppressed measles virus from lymph node biopsies. Science 173:840–841.

Best, J.M., Harcourt, G.C., Druce, A., Palmer, S.J., O'Shea, S., and Banatvala, J.E. 1980. Rubella immunity by four different techniques: Result of challenge studies. J. Med. Virol. 5:239–247.

Boger, W.P., 3d. 1980. Late ocular complications in congenital rubella syndrome. Ophthalmology (Rochester) 87:1244–1252.

Boger, W.P., 3d. 1981. Spontaneous absorption of the lens in the congenital rubella syndrome. Am. J. Ophthalmol. 99:433–434.

Boteler, W.L., Luipersheck, P.M., Fuccillo, D.A., and O'Beirne, A.J. 1983. Enzyme linked immunosorbent assay for detection of measles antibody. J. Clin. Microbiol. 17:814–818.

Buimovici-Klein, E., O'Beirne, A.J., Millian, S.J., and Cooper, L.Z. 1980. Low level of rubella immunity detected by ELISA and specific lymphocyte transformation. Arch. Virol. 66:321–327.

Buynak, E.B., and Hilleman, M.R. 1966. Live attenuated mumps virus vaccine. Proc. Soc. Exp. Biol. Med. 123:768–775.

Cantell, K. 1961. Mumps virus. Adv. Virus Res. 8:123–164.

Centers for Disease Control. 1981. Rubella Prevention. M.M.W.R. 30:37–47.

Centers for Disease Control. 1982a. Measles prevention. M.M.W.R. 31:217–24, 229–31.

Centers for Disease Control. 1982b. Countdown toward the elimination of measles in the U.S. M.M.W.R. 31:447–478.

Centers for Disease Control. 1982c. Mumps vaccine. M.M.W.R. 31:617–620.

Centers for Disease Control. 1983a. Rubella vaccination during pregnancy—United States, 1971–1982. M.M.W.R. 32:430–432.

Centers for Disease Control. 1983b. Rubella and congenital rubella—United States, 1980–83. M.M.W.R. 32:505–510.

Centers for Disease Control. 1983c. Efficacy of mumps vaccine. M.M.W.R. 32:391–398.

Centers for Disease Control. 1984. Rubella and congenital rubella—United States 1983. M.M.W.R. 33:237–247.

Cooper, L.Z. 1968. Rubella: A preventable cause of birth defects. In D. Bergsma (ed.), Birth Defects (Original Article Series). Vol 4. Intrauterine Infection. New York: The National Foundation.

DeJong, J.G., and Winkler, K.C. 1964. Survival of measles virus in air. Nature 201:1054–1055.

Desmond, M.M., Wilson, G.S., Melnick, J.L., Singer, D.B., Zion, T.E., Rudolph, A.J., Pineda, R.G., Ziai, M.-H., and Blattner, R.J. 1967. Congenital rubella encephalitis. Course and early sequelae. J. Pediatr. 71:311–331.

Desmond, M.M., Fisher, E.S., Vorderman, A.L., Schaffer, H.G., Andrew, L.P., Zion, T.E., and Catlin, F.I. 1978. The longitudinal course of congenital rubella encephalitis in non-retarded children. J. Pediatr. 93:584–591.

Enders, J.F., and Peebles, T.C. 1954. Propagation in tissue cultures of cytopathogenic agents from patients with measles 1954. Proc. Soc. Exp. Biol. Med. 86:277–286.

Floret, D., Rosenberg, D., Hage, G.N., and Monnet, P. 1980. Case report: Hyperthyroidism, diabetes mellitus and the congenital rubella syndrome. Acta Paediatr. Scand. 69:250–261.

Forrest, J.M., Menser, M.A., and Burgess, J.A. 1971. High frequency of diabetes mellitus in young adults with congenital rubella. Lancet i:332–334.

Fuccillo, D.A., and Sever, J.L. 1973. Viral teratology. Bact. Rev. 37:19–31.

Fuccillo, D.A., Steele, R.W., Henson, S.A., Vincent, M.M., Hardy, J.B., and Bellanti, J.A. 1974. Impaired cellular immunity to rubella virus in congenital rubella. Infect. Immunology 9:81–84.

Fulginiti, V.A., Eller, J.J., Downie, A.W., and Kempe, C.H. 1967. Altered reactivity to measles virus. J.A.M.A. 202:1075–1080.

Gershon, A., and Krupman, S. 1979. Measles Virus. In E.H. Lennette and N.J. Schmidt (eds.), Diagnostic Procedures for Viral, Rickettsial and Chlamydial Infections. Washington, D.C.: American Public Health Association.

Gregg, N. 1941. Congenital cataract following German measles in the mother. Trans. Ophthalmol. Soc. Aust. 3:35–46.

Habel, K. 1945. Cultivation of mumps virus in the developing chick embryo and its application to studies of immunity to mumps in man. Public Health Rep. 60:201–212.

Hall, W.W., and Martin, S.J. 1974. The biochemical and biological characteristics of the surface components of measles virus. J. Gen. Virol. 22:363–374.

Hardy, J.B., Sever, J.L., and Gilkeson, M.R. 1969. Declining antibody titers in children with congenital rubella. J. Pediatr. 75:213–220.

Herrmann, K.L. 1979. Rubella Virus. In E.H. Lennette and N.J. Schmidt (eds.), Diagnostic Procedures for Viral, Rickettsial and Chlamydial Infections. Washington, D.C.: American Public Health Association.

Hopps, H.E., and Parkman, P.D. 1979. In E.H. Lennette and N.J. Schmidt (eds.), Diagnostic Procedures for Viral, Rickettsial and Chlamydial Infections. Washington, D.C.: American Public Health Association.

Ingalls, T.H., Plotkin, S.A., Philbrook, F.R., and Thompson, R.F. 1970. Immunization of school children with rubella (RA 27/3) vaccine. Lancet i:99–101.

Johnson, S.C., and Goodpasture, E.W. 1934. An investigation of the etiology of mumps. J. Exp. Med. 59:1–19.

Katz, S.L., and Enders, J.F. 1969. Measles virus. In E.H. Lennette and N.J. Schmidt (eds.) Diagnostic Procedures for Viral and Rickettsial Infections, 4th ed. New York: American Public Health Association, 504–528.

Koplan, J.P., and Preblud, S.R. 1982. A benefit-cost analysis of mumps vaccine. Am. J. Dis. Child. 136:362–4.

Koplik, H. 1896. The diagnosis of the invasion of measles from a study of the exanthemata as it appears on the buccal mucus membrane. Arch. Pediatr. 13:918–922.

Lebon, P., and Lyon, G. 1974. Non-congenital rubella encephalitis (letter). Lancet, ii:468.

Leinikki, P.O., Shekarchi, I., Dorsett, P., and Sever, J.L. 1978. Determination of virus-specific IgM antibodies by using ELISA: Elimination of false-positive results protein A-Sepharose absorption and subsequent IgM antibody assay. J. Lab. Clin. Med. 92:849–857.

Leinikki, P.O., Shekarchi, I., Tzan, N., Madden, D.L., Sever, J.L., McLean, A., and Hilleman, M.R. 1979. Evaluation of enzyme-linked immunosorbent assay (ELISA) for mumps virus antibodies. Proc. Soc. Exp. Biol. Med. 160:363–367.

Lennette, D.A., Emmons, R.W., and Lennette, E.H. 1975. Rapid diagnoses of mumps virus infections by immunofluorescence methods. J. Clin. Microbiol. 2:81–84.

Maton, W.G. 1815. Some account of a rash liable to be mistaken for scarlatina. Medical Tr. Roy. Coll. Phys. London 5:149.

McAleer, W.J., Markus, H.Z., McLean, A.A., Buynak, E.B., and Hilleman, M.R. 1980. Stability on storage at various temperatures of live measles, mumps and rubella virus vaccines in new stabilizer. J. Biol. Stand. 8:281–287.

Menser, M.A., Dods, L., and Harley, J.D. 1967. A twenty-five year follow-up of congenital rubella. Lancet ii:1347–1350.

Menser, M.A., Forrest, J.M., Honeyman, M.C., and Burgess, J.A. 1974. Diabetes, HLA antigens, and congenital rubella. Lancet ii:1058–1059.

Meurman, O.H. 1978. Persistence of immunoglobulin G and immunoglobulin M antibodies after postnatal rubella infection determined by solid-phase radioimmunoassay. J. Clin. Microbiol. 7:34–38.

Meyer, H.M., Jr., Parkman, P.D., and Panos, T.C. 1966. Attenuated rubella virus: II. Production of an experimental live virus vaccine and clinical trial. N. Engl. J. Med. 275:575–580.

Musser, S.J., and Underwood, G.E. 1960. Studies on measles virus. II Physical properties and inactivation studies of measles virus. J. Immunol. 85:292–297.

Nicolai-Scholten, M.E., Ziegelmaier, R., Behrens, F., and Hopken, W. 1980. The enzyme-linked immunosorbent assay (ELISA) for determination of IgG and IgM antibodies after infection with mumps virus. Med. Microbiol. Immunol. 168:81–90.

Nommensen, F.E., and Dekkers, N.W. 1981. Detection of measles antigen in conjunctival epithelial lesions staining by lissamine green during measles virus infection. J. Med. Virol. 7:157–162.

Parkman, P.D., Buescher, E.L., and Artenstein, M.S. 1962. Recovery of rubella virus from Army recruits. Proc. Soc. Exp. Biol. Med. 111:225–230.

Parkman, P.D., McCown, Mundon, F.K., and Druzd, A.D. 1964. Studies of rubella. I. Properties of the virus. J. Immunol. 93:595–607.

Plotkin, S.A., and Kaye, R. 1970. Diabetes mellitus and congenital rubella. Pediatrics 46:650–651.

Plotkin, S.A., Farquhar, J.D., Katz, S., and Ingalls, T.H. 1967. Discussion of rubella vaccines. Pan. Am. Health Org. Sci. Publ. 147:405–408.

Popow-Kraupp, T. 1981. Enzyme-linked immunosorbent assay (ELISA) for mumps virus antibodies. J. Med. Virol. 8:79–88.

Preece, M.A., Kearney, P.J., and Marshall, W.C. 1977. Growth-hormone deficiency in congenital rubella. Lancet ii:842–844.

Rawls, W.E., and Melnick, J.L. 1966. Rubella

virus carrier cultures derived from congenitally infected infants. J. Exp. Med. 123:795–816.

Rice, G.P.A., Casali, P., and Oldstone, M.B.A. 1983. A new solid-phase enzyme-linked immunosorbent assay for specific antibodies to measles virus. J. Infect. Dis. 147:1055–1059.

Rosenberg, H.S., Oppenheimer, E.H., and Esterly, J.R. 1981. Congenital rubella syndrome: The late effects and their relation to early lesions. Perspectives in Pediatr. Pathol. 6:183–202.

St. Geme, J.W., Jr., Noren, G.R., and Adams, P. Jr. 1966. Proposed embryopathic relation between mumps virus and primary endocardial fibroelastosis. N. Engl. J. Med. 275:339–347.

Schiff, G.M., and Sever, J.L. 1966. Rubella: recent laboratory and clinical advances. Prog. Med. Virol. 8:30–61.

Schwartz, A.J.F. 1962. Preliminary tests of a highly attenuated measles vaccine. Am. J. Dis. Child. 103:386–389.

Sever, J.L. 1983. Virus as teratogens. In E.M. Johnson and D.M. Kochhar (eds.), Handbook of Experimental Pharmacology. Vol. 65. New York: Springer Verlag.

South, M.A., and Alford, C.A. 1980. The immunology of chronic intrauterine infections. In R. Stiehm and V.A. Fulginiti (eds.), Immunologic Disorders in Infants and Children, 2nd ed. Philadelphia: W.B. Saunders, 702–714.

Stewart, G.L., Parkman, P.D., Hopps, H.E., Douglas, R.D., Hamilton, J.P., and Meyer, H.M., Jr. 1967. Rubella-virus hemagglutination inhibition test. N. Engl. J. Med. 276:554–557.

Stokes, J., Jr., Weibel, R.E., Buynak, E.D., and Hilleman, M.R. 1967. Clinical and laboratory tests of Merck strain live attenuated rubella virus vaccine. Pan Am. Health Org. Sci. Publ. 147:402–405.

Sultz, H.A., Hart, B.A., Zielezny, M., and Schlesinger, E.R. 1975. Is mumps virus an etiologic factor in juvenile diabetes mellitus? J. Pediatr. 86:654–656.

Townsend, J.J., Baringer, J.R., Wolinsky, J.A., Malamud, N., Medick, J.P., Panirch, H.S., Scott, R.A.T., Oshiro, L.S., and Cremer, N.E. 1975. Progressive rubella panencephalitis. N. Engl. J. Med. 292:990–993.

Townsend, J.J., Stroop, W.G., Baringer, J.R., Wolinsky, J.H., McKerrow, J.H., and Berg, B.O. 1976. Neuropathy of progressive rubella panencephalitis after childhood rubella. Neurology 32:185–190.

Ukkonen, P., Vaisanen, O., and Penttinen, K. 1980. Enzyme-linked immunosorbent assay for mumps and parainfluenza type 1 immunoglobulin G and immunoglobulin M antibodies. J. Clin. Microbiol. 11:319–323.

Vejtorp, M. 1983. Serodiagnosis of postnatal rubella. Dan. Med. Bull. 30:(2)53–66.

Weil, M.L., Itabashi, H.H. Cremer, N.E., Oshiro, L.S., Lennette, E.H., and Carnay, L. 1975. Chronic progressive panencephalitis due to rubella virus simulating subacute sclerosing panencephalitis. N. Engl. J. Med. 292:994–998.

Weller, T.H., Alford, C.A., and Neva, F.A. 1964. Retrospective diagnosis by serological means of congenitally acquired rubella infection. N. Engl. J. Med. 270:1039–1041.

Weller, T.H., and Neva, F.A. 1962. Propagation in tissue culture of cytopathic agents from patients with rubella-like illness. Proc. Soc. Exp. Biol. Med. 111:215–225.

Wolinsky, J.S. 1978. Progressive rubella panencephalitis. In P.J. Vinken and G.W. Bruyn (eds.), Handbook of Clinical Neurology. Vol. 34. Amsterdam: North-Holland, 331–341.

Ziring, P.R., Fedun, B.A., and Cooper, L.A. 1975. Thyrotoxicosis in congenital rubella. (letter). J. Pediatr. 88:1002.

Ziring, P.R., Gallo, G., Finegold, M., Buimovici-Klein, E., and Ogra, P. 1977. Chronic lymphocytic thyroiditis: identification of rubella virus antigen in the thyroid of a child with congenital rubella. J. Pediatr. 90:419–420.

31

Human Retroviruses

Jörg Schüpbach, Fulvia diMarzo Veronese, and
Robert C. Gallo

31.1 INTRODUCTION

The search for human retroviruses started many years ago prompted by suggestive findings in animal model systems. It was well known that a variety of naturally occurring neoplasias and immune dysfunctions, especially immune suppression, in animals are induced by retroviruses. In contrast, evidence for an etiologic role of retroviruses in human neoplasia and lymphoproliferative disorders was long missing. In most animal leukemias, abundant retroviral replication was found and the virus could easily be isolated from these animals. However, viremia was not found in human leukemias. The inability to grow human cells of different lineages in vitro, and a poor understanding of what cell or tissue to select for virus isolation studies were two major problems encountered.

The discovery of human lymphotropic retroviruses called human T-cell leukemia/lymphoma viruses (HTLV) (Figure 31.1) was made possible by two principal achievements: (a) the ability to specifically detect the retroviral enzyme reverse transcriptase among a variety of cellular DNA polymerases (Sarngadharan et al., 1978), and (b) the ability to continuously grow human T cells in vitro with the aid of T-cell growth factor (TCGF) as originally described in this laboratory (Morgan et al., 1976; Ruscetti et al., 1977). To date the HTLV family members are the only known human retroviruses associated with a disease. In this chapter we briefly characterize these viruses and give some details about the pathology they produce and appropriate methods of virus detection.

31.2 HISTORY AND CHARACTERIZATION

31.2.1 Human T-Cell Lymphoma Virus Types I and II

As mentioned above, successful isolation of these viruses depended on the ability to grow hematopoietic cells of different lineages in vitro. After the discovery of TCGF, our laboratory developed a procedure for culturing normal human mature T lymphocytes (Morgan et al., 1976; Ruscetti et al., 1977). Using the same procedure, it was also possible to grow malignant T-cells obtained from patients with mature T-cell malignancies in suspension cultures (Poiesz et al., 1980a). Peripheral blood cells from these patients were separated on Ficoll–Hypaque, and grown in complete medium

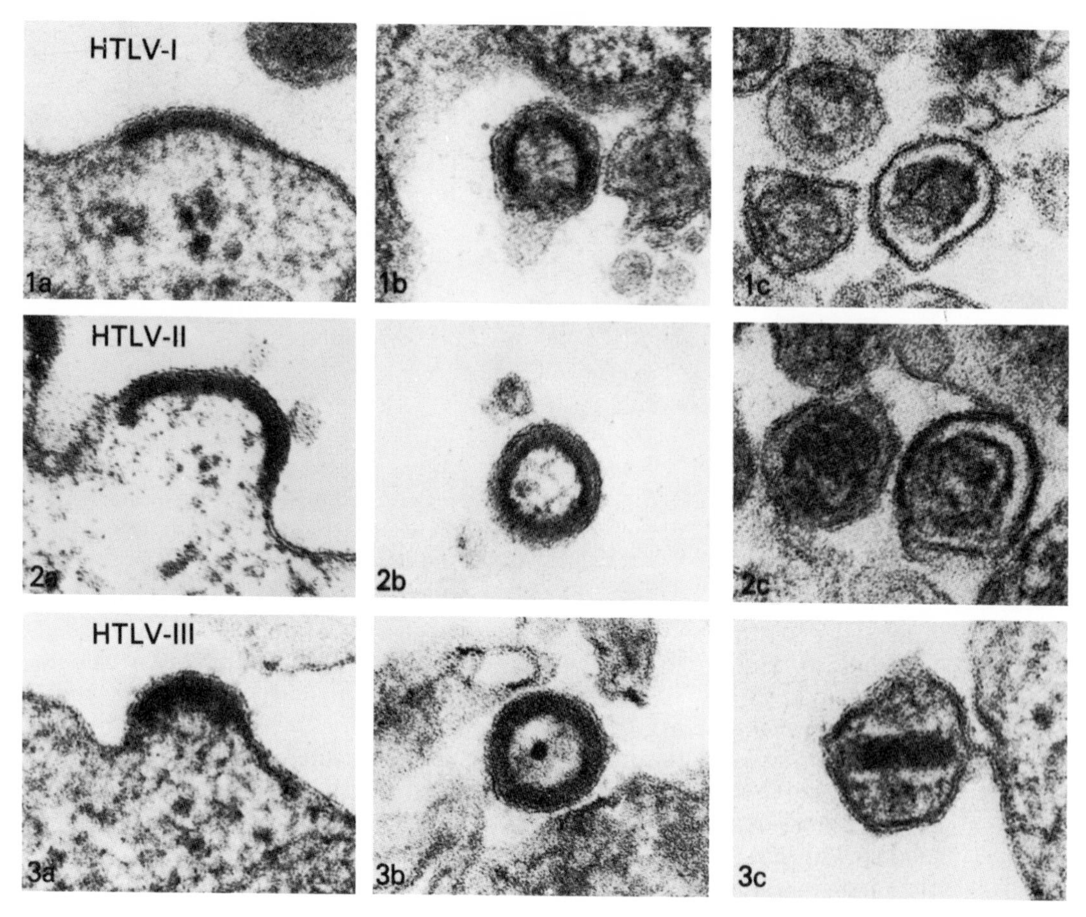

Figure 31.1. Electron microscopy of thin sections of cells producing HTLV-I, -II, or-III. *Top*: Cells producing HTLV-I; *middle*: Cells producing HTLV-II; *bottom*: Cells producing HTLV-III. (a) Virus particles budding from cell membrane. (b) Free, immature particles have separated from the cell membrane. (c) Mature particles, the particle in panel 3c is sectioned in a plane orthogonal to that of panel 3b. Note the rod-shaped core of HTLV-III. Bar indicates 100 μm. (Reproduced from Schüpbach et al., 1984b. Copyright 1985 by AAAS).

supplemented with 10% TCGF. The cultures were then monitored for changes in the morphology and production of virus by electron microscopy (EM) and by measuring viral reverse transcriptase activity in supernatant fluids. Several human T-cell lines were established from patients with mature T-cell malignancies and many of these released viruses that had the typical morphology of type C retroviruses by budding from the cell membrane. HTLV-I and -II are infectious agents that are exogenously introduced, because proviral genomes are absent from the DNA of normal human bulk T cells of HTLV-infected persons (Reitz et al., 1981). Moreover, HTLV sequences could not be detected in a B lymphoblastoid cell line established from a patient with an HTLV-positive T-cell lymphoma (Gallo et al., 1982).

Human T-cell lymphoma viruses were first isolated from cultured cells of two adult black U.S. patients with an aggressive form of T-cell malignancy (Poiesz et al., 1980b, 1981). These viruses were shown to be very closely related by nucleic acid and antigenic determinant analysis and were called HTLV type I. Human retrovirus isolates made from patients with malignancies of mature T cells were also independently reported from other laboratories. There are now more than 100 isolates of HTLV-I from several laboratories in the world, including the U.S., England, Holland, and Japan (Miyoshi et al., 1981; Catovsky et al., 1982; Yoshida et al., 1982; Haynes et al., 1983; Vyth-Dreese et al., 1983; Greaves et al., 1984). These isolates were obtained from T cells of patients with T-cell malignancies and, in some instances, from their normal family members (Sarin et al., 1983). In Japan, the isolates were at first called adult T-cell leukemia viruses from a distinct clinical entity, adult T-cell leukemia (ATL), which is present in southwest Japan and shows a pattern of geographic clustering suggestive of horizontal spreading of a transmittable agent. Comparative studies of the various isolates were performed on the basis of immunologic crossreactivities (Poiesz et al., 1981; Kalyanaraman et al., 1982a; Popovic et al., 1983b; Schüpbach et al., 1983b), and of sequence homology, as determined by molecular hybridization (Poiesz et al., 1981; Popovic et al., 1982), cleavage site maps of several restriction endonucleases (Wong-Staal et al., 1983), and nucleotide sequence analyses (Haseltine et al., 1984). The evidence from all these studies shows that all these isolates are closely related to or identical with the prototype HTLV-I isolates in the U.S.

In 1982 a retrovirus related to, but distinct from, HTLV-I was isolated in our laboratory from a cell line named MO that had been established by Saxon et al. from a patient with a T-cell variant of hairy T-cell leukemia (Kalyanaraman et al., 1982b). Accordingly, this virus was called HTLV-II_{MO}. Only one additional HTLV-II isolate was made since then (Hahn et al., 1984a). HTLV-I and -II isolates have the following common biological features: They are isolated from mature T cells, and have a special tropism for OKT4 T-cells. The morphologic appearance of budding particles and early forms of the virus are very similar by EM. In addition, HTLV-I and -II have a number of crossreactive immune determinants in the internal core and envelope proteins, some nucleotide sequence homology, and reverse transcriptases of similar size and biochemical features. Nevertheless, the two viruses are markedly different, as demonstrated by core protein (p24) competition radioimmunoassays, even though two mouse monoclonal antibodies directed against two different epitopes on HTLV-I and -II p24 do not discriminate between the two p24 (Sarngadharan et al., 1985c). Also markedly different are the envelope antigens as shown by biological assays, such as inhibition of syncytia induction and vesicular stomatitis virus pseudotype neutralization. Even though HTLV-II has been extensively characterized, its role in human diseases is still unclear.

31.2.2 Human T-Cell Lymphoma Virus Type III

The third type of human T-cell lymphotropic virus (HTLV-III) was detected and isolated from patients with acquired immunodeficiency syndrome (AIDS) or with signs and symptoms that frequently precede AIDS [AIDS-related complex (ARC)] (Gallo et al., 1984; Popovic et al., 1984b; Sarngadharan et al., 1984; Schüpbach et al., 1984b). What is now known to be the same virus was identified by Barré-Sinoussi et al. (1983) but not linked to AIDS. Acquired immunodeficiency syndrome has been recognized since 1981 as a distinct new disease entity and is characterized by severe immune suppression with depletion of the OKT4 subset of T lymphocytes, multiple opportunistic infections, and/or neoplasias (Gottlieb et al., 1981; Siegal et al., 1981). Several features of the disease suggested that it could be caused by a transmissible infectious agent, probably a virus. They included evidence from epidemiologic studies, especially the fact that it could be transmitted by blood transfusion or blood products (Curran et al., 1984), and evidence that the etiologic agent was filterable through devices that should have retained fungi or bacteria. Furthermore, it was well known that animal retroviruses, such as feline leukemia virus, are able to cause AIDS-like diseases in animals (Trainin et al., 1983).

Members of the HTLV family were first considered prime candidates for the cause of AIDS for the following reasons: (a) Sera from patients with AIDS frequently contained antibodies that reacted with antigens present on the membranes of HTLV-I–infected cells (Essex et al., 1983a,b). These antigens were later shown to be of viral origin and to represent the precursor of the envelope glycoproteins (Lee et al., 1984a; Schüpbach et al., 1984c). (b) HTLV are T lymphotropic and chiefly infect cells of the helper–inducer phenotype (OKT4/Leu 3+). Other than the members of the HTLV family few (if any) infectious agents, like bacteria and fungi or viruses, were known to have this kind of cell specificity. (c) HTLV were known to be transmitted from person to person by intimate contact or by blood or blood products. (d) In addition, HTLV-I and its variants are of probable African origin and it was suspected that AIDS may have originated in Africa (Gallo, 1984). (e) HTLV-I and -II had, indeed, been isolated from cultured T cells of some patients with AIDS (Gallo et al., 1983; Hahn et al., 1984a), and proviral DNA of HTLV-I had been identified in the cellular DNA of two AIDS patients (Gelmann, et al., 1983). However, these cases represented only rare instances.

Retroviruses belonging to a new subgroup, HTLV-III, have now been isolated at high frequency from patients with AIDS and ARC. These viruses share some properties with HTLV-I and -II, including preferred tropism for OKT4/Leu 3+ T lymphocytes, Mg^{2+}-dependent reverse transcriptase, a major core protein with a molecular weight of 24,000 (p24), some crossreactive antigenic determinants, the presence of a gene for transacting transcriptional activation, but only limited nucleic acid homology. The main difference involves the biological effects seen after transmission in vitro of the virus to primary human T cells. HTLV-I and -II can transform some of these cells and be propagated indefinitely in them. In contrast, HTLV-III is only transiently produced and has strong cytopathogenic effects, leading to cell death. The first HTLV-III isolate was obtained in our laboratory in November 1982, but detailed characterization of HTLV-III and serologic testing of large numbers of patients became possible only after the virus could be propagated in a human T-cell line, H9, which is not only resistant to the cytopathogenic effects of the virus, but also a good virus producer (Popovic et al., 1984b). Other retrovirus isolates from similar sources principally known as lymphoadenopathy-associated virus (LAV) (Barré-Sinoussi et al., 1983; Vilmer et al., 1984) and AIDS-related virus (ARV) (Levy et al., 1984) are closely related to the isolates of HTLV-III described above as evidenced by the nucleotide sequences of their proviral DNA and by endonuclease restriction maps (Muesing et al., 1985; Ratner et al., 1985; Sanchez-Pescador et al., 1985; Wain-Hobson et al., 1985).

31.3 NUCLEIC ACIDS AND PROTEINS

31.3.1 Human T-Cell Lymphoma Virus Types I and II

The genomes of HTLV-I and -II consist of a high-molecular weight polyadenylated RNA. Radioactive complementary DNA (cDNA) prepared by transcription of the viral RNA by the endogenous reverse transcriptase was used in extensive nucleic acid hybridization experiments. From these experiments came the demonstration that HTLV-I was not related to any of the previously known animal retroviruses and causes infection following its exogenous introduction in humans. The molecular cloning of HTLV-I and -II (Seiki et al., 1982; Manzari et al., 1983; Gelmann et al., 1984) and the complete nucleotide sequence of a DNA clone of HTLV-I (Seiki et al., 1983) allowed detailed molecular studies (see Figure 31.2). In analogy with the genomes of other retroviruses, the HTLV proviral genome has a long terminal repeat (LTR) sequence at both ends of the provirus. The portions of this sequence that are derived from unique 5′ RNA sequences and from sequences repeated at both ends of the genomic DNA are longer than those of most other retroviruses, resembling bovine leukemia virus in this respect. From the analysis of the deducted amino acid sequence of HTLV-I it was predicted that there are four coding genes in the HTLV genome: The location of *gag, pol*, and *env* is like in all other replication-competent retroviruses. In addition, there is a novel region between the *env* gene and the 3′ LTR that contains a gene coding for a 40-kilodalton (40 Kd) protein in HTLV-I and 37 Kd in HTLV-II (Lee et al., 1984b; Slamon et al., 1984). This region was initially designated pX, because no functional assignment could be made (Seiki et al., 1983). Sequence comparison of pX of HTLV-I and -II demonstrated that it can be further divided into a 5′ nonconserved region and a 3′ highly conserved region designated *lor* (Haseltine et al., 1984). The *lor* gene is believed to encode a transactivating factor that is involved in the transcriptional control of viral and, possibly, cellular genes (Sodroski et al., 1984b) and may be of importance in some biological aspects of the virus infection. The *gag* primary translational product is a precursor polyprotein Pr^{53gag}, composed of p19 at the aminoterminus, p24 in the middle, and p15 at the carboxyterminus. This precursor is then processed into the individual proteins. All three *gag* proteins were purified to homogeneity (Kalyanaraman et al., 1981;

Figure 31.2. Genetic organization of HTLVs. The *gag* gene codes for three proteins located in the viral core and associated with the viral RNA. *Pol* codes for the enzyme reverse transcriptase, *env* for the two proteins of the envelope, and *x-lor* for a protein involved in *trans* activation of viral and, possibly, cellular genes. Refer to text for details.

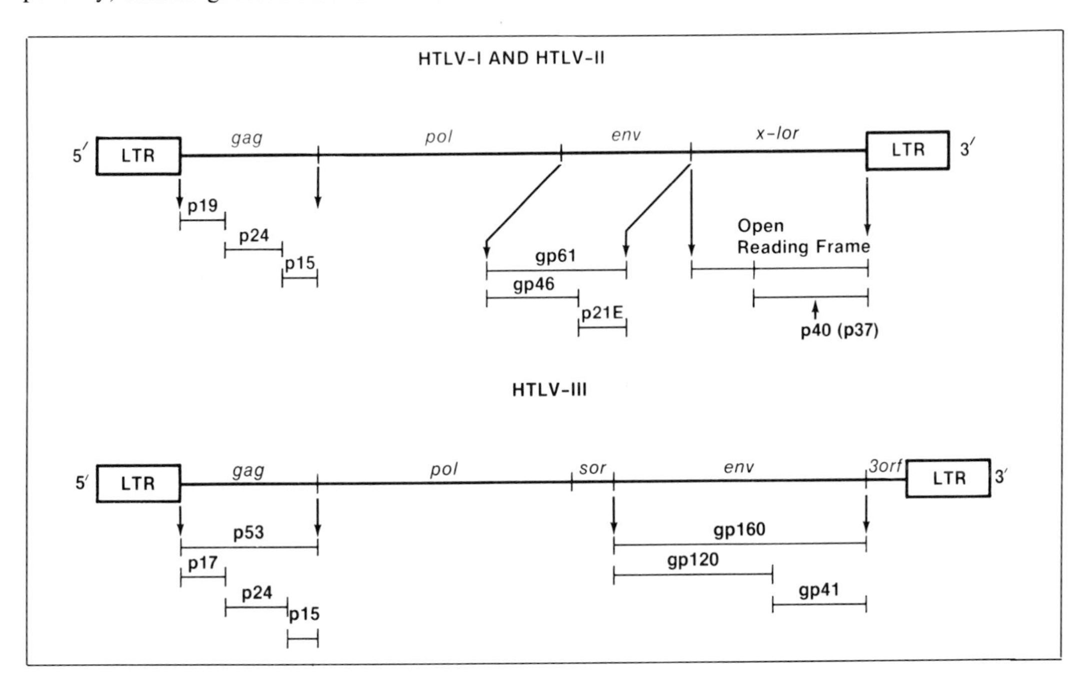

1984b). Their amino acid sequences were determined almost completely (Oroszlan et al., 1982; Copeland et al., 1983), and their immunologic properties were characterized (Kalyanaraman et al., 1981b, 1984b; Schüpbach et al., 1983a). Among the *gag* proteins of HTLV-I and -II, p24 seems to have the highest degree of antigenic crossreactivity. The *pol* gene codes for the reverse transcriptase, which is biochemically different from most mammalian type C retroviral reverse transcriptases. The HTLV enzymes are larger and prefer Mg^{2+} as divalent cation instead of Mn^{2+} (Rho et al., 1981). The *env* gene of HTLV-I codes for a primary translational glycosylated product of 61,000–68,000 daltons (gp61–68), the size being dependent on the cell line used. In the mature virion, gp61–68 is processed into the external glycoprotein gp46 and the transmembrane protein p21. The deduced amino acid sequence of the *env* precursor of HTLV-II showed great similarity with the corresponding protein of HTLV-I (Sodroski et al., 1984a).

31.3.2 Human T-Cell Lymphoma Virus Type III

The genome of HTLV-III has been cloned from the producer cell line, H9/HTLV-III, using several approaches. Results from the analysis of these clones coupled with studies that compared HTLV-III DNA from infected cell lines with that obtained from fresh tissues of AIDS patients indicate there is considerable polymorphism in the virion genome. The homology of HTLV-III to other subgroups of HTLV was analyzed by heteroduplex mapping and molecular hybridization (Hahn et al., 1984b; Gonda et al., 1985). This analysis showed short stretches of significant homology with HTLV-I in the *gag–pol* region and distant homology in a region at the 3′ end of the *env* gene, originally called the *lor* region. A greater extent of nucleotide sequence homology was found with visna virus, a non-oncogenic retrovirus of sheep, which is a member of the subfamily *Lentivirinae*. Five open reading frames were identified in the HTLV-III genome (Ratner et al., 1985), *gag, pol, sor, env, 3′ orf* but studies on the composition of the protein products are still going on. The major core protein derived from the *gag gene* of HTLV-III is p24. p17 is at the aminoterminus and p15, the nucleic acid binding protein, is located at the carboxyterminus (unpublished results with S. Orozlan). Studies using mouse monoclonal antibodies to HTLV-III p24, p17, and p15 indicate that these proteins are processed from the same precursor molecule $Pr53^{gag}$ (unpublished results). Another monoclonal antibody allowed the identification of the antigens of 65 and 51 Kd recognized by the majority of HTLV-III seropositive individuals in immunoblots as *pol* gene products (Veronese et al., 1986). The purified p66/51 was assayed for reverse transcriptase activity. The results clearly indicated that at least one of the two proteins is active reverse transcriptase of HTLV-III. HTLV-III infected cells contain a transacting factor which activates the expression of LTR-linked genes. The major functional domain of the transactivator gene of HTLV-III (*Tat-III*) is located in what was previously thought to be a noncoding region between the *sor* and *env* region. The *Tat-III* gene consists of three exons and its transcription into a functional messenger RNA involves double-splicing (Arya et al., 1985; Sodroski et al., 1985). The primary translational product of the HTLV-III *env* gene was identified as a glycoprotein of 160,000 daltons (gp160) which, in the mature virion, is cleaved into the external glycoprotein, gp120, located at the aminoterminus (Allan et al., 1985), and into the transmembrane protein gp41 at the carboxyterminus (Veronese et al., 1985).

31.4 PATHOGENESIS OF HUMAN T-CELL LYMPHOMA VIRUS TYPE I INFECTIONS

The infection in vitro of human T cells with HTLV-I may have two principal effects, namely, malignant transformation and alteration of immune function. Accordingly, the spectrum of human diseases found associated with HTLV-I includes a malignancy of T lymphocytes, adult T-cell leukemia/lymphoma. In some instances, HTLV-I is serologically associated with other types of human malignancies, such as some non-Hodgkin lymphomas, some chronic lymphatic leukemias, or certain types of cutaneous T-cell lymphomas. HTLV-I also may be linked to certain forms of immunodeficiency that predispose patients to infections with a variety of agents. This HTLV-I–associated immunodeficiency is not to be confused with the much more severe effect of HTLV-III.

31.4.1 Adult T-Cell Leukemia/Lymphoma

31.4.1.a Clustering of T-Cell Malignancies and Epidemiology of HTLV-I

The clustering of T-cell malignancies was originally recognized by Takatsuki et al. in a group of patients who were born in the southwestern region of the Japanese islands (Yodoi et al., 1974; Takatsuki et al., 1977; Uchijama et al., 1977). Serologic and other studies (Kalyanaraman et al., 1982a; Popovic et al., 1982; Robert-Guroff, et al., 1982b; Schüpbach et al., 1983b) subsequently linked these cancers to the newly detected HTLV-I (Poiesz et al., 1980b, 1981). Independently, Japanese investigators linked a retrovirus to adult T-cell leukemia/lymphoma (ATLL), which was first called ATLV (Hinuma et al., 1981; 1982; Miyoshi et al., 1981; Yoshida et al., 1982). Later, it was recognized that HTLV-I and ATLV were the same species of retrovirus (Watanabe et al., 1984). Other clusters of T-cell malignancies indistinguishable from Japanese ATLL were later identified in the Caribbean (Costello et al., 1980; Catovsky et al., 1982) and also linked to HTLV-I (Blattner et al., 1982; Schüpbach et al., 1983a). To date, additional clusters or cases of HTLV-I–associated malignancies are known in the U.S., especially in the southeastern regions, Central and South America, Africa, the Middle East, India, and some areas of the Far East outside Japan (Blayney et al., 1983a; Fleming et al., 1983; Hunsmann et al., 1983; Biggar et al., 1984; Merino et al., 1984; Saxinger et al., 1984a; Gallo, 1984; Gallo and Blattner, 1985). Recently, isolated cases of HTLV-I–associated malignancies were also reported from southern Italy (Manzari et al., 1984).

There is now solid evidence that HTLV-I is the etiologic agent of ATLL: Epidemiologic studies show that ATLL occurs only in regions where HTLV-I is endemic and that the incidence of ATLL in endemic regions is correlated with the prevalence of antibodies to HTLV-I (Robert-Guroff and Gallo, 1983). Infection by HTLV-I can be demonstrated in all patients with ATLL. It was further shown that the tumor cells of a given patient are of monoclonal origin and contain one or a few copies of the HTLV-I genome with a common (monoclonal) site of provirus integration (Wong-Staal et al., 1983; Yoshida et al., 1983).

In vitro models show tropism and transforming capacity of HTLV-I for lymphocytes bearing the T4 marker, which is the usual phenotype of the ATLL cells (Miyoshi et al., 1981; Yamamoto et al., 1982b; Popovic et al., 1983a). Furthermore, primates that were injected with infectious HTLV-I in 1981 after 4 years, are now coming down with the first signs of hematopoietic malignancies (P.D. Markham, personal communication). Thus, all four of Koch's postulates are satisfied. In addition, the molecular biological data indicate a direct causative role for ATLL by HTLV-I. Therefore, HTLV-I may be considered the etiologic agent of ATLL.

In all regions where clusters of ATLL are found, HTLV-I infection is also present among the healthy persons of an adequate control population. The amount of antibodies is closely associated with the regional incidence of ATLL. The antibodies and ATLL prevalence sometimes may vary greatly between places only short distances apart (Yamaguchi et al., 1983). Similarly, HTLV-I infection may be restricted to certain ethnic or social groups, for example, blacks emigrating from the Caribbean to European countries (Blattner et al., 1982; Greaves et al., 1984; Robert-Guroff et al., 1984; Schaffar-Deshaynes et al., 1984).

Family members of ATLL patients show a prevalence of antibodies to HTLV-I that is significantly higher than the unrelated control population (Robert-Guroff et al., 1983; Sarin et al., 1983; Schüpbach et al., 1983a,b; Miyamoto et al., 1985). The fact that HTLV-I is most frequently found in spouses, suggests that intimate contact, perhaps sexual intercourse, is an important means of virus transmission. The evidence so far favors a male-to-female transmission. Other routes of virus transmission may include blood transfusions (Saxinger and Gallo, 1982; Maeda et al., 1984; Miyamoto et al., 1984; Okochi et al., 1985) and mother-to-child transmission, possibly in utero or after birth (Tajima et al., 1982; Komuro et al., 1983). Some investigators also postulated a role for parasites or other vectors in the transmission of HTLV-I (Tajima et al., 1981, 1983; Merino et al., 1984).

It is estimated that about 1% of the antibody-positive persons eventually develop ATLL. Although the minimal incubation time for the development of ATLL is not known, circumstantial evidence suggests that it is in the range of years or decades. Cases of ATLL were reported to occur in individuals 20 years and longer after emigration from HTLV-I–endemic regions where they probably acquired their infection (Greaves et al., 1984).

31.4.1.b ATLL Clinical Course

Adult T-cell leukemia/lymphoma may manifest as leukemia (adult T-cell leukemia, ATL) or as a lymphoma. The disease usually is rapidly progressing and has a median survival time of less than 1 year, despite chemotherapy (Uchiyama et al., 1977; Hanoaka, 1982). A proportion of patients, however, have a subacute or chronic disease that eventually may progress to an acute clinical course. A few patients with subacute cases have gone into spontaneous remission (Kawano et al., 1984).

The malignant cells of ATLL display remarkable pleomorphism containing polylobulated nuclei with prominent nucleoli. They usually express the T4 lymphocyte phenotype, the transferrin receptor, and the receptor for TCGF [interleukin 2] (IL-2)] as demonstrated by the binding of the anti-"Tac" monoclonal antibody. The cells are terminal deoxytransferase-negative (Hattori et al., 1981; Lando et al., 1983; Popovic et al., 1983a;b; Waldmann et al., 1983). Although these cells are of the T4 phenotype, they have either suppressor or no detectable functions in vitro (Takatsuki et al., 1982; Waldmann et al., 1983; Yamada et al., 1983). The disease has a high frequency of lymphadenopathy and splenomegaly, and is frequently complicated by hypercalcemia, which is due to lytic bone lesions in the absence of tumor cell infiltration (Blayney et al., 1983b). Other frequent complications include lung or central nervous system involvement. Typically, there is no mediastinal involvement. The majority of patients show evidence in vitro and in vivo of immune dysfunction with frequent occurrence of opportunistic infections.

Five clinicopathologic patterns of HTLV-I–associated T-cell malignancies have been described (Table 31.1) (Clark et al., 1985): a) typical ATL usually shows a rapidly progressive clinical couse; b) smoldering ATL features an indolent clinical course with few circulating malignant cells, and skin involvement lymphadenopathy and/or hepatosplenomegaly may be present; c) chronic ATL has a high percentage of circulating malignant cells and is occasionally associated with skin involvement, chronic cough, mild lymphadenopathy or hepatosplenomegaly; d) the crisis phase of smoldering or chronic ATL represents the conversion to acute clinical dis-

Table 31.1. Features of Clinicopathologic Subgroups of ATLL

Clinicopathologic Subgroup	Leukemic Manifestation[a]	Skin Involvement	Lymphadenopathy	Splenomegaly	Hepatomegaly	Bone Marrow Infiltration	Hypercalcemia	Elevated Serum LDH	Elevated Serum Bilirubin
Classic ATL, or crisis phase of smoldering or chronic ATL	4[b]	2	4	3	3	3	3	4	3
Smoldering ATL	0	4	1	1	1	0	0	0	0
Chronic ATL	4	2	1	1	1	2	0	1	0
Mature T-cell NHL with integrated virus[c]	0	2	4	2	2	2	2	4	3

[a] >10,000 WBC/μl and abnormal lymphocytes with features typical of ATLL.
[b] The ratings indicate: 4, feature present in 76–100% of patients; 3, 51–75%; 2, 26–50%; 1, 1–25%, 0, absent.
[c] Aleukemic, "lymphoma-type" ATLL.
(Adapted from Clark et al., 1985)

ease with characteristic features of typical ATL; e) aleukemic ATLL finally represents cases of mature T-cell lymphomas in HTLV-I endemic areas that have integrated HTLV-I provirus but no leukemic manifestations. This group also includes the few cases of lymphomas that contain HTLV-I provirus in their tumor cells without having other features of ATLL.

Usually, the combination of clinical and pathologic features allows the distinction of ATLL from other malignancies of mature T cells, [i.e., mycosis fungoides (MF)/Sezary's syndrome and T-cell chronic lymphatic leukemia (T-CLL)]. However, the distinction may be difficult (Lennert et al., 1985). Most of the typical features of ATLL are uncommon for MF/Sezary's syndrome, with the exception of cutaneous involvement. In approximately 33% of ATLL cases, the leukemic infiltration is restricted to the subcutaneous tissue and the dermis, but does not affect the epidermis as usually seen in MF/Sezary's (Pautrier's microabscesses) (Jaffe et al., 1984). Moreover, the morphology of the polylobulated ATLL cells is different from typical Sezary cells. The nuclear convolutions of ATLL cells are usually less markedly indented than those of Sezary cells and have a more "lumpy" appearance. The characteristics that distinguish ATLL from T-CLL include the more aggressive clinical course of ATLL, the higher degree of nuclear pleomorphism, and the typical polylobulated nuclei of ATLL, compared with the more uniform T-CLL cells, and the absence of cytoplasmic granules in ATLL cells. In cases where a distinction on clinicopathologic criteria is not possible, HTLV-I serology, the demonstration of antigens in cultivated T cells, or the direct demonstration of the viral genome in uncultured leukemic cells will confirm or exclude the diagnosis of ATLL. In addition, the presence of the TCGF receptor (anti-Tac monoclonal) on malignant cells of the T4 phenotype strongly indicates an association with HTLV-I.

31.4.2 Human T-Cell Lymphoma Virus Type I-Related Immunodeficiency

An especially important feature of ATLL is the frequent opportunistic infections. They often include *Pneumocystis carinii* pneumonia, fungal infections with *Candida, Aspergillus* or *Cryptococcus, cytomegalovirus, herpes simplex* or *zoster*, and strongyloidiasis. Bacterial sepsis, especially with *Klebsiella,* is also frequent (Bunn et al., 1983; Clark et al., 1985). There is extensive evidence for abnormal immune functions in these patients, and in vitro studies have shown that HTLV-I–infected human T4 or T8 lymphocytes have impaired or abnormal function (Mitsuya et al., 1984; Popovic et al., 1984a; Suciu-Foca et al., 1984; Wainberg et al., 1985). HTLV-I infection appears to be three times more prevalent among patients hospitalized with infectious diseases than in the healthy control population (Essex et al., 1984). The biological properties of HTLV-I, therefore, are similar to those of certain animal retroviruses with T-cell tropism, most notably feline leukemia virus (FeLV). Feline leukemia virus causes profound immunosuppression in naturally infected cats. The manifestations include lymphopenia (Essex et al., 1975). thymic atrophy (Anderson et al., 1971), depressed cell-mediated (Perryman et al., 1972) and humoral immune responses (Trainin et al., 1983), frequent opportunistic infections (Essex et al., 1975), and an increased occurrence of certain cancers other than leukemia (Weijer et al., 1974).

31.4.3 Other Malignancies Serologically Linked with Human T-Cell Lymphoma Virus Type I

Apart from ATLL, where an etiologic relationship of HTLV-I is clearly established, a pathogenic role of HTLV-I is discussed in some other malignant disorders, based on the observation of HTLV-I–reactive antibodies in some percentage of patients.

A survey of lymphoid malignancies in Jamaica showed the presence of HTLV-I–reactive antibodies in 23% of patients with B-cell chronic lymphatic leukemia (B-CLL). These patients had never received blood transfusions. The antibody prevalence of these B-CLL patients is higher than that found among appropriate controls, suggesting an involvement of HTLV-I in B-CLL cases (Blattner and Gallo, 1985). In none of these cases was the viral genome found integrated in the malignant cell population, thus, ruling out a direct transforming role of the virus and suggesting an indirect mechanism. One possibility is that HTLV-I may be present in the T cells of these patients and that chronic antigenic stimulation of HTLV-I–reactive B-cell clones eventually may lead to an event of mutation causing malignant transformation. Such a mechanism was documented in at least one case (D. Mann et al., unpublished observations). Another possibility is that an underlying infection by HTLV-I causes immunodeficiency facilitating the development of lymphoid malignancies (see above). A last possibility is the presence of a different virus with antigenic properties similar to those of HTLV-I.

Other malignant disorders with some HTLV-I association were identified in the cutaneous T-cell lymphomas (CTCL) including MF and Sezary's syndrome. The first two isolates of HTLV-I were made from patients with a primary diagnosis of CTCL (Poiesz et al., 1980 and 1981), but subsequent revision of the clinicopathologic data established the final diagnosis of ATLL in both cases. Nevertheless, weak titers of HTLV-I–reactive antibodies were detected in 15% of Danish patients with early stages of CTCL (Saxinger et al., 1984b). The most likely explanation for these findings is the involvement of an agent that is distantly related to HTLV-I. This would explain the low titers of antibodies and their presence in only a small proportion of patients.

31.5 PATHOGENESIS OF HUMAN T-CELL LYMPHOMA VIRUS TYPE III INFECTIONS

31.5.1 Epidemiology and Transmission

The transmission of HTLV-III is governed by the same two basic rules as that of any other infectious agent. Thus, the likelihood of virus transmission is dependent on the closeness of contact with, and the frequency of exposure to, a source of infection. Similar to hepatitis B virus (HBV), HTLV-III is predominantly transmitted by sexual contact or parenteral inoculation (Goedert and Blattner, 1985). The most important ways of viral transmission and spread of HTLV-III infection are summarized in Figure 31.3. HTLV-III, so far has been isolated from blood, semen, saliva, lymph nodes, urine, bone marrow, cerebrospinal fluid, and neurologic tissues, (Gallo et al., 1984; Groopman et al., 1984; Ho et al., 1984; Popovic et al., 1984b; Zagury et al., 1984; Salahuddin et al., 1985; Ho et al., 1985). Mucous membrane contact with or inoculation of such and possibly other biological materials from an infected person, therefore, must be considered as exposure to the virus.

Male homosexuals account for 73% of all AIDS cases in the U.S. In San Francisco, HTLV-III infection (as indicated by seropositivity) was present in 65% of active homosexuals in 1984, and probably is still rising in this and other epicenters of the AIDS epidemic. Transmission of the virus among homosexuals, and possibly also among heterosexuals (Melbye et al., 1985), may be especially facilitated by receptive anal intercourse (Goedert et al., 1985).

Seventeen percent of the U.S. AIDS cases occurred so far in parenteral drug addicts. Exceedingly high prevalences of

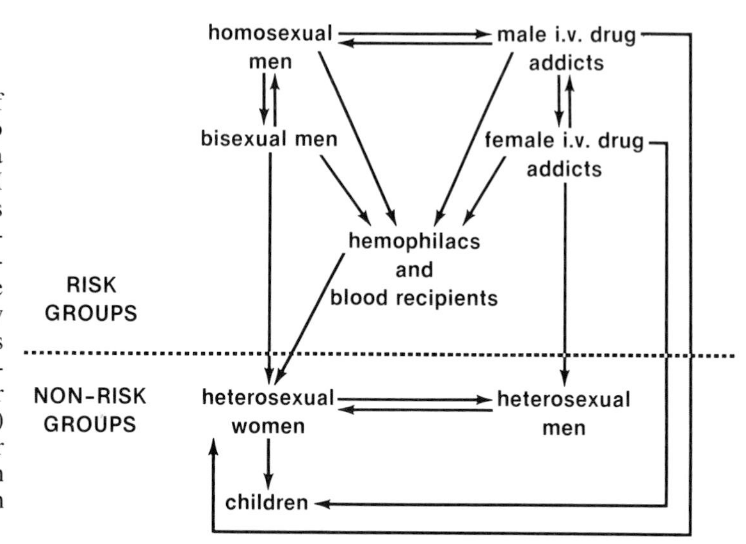

Figure 31.3. Spread of HTLV-III from risk groups to the general population. In a first phase of the HTLV-III epidemic, the virus was largely confined to homosexually active men and drug addicts. In a second step, the virus spread to heterosexually active members of risk groups (i.e., bisexual men, drug addicts, hemophiliacs, and other recipients of blood products) who now represent the major link of HTLV-III transmission to the heterosexual population at large.

HTLV-III antibodies are found in drug addicts. In Manhattan, 87% of drug addicts admitted to a Methadone clinic were antibody-positive in 1984. At the same time, prevalences around 50% were found among drug addicts in Newark and Jersey City, but also in certain European cities (Weiss S.H., et al., unpublished data; Schüpbach et al., 1985b). Possible contacts with homosexuals and sharing of blood-contaminated needles and syringes are considered the major ways of the spread of HTLV-III into and within this group.

In hemophiliacs and recipients of blood transfusions, the rate of infection is clearly linked to the intensity of treatment. In recipients of blood transfusions, the incidence of AIDS is linked to the number of units received (Hardy et al., 1985). In hemophiliacs, virtually all recipients of American-produced factor VIII concentrates (for the treatment of hemophilia A) were seropositive in 1984 (Kitchen et al., 1984; Melbye et al., 1984; Evatt et al., 1985; Koerper et al., 1985). These concentrates were made from the pooled plasma of thousands of commercial donors among whom some were members of high risk groups for AIDS. This is evidenced by the fact that the concentrates contain low amounts of HBV, which has a high prevalence among the risk groups for AIDS (Gerety and Aronson, 1982). Apparently, many of these concentrates harbored virulent HTLV-III as well, although it was not yet possible to isolate from or otherwise demonstrate virus in coagulation factor batches linked with the development of AIDS in recipients. However, it was reported that virulent HTLV-III added to human plasma could later be detected in traces in factor VIII concentrates (Koerper et al., 1985). Preliminary studies suggested that long-term heat inactivation of these coagulation factor preparations (e.g., pasteurization at 68°C for 72 hours) may be effective in destroying the virus without impairment to the clotting factor activity (Koerper et al., 1985; Mösseler et al., 1985). In contrast with factor VIII deficiency, hemophilia B is much less frequently associated with HTLV-III seropositivity. This may be due to the less intensive treatment required in hemophilia B, and possibly also to a lower ability of factor IX concentrates to harbor infectious virus. Similarly, the seropositivity is much lower in hemophilia A patients treated with cryoprecipitate prepared from single donors, fresh frozen plasma or, in some instances, with concentrates produced outside the USA (Melbye et al., 1984; Evatt et al., 1985; Schüpbach et al., unpublished observations).

Most cases of pediatric AIDS occur in offspring of parents belonging to AIDS risk groups. In these cases, the virus may be transmitted in utero (Lapointe et al., 1985) or in the postnatal period, possibly by milk (Ziegler et al., 1985). Some cases, however, are due to transfusions.

Although infections by HTLV-III were largely confined to homosexuals, bisexuals, and male drug addicts in the first phase of the epidemic, the virus has since spread to a considerable number of heterosexually active persons. Heterosexual contact has now been recognized as another risk factor for AIDS, especially when promiscuity and female prostitutes, who may be drug addicts, are involved (Redfield et al., 1985a;b). Heterosexual transmission of the virus could emerge as prevalent as homosexual transmission is now. Thus, the epidemiology of HTLV-III infection in the Western world, in the long run, may become similar to that in Africa, where HTLV-III has probably been around for more than a decade (Saxinger et al., 1985) and where HTLV-III seroreactivity and the incidence of AIDS are essentially evenly distributed among males and females (Biggar et al., 1985).

31.5.2 Principal Disorders Associated with Human T-Cell Lymphoma Virus Type III Infection

The development of serologic tests made it possible to identify and follow HTLV-III infections almost from the onset. Like most other retroviral infections, HTLV-III infections appear to be chronic and may persist for years if not for life. Infectious virus was isolated from a majority of asymptomatic seropositive individuals, up to 69 months after seroconversion (Feorino et al., 1985; Jaffe et al., 1985). A model depicting the course of HTLV-III infection and the diseases associated with it is shown in Figure 31.3. After an initial acute illness, the course of this chronic infection may be asymptomatic for a prolonged time (carrier status/latency period). Alternatively, the infection may lead to gradual impairment of immune functions, primarily the depletion of T4 lymphocytes, and clinical disorders such as ARC or full-blown AIDS may ensue. Preliminary observations suggest that 4–19% of seropositive individuals develop full-blown AIDS within observation periods ranging from 1 to 5 years. An additional 25% are estimated to develop early stages of AIDS-related illnesses (Landesman et al., 1985). Other studies, however, suggest that this rate may be much higher and that the immunologic manifestations usually start only in the third year after seroconversion (Eyster et al., 1985). It is not known how many of these persons will finally develop full-blown AIDS, but studies with observation periods of up to 3 years showed a progression rate of close to 20% (Metroka et al., 1983; Mathur-Wagh et al., 1984).

Independent of its attacks on the immune system, HTLV-III may infect cells of the nervous system and cause several neurologic syndromes. HTLV-III, thus, shows biological and some genetic similarity with certain members of the lentivirus subfamily of retroviruses, most notably with visna virus, which is the cause of slowly progressive, debilitating central nervous system infections in hoofed domestic animals (Gonda et al., 1985). Principally, four different clinical syndromes have been linked to HTLV-III so far (Figure 31.4).

31.5.2.a Acute Infectious Mononucleosis-Like Syndrome

An acute illness similar to infectious mononucleosis occurs in the majority of cases a few days to a few weeks after infection by HTLV-III (Anonymous, 1984; Cooper et al., 1985; Tucker et al., 1985). In one case, the incubation period was 6 days (Tucker et

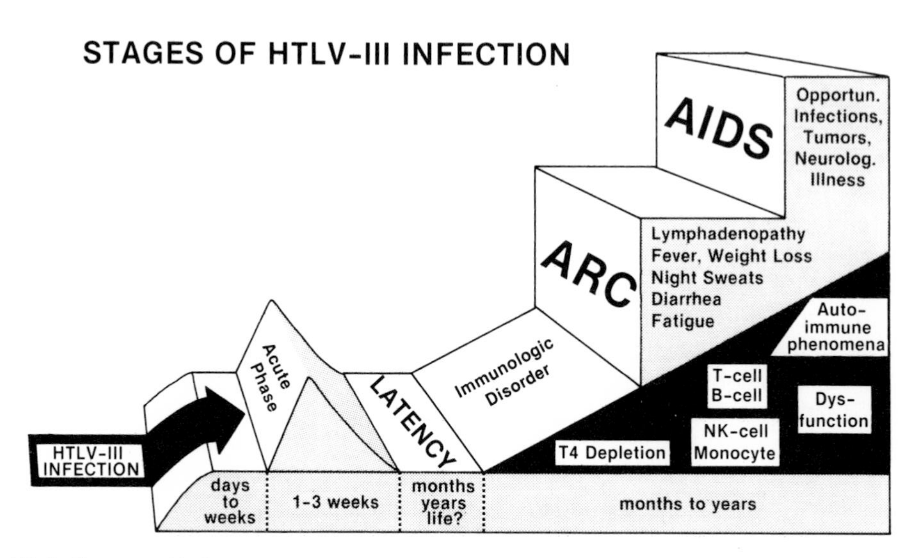

Figure 31.4. Stages of infection by HTLV-III. HTLV-III infection is usually followed by an acute clinical illness resembling infectious mononucleosis. After a latency period of variable length, the cytopathogenic effects of HTLV-III may lead to gradual depletion of T4 cells, followed by secondary effects in the immune functions that require T4 cell help. Autoimmune phenomena may further this process. Clinical manifestations of the immune disorder are subsumed in the syndromes of the AIDS-related complex and AIDS. Neurologic disease may be another manifestation. For details refer to text.

al., 1985). The illness is of sudden onset and lasts for 1 to 3 weeks. Symptoms frequently include fever, sweats, malaise, lethargy, anorexia, nausea, myalgia, arthralgia, headaches, sore throat, diarrhea, generalized lymphoadenopathy, splenomegaly, a macular erythematous truncal eruption, and thrombopenia. Desquamation of palms and soles occasionally may be seen. Immunologically, lymphopenia and an acute inversion of the T4/T8 lymphocyte ratio (<1.0) may be found, the latter of which appears in most cases to be due to an increase in the number of T8 cells (Cooper et al., 1985). In some cases, however, the inversion is due to a reduction of the T4 cells. Some residual symptoms may remain for considerable time after the acute phase before they subside, or they may eventually lead to the chronic complications of HTLV-III infection, ARC and AIDS. Seroconversion has been observed in a few cases after an interval ranging from 19 to 56 days from the onset of the acute illness (Cooper et al., 1985).

Probably, most persons infected will eventually develop antibodies to HTLV-III. There may be some delay in the response, however, and in the presence of viral antigenemia antibodies may become undetectable by the tests available. So far, virus has been isolated from three asymptomatic homosexuals and one man with lymphadenopathy who were seronegative by all available tests (Salahuddin et al., 1984a).

31.5.2.b AIDS-Related Complex

AIDS related complex is a milder form of AIDS or, in some cases, a prodrome to full-blown AIDS. According to the Centers for Disease Control (CDC), ARC is defined by any combination of two clinical and two laboratory findings from a list of abnormalities (Table 31.2). A major feature of ARC is the unexplained presence of generalized lymphadenopathy involving two or more

Table 31.2. Definition of AIDS-related Complex

Presence of any two clinical features plus any two laboratory anomalies from the following list: Clinical Features		Laboratory Abnormalities
Weight loss >10%, or 7 kg (15 lbs)		Helper T cells (T4) <400/mm^3
Fever >38°C (100°F)	>3 month duration	T4/T8 ratio <1.0
Lymphadenopathy (>2 extrainguinal sites)		Leukopenia
		Lymphopenia
Diarrhea		Thrombopenia
Fatigue		Anemia
Night sweats		Elevated serum immunoglobulins
		Increased circulating immune complexes
		Cutaneous anergy to multiple recall antigens
		Depressed in vitro T-cell blastogenesis

extrainguinal sites and lasting for at least 3 months. This condition also has been named lymphadenopathy syndrome (LAS). Biopsy of the soft, nontender, and mobile lymph nodes usually shows benign follicular hyperplasia, which is morphologically indistinguishable from that of other viral diseases. The usual OKT4/OKT8 (T4/T8) lymphocyte phenotype ratio in the lymph node is reversed in the center, mantle, and the interfollicular T-cell zone (Ziegler and Abrams, 1985). Molecular hybridization studies showed that only a small fraction of the cells are infected by HTLV-III (Gallo et al., 1985). This indicates that the hyperplasia is not directly caused by HTLV-III in the sense of a proliferative effect of the viral gene products, but is rather the result of immunoreactive mechanisms. Clinically, the lymphadenopathy is frequently accompanied by fatigue, flu-like illness, night sweats, fever, and marked weight loss. The patients frequently develop minor bacterial, fungal, and viral infections of the skin and mucous membranes, including seborrheic dermatitis, genital warts, and recurrent genital or oral herpes. Episodes of herpes zoster are also seen, as well as oral candidiasis and unexplained diarrhea. About 30% of the LAS patients, however, are asymptomatic (Ziegler and Abrams, 1985).

Lymphadenopathy usually involves the cervical and axillary lymph nodes. Splenomegaly is seen in about 30% of the patients. Routine laboratory findings are usually uncharacteristic. A minority of patients have anemia, leukopenia, or thrombopenia. The serum immunoglobulin levels (mainly IgG and IgA in adults, additionally IgM in children) are polyclonally increased in about 75% of patients. Immunologic studies reveal a lowered T4/T8 lymphocyte subset ratio in almost all patients and cutaneous anergy to a battery of recall skin antigens. Circulating immune complexes may be present in the serum, and the blastogenic responses to T- and B-cell mitogens are decreased.

HTLV-III specific antibodies are detected in about 90% of ARC cases (Sarngadharan et al., 1985b). Because the antibody titers in ARC generally are much higher than in AIDS where, nevertheless, 97% of the patients are positive, we feel that most of the 10% antibody-negative ARC cases are not associated with HTLV-III and have nothing to do with AIDS. In these cases, isolation of virus should be attempted and if negative, the diagnosis should be revised.

The prognosis of ARC and LAS remains uncertain. Very few patients with LAS were reported to have spontaneous regression (Metroka et al., 1983). As serologic testing for HTLV-III was not available at the time of those reports, it is well possible that the few cases showing regression were part of the 5–10% of patients with ARC or LAS who are seronegative for HTLV-III and whose disorders are probably not related to AIDS. Preliminary studies with observation periods of up to 3 years indicated that between 7% and 17% of the LAS patients progressed to AIDS (Metroka et al., 1983; Mathur-Wagh et al., 1984; Ziegler and Abrams, 1985). A "wasting syndrome," which is characterized by regression of lymphadenopathy, extreme fatigue, sweating, fever, and weight loss, frequently precedes the development of AIDS. A decrease in the titer of antibodies to the major core protein of HTLV-III, p24, in the presence of essentially unchanged titers of antibodies to the envelope component gp41 may represent another sign of unfavorable prognosis (Schüpbach et al., 1985a).

31.5.2.c Acquired Immunodeficiency Syndrome

AIDS is a highly lethal condition, initially described in 1981 (Gottlieb et al., 1981). Patients present with severe systemic opportunistic infections and/or certain types of tumors.

Up to 20% of HTLV-III seropositive persons develop AIDS within a period of 5 years (Landesman et al., 1985). The average incubation time for AIDS appears to be around 3 years. After this time, the AIDS incidence seems to decline. However, the long-term prospects of HTLV-III infection are still unknown (see below).

The current AIDS definition is still based solely on clinical criteria, (i.e., the secondary effects manifesting as opportunistic infections and tumors) (Table 31.3). The primary defect, however, resides in the

Table 31.3. AIDS—Surveillance Definition of the Centers for Disease Control

The occurrence of a disease that is at least moderately predictive of a defect in cell-mediated immunity, occurring in a person with no known cause for diminished resistance to that disease. These diseases include:

- Tumors
  - Kaposi's sarcoma in patients <60 years of age
  - Primary lymphoma of the CNS
- Infections
  - *Pneumocystis carinii* pneumonia
  - Unusually extensive mucocutaneous herpes simplex of >5 weeks' duration
  - *Cryptosporidium* enterocolitis >4 weeks' duration
  - Esophagitis due to *Candida albicans*, cytomegalovirus, or herpes simplex virus
  - Progressive multifocal leukoencephalopathy
  - Pneumonia, meningitis, or encephalitis due to one or more of the following: *Aspergillus, C. albicans, Cryptococcus neoformans,* cytomegalovirus, *Nocardia, Strongyloides, Toxoplasma gondii,* Zygomycosis, or atypical *Mycobacterium* species (excluding tuberculosis and lepra)

Pediatric AIDS—Provisional Surveillance Definition of the Centers for Disease Control

- Same as AIDS in adults, with the following provisions
  - Congenital infections that must be excluded are
    - *T. gondii* in patients <1 month of age
    - Herpes simplex virus in patients <1 month of age
    - Cytomegalovirus in patients <6 months of age
  - Specific conditions that must be excluded in a child are
    - Primary immunodeficiency disease: severe combined immunodeficiency, DiGeorge syndrome, Wiskott-Aldrich syndrome, ataxia-telangiectasia, graft-versus-host disease, neutropenia, neutrophil function abnormality, agammaglobulinemia, or hypogammaglobulinemia with raised IgM
    - Secondary immunodeficiency associated with immunosuppressive therapy, lymphoreticular malignancy or starvation

immune system and consists of the HTLV-III–mediated depletion of the T4 helper/inducer lymphocyte subset (Klatzman et al., 1984; Popovic et al., 1984b). T4 cells are crucial for many functions of the cellular and humoral immune system. They induce suppressor cells to inhibit the production of immunoglobulins by B cells. They also induce cytotoxic T cells to kill and, thus, are involved in the elimination of infected or otherwise altered cells. T4 cells also induce natural killer cells and monocyte functions, enhance the production of growth factors for other cells,, and suppress B-cell differentiation. It is clear that a loss of this important cell must result in a profound disturbance in the intricate network of immune functions (Bowen et al., 1985).

Most of the immunologic abnormalities and dysfunctions regularly seen in AIDS patients, thus, may be directly or indirectly explained by the deficiency of T4 cells ($<400/mm^3$). Among them are lymphopenia ($<1000/mm^3$), a lowered T4/T8 lymphocyte subset ratio (<1.0), a defect in delayed hypersensitivity manifesting in cutaneous anergy, elevated immunoglobulin levels usually due to IgG and IgA, and impaired blastogenic responses of both T and B cells in vitro.

Antibodies to HTLV-III are found in 97% of patients with AIDS (Sarngadharan et al., 1985b). The antibody titers usually decline with the advance of the disease and in some cases only antibodies to gp41 will remain, as demonstrated by Western blot (Safai et al., 1984; Schüpbach et al., 1985a).

31.5.2.d Neurologic Disorders

Apart from causing immunodeficiency, HTLV-III infection may be the cause of severe central nervous system (CNS) dysfunction (Jordan et al., 1985). Neurologic symptoms that are not due to secondary CNS infection or tumor are observed in at least 33% of patients with AIDS, and most of them have neuropathologic abnormalities at postmortem examination (Snider et al., 1983b). Spinal cord disease with signs and symptoms of paraparesis and ataxia, pathologically represented by a vacuolar myelopathy, afflicts approximately 20% of patients with AIDS (Petito et al., 1985). Another syndrome commonly seen is an encephalopathy associated with gross cerebral atrophy and microglial nodules that includes progressive dementia as the dominant feature (Snider et al., 1983b; Nielson et al., 1984). HTLV-III sequences were found to be integrated in brain tissue of patients with encephalopathy (Shaw et al., 1985), and infectious virus was transmitted from brain to chimpanzees (Gajdusek et al., 1985). HTLV-III also was isolated from cells present in cerebrospinal fluid (CSF) and in free form from CSF alone (Ho et al., 1985). Moreover, intra-blood–brain barrier synthesis of HTLV-III specific IgG could be demonstrated in all of nine ARC or AIDS patients, independent of the presence of neurologic symptoms. This suggests that CNS infection by HTLV-III occurs in the majority of patients with ARC or AIDS, irrespective of clinical symptomatology (Resnick et al., 1985a). In some cases, neurologic or psychiatric symptoms including acute meningitis, meningoencephalitis, or acute psychosis may even be the first manifestations of HTLV-III infection.

31.5.3 Possible Long-Term Effects of Human T-Cell Lymphoma Virus Type III Infection

Nothing is known about the long-term effects of HTLV-III infection. It is not known if persistent asymptomatic infection might not some day, by some unknown factors be activated and lead to ARC and AIDS. In addition, the spectrum of opportunistic agents might include a variety of putative oncogenic viruses whose effects will be seen only after considerable time. In animals, for example, most naturally occurring leukemias and lymphomas are caused by retroviruses. In humans, tumors that are proven or likely to be caused by viruses include adult T-cell leukemia, cutaneous T-cell lymphomas, some chronic lymphatic leukemias and B-cell lymphomas (all, as mentioned above, associated to various degrees with HTLV-I or similar retroviruses), Burkitt's lymphoma, nasopharyngeal carcinoma and Hodgkin's lymphoma (associated with Epstein–Barr virus), cervical carcinoma (certain types of papilloma virus), and hepatocellular carcinoma (hepatitis B virus). Also, Kaposi's sarcoma, which is one of the hallmarks of AIDS, is considered to be of viral origin (possibly cytomegalovirus). In addition, various disorders with aspects of autoimmunity may have a viral origin. It is of great concern that some of the tumors mentioned above are already officially recognized as AIDS manifestations or are now increasingly seen in connection with HTLV-III infection and symptoms of the ARC. So far, Burkitt-like lymphomas and other non-Hodgkin lymphomas, Hodgkin's disease, chronic lymphocytic leukemia, adenosquamous carcinoma of the lung, hepatocellular carcinoma, and carcinoma of the oropharynx have been reported in this connection (Doll and List, 1982; Lozada et al., 1982; Arlin et al., 1983; Snider et al., 1983a; Irwin et al., 1984; Levine et al., 1984; Ziegler et al., 1984).

31.6 LABORATORY DIAGNOSIS OF HUMAN RETROVIRUS INFECTIONS

A variety of tests were developed for the demonstration of infection by HTLV-I, -II, and -III. They include immunologic tests for the demonstration of virus-specific antibodies, and tests for the demonstration of viral proteins in HTLV-infected cells or tissues.

31.6.1 Tests for the Demonstration of Virus-Specific Antibodies

In dealing with immunologic tests, it should always be kept in mind that positive immunologic tests only indicate relatedness, but never identity of the antigens compared. Thus, the demonstration of antibodies reacting with proteins of HTLV-III may, in fact, indicate that a person has been infected with HTLV-III, especially if the antibody titers are high. However, a positive result may also merely mean that the person has been immunized with a crossreactive agent that has nothing to do with the causation of AIDS. This restriction is especially important in cases where the individuals tested are not members of risk groups and were not knowingly exposed to the agents tested.

Usually, the demonstration of antiviral antibodies indicates past but not necessarily ongoing viral infection. In the cases of HTLV-I and -III, however, the clear demonstration of specific antiviral antibodies may be considered adequate evidence for ongoing infection, as these retroviral infections generally persist for lifetime. Virus was isolated, or its presence demonstrated otherwise, in the vast majority of seropositive persons, irrespective of their clinical status (Gotoh et al., 1982; Miyoshi et al., 1982; Sarin et al., 1983; Feorino et al., 1984; Salahuddin et al., 1985).

31.6.2 Enzyme-Linked Immunosorbent Assays

Enzyme-linked immunosorbent assays (ELISA) were developed for both HTLV-I and -III (Saxinger and Gallo, 1983; Sarngadharan et al., 1984), and commercial kits are now available for both. In these assays, viral antigens (either purified whole virus, or single viral proteins) are bound to the wells of microtiter plates. Human serum or other body fluids to be tested for antibodies are diluted in buffer, added to the wells, and incubated for a period of time. The wells are then washed, and a secondary enzyme-conjugated antibody is added that is directed against human immunoglobulins. After a brief incubation, the wells are washed again. The enzyme's substrate and a color indicator are then added, and a color reaction develops in the wells where binding of human antibodies and, thus, of the enzyme-labeled secondary antibody, occurred.

The ELISA is a relatively inexpensive quick test that is ideally suited for large scale screening of blood conserves. In our experience, its sensitivity is somewhat lower than that of Western blotting or radioimmunoprecipitation/SDS-PAGE assays (Resnick et al., 1985; Schüpbach et al., 1985a). This is especially true with sera that contain low-titer antibodies against a single viral component only (frequently p24). An additional weakness of the ELISA is its high rate of false-positive results with sera that were not properly stored or transported (Schüpbach et al., 1985a). In our opinion, a positive ELISA result should be confirmed by an independent method, preferably Western blot or radioimmunoprecipitation, before a person is informed.

31.6.3 Western Blot

The advantages of a properly performed Western blot are its high sensitivity and its excellent specificity, which greatly surpasses that of the ELISA. Western blot allows simultaneous testing for antibodies to a variety of viral antigens, without a need for purifying them first. It is the complexity and expenditure of work and materials that prevent this excellent test from being used in a large scale.

The original procedure of Western blot (Towbin et al., 1979) was modified to allow the antibody testing of multiple serum samples (Schüpbach et al., 1984b,c, 1985a; Sarngadharan et al., 1984). Figure 31.5 illustrates the essential steps of the procedure.

1. In the first step, the proteins from sucrose gradient density-banded virus (250 μg/gel) were fractionated by electrophoresis on a 12% polyacrylamide slab gel in the presence of sodium dodecyl sulfate according to the method of Laemmli (1970)
2. The protein bands are then electrophoretically transferred (45V, overnight at 4°C) to a nitrocellulose sheet, as described by Towbin (1979); the sheet is saturated for 2 hours at 37°C in "BLOTTO" buffer, consisting of phosphate buffered saline (PBS) containing 5% nonfat dry milk and 0.001% Merthiolate (Johnson et al., 1984)
3. The sheet is rinsed with PBS containing 0.05% Tween 20 (PBS-T), placed between two layers of Parafilm (American Can Company), and cut to 3–5 mm wide strips; these can be used directly for testing or stored at −20°C
4. For testing, strips are freed of their Parafilm sheaths, placed in plastic tubes containing 2.4 ml BLOTTO and 100 μl normal goat serum (NGS), and incubated for 1 hour in a horizontal position on a shaker; test serum is then added in a dilution of 1:50 (50 μl) or 1:100 (25 μl) and the strips incubated overnight at 4°C; the strips are washed in three changes of PBS-T (5–10 ml, 5 minutes shaking for each change)
5. A secondary, labeled goat antibody with specificity to human immunoglobulins is added in BLOTTO containing 100 μl NGS; excellent results are obtained with affinity-purified and iodinated antibodies to IgG and IgM, which are used at an activity of 1.25×10^6 cpm per tube and with an incubation time of 30 minutes; after a final round of three washings with PBS-T, the strips are dried, mounted on filter paper, and processed by autoradiography. Alternative methods involve the use of biotinylated antibodies in combination with additional incubations with avidin D-horseradish peroxidase and, after another washing cycle, with an apppropriate substrate (e.g., a mixture consisting of 10 ml PBS, 2 ml of a solution of 0.3% 4-chloro-1-naphthol in methanol, and 4 μl of 30% H_2O_2).

This procedure is probably almost as sensitive as the use of radiolabeled antibodies,

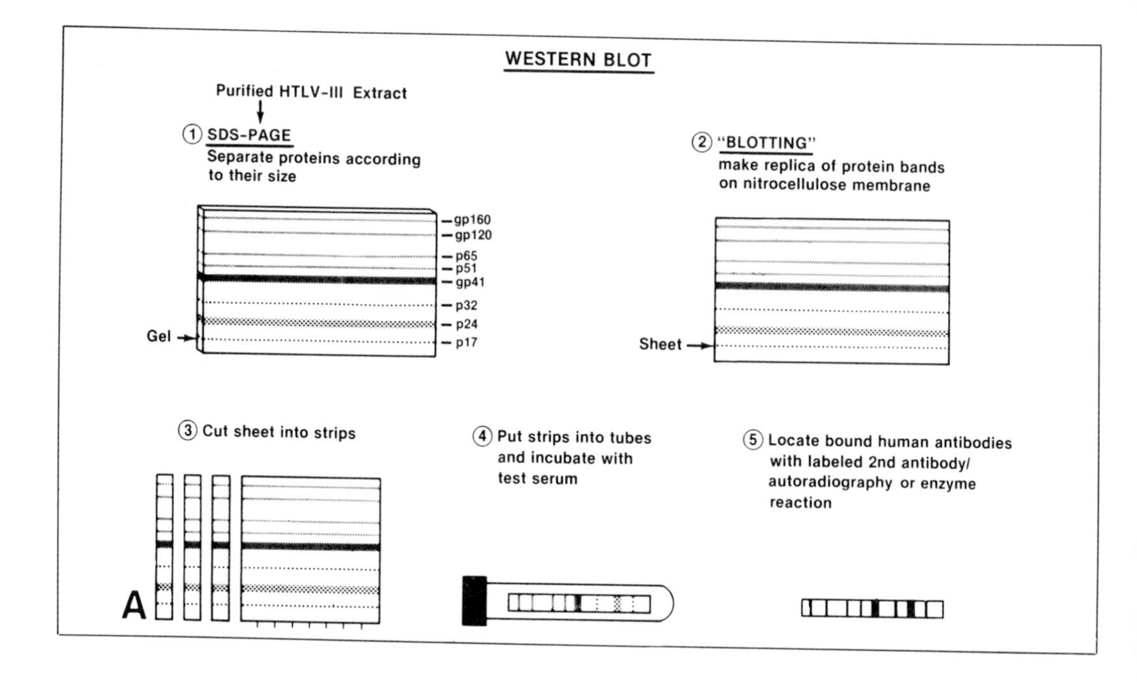

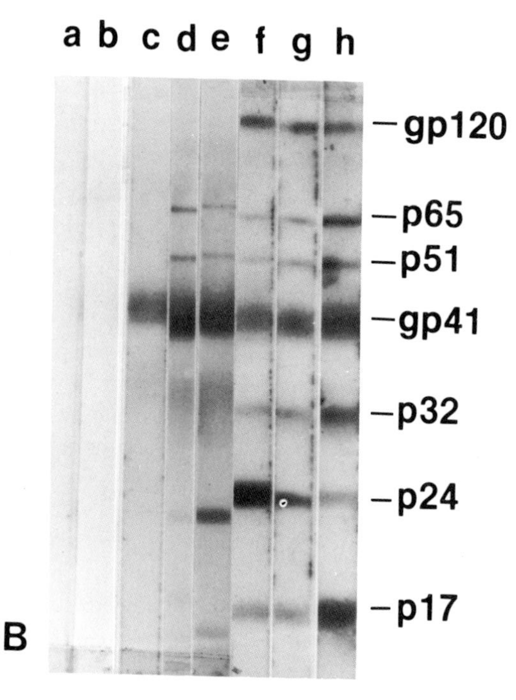

Figure 31.5. Western blotting with purified HTLV-III. (A) Principal steps; for details refer to text. (B) Representative strips of normal controls (a,b), patients with AIDS (c,d), a patient with ARC (e), and seropositive drug addicts (f–h).

and has the further advantage of shortening the test to a 24-hour period (strips can be prepared separately and stored frozen for several months). In contrast with the biotin/avidin system, the use of direct peroxidase-labeled antibodies is much less sensitive and should be avoided.

The interpretation of the results is easy in cases where antibodies against a variety of antigens are present and where the bands are strong. This is usually the case in patients with ARC. In advanced cases of AIDS, however, or at early stages of infection titers of antibodies may be low and only a single band may be detected. In addition, protein bands of cellular origin may be detected in positions slightly different from those of viral proteins. The quality of interpretation of Western blot, thus, depends largely on the experience of the interpreter. Proteins so far identified as coded for by HTLV-III include the *gag* proteins p24, p17, and p15, the *env* products gp160, gp120, and gp41 and the *pol* products p65/51. Antigen of unknown origin, but also associated with HTLV-III, is p32. Table 31.4 gives a summary of the distribution of some antibodies that are frequently detected by

Table 31.4. Distribution of Antibodies to Some HTLV-III–Associated Proteins Among Patients with AIDS or ARC, and Asymptomatically Infected Individuals

Group	Number Investigated	Prevalence of Antibodies (percentage) Directed at			
		p65	p51	gp41	p24
AIDS	23	74	70	100	61
ARC	46	91	89	98	98
Asymptomatic individuals	129	76	68	83	93

Western blot in the sera of patients with AIDS, ARC, and asymptomatically infected individuals. In patients with AIDS, antibodies to gp41 are thus present in virtually all cases, whereas, antibodies to p24 are found much less frequently. In ARC antibodies to gp41 and p24 are found at equally high frequency. Among asymptomatically infected individuals, antibodies to p24 are found at higher frequency than those to gp41.

The superiority of the Western blot was clearly demonstrated in an investigation of the presence of HTLV-III–specific antibodies in CSF of patients with AIDS or ARC (Resnick et al., 1985). The concentrations of virus-specific IgG antibodies in CSF are 100 to 1000 times less than in serum. The ELISA detected 91% of the CSF samples that were positive by Western blot and radioimmunoprecipitation. Less than 20% were positive by an indirect immunofluorescence test in which fixed HTLV-III producer cells were used as a target.

31.6.4 Radioimmunoassays

31.6.4.a Binding Assays (Solid Phase Radioimmunoassays)

The first generation of HTLV antibody test used were radioimmunoassays (RIA) (Posner et al., 1981; Robert-Guroff et al., 1982b). Whole virus proteins were coupled to wells of microtiter plates (Removawell Strips, Dynatech), as with the ELISA. An iodinated secondary antibody was used instead of an enzyme conjugate, and the activity of the detached wells was measured in a γ counter. The procedure appears to offer no advantage over the ELISA.

31.6.4.b Radioimmunoprecipitation Assays

These tests are usually done with single, purified proteins and, thus, are highly specific. Raidoimmunoprecipitation (RIP) assays were developed for p24 of HTLV-I (Kalyanaraman et al., 1981a,b, 1982a) and for the other *gag* proteins, p19 and p15 (Schüpbach et al., 1983a,b; Kalyanaraman et al., 1984b). Radioimmunoprecipitation assays were also developed for p24 of HTLV-III (Kalyanaraman et al., 1984a; Sarngadharan et al., 1985a).

For these tests, the purified proteins are iodinated to high activity (10,000–20,000 cpm/ng) and 8000–10,000 cpm are incubated with 10 μl of test serum. The total IgG is then precipitated by the addition of a predetermined amount of a goat antiserum to human IgG. The precipitate is washed and counted in a γ counter. In the presence of HTLV-specific antibodies in the test serum, labeled antigen will be bound in the precipitate.

The strength of the RIP assay is its specificity, which is partly due to the tedious work of antigen purification. Application of such tests in a larger scale might become feasible with genetically engineered viral proteins.

31.6.4.c RIP/SDS-PAGE Combination

In this test, metabolic labeling is used to label cellular or viral proteins. Cellular lysates are made and incubated with small amounts of test serum. The IgG fraction and any labeled antigens bound in immune complexes are then absorbed to protein A Sepharose and subjected to polyacrylamide gel electrophoresis in the presence of sodium dodecylsulfate (SDS-PAGE). Following autoradiography, the pattern of bands seen with the test serum is evaluated in comparison with the patterns obtained with positive and negative control sera.

The sensitivity of this test is very high and in the range of, or slightly higher than, that of Western blot (Resnick et al., 1985). The spectrum of antigens reactive in RIP/SDS-PAGE is different from that in Western blots performed with purified virus, because some proteins, especially the major envelope glycoprotein, tend to get lost during virus purification. Other antigens may become much more concentrated in the purified virus. Finally, a significant part of viral antigen may be present in the form of larger precursors in the cells, while the purified virus contains mostly their smaller end products. With HTLV-III, for example, the predominant antibodies seen by RIP/SDS-PAGE are usually directed at gp120, gp160, and gp41, and p24 is detected only in the presence of antibodies to these envelope antigens (Kitchen et al., 1984; Veronese et al., unpublished observations). In contrast, these larger antigens are only infrequently detected by Western blot with purified virus, and p24 may sometimes be the only antigen detected (Schüpbach et al., unpublished observations).

RIP/SDS-PAGE assays were used by some investigators as confirmatory tests after immunofluorescence or ELISA (Essex et al., 1983a,b; Schneider et al., 1985), but it is difficult to imagine how larger numbers of sera could be handled.

31.6.4.d Competition Radioimmunoassays

Competition RIA can be used for the detection of either antigens (see below) or antibodies. The principle of the test is that the interaction of a solid phase antigen with a corresponding radiolabeled antibody is competed by the previous binding of test serum antibodies. Any degree of antigen specificity may be obtained by the selective use of target antigens, the use of antisera or monoclonals of defined specificity, or by both. Tests based on this system were first developed for antibodies specific for p19 of HTLV-I (Robert-Guroff, 1982a), and later for antibodies to the total proteins of HTLV-I and -II (Tedder et al., 1984), and of HTLV-III (Cheinsong-Popov et al., 1984). These tests appear to be of high specificity, as demonstrated by the fact that none of more than 1000 unselected blood donors was positive in the HTLV-III assay (Cheinsong-Popov et al., 1984). The competition RIA may be quite sensitive, provided that the antibodies in the test serum recognize the same epitopes as the labeled reference antibody. HTLV-III antibody prevalences based on this test were in the same range as those obtained by Western blot 97% of the time. The competition RIA, thus, might be a good confirmatory test for the large scale screening of blood products.

31.6.5 Immunofluorescence Tests

Indirect immunofluorescence (IF) tests were some of the first tests used with human retrovirus infections. In fact, most of the serologic studies done in Japan were based on the demonstration of an ill-defined "ATLA" (adult T-cell leukemia antigen) (Hinuma et al., 1981), which only later was shown to be a conglomerate of viral or virus-associated proteins in HTLV-I–infected cells. Immunofluorescence tests, per se, are nonspecific tests in the sense that cells used as targets contain many other antigens that might possibly react with antibodies in any test serum. Even if uninfected cells of exactly the same type as the virus producer cells are available as negative controls (e.g., the uninfected H9 cell counterpart of the H9/HTLV-III producer), which is normally not the case, a positive result on H9/HTLV-III may still be nonspecific, because the reactive antibodies might be directed at a cellular component nonspecifically expressed in consequence of the virus infection. Immunofluorescence should only be used for screening, but never as a confirmatory test.

Indirect IF tests using live HTLV-infected cells (membrane IF) were also frequently used. These tests may be very sensitive due to the mechanism of patching and capping by which cell surface molecules crosslinked by antibodies are concentrated into relatively few spots or patches and then are moved to one cell pole. The specific interaction of relatively few antibody and antigen molecules may give a visible signal that, if dispersed over the whole cell surface, would be lost in the general background. Similar mechanisms are found elsewhere in nature, for example, in the color adaptation of certain fish or amphibia (flounders, frogs) to their environment. Other immunologic tests that take advantage of this principle are RIP assays and Western blot in which the concentration of antigens into defined areas (bands) is achieved by gel electrophoresis and, to a certain degree, fixed cell IF, because many antigens are located in spots or patches.

31.6.6 Large Scale Screening of Sera for Human T-Cell Lymphoma Virus Type III

Many blood banks and the plasma industries have now started to screen the donated blood by commercially available ELISA or HTLV-III. These assays will certainly eliminate the great majority of virus transmissions occurring by this route. Preliminary studies showed that about 60% of the sera of healthy blood donors that score—sometimes repeatedly—positive in these assays do not have antibodies detectable by any other method (Workshop on ELISA Test Kits for HTLV-III Antibody, NIH, Feb. 22, 1985). This proportion was even higher in studies done with noncommercial ELISA (Carlson et al., 1985; Schneider et al., 1985). The higher rate of positives by ELISA is certainly not due to a higher sensitivity of this test, because assays involving fractionation of viral and contaminant cellular proteins (Western blot and RIP/SDS-PAGE) that were shown to be of higher sensitivity than the ELISA (Resnick et al., 1985; and our personal observations) are negative. In view of the grave consequences a positive result may have, we recommend that positive ELISA results be confirmed by an independent assay before an individual is notified.

It has been proposed that an indirect fixed-cell IF test be used as a confirmatory assay in HTLV-III screening. We do not recommend IF for this purpose, for the following reasons: (a) IF per se is a nonspecific test that should only be used in combination with more specific assays; (b) fixed-cell IF is much less sensitive than the ELISA, Western blot, or RIP/SDS-PAGE (Resnick et al., 1985); (c) false-positive results may arise due to the circumstance that many AIDS or ARC patients may have T-cell specific autoantibodies (F. Jensen, personal communication). Such sera would not only score positive on the HTLV-III–infected cells, but also on noninfected control cells and, thus, be considered false-positive. An analogous situation may arise with the use of ELISA control plates that are coated with the extracts of uninfected H9 cells. At the moment, we only recommend Western blot, or RIP/SDS-PAGE for this purpose. Further work will show if these labor-intensive and expensive tests might be replaced by simpler assays, without loss of sensitivity and specificity.

31.6.7 Tests for Detection of Neutralizing Antibodies

Assays for the detection and quantitation of virus-neutralizing antibodies were developed for HTLV-I, -II, and -III. One system used was the construction of pseudotypes containing the genome and core proteins of vesicular stomatitis virus (VSV) and the envelope of HTLV-I, -II, or -III. The host cell specificity of these pseudotypes is determined by the HTLV envelope, but their other biological effects are still those of VSV; that is, they retain their lysing capabilities. Infection, thus, can be quantitated by plaque assays. HTLV-specific antibodies that, by their binding to the viral envelope, inhibit the binding of the particle to the virus receptor on the host cell membrane (virus-neutralizing antibodies) can simply be assayed by plaque inhibition (Clapham et al., 1984).

With this system, virus-neutralizing antibodies were readily detected in all of 12 patients infected by HTLV-I and in the one patient infected by HTLV-II (Clapham et al., 1984). The titers ranged from 50 to 30,000 in patients with ATLL and did not closely correlate with the titers to internal viral proteins. In HTLV-III infection, however, neutralizing activity was found in few individuals only and the geometric mean titer of 12 sera highly positive for nonneutralizing HTLV-III antibodies was less than 10. Only a small percentage of sera were capable of effective pseudotype neutralization, independent of whether they were from patients with AIDS, ARC, or from asymptomatic seropositive individuals. Similar antibody titers were found when the sera were tested with four different isolates of HTLV-III (Weiss et al., 1985).

Another assay used by the same group was the inhibition of human sera of HTLV-mediated syncytia formation (Nagy et al., 1983). The production of huge syncytia ("giant cells") containing many nuclei is a characteristic feature of infection by HTLV-I, -II, or -III (Popovic et al., 1983a, 1984b). Again, virus-neutralizing antibodies will block the interaction of viral envelope with cell membrane determinants and can thus be assayed on the base of syncytia inhibition.

Neutralizing activity was found in all of nine sera from ATLL patients, but the test was less sensitive and less specific than the pseudotype method. Hardly any activity was found when the test was applied to HTLV-III (Weiss et al., 1985).

In another type of assay, infectious virus is preincubated with test or control serum and the mixture then added to uninfected host cells. The appearance of viral proteins is inhibited or postponed in the presence of virus-neutralizing antibodies, and is monitored by indirect IF tests using HTLV-specific antisera or monoclonals, and by assays for reverse transcriptase in the supernatant fluids.

Using cell-free HTLV-III in combination with the H9 cell line and immunofluorescence of the cells on days 1 to 4 after infection, Robert-Guroff et al. detected neutralizing activity in 60% of 35 sera from AIDS patients and in 80% of the same number of ARC patients. The titers ranged from 10 to 560 with the mean geometric titers in ARC patients being twice as high as in the AIDS patients, and even higher in two healthy homosexual individuals (Robert-Guroff et al., 1985).

Using a similar system, but monitoring the cells for 2 weeks, Ho et al. detected neutralizing activity in four of four HTLV-III–positive healthy homosexuals, in three of three patients with ARC, and in three of five patients with AIDS. No activity was detected in uninfected normals, but the titers among the positives ranged only from 2 to 10 (Ho et al., 1985). Clavel et al., however, using their LAV isolate to infect fresh, phytohemagglutinin-stimulated normal peripheral blood lymphocytes and monitoring cytopathogenic effects and reverse transcriptase activity on day 7, found no neutralizing activity in seven of eight patients with AIDS or ARC. The only exception was the serum of the patient from whom LAV had originally been isolated. This serum neutralized 98% of the virus. All sera, however, inhibited LAV-mediated cell fusion to some degree higher than that of noninfected controls (Clavel et al., 1985).

Put together, the results of these tests indicate that virus-neutralizing activity is present in HTLV-I or -II–infected individuals. Whether or not these antibodies are of any importance with respect to the clinical course, remains to be determined. Only minor virus-neutralizing activity appears to be present in persons infected with HTLV-III. Whether or not this activity is due to antibodies directed against envelope components of the virus, remains to be determined. The finding of high neutralizing activity in the one patient from whom the virus used in the assay was isolated may indicate that, in view of the considerable heterogeneity found in the *env* products, neutralization assays using just one or a few closely related virus isolates may not be meaningful.

31.6.8 Tests for the Demonstration of Viral Proteins

31.6.8.a Demonstration of Antigens in Serum and Other Body Fluids

No reliable tests have been reported so far that would allow large scale screening of sera for the presence of antigens of HTLV-I, -II, or -III. As most persons infected with these viruses appear to have high titers of antiviral antibodies, any antigen present would be expected in the form of circulating immune complexes (CIC). Some of the antigen may be so well hidden in CIC that it is no longer detectable by the usual antigen assays. So far, CIC-bound viral antigen p24 was demonstrated by Western blot in consecutive serum samples of two patients with HTLV-I–associated lymphoma (Schüpbach et al., 1984a), and material compatible with CIC containing HTLV-I antigens was reportedly present in glomerular basal membrane deposits of a patient with ATLL (Miyoshi et al., 1983). The difficulties in developing reliable assays for these viruses, thus, may be partly due to this problem.

31.6.8.b Demonstration of Antigens in Cultured Cells

HTLV-I antigens usually are not detectable in fresh ATLL cells collected from leukemic blood, although virus-specific mRNA may be present in the cells (Franchini et al., 1984). The reason for this phenomenon is not known. After a short-term culture of one to a few days in the presence of TCGF, in most instances the cells start expressing viral proteins that can be detected by a variety of tests including assays for reverse transcriptase (see below), direct and indirect IF assays involving hyperimmune sera or monoclonal antibodies (Robert-Guroff et al., 1981; Palker et al., 1984), and competition RIP assays. In these, the precipitation of radiolabeled viral proteins by a reference antiserum is competed by preceding incubation of the antibody with test materials containing unlabeled antigen. The same test may be used to demonstrate crossreactivityy of antigens (Kalyanaraman et al., 1981b). As only a small minority of peripheral blood lymphocytes or lymph node cells of persons infected with HTLV-III contain the proviral genome (Gallo et al., 1985), reliable detection of antigen in fresh cells of HTLV-III–infected individuals is not possible, either. Currently, the demonstration of active infection can only be achieved by virus isolation methods.

31.6.8.c Reverse Transcriptase Assays

The activity of the retroviral enzyme reverse transcriptase can be used as a sensitive marker for infection by retroviruses. Even in the absence of demonstrable release of virus particles from cells, reverse transcriptase activity may suggest that retroviral information is expressed. Reverse transcriptase activity, however, is not specific in the sense that it would allow identification of the type of retrovirus involved.

For reverse transcriptase assays, it is essential to use virus preparations as free of contaminating cellular material as possible. The cell-free supernatants are subjected to polyethylene glycol 6000 precipitation (10% final concentration) in order to concentrate the virus. The samples are placed on ice for at least 2 hours and centrifuged at 1000 × *g* at 4°C for 45 minutes. The pellet is resuspended and the virus disrupted using a minimal volume of a buffer containing 25 m*M* Tris (pH 7.5), 5 m*M* dithiothreitol (DTT), 0.25 m*M* EDTA, 0.025% Triton X-100, 5 m*M* KCl, and 50% glycerol.

The reverse transcriptase assay measures the incorporation of tritiated deoxyribonucleotides into acid-insoluble material (Sarngadharan et al., 1978). The reaction mixture in a total volume of 100 μl contains 10 μl of the disrupted virus, 4 μl of 1 *M* Tris (pH 7.8), 4 μl of 0.2 *M* DTT, 5 μl of 0.2 *M* $MgCl_2$, 25 μl of ^3H-TTP (10–20 Ci/mmol), 72 μl of H_2O, and 5 μl of the appropriate template [$(dT)_{\sim 15} \cdot (A)_n$ and, as a control, $(dT)_{\sim 15} \cdot (dA)_n$; P.L. Biochemicals, Milwaukee, WI]. The reaction mixture is incubated at 37°C (for 1 hour) and the reaction is terminated by the addition of 50 μg of yeast tRNA and about 3 ml 10% TCA containing 0.02 *M* sodium pyrophosphate. The precipitate is kept on ice for at least 10 minutes, then collected on glass microfiber filters presoaked in 5% TCA containing 0.02 *M* sodium pyrophosphate. The filters are rinsed thoroughly with 5% TCA containing pyrophosphate, and finally with 70% ethanol. The washed filters are dried under a heat lamp. The incorporated radioactivity is determined by scintillation counting. Although RNA tumor viruses catalyze endogenous DNA synthesis in the absence of exogenous primer templates, the efficiency of the viral reverse transcriptase is amplified by the addition of several synthetic oligomer–homopolymer hybrid primer templates such as $(dT)_{\sim 15} \cdot (A)_n$ and $(dG)_{\sim 15} \cdot (C)_n$. The first of these is also used by certain cellular DNA polymerases like γ, but a combination of a high response with $(dT)_{\sim 15} \cdot (A)_n$ and a poor response with $(dT)_{\sim 15} \cdot (dA)_n$ in a duplicate sample is indicative of viral reverse transcriptase activity. A ratio of the two different activities of at least four represents a meaningful result. $(dG)_{\sim 15} \cdot (C)_n$ is specific for reverse transcriptase because DNA polymerases cannot use it, but is somehow less efficient than $(dT)_{\sim 15} \cdot (A)_n$, which is considered the primer-template of choice.

31.6.8.d Electron Microscopy

Electron microscopy does not play a role, so far, in the diagnosis of HTLV infections. This is due to the fact that the tumor cells of ATLL patients generally do not express the viral genome. Likewise, the fraction of peripheral blood or lymph node cells infected by HTLV-III is too small to allow detection of virus producing cells. Electron microscopy, thus, is restricted to scientific questions. Procedures have been published (Salahuddin et al., 1982) and may be found in detail elsewhere in this book.

31.6.9 Isolation Methods

31.6.9.a HTLV-I and -II

The discovery of TCGF or IL-2 allowed the establishment of procedures for long-term growth of mitogen- or antigen-stimulated normal T cells in vitro. When the same procedure is applied to grow T cells from peripheral blood or bone marrow from patients with various T-cell neoplasias, no prior stimulation with mitogen or antigen is required (Poiesz, 1980a). Heparinized peripheral blood or bone marrow samples are centrifuged at 1000 × *g* for 20 minutes and the cell pellets resuspended in RPMI 1640 medium. The mononuclear cells are banded in Ficoll–Hypaque, washed twice in PBS and resuspended in RPMI 1640 containing 1% L-glutamine, 1% penicillin/streptomycin mixture, 20% fetal calf serum, and 10% human TCGF. The cells are then placed into tissue culture flasks at a concentration of 10^6 cells/ml. The flasks must be checked twice weekly by monitoring cell number and viability. When the cells reach a concentration of 2 × 10^6 cells/ml, they are diluted with an equal volume of medium and continued in culture. When a continuous culture is established in the presence of TCGF, routine attempts are made to grow the cells without TCGF. Mitogen-stimulated normal peripheral blood lymphocytes are also grown as control cell lines following the same procedure. The cell cultures are also monitored at regular intervals for virus expression and release. Initially, virus expression is detected by release of viral reverse transcriptase activity into supernatant fluids and by EM examination; later specific reagents are used for the unequivocal detection and characterization of viral proteins.

Following this procedure, more than 100 isolates of HTLV-I were obtained and found comparable in terms of number of virus particles released. HTLV-I could be isolated only from two sources: peripheral blood and bone marrow mononuclear cells. An alternative, indirect method for virus isolation is by cocultivating normal umbilical cord blood lymphocytes with lethally irradiated lymphocytes from ATLL patients (Miyoshi et al., 1981). This procedure results in the productive infection and transformation of the cord blood T lymphocytes grown in suspension culture in the absence of exogenous TCGF (Markham et al., 1983). These transformed cord blood cells have morphologic and cytochemical properties similar to HTLV-positive fresh and cultured tumor cells, but are distinguishable from virus donor cells by HLA haplotype and chromosomal markers. They express HTLV proteins, release virus particles and contain surface receptors for TCGF. This method results in both TCGF independence and better virus production.

31.6.9.b HTLV-III

HTLV-III was isolated from cells of a large number of AIDS and ARC patients at all stages of the disease and from some clinically normal persons. The virus was isolated from several tissue sources including peripheral blood and bone marrow mononuclear cells, lymph nodes, saliva, urine, semen, cerebrospinal fluid and neural tissues and plasma (Gallo et al., 1984; Groopman et al., 1984; Ho et al., 1984; Popovic et al., 1984b; Zagury et al., 1984; Salahuddin et al., 1985; Ho et al., 1985). Isolation studies initiated in 1982 and procedures established for the isolation of HTLV-I and -II were modified for the isolation of HTLV-III (Popovic et al., 1984b). Heparinized peripheral blood and bone marrow mononuclear cells were collected and processed as mentioned above for HTLV-I and -II. Leukocytes from lymph nodes and brain biopsy specimens were prepared by mincing tissues to eliminate the stroma and banding on Ficoll–Hypaque before introduction into cell culture. Cell-free plasma was filtered and used directly or after pelleting of particulate material to infect normal peripheral blood mononuclear cells. Saliva samples were diluted to a final volume of 2 ml in complete growth medium, incubated for 2 hours at 37°C and centrifuged at 1000 × g for 10 minutes at 4°C. Supernatant fluids were filtered through a 0.45 μm filter and used for transmission experiments. Semen was obtained from AIDS patients and stored frozen. After thawing, mononuclear cells were banded on Ficoll–Hypaque and seeded in round bottom tissue culture clusters containing 200 μl of medium. The cells were activated for 24 hours with phytohemagglutinin using 0.1% (GIBCO, Grand Island, NY), and cultured in the presence of TCGF and a feeder layer containing 1.5 × 10^5 irradiated lymphoid cells. Proliferating cultures were used for transmission experiments.

Virus was directly isolated from peripheral blood and bone marrow mononuclear cells and from lymph node and brain cells established in culture. Virus was isolated from saliva and plasma by transmission or cocultivation of lymphocytes with mitogen-stimulated normal human lymphocytes (pretreated with polybrene or DEAE dextran) or permissive T-cell lines like subclones of the human T-cell line HT which, as mentioned before, is resistant to the cytopathic effect of HTLV-III and is a very good virus producer after infection. The tests employed to identify new HTLV-III isolates were: (a) release of particulate reverse transcriptase activity into cell culture supernatant fluids, (b) examination by EM to detect viral particles, (c) detection of HTLV-III proteins by indirect IF assay using HTLV-III–specific probes like monoclonal antibodies or hyperimmune sera, and (d) transmission of virus to cell cultures resulting in both cytopathic effect and virus release. Primary cells from patients usually start to produce virus within 2 to 3 weeks following establishment of the cultures. As soon as the cells start releasing virus, there is a coincidental reduction of viable cells, especially of those with the helper–inducer (OKT4/Leu 3+) phenotype. Nevertheless, a minor population of cells often survives and gives rise to another burst in virus production followed again by cell death. The same phenomenon occurs when virus collected from primary patient cells is used to infect fresh peripheral blood, bone marrow, or cord blood. However, the amount of virus released is higher and the destruction of the target cells more rapid. In this laboratory alone, more than 100 isolates have been obtained so far from individuals with AIDS, ARC, or healthy carriers of infection. The success rate of HTLV-III isolation varied depending on the condition and number of cells from different seropositive donors. The overall incidence of virus isolation was 50% for AIDS patients, 80% for ARC patients, and 30% for healthy individuals at risk with unknown serologic status. In contrast, presence or release of virus could not be demonstrated in more than 150 samples that were randomly selected from healthy donors and cultured the same way. A summary of the incidence and sources of HTLV-III isolation is given in Table 31.5. The higher efficiency of virus isolation from ARC patients than from AIDS patients reflects the relative availability of the cells carrying the virus.

Table 31.5. Incidence and Sources of HTLV-III Isolates

	Patient/Donor Diagnosis		
Tissue Source	Healthy Risk Group Individuals[a]	ARC	AIDS
Peripheral Blood	16/50	31/38	43/88
Bone Marrow	NT[b]	1/6	NT
Lymph Nodes	NT	4/4	NT
Brain Abscess	NT	NT	2/3
Plasma	NT	3/6	NT
Saliva	4/6	4/10	0/4
Semen	1/1	NT	2/2

[a] Serologic status unknown.
[b] NT, not tested.

31.7 STABILITY AND SAFETY PRECAUTIONS

Infection of cells with HTLV-I or -II requires direct contact of infected donor and uninfected target cells in most instances, thus, revealing low stability and infectivity of these viruses. Detailed studies were carried out on the stability of HTLV-III. To mimic some natural and clinical laboratory conditions, HTLV-III in medium supplemented with 50% human plasma was exposed to a dried state or incubated at different temperatures, such as room temperature, 37°C, and 56°C, and thereafter tested for infectivity. Complete inactivation of infectious virus was achieved only after 3 to 7 days in a dried state, 11 to 15 days after exposure at 37°C, and 3 to 5 hours at 56°C (Resnick et al., 1986). In contrast, a previous report described the inactivation of LAV after 30 minutes at 56°C (Spire et al., 1985). These results, however, were based merely on the determination of residual reverse transcriptase activity and did not take into account that infectious virus might still be present in the absence of detectable reverse transcriptase activity. Commonly used chemical disinfectants were also tested for their ability to inactivate HTLV-III. Sodium hypochlorite at a 0.5% final concentration (a 20-fold dilution of Chlorox) and ethanol at a 70% concentration completely inactivated the virus within 1 minute. Quaternary ammonium chloride (A-500) at a 15% concentration required at least 10 minutes to result in a complete inactivation. The effect of NP40 was also tested as an example of nonionic detergent used in the disruption of HTLV-III. Exposure of the virus to 0.5% NP40 resulted in complete inactivation of the virus within 1 minute. Inactivation of LAV by γ and ultraviolet irradiation was also studied (Spire et al., 1985). Lymphadenopathy-associated virus required exposure to 2.5×10^5 rad and 5×10^3 J/m^2 to become noninfectious. The latter result indicates that LAV is not inactivated by UV radiation exposures even higher then those normally used under laminar hoods or in laboratories.

These results underscore the need for special precautions in all dealings with this agent. Surfaces stained with possibly infected materials should be decontaminated immediately with 70% alcohol, sodium hypochlorite solution, or detergent. In general, biosafety level 2 practices, containment equipment, and facilities are recommended for all activities utilizing known or potentially infectious body fluids and tissues. Additional primary containment and personnel precautions, such as those described for biosafety level 3, may be indicated for activities with high potential for droplet or aerosol production and for activities involving production quantities or concentration of infectious materials. If these recommendations are followed, the risk of job-associated HTLV-III infection should be small. So far, only one case of needlestick transmission of HTLV-III has been documented (Anonymous, 1984) among hundreds of needlestick exposures reported to the CDC (Hirsch et al., 1985).

31.8 CONCLUSIONS

The isolation and characterization of human T-lymphotropic retroviruses and their etiologic association with ATLL and AIDS have brought the unfolding of a new chapter of retrovirology and medicine. Infection of T cells by viruses of the HTLV group may have two principal effects: augmentation or elimination. The consequences in vivo may include the development of various cancers and immune disorders, depending on the type of cell infected. A wide variety of cells might be affected, because HTLV were shown to infect cells other than the OKT4+ subset (e.g., OKT8+ suppressor/cytotoxic T cells, T cells neither expressing OKT4 nor OKT8, B cells, macrophages, fibroblasts, and glial cells of the CNS) (Yamamoto et al., 1982a; Clapham et al., 1983; Markham et al., 1984; Yoshikura et al., 1984; Shaw et al., 1985; and personal communications by M. Popovic and Z. Salahuddin). It is possible that other HTLV yet to be discovered may have preference for still other cell types. Consequences of retroviral infection also may be indirect. As many of the cells possibly infected are part of the immunologic network and may secrete a variety of lymphokines (Salahuddin et al., 1984b), the possible effects of retroviral infections in humans would include a wide range of disorders, such as leukemias or lymphomas of T, B, or "null" cell type, the macrophage/histiocyte lineage (Hodgkin's disease), the nonlymphocytic leukemias, disorders with aspects of autoimmunity, and degenerative disorders.

Diseases of known etiology are more amenable to prevention and treatment. For both ATLL and AIDS, the ultimate goal is to develop safe vaccines. This may be more difficult in AIDS, due to the apparent heterogeneity of HTLV-III. However, as evidence suggests that the destruction of the T4 cells in AIDS may be largely due to the cytopathogenic effects of HTLV-III, therapeutic approaches interfering with the spread of the virus at an early point of infection might prove valuable. Specific inhibitors for the various steps of the retroviral replication cycle, such as provirus transcription, attachment of virus particles to virus receptor and penetration, and reverse transcription, may prove successful.

REFERENCES

Allan, J.S., Coligan, J.E., Barin, F., et al. 1985. Major glycoprotein antigens that induce antibodies in AIDS patients are encoded by HTLV-III. Science 228:1091–1094.

Anderson, L.J., Jarret, W.F.H., Jarrett, O., and Laird, H.M. 1971. Feline leukemia virus infection of kittens: Mortality associated with atrophy of the thymus and lymphoid depletion. J. Natl. Cancer Inst. 47:807–817.

Anonymous 1984. Editorial: Needlestick transmission of HTLV-III from a patient infected in Africa. Lancet ii:1376–1377.

Arlin, Z.A., Mittelman, A., Gebhard, D., and Danieu, L. 1983. Chronic lymphocytic leukemia in a bisexual male. Cancer Invest. i:549–550.

Arya, S.K., Guo, C., Josephs, S.F., and Wong-Staal, F. 1985. Trans-activator gene of human T-lymphotropic virus type-III (HTLV-III). Science 229:69–73.

Barré-Sinoussi, F., Chermann, J.C., Rey, F., et al. 1983. Isolation of T-lymphotropic retrovirus from a patient at risk for acquired immune deficiency syndrome (AIDS). Science 220:868–871.

Biggar, R.J., Saxinger, C., Gardiner, C., et al. 1984. Type-I HTLV antibody in urban and rural Ghana, West Africa. Intl. J. Cancer 34:215–219.

Biggar, R.J., Melbye, M., Kestens, L., et al. 1985. Seroepidemiology of HTLV-III antibodies in a remote population of eastern Zaire. Br. Med. J. 290:808–810.

Blattner, W.A., Kalyanaraman, V.S., Robert-Guroff, M. et al. 1982. The human type-C retrovirus, HTLV, in blacks from the Caribbean, and relationship to adult T-cell leukemia/lymphoma. Int. J. Cancer 30:257–264.

Blattner, W.A., and Gallo, R.C. 1985. Human T-cell leukemia/lymphoma viruses: Clinical and epidemiologic features. Curr. Top. Microbiol. Immunol. 115:67–88.

Blayney, D.W., Blattner, W.A., Robert-Guroff, M., et al. 1983a. The human T-cell leukemia–lymphoma virus in the Southeastern United States. J. Am. Med. Assoc. 250:1048–1052.

Blayney, D.W., Jaffe, E.S., Fisher, R.I., et al. 1983b. The human T-cell leukemia/lymphoma virus, lymphoma, lytic bone lesions, and hypercalcemia. Ann. Intern. Med. 98:144–151.

Bowen, D.L., Lane, C.H., and Fauci, A.S. 1985. Immunologic features of AIDS. In V.T. DeVita, Jr., F. Hellman, S.A. Rosenberg (eds.), AIDS—Etiology, Diagnosis, Treatment and Prevention. Philadelphia: J.B. Lippincott Co., pp. 89–109.

Bunn, P.A., Schechter, G.P., Jaffe, E. et al. 1983. Clinical course of retrovirus-associated adult T-cell lymphoma in the United States. N. Engl. J. Med. 309:257–264.

Carlson, J., Hinrichs, S., Bryant, M. et al. 1985. HTLV-III antibody screening of blood bank donors. Lancet i:523–524.

Catovsky, D., Greaves, M.F., Rose, M. et al. 1982. Adult T-cell lymphoma–leukemia in blacks from the West Indies. Lancet i:639–643.

Cheingsong-Popov, R., Weiss, R.A., Dalgleish, A., et al. 1984. Prevalence of antibody to human T-lymphotropic virus type III in AIDS and AIDS-risk patients in Britain. Lancet ii:477–480.

Clapham, P., Nagy, K., Cheingsong-Popov, R., Exley, M., and Weiss, R.A. 1983. Productive infection and cell-free transmission of human T-cell leukemia virus in a nonlymphoid cell line. Science 222:1125–1127.

Clapham, P., Nagy, K., and Weiss, R.A. 1984. Pseudotypes of human T-cell leukemia virus types 1 and 2: Neutralization by patients' sera. Proc. Natl. Acad. Sci. USA 81:2886–2889.

Clark, J.W., Blattner, W.A., and Gallo, R.C. 1986. Human T-cell leukemia viruses and T-cell lymphoid malignancies. In P. Storf and J. Mendelsohn (eds.), Principles of Internal Medicine, 7th ed. New York: McGraw-Hill Co., in press.

Clavel, F., Klatzman, D., and Montagnier, L. 1985. Deficient LAV_1 neutralizing capacity of sera from patients with AIDS or related syndromes. Lancet i:879–880.

Cooper, D.A., Gold, J., Maclean, P. et al. 1985. Acute AIDS retrovirus infection. Definition of a clinical illness associated with seroconversion. Lancet i:537–540.

Copeland, T.D., Oroszlan, S., Kalyanaraman,

V.S., Sarngadharan, M.G. and Gallo, R.C. 1983. Complete amino acid sequence of human T-cell leukemia virus structural protein p15. FEBS Lett. 162:390–395.

Costello, C., Catovsky, D., O'Brien, M., Morilla, R., and Varadi, D. 1980. Chronic T-cell leukemias. I. Morphology, cytochemistry and ultrastructure. Leukemia Res. 4:463–476.

Curran, J.W., Lawrence, D.N., Jaffe, H., et al. 1984. Acquired immunodeficiency syndrome (AIDS) associated with transfusions. N. Engl. J. Med. 310:69–75.

Doll, D.C., and List, A.F. 1982. Burkitt's lymphoma in a homosexual man. Lancet i:1026–1027.

Essex, M., Hardy, W.D., Jr., Cotter, S.M., Jakowski, R.M., and Sliski, A.H. 1975. Naturally occurring persistent feline oncornavirus infections in the absence of disease. Infect. Immun. 11:470–475.

Essex, M., McLane, M.F., Lee, T.H., et al. 1983a. Antibodies to human T-cell leukemia virus membrane antigens (HTLV-MA) in hemophiliacs. Science 221:1061–1064.

Essex, M., McLane, M.F., Lee, T.H., et al. 1983b. Antibodies to cell membrane antigens associated with human T-cell leukemia virus in patients with AIDS. Science 220:859–862.

Essex, M.E., McLane, M.F., Tachibana, N., Francis, D.P., and Lee, T.H. 1984. Seroepidemiology of human T-cell leukemia virus in relation to immunosuppression and the acquired immunodeficiency syndrome. In R.C. Gallo, M. Essex, and L. Gross (eds.), Human T-Cell Leukemia/Lymphoma Virus. Cold Spring Harbor, N.Y.: Cold Spring Harbor Laboratory, pp. 355–362.

Evatt, B.L., Gromperts, E.D., McDougal, J.S., and Ramsey, R.B. 1984. Coincidental appearance of LAV/HTLV-III antibodies in hemophiliacs and the onset of the AIDS epidemic. N. Engl. J. Med. 312:483–486.

Eyster, M.E., Goedert, J.J., Sarngadharan, M.G., et al. 1985. Development and early natural history of HTLV-III antibodies in persons with hemophilia. J. Am. Med. Assoc. 253:2219–2224.

Feorino, P.M., Jaffe, H.W., Palmer, E. et al. 1985. Transfusion-associated acquired immunodeficiency syndrome. Evidence for persistent infection in blood donors. N. Engl. J. Med. 312:1293–1296.

Fleming, A.F., Yamamoto, N., Bhusnurmath, S.R., et al. 1983. Antibodies to ATLV (HTLV) in Nigerian blood donors and patients with chronic lymphatic leukemia or lymphoma. Lancet ii:334–335.

Franchini, G., Wong-Staal, F., and Gallo, R.C. 1984. Molecular studies of human T-cell leukemia virus and adult T-cell leukemia. J. Invest. Dermatol. 83:63s–66s.

Gallo, R.C. 1984. Human T-cell leukemia–lymphoma virus and T-cell malignancies in adults. In L.M. Franks, J. Wyke, and R.A. Weiss (eds.), Cancer Surveys, Vol. 3. Oxford: Oxford University Press, pp. 113–159.

Gallo, R.C., and Blattner, W.A. 1985. Human T-cell leukemia/lymphoma viruses: ATL and AIDS. In V.T. DeVita, Jr., S. Hellman, and S.A. Rosenberg (eds.), Important Advances in Oncology. Philadelphia: J.B. Lippincott Co., pp. 104–138.

Gallo, R.C., Mann, D., Broder, S., et al. 1982. Human T-cell leukemia–lymphoma virus (HTLV) is in T- but not B-lymphocytes from a patient with cutaneous T-cell lymphoma. Proc. Natl. Acad. Sci. USA 79:5680–5683.

Gallo, R.C., Salahuddin, S.Z., Popovic, M., et al. 1984. Frequent detection and isolation of cytopathic retroviruses (HTLV-III) from patients with AIDS and at risk for AIDS. Science 224:500–503.

Gallo, R.C., Sarin, P.S., Gelmann, E.P., et al. 1983. Isolation of human T-cell leukemia virus in acquired immune deficiency syndrome (AIDS). Science 220:865–867.

Gallo, R.C., Shaw, G.M., and Markham, P.D. 1985. The etiology of AIDS. In V.T. DeVita, Jr., F. Hellman, and S. Rosenberg (eds.), AIDS—Etiology, Diagnosis, Treatment and Prevention. Philadelphia: J.B. Lippincott Co., pp. 31–54.

Gajdusek, D.C., Amyx, H.L., Gibbs, C.J., et al. 1985. Infection of chimpanzees by human T-lymphotropic retroviruses from brain and other tissues from AIDS patients. Lancet i:55–56.

Gelmann, E.P., Franchini, G., Manzari, V., Wong-Staal, F., and Gallo, R.C. 1984. Molecular cloning of a new unique T-cell leukemia virus (HTLV-II_{MO}). Proc. Natl. Acad. Sci. USA 81:993–997.

Gelmann, E.P., Popovic, M., Blayney, D., et al. 1983. Proviral DNA of a retrovirus, human T-cell leukemia/lymphoma virus in two patients with AIDS. Science 220:862–865.

Gerety, R.J., and Aronson, D.L. 1982. Plasma derivatives and viral hepatitis. Transfusion 22:347–351.

Goedert, J.J., and Blattner, W.A. 1985. The epidemiology of AIDS and related conditions. In V.T. DeVita, Jr., S. Hellman, and S.A. Rosenberg (eds.), AIDS—Etiology, Diagnosis, Treatment and Prevention. Philadelphia: J.B. Lippincott Co., pp. 1–30.

Goedert, J.J., Sarngadharan, M.G., Biggar, R.J., et al. 1984. Determinants of retrovirus (HTLV-III) antibody and immunodeficiency conditions in homosexual men. Lancet ii:711–716.

Gonda, M.A., Wong-Staal, F., Gallo, R.C., Clements, J., Narayan, D., and Gilden, R.V. 1985. Sequence homology and morphologic similarity of HTLV-III and visna virus, a pathogenic lentivirus. Science 227:173–177.

Gotoh, Y.-I., Sagamura, K., and Hinuma, Y. 1982. Healthy carriers of a human retrovirus, adult T-cell leukemia virus (ATLV): Demonstration by clonal culture of ATLV-carrying T-cells from peripheral blood. Proc. Natl. Acad. Sci. USA 79:4780–4782.

Gottlieb, M.S., Schroff, R., Schanker, H.M. et al. 1981. *Pneumocystis carinii* pneumonia and mucosal candidiasis in previously healthy homosexual men. Evidence of a new acquired cellular immunodeficiency. N. Engl. J. Med. 305:1425–1431.

Greaves, M.F., Verbi, W., Tilley, R. et al. 1984. Human T-cell leukemia virus (HTLV) in the United Kingdom. Intl. J. Cancer 33:795–806.

Groopman, J.E., Salahuddin, S.Z., Sarngadharan, M.G. et al. 1984. HTLV-III in saliva of people with AIDS-related complex and healthy homosexual men at risk for AIDS. Science 226:447–449.

Hahn, B.H., Popovic, M., Kalyanaraman, V.S., et al. 1984a. Detection and characterization of an HTLV-II provirus in a patient with AIDS. In M.S. Gottlieb and J.E. Groopman (eds.), Acquired Immune Deficiency Syndrome. New York: Alan R. Liss, pp. 73–81.

Hahn, B.H., Shaw, G.M., Arya, S.K., Popovic, M., Gallo, R.C., Wong-Staal, F. 1984b. Molecular cloning and characterization of the virus associated with AIDS (HTLV-III). Nature 312:166–169.

Hanaoka, M. 1982. Progress in adult T-cell leukemia research. Acta Pathol. Jpn. 32(suppl.)171–185.

Hardy, A.M., Allen, J.R., Morgan, W.M., and Curran, J.W. 1985. The incidence of AIDS in selected population groups. J. Am. Med. Assoc. 253:215–220.

Haseltine, W.A., Sodroski, J., Patarca, R., Briggs, D., Perkins, D., and Wong-Staal, F. 1984. Structure of 3′-terminal region of type-II human T-lymphotropic virus: Evidence for new coding region. Science 225:419–421.

Hattori, T., Uchiyama, T., Toibana, T., et al. 1981. Surface phenotype of Japanese adult T-cell leukemia cells characterized by monoclonal antibodies. Blood 58:645–647.

Haynes, B.F., Miller, S.W., Palker, T.J., et al. 1983. Identification of human T-cell leukemia virus in a Japanese patient with adult T-cell leukemia and cutaneous lymphomatous vasculitis. Proc. Natl. Acad. Sci. USA 80:2054–2058.

Hinuma, Y., Nagata, K., Hanaoka, M., et al. 1981. Adult T-cell leukemia: Antigen in an ATL cell line and detection of antibodies to the antigens in human sera. Proc. Natl. Acad. Sci. USA 78:6476–6480.

Hinuma, Y., Komoda, H., Chosa, T. et al. 1982. Antibodies to adult T-cell leukemia-virus-associated antigens (ATLA) in sera from patients with ATL and controls in Japan: A nationwide sero-epidemiologic study. Intl. J. Cancer 29:631–635.

Hirsch, M.S., Wormser, G.P., Schooley, R.T., et al. 1985. Risk of nosocomial infection with human T-cell lymphotropic virus III (HTLV-III). N. Engl. J. Med. 312:1–4.

Ho, D.D., Schooley, R.T., Rota, T.R., et al. 1984. HTLV-III in the semen and blood of a healthy homosexual man. Science 226:451–453.

Ho, D.D., Rota, T.R., and Hirsch, M.S. 1985a. Antibody to lymphoadenopathy-associated virus in AIDS. N. Engl. J. Med. 312:649–650.

Ho, D.D., Rota, R.T., Schooley, R.T., Kaplan, J.C., Davis Allan, J., Groopman, J.E., Resnick, L., Felsenstein, D., Andrews, C.A., and Hirsch, M.S. 1985b. Isolation of HTLV-III from cerebrospinal fluid and neural tissues of patients with neurologic syndromes related to the acquired immunodeficiency syndrome. N. Engl. J. Med. 313:1493–1497.

Hunsmann, G., Schneider, J., Schmitt, J., and Yamamoto, N. 1983. Detection of serum antibodies to adult T-cell leukemia virus in non-human primates and in people from Africa. Intl. J. Cancer 32:329–332.

Irwin, L.E., Begandy, M.K., and Moore, T.M. 1984. Adenosquamous carcinoma of the lung and the acquired immunodeficiency syndrome. Ann. Intern. Med. 100:158.

Jaffe, E.S., Cossman, J., Blattner, W.A., et al. 1984. The pathologic spectrum of adult T-cell leukemia/lymphoma in the United States. Am. J. Surg. Pathol. 8:263–275.

Jaffe, H.W., Feorino, P.M., Darrow, W.W., et al. 1985. Persistent infection with HTLV-III/LAV in apparently healthy homosexual men. Ann. Intern. Med. 102:627–630.

Johnson, D.A., Gautsch, J.W., Sportsman, J.R., and Elder, J.H. 1984. Improved technique utilizing nonfat dry milk for analysis of proteins and nucleic acids transferred to nitrocellulose. Gene Anal. Techn. 1:3–8.

Jordan, B.D., Navia, B., Petito, C. et al. 1985. Neurological syndromes complicating AIDS. Front. Rad. Ther. Oncol. 19:82–87.

Kalyanaraman, V.S., Sarngadharan, M.G., Bunn, P.A., Minna, J.D., and Gallo, R.C. 1981a. Antibodies in human sera reactive against an internal structural protein of human T-cell lymphoma virus. Nature 294:271–273.

Kalyanaraman, V.S., Sarngadharan, M.G., Poiesz, B., Ruscetti, F.W., and Gallo, R.C. 1981b. Immunological properties of a type C retrovirus isolated from cultured human T-lymphoma cells and comparison to other mammalian retroviruses. J. Virol. 38:906–915.

Kalyanaraman, V.S., Sarngadharan, M.G., Nakao, Y. et al. 1982a. Natural antibodies to the structural core protein (p24) of the human T-cell leukemia (lymphoma) retrovirus (HTLV) found in sera of leukemic patients in Japan. Proc. Natl. Acad. Sci. USA. 79:1653–1657.

Kalyanaraman, V.S., Sarngadharan. M.G., Robert-Guroff, M., et al. 1982b. A new subtype of human T-cell leukemia virus (human T-leukemia virus-II) associated with a T-cell variant of hairy cell leukemia. Science 218:571–573.

Kalyanaraman, V.S., Cabradilla, C.D., Getchell, J.P., et al. 1984a. Antibodies to the core protein of lymphadenopathy-associated virus (LAV) in patients with AIDS. Science 225:321–323.

Kalyanaraman, V.S., Jarvis-Morar, M., Sarngadharan, MG., and Gallo, R.C. 1984b. Immunological characterization of the low molecular weight *gag* gene proteins p19 and p15 of human T-cell leukemia–lymphoma virus (HTLV), and demonstration of human natural antibodies to them. Virology 132:61–70.

Kawano, F., Tsuda, H., Yamaguchi, K. et al. 1984. Unusual clinical course of adult T-cell leukemia in siblings. Cancer 54:131–134.

Kitchen, L.W., Barin, F., Sullivan, J.L. et al. 1984. Aetiology of AIDS-antibodies to human T-cell leukaemia virus (type III) in haemophiliacs. Nature 312:367–369.

Klatzmann, D., Barré-Sinoussi, F., Nugeyre, M.T. et al. 1984. Selective tropism of lymphadenopathy associated virus (LAV) for helper–inducer T-lymphocytes. Science 225:59–63.

Koerper, M.A., Kaminsky, L.S., and Levy, J.A. 1985. Differential prevalence of antibody to AIDS-associated retrovirus in hemophiliacs treated with factor VIII concentrate versus cryoprecipitate: Recovery of infectious virus. Lancet i:275.

Komuro, A., Hayami, M, Fujii, H., Miyahara, S., and Hirayama, M. 1983. Vertical transmission of adult T-cell leukemia virus. Lancet i:240.

Laemmli, U.K. 1970. Cleavage of structural proteins during the assembly of the head of bacteriophage T4. Nature 227:680–685.

Landesman, S.H., Ginzburg, H.M., Weiss, S.H. 1985. The AIDS epidemic. N. Engl. J. Med. 312:521–524.

Lando, Z., Sarin, P.S., Megson, M. et al. 1983.

Association of human T-cell leukemia/lymphoma virus with the Tac antigen marker for the human T-cell growth factor receptor. Nature 305:733–736.

Lapointe, N., Michaud, J., Pekovic, D. et al. 1985. Transplacental transmission of HTLV-III virus. N. Engl. J. Med. 312:1325–1326.

Lee, T.H., Coligan, J.E., Homma, T., et al. 1984a. Human T-cell leukemia virus-associated cell membrane antigens: Identity of the major antigens recognized after virus infection. Proc. Natl. Acad. Sci. USA 81:3856–3860.

Lee, T.H., Coligan, J.E., Sodroski, J., et al. 1984b. Antigens encoded by the 3′-terminal region of human T-cell leukemia virus: Evidence for a functional gene. Science 226:57–61.

Lennert, K., Kikuchi, M., Sato, E., et al. 1985. HTLV-positive and -negative T-cell lymphomas. Morphological and immunohistochemical differences between European and HTLV-positive Japanese T-cell lymphomas. Intl. J. Cancer 35:65–72.

Levine, A.M., Meyer, P.R., Begandy, M.K. et al. 1984. Development of B-cell lymphoma in homosexual men. Ann. Intern. Med. 100:7–13.

Levy, J.A., Hoffman, A.D., Kramer, S.M., et al. 1984. Isolation of lymphocytopathic retroviruses from San Francisco patients with AIDS. Science 225:840–842.

Lozada, F., Silverman, S., Conant, M. 1982. New outbreaks of oral tumors, malignancies and infectious diseases strike young male homosexuals. Calif. Dent. J. 10:39.

Maeda, Y., Furukawa, M., Takehara, Y. et al. 1984. Prevalence of possible adult T-cell leukemia virus-carriers among volunteer blood donors in Japan: A nationwide study. Intl. J. Cancer 33:717–720.

Manzari, V., Wong-Staal, F., Franchini, G., et al. 1983. Human T-cell leukemia–lymphoma virus (HTLV): Cloning of an integrated defective provirus and flanking cellular sequences. Proc. Natl. Acad. Sci. USA 80:1574–1578.

Manzari, V., Fazio, V.M., Martinotti, S., et al. 1984. Human T-cell leukemia/lymphoma virus (HTLV-I) DNA: Detection in Italy in a lymphoma and in a Kaposi sarcoma patient. Intl. J. Cancer 34:891–892.

Markham, P.D. (personal communication).

Markham, P.D., Salahuddin, S.Z., Kalyanaraman, V.S., et al. 1983. Infection and transformation of fresh human umbilical cord blood cells by multiple sources of human T-cell leukemia–lymphoma virus (HTLV). Intl. J. Cancer 31:413–420.

Mathur-Wagh, U., Enlow, R.W., Spigland, I. et al. 1984. Longitudinal study of persistent generalized lymphadenopathy in homosexual men: Relation to acquired immunodeficiency syndrome. Lancet i:1033–1038.

Melbye, M., Froebel, K.S., Madhok, R., et al. 1984. HTLV-III seropositivity in European haemophiliacs exposed to factor VIII concentrate imported from the USA. Lancet ii:1444–1446.

Melbye, M., Ingerslev, J., Biggar, R.J., et al. 1985. Anal intercourse as a possible factor in heterosexual transmission of HTLV-III to spouses of hemophiliacs. N. Engl. J. Med. 312:857.

Merino, F., Robert-Guroff, M., Clark, J., et al. 1984. Natural antibodies to human T-cell leukemia/lymphoma virus in healthy Venezuelan populations. Intl. J. Cancer 34:501–506.

Metroka, C.E., Cunningham-Rundles, S., Pollack, M.S., et al. 1983. Generalized lymphadenopathy in homosexual men. Ann. Intern. Med. 99:585–591.

Mitsuya, H., Guo, H.-G., Megson, M. et al. 1984. Transformation and cytopathogenic effect in an immune human T-cell clone infected by HTLV-I. Science 223:1293–1296.

Miyamoto, K., Tomita, N., Ishii, A., et al. 1984. Transformation of ATLA-negative leukocytes by blood components from anti–ATLA-positive donors in vitro. Intl. J. Cancer 33:721–725.

Miyamoto, Y., Yamaguchi, K., Nishimura, H., et al. 1985. Familial adult T-cell leukemia. Cancer 55:181–185.

Miyoshi, I., Kubonishi, I., Yoshimoto, S., et al. 1981. Type-C virus particles in a cord T-cell line derived by cocultivating normal human cord leukocytes and human leukemic T-cells. Nature 294:770–771.

Miyoshi, I., Taguchi, H., Fujishita, M., et al.

1982. Asymptomatic type C virus carriers in the family of an adult T-cell leukemia patient. Gann 73:339–340.

Miyoshi, I., Yoshimoto, S., Ohtsuki, Y., et al. 1983. Adult T-cell leukemia antigen in renal glomerulus. Lancet i:768–769.

Morgan, D.A., Ruscetti, F.W., and Gallo, R.C. 1976. Selective in vitro growth of T-lymphocytes from normal human bone marrow. Science 193:1007–1008.

Mösseler, J., Schimpf, K., Auerswald, G., et al. 1985. Inability of pasteurized factor VIII preparations to induce antibodies to HTLV-III after long-term treatment. Lancet i:1111.

Muesing, M.A., Smith, D.H., Cabradilla, C.D., et al. 1985. Nucleic acid structure and expression of the human AIDS/lymphadenopathy retrovirus. Nature 313:450–458.

Nagy, K., Clapham, P., Cheingsong-Popov, R., and Weiss, R.A. 1983. Human T-cell leukemia virus type 1: Induction of syncytia and inhibition by patients' sera. Intl. J. Cancer 32:321–328.

National Institutes of Health. 1985. Workshop on ELISA Test Kits for HTLV-III Antibody.

Nielson, S., Petito, C.K., Urmacher, C.D., and Posner, J.B. 1984. Subacute encephalitis in acquired immune deficiency syndrome: A postmortem study. Am. J. Clin. Pathol. 82:678–682.

Okochi, K., Sato, H., and Hinuma, Y. 1986. A retrospective study on transmission of adult T-cell leukemia virus by blood transfusion: Seroconversion in recipients. Vox Sang (in press).

Oroszlan, S., Sarngadharan, M.G., Copeland, T.D., Kalyanaraman, V.S., Gilden, R.V., and Gallo, R.C. 1982. Primary structure analysis of the major internal protein p24 of human type-C T-cell leukemia virus. Proc. Natl. Acad. Sci. USA. 79:1291–1294.

Palker, T.J., Scearce, R.M., Miller, S.F., et al. 1984. Monoclonal antibodies against human T-cell leukemia–lymphoma virus (HTLV) p24 internal core protein. Use as diagnostic probes and cellular localization of HTLV. J. Exp. Med. 159:1117–1131.

Perryman, L.E., Hoover, E.A., and Yoh, D.S. 1972. Immunological reactivity of the cat: Immunosuppression in experimental feline leukemia. J. Natl. Cancer Inst. 49:1357–1365.

Petito, C.K., Navia, B.A., Cho, E.-S., et al. 1985. Vacuolar myelopathy pathologically resembling subacute combined degeneration in patients with the acquired immunodeficiency syndrome. N. Engl. J. Med. 312:874–879.

Poiesz, B.J., Ruscetti, F.W., Mier, J.W., Woods, A.M., and Gallo, R.C. 1980a. T-cell lines established from human T-lymphocyte neoplasias by direct response to T-cell growth factor. Proc. Natl. Acad. Sci. USA. 77:6815–6819.

Poiesz, B.J., Ruscetti, F.W., Gazdar, A.F., Bunn, P.A., Minna, J.D., and Gallo, R.C. 1980b. Detection and isolation of type-C retrovirus particles from fresh and cultured lymphocytes of a patient with cutaneous T-cell lymphoma. Proc. Natl. Acad. Sci. USA. 77:7415–7419.

Poiesz, B.J., Ruscetti, F.W., Reitz, M.S., Kalyanaraman, V.S., and Gallo, R.C. 1981. Isolation of a new type-C retrovirus (HTLV) in primary uncultured cells of a patient with Sezary T-cell leukemia. Nature 294:268–271.

Popovic, M., Reitz, M.S., Sarngadharan, M.G., et al. 1982. The virus of Japanese adult T-cell leukemia is a member of the human T-cell leukemia virus group. Nature 300:63–66.

Popovic, M., Lange-Wantzin, G., Sarin, P.S. et al. 1983a. Transformation of human umbilical cord blood T-cells by human T-cell leukemia/lymphoma virus. Proc. Natl. Acad. Sci. USA. 80:5402–5406.

Popovic, M., Sarin, P.S., Robert-Guroff, M., et al. 1983b. Isolation and transmission of human retrovirus (human T-cell leukemia virus). Science 219:856–859.

Popovic, M., Flomenberg, N., Volkman, D.J. et al. 1984a. Alteration of T-cell functions by infection with HTLV-I or HTLV-II. Science 226:459–462.

Popovic, M., Sarngadharan, M.G., Read, E., et al. 1984b. Detection, isolation and continuous production of cytopathic retroviruses (HTLV-III) from patients with AIDS and pre-AIDS. Science 224:497–500.

Posner, L.E., Robert-Guroff, M., Kalyanaraman, V.S., et al. 1981. Natural antibodies to the human T-cell lymphoma virus in patients with cutaneous T-cell lymphomas. J. Exp. Med. 154:333–346.

Ratner, L., Haseltine, W., Patarca, R., et al. 1985. Complete nucleotide sequence of the AIDS virus, HTLV-III. Nature 313:277–284.

Redfield, R.R., Markham, P.D., Salahuddin, S.Z. et al. 1985a. Frequent transmission of HTLV-III among spouses of patients with AIDS-related complex and AIDS. J. Am. Med. Assoc. 253:1571–1573.

Redfield, R.R., Markham, P.D., Salahuddin, S.Z., Wright, D.C., Sarngadharan, M.G., and Gallo, R.C. 1985b. Heterosexually acquired HTLV-III/LAV disease (AIDS-related complex and AIDS). J. Am. Med. Assoc. 254:2094–2096.

Reitz, M.S., Poiesz, B.J., Ruscetti, F.W., and Gallo, R.C. 1981. Characterization and distribution of nucleic acid sequences of a novel type C retrovirus isolated from neoplastic human T-lymphocytes. Proc. Natl. Acad. Sci. USA. 78:1878–1891.

Resnick, L., diMarzo-Veronese, F., Schüpbach, J., et al. 1985. Intra-blood–brain-barrier synthesis of HTLV-III specific IgG in patients with AIDS or AIDS-related complex N. Engl. J. Med. 313:1498–1503.

Resnick, L., Veren, K., Tondreau, S., Salahuddin, S.Z., and Markham, P.D. 1986. Stability and inactivation of HTLV-III/LAV under clinical and laboratory environments. J. Am. Med. Assoc. 255:1887–1891.

Rho, H.M., Poiesz, B., Ruscetti, F.W., and Gallo, R.C. 1981. Characterization of the reverse transcriptase from a new retrovirus (HTLV) produced by a human cutaneous T-cell lymphoma cell line. Virology 112:355–360.

Robert-Guroff, M., Ruscetti, F.W., Posner, L.E., Poiesz, B., and Gallo, R.C. 1981. Detection of the human T-cell lymphoma virus p19 in cells of some patients with cutaneous T-cell lymphoma and leukemia using a monoclonal antibody. J. Exp. Med. 154:1957–1964.

Robert-Guroff, M., Fahey, K.A., Maeda, M. et al. 1982a. Identification of HTLV p19 specific natural human antibodies by competition with monoclonal antibody. Virology 122:297–305.

Robert-Guroff, M., Nakao, Y., Notake, K. et al. 1982b. Natural antibodies to human retrovirus HTLV in a cluster of Japanese patients with adult T-cell leukemia. Science 215:975–978.

Robert-Guroff, M., and Gallo, R.C. 1983. Establishment of an etiologic relationship between the human T-cell leukemia/lymphoma virus (HTLV) and adult T-cell leukemia. Blut 47:1–12.

Robert-Guroff, M., Kalyanaraman, V.S., Blattner, W.A., et al. 1983. Evidence for human T-cell lymphoma–leukemia virus infection of family members of human T-cell lymphoma–leukemia virus positive T-cell leukemia–lymphoma patients. J. Exp. Med. 157:248–258.

Robert-Guroff, M., Coutinho, R.A., Zadelhoff, A.W., Vyth-Dreese, F.A., and Rümke, P. 1984. Prevalence of HTLV-specific antibodies in Surinam emigrants to the Netherlands. Leukemia Res. 8:501–504.

Robert-Guroff, M., Brown, M., and Gallo, R.C. 1985. HTLV-III neutralizing antibodies in patients with AIDS and ARC. Nature 316:72–74.

Ruscetti, F.W., Morgan, D.A., and Gallo, R.C. 1977. Functional and morphological characterization of human T-cells continuously grown in vitro. J. Immunol. 119:131–138.

Safai, B., Sarngadharan, M.G., Groopman, J.E., et al. 1984. Seroepidemiological studies of human T-lymphotropic retrovirus type III in acquired immunodeficiency syndrome. Lancet i:1438–1440.

Salahuddin, S.Z., Markham, P.D., and Gallo, R.C. 1982. Establishment of long-term monocyte suspension cultures from normal human peripheral blood. J. Exp. Med. 155:1842–1857.

Salahuddin, S.Z., Groopman, J.E., Markham, P.D., et al. 1984a. HTLV-III in symptom-free seronegative persons. Lancet ii:1418–1420.

Salahuddin, S.Z., Markham, P.D., Lindner, S.G., et al. 1984b. Lymphokine production of human T-cells transformed by human T-cell leukemia–lymphoma virus-I. Science 223:703–707.

Salahuddin, S.Z., Markham, P.D., Popovic, M., Sarngadharan, M.G., Orndorff, S., Fladagar, A., Patel, A., Gold, J., and Gallo, R.C. 1985. Isolation of infectious human T-cell leukemia/lymphotropic virus type III (HTLV-III) from patients with acquired immunodeficiency syndrome (AIDS) or AIDS-related complex (ARC) and from healthy carriers: A study of risk groups and tissue sources. Proc. Natl. Acad. Sci. USA. 82:5530–5534.

Sanchez-Pescador, R., Power, M.D., Barr, P.J., et al. 1985. Nucleotide sequence and expression of an AIDS-associated retrovirus (ARV-2). Science 227:484–492.

Sarin, P.S., Aoki, T., Shibata, A., et al. 1983. High incidence of human type-C retrovirus (HTLV) in family members of a HTLV-positive Japanese T-cell leukemia patient. Proc. Natl. Acad. Sci. USA. 80:2370–2374.

Sarngadharan, M.G., Robert-Guroff, M., and Gallo, R.C. 1978. DNA polymerases of normal and neoplastic mammalian cells. Biochim. Biophys. Acta 516:419–487.

Sarngadharan, M.G., Popovic, M., Bruch, L., Schüpbach, J., and Gallo, R.C. 1984. Antibodies reactive with a human T lymphotropic retrovirus (HTLV-III) in the sera of patients with acquired immune deficiency syndrome. Science 224:506–508.

Sarngadharan, M.G., Bruch, L., Popovic, M., and Gallo, R.C. 1985a. Immunological properties of HTLV-III core protein p24. Proc. Natl. Acad. Sci. USA 82:3481–3484.

Sarngadharan, M.G., Markham, P.D., and Gallo, R.C., 1985b. Human T-cell leukemia viruses. In B.N. Fields, D.M. Knipe, R.M. Chanock, J.L. Melnick, B. Roizman, and R.E. Shope (eds.), Virology. New York: Raven Press, pp. 1345–1371.

Sarngadharan, M.G., diMarzo-Veronese, F., Lee, S., and Gallo, R.C. 1985c. Immunological properties of HTLV-III antigens recognized by sera of patients with AIDS, ARC and asymptomatic carriers of HTLV-III infection. Cancer Res. 45:4574–4577.

Saxinger, W.C., and Gallo, R.C. 1982. Possible risk to recipients of blood from donors carrying serum markers of human T-cell leukemia virus. Lancet i:1074.

Saxinger, C., and Gallo, R.C. 1983. Methods in laboratory investigation. Application of the indirect enzyme-linked immunosorbent assay microtest to the detection and surveillance of human T-cell leukemia–lymphoma virus. Lab. Invest. 49:371–377.

Saxinger, W., Blattner, W.A., Levine, P.H. et al. 1984a. Human T-cell leukemia virus (HTLV-I) antibodies in Africa. Science 225:1473–1476.

Saxinger, W.C., Lange-Wantzin, G., Thomsen, K. et al. 1984b. Human T-cell leukemia virus: A diverse family of related exogenous retroviruses of humans and Old World primates. In R.C. Gallo, M. Essex, and L. Gross (eds.), Human T-cell Leukemia/Lymphoma Virus. Cold Spring Harbor, N.Y.: Cold Spring Harbor Laboratory, pp. 323–330.

Saxinger, W.C., Levine, P.H., Dean, A.G. et al. 1985. Evidence for exposure to HTLV-III in Uganda before 1973. Science 227:1036–1038.

Schaffar-Deshayes, L., Chavance, M., Monplaisir, N. et al. 1984. Antibodies to HTLV-I p24 in sera of blood donors, elderly people and patients with hematopoietic diseases in France and in French West Indies. Intl. J. Cancer 34:667–670.

Schneider, J., Bayer, H., Bienzle, U., Wernet, P., and Hunsmann, G. 1985. Antibodies against LAV/HTLV-III in German blood donors. Lancet i:275–276.

Schüpbach, J., Kalyanaraman, V.S., Sarngadharan, M.G., Blattner, W.A., and Gallo, R.C. 1983a. Antibodies against three purified proteins of the human type C retrovirus, human T-cell leukemia–lymphoma virus, in adult T-cell leukemia–lymphoma patients and healthy blacks from the Caribbean. Cancer Res. 43:886–891.

Schüpbach, J., Kalyanaraman, V.S., Sarngadharan, M.G., et al. 1983b. Antibodies against three purified structural proteins of the human type-C retrovirus. HTLV, in Japanese adult T-cell leukemia patients, healthy family members and unrelated normals. Intl. J. Cancer 32:583–590.

Schüpbach, J., Kalyanaraman, V.S., Sarngadharan, M.G., et al. 1984a. Demonstration of viral antigen p24 in circulating immune complexes of two patients with human T-cell leukaemia/lymphoma virus (HTLV) positive lymphoma. Lancet i:302–305.

Schüpbach, J., Popovic, M., Gilden, R.V., et al. 1984b. Serologic analysis of a subgroup of human T lymphotropic retroviruses (HTLV-III) associated with AIDS. Science 224:503–505.

Schüpbach, J., Sarngadharan, M.G., and Gallo, R.C. 1984c. Antigens on HTLV-infected cells recognized by leukemia and AIDS sera are related to HTLV viral glycoprotein. Science 224:607–610.

Schüpbach, J., Haller, O., Vogt, M., et al. 1985a. Antibodies to HTLV-III in Swiss patients with AIDS and pre-AIDS and in groups at risk for AIDS. N. Engl. J. Med. 312:265–270.

Schüpbach, J., Vogt, M., Bhushan, R. et al. 1985b. Prevalence of AIDS pathogen HTLV-III in Switzerland. Schweiz. Med. Wschr. 115:1048–1054.

Seiki, M., Hattori, S., and Yoshida, M. 1982. Human adult T-cell leukemia virus: Molecular cloning of the provirus DNA and the unique terminal structure. Proc. Natl. Acad. Sci. USA. 79:6899–6902.

Seiki, M., Hattori, S., Hirayama, Y. and Yoshida, M. 1983. Human adult T-cell leukemia virus: Complete nucleotide sequence of the provirus genome integrated in leukemia cell DNA. Proc. Natl. Acad. Sci. USA. 80:3618–3622.

Shaw, G.M., Harper, M.E., Hahn, B.H. et al. 1985. HTLV-III infection in brains of children and adults with AIDS encephalopathy. Science 227:177–182.

Siegal, F.P., Lopez, C., Hammer, G.S. et al. 1981. Severe acquired immunodeficiency in male homosexuals, manifested by chronic perianal ulcerative herpes simplex lesions. N. Engl. J. Med. 305:1439–1444.

Slamon, D.J., Shimotohno, K., Cline, M.J., et al. 1984. Identification of the putative transforming protein of the human T-cell leukemia viruses HTLV-I and HTLV-II. Science 226:61–65.

Snider, W.D., Simpson, D.M., Aronyk, K.E., and Nielsen, S.L. 1983a. Primary lymphoma of the central nervous system associated with the acquired immune deficiency syndrome. N. Engl. J. Med. 308:45.

Snider, W.D., Simpson, D.M., Nielsen, S. et al. 1983b. Neurological complications of acquired immune deficiency syndrome: Analysis of 50 patients. Ann. Neurol. 14:403–418.

Sodroski, J., Patarca, R., Perkins, D. et al. 1984a. Sequence of the envelope glycoprotein gene of Type II human T-lymphotropic virus. Science 225:421–424.

Sodroski, J.G., Rosen, C.A., and Haseltine, W.A. 1984b. *Trans*-acting transcriptional activation of the long terminal repeat of human T-lymphotropic virus. Science 225:381–385.

Sodroski, J., Patarca, R., Rosen, C., Wong-Staal, F., and Haseltine, W. 1985. Location of the trans-activating region on the genome of human T-cell lymphotropic virus type-III. Science 229:74–77.

Spire, B., Dormont, D., Barré-Sinoussi, F., Montagnier, L., and Chermann, J.L. 1985. Inactivation of lymphadenopathy-associated virus by heat, gamma rays, and ultraviolet light. Lancet i:188–189.

Suciu-Foca, N., Rubinstein, P., Popovic, M., Gallo, R.C., and King, D.W. 1984. Reactivity of HTLV-transformed human T-cell lines to MHC class II antigens. Nature 312:275–277.

Tajima, K., Tominaga, S., Shimizu, H. and Suchi, T. 1981. A hypothesis on the etiology of adult T-cell leukemia/lymphoma. Gann 72:684–691.

Tajima, K., Tominaga, S., and Suchi, T. 1982. Clinico-epidemiological analysis of adult T-cell leukemia. Gann Monogr. Cancer Res. 28:197–210.

Tajima, K., Fujita, K., Tsukidate, S. et al. 1983. Seroepidemiological studies on the effects of filarial parasites on infestation of adult T-cell leukemia virus in the Goto Islands, Japan. Gann 74:188–191.

Takatsuki, K., Uchiyama, J., Sagawa, K., and Yodoi, J. 1977. Adult T-cell leukemia in Japan. In S. Seno, F. Takaku, and S. Irino (eds.), Topics in Hematology. Amsterdam: Excerpta Medica, pp. 73–77.

Takatsuki, K., Uchiyama, T., Ueshima, Y., et al. 1982. Adult T-cell leukemia: Proposal as a new disease and cytogenetic, phenotypic, and functional studies of leukemic cells. Gann Monogr. Cancer Res. 28:13–22.

Tedder, R.S., Shanson, D.C., Jeffries, D.J., et al. 1984. Low prevalence in the UK of HTLV-I and HTLV-II infection in subjects with AIDS, with extended lymphadenopathy, and at risk of AIDS. Lancet ii:125–128.

Towbin, H., Staehelin, T., and Gordon, J. 1979. Electrophoretic transfer of proteins from polyacrylamide gels to nitrocellulose sheets: Procedure and some applications. Proc. Natl. Acad. Sci. USA. 76:4350–4354.

Trainin, Z., Wernicke, D., Ungar-Waron, H. and Essex, M. 1983. Suppression of the humoral antibody response in natural retrovirus infections. Science 220:858–859.

Tucker, J., Ludlam, C.A., Craig, A. et al. 1985. HTLV-III infection associated with glandular-fever-like illness in a haemophiliac. Lancet i:585.

Uchiyama, T., Yodoi, J., Sagawa, K., et al. 1977. Adult T-cell leukemia: Clinical and hematologic features of 16 cases. Blood 50:481–491.

Veronese, F., DeVico, A.L., Copeland, T.D., et al. 1985. Characterization of gp41 as the transmembrane protein coded by the HTLV-III envelope gene. Science 229:1402–1405.

Veronese di Marzo, F., Copeland, D.T., DeVico, A.L., Rahman, R., Oroszlan, S., Gallo, R.C., and Sarngadharan, M.G. 1986. Characterization of highly immunogenic p66/51 as the reverse transcriptase of HTLV-III/LAV. Science 231:1289–1291.

Vilmer, E., Barré-Sinoussi, F., Rouzioux, C. et al. 1984. Isolation of new lymphotropic retrovirus from two siblings with haemophilia B, one with AIDS. Lancet i:753–757.

Vyth-Dreese, F.A., and de Vries, J.E. 1983. Human T-cell leukemia virus in lymphocytes from a T-cell leukemia patient originating from Surinam. Lancet ii:993.

Wainberg, M.A., Spira, B., Boushira, M., and Margolese, R.G. 1985. Inhibition by human T-lymphotropic virus (HTLV-I) of T-lymphocyte mitogenesis: Failure of exogenous T-cell growth factor to restore responsiveness to lectin. Immunology 54:1–7.

Wain-Hobson, F., Sonigo, P., Danis, D., Cole, S., and Alizon, M. 1985. Nucleotide sequence of the AIDS virus, LAV. Cell 40:9–17.

Waldmann, T., Broder, S., Greene, W., et al. 1983. A comparison of the function and phenotype of Sezary T-cells with human T-cell leukemia/lymphoma virus (HTLV)-associated adult T-cell leukemia. Clin. Res. 31:5474–5480.

Watanabe, T., Seiki, M., and Yoshida, M. 1984 HTLV type 1 (U.S. isolate) and ATLV (Japanese isolate) are the same species of human retrovirus. Virology 133:238–241.

Weijer, K., Colofat, J., Daams, J.H., Hegeman, P.C., and Misdorp, W. 1974. Feline malignant mammary tumors. II. Immunologic and electron microscopic investigations into a possible viral etiology. J. Natl. Cancer Inst. 52:673–681.

Weiss, R.A., Clapham, P.R., Cheingsong-Popov, R. et al. 1985. Neutralization of human T-lymphotropic virus Type III by sera of AIDS and AIDS-risk patients. Nature 316:69–72.

Wong-Staal, F., Hahn, B., Manzari, V. et al., 1983. A survey of human leukemias for sequences of a human retrovirus. HTLV. Nature 302:626–628.

Yamada, Y. 1983. Phenotypic and functional analysis of leukemic cells from 16 patients with adult T-cell leukemia/lymphoma. Blood 61:192–199.

Yamaguchi, K., Nishimura, H., and Takatsuki, K. 1983. Clinical factors in malignant lymphoma and adult T-cell leukemia in Kumamamoto. Rinsho Ketsueki 24:1271–1276.

Yamamoto, N., Matsumoto, T., Koyanagi, Y., Tanaka, Y., and Hinuma, Y. 1982a. Unique cell lines harboring both Epstein–Barr virus and adult T-cell leukemia virus established from leukemic patients. Nature 299:367–369.

Yamamoto, N., Okada, M., Koyanagi, Y., Kannagi, M., and Hinuma, T. 1982b. Transformation of human leukocytes by cocultivation with an adult T-cell leukemia virus producer cell line. Science 217:737–739.

Yodoi, J., Takatsuki, K., Aoki, N., and Masuda, T. 1974. Chronic lymphocytic leukemia of T-cell origin: Demonstration in two cases by the use of antithymocyte membrane antiserum. Acta Haematol Jap. 37:289–292.

Yoshida, M., Miyoshi, I., and Hinuma, Y. 1982.

Isolation and characterization of retrovirus from cell lines of human adult T-cell leukemia and its implication in the disease. Proc. Natl. Acad. Sci. USA. 79:2031–2035.

Yoshida, M., Seiki, M., Yamaguchi, K., and Takatsuki, K. 1983. Monoclonal integration of HTLV provirus in primary tumors of adult T-cell leukemia suggests causative role of HTLV in disease. Proc. Natl. Acad. Sci. USA. 80:3618–3622.

Yoshikura, H., Nishida, J., Yoshida, M. et al. 1984. Isolation of HTLV derived from Japanese adult T-cell leukemia patients in human diploid fibroblast strain IMR90 and the biological characters of the infected cells. Intl. J. Cancer 33:745–749.

Zagury, D., Bernard, J., Leibowitch, J., et al. 1984. HTLV-III in cells cultured from semen of two patients with AIDS. Science 226:449–451.

Ziegler, J.L., and Abrams, D.I. 1985. The AIDS-related complex. In V.T. DeVita, F. Hellman, and S.A. Rosenberg (eds.), AIDS—Etiology, Diagnosis, Treatment and Prevention. Philadelphia: J.B. Lippincott Co., pp. 223–233.

Ziegler, J.L., Beckstead, J.A., and Volberding, P.A., et al. 1984. Non-Hodgkin's lymphoma in 90 homosexual men: Relation to generalized lymphoadenopathy and the acquired immunodeficiency syndrome. N. Engl. J. Med. 311:565–570.

Ziegler, J.B., Cooper, D.A., Johnson, R.O., and Gold, J. 1985. Postnatal transmission of AIDS-associated retrovirus from mother to infant. Lancet i:896–898.

Chlamydia

Julius Schachter

32.1 INTRODUCTION

Although the genus *Chlamydia* causes a number of human diseases, in a clinical virology setting the diagnosis most commonly called for is that of *Chlamydia trachomatis* oculogenital infections (Schachter, 1978). The lymphogranuloma venereum serovars of *C. trachomatis* cause a more extensive systemic sexually transmitted disease, which is relatively uncommon in the United States. *Chlamydia psittaci*, the other species within the genus, is a very common organism among domestic mammals and virtually ubiquitous in the avian kingdom, but affects humans only as a zoonosis and this diagnosis is usually established on serologic grounds (Schachter and Dawson, 1978). Therefore, this chapter will focus on the laboratory diagnosis and cultural methods for discovery of the organism.

Because chlamydial infections cause characteristic intracytoplasmic inclusions (reflecting the obligate intracellular nature of the organism's parasitism), the first demonstrations of the organism were cytologic (Schachter and Dawson, 1978). Initially, inclusions were demonstrated in the epithelium of experimentally infected subhuman primates. By 1910 similar studies using Giemsa stain of epithelial cell scrapings had demonstrated that *C. trachomatis* could infect the conjunctivae of adults and newborns, as well as cervical or urethral epithelial cells. A relationship with cervicitis and nongonococcal urethritis had been demonstrated.

The trachoma agent was first isolated by T'ang et al. in China, in 1957. The interest in sexually transmitted chlamydial infections was renewed by the report of Jones (1964) on recovery of the organism from the urethra and cervix of adults in England. The introduction of tissue culture procedures for the isolation of *Chlamydia* increased the clinical relevance of its detection, as the earlier yolk sac procedures were very time consuming (Gordon and Quan, 1965). The original isolation procedure, which involved irradiation of McCoy cells, has now been supplanted by treatment of the tissue culture cells with antimetabolites. Cultural diagnosis can be made in 2 to 7 days after processing of the specimen.

32.2 CHARACTERISTICS OF THE ORGANISM

The chlamydiae are among the more common pathogens throughout the animal kingdom (Storz, 1971; Page, 1972; Schachter,

1978; Schachter and Dawson, 1978). They are nonmotile, Gram-negative, obligate intracellular bacteria. Their unique developmental cycle differentiates them from all other microorganisms (Moulder et al., 1984). They replicate within the cytoplasm of host cells, forming characteristic intracellular inclusions that can be seen by light microscopy. They differ from the viruses by possessing both RNA and DNA and cell walls quite analogous in structure to those of Gram-negative bacteria. They are susceptible to many broad-spectrum antibiotics, possess a number of enzymes, and have a restricted metabolic capacity. None of these metabolic reactions results in the production of energy. Thus, they have been considered as energy parasites that use the adenosine 5′-triphosphate (ATP) produced by the host cell for their own requirements (Moulder, 1966).

32.2.1 Growth Cycle

Chlamydiae are phagocytosed by susceptible host cells (Byrne and Moulder, 1978). The phagocytic process is directly influenced by the chlamydiae and ingestion of organisms is specifically enhanced. Following attachment, at specific sites on the surface of the cell the elementary body enters the cell in a phagosome where the entire growth cycle is completed. The chlamydiae prevent phagolysosomal fusion. Once the elementary body (EM-diameter, 0.25–0.35 μm) has entered the cell, it reorganizes into a reticulate particle (initial body), which is larger (0.5–1.0 μm) and richer in RNA. After approximately 8 hours the initial body begins dividing by binary fission. Approximately 18 to 24 hours after infection, these initial bodies become elementary bodies by a poorly understood reorganization or condensation process. The elementary bodies are then released to initiate another cycle of infection. The elementary bodies are specifically adapted for extracellular survival and are the infectious form, whereas, the intracellular metabolically active and replicating form, the initial body, does not survive well outside the host cell and seems adapted for an intracellular milieu.

32.2.2 Taxonomy

Chlamydiae are presently placed in their own order, the *Chlamydiales*, family *Chlamydiaceae*, with one genus, *Chlamydia* (Moulder et al., 1984). There are two species, *C. trachomatis* and *C. psittaci*. *C. trachomatis* includes the organisms causing trachoma, inclusion conjunctivitis, lymphogranuloma venereum (LGV) and the other sexually transmitted infections, and some rodent pneumonia strains. *C. trachomatis* strains are sensitive to the action of sulfonamides and produce a glycogen-like material within the inclusion vacuole, which stains with iodine. *C. psittaci* strains infect many avian species and mammals, producing the diseases psittacosis, ornithosis, feline pneumonitis, bovine abortion, and others (Storz, 1971; Page, 1972). They are resistant to the action of sulfonamides and produce inclusions that do not stain with iodine.

32.2.3 Antigenic Relationships

The chlamydiae possess group (genus)-specific, species-specific, and type-specific antigens. Most of these are apparently located within the cell wall, but precise structural relationships are not known. The major outer membrane contains many of the subspecies specific antigens (Caldwell and Schachter, 1982). The group antigen, shared by all members of the genus, appears to be a lipopolysaccharide complex with a ketodeoxyoctanoic acid as the reactive moiety (Dhir et al., 1971). It may be analogous to the lipopolysaccharide (LPS) of certain Gram-negative bacteria (Nurminen et al., 1983). Species-specific protein antigens have been identified but have not been characterized (Caldwell and Kuo, 1977). Some type-specific antigens of *C. trachomatis* have been identified and appear to be proteins with an approximate molecular weight of 30,000 daltons (Sacks and MacDonald, 1979). Specific antigens of *C. psittaci* strains can be demonstrated by neutralization tests (Banks et al., 1970). The specific antigens of *C. trachomatis* are best recognized by a microimmunofluorescence (micro-IF) technique (Wang and Grayston, 1970), although these antigens are also associated with a toxic factor (large numbers of viable chlamydiae may kill mice in less than 24 hours after intravenous inoculation). One-way crossreactions have been reported between chlamydiae and some bacteria, but these do not appear to influence serodiagnosis.

32.3 PATHOGENESIS

C. trachomatis is almost exclusively a human pathogen (Grayston, 1975; Schachter, 1978). Serotypes within this species cause trachoma (serotypes A, B, Ba, and C have been associated with endemic trachoma, the most common preventable form of blindness, inclusion conjunctivitis, and LGV; serotypes L1, L2 and L3). Where sexual transmission of *C. trachomatis* strains other than LGV have been studied, serotypes D through K have been found to be the major identifiable cause of nongonococcal urethritis in men and may also cause epididymitis. Proctitis may occur in either sex. In women, cervicitis is a common result of chlamydial infection, and acute salpingitis may occur. The agent in the cervix may be transmitted to the neonate as it passes through the infected birth canal, and eye disease, inclusion conjunctivitis of the newborn, and a characteristic chlamydial pneumonia of infants may develop (Beem and Saxon, 1977). Vaginal infection and enteric infection in neonates are also recognized.

The organism is essentially a parasite of squamocolumnar epithelial cells (the LGV biovars are more invasive and involve lymphoid cells). Typical of the genus, *C. trachomatis* strains are capable of causing chronic and inapparent infections. Because their growth cycle is approximately 48 hours, the incubation periods are relatively long and are generally 1 to 3 weeks. *C. trachomatis* causes cell death as a result of its replicative cycle and, thus, is capable of producing some cell damage whenever it persists. However, because there are no toxic manifestations demonstrated, nor is there sufficient cell death as a result of replication, it is likely that the majority of the disease response is due to immunopathologic mechanisms or host response to the organism or its byproducts. In the absence of therapy chlamydial infections may persist for years, although symptoms usually abate.

32.4 INFECTED SITES AND METHODS OF COLLECTION

For cytologic studies, impression smears of involved tissues or scrapings of involved epithelial cell sites should be appropriately fixed (cold acetone for IF and methanol for Giemsa stain). As is also true for the isolation attempts, it is imperative that samples be collected from the involved epithelial cell sites by vigorous swabbing or scraping. Purulent discharges are inadequate and should be cleaned from the site prior to sampling.

For most *C. trachomatis* infections of humans, the involved mucous membranes should be vigorously swabbed or sampled by scraping. Thus, the conjunctiva for trachoma-inclusion conjunctivitis, the anterior urethra (several centimeters into the urethra), or the cervix at the endocervical canal would be tested. Because these strains appear to infect only columnar cells, cervical specimens must be collected at the transitional zone or within the os. Because the organism also can infect the urethra of the female, it may improve recovery rates if another sample is collected from the urethra and sent to the laboratory for testing in the same tube with the cervical sample. For women with salpingitis the samples may be collected by needle aspiration of the involved fallopian tube or endometrial specimens may yield the agent. Rectal mucosa, nasopharynx, and throat may also be sampled. For infants with pneumonia, swabs may be collected from the posterior nasopharynx or the throat, although nasopharyngeal or tracheobronchial aspirates collected by intubation appear to be a superior source of agent.

32.5 STABILITY, STORAGE, AND TRANSPORT

C. trachomatis is not a particularly labile organism and by providing reasonable care in the handling of specimens a minimal loss in infectivity is achieved by decreasing the time between specimen collection and processing in the laboratory.

Swabs, scrapings, and small tissue samples should be collected in a special transport medium. Because *Chlamydia* are bacteria, the selection of antibiotics to prevent other bacterial contamination is restricted. Broad-spectrum antibiotics such as tetracyclines, macrolides, or penicillin must be excluded. Aminoglycosides and fungicides are the mainstays. The chlamydial specimens should be refrigerated if they can be processed within 48 hours after collection. If not, they should be frozen at −60°C.

For isolation in cell culture, one suspending medium (2SP) consists of 0.2 *M* sucrose in 0.02 *M* phosphate buffer, pH 7.0–7.2 with 5% fetal calf serum and added antibiotics. A sucrose–phosphate–glutamate (SPG) medium also has been commonly used.

Chlamydial media

Growth medium

50 ml Eagle's minimum essential medium (EMEM) in Earle's salts 10x

50 ml fetal calf serum

5 ml of 200 m*M* solution L-glutamine

Up to 500 ml sterile distilled water

Adjust pH to 7.4 with 7.5% sodium bicarbonate

Isolation medium

Growth medium (above) with the following added (this medium may be used as a collection medium by doubling the concentrations of vancomycin and amphotericin B):

50 μg/ml vancomycin

10 μg/ml gentamicin

2 μg/ml amphotericin B

0.594 mg/ml glucose

1–2 μg/ml cycloheximide

Sucrose–phosphate transport medium (2SP)

68.46 g sucrose

2.088 g K_2HPO_4

1.088 g Na_2HPO_4

Distilled water to 1000 ml

Adjust pH to 7.0 and autoclave; add the following:

Bovine serum to 5%

50 μg/ml streptomycin

100 μg/ml vancomycin

25 U/ml nystatin

Sucrose–phosphate–glutamine transport medium (SPG)

75.00 g sucrose

0.52 g KH_2PO_4

1.22 g Na_2HPO_4

0.72 g glutamic acid

Distilled water to 1000 ml

Adjust pH to 7.4–7.6 and autoclave, add antibiotics as above (2SP).

It may be simpler to place the clinical specimen directly into standard tissue culture growth medium containing streptomycin (200 μg/ml) or gentamicin (10 μg/ml), vancomycin (100 μg/ml), and amphotericin B (4 μg/ml).

32.6 DIRECT CYTOLOGIC EXAMINATION

C. trachomatis infections of the conjunctiva, urethra, or cervix can be diagnosed by demonstrating typical intracytoplasmic inclusions, but cytologic procedures usually are less sensitive than isolation in tissue culture. The Giemsa stain was the method most often used previously, but IF procedures are more sensitive. These procedures are particularly useful in diagnosing acute, severe inclusion conjunctivitis of the newborn, and are less effective in diagnosing adult conjunctival and genital tract infections. With infants the ability to detect intracellular diplococci, if the child has gonococcal ophthalmia neonatorum, is another benefit and direct microscopy is obviously much faster than the isolation procedures. If isolation techniques are available, the use of cytology for genital tract disease is not recommended. Where diagnostic facilities are not available, however, direct microscopy of epithelial cell scrapings may be useful in diagnosing the chlamydial infections.

32.6.1 Fluorescent Antibody Technique

Most of the published experience with IF procedures reflects the use of polyclonal antibody, either in direct or indirect fluorescent antibody procedures. These represented efforts to detect typical chlamydial inclusions within epithelial cells. The procedures were more sensitive than other cytologic methods (Giemsa or iodine), but required an adequate source of appropriate antisera. There were no commercial sources and laboratories had to prepare their own reagents. The antisera were usually standardized against infected tissue culture monolayers to determine appropriate working dilutions. Trained microscopists were required to identify typical chlamydial intracytoplasmic inclusions. The procedure was less sensitive than isolation of the agent in cell culture.

More recently, fluorescein-conjugated monoclonal antibodies have been made available (Stephens et al., 1982). Although they are routinely used in some laboratories to identify *C. trachomatis* inclusions in cell culture (Stamm et al., 1983), the test may be applied directly to clinical specimens (Tam et al., 1984). There is inadequate experience with this procedure to recommend it at this writing. Anecdotal experience indicates that it will require trained microscopists and rigorous criteria for interpretation of the microscopic analysis of the smears. The procedure seems somewhat less sensitive than culture, but it is faster and less expensive and may represent an alternate method of diagnosing chlamydial infections in settings where culture is not available.

32.6.2 Iodine Staining Technique

Scrapings are air-dried, fixed in absolute methanol, and stained with Lugol iodine or 5% iodine in 10% potassium iodide for 3 to 5 minutes. Slides are examined as wet mounts. The matrix of inclusions may appear as a reddish-brown mass recognizable under low magnification. The slides may be decolorized with methanol and restained with Giemsa stain. This technique is the least sensitive cytologic procedure. It is not recommended for use with clinical specimens. Its speed and simplicity have made it the popular test for examining *C. trachomatis*-infected cell cultures.

32.6.3 Giemsa Staining Technique

The smear is air-dried, fixed with absolute methanol for at least 5 minutes, and dried again. It is then covered with the diluted Giemsa stain (freshly prepared the same day) for 1 hour. The slide is then rinsed rapidly in 95% ethyl alcohol to remove excess dye and to enhance differentiation; it is then dried and examined microscopically. Longer staining periods (1 to 5 hours) may be preferable with heavy tissue culture monolayers. Elementary bodies (EB) stain reddish-purple. The initial bodies are more basophilic, staining bluish, as do most bacteria.

32.7 METHODS FOR ISOLATION

The recommended procedures for primary isolation of chlamydiae are tissue culture techniques. The most common technique involves inoculation of clinical specimens into cycloheximide-treated McCoy cells (Ripa and Mardh, 1977). The basic principle involves centrifugation of the inoculum into the cell monolayer at approximately 2800 × *g* for 1 hour, incubation of monolayers for 48 to 72 hours, and then staining with iodine to detect the glycogen-positive inclusions. Fluorescent antibody staining may allow earlier detection of the inclusion (Thomas et al., 1977). Use of the abovementioned fluorescein-conjugated monoclonal antibodies represents the most sensitive method for detecting *C. trachomatis* inclusions in cell culture (Stamm et al., 1983). The procedure requires more attention to staining than the iodine technique and is more costly.

McCoy cells are plated onto 13-mm coverslips contained in 15-mm diameter (1 dram) disposable glass vials. Cell concentration (approximately 1×10^5 to 2×10^5) is selected to give a light, confluent monolayer after 24 to 48 hours of incubation at 37°C. For optimal results, the cells should be used within 24 to 72 hours after reaching confluency. If the laboratory is only passing cells on a sporadic basis, they may then be held at room temperature or in a low (2%) serum medium for at least 2 weeks prior to inoculation. The clinical specimen is inoculated into the cells, which are then washed, and medium containing 1–2 μg/ml cycloheximide (this must be titrated for each batch) is placed onto the cells 2 hours after centrifugation.

The clinical specimens should be shaken with glass beads prior to inoculation. This is safer and more convenient than sonication. Standard inoculation procedure involves removing medium from the cell monolayer and replacing it with the inoculum in a volume 0.1 to 1 ml. The specimen is then centrifuged onto the cell monolayer at approximately 3000 × *g* at 35°C for 1 hour. The vials should be held at 37°C for 2 hours before the cells are washed or the medium is changed. The cells are then incubated at 37°C for 48 to 72 hours, after which one coverslip is examined for inclusions by use of iodine, Giemsa, or IF staining. The use of IF can speed up the process, as inclusions can be seen clearly (although smaller) 24 hours postinfection, but this requires availability of immunologic reagents and uses more difficult microscopic procedures. Giemsa stain is more sensitive than iodine stain, but the microscopic evaluation is more difficult. Slide reading can be facilitated by examining the Giemsa stained coverslip by dark-field rather than bright-field microscopy (Darougar et al., 1971). The iodine stain is the simplest procedure and the one most commonly used, although it is less sensitive than either of the two.

If passage of positive material or blind passage of negative material is desired, the material should be passed at 72 to 96 hours postinoculation. The cell monolayer is disrupted by shaking with glass beads on a Vortex mixer; the material is centrifuged at low speed to remove cell debris, and the supernatant is inoculated as above. Approximately 90% of positive specimens are inclusion-positive in the first passage.

With trachoma, inclusion conjunctivitis, and the genital tract infections, the technique is as described above. In LGV, the aspirated bubo pus is diluted (10^{-1} and 10^{-2}) and treated as above. Second passages are always made because detritus from the inoculum may make it difficult to read the slides.

For laboratories processing large numbers of specimens, it may be convenient to use flat-bottomed 96-well microtiter plates rather than vials for the specimens (Yoder et al., 1981). Cells are plated onto coverslips or can be placed directly onto the plastic. Processing and incubation will be as above, but microscopy will be modified to use either long working objectives or inverted microscopes. This procedure is less sensitive than the vial technique but offers considerable savings in terms of reagents and time and may be suitable for settings where mostly symptomatic patients are being screened. These patients usually yield higher numbers of chlamydiae and, thus, minimize the impact of the decreased sensitivity of the test.

32.8 IDENTIFICATION

Because most laboratories will be using tissue culture isolation systems, the basic procedure for identification of chlamydiae involves demonstration of typical intracytoplasmic inclusions by appropriate (Giemsa or iodine) staining procedures. However, in laboratories initiating work with chlamydiae, it would be prudent to use at least one other parameter for identification of chlamydiae. Fluorescent antibody staining provides both a morphologic and an immunologic identification.

C. trachomatis strains may be serotyped by the micro-IF technique (Wang and Grayston, 1970). For this procedure antisera are produced by intravenous inoculation of mice at day 0, a booster is given at day 7, and exsanguination at day 11. The mouse antiserum is then tested in a titration against all serotypes, as well as the immunizing agent, and the serotype is identified presumptively by the pattern of reactivity and finally by appropriate box titration with the appropriate prototypic serotype.

32.9 SERODIAGNOSIS

The most widely used serologic test for diagnosing chlamydial infections is the complement fixation (CF) test. This is useful in diagnosing psittacosis, in which paired sera often show fourfold or greater increases in titer. It may also be useful in diagnosing LGV, in which single-point titers greater than 1:64 are highly supportive of this clinical diagnosis. With LGV it is difficult to demonstrate rising titers because the nature of the disease results in the patient being seen by the physician after the acute stage. Any titer above 1:16 is considered significant evidence of exposure to chlamydiae. The CF test is not particularly useful in diagnosing trachoma-inclusion conjunctivitis or the related genital tract infections, and it plays no role in diagnosing neonatal chlamydial infections.

The micro-IF method is a much more sensitive procedure for measuring anti-chlamydial antibodies. It may be used in diagnosing psittacosis, in which paired sera will show rising immunoglobulin G (IgG) titers (and often IgM antibody). With LGV, again, it is difficult to demonstrate rising titers, but single-point titers in active cases usually have relatively high levels of IgM antibody (>1:32) and IgG levels (≥1:2000). Trachoma, inclusion-conjunctivitis, and the genital tract infections may be diagnosed by the micro-IF technique if appropriately timed paired acute and convalescent sera can be obtained. It is often difficult to demonstrate rising antibody titers, however, particularly in sexually active populations. Many of these individuals will be seen for chronic or repeat infections. The background rate of seroreactors in venereal disease clinics is ≥60%, making it particularly difficult to demonstrate seroconversion. In general, first attacks of chlamydial urethritis have been regularly associated with seroconversion (Bowie et al., 1977). Individuals with systemic infection (epididymitis, salpingitis) usually have much higher antibody levels than those with superficial infections, and women tend to have higher antibody levels than men.

Serology is particularly useful in diagnosing chlamydial pneumonia in neonates.

In this case, high levels of IgM antibody are regularly found in association with disease (Schachter et al., 1982). IgG antibodies are less useful because the infants are being seen at a time when they have considerable levels of circulating maternal IgG because all these infections are acquired from the infected mother, who is always seropositive. It takes between 6 and 9 months for maternal antichlamydial antibodies to disappear. Infants older than 9 months may be tested for determination of prevalence of chlamydial infection without fear of confounding effects of maternal antibody. Infants with inclusion conjunctivitis or respiratory tract carriage of *Chlamydia* without pneumonia usually have very low levels of IgM antibodies. Thus, a single titer ≥1:64 may support the diagnosis of chlamydial pneumonia.

The micro-IF technique uses many serotypes of chlamydiae and the procedure as simplified by Wang et al. (1975) is recommended. Because serology is particularly useful in diagnosing neonatal infection, and the IgM antibody responses tend to be markedly specific, the use of single broadly reacting antigens will miss at least 15–25% of the infections that can be proven to be due to *Chlamydia* by other procedures, or that would be positive by a multiple antigen micro-IF. The single-antigen tests may involve either yolk sac suspensions of agent or identification of fluorescent inclusions in tissue monolayers (Thomas et al., 1976; Beem and Saxon, 1977; Richmond and Caul, 1977). Serotypes with the DEL serogroup are commonly chosen for this purpose.

Research workers should be warned that monotypic A seroreactions, at least in the U.S., are liable to be spurious. Long-term longitudinal studies on infants that are in progress in my laboratory suggest that the appearance of antibodies against type A (and to a lesser extent the cross-reacting CJI serotypes) may appear in response to nonchlamydial antigenic stimulus (Schachter et al., 1982). These antibodies are usually transient and do not result in the persistent high levels of IgG antibodies that usually follow chlamydial infections.

Enzyme immunoassay (EIA) techniques that measure antichlamydial antibodies have been described (Finn et al., 1983). Most of these procedures in measuring IgG antibody have been successful, albeit often less sensitive than the micro-IF test. They have not been successful in measuring IgM antibody. One test is commercially available. There is no published experience with that test. On the basis of published experience, it is likely that it will be less sensitive than the micro-IF, will miss some C complex reactors, and cannot be readily applied to IgM antibody. The procedure may be of some use in selected instances and for sero surveys in laboratories where micro-IF techniques are not available. A recently described EIA using major outer membrane protein appears to be promising (Puolakkainen et al., 1984).

32.9.1 Complement Fixation

The CF test may be performed in either the tube system or the microtiter system. Reagents should be standardized in the tube system, regardless of which test system will be used. The microtiter systems are most useful in screening large numbers of sera, but it is preferable to retest all positive sera in the tube system. Occasionally, sera giving titers in the 1:4 to 1:8 range in the microtiter system are positive at 1:16 (the significant level) in the tube system. The microtiter system uses standard plates and volumes one-tenth of those used in the tube test. The CF test is performed on serum specimens heated at 56°C for 30 minutes (preferably acute and convalescent paired sera tested together). In each test a positive control serum of high titer is included together with a known negative serum. The reagents for the CF test are standardized by the Kolmer technique and include special buffered saline, group antigen, antigen (normal yolk sac) control, the positive serum, the negative serum, guinea pig complement, rabbit anti-sheep hemolysin, and sheep erythrocytes. (The guinea pig complement should be carefully tested for chlamydial antibodies, because many herds are enzootically infected with chlamydiae, guinea pig inclusion-conjunctivitis agent.) The hemolytic system is titrated and the complement unitage is determined. The standard units used in the test are 4 units of antigen and 2 exact units of complement. The test may be performed using either a water bath at 37°C for 2 hours or using overnight incubation at 4°C; the former is preferable. Doubling dilutions of the serum (from 1:2) are made in a 0.25-ml volume of saline. The antigen is added at 4 units (0.25 ml), and 2 exact units of complement (0.5 ml) is added. Standard reagent controls are always included. The normal yolk sac control is used at the same dilution as the group antigen. The tubes are shaken well and incubated in a water bath at 37°C for 2 hours. Then 0.5 ml of sensitized sheep erythrocytes is added and the tubes are placed in a water bath for 1 hour, after which they are read for hemolysis on a 1+ to 4+ scale (roughly equivalent to 25–100% inhibition of erythrocyte lysis). The endpoint of the serum is considered the highest dilution inhibiting at least 50% (2+) hemolysis after a complete inhibition of hemolysis has been observed. It is good practice to shake the tubes to resuspend the settled cells, refrigerate them overnight, and recheck the results the next morning.

All reagents are available commercially, except for high-titered group antigen. This may be prepared as follows. Yolk sacs of 7-day embryonated eggs are inoculated with *Chlamydia* (e.g., psittacosis isolate 6BC) at a dose estimated to result in death of about 50% of inoculated eggs in 5 to 7 days. Eggs are candled daily, and those dying early are discarded. When the 50% death endpoint is approached, the remaining eggs (recently dead or live) are refrigerated for 3 to 24 hours. the yolk sacs are then harvested. If examination of random samples shows large numbers of particles, the yolk sacs are pooled. This preparation may be stored at −20°C until further processing. The yolk sacs are ground in a mortar with sterile sand. Beef heart broth (pH 7.0) is added to make a 20% suspension, and the material is cultured to determine if it is free of bacterial contamination. The suspension is placed in a flask containing sterile glass beads and stored at 4°C for 3 to 6 weeks with daily shaking. It is then centrifuged at about 500 × *g* to remove coarse particles, transferred to a heavy sterile flask, and steamed at 100°C or immersed in boiling water for 30 minutes. After it has cooled, liquefied phenol is added to 0.5%. The antigen should then be refrigerated for at least 1 week before use. It is stable for at least 1 year if not contaminated, and should have an antigen titer of 1:256 or greater. A similar preparation from uninfected yolk sacs must be included as one of the controls.

32.9.2 Microimmunofluorescence

The micro-IF test is usually performed against chlamydial organisms grown in yolk sac. Tissue culture grown agent may be used, but it may be necessary to concentrate the elementary bodies and to add some normal yolk sac to improve contrast for microscopy. The individual yolk sacs are selected for elementary body richness and pretitrated to give an even distribution of particles. It is generally found that a 1–3% yolk sac suspension (phosphate buffered saline at pH 7.0) is satisfactory. The antigens may be stored as frozen aliquots; after thawing, they are well mixed on a Vortex mixer before use. Antigen dots are placed on a slide in a specific pattern, using separate markings with a pen for each antigen. Each cluster of dots includes all the antigenic types to be tested. The antigen dots are air-dried and fixed on slides with acetone (15 minutes at room temperature). Slides may be stored frozen. When thawed for use, they may sweat, but they can be conveniently dried (as can the original antigen dots) with the cool air flow of a hair dryer. The slides have serial dilutions of serum (or tears or exudate) placed on the different antigen clusters. The clusters of dots are placed sufficiently separated to avoid the running of the serum from cluster to cluster. After the serum dilutions have been added, the slides are incubated for 0.5 to 1 hour in a moist chamber at 37°C. They are then placed in a buffered saline wash for 5 minutes, followed by a second 5-minute wash. The slides are then dried and stained with fluorescein-conjugated anti-human globulin. Conjugates are pretitrated in a known positive system to determine appropriate working dilutions. This reagent may be prepared against any class of globulin being considered (IgA or secretory piece for secretions, IgG or IgM). Counterstains, such as bovine serum albumin conjugated with rhodamine, may be included. The slides are then washed twice again, dried, and examined by standard fluorescence microscopy. Use of a monocular tube is recommended to allow greater precision in determining fluorescence of individual elementary body particles. The endpoints are read as the dilution giving bright fluorescence clearly associated with the well distributed elementary bodies throughout the antigen dot. Identification of the type-specific response is based on dilution differences reflected in the endpoints for different prototype antigens.

For each run of either CF or micro-IF, known positive and negative sera always should be included. These sera should always duplicate their titers as previously observed within the experimental (±1 dilution) error of the system.

REFERENCES

Banks, J., Eddie, B., Sung, M., Sugg, N., Schachter, J., and Meyer, K.F. 1970. Plaque reduction technique for demonstrating neutralizing antibodies for *Chlamydia*. Infect. Immun. 2:443–447.

Beem, M.O., and Saxon, E.M. 1977. Respiratory-tract colonization and a distinctive pneumonia syndrome in infants infected with *Chlamydia trachomatis*. N. Engl. J. Med. 296:306–310.

Bowie, W.R., Wang, S.-P., Alexander, E.R., Floyd, J., Forsyth, P., Pollock, H., Tin, J.-S., Buchanan, T., and Holmes, K.K. 1977. Etiology of nongonococcal urethritis: Evidence for *Chlamydia trachomatis* and *Ureaplasma urealyticum*. J. Clin. Invest. 59:735–742.

Byrne, G.I., and Moulder, J.W. 1978. Parasite-specified phagocytosis of *Chlamydia psittaci* and *Chlamydia trachomatis* by L and HeLa cells. Infect. Immun. 19:598–606.

Caldwell, H.D., and Kuo, C.C. 1977. Serologic diagnosis of lymphogranuloma venereum by counter immunoelectrophoresis with a *Chlamydia trachomatis* protein antigen. J. Immunol. 118:442–445.

Caldwell, H.D., and Schachter, J. 1982. Antigenic analysis of the major outer membrane protein of *Chlamydia* spp. Infect. Immun. 35:1024–1031.

Darougar, S., Kinnison, J.R., and Jones, B.R. 1971. Simplified irradiated McCoy cell culture for isolation of chlamydiae. In R.L. Nichols (ed.), Trachoma and Related Disorders Caused by Chlamydial Agents. Amsterdam: Excerpta Medica, pp. 63–70.

Dhir, S.P., Kenny, G.E., and Grayston, J.T. 1971. Characterization of the group antigen of *Chlamydia trachomatis*. Infect. Immun. 4:725–730.

Finn, M.P., Ohlin, A., and Schachter, J. 1983. Enzyme-linked immunosorbent assay for immunoglobulin G and M antibodies to *Chlamydia trachomatis* in human sera. J. Clin. Microbiol. 17:848–852.

Gordon, F., and Quan, A.L. 1965. Isolation of the trachoma agent in cell culture. Proc. Soc. Exp. Biol. 118:354–359.

Grayston, J.T., and Wang, S.-P. 1975. New knowledge of chlamydiae and the diseases they cause. J. Infect. Dis. 132:87–105.

Jones, B.R. 1964. Ocular syndromes of TRIC

virus infection and their possible genital significance. Br. J. Vener. Dis. 40:3–18.

Moulder, J.W. 1966. The relation of the psittacosis group (chlamydiae) to bacteria and viruses. Ann. Rev. Microbiol. 20:107–130.

Moulder, J.W., Hatch, T.P., Kuo, C.C., Schachter, J., and Storz, J. 1984. Order II. Chlamydiales Storz and Page 1971, 334. In N.R. Krieg and J.G. Holt (eds.), Bergey's Manual of Systematic Bacteriology. Baltimore, MD: Williams and Wilkins, pp. 729–739.

Nurminen, M., Leinonen, M., Saikku, P., and Makela, P.H. 1983. The genus-specific antigen of *Chlamydia:* Resemblance to the lipopolysaccharide of enteric bacteria. Science 220:1279–1281.

Page, L.A. 1972. Chlamydiosis (ornithosis). In M.S. Hofstad (ed.), Diseases of Poultry. Ames, IO: Iowa State University Press, pp. 414–417.

Puolakkainen, M., Saikku, P., Leinonen, M., Nurminen, M., Vaananen, P., and Makela, P.H. 1984. Chlamydial pneumonitis and its serodiagnosis in infants. J. Infect. Dis. 149:598–604.

Richmond, S.J., and Caul, E.O. 1977. Single-antigen indirect immunofluorescence test for screening venereal disease clinical populations for chlamydial antibodies. In K.K. Holmes and D. Hobson (eds.), Nongonococcal Urethritis and Related Infections. Washington, D.C.: American Society for Microbiology, pp. 259–265.

Ripa, K.T., and Mardh, P.-A. 1977. New simplified culture technique for *Chlamydia trachomatis*. In K.K. Holmes and D. Hobson (eds.), Nongonococcal Urethritis and Related Infections. Washington, D.C.: American Society for Microbiology, pp. 323–327.

Sacks, D.L., and MacDonald, A.B. 1979. Isolation of type-specific antigen from *Chlamydia trachomatis* by sodium dodecyl sulfate-polyacrylamide gel electrophoresis. J. Immunol. 122:136–139.

Schachter, J. 1978. Chlamydial infections. N. Engl. J. Med. 298:428–435, 490–495, 540–549.

Schachter, J., and Dawson, C.R. 1978. Human Chlamydial Infections. Littleton: Publishing Sciences Group.

Schachter, J., Grossman, M., and Azimi, P.H. 1982. Serology of *Chlamydia trachomatis* in infants. J. Infect. Dis. 146:530–535.

Stamm, W.E., Tam, M., Koester, M., and Cles, L. 1983. Detection of *Chlamydia trachomatis* inclusions in McCoy cell cultures with fluorescein-conjugated monoclonal antibodies. J. Clin. Microbiol. 17:666–668.

Stephens, R.S., Tam, M.R., Kuo, C.-C., and Nowinski, R.C. 1982. Monoclonal antibodies to *Chlamydia trachomatis:* Antibody specificities and antigen characterization. J. Immunol. 128:1083–1089.

Storz, J. 1971. Chlamydia and Chlamydia-induced Diseases. Springfield: Charles C. Thomas.

Tam, M.R., Stamm, W.E., Handsfield, H.H., Stephens, R., Kuo, C.-C., Holmes, K.K., Ditzenberger, K., Crieger, M., and Nowinski, R.C. 1984. Culture-independent diagnosis of *Chlamydia trachomatis* using monoclonal antibodies. N. Engl. J. Med. 310:1146–1150.

Thomas, B.J., Evans, R.T., Hutchinson, G.R., and Taylor-Robinson, D. 1977. Early detection of chlamydial inclusions combining the use of cycloheximide-treated McCoy cells and immunofluorescence staining. J. Clin. Microbiol. 6:285–292.

Thomas, B.J., Reeve, P., and Oriel, J.D. 1976. Simplified serological test for antibodies to *Chlamydia trachomatis*. J. Clin. Microbiol. 4:6–10.

Wang, S.-P., and Grayston, J.T. 1970. Immunologic relationship between genital TRIC, lymphogranuloma venereum and related organisms in a new microtiter indirect immunofluorescence test. Am. J. Ophthalmol. 70:367–374.

Wang, S.-P., Grayston, J.T., Alexander, E.R., and Holmes, K.K. 1975. Simplified microimmunofluorescence test with trachoma–lymphogranuloma venereum (*Chlamydia trachomatis*) antigens for use as a screening test for antibody. J. Clin. Microbiol. 1:250–255.

Yoder, B.L., Stamm, W.E., Koester, M.C., and Alexander, E.R. 1981. Microtest procedure for isolation of *Chlamydia trachomatis*. J. Clin. Microbiol. 13:1036–1039.

Section 3

Reference Laboratories

33

Virology Services Offered by the Federal Reference Laboratory—Centers for Disease Control

33.1 INTRODUCTION

A strong collaborative effort among local, state, and federal laboratories provides the foundation for a successful nationwide program for the surveillance, prevention, and control of infectious diseases. Laboratories at each level have distinct responsibilities. As the state public health laboratories provide reference and disease surveillance at the state level, so the Centers for Disease Control (CDC) provide reference and disease surveillance at the national level. The effectiveness and timeliness with which each laboratory fulfills its responsibilities greatly influence the success of the nationwide program.

Providing reference and disease surveillance (RDS) in microbiology, hematology, histopathology, and immunology at CDC is the responsibility of the Center for Infectious Diseases (CID). This program, in collaboration with state and other qualified laboratories, constitutes an important segment of the mission of the CID. *All* RDS specimens, with proper justification and a completed request form, must be submitted to CID by or through the state public health laboratory or with the knowledge and consent of the state laboratory director or designee. CID will provide RDS on the following:

- Cultures, serum, or cerebrospinal fluid (CSF) samples, transudates, exudates, tissues, or histologic specimens from patients suspected of having an unusual infectious disease and/or other kinds of specimens (vectors, foods, liquids, etc.) that aid in the diagnosis of life-threatening, unusual, or exotic infectious diseases
- Cultures or serum specimens obtained from patients who have infectious diseases that occur only sporadically or who are involved in outbreaks of diseases caused by organisms for which satisfactory diagnostic reagents are not commercially or widely available
- Organisms suspected of being unusual pathogens or that are associated with hospital-acquired infections
- Specimens that are clinically important and are forwarded to CDC for confirmation because earlier test results were difficult to interpret or require confirmation for quality assurance of test performance

Material in this chapter is reprinted from Reference and Disease Surveillance, 1985, by the Centers for Disease Control, Atlanta, Georgia, with permission.

- Serum specimens or cultures that are clinically important and are sent to CID for confirmation because the results in state laboratories were atypical, aberrant, or difficult to interpret, or difficulties were encountered with the reagents used
- Arthropod and vertebrate specimens necessary for confirmation of zoonotic diseases

In assigning reference priorities strong emphasis is placed on the quality of both the specimen and the accompanying information. Prior consultation on urgent or unusual specimens will enable CID to be more responsive and efficient. Protecting our nation's health through an effective reference and disease surveillance program requires teamwork, cooperation, and good communication.

33.2 CID ORGANIZATION

The Table of Organization shows the structure of the Center for Infectious Diseases (CID). The virology laboratories are located in the CID Divisions or Programs as shown below. The staff members to contact for consultation are listed by organization on the following page.

Bacteriology Laboratories
- Division of Bacterial Diseases (DBD)
- Hospital Infections Program (HIP)
- Sexually Transmitted Diseases Laboratory Program (STDLP)
- Plague Branch (see DVBVD)

Mycology Laboratories
- Division of Mycotic Diseases (DMD)

Parasitology Laboratories
- Division of Parasitic Diseases (DPD)

Virology Laboratories
- Division of Vector-Borne Viral Diseases (DVBVD) (includes plague lab)
- Division of Viral Diseases (DVD) (includes Rickettsia lab)

Hematology, Histopathology, and Immunology Laboratories
- Division of Host Factors (DHF)

33.3 REFERENCE AND DISEASE SURVEILLANCE

The following tables list the virology RDS activities of CID. These tables are organized by disease category and alphabetized for easy referral. Use Tables 33.1–33.3 and the footnoted instructions along with the information in section 33.4 for the shortest possible response time from CID laboratories.

Reference and disease surveillance activities are not undertaken for diseases for which satisfactory reagents or tests are commercially available or for which there is no public health need. The Assistant Director for Laboratory Science, CID, is available to consult with state health department laboratory directors about firms producing reagents or offering interstate diagnostic services.

33.4 REQUIREMENTS FOR ALL SPECIMENS

33.4.1 General Information

1. Etiologic agents should be cultivated and shipped in a medium that will protect and ensure the viability of the microorganism during transit
2. Only pure cultures of etiologic agents should be sent; mixed cultures cannot be accepted without written justification
3. Optimum containers for different groups of etiologic agents vary depending on the agent and the distance involved in shipment. In all instances the primary container should be of a durable material that, when properly packaged, is leakproof and can withstand the temperature and pressure variations likely to occur in the air and on the ground during shipment to CDC.

(Continued on page 514)

Table 33.1. Laboratories of the CDC Handling Virus Specimens

Division		Office	Commercial	Federal Telephone System FTS
VECTOR-BORNE VIRAL DISEASES				
Director	Thomas P. Monath, M.D.	Ft. Collins, CO	303 221-6428	330-6428
Asst. Dir. Med. Sc.	Jack D. Poland, M.D.	Ft. Collins, CO	303 221-6429	330-6429
Asst. Dir. Lab. Sc.	Dennis W. Trent, Ph.D.	Ft. Collins, CO	303 221-6420	330-6420
ARBOVIRUS ECOLOGY BRANCH				
Chief	D. Bruce Francy, Ph.D.	Ft. Collins, CO	303 221-6432	330-6428
ARBOVIRUS REFERENCE BRANCH				
Chief	Charles H. Calisher, Ph.D.	Ft. Collins, CO	303 221-6428	330-6459
DENGUE BRANCH				
Chief	Duane J. Gubler, Ph.D.	San Juan, PR	809 781-3636	
VIRAL DISEASES				
Director	Frederick A. Murphy, D.V.M., Ph.D.	Bldg. 6, Rm. 110	404 329-3574	236-3574
Asst. Dir. Med. Sc.	James E. Maynard, M.D.	Bldg. 6, Rm. 154A	404 329-3491	236-3491
Asst. Dir. Lab. Sc.	Donald P. Francis, M.D., D.Sc.	Bldg. 7, Rm. SB16	404 329-3577	236-3577
Asst. Dir. Lab. Sc.	Kenneth L. Herrmann, M.D.	Bldg. 6, Rm. 110	404 329-3574	236-3574
EPIDEMIOLOGY OFFICE				
Chief	Lawrence B. Schonberger, M.D.	Bldg. 6, Rm. 128	404 329-3091	236-3091
AIDS BRANCH				
Chief	James W. Curran, M.D., M.P.H.	Bldg. 6, Rm. 292	404 329-2891	236-2891
HEPATITIS BRANCH				
Act. Chief	James E. Maynard, M.D., Ph.D.	Bldg. 6, Rm. 154A	404 321-2339	236-2339
INFLUENZA BRANCH				
Chief	Alan P. Kendal, Ph.D.	Bldg. 7, Rm. 111	404 329-3591	236-3591
RESPIRATORY AND ENTEROVIRUS BRANCH				
Chief	Larry J. Anderson, M.D.	Bldg. 7, Rm. 144	404 329-3596	236-3596
SPECIAL PATHOGENS BRANCH				
Chief	Joseph B. McCormick, M.D.	Bldg. 7, Rm. SSB8	404 329-3308	236-3308
VIRAL EXANTHEMS AND HERPES VIRUS BRANCH				
Chief	Carlos Lopez, Ph.D.	Bldg. 7, Rm. 206	404 329-3532	236-3532
VIRAL AND RICKETTSIAL ZOONOSES BRANCH				
Chief	Joseph E. McDade, Ph.D.	Bldg. 7, Rm. B	404 329-3095	236-3095
STATISTICAL SERVICES AND DATA PROCESSING ACTIVITY				
Chief	Dennis J. Bregman	Bldg. 6, Rm. 113	404 329-3490	236-3490

Table 33.2. Viral Diseases for which Testing is Available at CDC

Disease or Agent[b]	Organizational Unit	Serology	Isolation Specimens	Antigen Detection	Specimens
Arboviruses[a]					
California encephalitis (LaCrosse)	DVBVD	EIA, HI, CF, NT	Brain, mosquitoes	IIF, EIA	organs, mosquitoes
Colorado tick fever	DVBVD	EIA, CF, NT	Blood (unfrozen)[c]	IIF, EIA	blood
Dengue 1–4	DVBVD	EIA, HI, CF, NT	Serum, mosquitoes	IIF, EIA	organs, mosquitoes
Eastern equine encephalitis	DVBVD	EIA, HI, CF, NT	Brain, serum, mosquitoes	IIF, EIA	organs, mosquitoes
Japanese encephalitis	DBVVD	EIA, HI, CF, NT	Brain, serum, mosquitoes	IIF, EIA	organs, mosquitoes
Murray Valley encephalitis	DVBVD	EIA, HI, CF, NT	Brain, serum, mosquitoes	IIF, EIA	organs, mosquitoes
Powassan	DVBVD	EIA, HI, CF, NT	Brain, serum mosquitoes	IIF, EIA	organs, mosquitoes
St. Louis encephalitis	DVBVD	EIA, HI, CF, NT	Brain, serum, mosquitoes	IIF, EIA	organs, mosquitoes, body fluids
Venezuelan equine encephalitis	DVBVD	EIA, HI, CF, NT	Brain, serum, mosquitoes	IIF, EIA	organs, mosquitoes
Western equine encephalitis	DVBVD	EIA, HI, CF, NT	Brain, serum, mosquitoes	IIF, EIA	organs, mosquitoes
Yellow fever	DVBVD	EIA, HI, CF, NT	Serum, liver, mosquitoes	IIF, EIA	blood, liver, mosquitoes
Other arboviruses	DVBVD	EIA, HI, CF, NT	Serum, mosquitoes	IIF, EIA	organs, mosquitoes
Crimean hemorrhagic fever (Congo)[d]	DVD	EIA, IIF, ELISA	Serum, liver		
Kyasanur Forest disease[d]	DVD				
Omsk hemorrhagic fever[d]	DVD				
Tick-borne encephalitis[d]	DVD				
Hantaan (Korean hemorrhagic fever)[d]	DVD				

Arenaviruses			
Junin[d]	DVD		
Lassa[d]	DVD		
Lymphocytic choriomeningitis	DVD	IIF	
Machupo[d]	DVD		
Enteroviruses			
Coxackie Viruses[e,f]	DVD		
Echoviruses[e,f]	DVD		
Enterovirus 70[e,f]	DVD		
Polioviruses 1–3[e,f]	DVD		
Respiratory viruses			
Adenoviruses[e]	DVD		
Coronaviruses[e]	DVD		
Influenza[h]	DVD		
Mumps[g]	DVD		
Parainfluenza[e]	DVD		
Parvovirus, human[e]	DVD		
Resp. syncytial virus[e]	DVD		
Picornaviruses[e]	DVD		
Herpesviruses			
Cytomegalovirus[e]	DVD		
Herpes simplex[e]	DVD		
Infectious mononucleosis (Epstein–Barr)	DVD	IIF, OCH	
Varicella-zoster[g]	DVD	IIF	Vesicular fluid, scabs[j]
Exanthematous viruses			
Measles (rubeola)[g]	DVD	HI	
Rubella[g]	DVD	HI, EIA	
Orf-paravaccinia	DVD	IIF	Vesicular fluid, scabs
Variola/Vaccinia[i]	DVD	HI, IIF, EIA	Vesicular fluid, scabs, brain, saliva[j]

Table 33.2. *(continued)*

Disease or Agent[b]	Organizational Unit	Serology	Isolation Specimens	Antigen Detection	Specimens
Miscellaneous viruses					
Ebola[d]	DVD				
Marburg[d]	DVD				
Rabies	DVD	RFFIT[j], DFA	Brain or skin biopsy		
Retroviruses[k]					
HTLV-I	DVD	RIP			
HTLV-II	DVD	RIP			
HTLV-III/LAV	DVD	RIP, ELISA, WIB			
Hepatitis A					
Hepatitis A (anti-HAV)	DVD	EIA[j]			
IgM-anti HAV	DVD	EIA[j]			
Hepatitis B					
HBsAg	DVD	EIA[j]			
Anti HBs	DVD	EIA[j]			
Anti HBc	DVD	EIA[j]			
IgM anti HBc	DVD	EIA[j]			
HBeAg	DVD	EIA[j]			
Anti HBe	DVD	EIA[j]			
Subtyping HBsAg	DVD	PPT[j]			
Δ Hepatitis					
Anti Δ	DVD	EIA[j]			
IgM anti Δ	DVD	EIA[j]			
Δ antigen	DVD	EIA[j]			
Viral gastroenteritis					
Rotavirus, human[e]	DVD	ELISA			
Norwalk virus[e]	DVD	RIA[j]			
Adenovirus[e]	DVD	ELISA			

Abbreviations: CF, complement fixation; NT, neutralization; ELISA, enzyme-linked immunosorbent assay; HI, hemagglutination inhibition; OCH, ox cell hemolysin; RFFIT, rapid fluorescent focus inhibition test; IIF, indirect immune fluorescence; PPT, precipitation test; DVBVD, Division of Vector Borne Viral Diseases, Ft. Collins, CO; RIP, radioimmunoprecipitation; EIA, enzyme immunoassay; DVD, Division of Viral Diseases, Atlanta, GA; RIA, radioimmunoassay; WIB, Western immunoblot; DFA, direct fluorescent antibody test.

[a] Antibody titers for the arboviral diseases or agents listed may be detected in serum, CSF or both.

[b] Other tests or tests for agents not listed may be made by prior consultation and arrangement. All serum specimens for serologic tests should be sterile and shipped frozen or iced. Serum specimens may be sent without refrigeration if delivery is assured within 48 hours. In most instances paired (acute and convalescent) serum specimens are required for serologic diagnosis. Specimens for viral isolation should be shipped frozen on dry ice if they will not arrive within 48 hours of collection; if delivery to CDC will take less than 48 hours specimens for viral isolation should be shipped chilled by cold packs or wet ice.

[c] Send specimens on wet ice or cold packs; *do not* freeze.

[d] This is one of the most highly pathogenic viruses known. Infections (and deaths) may occur in medical and supportive personnel through close contact with patients and their exudates. Contact Special Pathogens Branch, Division of Viral Diseases, CID, CDC, for special instructions before taking, packing, and shipping specimens suspected of containing these agents. *Do not submit specimens for testing without making prior arrangements.*

[e] Because of the ubiquitous nature of these viruses, primary diagnosis by serologic testing or virus isolation cannot routinely be offered. These services may be provided by prior consultation and arrangement for outbreaks or cases of unusual public health significance.

[f] NT tests are performed by prior arrangement only when virus is isolated from patient with enterovirus/poliovirus.

[g] Immune status testing will be performed only by prior consultation and arrangement; indicate on request form that such arrangements were made.

[h] Because of the epidemic nature of the virus, primary diagnosis by serologic testing or virus isolation cannot routinely be offered. These services may be provided by individual agreement for outbreaks or cases of unusual public health significance. Reference antigenic analysis of representative or unusual isolates ("strain comparison") is organized and conducted yearly by the WHO Collaborating Center for Influenza at CDC. Isolates should be sent directly to that Center (Bldg. 7, Room 112). The completed WHO data form, *Not* CDC form 50.34, is required. This form may be obtained from the Influenza Branch, DVD, CID, CDC.

[i] For information on collecting, packing, and shipping specimens from a person with suspected pox disease, contact Viral Exanthems and Herpesvirus Branch, DVD, CID, CDC.

[j] Prior consultation arrangements with appropriate Branch are required before submitting specimens; indicate on request form that such arrangements were made.

[k] Laboratory support and collaboration for AIDS and lymphatic malignancies available through prior arrangements with Retrovirus Laboratory, AIDS Branch, DVD, CID, CDC.

Table 33.3. Chlamydial Diseases for which Testing is Available at CDC

Disease or Agent	Organizational Unit	Identification	Typing	Antimicrobial Susceptibilities	Serology[a]	Isolation
Chlamydia psittaci	STDLP	X			CF	*
Chlamydia trachomatis infections	STDLP	X	X			
LGV	STDLP	X	*	*	CF	*
Genital non-LGV	STDLP	X	*	*	IIF*	*
Trachoma-inclusion conjunctivitis	STDLP	X	*	*	IIF	*
Pneumonia	STDLP	X	*	*	IIF*	*

Symbols and abbreviations: *, Prior consultation and arrangements are required before submitting specimens; these arrangements should be indicated on submitted request form. X, Provided only for pure isolates unless otherwise indicated; *environmental or animal isolates not accepted unless sufficient evidence of public health significance is presented.* CF, complement fixation; IIF, indirect immune fluorescence; LGV, lymphogranuloma venereum; STDLP, Sexually Transmitted Diseases Laboratory Program.

[a] Paired serum specimens required.

V. U.S. DEPARTMENT OF HEALTH AND HUMAN SERVICES
PUBLIC HEALTH SERVICE
CENTERS FOR DISEASE CONTROL
ATLANTA, GEORGIA 30333
Telephone: (404) 329-3883 (Commercial)
236-3883 (FTS)

42 CFR Part 72 – Interstate Shipment of Etiologic Agents

PART 72—INTERSTATE SHIPMENT OF ETIOLOGIC AGENTS[1]

Sec.
72.1 Definitions.
72.2 Transportation of diagnostic specimens, biological products, and other materials; minimum packaging requirements.
72.3 Transportation of materials containing certain etiologic agents; minimum packaging requirements.
72.4 Notice of delivery; failure to receive.
72.5 Requirements; variations.

Authority: Sec. 215, 58 Stat. 690, as amended, 42 U.S.C. 216; sec. 361, 58 Stat. 703, (42 U.S.C. 264)

§ 72.1 Definitions.

As used in this part:

"Biological product" means a biological product prepared and manufactured in accordance with the provisions of 9 CFR Parts 102–104 and 21 CFR Parts 312 and 600–680 and which, in accordance with such provisions, may be shipped in interstate traffic.

"Diagnostic specimen" means any human or animal material including, but not limited to, excreta, secreta, blood and its components, tissue, and tissue fluids being shipped for purposes of diagnosis.

"Etiologic agent" means a viable microorganism or its toxin which causes, or may cause, human disease.

"Interstate traffic" means the movement of any conveyance or the transportation of persons or property, including any portion of such movement or transportation which is entirely within a State or possession, (a) from a point of origin in any State or possession to a point of destination in any other State or possession, or (b) between a point of origin and a point of destination in the same State or possession but through any other State, possession, or contiguous foreign country.

§ 72.2 Transportation of diagnostic specimens, biological products, and other materials; minimum packaging requirements.

No person may knowingly transport or cause to be transported in interstate traffic, directly or indirectly, any material including, but not limited to, diagnostic specimens and biological products which such person reasonably believes may contain an etiologic agent unless such material is packaged to withstand leakage of contents, shocks, pressure changes, and other conditions incident to ordinary handling in transportation.

§ 72.3 Transportation of materials containing certain etiologic agents; minimum packaging requirements.

Notwithstanding the provisions of § 72.2, no person may knowingly transport or cause to be transported in interstate traffic, directly or indirectly, any material (other than biological products) known to contain, or reasonably believed by such person to contain, one or more of the following etiologic agents unless such material is packaged, labeled, and shipped in accordance with the requirements specified in paragraphs (a)–(f) of this section:

Measles virus.
Mumps virus.
Parainfluenza viruses—all types.
Polioviruses—all types.
Poxviruses—all members.
Rabies virus—all strains.
Reoviruses—all types.
Respiratory syncytial virus.
Rhinoviruses—all types.
Rickettsia—all species.
Rochalimaea quintana.
Rotaviruses—all types.
Rubella virus.
Simian virus 40.
Tick-borne encephalitis virus complex, including Russian spring-summer encephalitis, Kyasanur forest disease, Omsk hemorrhagic fever, and Central European encephalitis viruses.
Vaccinia virus.
Varicella virus.
Variola major and Variola minor viruses.
Vesicular stomatis viruses—all types.
White pox viruses.
Yellow fever virus.[2]

(a) *Volume not exceeding 50 ml.* Material shall be placed in a securely closed, watertight container (primary container (test tube, vial, etc.)) which shall be enclosed in a second, durable watertight container (secondary container). Several primary containers may be enclosed in a single secondary container, if the total volume of all the primary containers so enclosed does not exceed 50 ml. The space at the top, bottom, and sides between the primary and secondary containers shall contain sufficient nonparticulate absorbent material (e.g., paper towel) to absorb the entire contents of the primary container(s) in case of breakage or leakage. Each set of primary and secondary containers shall then be enclosed in an outer shipping container constructed of corrugated fiberboard, cardboard, wood, or other material of equivalent strength.

(b) *Volume greater than 50 ml.* Packaging of material in volumes of 50 ml. or more shall comply with requirements specified in paragraph (a) of this section. In addition, a shock absorbent material, in volume at least equal to that of the absorbent material between the primary and secondary containers, shall be placed at the top, bottom, and sides between the secondary container and the outer shipping container. Single primary containers shall not contain more than 1,000 ml of material. However, two or more primary containers whose combined volumes do not exceed 1,000 ml may be placed in a single, secondary container. The maximum amount of etiologic agent which may be enclosed within a single outer shipping container shall not exceed 4,000 ml.

(c) *Dry ice.* If dry ice is used as a refrigerant, it must be placed outside the secondary container(s). If dry ice is used between the secondary container and the outer shipping container, the shock absorbent material shall be placed so that the secondary container does not become loose inside the outer shipping container as the dry ice sublimates.

(d)(1) The outer shipping container of all materials containing etiologic agents transported in interstate traffic must bear a label as illustrated and described below:

(2) The color of material on which the label is printed must be white, the symbol red, and the printing in red or white as illustrated.

(3) The label must be a rectangle measuring 51 millimeters (mm) (2 inches) high by 102.5 mm (4 inches) long.

(4) The red symbol measuring 38 mm (1½ inches) in diameter must be centered in a white square measuring 51 mm (2 inches) on each side.

(5) Type size of the letters of label shall be as follows:

Etiologic agents—10 pt. rev.
Biomedical material—14 pt.
In case of damage or leakage—10 pt. rev.
Notify Director CDC, Atlanta, Georgia—8 pt. rev.
404–633–5313—10 pt. rev.

[1] The requirements of this part are in addition to and not in lieu of any other packaging or other requirements for the transportation of etiologic agents in interstate traffic prescribed by the Department of Transportation and other agencies of the Federal Government.

[2] This list may be revised from time to time by Notice published in the Federal Register to identify additional agents which must be packaged in accordance with the requirements contained in this part.

(e) *Damaged packages.* The carrier shall promptly, upon discovery of evidence of leakage or any other damage to packages bearing an Etiologic Agents/Biomedical Material label, isolate the package and notify the Director, Center for Disease Control, 1600 Clifton Road, NE., Atlanta, GA 30333, by telephone: (404) 633-5313. The carrier shall also notify the sender.

(f) *Registered mail or equivalent system.* Transportation of the following etiologic agents shall be by registered mail or an equivalent system which requires or provides for sending notification of receipt to the sender immediately upon delivery:

Hemorrhagic fever agents including, but not limited to, Crimean hemorrhagic fever (Congo), Junin, Machupo viruses, and Korean hemorrhagic fever viruses.
Herpesvirus simiae (B virus).
Lassa virus.
Marburg virus.
Tick-borne encephalitis virus complex including, but not limited to, Russian spring-summer encephalitis, Kyasanur forest disease, Omsk Hemorrhagic fever, and Central European encephalitis viruses, Variola minor, and Variola major.
Variola major, Variola minor, and Whitepox viruses.

§ 72.4 Notice of delivery; failure to receive.

When notice of delivery of materials known to contain or reasonably believed to contain etiologic agents listed in § 72.3(f) is not received by the sender within 5 days following anticipated delivery of the package, the sender shall notify the Director, Center for Disease Control, 1600 Clifton Road, NE., Atlanta, GA 30333 (telephone (404) 633-5313).

§ 72.5 Requirements; variations.

The Director, Center for Disease Control, may approve variations from the requirements of this section if, upon review and evaluation, it is found that such variations provide protection at least equivalent to that provided by compliance with the requirements specified in this section and such findings are made a matter of official record.

[FR Doc. 80-21757 Filed 7-18-80; 8:45 am]
BILLING CODE 4110-86-M

Effective August 20, 1980

FEDERAL REGISTER, VOL. 45, NO. 141-- MONDAY, July 21. 1980

4. Serum specimens for serology should be aseptically separated from whole blood that was collected using aseptic technique. Contaminated serum specimens are unsuited for almost all purposes. Paired serum specimens are preferred, and in many cases required. The first specimen should be obtained as soon after the onset of illness as possible and refrigerated. The second specimen should be collected 2 to 4 weeks later. The optimal interval for collecting the second serum specimen will vary with different infectious diseases.
5. When whole blood is sent for isolation of certain viral agents, the blood should be kept cold but not frozen prior to shipment and shipped in wet ice, *not* dry ice; however, whole blood submitted for rickettsial isolation *must* be packed in dry ice and shipped frozen.
6. Slides with tissue sections, blood films, and smears of clinical material should be dry, free of immersion oil, properly labeled, and carefully packed in a slide mailing container.

33.4.2 Additional Suggestions for Viral and Chlamydial Specimens

33.4.2.a General

When shipping specimens in tubes or vials, use only tight fitting soft rubber stoppers or leak-proof screw caps; seal well with waterproof tape. Avoid direct contact between specimen container and dry ice to prevent breakage. To prevent thawing of specimens in the event of transit delays, use enough dry ice to last 48 hours beyond expected arrival time.

33.4.2.b Specimens for Isolation of Etiologic Agents

Select specimens during the acute, febrile phase of illness. Depending on disease suspected, the specimen submitted may be nasal or throat washings, sputum, urine, feces, CSF, skin scrapings, aspirate from lesions, blood, serum, or various tissues. Throat, nasal, or rectal swabs must be immersed in an appropriate virus transport fluid to prevent drying. Blood, CSF, and tissues should be handled aseptically. Autopsy tissues from several organs should be placed in separate containers, not pooled. Most specimens deliverable to the laboratory within 48 hours of collection may be sent chilled using frozen ice packs to maintain the initial condition. If more than 48 hours are required, the specimens should be frozen and kept frozen during shipment (specimens suspected of containing cytomegalovirus (CMV) should not be frozen but sent chilled). When shipment of frozen specimens is not possible, some types of specimens can be preserved in buffered glycerin, but partial loss of virus generally occurs (complete rickettsial and chlamydial loss will occur). Consultation is required before buffered glycerin is used for transport of specific virus specimens.

33.4.2.c Specimens for Serology

Generally, serum may be shipped unfrozen if drawn aseptically and delivery within 48 hours is assured. *Never* add a preservative to serum for serology.

33.4.3 Arthropod Specimens (Viral Isolation)

Arthropod specimens must be collected alive, killed by freezing or exposure to ether vapor, sealed in ampules, and shipped on dry ice. Cyanide or chloroform must not be used because these chemicals inactivate viruses and rickettsia.

33.4.4 Instructions for Identifying Specimens

1. Enclose the specimen in a tight fitting rubber stopper or screwcap tube or vial (primary container). Seal the tube or vial closure with waterproof tape; plastic tubes or vials are not recommended for routine shipment of specimens
2. Identify individual specimen tubes or vials by encircling them with adhesive tape with typed or pencilled name and other identifying information
3. Print patient's complete name and/or other information to identify with request form, type of specimen, and date the specimen was collected
4. *Do not* use ball point pens, wax, indelible pencils, or other writing instruments that tend to smear

33.4.5 Instructions for Packing Specimens

1. Package specimens properly to protect the material while in transit and the personnel who handle the packages:
   - *Never* mail clinical specimens or cultures in glass or plastic Petri plates or similar containers
   - *Never* enclose dry ice in hermetically sealed containers
2. Place the tube or vial in a watertight shipping container (secondary container); pack absorbent cotton or other suitable absorbent material around the tube to absorb shock and possible leakage; do not use particulate material for this purpose; if several tubes are to be packed in the same can, cushion them by wrapping each individually in paper towels or cotton; *do not* wrap form CDC 50.34 around the tube; put it around the secondary container (Figure 33.1)
3. Pack the secondary container in a cardboard outer shipping container with crumpled newspaper or other shock absorbing insulating material; CDC strongly recommends using special commercially available styrofoam-lined cardboard shipping containers rather than ordinary cardboard containers (Figure 33.2); seal the outer shipping container securely, and affix a properly completed address label

33.4.6 Instructions for Shipping Specimens

1. When CO_2 is used as a refrigerant, the outer shipping container should be marked in accordance with applicable Department of Transportation regulations and Air Transport Association tariffs
2. The Etiologic Agent/Biological Materials label should be affixed to all shipments of etiologic agents (cultures, virus suspensions, etc.) in accordance with Title 42, Code of Federal Regulations, Section 72.25 (Figure 33.3)
3. If specimens are to be shipped for long distances, send them by an expedited package service to assure prompt arrival; avoid air freight because it is often delayed
4. Avoid delay at CDC by addressing shipments as follows:

Data and Specimen Handling Section
Center for Infectious Diseases
Centers for Disease Control
Atlanta, GA 30333

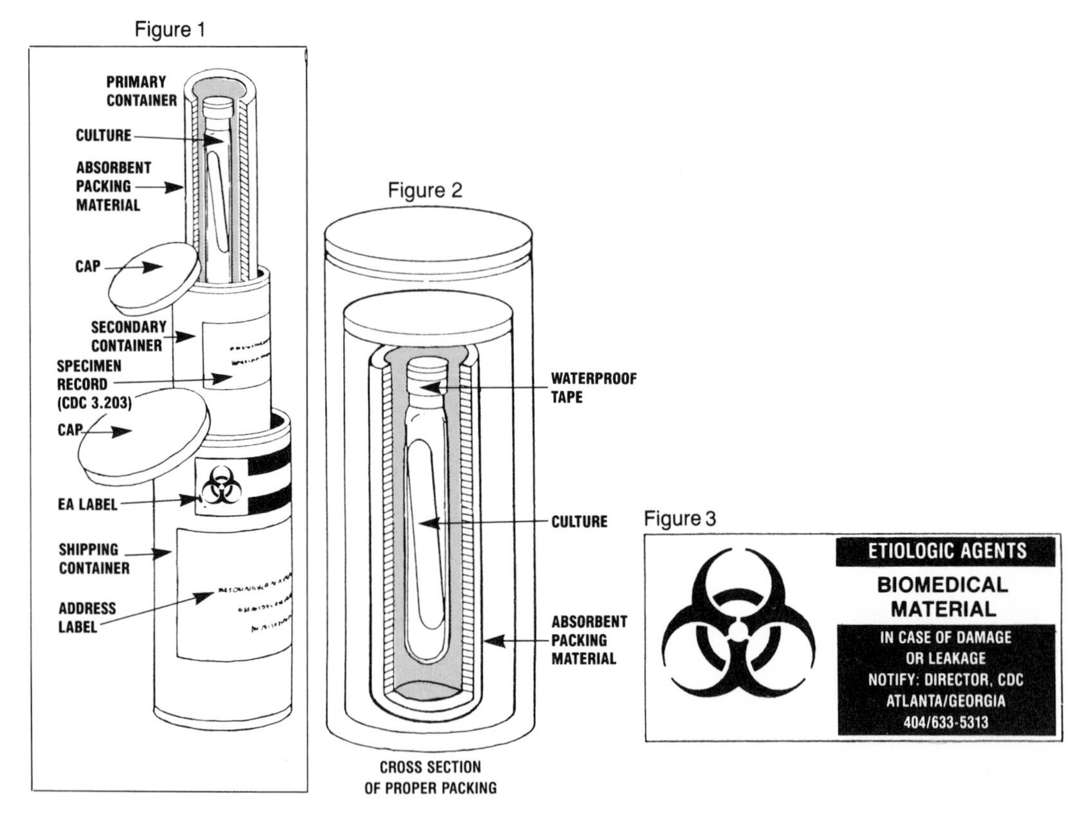

Figures 33.1, 33.2, 33.3. Figures 1 and 2 diagram the packaging and labeling of etiologic agents in volumes of less than 50 ml. in accordance with the provisions of subparagraph 72.3 (a) of the cited regulation. Figure 3 illustrates the label, described in subparagraph 72.3 (d) (1–5) of the regulations, which shall be affixed to all shipments of etiologic agents.

5. Whenever possible, time the shipment to arrive at CDC at the beginning or middle of the workweek *not* just before or on a weekend or holiday. Section 72.25, Part 72, Title 42, Code of Federal Regulations, governs the interstate shipment of etiologic agents and diagnostic specimens and contains packing and labeling requirements

33.4.7 Instructions for Completing Request Form for Reference and Disease Surveillance Support CDC 50.34 REV 8-84

The CDC form is a combined request, specimen information, patient history, and results report form. The upper third of the front of the form is for the required identifying information used to start a computer record on each specimen, to track a specimen while being tested in CID laboratories, and to direct the reports to the proper addressee. *All* of the information requested on this *front third* must be provided. The results of the CID laboratories are affixed to the lower portion of the front and constitute the report that is returned to the state health department laboratory or other authorized sender. The back of the form is to provide CID laboratories with certain essential information about the specimen, the assistance requested, or the patient. The required information on the back of the form depends on the type of specimen submitted and the assistance requested. Detailed instructions for completing the form are on the following pages.

33.5 SPECIAL REQUIREMENT FOR VIRAL DISEASES

Considerable research and development in the rapid diagnosis of viral diseases, together with the recognition of many new viral agents responsible for infectious disease, prompts CID to encourage state laboratories to request consultation and reference testing for both those diseases/agents listed, as well as new agents that may not be included in the tables.

The Division of Viral Diseases will provide confirmation of serologic and virologic diagnosis when deemed essential by the state laboratories (after that laboratory has consulted with the Division of Viral Diseases specialty laboratory). Highest priority will be given to the diagnosis of a life-threatening condition, or to an outbreak of illness of public health importance. Infection with a highly pathogenic virus, such as Lassa and Marburg, is a major health hazard not only for the individual patient, but for the medical and support personnel involved. In such cases, prompt contact with the Special Pathogens Branch, Division of Viral Diseases, is important. Laboratory support is also available to evaluate problem specimens and to study the complications of viral vaccines.

The Division of Vector Borne Viral Diseases (DVBVD) is concerned with surveillance of arthropod-borne viral diseases in the U.S. and offers clinical and public health laboratory support in a) primary or confirmatory diagnosis of arboviral infections; b) identification of arbovirus isolates; and c) serologic and virologic surveillance of arthropods and vertebrates.

Clinical specimens from suspected arboviral diseases *must* be accompanied by the following items of clinical and epidemiologic history: dates and places of travel; yellow fever, Japanese encephalitis or other relevant immunizations (WEE, EEE, VEE virus vaccine in horses and avians); and dates of onset and specimen collections.

33.5.1 Sera and Other Body Fluids

Paired sera (acute and convalescent phase) are necessary for serologic diagnosis but in some cases of arboviral diseases, a presumptive diagnosis is possible if only a single convalescent serum is available.

The DVBVD requests CSF from encephalitis patients. Newly introduced antibody capture tests of antibody in CSF offer a rapid and sensitive means of specific diagnosis. Detection of viral antigen in other body fluids is under investigation and approved fluids are requested for testing. Prearrangement for these tests should be made with the DVBVD.

33.5.2 Tissues

Biopsy or autopsy specimens from patients with arboviral diseases may be submitted for viral isolation and/or examination by immunofluorescence (IF) for viral antigen. Tissues submitted for IF examination should be embedded in appropriate medium for frozen sections, or sent cold for processing in Fort Collins, CO. Prearrangement for these procedures is requested.

33.5.3 Isolates

Viral isolates referred to the DVBVD for identification and/or characterization should be accompanied by as complete an isolation history as possible. Source of the isolate, date of isolation, host system used for isolation, success of reisolation and passage level, and titer of the material being sent should be given. If available, information concerning sensitivity to lipid solvents, spectrum of sensitive host systems, presence or absence of HA activity, and antigenic relationship to other established viruses also should be offered. Prearrangements for these studies must be made with the DVBVD.

33.5.4 Arthropod, Avian Specimens

In the event of a known or presumed arthropod-borne viral encephalitis outbreak, arthropods may be submitted for virus isolation. Contact with DVBVD personnel is required to arrange for testing prior to submission of specimens. Specimens should be carefully put into lipless glass vials, tightly stoppered, and shipped on dry ice. The date(s), method(s), and location(s) of collection should accompany specimens. Wild avian sera may also be submitted for antibody tests under similar conditions. Sera should be collected from abundant passerine bird species occurring in the potential outbreak area. Birds may be bled from the jugular vein (0.2 ml) and the blood added to 0.9 ml diluent [buffered saline pH 7.4–7.8, preferably with protein stabilizer such as fetal calf serum (FCS) or bovine albumin (BA)]. Specimens should be centrifuged to remove cells and debris, supernatant fluid removed, frozen, and shipped on dry ice. As for arthropods, prearrangements for testing are required.

33.6 FRONT OF FORM

Name, address, and phone number of requesting physician or organization: Print or type the name and address and telephone number of physician, microbiologist, or organization where specimen originated (person or institution to contact if additional information is needed and to whom final report will be forwarded by the State laboratory)

State Health Department address: Print or stamp the address of the state health department laboratory or agency sending specimen (this will ensure that reports are directed through the correct state health department)

State Health Department number: Print state health department laboratory number assigned to the specimen, if any (used as a crossreference for specimen identification)

Date sent to CDC: Print date (month, day, and year) that specimen is shipped to CDC

Hospital number: For those specimens originating from a hospital, print patient's hospital number (not required by CDC but is helpful information for the hospital record office to match report to patient's record)

Name: Print last name, first name, and middle initial of patient or other equally appropriate specimen identification (required to track and locate specimens)

Birthdate: Print date (month, day, and year) of patient's birth; age in years, or months if an infant, is acceptable

Sex: Print an "M" (male) or "F" (female)

Clinical diagnosis: Print patient's clinical diagnosis; if none has been made, indicate why assistance is requested (e.g., possible outbreak, exotic isolate, or disease)

Associated illness: Print patient's associated or underlying illness, such as cancer, arthritis, hypertension, immunocompromised, or enter major symptoms

Date of onset: Print date (month, day, and year) illness started (this date is critical for the interpretation of serologic results; if uncertain, give approximate date)

Fatal: Check 1 box only (this element has epidemiologic significance)

Type of specimen: Print type of specimen, such as serum, CSF, fungus culture, etc.

33.7 BACK OF FORM

The types of specimens usually sent to CDC laboratories are serum specimens, reference cultures, or clinical specimens. To assist state health department laboratories and others in obtaining the information on the back of the request form that CID requires, the following tabulation for each of

*Justification must be completed by State health department laboratory before specimen can be accepted by CDC. Please check the first applicable statement and when appropriate complete the statement with the *.*

1. Disease suspected to be of public health importance. Specimen is:
   - (a) ☐ from an outbreak.
   - (b) ☐ from uncommon or exotic disease.
   - (c) ☐ an isolate that cannot be identified, is atypical, shows multiple antibiotic resistance, or from a normally sterile site(s).
   - (d) ☐ from a disease for which reliable diagnostic reagents or expertise are unavailable in State.
2. ☐ Ongoing collaborative CDC/State project.
3. ☐ Confirmation of results requested for quality assurance.

*Prior arrangement for testing has been made. Please bring to the attention of:

(Name) ______

Completed by: ______

Date: ______

Name, Address and Phone Number of Physician or Organization:

STATE HEALTH DEPARTMENT LABORATORY ADDRESS:

STATE HEALTH DEPT. NO.:

DATE SENT TO CDC: Month Day Year

PATIENT IDENTIFICATION **Hospital No.:**

NAME: Last (18-37) First (38-47) Middle Initial (48)

BIRTHDATE: (49-54) Month Day Year

SEX: (55)

CLINICAL DIAGNOSIS: (56-57)

ASSOCIATED ILLNESS: (58-59)

DATE OF ONSET (Mo. Da. Yr.) (60-65)

FATAL? (66) ☐YES ☐NO

(For CDC Use Only) **CDC NUMBER**

UNIT | FY 3-4) | NUMBER (5-10) | SUF (11)

DATE RECEIVED (12-17) Month Day Year

REVERSE SIDE OF THIS FORM MUST BE COMPLETED

Type Specimen

THIS FORM MUST BE EITHER PRINTED OR TYPED

PLEASE PREPARE A SEPARATE FORM FOR EACH SPECIMEN

D.A.S.H.

0	3

(12-13)

Date Reported

Mo. Day Yr.

(14-19)

Comments:

(40-41)

D	6	5

(198-200)

DEPARTMENT OF HEALTH AND HUMAN SERVICES
PUBLIC HEALTH SERVICE
CENTERS FOR DISEASE CONTROL
Center for Infectious Diseases
Atlanta, Georgia 30333

The information collected on the above form for the use of CDC researchers is in accordance with the Public Health Service Act, Section 301. Furnishing the information is voluntary, but the completeness of the information accompanying the sample will influence the assignment of reference priorities. The individually identified data may be shared with public health or cooperating medical authorities in conjunction with CDC's ongoing program of providing reference laboratory tests in support of public health activities. An accounting of such disclosures may be made available to the subject individual upon request.

CDC 50.34 REV. 7-85
(Formerly 3.203)

LABORATORY EXAMINATION(S) REQUESTED (31-36)
☐ ANtimicrobial Susceptibility ☐ IDentification ☐ SErology (Specify Test) ____
☐ HIstology ☐ ISolation ☐ OTher (Specify) ____

CATEGORY OF AGENT SUSPECTED (37) ☐ Bacterial ☐ Viral ☐ Fungal ☐ Rickettsial ☐ Parasitic ☐ Other (Specify) ____

SPECIFIC AGENT SUSPECTED: ____ (38-40) | **OTHER ORGANISM(S) FOUND:** ____ (41-46)

ISOLATION ATTEMPTED? (47) ☐ Yes ☐ No **NO. TIMES ISOLATED** (48-49) ____ **NO. TIMES PASSED** (50-51) ____

SPECIMEN SUBMITTED IS (52): ☐ Original Material ☐ Pure Isolate ☐ Mixed Isolate

DATE SPECIMEN TAKEN (53-58) Mo. Da. Year
ORIGIN (59-60) ☐ HUman ☐ FOod ☐ SOil ☐ ANimal (specify) ____ ☐ OTher (Specify) ____

SOURCE OF SPECIMEN (61-62): ☐ BLood ☐ GAstric ☐ SErum ☐ SPutum ☐ URine ☐ CSF ☐ HAir ☐ SKin ☐ STool ☐ THroat
☐ WOund (Site) ____ ☐ TIssue (Specify) ____
☐ EXudate (Site) ____ ☐ OTher (Specify) ____

SUBMITTED ON (63-64): ☐ MEdium (Specify) ____ ☐ EGg ☐ TIssue Culture (Type) ____
☐ ANimal (Specify) ____ ☐ OTher (Specify) ____

SERUM INFORMATION: Mo. Da. Yr.
(65-72) ☐ ACute
(73-80) ☐ COnvalescent
(81-88) ☐ S3
(89-96) ☐ S4
(97-104) ☐ S5

IMMUNIZATIONS: Mo. Yr.
____ (105-110)
____ (111-116)
____ (117-122)
____ (123-128)

TREATMENT: Drugs Used ☐ None (129) Date Begun Mo. Da. Yr. Date Completed Mo. Da. Yr.
____ (130-143)
____ (144-157)
____ (158-171)

EPIDEMIOLOGICAL DATA: (172-173)
☐ SIngle Case ☐ SPoradic ☐ COntact ☐ EPidemic ☐ CArrier
Family Illness ____ (174-175)
Community Illness ____ (176-177)
Travel and Residence (Location) Mo. Yr.
☐ **Foreign** ____ (178-183)
☐ **USA** ____ (184-189)
Animal Contacts (Species) ____ (190-191)
Arthropod Contacts: (192) ☐ None ☐ Exposure Only ☐ Bite
Type of Arthropod ____ (193-194)
Suspected Source of Infection ____ (195-196)

PREVIOUS LABORATORY RESULTS/OTHER CLINICAL INFORMATION:
(Information supplied should be related to this case and/or specimen (s) and relative to the test (s) requested.

CLINICAL TEST RESULTS: (12-13) 0 2
Sputum and Histological Findings ____
Blood Counts ____ Urine Exams ____
Type Skin Tests Performed — Date Mo. Da. Yr. — Strength — Pos. — Neg.
____ (14-21) ____ (22)
____ (23-30) ____ (31)
____ (32-39) ____ (40)

SIGNS AND SYMPTOMS
(48-49) ☐ FEver
Maximum Temperature: ____ (50-53)
Duration: ____ Days (54-55)
(56-57) ☐ CHills

RASH:
(58-59) ☐ MAculopapular
(60-61) ☐ HEmorrhagic
(62-63) ☐ VEsicular
(64-65) ☐ Erythema Nodosum
(66-67) ☐ Erythema Marginatum
(68-69) ☐ OTher ____

RESPIRATORY:
(70-71) ☐ RHinitis
(72-73) ☐ PUlmonary
(74-75) ☐ PHaryngitis
(76-77) ☐ CAlcifications
(78-79) ☐ PNeumonia (type) ____
(80-81) ☐ OTher ____

CARDIOVASCULAR:
(82-83) ☐ MYocarditis
(84-85) ☐ PEricarditis
(86-87) ☐ ENdocarditis
(88-89) ☐ OTher ____

GASTROINTESTINAL:
(90-91) ☐ DIarrhea
(92-93) ☐ BLood
(94-95) ☐ MUcous
(96-97) ☐ COnstipation
(98-99) ☐ ABdominal pain
(100-101) ☐ VOmiting
(102-103) ☐ OTher ____

CENTRAL NERVOUS SYSTEM:
(104-105) ☐ HEadache
(106-107) ☐ MEningismus
(108-109) ☐ MIcrocephalus
(110-111) ☐ HYdrocephalus
(112-113) ☐ SEizures
(114-115) ☐ CErebral Calcification
(116-117) ☐ CHorea
(118-119) ☐ PAralysis
(120-121) ☐ OTher ____

MISCELLANEOUS:
(122-123) ☐ JAundice
(124-125) ☐ MYalgia
(126-127) ☐ PLeurodynia
(128-129) ☐ COnjunctivitis
(130-131) ☐ CHorioretinitis
(132-133) ☐ SPlenomegaly
(134-135) ☐ HEpatomegaly
(136-137) ☐ LIver Abscess
(138-139) ☐ LYmphadenopathy
(140-141) ☐ MUcous Membrane Lesions
(142-143) ☐ OTher ____

STATE OF ILLNESS:
(144-145) ☐ SYmptomatic
(146-147) ☐ ASymptomatic
(148-149) ☐ SUbacute
(150-151) ☐ CHronic
(152-153) ☐ DIsseminated
(154-155) ☐ LOcalized
(156-157) ☐ INtraintestinal
(158-159) ☐ EXtraintestinal
(160-161) ☐ OTher ____

FOR CDC USE ONLY 0 1 (12-13) No. Specimens: (16-20) ____ No. Tests: (21-25) ____

TYPE SERVICE: (14-15)
01-Reference
02-Epid. Aid
03-Proficiency Testing
04-Special Projects
____ - Other

LOCATION CODE: (26-27)

AR	Argentina	CM	Cameroon	GT	Guatemala	NU	Nicaragua	SZ	Switzerland		
AS	Australia	CO	Colombia	HA	Haiti	NZ	New Zeland	TD	Trinidad-Tobago		
AU	Austria	CS	Costa Rica	HO	Honduras	PA	Paraguay	TH	Thailand		
BC	Bermuda	CY	Cyprus	IN	India	PE	Peru	TW	Taiwan		
BE	Belgium	DR	Dominican Rep.	IS	Israel	PK	Pakistan	UK	United Kingdom		
BH	British Honduras	EC	Ecuador	IT	Italy	PL	Poland	UR	Soviet Union		
BL	Bolivia	ES	El Salvador	IV	Ivory Coast	PN	Panama	UY	Uruguay		
BR	Brazil	ET	Ethiopia	JM	Jamaica	PP	New Guinea	VE	Venezuela		
CA	Canada	FR	France	MX	Mexico	RP	Philippines	VN	Vietnam		
CB	Cambodia	GE	Germany	MY	Malaysia	RQ	Puerto Rico	VQ	Virgin Islands		
CI	Chile	GQ	Guam	NI	Nigeria	SL	Sierra Leone	____	Other ____		
						SP	Spain				

SPECIMEN SUBMITTED BY: (28-30)
100-Health Dept.
200-CDC Clinic
202-Biological Reagents
205-Proficiency Testing
214-S.E.P.
301-Army
302-Navy
303-Air Force
307-V.A. Hosp.
310-U.S.D.A.
321-F.D.A.
323-Indian Hosp.
324-PHS Hosp.
550-University
606-Physician/Clinic
____ - Other

CDC 50.34 REV. 7-85 (BACK)
(Formerly 3.203)

CDC Number Unit FY Number Suf.

the three types of specimens should serve as a guide.

Serum Specimens

Required

Laboratory exam requested
Specific agent suspected
Serum information[1]
Immunization[1]
Treatment[1]
Epidemiologic data[1]
Previous lab results

Useful

Clinical information
Signs, symptoms, etc.

Reference Cultures

Required

Laboratory exam requested
Category of agent suspected
Specific agent suspected
Kind of specimen
Origin of specimen
Source of specimen
Submitted on what medium
Previous lab results

Useful

Isolation attempted
Date specimen taken/number times isolated
Other clinical information
Clinical test results
Signs, symptoms, etc.
Other organisms found
Epidemiologic data[1]
Treatment[1]

Clinical Specimens

Required

Laboratory exam requested
Category of agent suspected
Specific agent suspected
Specimen submitted is
Date specimen taken
Source of specimen
Epidemiologic data[1]
Previous lab results

Useful

Other clinical information
Clinical test results
Signs, symptoms, etc.

ACKNOWLEDGMENT

The editors acknowledge with gratitude Dr. Albert Balows, Assistant Director for Laboratory Science, Center for Infectious Diseases, CDC, Atlanta, GA, who was responsible for supplying the material for this chapter and guiding us through its organization.

---

[1] *Exercise good judgment to determine the relevance of these items.* Paired sera are required for viral and bacterial disease serology, a single serum is required for mycotic and parasitic diseases and for syphilis serology (congenital syphilis excepted). In all instances the date(s) of collection of serum specimens *must* be provided. Immunization history is required when such information relates to the serology requested (i.e., required for polio, measles, etc.). Information on treatment, such as administration of immune serum or globulin, or antibiotics, is often of great benefit when doing serology or identifying reference cultures. As much relevant epidemiologic data as can be obtained should be provided. History of travel and animal or arthropod contacts are required for those RDS in which this kind of information is clearly necessary. If any required item of information is not available after efforts have been made to obtain it, please so indicate.

34

State Laboratory Virology Services

Steven Specter and Gerald J. Lancz

34.1 INTRODUCTION

State public health laboratories, much like the federal laboratories at the Centers for Disease Control (CDC), are charged with providing laboratory diagnosis of viral infections when local services are not available. Thus, most states do not encourage routine use of the state virology laboratories as a primary diagnostic laboratory, provided a local laboratory service is available. Limited services by state laboratories are often necessitated due to the limited financial support. Therefore, whenever possible, it is recommended that private hospital, commercial, or local public health laboratories be used for routine primary diagnostic virology services. However, the state virus laboratory should be utilized for diagnostic problems that go beyond the scope or capability of local laboratories, especially for viruses for which statewide surveillance is performed (e.g., influenza, arboviruses) as well as viruses of epidemiologic significance. Although many state laboratories will accept specimens for routine primary isolation, their budget may be adversely affected if the laboratories are confronted with large numbers of specimens. This may impinge unfavorably on their ability to perform a key function as a center to collate and send information regarding viral diseases to the CDC. Within this context we have listed some of the functions and viral diagnostic services available in state and U.S. territorial public health laboratories.

34.2 SUBMISSION OF SPECIMENS

All but a handful of the states have laboratories that accept specimens for the diagnosis of viral diseases (Table 34.1). In most cases the submission of specimens to the state laboratory is via local public health laboratories, when they are available. Each state with laboratories that accept specimens for virus isolation and/or identification has its own set of requirements for shipping, type of specimen, etc. In most cases these are described in detail in the literature provided by the appropriate state authority. In many instances, conditions for specimen submission are similar for many states; however, specific requirements are imposed by other states. It is advised that anyone who desires to submit a specimen to their state clinical virology laboratory, contact the laboratory head to determine their requirements for submitting clinical material. Some generally accepted require-

Table 34.1. Virology Services Available in State and Territorial Public Health Laboratory[a]

	Serology	Isolation	Refer Specimens to CDC[b]	Regulate Primary Labs in State	Arbovirus Surveillance	Influenza Surveillance	Rabies Detection	Special Epidemiologic Services
1. Alabama	+	No	+	No	+	NI	NI	Hepatitis testing
2. Alaska	+	+	No	NA	+	+	+	
3. Arizona	+	+	No[i]	No	+	+	+	
4. Arkansas	+	No	+	No	+	NI	NI	
5. California	+	+	+	+	+	+	+	Many others
6. Colorado	+	+[c]	+	No	NI	NI	+	Upon request
7. Connecticut	+	+	+	+	NI	NI	+	Many others
8. Delaware	+	+	+	+	+	+	NI	
9. District of Columbia	+	No	NI	+	NI	NI	+	
10. Florida	+	+	+	+[d]	+	NI	+	
11. Georgia	+	+	+	+[d]	NI	NI	NI	HBV screening-high risk groups
12. Guam	+[j]	No	+	No	NI	NI	NI	
13. Hawaii	+	+	+	No	NI	+	+	Respiratory & Enterovirus Surveillance
14. Idaho	+[g]	+[g]	No	NI	NI	NI		
15. Illinois	+	+	+	+	+	+	NI	Measles
16. Indiana	+	+	+	No	+	NI	NI	
17. Iowa	+	+	+	No	+	+	NI	Enterovirus Surveillance Mycoplasma monitoring of cell cultures
18. Kansas	+	+	+	No	NI	+	NI	Enterovirus, Rotavirus, Resp. Virus Surveillance

19. Kentucky	+	+	+	No	NI	+	+	
20. Louisiana	+	+	+	No	+	NI	+	
21. Maine	+	+	+	+	NI	NI	NI	
22. Maryland	+	+	+	+	+	+	+	Vaccine Preventable Dis. and Hepatitis Surveillance
23. Massachusetts	+	+	+	+[d]	+	NI	+	
24. Michigan	+	+	+	+	+	NI	+	Hepatitis B Measles and Rubella IgM
25. Minnesota	+	+	+	No	NI	NI	+	
26. Mississippi	+[e]	No	+	No	NI	NI	+	
27. Missouri	+	+	No	No	+	+	NI	Many others
28. Montana	+	+	+	+	+	+	NI	
29. Nebraska	NI	NI	NI	NI	NI	NI	NI	
30. Nevada	No	No	No	+[d]	NI	NI	NI	
31. New Hampshire	No[f]	No[f]	NI	NI	NI	NI	NI	
32. New Jersey	+	+	No	No	+	NI	+	
33. New Mexico					+	+	+	Enterovirus, adenovirus, Rubella IgM
34. New York	+	+	+	+	+	NI	+	
35. North Carolina	+	+	+	No	+	+	+	
36. North Dakota	+	+			NI	NI	+	
37. Ohio	+	+	+	No	+	NI	+	
38. Oklahoma	+	+	+	No	NI	+	+	
39. Oregon	+	+	+	No	NI	NI	NI	Hepatitis B screening certain high risk groups
40. Pennsylvania	+	+	+	+	+	+	+	
41. Puerto Rico	+[e,j]	No	No	No	NI	NI	+	
42. Rhode Island	+	No	+	No	NI	NI	+	
43. South Carolina	+	+	+	No	NI	NI	+	
44. South Dakota	+	+	+	No	+	+	+	

Table 34.1. (*continued*)

	Serology	Isolation	Refer Specimens to CDC[b]	Regulate Primary Labs in State	Arbovirus Surveillance	Influenza Surveillance	Rabies Detection	Special Epidemiologic Services
45. Tennesee	+	+	+	No	NI	NI	+	
46. Texas	+	+	No	No	+	NI	+	Herpes simplex screening 33rd week pregnancy Rubella screening
47. Utah	+	+	+	No	+	+	+	Some special projects
48. Vermont	+	+[h]	+	No	NI	+	NI	
49. Virginia	NI	NI	NI	No				
50. Virgin Islands	No	No	+	No	NI	NI	NI	
51. Washington	+	+	+	No	+	+	+	
52. West Virginia	+	No	NI	NI	NI	NI	+	
53. Wisconsin	+	+	+	+	+	+	NI	
54. Wyoming	No	No	+	No	NI	NI	NI	

Abbreviations: NA, not applicable; NI, no information provided; +, yes.
[a] Information is based on responses to a questionnaire sent to the state laboratories in fall and winter 1983 to 1984.
[b] Specimens submitted to CDC refer to those not normally tested in the state laboratories.
[c] Rabies virus isolation only is available.
[d] Regulation by state authorities but not through the Public Health Laboratory.
[e] Rabies direct immunofluorescence or animal brain only.
[f] Laboratory is presently in developmental stage.
[g] Herpes simplex virus and influenza only.
[h] Influenza only.
[i] Some seriologic requests go to CDC.
[j] Rubella only.

ments include:

1. A good clinical history with a listing of the patient's name and age, date of specimen collection, date of onset of illness, site and type of specimen collected, major clinical symptoms, relevant immunization history, virus(es) for which specimens are to be tested, and physician's name, address, and phone number.
2. Serologic testing for antiviral antibodies generally requires submission of acute and convalescent sera, except for special screening studies (e.g., rubella or varicella-zoster immune status)
3. Neonatal serum should be accompanied by a maternal serum

Many state laboratories request that during an epidemic, only a limited number of specimens be submitted to allow determination of the causative agents, rather than the submission of specimens for each patient seen in the course of the epidemic. Shipping requirements may include both conditions for handling and packaging specimens as well as prepayment of shipping costs. Most states do not require a fee for testing; however, a few states do have minimal fees (e.g., Missouri, $3 per specimen; Georgia, $8 minimum; Minnesota, unspecified handling fee). Submission of specimens may be limited to only certain individuals. For example, Georgia, Maryland, and New York have requirements that limit submission of specimens to physicians and/or patients who reside in their state. Several states limit specimen submission to licensed physicians, whereas, others also accept specimens from veterinarians and other legitimate public health services, including hospitals, state agencies, public or community health laboratories.

Specimens for the detection of rabies virus often have additional requirements for handling, shipping, and clinical history. Many states provide specific instructions as well as a Rabies Investigation Report Form to accompany such specimens.

34.3 SCOPE OF SERVICES

An overview of the services available in each state laboratory is provided in Table 34.1. The diversity of services offered by different states is apparent. These range from laboratories that offer only serology for select viruses, to those that provide extensive serology and isolation services for virtually any virus. There are laboratories that a) exclude class IV agents (e.g., lassa fever virus); b) exclude viruses that could be easily tested for in-hospital or private laboratories (e.g., herpes simplex); c) include only viruses that have epidemiologic importance (e.g., influenza, arboviruses). A listing of such specific services by state has been avoided here because these services no doubt change periodically. Again, it is recommended that you refer to your state laboratory to determine the extent of their services. The addresses and telephone numbers for the state laboratories, as of March 1984, are provided in Table 34.2.

Most laboratories will send specimens to the CDC if they do not handle them on site. This may be limited to class IV specimens or may cover a broad range of viral agents. Also listed in Table 34.1 are some special services offered by the laboratories, including participation in national or international surveillance programs for arboviruses and influenza; detection services for rabies virus, which usually results in rapid reporting in suspected rabies exposure cases and special epidemiologic services. However, a service that is considered to be special by one state laboratory may be part of the normal service offered by another state laboratory. The listing of special services placed in Table 34.1 is based on information supplied by the laboratory directors.

Regulation of licensure of virology laboratories by the state laboratories is not a common practice. Approximately 22% of the state laboratories indicated that they were involved in regulating primary laboratories. Four additional state laboratories

Table 34.2. State and Territorial Public Health Laboratories

ALABAMA
Director
Clinical Laboratory Administration
State Department of Public Health
University Drive
Montgomery, AL 36130
FTS Direct and Commercial: (205) 277-8660, ext. 215

ALASKA
Chief
Section of Laboratories
Alaska Division of Public Health
Department of Health and Social Services
Pouch H-06-D
Juneau, AK 99811
FTS Direct and Commercial: (907) 465-3140

ARIZONA
Chief
Bureau of Laboratory Services
Arizona Department of Health Services
1520 West Adams Street
Phoenix, AZ 85007
FTS Direct and Commercial: (602) 255-1188

ARKANSAS
Director
Division of Public Health Laboratories
4815 West Markham Street
Little Rock, AR 72201
FTS Operator: 740-5011
Commercial: (501) 661-2217

CALIFORNIA
Chief
Laboratory Services Branch
California Department of Health
2151 Berkeley Way
Berkeley, CA 94704
FTS Direct and Commercial: (415) 540-2408

COLORADO
Director
Division of Laboratories
Department of Public Health
4210 East 11th Avenue
Denver, CO 80220
FTS Direct and Commercial: (303) 320-1166

CONNECTICUT
Director of Laboratories
State Department of Health
P.O. Box 1689
Hartford, CT 06101
FTS Direct: 641-5063
Commercial: (203) 566-5063

DELAWARE
Director
Division of Public Health Laboratories
Jesse S. Cooper Memorial Building
Capitol Square
Dover, DE 19901
FTS Direct and Commercial: (302) 736-4734

DISTRICT OF COLUMBIA
Acting Director
Bureau of Laboratories
Department of Human Services
300 Indiana Avenue, NW, Room 6154
Washington, DC 20001
FTS Direct and Commercial: (202) 727-0557

FLORIDA
Director
Office of Laboratory Services
Department of Health and Rehabilitative Services
P.O. Box 210 (1217 Pearl Street)
Jacksonville, FL 32231
FTS Direct and Commercial: (904) 354-3961

GEORGIA
Director of Laboratories
Georgia Department of Human Resources
47 Trinity Avenue
Atlanta, GA 30334
Commercial: (404) 656-4852

GUAM
Laboratory Director
Public Health and Social Services
P.O. Box 2816
Agana, Guam 96910

HAWAII
Chief
Laboratories Branch
State Department of Health
P.O. Box 3378

Table 34.2. (*continued*)

Honolulu, HI 96801
FTS Direct and Commercial: (808) 548-6324

IDAHO
Chief
Bureau of Laboratories
Department of Health and Welfare
2220 Old Penitentiary Road
Boise, ID 83701
FTS Direct: 554-2235
Commercial: (208) 334-2235
Evaluation and Specimens:
Box 640
Boise, ID 83701

ILLINOIS
Chief
Division of Laboratories
Illinois Department of Public Health
535 W. Jefferson, 4th Floor
Springfield, IL 62761
FTS Direct and Commercial: (217) 782-4977

INDIANA
Director
Bureau of Laboratories
State Board of Health
1330 West Michigan Street
Indianapolis, IN 46206
FTS Direct and Commercial: (317) 633-0376

IOWA
Director
University Hygienic Laboratory
University of Iowa
Iowa City, IA 52242
FTS Direct and Commercial: (319) 353-5990

KANSAS
Director
Office of Laboratories and Research
Department of Health and Environment
Forbes Building, #740
Topeka, KS 66620
FTS Direct and Commercial: (913) 862-9360

KENTUCKY
Director
Division of Laboratory Services
Department for Health Services
Cabinet for Human Resources
275 East Main Street
Frankfort, KY 40621
FTS Direct and Commercial: (502) 564-4446

LOUISIANA
Director
Bureau of Laboratories
Office of Health Services and Environmental Quality
Louisiana State Department of Health
325 Loyola Avenue, 7th Floor
New Orleans, LA 70112
FTS Direct and Commercial: (504) 568-5373

MAINE
Director
Public Health Laboratory
Department of Human Services
State House - Station No. 12
Augusta, ME 04333
FTS Direct: 868-2727
Commercial: (207) 289-2727

MARYLAND
Director
Laboratories Administration
State Department of Health and Mental Hygiene
P.O. Box 2355
Baltimore, MD 21203
FTS Direct: 932-2880
Commercial: (301) 383-2880

MASSACHUSETTS
Director
State Laboratory Institute
Department of Public Health
305 South Street
Jamaica Plain, MA 02130
FTS Direct and Commercial: (617) 522-3700

MICHIGAN
Laboratory Director
Laboratory and Epidemiological Services Administration
Michigan Department of Public Health
P.O. Box 30035 - 3500 N. Logan
Lansing, MI 48909
FTS Direct: 253-1381
Commercial: (517) 373-1381

Table 34.2. *(continued)*

MINNESOTA
Director
Division of Medical Laboratories
Minnesota Department of Health
P.O. Box 9441
Minneapolis, MN 55440
FTS Direct and Commercial: (612) 623-5210

MISSISSIPPI
Director
Public Health Laboratories
State Board of Health
P.O. Box 1700
Jackson, MS 39205
FTS Direct and Commercial: (601) 354-6672

MISSOURI
Director
Bureau of Laboratory Services
Missouri Division of Health
307 W. McCarty
Jefferson City, MO 65101
FTS Direct and Commercial: (314) 751-3334

MONTANA
Chief
State Microbiology Laboratory
State Department of Health and Environmental Sciences
Cogswell Building
Helena, MT 59620
FTS Direct: 587-2642
Commercial: (406) 449-2642

NEBRASKA
Director of Laboratories
State Department of Health
P.O. Box 2755
Lincoln, NE 68502
FTS Direct: 541-2122
Commercial: (402) 471-2122

NEVADA
Administrator
Nevada State Health Laboratory
Department of Human Resources
1660 N. Virginia Street
Reno, NV 89503
FTS Operator: 598-6011
Commercial: (702) 885-4475

NEW HAMPSHIRE
Acting Director
Diagnostic Laboratories
Division of Public Health
State Laboratory Building
Hazen Drive
Concord, NH 03301
FTS Direct: 842-1110, ext. 4657
Commercial: (603) 271-4657

NEW JERSEY
Director
Public Health and Environmental Laboratories
State Department of Health
P.O. Box 1540-CN 360
Trenton, NJ 08625
FTS Direct and Commercial: (609) 292-5605

NEW MEXICO
Director
Scientific Laboratory Division
700 Camino de Salud, NE
Albuquerque, NM 87106
FTS Direct and Commercial: (505) 841-2500

NEW YORK
Director
Division of Laboratories and Research
State Department of Health
Tower Building, Empire State Plaza
Albany, NY 12201
FTS Direct and Commercial: (518) 474-4170, ext. 4180

NORTH CAROLINA
Director
Public Health Laboratory
State Board of Health
P.O. Box 28047
Raleigh, NC 27611
FTS Direct and Commercial: (919) 733-7834

NORTH DAKOTA
Chief
Laboratory Services Section
State Department of Health
Box 1618
Bismarck, ND 58502-1618
FTS Operator: 783-4011
Commercial: (701) 224-2384

Table 34.2. (*continued*)

OHIO
Chief
Division of Public Health Laboratories
State Department of Health
P.O. Box 2568
Columbus, OH 43216
FTS Direct and Commercial: (614) 421-1078

OKLAHOMA
Chief
Public Health Laboratory Service
State Department of Health
P.O. Box 24106
Oklahoma City, OK 73124
FTS Operator: 736-4011
Commercial: (405) 271-5070

OREGON
Manager-Director
Public Health Laboratory
Department of Human Resources
1717 SW 10th Avenue
Portland, OR 97201
FTS Direct and Commercial: (503) 229-5884

PENNSYLVANIA
Director
Bureau of Laboratories
Pennsylvania Department of Health
Pickering Way and Welsh Pool Road
Lionville, PA 19353
FTS Direct and Commercial: (215) 363-8500

PUERTO RICO
Director
Institute of Health Laboratories
Department of Health
Building A - Call Box 70184
San Juan, PR 00922
FTS Direct and Commercial: (809) 767-2014

RHODE ISLAND
Associate Director
Division of Laboratories
Health Laboratory Building
50 Orms Street
Providence, RI 02904
FTS Operator: 838-1000
Commercial: (401) 274-1011

SOUTH CAROLINA
Chief
Bureau of Laboratories
Department of Health and Environmental Control
P.O. Box 2202
Columbia, SC 29202
FTS Direct and Commercial: (803) 758-4491

SOUTH DAKOTA
Director
State Health Laboratory
Laboratory Building
Pierre, SD 57501
FTS Operator: 782-7000
Commercial: (605) 773-3368

TENNESSEE
Director
Division of Laboratory Services
Tennesse Department of Public Health
Cordell Hull Building, Room 425
Nashville, TN 37219
FTS Direct and Commercial: (615) 741-3596

TEXAS
Chief
Bureau of Laboratories
Texas Department of Health
1100 West 49th Street
Austin, TX 78756
FTS Operator: 729-4011
Commercial: (512) 458-7318

UTAH
Director
Utah State Health Laboratory
44 Medical Drive, Room 207
Salt Lake City, UT 84113
FTS Direct and Commercial: (801) 533-6131

VERMONT
Director
State Public Health Laboratory
State Department of Health
115 Colchester Avenue
Burlington, VT 05401
FTS Operator: 832-6501
Commercial: (802) 862-5701

Table 34.2. *(continued)*

VIRGINIA Director Bureau of Microbiological Science Division of Consolidated Laboratory Services Department of General Services Commonwealth of Virginia Box 1877 Richmond, VA 23215 FTS Direct: 936-3756 Commercial: (804) 786-3756	WEST VIRGINIA Director State Hygienic Laboratory 167 11th Avenue South Charleston, WV 25303 FTS Direct: 885-3530 Commercial: (304) 348-3530
VIRGIN ISLANDS Director of Public Health Laboratory P.O. Box 8585 St. Thomas, VI 00801 FTS Direct and Commercial: (809) 774-5955	WISCONSIN Director State Laboratory of Hygiene William D. Stovall Building 465 Henry Mall Madison, WI 53706 FTS Direct and Commercial: (608) 262-1293
WASHINGTON Chief Laboratory Section State Department of Social and Health Services 1409 Smith Tower Seattle, WA 98104 FTS Direct and Commercial: (206) 464-6461	WYOMING Director Public Health Laboratory Services Division of Health and Medical Services State Office Building Cheyenne, WY 82001 FTS Operator: 328-1110 Commercial: (307) 777-7431

indicated that they are not involved in regulation of primary laboratory licensure but that another stage agency performed this function. In some states there is no State Virology Laboratory, whereas, in other states there are no practicing primary virology laboratories that require regulation. It would seem that some regulation is desirable in all states to ensure that standard, accepted practices are used to obtain reliable results. The establishment of some level of regulation within states would promote this standardization.

34.4 TURNAROUND TIME FOR RESULT REPORTING

The bane of viral diagnosis by state laboratories in the past has been the long turnaround time from submission of a specimen by the physician until a report is returned to that physician. Frequently, this was a matter of months in all but emergency cases, as with exposure to rabies virus. Today, the turnaround time for diagnosis of many viral diseases is no longer a significant problem. Based on the response to our questionnaire, it appears that most laboratories send out a report upon identification of a virus, in some states this may be a telephone report.

Virtually all laboratories indicated that the length of time for reporting results was variable and dependent on the type of testing to be performed, as well as whether a specimen is positive or negative. Serology results are frequently reported within a few days to 1 week of receipt of the acute and convalescent sera, however, a few laboratories indicate this may be as long as 2 to 4 weeks. Testing for immune status against a particular virus is reported from a few days to 4 weeks after the receipt of a single serum. Positive isolation of many viruses is reported out within 72 hours, whereas, some isolation (as in the case of cytomegalovirus) may take up to 2 weeks. Reports on specimens that are negative for virus isolation may be sent out as soon as 2 weeks or not until 6 weeks after receipt of the specimen.

ACKNOWLEDGMENT

We thank the directors of the various state laboratories who supplied information for this chapter.

35

Laboratories Offering Viral Diagnostic Services

Steven Specter and Gerald J.Lancz

35.1 INTRODUCTION

This chapter identifies local laboratories that offer viral diagnostic services, including hospital, university and commercial laboratories (Table 35.1). We have not included local public health laboratories, because the services they provide are generally limited to epidemiology. Because diagnostic services offered by any individual laboratory may change, we have not attempted to indicate the extent of services available. Additionally, laboratories that provide viral diagnostic services may open, close or discontinue certain services at any time. This must be considered when reference is made to the information compiled in Table 35.1 for possible submission of specimens, utilization of services, etc.

The information presented was supplied by state public health laboratories and individual clinical virologists, who provided names of laboratories with which they were familiar. For some states, the state laboratory personnel were unable to supply such a list and it was not possible for us to identify individuals with knowledge of viral diagnostic services. Thus, a failure to list services available in a particular state may reflect either our inability to identify these laboratories or that laboratory services are not offered in that state.

Some commercial laboratories have a national network to process clinical specimens. These laboratories fill the void when there is no local laboratory that performs these services (Table 35.2).

The extent of services available from viral diagnostic laboratories varies widely. Some laboratories offer viral diagnostic and serology services for virtually all common human pathogenic viruses, while others may perform a limited number of tests (e.g., herpes simplex virus isolation or rubella serology) only. Individuals should contact the laboratory(ies) listed in their locale to determine how their needs can be served best.

Table 35.1. Laboratories Performing Viral Diagnosis

Alabama
- Birmingham
  - University of Alabama Medical Center

Alaska
- Anchorage
  - Providence Hospital

Arizona
- Glendale
  - Automated Pathology Services-Thunderbird Med Labs
- Mesa
  - Sonora Lab Sciences, Inc.
- Phoenix
  - Boland (Bolin) Laboratories, Inc.
- Tempe
  - Mobile Microbiology Services
- Tucson
  - University of Arizona Health Sciences Center

Arkansas
- Little Rock
  - Arkansas Children's Hospital
  - University of Arkansas Medical Services Center

California
- Emeryville
  - Virolab, Inc.
- La Mirada
  - Immunology Consultants
- Long Beach
  - Medical Reference Laboratory
  - Memorial Hospital Medical Center
- Los Angeles
  - Cedars Sinai Medical Center
  - CLMG, Inc.
  - Specialty Labs
  - UCLA
  - USC/LAC General Hospital
- Newberg Park
  - Reference Laboratory
- North Hollywood
  - Kaiser-Permanente
- Orange
  - University of California, Irvine
- San Diego
  - University of California, San Diego
- San Francisco
  - Mt. Zion Hospital
- Stanford
  - Children's Hospital at Stanford
- Torrance
  - Harbor General Hospital
- Woodland Hills
  - Smith Kline Clinical Laboratories, Inc.

Colorado
- None identified

Connecticut
- Danbury
  - Danbury Hospital
- Farmington
  - University of Connecticut Medical School
- Hartford
  - Hartford Hospital
  - St. Francis Hospital
- New Haven
  - Yale University
- West Haven
  - Veterans Hospital
- Misc.
  - CLN Medical Laboratory
  - Columbia Medical Laboratory
  - Virus Serology Laboratory

Delaware
- None Identified

District of Columbia
  - Children's Hospital National Medical Center
  - Georgetown University
  - Walter Reed Army Medical Center

Florida
- Miami
  - Jackson Memorial Hospital
- Tampa
  - Smith Kline Clinical Laboratories
  - St. Joseph's Hospital
  - University of South Florida Medical Center

Georgia
- Atlanta
  - Emory University Laboratory
  - Smith Kline Clinical Laboratories
- Tucker
  - PSPA

Table 35.1. (*continued*)

- Hawaii
  - None Identified
- Idaho
  - Boise
    - St. Luke's Hospital
- Illinois
  - Chicago
    - Columbus Hospital
    - Illinois Masonic Medical Center Hospital
    - Michael Reese Hospital
    - Mt. Sinai Hospital
    - Northwestern University Medical Center
    - Rush-Presbyterian St. Luke's Hospital
  - Elmhurst
    - Memorial Hospital of DuPage County
  - Maywood
    - Loyola Medical Center Hospital
  - Park Ridge
    - Victoria Clinic Reference Laboratory
  - Peoria
    - Mobilab, Inc.
  - Rockford
    - Rockford School of Medicine
  - Springfield
    - Memorial Medical Center
- Indiana
  - South Bend
    - Notre Dame University
  - Indianapolis
    - Indiana Medical Center
- Iowa
  - Des Moines
    - Iowa Methodist Medical Center
  - Iowa City
    - University of Iowa Hospitals and Clinics
  - Sioux City
    - St. Luke's Hospital
- Kansas
  - Kansas City
    - Kansas University Medical Center
  - Wichita
    - Consolidated Biological Laboratory
    - St. Francis Hospital
    - Wesley Medical Center
- Kentucky
  - Lexington
    - Univeristy of Kentucky Medical Center
  - Louisville
    - Jewish Hospital
    - Pediatric Virology Laboratory
- Louisiana
  - New Orleans
    - New Orleans Charity Hospital
- Maine
  - Portland
    - Maine Medical Center
- Maryland
  - Baltimore
    - Johns Hopkins Hospital
    - Sinai Hospital
    - University of Maryland Hospital
  - Fort Meade
    - Kimbrough Hospital
  - Rockville
    - Bionetics Laboratory
    - NIH Clinical Center
  - Walkersville
    - MA Bioproducts Laboratory
  - Misc.
    - Maryland Medical Laboratory
- Massachusetts
  - Boston
    - Children's Hospital
    - Damon Laboratories
    - Massachusetts General Hospital
    - Smith Kline Clinical Laboratories
    - Tufts-New England Medical Center
- Michigan
  - Annapolis
    - Peoples Community Hospital Association
  - Ann Arbor
    - University of Michigan Hospital
  - Detroit
    - Children's Hospital
    - Henry Ford Hospital
    - William Beaumont Hospital
  - Farmington Hills
    - Bio-Science Lab
  - Grand Rapids
    - Continental Clinical Biochemical
    - St. Mary's Hospital
  - Lansing
    - Edward Sparrow Hospital

Table 35.1. (*continued*)

- Livonia
  - Roche Bio-Medical Laboratory

Minnesota
- Duluth
  - University of Minnesota Medical School
- Minneapolis
  - Hennepin County Metropolitan Medical Center
  - Lufkin Laboratories
  - University of Minnesota Hospitals
  - Veteran's Hospital
  - Viromed
- Rochester
  - Mayo Clinic
- St. Cloud
  - North Central Laboratories
- St. Paul
  - Children's Hospital
  - Ramsey Hospital

Mississippi
- Jackson
  - University Medical Center

Missouri
- Columbia
  - Boyce and Bynum Laboratories
  - Veteran's Hospital
- Kansas City
  - Children's Mercy Hospital
  - North Kansas City Memorial Hospital
- Springfield
  - Cox Medical Center
  - St. John's Hospital
- St. Louis
  - Cardinal Glennon Hospital
  - Smith Kline Clinical Laboratories
  - St. Louis Children's Hospital

Montana
- Missoula
  - Community Hospital

Nebraska
- None Identified

Nevada
- Las Vegas
  - Associated Pathologist's Hospital
  - Sunrise Hospital Laboratory
- Reno
  - Sierra Nevada Laboratory
  - Washoe Medical Center

New Hampshire
- None Identified

New Jersey
- Hackensack
  - Hackensack Hospital
- Newark
  - St. Michael's Medical Center
- New Brunswick
  - Middlesex Hospital
- Teterboro
  - Metpath

New Mexico
- Albuquerque
  - SED Medical Laboratories
  - University of New Mexico Medical School

New York
- Buffalo
  - Children's Hospital
- New York City
  - Bellevue Hospital
  - Beth Israel Medical Center
  - Columbia Presbyterian Medical Center
  - Lenox Hill Hospital
  - Memorial Hospital
  - Montefiore Medical Center
  - Mt. Sinai Hospital
  - New York Hospital Pathology Laboratory
  - St. Luke's Roosevelt Hospital Center
- Staten Island
  - IBR-Consolidated Clinical Laboratory

North Carolina
- Burlington
  - Biomedical Reference Laboratories
- Chapel Hill
  - Frank Porter Graham Child Development Center
  - North Carolina Memorial Hospital
- Charlotte
  - Presbyterian Hospital
- Durham
  - Duke University Medical Center
  - Veteran's Hospital

Table 35.1. (*continued*)

Greensboro
Moses Cone Memorial Hospital
Greenville
East Carolina University Virology Laboratory
Raleigh
Rex Hospital
Winston-Salem
North Baptist Hospital

North Dakota
None Identified

Ohio
Akron
Akron City Hospital
Children's Hospital
Canton
Aultman Hospital
Timkin Mercy Medical Hospital
Cincinnati
Children's Hospital
University of Cincinnati Hospital
Cleveland
Cleveland Metropolitan General Hospital
St. Luke's Hospital
Columbus
Children's Hospital
Doctor's Hospital
Mercy Medical Center
Ohio State University Hospital
Dayton
Children's Hospital Medical Center
Rootstown
North East Ohio Universities
Springfield
Springfield Community Hospital
Toledo
Medical College of Ohio
Youngstown
Youngstown Hospital
Misc.
Diagnostic Virology Service

Oklahoma
Oklahoma City
Oklahoma Children's Memorial Hospital
Presbyterian Hospital
St. Anthony Hospital
Tulsa
St. Francis Hospital

Oregon
Coos Bay
Bay Area Hospital Laboratory
Dr. D. H. McGowan, Medical Laboratory
Corvallis
Good Samaritan Hospital Laboratory
Eugene
Pathology Consultants
Sacred Heart
Medford
Rouge Valley Medical Center
Oregon City
Drs. Haug and Hoffman
Pendleton
Interpath Laboratory, PC
Portland
Good Samaritan Hospital Laboratory
Medical Lab-Pathologists Central Laboratory
Oregon Health Science University Clinical Laboratory
Physicians Medical Laboratory
Providence Medical Center
St. Vincent Hospital Laboratory
Reedport
Lower Umpqua District Hospital
Roseburg
Douglas Community Hospital
Salem
Capitol Medical Laboratory
The Dalles
Mid Columbia Medical Center

Pennsylvania
Harrisburg
Harrisburg Hospital
Johnstown
Conemaugh Valley Memorial Hospital
King of Prussia
Smith Kline Clinical Laboratories
Philadelphia
Albert Einstein Medical Center
Children's Hospital of Pennsylvania
St. Christopher's Hospital for Children
Temple University Hospital
Thomas Jefferson University Hospital

Table 35.1. *(continued)*

Pittsburgh
- Allegheny General Hospital Singer Memorial
- Children's Hospital of Pittsburgh
- Mercy Hospital 3
- Presbyterian University Hospital

Sayre
- Robert Packer Hospital

Rhode Island

Providence
- Brown University Medical Center
- Rhode Island Hospital

South Carolina

Charleston
- Medical University of South Carolina

South Dakota

Sioux Falls
- Veteran's Medical Center

Vermillion
- University of South Dakota Virology Laboratory

Tennessee

Johnson City
- East Tennessee State Medical School

Memphis
- Baptist Hospital

Nashville
- Vanderbilt Hospital

Texas

Dallas
- Parkland Memorial Hospital

Houston
- Influenza Research Center, Baylor College of Medicine
- Methodist Hospital

Utah

Salt Lake City
- University of Utah Medical Center

Vermont

Burlington
- Medical Center Hospital of Vermont

Virginia

Richmond
- Medical College of Virginia

Washington

Seattle
- Children's Orthopedic Hospital

Spokane
- Sacred Heart Medical Center

West Virginia

Charleston
- Charleston Area Medical Center

Huntington
- Marshall University School of Medicine

Wisconsin

Green Bay
- Bellin Memorial Hospital

La Crosse
- Gunderson Clinic

Marshfield
- Marshfield Clinic

Wyoming

Jackson
- Intermountain Virology Laboratory

Table 35.2. Commercial Laboratories Which Offer Viral Diagnostic Services Nationally

Laboratory	Location
Bionetics Lab	Kensington, MD
Bio-Science Lab	Van Nuys, CA
MA Bioproducts	Walkersville, MD
Metpath	Teterboro, NJ
Roche Bio-Medical Lab	Columbus, OH
Smith Kline Clinical Labs	King of Prussia, PA
Specialty Labs.	Los Angeles, CA
Virolab, Inc.	Emeryville, CA
Viromed	Minneapolis, MN

Index

D

E